Business Analytics

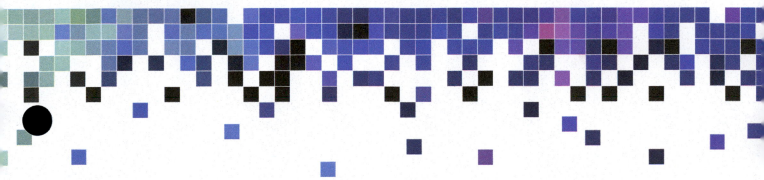

D0713707

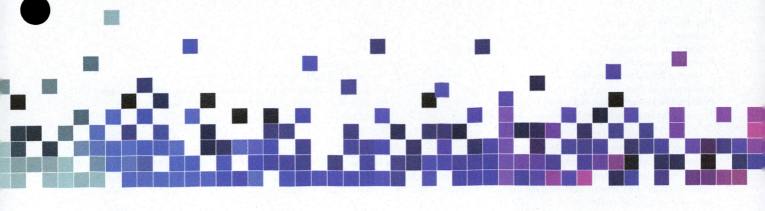

Business Analytics

Methods, Models, and Decisions

James R. Evans | University of Cincinnati

THIRD EDITION

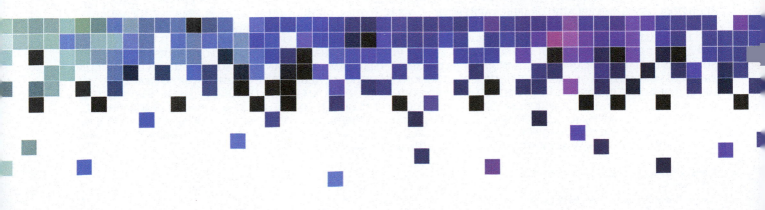

Director, Portfolio Management: Deirdre Lynch
Courseware Portfolio Manager: Patrick Barbera
Courseware Portfolio Management Assistan: Morgan Danna
Content Producer: Angela Montoya
Managing Producer: Karen Wernholm
Producer: Jean Choe
Product Marketing Manager: Kaylee Carlson
Product Marketing Assistant: Marianela Silvestri

Field Marketing Manager: Thomas Hayward
Field Marketing Assistant: Derrica Moser
Senior Author Support/Technology Specialist: Joe Vetere
Manager, Rights and Permissions: Gina Cheselka
Text and Cover Design: Jerilyn Bockorick
Production Coordination, Composition, and Illustrations: Pearson CSC
Manufacturing Buyer: Carol Melville, LSC Communications
Cover Image: shuoshu/Getty Images

Copyright © 2020, 2016, 2013 by Pearson Education, Inc. 221 River Street, Hoboken, NJ 07030. All Rights Reserved. Printed in the United States of America. This publication is protected by copyright, and permission should be obtained from the publisher prior to any prohibited reproduction, storage in a retrieval system, or transmission in any form or by any means, electronic, mechanical, photocopying, recording, or otherwise. For information regarding permissions, request forms and the appropriate contacts within the Pearson Education Global Rights & Permissions department, please visit www.pearsoned.com/permissions/.

PEARSON, ALWAYS LEARNING, and MYLAB are exclusive trademarks owned by Pearson Education, Inc. or its affiliates in the U.S. and/or other countries.

Unless otherwise indicated herein, any third-party trademarks that may appear in this work are the property of their respective owners and any references to third-party trademarks, logos or other trade dress are for demonstrative or descriptive purposes only. Such references are not intended to imply any sponsorship, endorsement, authorization, or promotion of Pearson's products by the owners of such marks, or any relationship between the owner and Pearson Education, Inc. or its affiliates, authors, licensees or distributors.

[For instructor editions: This work is solely for the use of instructors and administrators for the purpose of teaching courses and assessing student learning. Unauthorized dissemination, publication or sale of the work, in whole or in part (including posting on the internet) will destroy the integrity of the work and is strictly prohibited.]

MICROSOFT SCREENSHOTS IN THIS TEXT HAVE BEEN TAKEN FROM EXCEL 2016. THE CONCEPTS AND THEORY BEING TAUGHT ARE APPLICABLE IN OTHER, SIMILAR SOFTWARE.

MICROSOFT® AND WINDOWS® ARE REGISTERED TRADEMARKS OF THE MICROSOFT CORPORATION IN THE U.S.A. AND OTHER COUNTRIES. SCREEN SHOTS AND ICONS REPRINTED WITH PERMISSION FROM THE MICROSOFT CORPORATION. THIS BOOK IS NOT SPONSORED OR ENDORSED BY OR AFFILIATED WITH THE MICROSOFT CORPORATION.

MICROSOFT AND/OR ITS RESPECTIVE SUPPLIERS MAKE NO REPRESENTATIONS ABOUT THE SUITABILITY OF THE INFORMATION CONTAINED IN THE DOCUMENTS AND RELATED GRAPHICS PUBLISHED AS PART OF THE SERVICES FOR ANY PURPOSE. ALL SUCH DOCUMENTS AND RELATED GRAPHICS ARE PROVIDED "AS IS" WITHOUT WARRANTY OF ANY KIND. MICROSOFT AND/ OR ITS RESPECTIVE SUPPLIERS HEREBY DISCLAIM ALL WARRANTIES AND CONDITIONS WITH REGARD TO THIS INFORMATION, INCLUDING ALL WARRANTIES AND CONDITIONS OF MERCHANTABILITY, WHETHER EXPRESS, IMPLIED OR STATUTORY, FITNESS FOR A PARTICULAR PURPOSE, TITLE AND NON-INFRINGEMENT. IN NO EVENT SHALL MICROSOFT AND/OR ITS RESPECTIVE SUPPLIERS BE LIABLE FOR ANY SPECIAL, INDIRECT OR CONSEQUENTIAL DAMAGES OR ANY DAMAGES WHATSOEVER RESULTING FROM LOSS OF USE, DATA OR PROFITS, WHETHER IN AN ACTION OF CONTRACT, NEGLIGENCE OR OTHER TORTIOUS ACTION, ARISING OUT OF OR IN CONNECTION WITH THE USE OR PERFORMANCE OF INFORMATION AVAILABLE FROM THE SERVICES. THE DOCUMENTS AND RELATED GRAPHICS CONTAINED HEREIN COULD INCLUDE TECHNICAL INACCURACIES OR TYPOGRAPHICAL ERRORS. CHANGES ARE PERIODICALLY ADDED TO THE INFORMATION HEREIN. MICROSOFT AND/OR ITS RESPECTIVE SUPPLIERS MAY MAKE IMPROVEMENTS AND/OR CHANGES IN THE PRODUCT(S) AND/OR THE PROGRAM(S) DESCRIBED HEREIN AT ANY TIME. PARTIAL SCREEN SHOTS MAY BE VIEWED IN FULL WITHIN THE SOFTWARE VERSION SPECIFIED.

Library of Congress Cataloging-in-Publication Data

Names: Evans, James R. (James Robert), 1950- author.
Title: Business analytics : methods, models, and decisions / James R. Evans.
Description: [Third edition] | Boston, MA : [Pearson, 2018]
Identifiers: LCCN 2018042697| ISBN 9780135231678 | ISBN 0135231671
Subjects: LCSH: Business planning. | Strategic planning. | Industrial
    management--Statistical methods.
Classification: LCC HD30.28 .E824 2018 | DDC 658.4/01--dc23 LC record available at https://lccn.loc.gov/2018042697

ISBN 13: 978-0-13-523167-8
ISBN 10: 0-13-523167-1

23 2022

NC 06.15.2022 1440

Brief Contents

Contents

Online Supplements: Information about how to access and use Analytic Solver Basic are available for download at www.pearsonhighered.com/evans.

Getting Started with Analytic Solver

Using Advanced Regression Techniques in Analytic Solver

Using Forecasting Techniques in Analytic Solver

Using Data Mining in Analytic Solver

Model Analysis in Analytic Solver

Using Monte Carlo Simulation in Analytic Solver

Using Linear Optimization in Analytic Solver

Using Integer and Nonlinear Optimization in Analytic Solver

Using Optimization Parameter Analysis in Analytic Solver

Using Decision Trees in Analytic Solver

Preface

In 2007, Thomas H. Davenport and Jeanne G. Harris wrote a groundbreaking book, *Competing on Analytics: The New Science of Winning* (Boston: Harvard Business School Press). They described how many organizations are using analytics strategically to make better decisions and improve customer and shareholder value. Over the past several years, we have seen remarkable growth in analytics among all types of organizations. The Institute for Operations Research and the Management Sciences (INFORMS) noted that analytics software as a service is predicted to grow at three times the rate of other business segments in upcoming years.[1] In addition, the *MIT Sloan Management Review* in collaboration with the IBM Institute for Business Value surveyed a global sample of nearly 3,000 executives, managers, and analysts.[2] This study concluded that top-performing organizations use analytics five times more than lower performers, that improvement of information and analytics was a top priority in these organizations, and that many organizations felt they were under significant pressure to adopt advanced information and analytics approaches. Since these reports were published, the interest in and the use of analytics has grown dramatically.

In reality, business analytics has been around for more than a half-century. Business schools have long taught many of the core topics in business analytics—statistics, data analysis, information and decision support systems, and management science. However, these topics have traditionally been presented in separate and independent courses and supported by textbooks with little topical integration. This book is uniquely designed to present the emerging discipline of business analytics in a unified fashion consistent with the contemporary definition of the field.

About the Book

This book provides undergraduate business students and introductory graduate students with the fundamental concepts and tools needed to understand the role of modern business analytics in organizations, to apply basic business analytics tools in a spreadsheet environment, and to communicate with analytics professionals to effectively use and interpret analytic models and results for making better business decisions. We take a balanced, holistic approach in viewing business analytics from descriptive, predictive, and prescriptive perspectives that define the discipline.

---

[1]Anne Robinson, Jack Levis, and Gary Bennett, INFORMS News: INFORMS to Officially Join Analytics Movement. http://www.informs.org/ORMS-Today/Public-Articles/October-Volume-37-Number-5/INFORMS-News-INFORMS-to-Officially-Join-Analytics-Movement.
[2]"Analytics: The New Path to Value," *MIT Sloan Management Review* Research Report, Fall 2010.

This book is organized in five parts.

1. Foundations of Business Analytics

    The first two chapters provide the basic foundations needed to understand business analytics and to manipulate data using Microsoft Excel. Chapter 1 provides an introduction to business analytics and its key concepts and terminology, and includes an appendix that reviews basic Excel skills. Chapter 2, Database Analytics, is a unique chapter that covers intermediate Excel skills, Excel template design, and PivotTables.

2. Descriptive Analytics

    Chapters 3 through 7 cover fundamental tools and methods of data analysis and statistics. These chapters focus on data visualization, descriptive statistical measures, probability distributions and data modeling, sampling and estimation, and statistical inference. We subscribe to the American Statistical Association's recommendations for teaching introductory statistics, which include emphasizing statistical literacy and developing statistical thinking, stressing conceptual understanding rather than mere knowledge of procedures, and using technology for developing conceptual understanding and analyzing data. We believe these goals can be accomplished without introducing every conceivable technique into an 800–1,000 page book as many mainstream books currently do. In fact, we cover all essential content that the state of Ohio has mandated for undergraduate business statistics across all public colleges and universities.

3. Predictive Analytics

    In this section, Chapters 8 through 12 develop approaches for applying trendlines and regression analysis, forecasting, introductory data mining techniques, building and analyzing models on spreadsheets, and simulation and risk analysis.

4. Prescriptive Analytics

    Chapters 13 and 14 explore linear, integer, and nonlinear optimization models and applications. Chapter 15, Optimization Analytics, focuses on what-if and sensitivity analysis in optimization, and visualization of Solver reports.

5. Making Decisions

    Chapter 16 focuses on philosophies, tools, and techniques of decision analysis.

Changes to the Third Edition

The third edition represents a comprehensive revision that includes many significant changes. The book now relies only on native Excel, and is independent of platforms, allowing it to be used easily by students with either PC or Mac computers. These changes provide students with enhanced Excel skills and basic understanding of fundamental concepts. *Analytic Solver* is no longer integrated directly in the book, but is illustrated in online supplements to facilitate revision as new software updates may occur. These supplements plus information regarding how to access *Analytic Solver* may be accessed at http://pearsonhighered.com/evans.

Key changes to this edition are as follows:

■   This edition is paired with MyLab Statistics, a teaching and learning platform that empowers you to reach every student. By combining trusted author content with digital tools and a flexible platform, MyLab personalizes the learning experience and improves results for each student. For example, new Excel and StatCrunch Projects help students develop business decision-making skills.

■ Each chapter now includes a short section called *Technology Help*, which provides useful summaries of key Excel functions and procedures, and the use of supplemental software including *StatCrunch* and *Analytic Solver Basic.*

■ Chapter 1 includes an Appendix reviewing basic Excel skills, which will be used throughout the book.

■ Chapter 2, Database Analytics, is a new chapter derived from the second edition that focuses on applications of Excel functions and techniques for dealing with spreadsheet data, including a new section on Excel template design.

■ Chapter 3, Data Visualization, includes a new Appendix illustrating Excel tools for Windows and a brief overview of Tableau.

■ Chapter 5, Probability Distributions and Data Modeling, includes a new section on Combinations and Permutations.

■ Chapter 6, Sampling and Estimation, provides a discussion of using data visualization for confidence interval comparison.

■ Chapter 9, Forecasting Techniques, now includes Excel approaches for double exponential smoothing and Holt-Winters models for seasonality and trend.

■ Chapter 10, Introduction to Data Mining, has been completely rewritten to illustrate simple data mining techniques that can be implemented on spreadsheets using Excel.

■ Chapter 11, Spreadsheet Modeling and Analysis, is now organized along the analytic classification of descriptive, predictive, and prescriptive modeling.

■ Chapter 12 has been rewritten to apply Monte-Carlo simulation using only Excel, with an additional section of systems simulation concepts and approaches.

■ Optimization topics have been reorganized into two chapters—Chapter 13, Linear Optimization, and Chapter 14, Integer and Nonlinear Optimization, which rely only on the Excel-supplied *Solver.*

■ Chapter 15 is a new chapter called Optimization Analytics, which focuses on what-if and sensitivity analysis, and visualization of *Solver* reports; it also includes a discussion of how *Solver* handles models with bounded variables.

In addition, we have carefully checked, and revised as necessary, the text and problems for additional clarity. We use major section headings in each chapter and tie these clearly to the problems and exercises, which have been revised and updated throughout the book. At the end of each section we added several "Check Your Understanding" questions that provide a basic review of fundamental concepts to improve student learning. Finally, new Analytics in Practice features have been incorporated into several chapters.

Features of the Book

■ **Chapter Section Headings**—with "Check Your Understanding" questions that provide a means to review fundamental concepts.

■ **Numbered Examples**—numerous, short examples throughout all chapters illustrate concepts and techniques and help students learn to apply the techniques and understand the results.

■ **"Analytics in Practice"**—at least one per chapter, this feature describes real applications in business.

■ **Learning Objectives**—lists the goals the students should be able to achieve after studying the chapter.

- **Key Terms**—bolded within the text and listed at the end of each chapter, these words will assist students as they review the chapter and study for exams. Key terms and their definitions are contained in the glossary at the end of the book.
- **End-of-Chapter Problems and Exercises**—clearly tied to sections in each chapter, these help to reinforce the material covered through the chapter.
- **Integrated Cases**—allow students to think independently and apply the relevant tools at a higher level of learning.
- **Data Sets and Excel Models**—used in examples and problems and are available to students at www.pearsonhighered.com/evans.

Software Support

Technology Help sections in each chapter provide additional support to students for using Excel functions and tools, Tableau, and StatCrunch.

Online supplements provide detailed information and examples for using *Analytic Solver Basic*, which provides more powerful tools for data mining, Monte-Carlo simulation, optimization, and decision analysis. These can be used at the instructor's discretion, but are not necessary to learn the fundamental concepts that are implemented using Excel. Instructions for obtaining licenses for *Analytic Solver Basic* can be found on the book's website, http://pearsonhighered.com/evans.

To the Students

To get the most out of this book, you need to do much more than simply read it! Many examples describe in detail how to use and apply various Excel tools or add-ins. We highly recommend that you *work through these examples* on your computer to replicate the outputs and results shown in the text. You should also *compare mathematical formulas* with spreadsheet formulas and *work through basic numerical calculations by hand*. Only in this fashion will you learn how to use the tools and techniques effectively, gain a better understanding of the underlying concepts of business analytics, and increase your proficiency in using Microsoft Excel, which will serve you well in your future career.

Visit the companion Web site (www.pearsonhighered.com/evans) for access to the following:

- **Online Files:** Data Sets and Excel Models—files for use with the numbered examples and the end-of-chapter problems. (For easy reference, the relevant file names are italicized and clearly stated when used in examples.)
- **Online Supplements for *Analytic Solver Basic*:** Online supplements describing the use of *Analytic Solver* that your instructor might use with selected chapters.

To the Instructors

MyLab Statistics is now available with Evans "Business Analytics" 3e: MyLab™ Statistics is the teaching and learning platform that empowers instructors to reach every student. Teach your course your way with a flexible platform. Collect, crunch, and communicate with data in StatCrunch®, an integrated Web-based statistical software. Empower each learner with personalized and interactive practice. Tailor your course to your students' needs with enhanced reporting features. Available with the complete eText, accessible anywhere with the Pearson eText app.

Instructor's Resource Center—Reached through a link at www.pearsonhighered.com/evans, the Instructor's Resource Center contains the electronic files for the complete Instructor's Solutions Manual, PowerPoint lecture presentations, and the Test Item File.

- **Register, redeem, log in at www.pearsonhighered.com/irc:** instructors can access a variety of print, media, and presentation resources that are available with this book in downloadable digital format. Resources are also available for course management platforms such as Blackboard, WebCT, and CourseCompass.
- *Instructor's Solutions Manual*—The Instructor's Solutions Manual, updated and revised for the second edition by the author, includes Excel-based solutions for all end-of-chapter problems, exercises, and cases. The Instructor's Solutions Manual is available for download by visiting www.pearsonhighered.com/evans and clicking on the Instructor Resources link.
- *PowerPoint presentations*—The PowerPoint slides, revised and updated by the author, are available for download by visiting www.pearsonhighered.com/ evans and clicking on the Instructor Resources link. The PowerPoint slides provide an instructor with individual lecture outlines to accompany the text. The slides include nearly all of the figures, tables, and examples from the text. Instructors can use these lecture notes as they are or can easily modify the notes to reflect specific presentation needs.
- *Test Bank*—The TestBank, prepared by Paolo Catasti from Virginia Commonwealth University, is available for download by visiting www.pearsonhighered.com/evans and clicking on the Instructor Resources link.
- **Need help?** Pearson Education's dedicated technical support team is ready to assist instructors with questions about the media supplements that accompany this text. The supplements are available to adopting instructors. Detailed descriptions are provided at the Instructor's Resource Center.

Acknowledgments

I would like to thank the staff at Pearson Education for their professionalism and dedication to making this book a reality. In particular, I want to thank Angela Montoya, Kathleen Manley, Karen Wernholm, Kaylee Carlson, Jean Choe, Bob Carroll, and Patrick Barbera. I would also like to thank Gowri Duraiswamy at SPI, and accuracy and solutions checker Jennifer Blue for their outstanding contributions to producing this book. I also want to acknowledge Daniel Fylstra and his staff at Frontline Systems for working closely with me on providing *Analytic Solver Basic* as a supplement with this book. If you have any suggestions or corrections, please contact the author via email at james.evans@uc.edu.

James R. Evans
Department of Operations, Business Analytics, and Information Systems
University of Cincinnati
Cincinnati, Ohio

Get the Most Out of
MyLab Statistics

Statistics courses are continuously evolving to help today's students succeed. It's more challenging than ever to support students with a wide range of backgrounds, learner styles, and math anxieties. The flexibility to build a course that fits instructors' individual course formats—with a variety of content options and multimedia resources all in one place—has made MyLab Statistics the market-leading solution for teaching and learning mathematics since its inception.

78% of students say MyLab Statistics helped them learn their course content.*

Teach your course with a consistent author voice

With market-leading author content options, your course can fit your style. Pearson offers the widest variety of content options, addressing a range of approaches and learning styles, authored by thought leaders across the business and math curriculum. MyLab™ Statistics is tightly integrated with each author's style, offering a range of author-created multimedia resources, so your students have a consistent experience.

Thanks to feedback from instructors and students from more than 10,000 institutions, MyLab Statistics continues to transform—delivering new content, innovative learning resources, and platform updates to support students and instructors, today and in the future.

*Source: 2018 Student Survey, n 31,721

pearson.com/mylab/statistics

Resources for Success

MyLab Statistics Online Course for Business Analytics

by James R. Evans

MyLab™ Statistics is available to accompany Pearson's market leading text offerings. To give students a consistent tone, voice, and teaching method each text's flavor and approach is tightly integrated throughout the accompanying MyLab Statistics course, making learning the material as seamless as possible.

Enjoy hands off grading with Excel Projects

Using proven, field-tested technology, auto-graded Excel Projects let instructors seamlessly integrate Microsoft Excel content into the course without manually grading spreadsheets. Students can practice important statistical skills in Excel, helping them master key concepts and gain proficiency with the program.

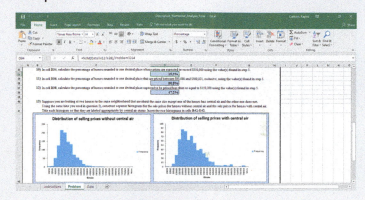

StatCrunch

StatCrunch, Pearson's powerful web-based statistical software, instructors and students can access tens of thousands of data sets including those from the textbook, perform complex analyses, and generate compelling reports. StatCrunch is integrated directly into MyLab Statistics or available as a standalone product. To learn more, go to www.statcrunch.com on any laptop, tablet, or smartphone.

Technology Tutorials and Study Cards

MyLab makes learning and using a variety of statistical software programs as seamless and intuitive as possible. Download data sets from the text and MyLab exercises directly into Excel. Students can also access instructional support tools including tutorial videos, study cards, and manuals for a variety of statistical software programs including StatCrunch, Excel, Minitab, JMP, R, SPSS, and TI83/84 calculators.

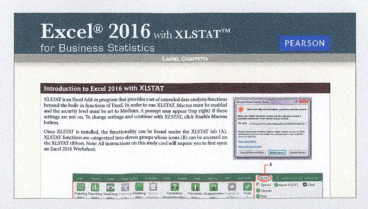

pearson.com/mylab/statistics

About the Author

James R. Evans
Professor Emeritus, University of Cincinnati, Lindner College of Business

James R. Evans is Professor Emeritus in the Department of Operations, Business Analytics, and Information Systems in the College of Business at the University of Cincinnati. He holds BSIE and MSIE degrees from Purdue and a PhD in Industrial and Systems Engineering from Georgia Tech.

Dr. Evans has published numerous textbooks in a variety of business disciplines, including statistics, decision models, and analytics, simulation and risk analysis, network optimization, operations management, quality management, and creative thinking. He has published 100 papers in journals such as *Management Science, IIE Transactions, Decision Sciences, Interfaces*, the *Journal of Operations Management, the Quality Management Journal*, and many others, and wrote a series of columns in *Interfaces* on creativity in management science and operations research during the 1990s. He has also served on numerous journal editorial boards and is a past-president and Fellow of the Decision Sciences Institute. In 1996, he was an INFORMS Edelman Award Finalist as part of a project in supply chain optimization with Procter & Gamble that was credited with helping P&G save over $250,000,000 annually in their North American supply chain, and consulted on risk analysis modeling for Cincinnati 2012's Olympic Games bid proposal.

A recognized international expert on quality management, he served on the Board of Examiners and the Panel of Judges for the Malcolm Baldrige National Quality Award. Much of his research has focused on organizational performance excellence and measurement practices.

Credits

Text Credits

Chapter 3   *page 101* Prem Thomas, MD and Seth Powsner, MD "Data Presentation for Quality Improvement", 2005, AMIA.

Appendix A   *page 633–635* National Institute of Standards and Technology.

Photo Credits

Chapter 1   *page 1* NAN728/Shutterstock *page 28* hans12/Fotolia

Chapter 2   *page 47* NAN728/Shutterstock *page 58* 2jenn/Shutterstock

Chapter 3   *page 85* ESB Professional/Shutterstock

Chapter 4   *page 115* Nataliiap/Shutterstock *page 163* langstrup/123RF

Chapter 5   *page 173* PeterVrabel/Shutterstock *page 195* Fantasista/Fotolia *page 209* Victor Correia/Shutterstock

Chapter 6   *page 219* Robert Brown Stock/Shutterstock *page 224* Stephen Finn/Shutterstock

Chapter 7   *page 247* Jirsak/Shutterstock *page 273* Hurst Photo/Shutterstock

Chapter 8   *page 283* Luca Bertolli/123RF *page 305* Gunnar Pippel/Shutterstock *page 305* Vector Illustration/Shutterstock *page 305* Claudio Divizia/Shutterstock *page 305* Natykach Nataliia/Shutterstock

Chapter 9   *page 325* rawpixel/123RF *page 351* Sean Pavone/Shutterstock

Chapter 10   *page 355* Laborant/Shutterstock *page 374* Helder Almeida/Shutterstock

Chapter 11   *page 377* marekuliasz/Shutterstock *page 388* Bryan Busovicki/Shutterstock *page 393* Poprotskiy Alexey/Shutterstock

Chapter 12   *page 423* Stephen Rees/Shutterstock *page 447* Vladitto/Shutterstock

Chapter 13   *page 465* Pinon Road/Shutterstock *page 468* bizoo_n/Fotolia *page 509* 2jenn/Shutterstock

Chapter 14   *page 523* Jirsak/Shutterstock *page 539* Kheng Guan Toh/Shutterstock

Chapter 15   *page 565* Alexander Orlov/Shutterstock

Chapter 16   *page 603* marekuliasz/Shutterstock *page 627* SSokolov/Shutterstock

Front Matter   James R. Evans

Introduction to Business Analytics

NAN728/Shutterstock

LEARNING OBJECTIVES   After studying this chapter, you will be able to:

- Define business analytics.
- Explain why analytics is important in today's business environment.
- State some typical examples of business applications in which analytics would be beneficial.
- Summarize the evolution of business analytics and explain the concepts of business intelligence, operations research and management science, and decision support systems.
- Explain the difference between descriptive, predictive, and prescriptive analytics.
- State examples of how data are used in business.

- Explain the concept of a model and various ways a model can be characterized.
- Define and list the elements of a decision model.
- Illustrate examples of descriptive, predictive, and prescriptive models.
- Explain the difference between uncertainty and risk.
- Define the terms *optimization*, *objective function*, and *optimal solution*.
- Explain the difference between a deterministic and stochastic decision model.
- List and explain the steps in the problem-solving process.

The purpose of this book is to provide you with a basic introduction to the concepts, methods, and models used in business analytics so that you will develop an appreciation not only for its capabilities to support and enhance business decisions, but also for the ability to use business analytics at an elementary level in your work. In this chapter, we introduce you to the field of business analytics and set the foundation for many of the concepts and techniques that you will learn. Let's start with a rather innovative example.

Most of you have likely been to a zoo, seen the animals, had something to eat, and bought some souvenirs. You probably wouldn't think that managing a zoo is very difficult; after all, it's just feeding and taking care of the animals, right? A zoo might be the last place that you would expect to find business analytics being used, but not anymore. The Cincinnati Zoo & Botanical Garden has been an "early adopter" and one of the first organizations of its kind to exploit business analytics.[1]

Despite generating more than two-thirds of its budget through its own fund-raising efforts, the zoo wanted to reduce its reliance on local tax subsidies even further by increasing visitor attendance and revenues from secondary sources such as membership, food, and retail outlets. The zoo's senior management surmised that the best way to realize more value from each visit was to offer visitors a truly transformed customer experience. By using business analytics to gain greater insight into visitors' behavior and tailoring operations to their preferences, the zoo expected to increase attendance, boost membership, and maximize sales.

The project team—which consisted of consultants from IBM and Brightstar Partners, as well as senior executives from the zoo—began translating the organization's goals into technical solutions. The zoo worked to create a business analytics platform that was capable of delivering the desired goals by combining data from ticketing and point-of-sale systems throughout the zoo with membership information and geographical data gathered from the ZIP codes of all visitors. This enabled the creation of reports and dashboards that gave everyone from senior managers to zoo staff access to real-time information that helped them optimize operational management and transform the customer experience.

By integrating weather forecast data, the zoo is now able to compare current forecasts with historic attendance and sales data, supporting better decision making for labor scheduling and inventory planning. Another area where the solution delivers new insight is food service. By opening food outlets at specific times of day when demand is highest (for example, keeping ice cream kiosks open in the

[1]IBM Software Business Analtyics, "Cincinnati Zoo transforms customer experience and boosts profits," © IBM Corporation 2012.

final hour before the zoo closes), the zoo has been able to increase sales significantly. In addition, attendance and revenues have dramatically increased, resulting in annual return on investment of 411%. The business analytics initiative paid for itself within three months and delivers, on average, benefits of $738,212 per year. Specifically,

- The zoo has seen a 4.2% rise in ticket sales by targeting potential visitors who live in specific ZIP codes.
- Food revenues increased 25% by optimizing the mix of products on sale and adapting selling practices to match peak purchase times.
- Eliminating slow-selling products and targeting visitors with specific promotions enabled an 18% increase in merchandise sales.
- The zoo was able to cut its marketing expenditure, saving $40,000 in the first year, and reduce advertising expenditure by 43% by eliminating ineffective campaigns and segmenting customers for more targeted marketing.

Because of the zoo's success, other organizations such as Point Defiance Zoo & Aquarium in Tacoma, Washington, and History Colorado Center, a museum in Denver, have embarked on similar initiatives.

 ## What Is Business Analytics?

Everyone makes decisions. Individuals face personal decisions such as choosing a college or graduate program, making product purchases, selecting a mortgage instrument, and investing for retirement. Managers in business organizations make numerous decisions every day. Some of these decisions include what products to make and how to price them, where to locate facilities, how many people to hire, where to allocate advertising budgets, whether or not to outsource a business function or make a capital investment, and how to schedule production. Many of these decisions have significant economic consequences; moreover, they are difficult to make because of uncertain data and imperfect information about the future.

Managers today no longer make decisions based on pure judgment and experience; they rely on factual data and the ability to manipulate and analyze data to supplement their intuition and experience, and to justify their decisions. What makes business decisions complicated today is the overwhelming amount of available data and information. Data to support business decisions—including those specifically collected by firms as well as through the Internet and social media such as Facebook—are growing exponentially and becoming increasingly difficult to understand and use. As a result, many companies have recently established analytics departments; for instance, IBM reorganized its consulting business and established a new 4,000-person organization focusing on analytics. Companies are increasingly seeking business graduates with the ability to understand and use analytics. The demand for professionals with analytics expertise has skyrocketed, and many universities now have programs in analytics.[2]

[2]Matthew J. Liberatore and Wenhong Luo, "The Analytics Movement: Implications for Operations Research," Interfaces, 40, 4 (July–August 2010): 313–324.

Business analytics, or simply **analytics**, is the use of data, information technology, statistical analysis, quantitative methods, and mathematical or computer-based models to help managers gain improved insight about their business operations and make better, fact-based decisions. Business analytics is "a process of transforming data into actions through analysis and insights in the context of organizational decision making and problem solving."[3] Business analytics is supported by various tools such as Microsoft Excel and various Excel add-ins, commercial statistical software packages such as SAS or Minitab, and more complex business intelligence suites that integrate data with analytical software.

Using Business Analytics

Tools and techniques of business analytics are used across many areas in a wide variety of organizations to improve the management of customer relationships, financial and marketing activities, human capital, supply chains, and many other areas. Leading banks use analytics to predict and prevent credit fraud. Investment firms use analytics to select the best client portfolios to manage risk and optimize return. Manufacturers use analytics for production planning, purchasing, and inventory management. Retailers use analytics to recommend products to customers and optimize marketing promotions. Pharmaceutical firms use analytics to get life-saving drugs to market more quickly. The leisure and vacation industries use analytics to analyze historical sales data, understand customer behavior, improve Web site design, and optimize schedules and bookings. Airlines and hotels use analytics to dynamically set prices over time to maximize revenue. Even sports teams are using business analytics to determine both game strategy and optimal ticket prices.[4] For example, teams use analytics to decide on ticket pricing, who to recruit and trade, what combinations of players work best, and what plays to run under different situations.

Among the many organizations that use analytics to make strategic decisions and manage day-to-day operations are Caesars Entertainment, the Cleveland Indians baseball, Phoenix Suns basketball, and New England Patriots football teams, Amazon.com, Procter & Gamble, United Parcel Service (UPS), and Capital One bank. It was reported that nearly all firms with revenues of more than $100 million are using some form of business analytics.

Some common types of business decisions that can be enhanced by using analytics include

- pricing (for example, setting prices for consumer and industrial goods, government contracts, and maintenance contracts),
- customer segmentation (for example, identifying and targeting key customer groups in retail, insurance, and credit card industries),
- merchandising (for example, determining brands to buy, quantities, and allocations),
- location (for example, finding the best location for bank branches and ATMs, or where to service industrial equipment),
- supply chain design (for example, determining the best sourcing and transportation options and finding the best delivery routes),

---

[3]Liberatore and Luo, "The Analytics Movement".

[4]Jim Davis, "8 Essentials of Business Analytics," in "Brain Trust—Enabling the Confident Enterprise with Business Analytics" (Cary, NC: SAS Institute, Inc., 2010): 27–29. www.sas.com/bareport

- staffing (for example, ensuring the appropriate staffing levels and capabilities and hiring the right people—sometimes referred to as "people analytics"),
- health care (for example, scheduling operating rooms to improve utilization, improving patient flow and waiting times, purchasing supplies, predicting health risk factors),

and many others in operations management, finance, marketing, and human resources—in fact, in every discipline of business.[5]

Various research studies have discovered strong relationships between a company's performance in terms of profitability, revenue, and shareholder return and its use of analytics. Top-performing organizations (those that outperform their competitors) are three times more likely to be sophisticated in their use of analytics than lower performers and are more likely to state that their use of analytics differentiates them from competitors.[6] However, research has also suggested that organizations are overwhelmed by data and struggle to understand how to use data to achieve business results and that most organizations simply don't understand how to use analytics to improve their businesses. Thus, understanding the capabilities and techniques of analytics is vital to managing in today's business environment.

So, no matter what your job position in an organization is or will be, the study of analytics will be quite important to your future success. You may find many uses in your everyday work for the Excel-based tools that we will study. You may not be skilled in all the technical nuances of analytics and supporting software, but you will, at the very least, be a consumer of analytics and work with analytics professionals to support your analyses and decisions. For example, you might find yourself on project teams with managers who know very little about analytics and analytics experts such as statisticians, programmers, and economists. Your role might be to ensure that analytics is used properly to solve important business problems.

Impacts and Challenges

The benefits of applying business analytics can be significant. Companies report reduced costs, better risk management, faster decisions, better productivity, and enhanced bottom-line performance such as profitability and customer satisfaction. For example, 1-800-Flowers.com used analytic software to target print and online promotions with greater accuracy; change prices and offerings on its Web site (sometimes hourly); and optimize its marketing, shipping, distribution, and manufacturing operations, resulting in a $50 million cost savings in one year.[7]

Business analytics is changing how managers make decisions.[8] To thrive in today's business world, organizations must continually innovate to differentiate themselves from competitors, seek ways to grow revenue and market share, reduce costs, retain existing customers and acquire new ones, and become faster and leaner. IBM suggests that traditional management

[5]Thomas H. Davenport, "How Organizations Make Better Decisions," edited excerpt of an article distributed by the International Institute for Analytics published in "Brain Trust—Enabling the Confident Enterprise with Business Analytics" (Cary, NC: SAS Institute, Inc., 2010): 8–11. www.sas.com/bareport
[6]Thomas H. Davenport and Jeanne G. Harris, *Competing on Analytics* (Boston: Harvard Business School Press, 2007): 46; Michael S. Hopkins, Steve LaValle, Fred Balboni, Nina Kruschwitz, and Rebecca Shockley, "10 Data Points: Information and Analytics at Work," *MIT Sloan Management Review*, 52, 1 (Fall 2010): 27–31.
[7]Jim Goodnight, "The Impact of Business Analytics on Performance and Profitability," in "Brain Trust—Enabling the Confident Enterprise with Business Analytics" (Cary, NC: SAS Institute, Inc., 2010): 4–7. www.sas.com/bareport
[8]*Analytics: The New Path to Value*, a joint MIT Sloan Management Review and IBM Institute for Business Value study.

approaches are evolving in today's analytics-driven environment to include more fact-based decisions as opposed to judgment and intuition, more prediction rather than reactive decisions, and the use of analytics by everyone at the point where decisions are made rather than relying on skilled experts in a consulting group.[9] Nevertheless, organizations face many challenges in developing analytics capabilities, including lack of understanding of how to use analytics, competing business priorities, insufficient analytical skills, difficulty in getting good data and sharing information, and not understanding the benefits versus perceived costs of analytics studies. Successful application of analytics requires more than just knowing the tools; it requires a high-level understanding of how analytics supports an organization's competitive strategy and effective execution that crosses multiple disciplines and managerial levels.

In 2011, a survey by Bloomberg Businessweek Research Services and SAS concluded that business analytics was still in the "emerging stage" and was used only narrowly within business units, not across entire organizations. The study also noted that many organizations lacked analytical talent, and those that did have analytical talent often didn't know how to apply the results properly. While analytics was used as part of the decision-making process in many organizations, most business decisions are still based on intuition.[10] Today, business analytics has matured in many organizations, but many more opportunities still exist. These opportunities are reflected in the job market for analytics professionals, or "data scientists," as some call them. McKinsey & Company suggested that there is a shortage of qualified data scientists.[11]

CHECK YOUR UNDERSTANDING

1. Explain why analytics is important in today's business environment.

2. Define business analytics.

3. State three examples of how business analytics is used in organizations.

4. What are the key benefits of using business analytics?

5. What challenges do organizations face in using analytics?

Evolution of Business Analytics

Analytical methods, in one form or another, have been used in business for more than a century. The core of business analytics consists of three disciplines: business intelligence and information systems, statistics, and modeling and optimization.

Analytic Foundations

The modern evolution of analytics began with the introduction of computers in the late 1940s and their development through the 1960s and beyond. Early computers provided the ability to store and analyze data in ways that were either very difficult or impossible to do manually. This facilitated the collection, management, analysis, and reporting of data, which

---

[9]"Business Analytics and Optimization for the Intelligent Enterprise" (April 2009). www.ibm.com/qbs/intelligent-enterprise

[10]Bloomberg Businessweek Research Services and SAS, "The Current State of Business Analytics: Where Do We Go From Here?" (2011).

[11]Andrew Jennings, "What Makes a Good Data Scientist?" *Analytics Magazine* (July–August 2013): 8–13. www.analytics-magazine.org

is often called **business intelligence (BI)**, a term that was coined in 1958 by an IBM researcher, Hans Peter Luhn.[12] Business intelligence software can answer basic questions such as "How many units did we sell last month?" "What products did customers buy and how much did they spend?" "How many credit card transactions were completed yesterday?" Using BI, we can create simple rules to flag exceptions automatically; for example, a bank can easily identify transactions greater than $10,000 to report to the Internal Revenue Service.[13] BI has evolved into the modern discipline we now call **information systems (IS)**.

Statistics has a long and rich history, yet only rather recently has it been recognized as an important element of business, driven to a large extent by the massive growth of data in today's world. Google's chief economist noted that statisticians surely have one of the best jobs.[14] Statistical methods allow us to gain a richer understanding of data that goes beyond business intelligence reporting by not only summarizing data succinctly but also finding unknown and interesting relationships among the data. Statistical methods include the basic tools of description, exploration, estimation, and inference, as well as more advanced techniques like regression, forecasting, and data mining.

Much of modern business analytics stems from the analysis and solution of complex decision problems using mathematical or computer-based models—a discipline known as operations research, or management science. *Operations research (OR)* was born from efforts to improve military operations prior to and during World War II. After the war, scientists recognized that the mathematical tools and techniques developed for military applications could be applied successfully to problems in business and industry. A significant amount of research was carried on in public and private think tanks during the late 1940s and through the 1950s. As the focus on business applications expanded, the term *management science (MS)* became more prevalent. Many people use the terms *operations research* and *management science* interchangeably, so the field became known as **Operations Research/Management Science (OR/MS)**. Many OR/MS applications use **modeling and optimization**—techniques for translating real problems into mathematics, spreadsheets, or various computer languages, and using them to find the best ("optimal") solutions and decisions. INFORMS, the Institute for Operations Research and the Management Sciences, is the leading professional society devoted to OR/MS and analytics and publishes a bimonthly magazine called *Analytics* (http://analytics-magazine.org/). Digital subscriptions may be obtained free of charge at the Web site.

Modern Business Analytics

Modern business analytics can be viewed as an integration of BI/IS, statistics, and modeling and optimization, as illustrated in Figure 1.1. While these core topics are traditional and have been used for decades, the uniqueness lies in their intersections. For example, **data mining** is focused on better understanding characteristics and patterns among variables in large databases using a variety of statistical and analytical tools. Many standard statistical tools as well as more advanced ones are used extensively in data mining. **Simulation and risk analysis** relies on spreadsheet models and statistical analysis to examine the impacts of uncertainty in estimates and their potential interaction with one another on the output variable of interest.

---

[12] H. P. Luhn, "A Business Intelligence System." *IBM Journal* (October 1958).

[13] Jim Davis, "Business Analytics: Helping You Put an Informed Foot Forward," in "Brain Trust—Enabling the Confident Enterprise with Business Analytics," (Cary, NC: SAS Institute, Inc., 2010): 4–7. www.sas.com/bareport

[14] James J. Swain, "Statistical Software in the Age of the Geek," *Analytics Magazine* (March -April 2013): 48–55.

▶ **Figure 1.1**

A Visual Perspective of Business Analytics

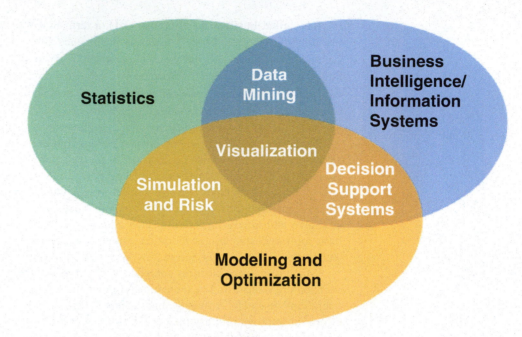

Decision support systems (DSSs) began to evolve in the 1960s by combining business intelligence concepts with OR/MS models to create analytical-based computer systems to support decision making. DSSs include three components:

1. *Data management.* The data management component includes databases for storing data and allows the user to input, retrieve, update, and manipulate data.
2. *Model management.* The model management component consists of various statistical tools and management science models and allows the user to easily build, manipulate, analyze, and solve models.
3. *Communication system.* The communication system component provides the interface necessary for the user to interact with the data and model management components.[15]

DSSs have been used for many applications, including pension fund management, portfolio management, work-shift scheduling, global manufacturing and facility location, advertising-budget allocation, media planning, distribution planning, airline operations planning, inventory control, library management, classroom assignment, nurse scheduling, blood distribution, water pollution control, ski-area design, police-beat design, and energy planning.[16]

A key feature of a DSS is the ability to perform **what-if analysis**—how specific combinations of inputs that reflect key assumptions will affect model outputs. What-if analysis is also used to assess the sensitivity of optimization models to changes in data inputs and provide better insight for making good decisions.

Perhaps the most useful component of business analytics, which makes it truly unique, is the center of Figure 1.1—**visualization**. Visualizing data and results of analyses provides a way of easily communicating data at all levels of a business and can reveal surprising patterns and relationships. Software such as IBM's Cognos system exploits data visualization

---

[15]William E. Leigh and Michael E. Doherty, *Decision Support and Expert Systems* (Cincinnati, OH: South-Western Publishing Co., 1986).
[16]H. B. Eom and S. M. Lee, "A Survey of Decision Support System Applications (1971–April 1988)," *Interfaces*, 20, 3 (May–June 1990): 65–79.

for query and reporting, data analysis, dashboard presentations, and scorecards linking strategy to operations. The Cincinnati Zoo, for example, has used this on an iPad to display hourly, daily, and monthly reports of attendance, food and retail location revenues and sales, and other metrics for prediction and marketing strategies. UPS uses telematics to capture vehicle data and display them to help make decisions to improve efficiency and performance. You may have seen a **tag cloud** (see the graphic at the beginning of this chapter), which is a visualization of text that shows words that appear more frequently with larger fonts.

Software Support and Spreadsheet Technology

Many companies, such as IBM, SAS, and Tableau Software, have developed a variety of software and hardware solutions to support business analytics. For example, IBM's Cognos Express, an integrated business intelligence and planning solution designed to meet the needs of midsize companies, provides reporting, analysis, dashboard, scorecard, planning, budgeting, and forecasting capabilities. It is made up of several modules, including Cognos Express Reporter, for self-service reporting and ad hoc query; Cognos Express Advisor, for analysis and visualization; and Cognos Express Xcelerator, for Excel-based planning and business analysis. Information is presented to users in a context that makes it easy to understand; with an easy-to-use interface, users can quickly gain the insight they need from their data to make the right decisions and then take action for effective and efficient business optimization and outcome. SAS provides a variety of software that integrate data management, business intelligence, and analytics tools. SAS Analytics covers a wide range of capabilities, including predictive modeling and data mining, visualization, forecasting, optimization and model management, statistical analysis, text analytics, and more. Tableau Software provides simple drag and drop tools for visualizing data from spreadsheets and other databases. We encourage you to explore many of these products as you learn the basic principles of business analytics in this book.

Although commercial software often have powerful features and capabilities, they can be expensive, generally require advanced training to understand and apply, and may work only on specific computer platforms. Spreadsheet software, on the other hand, is widely used across all areas of business and used by nearly everyone. Spreadsheets are an effective platform for manipulating data and developing and solving models; they support powerful commercial add-ins and facilitate communication of results. Spreadsheets provide a flexible modeling environment and are particularly useful when the end user is not the designer of the model. Teams can easily use spreadsheets and understand the logic upon which they are built. Information in spreadsheets can easily be copied from spreadsheets into other documents and presentations. A recent survey identified more than 180 commercial spreadsheet products that support analytics efforts, including data management and reporting, data- and model-driven analytical techniques, and implementation.[17] Many organizations have used spreadsheets extremely effectively to support decision making in marketing, finance, and operations. Some illustrative applications include the following:[18]

- Analyzing supply chains (Hewlett-Packard)
- Determining optimal inventory levels to meet customer service objectives (Procter & Gamble)

---

[17]Thomas A. Grossman, "Resources for Spreadsheet Analysts," *Analytics Magazine* (May/June 2010): 8. www.analytics-magazine.org
[18]Larry J. LeBlanc and Thomas A. Grossman, "Introduction: The Use of Spreadsheet Software in the Application of Management Science and Operations Research," *Interfaces*, 38, 4 (July–August 2008): 225–227.

- Selecting internal projects (Lockheed Martin Space Systems)
- Planning for emergency clinics in response to a sudden epidemic or bioterrorism attack (Centers for Disease Control)
- Analyzing the default risk of a portfolio of real estate loans (Hypo International)
- Assigning medical residents to on-call and emergency rotations (University of Vermont College of Medicine)
- Performance measurement and evaluation (American Red Cross)

Some optional software packages for statistical applications that your instructor might use are SAS, Minitab, *XLSTAT* and *StatCrunch*. These provide many powerful procedures as alternatives or supplements to Excel.

Spreadsheet technology has been influential in promoting the use and acceptance of business analytics. Spreadsheets provide a convenient way to manage data, calculations, and visual graphics simultaneously, using intuitive representations instead of abstract mathematical notation. Although the early applications of spreadsheets were primarily in accounting and finance, spreadsheets have developed into powerful general-purpose managerial tools for applying techniques of business analytics. The power of analytics in a personal computing environment was noted decades ago by business consultants Michael Hammer and James Champy, who said, "When accessible data is combined with easy-to-use analysis and modeling tools, frontline workers—when properly trained—suddenly have sophisticated decision-making capabilities."[19]

ANALYTICS IN PRACTICE: Social Media Analytics

One of the emerging applications of analytics is helping businesses learn from social media and exploit social media data for strategic advantage.[20] Using analytics, firms can integrate social media data with traditional data sources such as customer surveys, focus groups, and sales data; understand trends and customer perceptions of their products; and create informative reports to assist marketing managers and product designers.

Social media analytics is useful in decision making in many business domains to understand how user-generated content spreads and influences user interactions, how information is transmitted, and how it influences decisions. A review of research published in social media analytics provides numerous examples:[21]

- The analysis of public responses from social media before, during, and after disasters, such as the 2010 Haiti earthquake and Hurricane Sandy in New York City

in 2012, has the potential to improve situational knowledge in emergency and disaster management practices.
- Social media platforms enable citizens' engagement with politicians, governments, and other citizens. Studies have examined how voters discuss the candidates during an election, how candidates are adopting Twitter for campaigning and influencing conversations in the public space, and how presidential candidates in the United States used Twitter to engage people and identify the topics mentioned by candidates during their campaigns. Others have used analytics to track political preference by monitoring online popularity.
- In the entertainment industry, one study analyzed viewer ratings to predict the impact on revenue for upcoming movies. Another developed a web intelligence application to aggregate the news about popular TV serials and identify emerging storylines.

---

[19]Michael Hammer and James Champy, *Reengineering the Corporation* (New York: HarperBusiness, 1993): 96.

[20]Jim Davis, "Convergence—Taking Social Media from Talk to Action," SASCOM (First Quarter 2011): 17.

[21]Ashish K. Rathore, Arpan K. Kar, and P. Vigneswara Ilavarasana, "Social Media Analytics: Literature Review and Directions for Future Research," *Decision Analysis*, 14, 4 (December 2017): 229–249.

■ Retail organizations monitor and analyze social media data about their own products and services and also about their competitors' products and services to stay competitive. For instance, one study analyzed different product features based on rankings from users' online reviews.

■ The integration of social media application and health care leads to better patient management and empowerment. One researcher classified various online health communities, such as a diabetes patients' community, using posts from WebMD.com. Another analyzed physical activity–related tweets for a better understanding of physical activity behaviors. To predict the spread of influenza, one researcher developed a forecasting approach using flu-related tweets.

In this book, we use Microsoft Excel as the primary platform for implementing analytics. In the Chapter 1 Appendix, we review some key Excel skills that you should have before moving forward in this book.

The main chapters in this book are designed using Excel 2016 for Windows or Excel 2016 for Mac. Earlier versions of Excel do not have all the capabilities that we use in this book. In addition, some key differences exist between Windows and Mac versions that we will occasionally point out. Thus, some Excel tools that we will describe in chapter appendixes require you to use Excel for Windows, Office 365, or Google Sheets, and will not run on Excel for Mac; these are optional to learn, and are not required for any examples or problems. Your instructor may use optional software, such as XLSTAT and StatCrunch, which are provided by the publisher (Pearson), or Analytic Solver, which is described in online supplements to this book.

CHECK YOUR UNDERSTANDING

1. Provide two examples of questions that business intelligence can address.

2. How do statistical methods enhance business intelligence reporting?

3. What is operations research/management science?

4. How does modern business analytics integrate traditional disciplines of BI, statistics, and modeling/optimization?

5. What are the components of a decision support system?

Descriptive, Predictive, and Prescriptive Analytics

Business analytics begins with the collection, organization, and manipulation of data and is supported by three major components:[22]

1. *Descriptive analytics.* Most businesses start with **descriptive analytics**—the use of data to understand past and current business performance and make informed decisions. Descriptive analytics is the most commonly used and most well-understood type of analytics. These techniques categorize, characterize, consolidate, and classify data to convert them into useful information for the purposes of understanding and analyzing business performance. Descriptive

---

[22]Parts of this section are adapted from Irv Lustig, Brenda Dietric, Christer Johnson, and Christopher Dziekan, "The Analytics Journey," *Analytics* (November/December 2010). http://analytics-magazine.org/novemberdecember-2010-table-of-contents/

analytics summarizes data into meaningful charts and reports, for example, about budgets, sales, revenues, or cost. This process allows managers to obtain standard and customized reports and then drill down into the data and make queries to understand the impact of an advertising campaign, such as reviewing business performance to find problems or areas of opportunity, and identifying patterns and trends in data. Typical questions that descriptive analytics helps answer are "How much did we sell in each region?" "What was our revenue and profit last quarter?" "How many and what types of complaints did we resolve?" "Which factory has the lowest productivity?" Descriptive analytics also helps companies to classify customers into different segments, which enables them to develop specific marketing campaigns and advertising strategies.

2. *Predictive analytics*. **Predictive analytics** seeks to predict the future by examining historical data, detecting patterns or relationships in these data, and then extrapolating these relationships forward in time. For example, a marketer might wish to predict the response of different customer segments to an advertising campaign, a commodities trader might wish to predict short-term movements in commodities prices, or a skiwear manufacturer might want to predict next season's demand for skiwear of a specific color and size. Predictive analytics can predict risk and find relationships in data not readily apparent with traditional analyses. Using advanced techniques, predictive analytics can help detect hidden patterns in large quantities of data, and segment and group data into coherent sets to predict behavior and detect trends. For instance, a bank manager might want to identify the most profitable customers, predict the chances that a loan applicant will default, or alert a credit card customer to a potential fraudulent charge. Predictive analytics helps to answer questions such as "What will happen if demand falls by 10% or if supplier prices go up 5%?" "What do we expect to pay for fuel over the next several months?" "What is the risk of losing money in a new business venture?"

3. *Prescriptive analytics*. Many problems, such as aircraft or employee scheduling and supply chain design, simply involve too many choices or alternatives for a human decision maker to effectively consider. **Prescriptive analytics** uses optimization to identify the best alternatives to minimize or maximize some objective. Prescriptive analytics is used in many areas of business, including operations, marketing, and finance. For example, we may determine the best pricing and advertising strategy to maximize revenue, the optimal amount of cash to store in ATMs, or the best mix of investments in a retirement portfolio to manage risk. Prescriptive analytics addresses questions such as "How much should we produce to maximize profit?" "What is the best way of shipping goods from our factories to minimize costs?" "Should we change our plans if a natural disaster closes a supplier's factory, and if so, by how much?" The mathematical and statistical techniques of predictive analytics can also be combined with prescriptive analytics to make decisions that take into account the uncertainty in the data.

A wide variety of tools are used to support business analytics. These include

- Database queries and analysis
- "Dashboards" to report key performance measures
- Data visualization
- Statistical methods
- Spreadsheets and predictive models

ANALYTICS IN PRACTICE: Analytics in the Home Lending and Mortgage Industry[23]

Sometime during their lives, most Americans will receive a mortgage loan for a house or condominium. The process starts with an application. The application contains all pertinent information about the borrower that the lender will need. The bank or mortgage company then initiates a process that leads to a loan decision. It is here that key information about the borrower is provided by third-party providers. This information includes a credit report, verification of income, verification of assets, verification of employment, and an appraisal of the property. The result of the processing function is a complete loan file that contains all the information and documents needed to underwrite the loan, which is the next step in the process. Underwriting is where the loan application is evaluated for its risk. Underwriters evaluate whether the borrower can make payments on time, can afford to pay back the loan, and has sufficient collateral in the property to back up the loan. In the event the borrower defaults on their loan, the lender can sell the property to recover the amount of the loan. But if the amount of the loan is greater than the value of the property, then the lender cannot recoup their money. If the underwriting process indicates that the borrower is creditworthy and has the capacity to repay the loan and the value of the property in question is greater than the loan amount, then the loan is approved and will move to closing. Closing is the step where the borrower signs all the appropriate papers, agreeing to the terms of the loan.

In reality, lenders have a lot of other work to do. First, they must perform a quality control review on a sample of the loan files that involves a manual examination of all the documents and information gathered. This process is designed to identify any mistakes that may have been made or information that is missing from the loan file. Because lenders do not have unlimited money to lend to borrowers, they frequently sell the loan to a third party so that they have fresh capital to lend to others. This occurs in what is called the secondary market. Freddie Mac and Fannie Mae are the two largest purchasers of mortgages in the secondary market. The final step in the process is servicing. Servicing includes all the activities associated with providing the customer service on the loan, like processing payments, managing property taxes held in escrow, and answering questions about the loan.

In addition, the institution collects various operational data on the process to track its performance and efficiency, including the number of applications, loan types and amounts, cycle times (time to close the loan), bottlenecks in the process, and so on. Many different types of analytics are used:

Descriptive analytics—This focuses on historical reporting, addressing such questions as

- How many loan applications were taken in each of the past 12 months?
- What was the total cycle time from application to close?
- What was the distribution of loan profitability by credit score and loan-to-value (LTV), which is the mortgage amount divided by the appraised value of the property?

Predictive analytics—Predictive modeling uses mathematical, spreadsheet, and statistical models and addresses questions such as

- What impact on loan volume will a given marketing program have?
- How many processors or underwriters are needed for a given loan volume?
- Will a given process change reduce cycle time?

Prescriptive analytics—This involves the use of simulation or optimization to drive decisions. Typical questions include

- What is the optimal staffing to achieve a given profitability constrained by a fixed cycle time?
- What is the optimal product mix to maximize profit constrained by fixed staffing?

The mortgage market has become much more dynamic in recent years due to rising home values, falling interest rates, new loan products, and an increased desire by home owners to utilize the equity in their homes as a financial resource. This has increased the complexity and variability of the mortgage process and created an opportunity for lenders to proactively use the data that are available to them as a tool for managing their business. To ensure that the process is efficient, effective, and performed with quality, data and analytics are used every day to track what is done, who is doing it, and how long it takes.

[23]Contributed by Craig Zielazny, BlueNote Analytics, LLC.

- Scenario and "what-if" analyses
- Simulation
- Forecasting
- Data and text mining
- Optimization
- Social media, Web, and text analytics

Although the tools used in descriptive, predictive, and prescriptive analytics are different, many applications involve all three. Here is a typical example in retail operations.

EXAMPLE 1.1   **Retail Markdown Decisions**[24]

As you probably know from your shopping experiences, most department stores and fashion retailers clear their seasonal inventory by reducing prices. The key question they face is what prices should they set—and when should they set them—to meet inventory goals and maximize revenue? For example, suppose that a store has 100 bathing suits of a certain style that go on sale on April 1 and wants to sell all of them by the end of June. Over each week of the 12-week selling season, they can make a decision to discount the price. They face two decisions: When to reduce the price, and by how much. This results in 24 decisions to make. For a major national chain that may carry thousands of products, this can easily result in millions of decisions that store managers have to make. Descriptive analytics can be used to examine historical data for similar products, such as the number of units sold, price at each point of sale, starting and ending inventories, and special promotions, newspaper ads, direct marketing ads, and so on, to understand what the results of past decisions achieved. Predictive analytics can be used to predict sales based on pricing decisions. Finally, prescriptive analytics can be applied to find the best set of pricing decisions to maximize the total revenue.

 CHECK YOUR UNDERSTANDING

1. Define descriptive analytics and provide two examples.
2. Define predictive analytics and provide two examples.
3. Define prescriptive analytics and provide two examples.

Data for Business Analytics

Since the dawn of the electronic age and the Internet, both individuals and organizations have had access to an enormous wealth of data and information. Most data are collected through some type of measurement process, and consist of numbers (e.g., sales revenues) or textual data (e.g., customer demographics such as gender). Other data might be extracted from social media, online reviews, and even audio and video files. *Information* comes from analyzing data—that is, extracting meaning from data to support evaluation and decision making.

Data are used in virtually every major function in a business. Modern organizations—which include not only for-profit businesses but also nonprofit organizations—need good data to support a variety of company purposes, such as planning, reviewing company performance, improving operations, and comparing company performance with competitors'

---

[24]Inspired by a presentation by Radhika Kulkarni, SAS Institute, "Data-Driven Decisions: Role of Operations Research in Business Analytics," INFORMS Conference on Business Analytics and Operations Research, April 10–12, 2011.

or best-practice benchmarks. Some examples of how data are used in business include the following:

- Annual reports summarize data about companies' profitability and market share both in numerical form and in charts and graphs to communicate with shareholders.
- Accountants conduct audits to determine whether figures reported on a firm's balance sheet fairly represent the actual data by examining samples (that is, subsets) of accounting data, such as accounts receivable.
- Financial analysts collect and analyze a variety of data to understand the contribution that a business provides to its shareholders. These typically include profitability, revenue growth, return on investment, asset utilization, operating margins, earnings per share, economic value added (EVA), shareholder value, and other relevant measures.
- Economists use data to help companies understand and predict population trends, interest rates, industry performance, consumer spending, and international trade. Such data are often obtained from external sources such as Standard & Poor's Compustat data sets, industry trade associations, or government databases.
- Marketing researchers collect and analyze extensive customer data. These data often consist of demographics, preferences and opinions, transaction and payment history, shopping behavior, and much more. Such data may be collected by surveys, personal interviews, or focus groups, or from shopper loyalty cards.
- Operations managers use data on production performance, manufacturing quality, delivery times, order accuracy, supplier performance, productivity, costs, and environmental compliance to manage their operations.
- Human resource managers measure employee satisfaction, training costs, turnover, market innovation, training effectiveness, and skills development.

Data may be gathered from primary sources such as internal company records and business transactions, automated data-capturing equipment, and customer market surveys and from secondary sources such as government and commercial data sources, custom research providers, and online research.

Perhaps the most important source of data today is data obtained from the Web. With today's technology, marketers collect extensive information about Web behaviors, such as the number of page views, visitor's country, time of view, length of time, origin and destination paths, products they searched for and viewed, products purchased, and what reviews they read. Using analytics, marketers can learn what content is being viewed most often, what ads were clicked on, who the most frequent visitors are, and what types of visitors browse but don't buy. Not only can marketers understand what customers have done, but they can better predict what they intend to do in the future. For example, if a bank knows that a customer has browsed for mortgage rates and homeowner's insurance, they can target the customer with homeowner loans rather than credit cards or automobile loans. Traditional Web data are now being enhanced with social media data from Facebook, cell phones, and even Internet-connected gaming devices.

As one example, a home furnishings retailer wanted to increase the rate of sales for customers who browsed their Web site. They developed a large data set that covered more than 7,000 demographic, Web, catalog, and retail behavioral attributes for each customer. They used predictive analytics to determine how well a customer would respond to different e-mail marketing offers and customized promotions to individual customers. This not only helped them to determine where to most effectively spend marketing resources but also doubled the response rate compared to previous marketing campaigns, with a projected and multimillion dollar increase in sales.[25]

Big Data

Today, nearly all data are captured digitally. As a result, data have been growing at an overwhelming rate, being measured by terabytes (10^{12} bytes), petabytes (10^{15} bytes), exabytes (10^{18} bytes), and even by higher-dimensional terms. Just think of the amount of data stored on Facebook, Twitter, or Amazon servers, or the amount of data acquired daily from scanning items at a national grocery chain such as Kroger and its affiliates. Walmart, for instance, has over one million transactions each hour, yielding more than 2.5 petabytes of data. Analytics professionals have coined the term **big data** to refer to massive amounts of business data from a wide variety of sources, much of which is available in real time. IBM calls these characteristics *volume, variety,* and *velocity.* Most often, big data revolve around customer behavior and customer experiences. Big data provide an opportunity for organizations to gain a competitive advantage—if the data can be understood and analyzed effectively to make better business decisions.

The volume of data continues to increase; what is considered "big" today will be even bigger tomorrow. In one study of information technology (IT) professionals in 2010, nearly half of survey respondents ranked data growth among their top three challenges. Big data are captured using sensors (for example, supermarket scanners), click streams from the Web, customer transactions, e-mails, tweets and social media, and other ways. Big data sets are unstructured and messy, requiring sophisticated analytics to integrate and process the data and understand the information contained in them. Because much big data are being captured in real time, they must be incorporated into business decisions at a faster rate. Processes such as fraud detection must be analyzed quickly to have value. In addition to *volume, variety,* and *velocity,* IBM proposed a fourth dimension: *veracity*—the level of reliability associated with data. Having high-quality data and understanding the uncertainty in data are essential for good decision making. Data veracity is an important role for statistical methods.

Big data can help organizations better understand and predict customer behavior and improve customer service. A study by the McKinsey Global Institute noted that, "The effective use of big data has the potential to transform economies, delivering a new wave of productivity growth and consumer surplus. Using big data will become a key basis of competition for existing companies, and will create new competitors who are able to attract employees that have the critical skills for a big data world."[26] However, understanding big data requires advanced analytics tools such as data mining and text analytics, and new technologies such as cloud computing, faster multi-core processors, large memory spaces, and solid-state drives.

Data Reliability and Validity

Poor data can result in poor decisions. In one situation, a distribution system design model relied on data obtained from the corporate finance department. Transportation costs were

---

[25]Based on a presentation by Bill Franks of Teradata, "Optimizing Customer Analytics: How Customer Level Web Data Can Help," INFORMS Conference on Business Analytics and Operations Research, April 10–12, 2011.

[26]James Manyika, Michael Chui, Brad Brown, Jacques Bughin, Richard Dobbs, Charles Roxburgh, and Angela Hung Byers, "Big Data: The Next Frontier for Innovation, Competition, and Productivity," McKinsey & Company May 2011.

determined using a formula based on the latitude and longitude of the locations of plants and customers. But when the solution was represented on a geographic information system (GIS) mapping program, one of the customers was located in the Atlantic Ocean.

Thus, data used in business decisions need to be reliable and valid. **Reliability** means that data are accurate and consistent. **Validity** means that data correctly measure what they are supposed to measure. For example, a tire pressure gauge that consistently reads several pounds of pressure below the true value is not reliable, although it is valid because it does measure tire pressure. The number of calls to a customer service desk might be counted correctly each day (and thus is a reliable measure), but it is not valid if it is used to assess customer dissatisfaction, as many calls may be simple queries. Finally, a survey question that asks a customer to rate the quality of the food in a restaurant may be neither reliable (because different customers may have conflicting perceptions) nor valid (if the intent is to measure customer satisfaction, as satisfaction generally includes other elements of service besides food).

CHECK YOUR UNDERSTANDING

1. State three examples of how data are used in different business functions.

2. How are data obtained from the Web used in marketing and business?

3. Define big data and list the four characteristics of big data.

4. Explain the concepts of data reliability and validity.

Models in Business Analytics

To make an informed decision, we must be able to specify the decision alternatives that represent the choices that can be made and criteria for evaluating the alternatives. Specifying decision alternatives might be very simple; for example, you might need to choose one of three corporate health plan options. Other situations can be more complex; for example, in locating a new distribution center, it might not be possible to list just a small number of alternatives. The set of potential locations might be anywhere in the United States or even across the globe. Decision criteria might be to maximize discounted net profits, customer satisfaction, or social benefits or to minimize costs, environmental impact, or some measure of loss.

Many decision problems can be formalized using a model. A **model** is an abstraction or representation of a real system, idea, or object. Models capture the most important features of a problem and present them in a form that is easy to interpret. A model can be as simple as a written or verbal description of some phenomenon, a visual representation such as a graph or a flowchart, or a mathematical or spreadsheet representation. Example 1.2 illustrates three ways to express a model.

A **decision model** is a logical or mathematical representation of a problem or business situation that can be used to understand, analyze, or facilitate making a decision. Decision models can be represented in various ways, most typically with mathematical functions and spreadsheets. Spreadsheets are ideal vehicles for implementing decision models because of their versatility in managing data, evaluating different scenarios, and presenting results in a meaningful fashion. We will focus on spreadsheet models beginning with Chapter 11.

EXAMPLE 1.2   Three Forms of a Model

Models are usually developed from theory or observation and establish relationships between actions that decision makers might take and results that they might expect, thereby allowing the decision makers to evaluate scenarios or to predict what might happen. For example, the sales of a new product, such as a first-generation iPad, Android phone, or 3-D television, often follow a common pattern. We might represent this in one of the three following ways:

1. A simple verbal description of sales might be: *The rate of sales starts small as early adopters begin to evaluate a new product and then begins to grow at an increasing rate over time as positive customer feedback spreads.*

Eventually, the market begins to become saturated, and the rate of sales begins to decrease.

2. A sketch of sales as an S-shaped curve over time, as shown in Figure 1.2, is a visual model that conveys this phenomenon.

3. Finally, analysts might identify a mathematical model that characterizes this curve. Several different mathematical functions do this; one is called a Gompertz curve and has the formula: $S = ae^{be^{ct}}$, where S = sales, t = time, e is the base of natural logarithms, and a, b, and c are constants. Of course, you would not be expected to know this; that's what analytics professionals do.

Decision models typically have three types of input:

1. *Data*, which are assumed to be constant for purposes of the model. Some examples are costs, machine capacities, and intercity distances.
2. *Uncontrollable inputs*, which are quantities that can change but cannot be directly controlled by the decision maker. Some examples are customer demand, inflation rates, and investment returns. Often, these variables are uncertain.
3. *Decision options*, which are controllable and can be selected at the discretion of the decision maker. Some examples are production quantities, staffing levels, and investment allocations. Decision options are often called **decision variables**.

Decision models characterize the relationships among these inputs and the outputs of interest to the decision maker (see Figure 1.3). In this way, the user can manipulate the decision options and understand how they influence outputs, make predictions for the future, or use analytical tools to find the best decisions. Thus, decision models can be descriptive, predictive, or prescriptive and therefore are used in a wide variety of business analytics applications.

▶ **Figure 1.2**

New Product Sales over Time

▶ **Figure 1.3**

The Nature of Decision Models

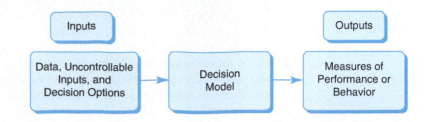

Decision models complement decision makers' intuition and often provide insights that intuition cannot. For example, one early application of analytics in marketing involved a study of sales operations. Sales representatives had to divide their time between large and small customers and between acquiring new customers and keeping old ones. The problem was to determine how the representatives should best allocate their time. Intuition suggested that they should concentrate on large customers and that it was much harder to acquire a new customer than to keep an old one. However, intuition could not tell whether they should concentrate on the 100 largest or the 1,000 largest customers, or how much effort to spend on acquiring new customers. Models of sales force effectiveness and customer response patterns provided the insight to make these decisions. However, it is important to understand that all models are only representations of the real world and, as such, cannot capture every nuance that decision makers face in reality. Decision makers must often modify the policies that models suggest to account for intangible factors that they might not have been able to incorporate into the model.

Descriptive Models

Descriptive models explain behavior and allow users to evaluate potential decisions by asking "what-if?" questions. The following example illustrates a simple descriptive mathematical model.

EXAMPLE 1.3   **Gasoline Usage Model**

Automobiles have different fuel economies (miles per gallon), and commuters drive different distances to work or school. Suppose that a state Department of Transportation (DOT) is interested in measuring the average monthly fuel consumption of commuters in a certain city. The DOT might sample a group of commuters and collect information on the number of miles driven per day, the number of driving days per month, the fuel economy of their vehicles, and additional miles driven per month for leisure and household activities. We may develop a simple descriptive model for calculating the amount of gasoline consumed, using the following symbols for the data:

G = gallons of fuel consumed per month
m = miles driven per day to and from work or school
d = number of driving days per month
f = fuel economy in miles per gallon (mpg)
a = additional miles for leisure and household activities per month

When developing mathematical models, it is very important to use the dimensions of the variables to ensure logical consistency. In this example, we see that

$$(m \text{ miles/day}) \times (d \text{ days/month}) = m \times d \text{ miles/month}$$

Thus, the total number of miles driven per month = $m \times d + a$. If the vehicle gets f miles/gallon, then the total number of gallons consumed per month is

$$G = (m \times d + a \text{ miles/month})/(f \text{ miles/gallon})$$
$$= (m \times d + a)/f \text{ gallons/month} \qquad \textbf{(1.1)}$$

Suppose that a commuter drives 30 miles round trip to work for 20 days each month, achieves a fuel economy of 34 mpg, and drives an additional 250 miles each month. Using formula (1.1), the number of gallons consumed is

$$G = (30 \times 20 + 250)/34 = 25.0 \text{ gallons/month}$$

In the previous example, we have no decision options; the model is purely descriptive, but allows us to evaluate "what-if?" questions, for example, "What if we purchase a hybrid vehicle with a fuel economy of 45 miles/gallon?" "What if leisure and household activity driving increases to 400 miles/month?" Most of the models we will be using include decision options. As an example, suppose that a manufacturer has the option of producing a part in house or outsourcing it from a supplier (the decision options). Should the firm produce the part or outsource it? The decision depends on the costs of manufacturing and outsourcing, as well as the anticipated volume of demand (the uncontrollable inputs). By developing a model to evaluate the total cost of both alternatives (the outputs), the best decision can be made.

EXAMPLE 1.4   An Outsourcing Decision Model

Suppose that a manufacturer can produce a part for $125/unit with a fixed cost of $50,000. The alternative is to outsource production to a supplier at a unit cost of $175. The total manufacturing and outsourcing costs can be expressed by simple mathematical formulas, where Q is the production volume:

$$TC \text{ (manufacturing)} = \$50,000 + \$125 \times Q \quad \textbf{(1.2)}$$
$$TC \text{ (outsourcing)} = \$175 \times Q \quad \textbf{(1.3)}$$

These formulas comprise the decision model, which simply describes what the costs of manufacturing and outsourcing are for any level of production volume. Thus, if the anticipated production volume is 1,500 units, the cost of manufacturing will be $50,000 + $125 ×1,500 = $237,500, and the cost of outsourcing would be $175 ×1,500 = $262,500;

therefore, manufacturing would be the best decision. On the other hand, if the anticipated production volume is only 800 units, the cost of manufacturing will be $50,000 + $125 ×800 = $150,000 and the cost of outsourcing would be $175 ×800 = $140,000, and the best decision would be to outsource. If we graph the two total cost formulas, we can easily see how the costs compare for different values of Q. This is shown graphically in Figure 1.4. The point at which the total costs of manufacturing and outsourcing are equal is called the break-even volume. This can easily be found by setting TC (manufacturing) = TC (outsourcing) and solving for Q:

$$\$50,000 + \$125 \times Q = \$175 \times Q$$
$$\$50,000 = 50 \times Q$$
$$Q = 1,000$$

▶ **Figure 1.4**

Graphical Illustration of Break-Even Analysis

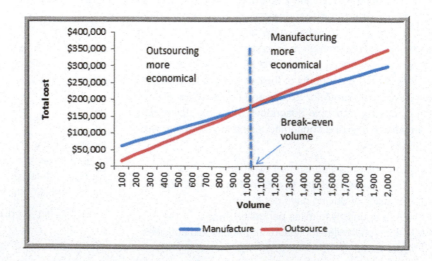

Predictive Models

Predictive models focus on what will happen in the future. Many predictive models are developed by analyzing historical data and assuming that the past is representative of the future. Example 1.5 shows how historical data might be used to develop a model that can be used to predict the impact of pricing and promotional strategies in the grocery industry.[27]

EXAMPLE 1.5 A Predictive Sales-Promotion Model

In the grocery industry, managers typically need to know how best to use pricing, coupons, and advertising strategies to influence sales. Grocers often study the relationship of sales volume to these strategies by conducting controlled experiments. That is, they implement different combinations of pricing, coupons, and advertising, observe the sales that result, and use analytics to develop predictive models of sales as a function of these decision strategies.

For example, suppose that a grocer who operates three stores in a small city varied the price, coupons (yes = 1, no = 0), and advertising expenditures in a local newspaper over a 16-week period and observed the following sales:

Week	Price ($)	Coupon (0,1)	Advertising ($)	Store1 Sales (Units)	Store 2 Sales (Units)	Store 3 Sales (Units)
1	6.99	0	0	501	510	481
2	6.99	0	150	772	748	775
3	6.99	1	0	554	528	506
4	6.99	1	150	838	785	834
5	6.49	0	0	521	519	500
6	6.49	0	150	723	790	723
7	6.49	1	0	510	556	520
8	6.49	1	150	818	773	800
9	7.59	0	0	479	491	486
10	7.59	0	150	825	822	757
11	7.59	1	0	533	513	540
12	7.59	1	150	839	791	832
13	5.49	0	0	484	480	508
14	5.49	0	150	686	683	708
15	5.49	1	0	543	531	530
16	5.49	1	150	767	743	779

To better understand the relationships among price, coupons, and advertising, an analyst might have developed the following model using business analytics tools (we will see how to do this in Chapter 8):

$$\text{Total Sales} = 1105.55 + 56.18 \times \text{Price} + 123.88 \times \text{Coupon} + 5.24 \times \text{Advertising} \quad \textbf{(1.4)}$$

In this example, the uncontrollable inputs are the sales at each store. The decision options are price, coupons, and advertising. The numerical values in the model are estimated from the data obtained from the experiment. They reflect the impact on sales of changing the decision options. For example, an increase in price of $1 results in a 56.18-unit

(continued)

---

[27]Roger J. Calantone, Cornelia Droge, David S. Litvack, and C. Anthony di Benedetto. "Flanking in a Price War," *Interfaces*, 19, 2 (1989): 1–12.

increase in weekly sales; using coupons (that is, setting Coupon = 1 in the model) results in a 123.88-unit increase in weekly sales. The output of the model is the predicted total sales units of the product. For example, if the price is $6.99, no coupons are offered, and no advertising is done (the experiment corresponding to week 1), the model estimates sales as

$$\text{Total Sales} = 1{,}105.55 + 56.18 \times 6.99 + 123.88 \times 0 \\ + 5.24 \times 0 = 1{,}498.25 \text{ units}$$

We see that the actual total sales in the three stores for week 1 was 1,492. Thus, this model appears to provide good estimates for sales using the historical data. We would hope that this model would also provide good predictions of future sales. So if the grocer decides to set the price at $5.99, does not use coupons, and spends $100 in advertising, the model would predict sales to be

$$\text{Total Sales} = 1{,}105.55 + 56.18 \times 5.99 + 123.88 \times 0 \\ + 5.24 \times 100 = 1{,}966.07 \text{ units}$$

Prescriptive Models

A prescriptive decision model helps decision makers to identify the best solution to a decision problem. **Optimization** is the process of finding a set of values for decision options that minimize or maximize some quantity of interest—profit, revenue, cost, time, and so on—called the **objective function**. Any set of decision options that optimizes the objective function is called an **optimal solution**. In a highly competitive world, where one percentage point can mean a difference of hundreds of thousands of dollars or more, knowing the best solution can mean the difference between success and failure.

EXAMPLE 1.6 A Prescriptive Model for Pricing

To illustrate an example of a prescriptive model, suppose that a firm wishes to determine the best pricing for one of its products to maximize revenue over the next year. A market research study has collected data that estimate the expected annual sales for different levels of pricing. Analysts determined that sales can be expressed by the following model:

$$\text{Sales} = -2.9485 \times \text{Price} + 3{,}240.9 \qquad (1.5)$$

Because revenue equals price × sales, a model for total revenue is

$$\text{Total Revenue} = \text{Price} \times \text{Sales} \\ = \text{Price} \times (-2.9485 \times \text{Price} + 3{,}240.9) \\ = -2.9485 \times \text{Price}^2 + 3{,}240.9 \times \text{Price} \qquad (1.6)$$

The firm would like to identify the price that maximizes the total revenue. One way to do this would be to try different prices and search for the one that yields the highest total revenue. This would be quite tedious to do by hand or even with a calculator; however, as we will see in later chapters, spreadsheet models make this much easier.

Although the pricing model did not, most optimization models have **constraints**—limitations, requirements, or other restrictions that are imposed on any solution, such as "Do not exceed the allowable budget" or "Ensure that all demand is met." For instance, a consumer products company manager would probably want to ensure that a specified level of customer service is achieved with the redesign of the distribution system. The presence of constraints makes modeling and solving optimization problems more challenging; we address constrained optimization problems later in this book, starting in Chapter 13.

For some prescriptive models, analytical solutions—closed-form mathematical expressions or simple formulas—can be obtained using such techniques as calculus or other types of mathematical analyses. In most cases, however, some type of computer-based procedure is needed to find an optimal solution. An **algorithm** is a systematic procedure that finds a solution to a problem. Researchers have developed effective algorithms to solve many types of optimization problems. For example, Microsoft Excel has a built-in add-in called *Solver* that allows you to find optimal solutions to optimization problems formulated as spreadsheet models. We use *Solver* in later chapters. However, we will not be concerned with the detailed mechanics of these algorithms; our focus will be on the use of the algorithms to solve and analyze the models we develop.

If possible, we would like to ensure that an algorithm such as the one *Solver* uses finds the best solution. However, some models are so complex that it is impossible to solve them optimally in a reasonable amount of computer time because of the extremely large number of computations that may be required or because they are so complex that finding the best solution cannot be guaranteed. In these cases, analysts use **search algorithms**—solution procedures that generally find good solutions without guarantees of finding the best one. Powerful search algorithms exist to obtain good solutions to extremely difficult optimization problems. One of these is discussed in Chapter 14.

Model Assumptions

All models are based on assumptions that reflect the modeler's view of the "real world." Some assumptions are made to simplify the model and make it more tractable, that is, able to be easily analyzed or solved. Other assumptions might be made to better characterize historical data or past observations. The task of the modeler is to select or build an appropriate model that best represents the behavior of the real situation. For example, economic theory tells us that demand for a product is negatively related to its price. Thus, as prices increase, demand falls, and vice versa (a phenomenon that you may recognize as price elasticity—the ratio of the percentage change in demand to the percentage change in price). Different mathematical models can describe this phenomenon. In the following examples, we illustrate two of them.

EXAMPLE 1.7   **A Linear Demand Prediction Model**

A simple model to predict demand as a function of price is the linear model

$$D = a - bP \qquad (1.7)$$

where D is the demand, P is the unit price, a is a constant that estimates the demand when the price is zero, and b is the slope of the demand function. This model is most applicable when we want to predict the effect of small changes around the current price. For example, suppose we know that when the price is $100, demand is 19,000 units and that demand falls by 10 for each dollar of price increase. Using simple algebra, we can determine that $a = 20,000$ and $b = 10$. Thus, if the price is $80, the predicted demand is

$$D = 20,000 - 10(80) = 19,200 \text{ units}$$

If the price increases to $90, the model predicts demand as

$$D = 20,000 - 10(90) = 19,100 \text{ units}$$

If the price is $100, demand would be

$$D = 20,000 - 10(100) = 19,000 \text{ units}$$

and so on. A graph of demand as a function of price is shown in Figure 1.5 as price varies between $80 and $120. We see that there is a constant decrease in demand for each $10 increase in price, a characteristic of a linear model.

▶ **Figure 1.5**

Graph of Linear Demand Model $D = a - bP$

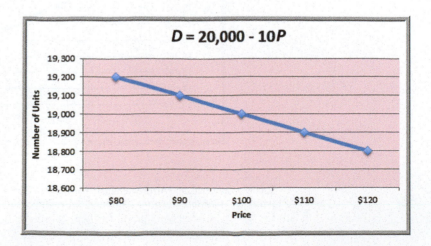

EXAMPLE 1.8 A Nonlinear Demand Prediction Model

An alternative model assumes that price elasticity is constant. In this case, the appropriate model is

$$D = cP^{-d} \qquad (1.8)$$

where c is the demand when the price is 0 and $d > 0$ is the price elasticity. To be consistent with Example 1.7, we assume that when the price is zero, demand is 20,000. Therefore, $c = 20,000$. We will also, as in Example 1.7, assume that when the price is $100, $D = 19,000$.

Using these values in equation (1.8), we can determine the value for d as 0.0111382 (we can do this mathematically using logarithms, but we'll see how to do this very easily using Excel in Chapter 11). Thus, if the price is $80, then the predicted demand is

$$D = 20,000(80)^{-0.0111382} = 19,047$$

If the price is 90, the demand would be

$$D = 20,000(90)^{-0.0111382} = 19,022$$

If the price is 100, demand is

$$D = 20,000(100)^{-0.0111382} = 19,000$$

A graph of demand as a function of price is shown in Figure 1.6. The predicted demand falls in a slight nonlinear fashion as price increases. For example, demand decreases by 25 units when the price increases from $80 to $90, but only by 22 units when the price increases from $90 to $100. If the price increases to $110, you would see a smaller decrease in demand. Therefore, we see a nonlinear relationship, in contrast to Example 1.7.

Both models in Examples 1.7 and 1.8 make different predictions of demand for different prices (other than $90). Which model is best? The answer may be neither. First of all, the development of realistic models requires many price point changes within a carefully designed experiment. Second, it should also include data on competition and customer disposable income, both of which are hard to determine. Nevertheless, it is possible to develop price elasticity models with limited price ranges and narrow customer segments. A good starting point would be to create a historical database with detailed information on all past pricing actions. Unfortunately, practitioners have observed that such models are not widely used in retail marketing, suggesting ample opportunity to apply business analytics.[28]

▶ **Figure 1.6**

Graph of Nonlinear Demand Model $D = cP^{-d}$

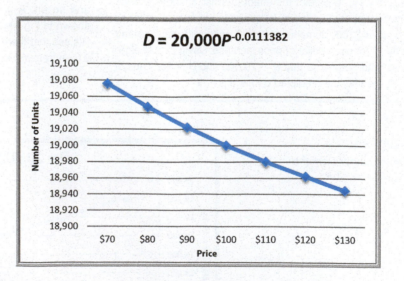

[28]Zhang, Clay Duan, and Arun Muthupalaniappan, "Analytics Applications in Consumer Credit and Retail Marketing," *Analytics Magazine* (November - December 2011): 27–33.

Uncertainty and Risk

As we all know, the future is always uncertain. Thus, many predictive models incorporate uncertainty and help decision makers analyze the risks associated with their decisions. **Uncertainty** is imperfect knowledge of what will happen; **risk** is associated with the consequences and likelihood of what might happen. For example, the change in the stock price of Apple on the next day of trading is uncertain. If you own Apple stock, then you face the risk of losing money if the stock price falls. If you don't own any stock, the price is still uncertain, although you would not have any risk. Risk is evaluated by the magnitude of the consequences and the likelihood that they would occur. For example, a 10% drop in a stock price would incur a higher risk if you own $1 million worth of that stock than if you only owned $1,000 worth of that stock. Similarly, if the chances of a 10% drop were 1 in 5, the risk would be higher than if the chances were only 1 in 100.

The importance of risk in business has long been recognized. The renowned management writer Peter Drucker observed in 1974:

> *To try to eliminate risk in business enterprise is futile. Risk is inherent in the commitment of present resources to future expectations. Indeed, economic progress can be defined as the ability to take greater risks. The attempt to eliminate risks, even the attempt to minimize them, can only make them irrational and unbearable. It can only result in the greatest risk of all: rigidity.*[29]

Consideration of risk is a vital element of decision making. For instance, you would probably not choose an investment simply on the basis of the return you might expect because, typically, higher returns are associated with higher risk. Therefore, you have to make a trade-off between the benefits of greater rewards and the risks of potential losses. Analytic models can help assess risk. A model in which some of the model input information is uncertain is often called a **stochastic**, or **probabilistic**, **model**. In contrast, a **deterministic model** is one in which all model input information is either known or assumed to be known with certainty. For instance, suppose that customer demand is an important element of some model. We can make the assumption that the demand is known with certainty; say, 5,000 units per month. In this case, we would be dealing with a deterministic model. On the other hand, suppose we have evidence to indicate that demand is uncertain, with an average value of 5,000 units per month, but which typically varies between 3,200 and 6,800 units. If we make this assumption, we would be dealing with a stochastic model. Stochastic models are useful in analyzing uncertainty in real-world situations, and we will discuss these later in this book.

CHECK YOUR UNDERSTANDING

1. Define a model and state three common forms of a model.

2. Explain the elements of a decision model.

3. Explain how decision models are used for descriptive, predictive, and prescriptive applications.

4. Define optimization and the characteristics of optimization models.

5. Explain the importance of assumptions in building decision models.

6. What is the difference between uncertainty and risk?

---

[29]P. F. Drucker, *The Manager and the Management Sciences in Management: Tasks, Responsibilities, Practices* (London: Harper and Row, 1974).

Problem Solving with Analytics

The fundamental purpose of analytics is to help managers solve problems and make decisions. The techniques of analytics represent only a portion of the overall problem-solving and decision-making process. Problem solving is the activity associated with defining, analyzing, and solving a problem and selecting an appropriate solution that solves a problem.

Problem solving consists of several phases:

1. Recognizing a problem
2. Defining the problem
3. Structuring the problem
4. Analyzing the problem
5. Interpreting results and making a decision
6. Implementing the solution

Recognizing a Problem

Managers at different organizational levels face different types of problems. In a manufacturing firm, for instance, top managers face decisions regarding allocating financial resources, building or expanding facilities, determining product mix, and strategically sourcing production. Middle managers in operations develop distribution plans, production and inventory schedules, and staffing plans. Finance managers analyze risks, determine investment strategies, and make pricing decisions. Marketing managers develop advertising plans and make sales force allocation decisions. In manufacturing operations, problems involve the size of daily production runs, individual machine schedules, and worker assignments. Whatever the problem, the first step is to realize that it exists.

How are problems recognized? Problems exist when there is a gap between what is happening and what we think should be happening. For example, a consumer products manager might feel that distribution costs are too high. This recognition might result from comparing performance with that of a competitor, or observing an increasing trend compared to previous years.

Defining the Problem

The second step in the problem-solving process is to clearly define the problem. Finding the real problem and distinguishing it from symptoms that are observed is a critical step. For example, high distribution costs might stem from inefficiencies in routing trucks, poor location of distribution centers, or external factors such as increasing fuel costs. The problem might be defined as improving the routing process, redesigning the entire distribution system, or optimally hedging fuel purchases.

Defining problems is not a trivial task. The complexity of a problem increases when the following occur:

- The number of potential courses of action is large.
- The problem belongs to a group rather than to an individual.
- The problem solver has several competing objectives.
- External groups or individuals are affected by the problem.
- The problem solver and the true owner of the problem—the person who experiences the problem and is responsible for getting it solved—are not the same.
- Time limitations are important.

These factors make it difficult to develop meaningful objectives and characterize the range of potential decisions. In defining problems, it is important to involve all people who make the decisions or who may be affected by them.

Structuring the Problem

This usually involves stating goals and objectives, characterizing the possible decisions, and identifying any constraints or restrictions. For example, if the problem is to redesign a distribution system, decisions might involve new locations for manufacturing plants and warehouses (where?), new assignments of products to plants (which ones?), and the amount of each product to ship from different warehouses to customers (how much?). The goal of cost reduction might be measured by the total delivered cost of the product. The manager would probably want to ensure that a specified level of customer service—for instance, being able to deliver orders within 48 hours—is achieved with the redesign. This is an example of a constraint. Structuring a problem often involves developing a formal model.

Analyzing the Problem

Here is where analytics plays a major role. Analysis involves some sort of experimentation or solution process, such as evaluating different scenarios, analyzing risks associated with various decision alternatives, finding a solution that meets certain goals, or determining an optimal solution. Analytics professionals have spent decades developing and refining a variety of approaches to address different types of problems. Much of this book is devoted to helping you understand these techniques and gain a basic facility in using them.

Interpreting Results and Making a Decision

Interpreting the results from the analysis phase is crucial in making good decisions. Models cannot capture every detail of the real problem, and managers must understand the limitations of models and their underlying assumptions and often incorporate judgment into making a decision. For example, in locating a facility, we might use an analytical procedure to find a "central" location; however, many other considerations must be included in the decision, such as highway access, labor supply, and facility cost. Thus, the location specified by an analytical solution might not be the exact location the company actually chooses.

Implementing the Solution

This simply means making the solution work in the organization, or translating the results of a model back to the real world. This generally requires providing adequate resources, motivating employees, eliminating resistance to change, modifying organizational policies, and developing trust. Problems and their solutions affect people: customers, suppliers, and employees. All must be an important part of the problem-solving process. Sensitivity to political and organizational issues is an important skill that managers and analytical professionals alike must possess when solving problems.

   In each of these steps, good communication is vital. Analytics professionals need to be able to communicate with managers and clients to understand the business context of the problem and be able to explain results clearly and effectively. Such skills as constructing good visual charts and spreadsheets that are easy to understand are vital to users of analytics. We emphasize these skills throughout this book.

ANALYTICS IN PRACTICE: Developing Effective Analytical Tools at Hewlett-Packard[30]

Hewlett-Packard (HP) uses analytics extensively. Many applications are used by managers with little knowledge of analytics. These require that analytical tools be easily understood. Based on years of experience, HP analysts compiled some key lessons. Before creating an analytical decision tool, HP asks three questions:

1. *Will analytics solve the problem?* Will the tool enable a better solution? Should other, nonanalytical solutions be used? Are there organizational or other issues that must be resolved? Often, what may appear to be an analytical problem may actually be rooted in problems of incentive misalignment, unclear ownership and accountability, or business strategy.
2. *Can we leverage an existing solution?* Before "reinventing the wheel," can existing solutions address the problem? What are the costs and benefits?
3. *Is a decision model really needed?* Can simple decision guidelines be used instead of a formal decision tool?

Once a decision is made to develop an analytical tool, several guidelines are used to increase the chances of successful implementation:

- *Use prototyping*, a quick working version of the tool designed to test its features and gather feedback.
- *Build insight, not black boxes.* A "black box" tool is one that generates an answer, but may not provide confidence to the user. Interactive tools that create insights to support a decision provide better information.
- *Remove unneeded complexity.* Simpler is better. A good tool can be used without expert support.

hans12/Fotolia

- *Partner with end users in discovery and design.* Decision makers who will actually use the tool should be involved in its development.
- *Develop an analytic champion.* Someone (ideally, the actual decision maker) who is knowledgeable about the solution and close to it must champion the process.

CHECK YOUR UNDERSTANDING

1. List the major phases of problem solving and explain each.
2. What lessons did Hewlett-Packard learn about using analytics?

---

[30]Based on Thomas Olavson and Chris Fry, "Spreadsheet Decision-Support Tools: Lessons Learned at Hewlett-Packard," *Interfaces*, 38, 4, July–August 2008: 300–310.

KEY TERMS

Algorithm	Optimal solution
Big data	Optimization
Business analytics (analytics)	Predictive analytics
Business intelligence (BI)	Prescriptive analytics
Constraint	Price elasticity
Data mining	Problem solving
Decision model	Reliability
Decision options (decision variables)	Risk
Decision support systems (DSS)	Search algorithm
Descriptive analytics	Simulation and risk analysis
Deterministic model	Statistics
Information systems (IS)	Stochastic (probabilistic) model
Model	Tag cloud
Modeling and optimization	Uncertainty
Objective function	Validity
Operations Research/Management Science (OR/MS)	Visualization
	What-if analysis

CHAPTER 1 TECHNOLOGY HELP

Useful Excel Functions (*see Appendix A1*)

MIN(*range*) Finds the smallest value in a range of cells.

MAX(*range*) Finds the largest value in a range of cells.

SUM(*range*) Finds the sum of values in a range of cells.

AVERAGE(*range*) Finds the average of the values in a range of cells.

COUNT(*range*) Finds the number of cells in a range that contain numbers.

COUNTIF(*range, criteria*) Finds the number of cells within a range that meet a specified criterion.

NPV(*rate, value1, value2, . . .*) Calculates the net present value of an investment by using a discount rate and a series of future payments (negative values) and income (positive values).

DATEDIF(*startdate*, *enddate*, *time unit*) Computes the number of whole years, months, or days between two dates.

PROBLEMS AND EXERCISES

What Is Business Analytics?

1. Discuss how you might use business analytics in your personal life, such as in managing your grocery purchasing, automobile maintenance, budgeting, or sports. Be creative in identifying opportunities!

2. How might analytics be used in the following situations?
   a. Making and marketing a new product
   b. Deciding where to locate a new plant
   c. Determining a personal investment plan

3. A supermarket has been experiencing long lines during peak periods of the day. The problem is noticeably worse on certain days of the week, and the peak periods sometimes differ according to the day of the week. There are usually enough workers on the job to open all cash registers. The problem the supermarket manager faces is knowing when to call some of the workers who are stocking shelves up to the front to work the checkout counters. How might business analytics help the supermarket manager? What data would be needed to facilitate good decisions?

Descriptive, Predictive, and Prescriptive Analytics

4. For each of the following scenarios, state whether descriptive, predictive, or prescriptive analytics tools would most likely be used.

   a. The chief financial officer for a small manufacturing firm would like to estimate the net profit that the firm could expect over the next three years.

   b. A human resource manager needs to understand whether the company's current employee mix has the skills and capabilities needed to achieve the goals laid out by a new strategic plan.

   c. A financial advisor would like to develop the best mix of stocks, bonds, and other investments for a client to achieve a comfortable level of risk.

   d. A large service firm wishes to determine how to invest the cash received from its financial product to achieve the best return.

   e. A logistics company wants to better understand the relative profitability of its numerous customers over the past three years.

   f. A disaster relief agency needs to allocate its budget for the next year among various relief efforts and programs.

   g. An automobile company would like to determine the number of vehicles it could sell next year based on the proposed price.

   h. A baseball team would like to set ticket prices for different sections in its stadium to attract the highest number of fans throughout the season.

Models in Business Analytics

5. Suppose that a manufacturer can produce a part for $10.00 with a fixed cost of $5,000. Alternately, the manufacturer could contract with a supplier in Asia to purchase the part at a cost of $12.00, which includes transportation.

   a. If the anticipated production volume is 1,200 units, compute the total cost of manufacturing and the total cost of outsourcing.

   b. What is the best decision?

6. Use the model developed in Example 1.5 to predict the total sales for weeks 2 through 16, and compare the results to the observed sales. Does the accuracy of the model seem to be different when coupons are used or not? When advertising is used or not?

7. A bank developed a model for predicting the average checking and savings account balance as balance $= -17{,}732 + 367 \times$ age $+ 1{,}300 \times$ years education $+ 0.116 \times$ household wealth.

   a. Explain how to interpret the numbers in this model.

   b. Suppose that a customer is 32 years old, is a college graduate (so that years education $= 16$), and has a household wealth of $150,000. What is the predicted bank balance?

8. Four key marketing decision options are price (P), advertising (A), transportation (T), and product quality (Q). Consumer demand (D) is influenced by these variables. The simplest model for describing demand in terms of these variables is

   $$D = k - pP + aA + tT + qQ$$

   where k, p, a, t, and q are positive constants.

   a. How does a change in each variable affect demand?

   b. How do the variables influence each other?

   c. What limitations might this model have? Can you think of how this model might be made more realistic?

9. Total marketing effort is a term used to describe the critical decision factors that affect demand: price, advertising, distribution, and product quality. Let the variable x represent total marketing effort. A typical model that is used to predict demand as a function of total marketing effort is

   $$D = ax^b$$

   Suppose that a is a positive number. Different models result from varying the constant b. Sketch the graphs of these models for $b = 0$, $b = 1$, $0 < b < 1$, $b < 0$, and $b > 1$. What does each model tell you about the relationship between demand and marketing effort? What assumptions are implied? Are they reasonable? How would you go about selecting the appropriate model?

10. A manufacturer of headphones is preparing to set the price on a new design. Demand is thought to depend on the price and is represented by the model

    $$D = 2{,}500 - 3P$$

    The accounting department estimates that the total costs can be represented by

    $$C = 5{,}000 + 5D$$

    Develop a model for the total profit in terms of the price, P.

Problem Solving with Analytics

11. In this chapter, we noted the importance of defining and analyzing a problem prior to attempting to find a solution. Consider this example: One of the earliest operations research groups during World War II was conducting a study on the optimum utilization of Spitfire and Hurricane aircraft during the Battle of Britain. Whenever one of these planes returned from battle, the locations of the bullet holes on it were carefully plotted. By repeatedly recording these data over time, and studying the clusters of data, the group was able to estimate the regions of the aircraft most likely to be hit by enemy gunfire, with the objective of reinforcing these regions with special armor. What difficulties do you see here?

12. John Toczek, a practicing analytics professional, maintains a Web site called the PuzzlOR (OR standing for "operations research") at www.puzzlor.com. Each month he posts a new puzzle. Choose one of them and work through the problem-solving process, focusing on steps two through four. A good one to start with is "SurvivOR" from June 2010.

CASE: PERFORMANCE LAWN EQUIPMENT

In each chapter of this book, we use a a fictitious company, Performance Lawn Equipment (PLE), within a case exercise for applying the tools and techniques introduced in the chapter.[31] To put the case in perspective, we first provide some background about the company, so that the applications of business analytic tools will be more meaningful.

PLE, headquartered in St. Louis, Missouri, is a privately owned designer and producer of traditional lawn mowers used by homeowners. In the past ten years, PLE has added another key product, a medium-size diesel power lawn tractor with front and rear power takeoffs, Class I three-point hitches, four-wheel drive, power steering, and full hydraulics. This equipment is built primarily for a niche market consisting of large estates, including golf and country clubs, resorts, private estates, city parks, large commercial complexes, lawn care service providers, private homeowners with five or more acres, and government (federal, state, and local) parks, building complexes, and military bases. PLE provides most of the products to dealerships, which, in turn, sell directly to end users. PLE employs 1,660 people worldwide. About half the workforce is based in St. Louis; the remainder is split among their manufacturing plants.

In the United States, the focus of sales is on the eastern seaboard, California, the Southeast, and the south central states, which have the greatest concentration of customers. Outside the United States, PLE's sales include a European market, a growing South American market, and developing markets in the Pacific Rim and China.

Both end users and dealers have been established as important customers for PLE. Collection and analysis of end-user data showed that satisfaction with the products depends on high quality, easy attachment/dismount of implements, low maintenance, price value, and service.

For dealers, key requirements are high quality, parts and feature availability, rapid restock, discounts, and timeliness of support.

PLE has several key suppliers: Mitsitsiu, Inc., the sole source of all diesel engines; LANTO Axles, Inc., which provides tractor axles; Schorst Fabrication, which provides subassemblies; Cuberillo, Inc, supplier of transmissions; and Specialty Machining, Inc., a supplier of precision machine parts.

To help manage the company, PLE managers have developed a "balanced scorecard" of measures. These data, which are summarized shortly, are stored in the form of a Microsoft Excel workbook (*Performance Lawn Equipment*) accompanying this book. The database contains various measures captured on a monthly or quarterly basis and is used by various managers to evaluate business performance. Data for each of the key measures are stored in a separate worksheet. A summary of these worksheets is given next:

- *Dealer Satisfaction*, measured on a scale of 1–5 (1 = poor, 2 = less than average, 3 = average, 4 = above average, and 5 = excellent). Each year, dealers in each region are surveyed about their overall satisfaction with PLE. The worksheet contains summary data from surveys for the past five years.
- *End-User Satisfaction*, measured on the same scale as dealers. Each year, 100 users from each region are surveyed. The worksheet contains summary data for the past five years.
- *Customer Survey*, results from a survey for customer ratings of specific attributes of PLE tractors: quality, ease of use, price, and service on the same 1–5 scale. This sheet contains 200 observations of customer ratings.

[31]The case scenario was based on *Gateway Estate Lawn Equipment Co. Case Study*, used for the 1997 Malcolm Baldrige National Quality Award Examiner Training course. This material is in the public domain. The database, however, was developed by the author.

■ *Complaints*, which shows the number of complaints registered by all customers each month in each of PLE's five regions (North America, South America, Europe, the Pacific, and China).

■ *Mower Unit Sales* and *Tractor Unit Sales*, which provide sales by product by region on a monthly basis. Unit sales for each region are aggregated to obtain world sales figures.

■ *Industry Mower Total Sales* and *Industry Tractor Total Sales*, which list the number of units sold by all producers by region.

■ *Unit Production Costs*, which provides monthly accounting estimates of the variable cost per unit for manufacturing tractors and mowers over the past five years.

■ *Operating and Interest Expenses*, which provides monthly administrative, depreciation, and interest expenses at the corporate level.

■ *On-Time Delivery*, which provides the number of deliveries made each month from each of PLE's major suppliers, the number on time, and the percent on time.

■ *Defects After Delivery*, which shows the number of defects in supplier-provided material found in all shipments received from suppliers.

■ *Time to Pay Suppliers*, which provides measurements in days from the time the invoice is received until payment is sent.

■ *Response Time*, which gives samples of the times taken by PLE customer-service personnel to respond to service calls by quarter over the past two years.

■ *Employee Satisfaction*, which provides data for the past four years of internal surveys of employees to determine their overall satisfaction with their jobs, using the same scale used for customers. Employees are surveyed quarterly, and results are stratified by employee category: design and production, managerial, and sales/administrative support.

In addition to these business measures, the PLE database contains worksheets with data from special studies:

■ *Engines*, which lists 50 samples of the time required to produce a lawn mower blade using a new technology.

■ *Transmission Costs*, which provides the results of 30 samples each for the current process used to produce tractor transmissions and two proposed new processes.

■ *Blade Weight*, which provides samples of mower blade weights to evaluate the consistency of the production process.

■ *Mower Test*, which lists test results of mower functional performance after assembly for 30 samples of 100 units each.

■ *Employee Retention*, data from a study of employee duration (length of hire) with PLE. The 40 subjects were identified by reviewing hires from ten years prior and identifying those who were involved in managerial positions (either hired into management or promoted into management) at some time in this ten-year period.

■ *Shipping Cost*, which gives the unit shipping cost for mowers and tractors from existing and proposed plants for a supply chain design study.

■ *Fixed Cost*, which lists the fixed cost to expand existing plants or build new facilities, also as part of the supply chain design study.

■ *Purchasing Survey*, which provides data obtained from a third-party survey of purchasing managers of customers of Performance Lawn Care.

Elizabeth Burke has recently joined the PLE management team to oversee production operations. She has reviewed the types of data that the company collects and has assigned you the responsibility to be her chief analyst in the coming weeks. She has asked you to do some preliminary analysis of the data for the company.

1. First, she would like you to edit the worksheets *Dealer Satisfaction* and *End-User Satisfaction* to display the total number of responses to each level of the survey scale across all regions for each year.

2. Second, she wants a count of the number of failures in the worksheet *Mower Test*.

3. Next, Elizabeth has provided you with prices for PLE products for the past five years:

Year	Mower Price	Tractor Price
2014	$150	$3,250
2015	$175	$3,400
2016	$180	$3,600
2017	$185	$3,700
2018	$190	$3,800

Create a new worksheet to compute gross revenues by month and region, as well as worldwide totals, for each product using the data in *Mower Unit Sales* and *Tractor Unit Sales*.

4. Finally, she wants to know the market share for each product and region by month based on the PLE and industry sales data, and the average market share by region over the five years.

Summarize all your findings in a report to Ms. Burke.

Basic Excel Skills

To be able to apply the procedures and techniques that you will learn in this book, it is necessary for you to be relatively proficient in using Excel. We assume that you are familiar with the most elementary spreadsheet concepts and procedures, such as

- opening, saving, and printing files;
- using workbooks and worksheets;
- moving around a spreadsheet;
- selecting cells and ranges;
- inserting/deleting rows and columns;
- entering and editing text, numerical data, and formulas in cells;
- formatting data (number, currency, decimal places, etc.);
- working with text strings;
- formatting data and text; and
- modifying the appearance of the spreadsheet using borders, shading, and so on.

Menus and commands in Excel reside in the "ribbon" shown in Figure A1.1. All Excel discussions in this book will be based on Excel 2016 for Windows; if you use Excel 2016 for Mac, some differences may exist, and we may point these out as appropriate. Menus and commands are arranged in logical *groups* under different *tabs* (*Home, Insert, Formulas,* and so on); small triangles pointing downward indicate *menus* of additional choices. We often refer to certain commands or options and where they may be found in the ribbon. For instance, in the Mac version, groups are not specified.

Excel provides an add-in called the *Analysis Toolpak*, which contains a variety of tools for statistical computation, and *Solver*, which is used for optimization. They will be found in the *Data* tab ribbon; you should ensure that these are activated. To activate them in Windows, click the *File* tab and then *Options* in the left column. Choose *Add-Ins* from the left column. At the bottom of the dialog, make sure *Excel Add-ins* is selected in the *Manage:* box and click *Go*. In the *Add-Ins* dialog, if *Analysis Toolpak, Analysis Toolpak VBA,* and *Solver Add-in* are not checked, simply check the boxes and click *OK*. You will not have to repeat this procedure every time you run Excel in the future. On Excel 2016 for Mac, go to *Tools > Excel Add-ins* and select both *Analysis Toolpak* and *Solver*.

▲ **Figure A1.1**

Excel Ribbons for Windows and Mac

Excel Formulas and Addressing

Formulas in Excel use common mathematical operators:

- addition (+)
- subtraction (−)
- multiplication (*)
- division (/)

Exponentiation uses the ^ symbol; for example, 2^5 is written as 2^5 in an Excel formula.

Cell references in formulas can be written either with *relative addresses* or *absolute addresses*. A **relative address** uses just the row and column label in the cell reference (for example, A4 or C21); an **absolute address** uses a dollar sign ($ sign) before either the row or column label or both (for example, $A2, C$21, or $B$15). Which one we choose makes a critical difference if you copy the cell formulas. If only relative addressing is used, then copying a formula to another cell changes the cell references by the number of rows or columns in the direction that the formula is copied. So, for instance, if we would use a formula in cell B8, =B4−B5*A8, and copy it to cell C9 (one column to the right and one row down), all the cell references are increased by one and the formula would be changed to =C5−C6*B9.

Using a $ sign before a row label (for example, B$4) keeps the reference fixed to row 4 but allows the column reference to change if the formula is copied to another cell. Similarly, using a $ sign before a column label (for example, $B4) keeps the reference to column B fixed but allows the row reference to change. Finally, using a $ sign before both the row and column labels (for example, $B$4) keeps the reference to cell B4 fixed no

---

EXAMPLE A1.1 | **Implementing Price-Demand Models in Excel**

In Chapter 1, we described two models for predicting demand as a function of price:

$$D = a - bP$$

and

$$D = cP^{-d}$$

Figure A1.2 shows a spreadsheet (Excel file *Demand Prediction Models*) for calculating demand for different prices using each of these models. For example, to calculate the demand in cell B8 for the linear model, we use the formula

$$=\$B\$4 - \$B\$5*A8$$

To calculate the demand in cell E8 for the nonlinear model, we use the formula

$$=\$E\$4*D8\text{\textasciicircum} - \$E\$5$$

Note how the absolute addresses are used so that as these formulas are copied down, the demand is computed correctly.

▶ **Figure A1.2**

Excel Models for Demand Prediction

	A	B	C	D	E
1	**Demand Prediction Models**				
2					
3	**Linear Model**			**Nonlinear Model**	
4	a	20,000		c	20,000
5	b	10		d	0.0111382
6					
7	**Price**	**Demand**		**Price**	**Demand**
8	$80.00	$19,200		$70.00	$19,075.63
9	$90.00	$19,100		$80.00	$19,047.28
10	$100.00	$19,000		$90.00	$19,022.31
11	$110.00	$18,900		$100.00	$19,000.00
12	$120.00	$18,800		$110.00	$18,979.84
13				$120.00	$18,961.45
14				$130.00	$18,944.56

matter where the formula is copied. You should be very careful to use relative and absolute addressing appropriately in your models, especially when copying formulas.

Copying Formulas

Excel provides several ways of copying formulas to different cells. This is extremely useful in building decision models, because many models require replication of formulas for different periods of time, similar products, and so on. The easiest way is to select the cell with the formula to be copied, and then press Ctrl-C on your Windows keyboard or Command-C on a Mac, click on the cell you wish to copy to, and then press Ctrl-V in Windows or Command-V on a Mac. You may also enter a formula directly in a range of cells without copying and pasting by selecting the range, typing in the formula, and pressing Ctrl-Enter in Windows or Command-Enter on a Mac.

To copy a formula from a single cell or range of cells down a column or across a row, first select the cell or range, click and hold the mouse on the small square in the lower right-hand corner of the cell (the "fill handle"), and drag the formula to the "target" cells to which you wish to copy.

 **Useful Excel Tips**

- **Split Screen.** You may split the worksheet horizontally and/or vertically to view different parts of the worksheet at the same time. The vertical splitter bar is just to the right of the bottom scroll bar, and the horizontal splitter bar is just above the right-hand scroll bar. Position your cursor over one of these until it changes shape, click, and drag the splitter bar to the left or down.
- **Column and Row Widths.** Many times a cell contains a number that is too large to display properly because the column width is too small. You may change the column width to fit the largest value or text string anywhere in the column by positioning the cursor to the right of the column label so that it changes to a cross with horizontal arrows and then double-clicking. You may also move the arrow to the left or right to manually change the column width. You may change the row heights in a similar fashion by moving the cursor below the row number label. This can be especially useful if you have a very long formula to display. To break a formula within a cell, position the cursor at the break point in the formula bar and press Alt-Enter.
- **Displaying Formulas in Worksheets.** Choose *Show Formulas* in the *Formulas* tab. You may also press Ctrl ~ in either Windows or Mac to toggle formulas on and off. You often need to change the column width to display the formulas properly.
- **Displaying Grid Lines and Row and Column Headers for Printing.** Check the *Print* boxes for gridlines and headings in the *Sheet Options* group under the *Page Layout* tab. Note that the *Print* command can be found by clicking on the *Office* button in Windows or under the *File* menu in Mac.
- **Filling a Range with a Series of Numbers.** Suppose you want to build a worksheet for entering 100 data values. It would be tedious to have to enter the numbers from 1 to 100 one at a time. Simply fill in the first few values in the series and highlight them. Then click and drag the small square (fill handle) in the lower right-hand corner down (Excel will show a small pop-up window that tells you the last value in the range) until you have filled in the column to 100; then release the mouse.

Excel Functions

Functions are used to perform special calculations in cells and are used extensively in business analytics applications. All Excel functions require an equal sign and a function name followed by parentheses, in which you specify arguments for the function.

Basic Excel Functions

Some of the more common functions that we will use in applications include the following:

> MIN(*range*)—finds the smallest value in a range of cells
> MAX(*range*)—finds the largest value in a range of cells
> SUM(*range*)—finds the sum of values in a range of cells
> AVERAGE(*range*)—finds the average of the values in a range of cells
> COUNT(*range*)—finds the number of cells in a range that contain numbers
> COUNTIF(*range, criteria*)—finds the number of cells within a range that meet a specified criterion

Logical functions, such as IF, AND, OR, and VLOOKUP will be discussed in Chapter 2.

The COUNTIF function counts the number of cells within a range that meet a criterion you specify. For example, you can count all the cells that start with a certain letter, or you can count all the cells that contain a number that is larger or smaller than a number you specify. Examples of criteria are 100, ">100", a cell reference such as A4, and a text string such as "Facebook." Note that text and logical formulas must be enclosed in quotes. See Excel Help for other examples.

Excel has other useful COUNT-type functions: COUNTA counts the number of non-blank cells in a range, and COUNTBLANK counts the number of blank cells in a range. In addition, COUNTIFS(*range1, criterion1, range2, criterion2, . . . , range_n, criterion_n*) finds the number of cells within multiple ranges that meet specific criteria for each range.

We illustrate these functions using the *Purchase Orders* data set in Example A1.2.

EXAMPLE A1.2   **Using Basic Excel Functions**

In the *Purchase Orders* data set, we will find the following:

- Smallest and largest quantity of any item ordered
- Total order costs
- Average number of months per order for accounts payable
- Number of purchase orders placed
- Number of orders placed for O-rings
- Number of orders with A/P (accounts payable) terms shorter than 30 months
- Number of O-ring orders from Spacetime Technologies
- Total cost of all airframe fasteners
- Total cost of airframe fasteners purchased from Alum Sheeting

The results are shown in Figure A1.3. In this figure, we used the split-screen feature in Excel to reduce the number of rows shown in the spreadsheet. To find the smallest and largest quantity of any item ordered, we use the MIN and MAX functions for the data in column F. Thus, the formula in cell B99 is =MIN(F4:F97) and the formula in cell B100 is =MAX(F4:F97). To find the total order costs,

we sum the data in column G using the SUM function: =SUM(G4:G97); this is the formula in cell B101. To find the average number of A/P months, we use the AVERAGE function for the data in column H. The formula in cell B102 is =AVERAGE(H4:H97). To find the number of purchase orders placed, use the COUNT function. Note that the COUNT function counts only the number of cells in a range that contain numbers, so we could not use it in columns A, B, or D; however, any other column would be acceptable. Using the item numbers in column C, the formula in cell B103 is =COUNT(C4:C97). To find the number of orders placed for O-rings, we use the COUNTIF function. For this example, the formula used in cell B104 is =COUNTIF(D4:D97, "O-Ring"). We could have also used the cell reference for any cell containing the text O-Ring, such as =COUNTIF(D4:D97, D12). To find the number of orders with A/P terms less than 30 months, we use the formula =COUNTIF(H4:H97, "<30") in cell B105. Finally, to count the number of O-ring orders for Spacetime Technologies, we use =COUNTIFS(D4:D97, "O-Ring", A4:A97, "Spacetime Technologies").

IF-type functions are also available for other calculations. For example, the functions SUMIF, AVERAGEIF, SUMIFS, and AVERAGEIFS can be used to embed IF logic within mathematical functions. For instance, the syntax of SUMIF is SUMIF(*range, criterion, [sum range]*); *sum range* is an optional argument that allows you to add cells in a different range. Thus, in the *Purchase Orders* database, to find the total cost of all airframe fasteners, we would use

$$=\text{SUMIF(D4:D97, ``Airframe fasteners'', G4:G97)}$$

This function looks for airframe fasteners in the range D4:D97, but then sums the associated values in column G (cost per order). The arguments for SUMIFS and AVERAGEIFS are *(sumrange, range1, criterion1, range2, criterion2, . . . ,rangeN, criterionN).* For example, the function

$$=\text{SUMIFS(F4:F97,A4:A97,``Alum Sheeting'', D4:D97, ``Airframe fasteners'')}$$

will find the total quantity (from the *sumrange* in column F) of all airframe fasteners purchased from Alum Sheeting.

Functions for Specific Applications

Excel has a wide variety of other functions for statistical, financial, and other applications, many of which we introduce and use throughout the text. For instance, some financial models that we develop require the calculation of net present value (NPV). **Net present value** (also called **discounted cash flow**) measures the worth of a stream of cash flows, taking into account the time value of money. That is, a cash flow of F dollars t time periods in the future is worth $F/(1 + i)^t$ dollars today, where i is the **discount rate.** The discount rate reflects the opportunity costs of spending funds now versus achieving a return through another investment, as well as the risks associated with not receiving returns until a later

	A	B	C	D	E	F	G	H	I	J
1	Purchase Orders									
2										
3	Supplier	Order No.	Item No.	Item Description	Item Cost	Quantity	Cost per order	A/P Terms (Months)	Order Date	Arrival Date
4	Hulkey Fasteners	Aug11001	1122	Airframe fasteners	$ 4.25	19,500	$ 82,875.00	30	08/05/11	08/13/11
5	Alum Sheeting	Aug11002	1243	Airframe fasteners	$ 4.25	10,000	$ 42,500.00	30	08/08/11	08/14/11
6	Fast-Tie Aerospace	Aug11003	5462	Shielded Cable/ft.	$ 1.05	23,000	$ 24,150.00	30	08/10/11	08/15/11
7	Fast-Tie Aerospace	Aug11004	5462	Shielded Cable/ft.	$ 1.05	21,500	$ 22,575.00	30	08/15/11	08/22/11
8	Steelpin Inc.	Aug11005	5319	Shielded Cable/ft.	$ 1.10	17,500	$ 19,250.00	30	08/20/11	08/31/11
9	Fast-Tie Aerospace	Aug11006	5462	Shielded Cable/ft.	$ 1.05	22,500	$ 23,625.00	30	08/20/11	08/26/11
10	Steelpin Inc.	Aug11007	4312	Bolt-nut package	$ 3.75	4,250	$ 15,937.50	30	08/25/11	09/01/11
11	Durrable Products	Aug11008	7258	Pressure Gauge	$ 90.00	100	$ 9,000.00	45	08/25/11	08/28/11
12	Fast-Tie Aerospace	Aug11009	6321	O-Ring	$ 2.45	1,300	$ 3,185.00	30	08/25/11	09/04/11
96	Steelpin Inc.	Nov11009	5677	Side Panel	$ 195.00	110	$ 21,450.00	30	11/05/11	11/17/11
97	Manley Valve	Nov11010	9955	Door Decal	$ 0.55	125	$ 68.75	30	11/05/11	11/10/11
98										
99	Minimum Quantity	90								
100	Maximum Quantity	25,000								
101	Total Order Costs	$ 2,471,760.00								
102	Average Number of A/P Months	30.63829787								
103	Number of Purchase Orders	94								
104	Number of O-ring Orders	12								
105	Number of A/P Terms < 30	17								
106	Number of O-ring Orders Spacetime	3								

▲ Figure A1.3

Application of Basic Excel Functions to Purchase Orders Data

time. The sum of the present values of all cash flows over a stated time horizon is the net present value:

$$\text{NPV} = \sum_{t=0}^{n} \frac{F_t}{(1 + i)^t} \qquad (A1.1)$$

where F_t = cash flow in period t. A positive NPV means that the investment will provide added value because the projected return exceeds the discount rate.

The Excel function NPV(*rate, value1, value2, . . .*) calculates the net present value of an investment by using a discount rate and a series of future payments (negative values) and income (positive values). *Rate* is the value of the discount rate i over the length of one period, and *value1, value2, . . .* are 1 to 29 arguments representing the payments and income for each period. The values must be equally spaced in time and are assumed to occur at the end of each period. The NPV investment begins one period before the date of the *value1* cash flow and ends with the last cash flow in the list. The NPV calculation is based on future cash flows. If the first cash flow (such as an initial investment or fixed cost) occurs at the beginning of the first period, then it must be added to the NPV result and *not* included in the function arguments.

EXAMPLE A1.3 **Using the NPV Function**

A company is introducing a new product. The fixed cost for marketing and distribution is $25,000 and is incurred just prior to launch. The forecasted net sales revenues for the first six months are shown in Figure A1.4. The formula in cell B8 computes the net present value of these cash flows as =NPV(B6, C4:H4)−B5. Note that the fixed cost is not a future cash flow and is not included in the NPV function arguments.

Insert Function

The easiest way to locate a particular function is to select a cell and click on the *Insert Function* button $[f_x]$, which can be found under the ribbon next to the formula bar and also in the *Function Library* group in the *Formulas* tab. You may either type in a description in the search field, such as "net present value," or select a category, such as "financial," from the drop-down box.

This feature is particularly useful if you know what function to use but are not sure of what arguments to enter because it will guide you in entering the appropriate data for the function arguments. Figure A1.5 shows the dialog from which you may select the function you wish to use. For example, if we would choose the COUNTIF function, the dialog in Figure A1.6 appears. When you click in an input cell, a description of the argument is shown. Thus, if you are not sure what to enter for the range, the explanation in Figure A1.6

▶ **Figure A1.4**

Net Present Value Calculation

	A	B	C	D	E	F	G	H
1	Net Present Value							
2								
3		Month	January	February	March	April	May	June
4		Sales Revenue Forecast	$2,500	$4,000	$5,000	$8,000	$10,000	$12,500
5	Fixed Cost	$25,000.00						
6	Discount Rate	3%						
7								
8	NPV	$11,975.81						

▶ **Figure A1.5**

Insert Function Dialog

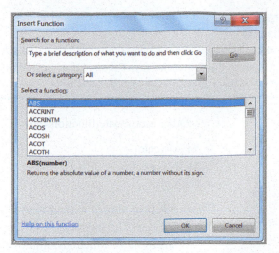

▶ **Figure A1.6**

Function Arguments Dialog for COUNTIF

will help you. For further information, you could click on the *Help* button in the lower left-hand corner.

Date and Time Functions

In many analytics applications, a database might contain dates, such as when an order is placed or when an employee was hired. Excel can display a date in a variety of formats, such as 2/14/17 or 14-Feb-17. You may choose the standard date format (for example, 2/14/17) by selecting *Date* in the *Number* formatting box or select a custom format by selecting *Custom* in the *Number* box. Excel stores dates in a serial format. January 1, 1900 is day 1, and each subsequent day is numbered sequentially. Both the current date and January 1, 1900 are included in the count. So 2/14/17 is 42,780 when expressed in this format. This means there are 42,780 days between January 1, 1900 and February 14, 2017 (including both days). Therefore, to determine the number of days between two dates, you can simply subtract them.

Another useful date function is DATEDIF (which surprisingly doesn't appear in the *Insert Function* list!), which can compute the number of whole years, months, or days between two dates. The syntax of DATEDIF is

DATEDIF(*startdate, enddate, time unit*)

The time unit can be "y," "m," or "d." For instance, DATEDIF(4/26/89, 2/14/17, "y") will return 27 (years), while DATEDIF(4/26/89, 2/14/17, "m") will return 333 (months).

Computing Lead Times

In the *Purchase Orders* database, we will compute the lead time for each order, that is, the number of days between the order date and arrival date. Figure A1.7 shows the use of the DATEDIF function. Alternatively, we could have simply subtracted the values: for example, in cell K4, use = J4−I4.

Other useful date functions are

- YEAR(date)
- MONTH(date)
- DAY(date)

These functions simply extract the year, month, and day from a date or cell reference that contains a date. The function TODAY( ) displays the current date.

Similar to date functions, times can be formatted in a variety of ways, such as 12:26 PM, Hours:Minutes:Seconds, or in military time. The function NOW( ) displays the current time and date.

Miscellaneous Excel Functions and Tools

In this section, we will illustrate a few miscellaneous Excel functions and tools that support analytics applications.

Range Names

A range name is a descriptive label assigned to a cell or range of cells. Range names can help to facilitate building models on spreadsheets and understanding the formulas on which models are based. There are several ways to create range names in Excel.

Using the *Name* Box to Create a Range Name

Suppose that we create a simple spreadsheet for computing total cost (which is the fixed cost plus unit variable cost times quantity produced) shown in Figure A1.8. We will define range names for each of the numerical cells that correspond to the labels on the left. That is, we will name cell B3 Fixed cost, cell B4 Unit variable cost, and so on. Click on cell B3; in the *Name* box, type the name Fixed_cost (note the underscore), and then press Enter. Figure A1.8 shows that the name for cell B3 is displayed in the *Name* box. Repeat this process for each of the other numerical cells.

K4			f_x	=DATEDIF(I4,J4,"d")							
	A	B	C	D	E	F	G	H	I	J	K
1	Purchase Orders										
2											
3	Supplier	Order No.	Item No.	Item Description	Item Cost	Quantity	Cost per order	A/P Terms (Months)	Order Date	Arrival Date	Lead Time
4	Hulkey Fasteners	Aug11001	1122	Airframe fasteners	$ 4.25	19,500	$ 82,875.00	30	08/05/11	08/13/11	8
5	Alum Sheeting	Aug11002	1243	Airframe fasteners	$ 4.25	10,000	$ 42,500.00	30	08/08/11	08/14/11	6
6	Fast-Tie Aerospace	Aug11003	5462	Shielded Cable/ft.	$ 1.05	23,000	$ 24,150.00	30	08/10/11	08/15/11	5
7	Fast-Tie Aerospace	Aug11004	5462	Shielded Cable/ft.	$ 1.05	21,500	$ 22,575.00	30	08/15/11	08/22/11	7
8	Steelpin Inc.	Aug11005	5319	Shielded Cable/ft.	$ 1.10	17,500	$ 19,250.00	30	08/20/11	08/31/11	11
9	Fast-Tie Aerospace	Aug11006	5462	Shielded Cable/ft.	$ 1.05	22,500	$ 23,625.00	30	08/20/11	08/26/11	6

▲ Figure A1.7

Using DATEDIF to Compute Lead Times

▶ **Figure A1.8**

Defined Range Name for Cell B3 Using the Name *Box*

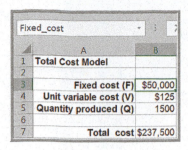

1. Use the *Name* box. The name box is at the upper left of a spreadsheet between the ribbon and the column headers. Usually, it displays the cell reference that is selected. To name a range, first select a cell or range and enter the range name in the *Name* box. Range names cannot contain any spaces, so it is common to use an underscore between words.

2. Use *Create from Selection*. This option is particularly useful when the names you want to use are listed in the column immediately to the right or left of the cell, or in the row immediately above or below the cell or range. The next example illustrates this procedure.

| EXAMPLE A1.6 | **Using *Create from Selection* to Define Range Names** |

In the *Total Cost Model* spreadsheet, we will use the text labels to the left of the numerical inputs as the range names. First, highlight the range A3:B5. Then, on the *Formulas* tab, choose *Create from Selection*. The box for the left column will automatically be checked. Click OK. If you select any numerical cell, you will see the range name in the *Name* box as shown in Figure A1.9.

3. Use *Define Name*. This allows you to enter the name but also provides an option to restrict the name to the current worksheet or allow it to be used in any worksheet of the workbook.

| EXAMPLE A1.7 | **Using *Define Name* to Create a Range Name** |

In the *Total Cost Model* spreadsheet, select cell B3. Click *Define Name* on the *Formulas* tab. This will bring up a dialog that allows you to enter a range name. Figure A1.10 illustrates this. Click OK.

▶ **Figure A1.9**

Defined Range Name for Cell B4 Using Create from Selection

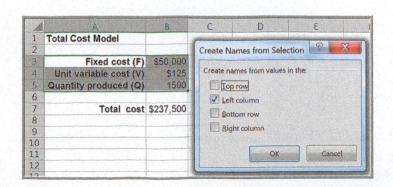

► **Figure A1.10**

Defined Range Name for Cell B3 Using Define Name

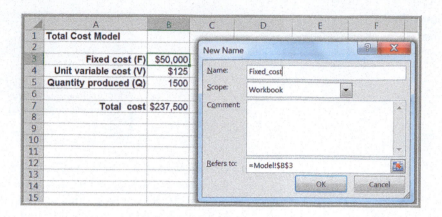

After range names have been defined, we can display a summary by clicking the *Name Manager* in the *Formulas* tab. Figure A1.11 shows this. (Note: The *Name Manager* button in the *Formulas* tab is only available in Windows. On a Mac, you can click *Define Name* and see a list of your range names, and then modify, add, or delete them.) This allows you to easily add, edit, or delete range names.

Finally, you can apply range names to the formulas in the spreadsheet. This replaces the cell references by the names, making it easy to understand the formulas. To do this, click on the drop-down arrow next to *Define Name* and select *Apply Names....* In the dialog box, select all the names you wish to use and click OK. Figure A1.12 shows the result for the total cost cell, B7. We see that the original formula=B3+B4*B5 now displays the names.

► **Figure A1.11**

Name Manager *Dialog (Windows)*

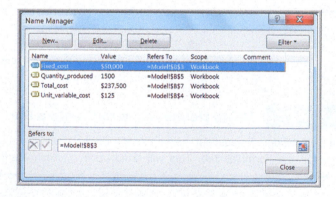

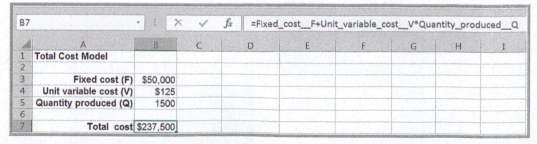

▲ **Figure A1.12**

Using Range Names in Formulas

VALUE Function

It's not unusual to download data from the Web that looks numerical but is actually expressed as text; this is often true of data from U.S. government Web sites. One way to tell is that text is usually left justified, while numbers are right justified. The function VALUE(*text*) can be used to convert a text string that represents a number to a number.

Paste Special

If you cut and paste a cell or a range, Excel will copy any formulas that are contained within the cell or range (using the appropriate relative or absolute addressing). In some cases, you may only want to paste the numerical values, not the formulas. This can be accomplished using the *Paste Special* command. From the *Edit* group, click the down arrow on the *Paste* icon to display the options and choose *Paste Special*.

Figure A1.13 shows the Excel *Paste Special* dialog. There are many options, but we will only illustrate three of them. First, copy the range of cells of interest and click on the cell where you want to paste the results.

- To paste only the values in cells (not the formulas), select *Values* and then click OK.
- To transpose data in columns to rows and vice versa, check the *Transpose* box and then click OK. (Make sure you won't overwrite any existing data!)

The *Operation* choices allow you to transform a range of numbers by adding, subtracting, dividing, or multiplying each number in the range by a constant. We illustrate this in the following example.

EXAMPLE A1.8   **Currency Conversion Using *Paste Special***

Figure A1.14 shows sales data in column B expressed in euros. Assume that the current conversion factor to U.S. dollars is 1 euro = $1.117. To convert these into U.S. dollars, first copy the data to column C (if you don't, the procedure will convert the data in column B to dollars, but you will no longer have the original data). Select cell C3 corresponding to the conversion factor. Next, select the range of data in column C and open the *Paste Special* dialog. Select *Multiply* and then press OK. The original data in euros will be converted into dollars, as shown in the figure (you will need to reformat the data as dollars).

► **Figure A1.13**

Paste Special *Dialog*

	A	B	C
1	European Sales		
2			
3	Euro-Dollar conversion factor		1.117
4			
5			
6	Month	Sales (Euros)	Sales (Dollars)
7	January	€ 24,169.00	$26,996.77
8	February	€ 30,472.00	$34,037.22
9	March	€ 29,547.00	$33,004.00
10	April	€ 25,695.00	$28,701.32
11	May	€ 27,580.00	$30,806.86
12	June	€ 27,963.00	$31,234.67
13	July	€ 29,647.00	$33,115.70
14	August	€ 32,513.00	$36,317.02
15	September	€ 35,176.00	$39,291.59
16	October	€ 31,468.00	$35,149.76
17	November	€ 30,274.00	$33,816.06
18	December	€ 27,486.00	$30,701.86

Concatenation

To concatenate means to join. In many applications, you might wish to take text data that are in different columns and join them together (for example, first and last names). The Excel function CONCATENATE(*text1, text2, . . . , text30*) can be used to join up to 30 text strings into a single string. For example, suppose that you have a database of names, with last names in column A and first names in column B. Suppose that cell A1 contains the last name Smith, and cell B1 contains the first name John. Then CONCATENATE(B1, " ", A1) will result in the text string John Smith. Note that the cells are separated by a field of open quotes around a blank space; this provides a space between the first and last name; otherwise, the function would have produced JohnSmith.

We may also perform concatenation using Excel formulas and the operator &. For example, to concatenate the text in cells B1 and A1, enter the formula =B1&A1, which would result in JohnSmith. If we use the formula =B1&" "&A1, then we are simply inserting a blank between the names: John Smith. Any text string can be used; for example, ="Dr. "&B1&" "&A1 would result in Dr. John Smith.

Error Values

Excel displays various error values if it cannot compute a formula. These error values are the following:

- #DIV/0!—A formula is trying to divide by zero.
- #N/A—"Not available," meaning that the formula could not return a result.
- #NAME?—An invalid name is used in a formula.
- #NUM!—An invalid argument is used in a function, such as a negative number in SQRT.
- #REF!—A formula contains an invalid cell reference.
- #VALUE!—Excel cannot compute a function because of an invalid argument.

Sometimes these are not user errors, but quirks in the data. For instance, you might have missing data in a column that is used in a denominator of a formula and obtain a #DIV/0! error. You might use the Excel function IFERROR(*value, value_if_error*) to display a specific value if an error is present. For example, if you are computing A1/B1, then if B1 is zero or blank, the function =IFERROR(A1/B1, " ") will display a blank cell instead of #DIV/0!; or =IFERROR(A1/B1, "Check the data in cell B1!") will provide better guidance.

PROBLEMS AND EXERCISES APPENDIX A1

Excel Formulas and Addressing

1. Develop a spreadsheet for computing the demand for any values of the input variables in the linear demand and nonlinear demand prediction models in Examples 1.7 and 1.8 in the chapter.

2. The Excel file *Science and Engineering Jobs* shows the number of jobs in thousands in the year 2000 and projections made for 2010 from a government study. Use the Excel file to compute the projected increase from the 2000 baseline and the percentage increase for each occupational category.

3. A new graduate has taken a job with an annual salary of $60,000. She expects her salary to go up by 2.5% each year for the first five years. Her starting salary is stored in cell A4 of an Excel worksheet, and the salary increase rate is stored in cell B4. Construct a table with years 1 through 5 in cells A6:A10 and her salary in cells B6:B10. Write the formula for her salary in year 2 (in cell B7) that can be copied and pasted correctly in cells B8 through B10.

4. Example 1.2 in the chapter described a scenario for new product sales that can be characterized by a formula called a Gompertz curve:

$$S = ae^{be^{ct}}$$

Develop a spreadsheet for calculating sales using this formula for $t = 0$ to 160 in increments of 10 when $a = 15{,}000$, $b = -8$, and $c = -0.05$.

5. Return on investment (ROI) is profit divided by investment. In marketing, ROI is determined as incremental sales times gross margin minus marketing investment, all divided by marketing investment. Suppose that a company plans to spend $3 million to place search engine ads and expects $15 million in incremental sales. Its gross margin is estimated to be 45%.

   a. Develop a spreadsheet to compute the marketing ROI.

   b. Use the spreadsheet to predict how ROI will change if the incremental sales estimate is wrong (consider a range of values above and below the expected sales).

Excel Functions

6. In the *Accounting Professionals* database, use Excel functions to find

   a. the maximum number of years of service.

   b. the average number of years of service.

   c. the number of male employees in the database.

   d. the number of female employees who have a CPA.

7. A company forecasts its net revenue for the next three years as $172,800, $213,580, and $293,985. Find the net present value of these cash flows, assuming a discount rate of 4.2%

8. A pharmaceutical manufacturer has projected net profits for a new drug that is being released to the market over the next five years:

Year	Net Profit
1	($300,000,000)
2	($145,000,000)
3	$50,000,000
4	$125,000,000
5	$530,000,000

A fixed cost of $80,000,000 has been incurred for research and development (in year 0). Use a spreadsheet to find the net present value of these cash flows for a discount rate of 3%.

9. The worksheet *Base Data* in the Excel file *Credit Risk Data* provides information about 425 bank customers who had applied for loans. The data include the purpose of the loan, checking and savings account balances, number of months as a customer of the bank, months employed, gender, marital status, age, housing status and number of years at current residence, job type, and credit-risk classification by the bank.[1] Use the COUNTIF function to determine (1) how many customers applied for new-car, used-car, business, education, small-appliance, and furniture loans and (2) the number of customers with checking account balances less than $500.

---

[1]Based on Efraim Turban, Ranesh Sharda, Dursun Delen, and David King, *Business Intelligence: A Managerial Approach*, 2nd ed. (Upper Saddle River NJ: Prentice Hall, 2011).

10. The Excel file *Store and Regional Sales Database* provides sales data for computers and peripherals showing the store identification number, sales region, item number, item description, unit price, units sold, and month when the sales were made during the fourth quarter of last year.[2] Modify the spreadsheet to calculate the total sales revenue for each of the eight stores as well as each of the three sales regions.

Miscellaneous Excel Functions and Tools

11. Define range names for all the data and model entities in the *Break-Even Decision Model* spreadsheet and apply them to the formulas in the model.

12. Define range names for all the entities in the *Crebo Manufacturing Model* spreadsheet and apply them to the formulas in the model.

13. Define range names for all the entities in the *Hotel Overbooking Model* spreadsheet and apply them to the formulas in the model.

---

[2]Based on Kenneth C. Laudon and Jane P. Laudon, *Essentials of Management Information Systems*, 9th ed. (Upper Saddle River, NJ: Prentice Hall, 2011).

Database Analytics

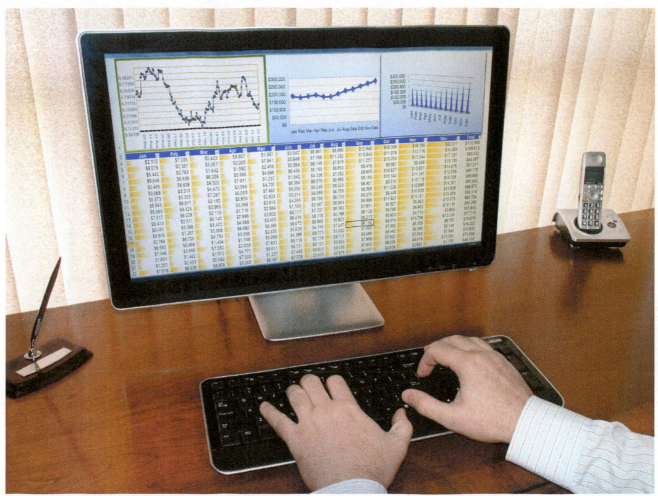

NAN728/Shutterstock

LEARNING OBJECTIVES After studying this chapter, you will be able to:

- Explain the difference between a data set and a database and apply Excel range names in data files.
- Construct Excel tables and be able to sort and filter data.
- Apply the Pareto principle to analyze data.
- Use database functions to extract records.

- Apply logical functions in Excel formulas.
- Use Excel lookup functions to make database queries.
- Design simple Excel templates for descriptive analytics.
- Use PivotTables to analyze and gain insight from data.

Most organizations begin their analytics journey using descriptive analytics tools to understand past and current performance to support customer service representatives, technical support, manufacturing, and other needs. This typically involves database analytics—querying databases and "drilling down" to better understand relationships within data, developing summarized reports, sorting data, and creating charts to visualize data (discussed in the next chapter)—techniques most commonly associated with business intelligence/information systems. These techniques provide a means of translating data into meaningful information. Data are raw numbers and facts, for example, information about customers, sales, costs, revenue, and so on. When data are manipulated, summarized, or processed in some way to provide insight and understanding, information results, which allows managers to make informed decisions.

In your career, you will likely need to extract or manipulate information from a database. Typical questions that you might address or be asked to determine are the following:

- How many units did we sell last quarter in each store?
- What was our total cost of materials last month?
- How much profit did we make last month?
- What percentage of customers responded that they are highly likely to recommend us?
- How many defective items did each factory make?
- How many orders were delivered on time?
- How does employee satisfaction differ by job classification and region or factory?

You might also use database analytics in your personal life, for example:

- How did my investment portfolio perform this year as compared to the previous year?
- In what stocks or mutual funds should I invest?
- Which of my fantasy sports players ranked in the top 10% of performance?
- How did we spend our money last year as compared to our budget?
- How can I compare the cost of living among various cities where I am interviewing for a job?
- What foods should I eat to meet my health goals or nutritional requirements?

The applications are endless! All you need is data, and data are everywhere.

Many sophisticated tools, such as Microsoft Access, SQL (structured query language), and OLAP (online analytical processing) software, are available to organize and query data to answer routine business questions, and to process data to create meaningful information. However, in this book, we won't be delving deeply into databases or database management systems but will work with individual database files or simple data sets. Because spreadsheets are convenient tools for storing and manipulating data sets and database files, we will use them for all examples and problems. Thus, in this chapter, we focus only on basic concepts and approaches that can be implemented easily using Excel. We will introduce many useful advanced functions available in Excel and illustrate how they can be used for descriptive analytics. These functions are also useful in spreadsheet modeling, which we will study later in this book.

Data Sets and Databases

A **data set** is simply a collection of data. Marketing survey responses, a table of historical stock prices, and a collection of measurements of dimensions of a manufactured item are examples of data sets. As we embark on statistical analysis procedures in subsequent chapters, we will see many examples of **empirical data**—data that come from experimentation and observation.

Most data that managers work with are stored in databases. A **database** is a collection of related files containing records on people, places, or things. Databases provide structure to data. The people, places, or things for which we store and maintain information are called *entities*.[1] A database is usually organized in a two-dimensional table, where the columns correspond to each individual element of data (called *fields*, or *attributes*), and the rows represent records of related data elements.

Empirical data, such as those obtained from questionnaires and surveys, are often organized as databases, and all the database tools that we will learn can be used. The Drout

EXAMPLE 2.1	A Purchase Order Database[2]

Figure 2.1 shows a portion of the Excel file *Purchase Orders* that shows a list of orders for parts used in airplane assemblies. The fields are listed in row 3 of the database and consist of the supplier name, order number, cost, quantity, A/P (accounts payable) terms, and so on. Each record starting in row 4 provides the data for each of these fields for one order.

	A	B	C	D	E	F	G	H	I	J
1	Purchase Orders									
2										
3	Supplier	Order No.	Item No.	Item Description	Item Cost	Quantity	Cost per order	A/P Terms (Months)	Order Date	Arrival Date
4	Hulkey Fasteners	Aug11001	1122	Airframe fasteners	$ 4.25	19,500	$ 82,875.00	30	08/05/11	08/13/11
5	Alum Sheeting	Aug11002	1243	Airframe fasteners	$ 4.25	10,000	$ 42,500.00	30	08/08/11	08/14/11
6	Fast-Tie Aerospace	Aug11003	5462	Shielded Cable/ft.	$ 1.05	23,000	$ 24,150.00	30	08/10/11	08/15/11
7	Fast-Tie Aerospace	Aug11004	5462	Shielded Cable/ft.	$ 1.05	21,500	$ 22,575.00	30	08/15/11	08/22/11
8	Steelpin Inc.	Aug11005	5319	Shielded Cable/ft.	$ 1.10	17,500	$ 19,250.00	30	08/20/11	08/31/11
9	Fast-Tie Aerospace	Aug11006	5462	Shielded Cable/ft.	$ 1.05	22,500	$ 23,625.00	30	08/20/11	08/26/11
10	Steelpin Inc.	Aug11007	4312	Bolt-nut package	$ 3.75	4,250	$ 15,937.50	30	08/25/11	09/01/11

▶ **Figure 2.1**

A Portion of Excel File Purchase Orders

---

[1]Kenneth C. Laudon and Jane P. Laudon, *Essentials of Management Information Systems*, 9th ed. (Upper Saddle River, NJ: Prentice Hall, 2011): 159.

[2]Based on Laudon and Loudon, ibid.

Advertising case at the end of this chapter is one example where you can apply database tools to empirical data. Although the databases we will use in this book are "clean," you must be aware that many encountered in practice have typos, missing values, or other errors, which must be corrected in order to use them effectively.

Using Range Names in Databases

We introduced the concept of range names in spreadsheets in the Appendix to Chapter 1. Using range names for data arrays (that is, rows, columns, or rectangular matrix arrays) can greatly simplify many database calculations. The following example shows how to do this.

EXAMPLE 2.2   **Defining and Using Range Names for a Database**

Figure 2.2 shows a simple database for monthly sales of five products for one year (Excel file *Monthly Product Sales*). We will define names for each row (months) and column (product name) of data. Any of the techniques described in the Appendix to Chapter 1 can be used (we suggest that you review this material first). For instance, select the range B4:B15, which consists of the sales of Product A. In the *Name* box, enter Product_A and press *Enter*. Alternatively,

you can use *Define Name* to do this. Finally, you can select the entire range of data including the product names and use the *Create from Selection* option to define range names for all products at the same time.

You may use range names to simplify calculations. For example, to find the total sales, you can use the function =SUM(Product_A) instead of =SUM(B4:B15).

Excel also provides a convenient way to include all data in a row or column in a function. For example, the range for column B is expressed as B:B; the range for row 4 is expressed as 4:4, and so on. Therefore, =SUM(B:B) will calculate the sum of all data in column B; in Figure 2.2, this will be the total sales of Product A. The function =SUM(4:4) will calculate the sum of all data in row 4; in Figure 2.2, this will be the sum of January sales for all products. Using these ranges is particularly helpful if you add new data to a spreadsheet; you would not have to adjust any defined range names.

CHECK YOUR UNDERSTANDING

1. Explain the difference between a data set and a database.

2. For the *Monthly Product Sales* Excel file, use the *Create from Selection* option to define range names for all products (column data) and all months (row data).

3. For the *Monthly Product Sales* Excel file, use the function SUM(B:B) to calculate the total sales for Product A. What happens if you enter this function in column B?

▶ **Figure 2.2**

Monthly Product Sales *Data*

	A	B	C	D	E	F
1	Sales Units					
2						
3	Month	Product A	Product B	Product C	Product D	Product E
4	January	7792	5554	3105	3168	10350
5	February	7268	3024	3228	3751	8965
6	March	7049	5543	2147	3319	6827
7	April	7560	5232	2636	4057	8544
8	May	8233	5450	2726	3837	7535
9	June	8629	3943	2705	4664	9070
10	July	8702	5991	2891	5418	8389
11	August	9215	3920	2782	4085	7367
12	September	8986	4753	2524	5575	5377
13	October	8654	4746	3258	5333	7645
14	November	8315	3566	2144	4924	8173
15	December	7978	5670	3071	6563	6088

ANALYTICS IN PRACTICE: Using Big Data to Monitor Water Usage in Cary, North Carolina[3]

When the Town of Cary installed wireless meters for 60,000 customers in 2010, it knew the new technology wouldn't just save money by eliminating manual monthly readings; the town also realized it would get more accurate and timely information about water consumption. The Aquastar wireless system reads meters once an hour—that's 8,760 data points per customer each year instead of 12 monthly readings. The data had tremendous potential, if it could be easily consumed. The challenge was to analyze half-a-billion data points on water usage and make them available, visually, to all customers, and gain a big-picture view of water consumption to better plan future water plant expansions and promote targeted conservation efforts.

Cary used advanced analytics software from SAS Institute. The ability to visually look at data by household or commercial customer, by hour, has led to some very practical applications:

- The town can notify customers of potential leaks within days.
- Customers can set alerts that notify them within hours if there is a spike in water usage.
- Customers can track their water usage online, helping them to be more proactive in conserving water.

The town estimates that by just removing the need for manual readings, the Aquastar system will save more than $10 million after the cost of the project. But the analytics component could provide even bigger savings. Already, both the town and individual citizens have saved money by catching water leaks early. As the Town of Cary continues to plan its future infrastructure needs, having accurate information on water usage will help it invest in the right amount of infrastructure at the right time. Additionally, understanding water usage will help the town if it experiences something detrimental like a drought.

Data Queries: Tables, Sorting, and Filtering

Managers make numerous queries about data. For example, Figure 2.3 shows a portion of the *Credit Risk Data* Excel file, which summarizes loan application data and credit risk assessment. This database contains the purpose of the loan, the balance of checking and savings accounts, how long the applicant has been a customer and length of employment, other demographic information, and an assessment of credit risk. A manager might be interested in comparing the amount of financial assets as related to the number of months employed, finding all loans of a certain type, or the proportion of low credit risk individuals who own a home. To address these queries, we need to sort the data in some way. Extracting a set of records having certain characteristics is called *filtering* the data. Excel provides a convenient way of formatting databases to facilitate such types of analyses, called *Tables*.

An Excel table allows you to use table references to perform basic calculations, as the next example illustrates.

▶ **Figure 2.3**

Portion of Excel File Credit Risk Data (Based on Kenneth C. Laudon and Jane P. Laudon, Essentials of Management Information Systems, 9th ed. Upper Saddle River, NJ: Prentice Hall, 2011).

	A	B	C	D	E	F	G	H	I	J	K	L
1	Credit Risk Data											
2												
3	Loan Purpose	Checking	Savings	Months Customer	Months Employed	Gender	Marital Status	Age	Housing	Years	Job	Credit Risk
4	Small Appliance	$0	$739	13	12	M	Single	23	Own	3	Unskilled	Low
5	Furniture	$0	$1,230	25	0	M	Divorced	32	Own	1	Skilled	High
6	New Car	$0	$389	19	119	M	Single	38	Own	4	Management	High
7	Furniture	$638	$347	13	14	M	Single	36	Own	2	Unskilled	High
8	Education	$963	$4,754	40	45	M	Single	31	Rent	3	Skilled	Low
9	Furniture	$2,827	$0	11	13	M	Married	25	Own	1	Skilled	Low
10	New Car	$0	$229	13	16	M	Married	26	Own	3	Unskilled	Low

[3]"Municipality puts wireless water meter-reading data to work," Copyright 2016 SAS Institute Inc. Carey, NC USA. https://www.sas.com/en_us/customers/townofcary-aquastar.html

	Loan Purpo ⌄	Checkin ⌄	Savin ⌄	Months Customer ⌄	Months Employ ⌄	Gend ⌄	Marital Stat ⌄	Age ⌄	Housi ⌄	Years ⌄	J ⌄	Credit Ri ⌄
1	Credit Risk Data											
2												
4	Small Appliance	$0	$739	13	12	M	Single	23	Own	3	Unskilled	Low
5	Furniture	$0	$1,230	25	0	M	Divorced	32	Own	1	Skilled	High
6	New Car	$0	$389	19	119	M	Single	38	Own	4	Management	High
7	Furniture	$638	$347	13	14	M	Single	36	Own	2	Unskilled	High
8	Education	$963	$4,754	40	45	M	Single	31	Rent	3	Skilled	Low
9	Furniture	$2,827	$0	11	13	M	Married	25	Own	1	Skilled	Low
10	New Car	$0	$229	13	16	M	Married	26	Own	3	Unskilled	Low
11	Business	$0	$533	14	2	M	Single	27	Own	1	Unskilled	Low
12	Small Appliance	$6,509	$493	37	9	M	Single	25	Own	2	Skilled	High
13	Small Appliance	$966	$0	25	4	F	Divorced	43	Own	1	Skilled	High
14	Business	$0	$989	49	0	M	Single	32	Rent	2	Management	High

▶ **Figure 2.4**

Portion of Credit Risk Data
Formatted as an Excel Table

EXAMPLE 2.3 Creating an Excel Table

We will use the *Credit Risk Data* file to illustrate an **Excel table**. First, select the range of the data, including headers (a useful shortcut is to select the first cell in the upper left corner, then click Ctrl+Shift+down arrow, and then Ctrl+Shift+right arrow; on a Mac, use the Command key instead of Ctrl). Next, click *Table* from the *Tables* group on the *Insert* tab and make sure that the box for *My Table Has Headers* is checked. (You may also just select a cell within the table and then click on *Table* from the *Insert* menu. Excel will choose the table range for you to verify.) The table range will now be formatted and will continue automatically when new data are entered. Figure 2.4 shows a portion of the result. Note that the rows are shaded and that each column header has a drop-down arrow to filter the data (we'll discuss this shortly). If you click within a table, the *Table Tools Design* tab will appear in the ribbon, allowing you to do a variety of things, such as change the color scheme, remove duplicates, and change the formatting.

A useful option in the *Design* tab is to add a total row. In the *Table Style Options* group, check the box for *Total Row*, and Excel will add a new row at the bottom of the table. If you click on any cell in the total row, you can select the type of calculation you want from the drop-down box; this includes SUM, AVERAGE, COUNT, MAX, and MIN.

EXAMPLE 2.4 Table-Based Calculations

Suppose that in the *Credit Risk Data* table, we wish to calculate the total amount of savings in column C. We could, of course, simply use the function =SUM(C4:C428). However, with a table, we could use the formula =SUM(Table1[Savings]). The table name, Table1, can be found (and changed) in the *Properties* group of the *Table Tools Design* tab in Windows or in the *Table* tab on a Mac. Note that Savings is the name of the header in column C. One of the advantages of doing this is that if we add new records to the table, the calculation will be updated automatically, and we don't have to change the range in the formula or get a wrong result if we forget to. As another example, we could find the number of home owners using the function =COUNTIF(Table1[Housing], "Own").

If you add additional records at the end of the table, they will automatically be included and formatted, and if you create a chart based on the data, the chart will automatically be updated if you add new records.

Sorting Data in Excel

Excel provides many ways to sort lists by rows or column or in ascending or descending order and using custom sorting schemes. The sort buttons in Excel can be found under the *Data* tab (see Figure 2.5; the Mac *Data* tab ribbon is similar). Select a single cell in the column you want to sort on and click the "AZ down arrow" button to sort from smallest to largest or the "AZ up arrow" button to sort from largest to smallest. You may also click the *Sort* button to specify criteria for more advanced sorting capabilities.

▶ **Figure 2.5**

Excel Data *Tab Ribbon*

EXAMPLE 2.5   **Sorting Data in the *Purchase Orders* Database**

In the *Purchase Orders* database, suppose we wish to sort the data by supplier. Click on any cell in column A of the data (but not the header cell A3) and then the "*AZ down*" button in the *Data* tab. Excel will select the entire range of the data and sort by name of supplier in column A, a portion of which is shown in Figure 2.6. This allows you to easily identify the records that correspond to all orders from a particular supplier.

Pareto Analysis

Pareto analysis is a term named after an Italian economist, Vilfredo Pareto, who, in 1906, observed that a large proportion of the wealth in Italy was owned by a relatively small proportion of the people. The Pareto principle is often seen in many business situations. For example, a large percentage of sales usually comes from a small percentage of customers, a large percentage of quality defects stems from just a couple of sources, or a large percentage of inventory value corresponds to a small percentage of items. As a result, the Pareto principle is also often called the "80–20 rule," referring to the generic situation in which 80% of some output comes from 20% of some input. A Pareto analysis relies on sorting data and calculating the cumulative percentage of the characteristic of interest.

EXAMPLE 2.6   **Applying the Pareto Principle**

The Excel file *Bicycle Inventory* lists the inventory of bicycle models in a sporting goods store.[4] To conduct a Pareto analysis, we first compute the inventory value of each product by multiplying the quantity on hand by the purchase cost; this is the amount invested in the items that are currently in stock. Then we sort the data in decreasing order of inventory value and compute the percentage of the total inventory value for each product and the cumulative percentage. See columns G through I in Figure 2.7. We see that about 75% of the inventory value is accounted for by less than 40% (9 of 24) of the items. If these high-value inventories aren't selling well, the store manager may wish to keep fewer in stock.

	A	B	C	D	E	F	G	H	I	J
1	Purchase Orders									
2										
3	**Supplier**	**Order No.**	**Item No.**	**Item Description**	**Item Cost**	**Quantity**	**Cost per order**	**A/P Terms (Months)**	**Order Date**	**Arrival Date**
4	Alum Sheeting	Aug11002	1243	Airframe fasteners	$ 4.25	10,000	$ 42,500.00	30	08/08/11	08/14/11
5	Alum Sheeting	Sep11002	5417	Control Panel	$ 255.00	406	$ 103,530.00	30	09/01/11	09/10/11
6	Alum Sheeting	Sep11008	1243	Airframe fasteners	$ 4.25	9,000	$ 38,250.00	30	09/05/11	09/12/11
7	Alum Sheeting	Oct11016	1243	Airframe fasteners	$ 4.25	10,500	$ 44,625.00	30	10/10/11	10/17/11
8	Alum Sheeting	Oct11022	4224	Bolt-nut package	$ 3.95	4,500	$ 17,775.00	30	10/15/11	10/20/11
9	Alum Sheeting	Oct11026	5417	Control Panel	$ 255.00	500	$ 127,500.00	30	10/20/11	10/27/11
10	Alum Sheeting	Oct11028	5634	Side Panel	$ 185.00	150	$ 27,750.00	30	10/25/11	11/03/11
11	Alum Sheeting	Oct11036	5634	Side Panel	$ 185.00	140	$ 25,900.00	30	10/29/11	11/04/11
12	Durrable Products	Aug11008	7258	Pressure Gauge	$ 90.00	100	$ 9,000.00	45	08/25/11	08/28/11
13	Durrable Products	Sep11009	7258	Pressure Gauge	$ 90.00	120	$ 10,800.00	45	09/05/11	09/09/11
14	Durrable Products	Sep11027	1369	Airframe fasteners	$ 4.20	15,000	$ 63,000.00	45	09/25/11	09/30/11
15	Durrable Products	Sep11031	1369	Airframe fasteners	$ 4.20	14,000	$ 58,800.00	45	09/27/11	10/03/11

▶ **Figure 2.6**

Portion of Purchase Orders *Database Sorted by Supplier Name*

---

[4]Based on Kenneth C. Laudon and Jane P. Laudon, *Essentials of Management Information Systems*, 9th ed. (Upper Saddle River, NJ: Prentice Hall, 2011).

	A	B	C	D	E	F	G	H	I
1	Bicycle Inventory								
2									
3	Product Category	Product Name	Purchase Cost	Selling Price	Supplier	Quantity on Hand	Inventory Value	Percentage	Cumulative %
4	Road	Runroad 5000	$450.95	$599.99	Run-Up Bikes	5	$ 2,254.75	11.2%	11.2%
5	Road	Runroad 1000	$250.95	$350.99	Run-Up Bikes	8	$ 2,007.60	10.0%	21.1%
6	Road	Elegant 210	$281.52	$394.13	Bicyclist's Choice	7	$ 1,970.64	9.8%	30.9%
7	Road	Runroad 4000	$390.95	$495.99	Run-Up Bikes	5	$ 1,954.75	9.7%	40.6%
8	Mtn.	Eagle 3	$350.52	$490.73	Bike-One	5	$ 1,752.60	8.7%	49.3%
9	Road	Classic 109	$207.49	$290.49	Bicyclist's Choice	7	$ 1,452.43	7.2%	56.5%
10	Hybrid	Eagle 7	$150.89	$211.46	Bike-One	9	$ 1,358.01	6.7%	63.3%
11	Hybrid	Tea for Two	$429.02	$609.00	Simpson's Bike Supply	3	$ 1,287.06	6.4%	69.7%
12	Mtn.	Bluff Breaker	$375.00	$495.00	The Bike Path	3	$ 1,125.00	5.6%	75.2%
13	Mtn.	Eagle 2	$401.11	$561.54	Bike-One	2	$ 802.22	4.0%	79.2%
14	Leisure	Breeze LE	$109.95	$149.95	The Bike Path	5	$ 549.75	2.7%	81.9%
15	Children	Runkidder 100	$50.95	$75.99	Run-Up Bikes	10	$ 509.50	2.5%	84.5%
16	Mtn.	Jetty Breaker	$455.95	$649.95	The Bike Path	1	$ 455.95	2.3%	86.7%
17	Leisure	Runcool 3000	$85.95	$135.99	Run-Up Bikes	5	$ 429.75	2.1%	88.9%
18	Children	Coolest 100	$69.99	$97.98	Bicyclist's Choice	6	$ 419.94	2.1%	91.0%
19	Mtn.	Eagle 1	$410.01	$574.01	Bike-One	1	$ 410.01	2.0%	93.0%
20	Children	Green Rider	$95.47	$133.66	Simpson's Bike Supply	4	$ 381.88	1.9%	94.9%
21	Leisure	Breeze	$89.95	$130.95	The Bike Path	4	$ 359.80	1.8%	96.7%
22	Leisure	Blue Moon	$75.29	$105.41	Simpson's Bike Supply	4	$ 301.16	1.5%	98.2%
23	Leisure	Supreme 350	$50.00	$70.00	Bicyclist's Choice	3	$ 150.00	0.7%	98.9%
24	Children	Red Rider	$15.00	$25.50	Simpson's Bike Supply	8	$ 120.00	0.6%	99.5%
25	Leisure	Starlight	$100.47	$140.66	Simpson's Bike Supply	1	$ 100.47	0.5%	100.0%
26	Hybrid	Runblend 2000	$180.95	$255.99	Run-Up Bikes	0	$ -	0.0%	100.0%
27	Road	Twist & Shout	$490.50	$635.70	Simpson's Bike Supply	0	$ -	0.0%	100.0%
28						Total	$ 20,153.27		

▶ **Figure 2.7**

Pareto Analysis of Bicycle Inventory

Filtering Data

For large data files, finding a particular subset of records that meet certain characteristics by sorting can be tedious. **Filtering** simplifies this process. Excel provides two filtering tools: *AutoFilter* for simple criteria and *Advanced Filter* for more complex criteria. These tools are best understood by working through some examples.

EXAMPLE 2.7 **Filtering Records by Item Description**

In the *Purchase Orders* database, suppose we are interested in extracting all records corresponding to the item bolt-nut package. First, select any cell within the database. Then, from the Excel *Data* tab, click on *Filter*. A drop-down arrow will then be displayed on the right side of each header column. Clicking on one of these will display a drop-down box. These are the options for filtering on that column of data. Click the one next to the Item Description header. Uncheck the box for *Select All* and then check the box corresponding to the bolt-nut package, as shown in Figure 2.8. Click the *OK* button, and the *Filter* tool will display only those orders for this item (Figure 2.9). To restore the original data file, click on the drop-down arrow again and then click *Clear filter from "Item Description."*

▶ **Figure 2.8**

Selecting Records for Bolt-Nut Package

	A	B	C	D	E
1	Purchase Orders				
2					
3	Supplier	Order N	Item N	Item Description	Item Co
4	Hulkey Fasteners	Aug1	↓ Sort A to Z		$4.25
5	Alum Sheeting	Aug1	↓ Sort Z to A		$4.25
6	Fast-Tie Aerospace	Aug1	Sort by Color ▶		$1.05
7	Fast-Tie Aerospace	Aug1			$1.05
8	Steelpin Inc.	Aug1	Clear Filter From "Item Description"		$1.10
9	Fast-Tie Aerospace	Aug1	Filter by Color ▶		$1.05
10	Steelpin Inc.	Aug1	Text Filters ▶		$3.75
11	Durrable Products	Aug1	Search		$90.00
12	Fast-Tie Aerospace	Aug1	☑ (Select All)		$2.45
13	Fast-Tie Aerospace	Aug1	☐ Airframe fasteners		$1.05
14	Steelpin Inc.	Aug1	☑ Bolt-nut package		$1.10
15	Hulkey Fasteners	Aug1	☐ Control Panel		$1.25
16	Hulkey Fasteners	Aug1	☐ Door Decal		$0.75
17	Steelpin Inc.	Aug1	☐ Electrical Connector		$1.65
18	Steelpin Inc.	Sep1	☐ Gasket		$3.75
19	Alum Sheeting	Sep1	☐ Hatch Decal		$255.00
20	Hulkey Fasteners	Sep1	☐ Machined Valve		$1.25
21	Steelpin Inc.	Sep1	☐ O-Ring		$1.65
22	Steelpin Inc.	Sep1	OK Cancel		$3.75
23	Hulkey Fasteners	Sep1			$4.25

	A	B	C	D	E	F	G	H	I	J
1	Purchase Orders									
2										
3	Supplier	Order N	Item N	Item Description	Item Co	Quanti	Cost per orde	A/P Terms (Months	Order Da	Arrival Da
10	Steelpin Inc.	Aug11007	4312	Bolt-nut package	$3.75	4,250	$15,937.50	30	08/25/11	09/01/11
18	Steelpin Inc.	Sep11001	4312	Bolt-nut package	$3.75	4,200	$15,750.00	30	09/01/11	09/10/11
22	Steelpin Inc.	Sep11005	4312	Bolt-nut package	$3.75	4,150	$15,562.50	30	09/03/11	09/11/11
24	Spacetime Technologies	Sep11007	4111	Bolt-nut package	$3.55	4,800	$17,040.00	25	09/05/11	09/20/11
32	Spacetime Technologies	Sep11015	4111	Bolt-nut package	$3.55	4,585	$16,276.75	25	09/10/11	09/30/11
36	Spacetime Technologies	Sep11019	4111	Bolt-nut package	$3.55	4,200	$14,910.00	25	09/15/11	10/15/11
39	Spacetime Technologies	Sep11022	4111	Bolt-nut package	$3.55	4,250	$15,087.50	25	09/20/11	10/10/11
43	Spacetime Technologies	Sep11026	4111	Bolt-nut package	$3.55	4,200	$14,910.00	25	09/25/11	10/25/11
61	Spacetime Technologies	Oct11010	4111	Bolt-nut package	$3.55	4,600	$16,330.00	25	10/05/11	10/19/11
62	Durrable Products	Oct11011	4569	Bolt-nut package	$3.50	3,900	$13,650.00	45	10/05/11	10/15/11
73	Alum Sheeting	Oct11022	4224	Bolt-nut package	$3.95	4,500	$17,775.00	30	10/15/11	10/20/11

▶ **Figure 2.9**

Filter Results for Bolt-Nut Package

EXAMPLE 2.8	**Filtering Records by Item Cost**

In this example, suppose we wish to identify all records in the *Purchase Orders* database whose item cost is at least $200. First, click on the drop-down arrow in the Item Cost column and position the cursor over *Numbers Filter*. This displays a list of options, as shown in Figure 2.10. Select *Greater Than Or Equal To . . .* from the list. (The Mac interface is slightly different, but has the same capabilities.) This brings up a *Custom AutoFilter* dialog (Figure 2.11) that allows you to specify up to two specific criteria using "and" and "or" logic. Enter 200 in the box as shown in Figure 2.11 and then click *OK*. The tool will display all records having an item cost of $200 or more.

AutoFilter creates filtering criteria based on the type of data being filtered. For instance, in Figure 2.10 we see that the *Number Filters* menu list includes numerical criteria such as "equals," "does not equal," and so on. If you choose to filter on Order Date or Arrival Date, the *AutoFilter* tools will display a different *Date Filters* menu list for filtering that includes "tomorrow," "next week," "year to date," and so on.

AutoFilter can be used sequentially to "drill down" into the data. For example, after filtering the results by bolt-nut package in Figure 2.9, we could then filter by order date and select all orders processed in September.

▶ **Figure 2.10**

Selecting Records for Item Cost Filtering

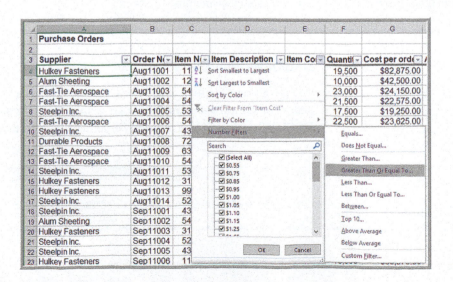

▶ **Figure 2.11**

Custom AutoFilter *Dialog*

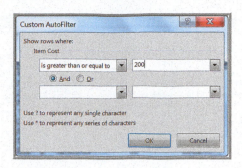

▶ **Figure 2.11**

Custom AutoFilter *Dialog*

Caution! The *Filter* tool does not *extract* the records; it simply *hides* the records that don't match the criteria. This means that if you highlight a range of filtered data to compute a sum or average, for example, you will get the results for *all* records, including those that are hidden from view. However, you can copy and paste the filtered data to another range or Excel worksheet, which then uses only the filtered records, and get the correct results.

The *Advanced Filter* provides a way of explicitly defining criteria by which to filter a database. To do this, first copy the headers from the database to an open location in the worksheet. Under the headers, specify the criteria that you want to use to filter the data. Multiple criteria in the same row are logically joined by "and," while criteria in rows are joined by "or." In general, criteria for numerical values can include $=$, $>$, $<$, $<>$ (not equal), $>=$, or $<=$.

EXAMPLE 2.9   **Using the *Advanced Filter***

We will use the *Purchase Orders* database. Figure 2.12 shows the criteria by which we want to filter the records. These criteria will find all the records for Hulkey Fasteners having order quantities that exceed 5,000 and order dates before 9/1/11, as well as all records for Steelpin Inc. with order quantities less than 5,000 and order dates before 9/1/11. You need not include columns that have no criteria; in this case, we could have simply listed the columns Supplier, Quantity, and Order Date along with the specified criteria.

To use the *Advanced Filter*, choose *Advanced* next to the *Filter* in the *Data* tab. In the dialog box, enter the list range of the database and the criteria range (see Figure 2.13). Figure 2.14 shows the results; note that only the records meeting the criteria are displayed.

Database Functions

You are already familiar with basic Excel functions such as SUM, AVERAGE, COUNT, and so on. **Database functions** start with a "D" (for example, DSUM, DAVERAGE, DCOUNT) and allow you to specify criteria that limit the calculations to a subset of records

	A	B	C	D	E	F	G	H	I	J
1	Purchase Orders									
2										
3	Supplier	Order No.	Item No.	Item Description	Item Cost	Quantity	Cost per order	A/P Terms (Months)	Order Date	Arrival Date
4	Hulkey Fasteners					>5000			<9/1/11	
5	Steelpin Inc.					<5000			<9/1/11	

▶ **Figure 2.12**

Criteria for Using the Advanced Filter *for the* Purchase Orders *Database*

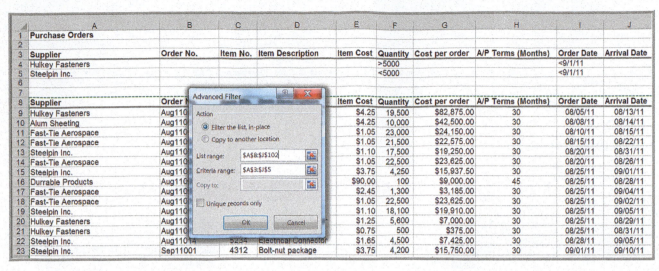

	A	B	C	D	E	F	G	H	I	J
1	Purchase Orders									
2										
3	Supplier	Order No.	Item No.	Item Description	Item Cost	Quantity	Cost per order	A/P Terms (Months)	Order Date	Arrival Date
4	Hulkey Fasteners					>5000			<9/1/11	
5	Steelpin Inc.					<5000			<9/1/11	
6										
7										
8	Supplier	Order N			Item Cost	Quantity	Cost per order	A/P Terms (Months)	Order Date	Arrival Date
9	Hulkey Fasteners	Aug110			$4.25	19,500	$82,875.00	30	08/05/11	08/13/11
10	Alum Sheeting	Aug110			$4.25	10,000	$42,500.00	30	08/08/11	08/14/11
11	Fast-Tie Aerospace	Aug110			$1.05	23,000	$24,150.00	30	08/10/11	08/15/11
12	Fast-Tie Aerospace	Aug110			$1.05	21,500	$22,575.00	30	08/15/11	08/22/11
13	Steelpin Inc.	Aug110			$1.10	17,500	$19,250.00	30	08/20/11	08/31/11
14	Fast-Tie Aerospace	Aug110			$1.05	22,500	$23,625.00	30	08/20/11	08/26/11
15	Steelpin Inc.	Aug110			$3.75	4,250	$15,937.50	30	08/25/11	09/01/11
16	Durrable Products	Aug110			$90.00	100	$9,000.00	45	08/25/11	08/28/11
17	Fast-Tie Aerospace	Aug110			$2.45	1,300	$3,185.00	30	08/25/11	09/04/11
18	Fast-Tie Aerospace	Aug110			$1.05	22,500	$23,625.00	30	08/25/11	09/02/11
19	Steelpin Inc.	Aug110			$1.10	18,100	$19,910.00	30	08/25/11	09/05/11
20	Hulkey Fasteners	Aug110			$1.25	5,600	$7,000.00	30	08/25/11	08/29/11
21	Hulkey Fasteners	Aug110			$0.75	500	$375.00	30	08/25/11	08/31/11
22	Steelpin Inc.	Aug11014	5234	Electrical Connector	$1.65	4,500	$7,425.00	30	08/28/11	09/05/11
23	Steelpin Inc.	Sep11001	4312	Bolt-nut package	$3.75	4,200	$15,750.00	30	09/01/11	09/10/11

Advanced Filter dialog box:

Action
- Filter the list, in-place
- Copy to another location

List range: $A$8:$J$102

Criteria range: $A$3:$J$5

Copy to:

☐ Unique records only

OK   Cancel

▶ **Figure 2.13**

Advanced Filter *Dialog*

	A	B	C	D	E	F	G	H	I	J
1	Purchase Orders									
2										
3	Supplier	Order No.	Item No.	Item Description	Item Cost	Quantity	Cost per order	A/P Terms (Months)	Order Date	Arrival Date
4	Hulkey Fasteners					>5000			<9/1/11	
5	Steelpin Inc.					<5000			<9/1/11	
6										
7										
8	Supplier	Order No.	Item No.	Item Description	Item Cost	Quantity	Cost per order	A/P Terms (Months)	Order Date	Arrival Date
9	Hulkey Fasteners	Aug11001	1122	Airframe fasteners	$4.25	19,500	$82,875.00	30	08/05/11	08/13/11
15	Steelpin Inc.	Aug11007	4312	Bolt-nut package	$3.75	4,250	$15,937.50	30	08/25/11	09/01/11
20	Hulkey Fasteners	Aug11012	3166	Electrical Connector	$1.25	5,600	$7,000.00	30	08/25/11	08/29/11
22	Steelpin Inc.	Aug11014	5234	Electrical Connector	$1.65	4,500	$7,425.00	30	08/28/11	09/05/11

▶ **Figure 2.14**

Advanced Filter *Results for the* Purchase Orders *Database*

in a database using the same format as the *Advanced Filter*. You can find all database functions by selecting the category *Database* in the *Insert Function* dialog. For example, the syntax for the DSUM function is DSUM(*database, field, criteria*). *Database* is the range that includes the column labels; *field* is the column name that contains the values to sum, enclosed in quotation marks, or a reference to the column name; and *criteria* is the range that specifies the records you want to sum (the same format as in Figure 2.12).

EXAMPLE 2.10   **Using a Database Function**

In the *Purchase Orders* database, suppose that we wish to find the total cost of all orders that meet the criteria specified in Figure 2.12. Figure 2.15 shows the application of the DSUM function. In this function, Criteria is a range name that corresponds to the criteria range A3:J5, in the

worksheet. Note that the function references cell G3 for the *field*, which corresponds to cost per order (we might also have created a range name for the entire database). Alternatively, we could have used the function =DSUM(A8:J102, "Cost per order", A3:J5).

| G6 | | | ⋮ | × | ✓ | f_x | =DSUM(A8:J102,G3,Criteria) | | | | |

⊿	A	B	C	D	E	F	G	H	I	J
1	Purchase Orders									
2										
3	Supplier	Order No.	Item No.	Item Description	Item Cost	Quantity	Cost per order	A/P Terms (Months)	Order Date	Arrival Date
4	Hulkey Fasteners					>5000			<9/1/11	
5	Steelpin Inc.					<5000			<9/1/11	
6						Total Cost	$113,237.50			

▶ **Figure 2.15**

Using the DSUM Function

CHECK YOUR UNDERSTANDING

1. What does the term *filtering* mean in the context of a database?

2. Convert the *Credit Risk* database to an Excel table, and use a table-based function to count the number of records designated as having high credit risk.

3. Sort the data in the *Purchase Orders* database from lowest cost per order to highest.

4. Use both the *AutoFilter* and the *Advanced Filter* to identify all orders for bolt-nut packages that are processed in September in the *Purchase Orders* database.

ANALYTICS IN PRACTICE: Discovering the Value of Database Analytics at Allders International[5]

Allders International specializes in duty-free operations with 82 tax-free retail outlets throughout Europe, including shops in airports and seaports and on cross-channel ferries. Like most retail outlets, Allders International must track masses of point-of-sale data to assist in inventory and product-mix decisions. Which items to stock at each of its outlets can have a significant impact on the firm's profitability. To assist them, they implemented a computer-based data warehouse to maintain the data. Prior to doing this, they had to analyze large quantities of paper-based data. This manual process was so overwhelming and time-consuming that the analyses were often too late to provide useful information for their decisions. The data warehouse allowed the company to make simple queries, such as finding the performance of a particular item across all retail outlets or the financial performance of a particular outlet, quickly and easily. This allowed them to identify which inventory items or outlets were underperforming. For instance, a Pareto analysis of its product lines (groups of similar items)

2jenn/Shutterstock

found that about 20% of the product lines were generating 80% of the profits. This allowed them to selectively eliminate some of the items from the other 80% of the product lines, which freed up shelf space for more profitable items and reduced inventory and supplier costs.

[5]Based on Stephen Pass, "Discovering Value in a Mountain of Data," *OR/MS Today*, 24, 5 (December 1997): 24–28. (*OR/MS Today* was the predecessor of *Analytics* magazine.)

Logical Functions

Logical functions depend on whether one or more conditions are true or false. A *condition* is a statement about the value of a cell, either numeric or text. Three useful logical functions in business analytics applications are the following:

> IF(*condition, value if true, value if false*)—a logical function that returns one value if the condition is true and another if the condition is false
>
> AND(*condition 1, condition 2, . . .*)—a logical function that returns TRUE if all conditions are true and FALSE if not
>
> OR(*condition 1, condition 2, . . .*)—a logical function that returns TRUE if any condition is true and FALSE if not

The IF function, IF(*condition, value if true, value if false*), allows you to choose one of two values to enter into a cell. If the specified *condition* is true, *value if true* will be put in the cell. If the condition is false, *value if false* will be entered. *Value if true* and *value if false* can be a number or a text string enclosed in quotes. For example, if cell C2 contains the function =IF(A8 = 2,7,12), it states that if the value in cell A8 is 2, the number 7 will be assigned to cell C2; if the value in cell A8 is not 2, the number 12 will be assigned to cell C2. Conditions may include the following:

> = equal to
> \> greater than
> < less than
> >= greater than or equal to
> <= less than or equal to
> <> not equal to

Note that if a blank is used between quotes, " ", then the result will simply be a blank cell. This is often useful to create a clean spreadsheet.

AND and OR functions simply return the values of *true* or *false* if all or at least one of multiple conditions are met, respectively. You may use AND and OR functions as the *condition* within an IF function; for example, =IF(AND(B1=3,C1=5),12,22). Here, if cell B1=3 and cell C1=5, then the value of the function is 12; otherwise, it is 22.

| EXAMPLE 2.11 | Using the IF Function |

In the *Purchase Orders* database, suppose that the aircraft component manufacturer considers any order of 10,000 units or more to be large, whereas any other order size is considered to be small. We may use the IF function to classify the orders. First, create a new column in the spreadsheet for the order size, say, column K. In cell K4, use the formula

> =IF(F4 >= 10000, "Large", "Small")

This function will return the value *Large* in cell K4 if the order size in cell F4 is 10,000 or more; otherwise, it returns the value *Small*. Further, suppose that large orders with a total cost of at least $25,000 are considered critical. We may flag these orders as critical by using the function in cell L4:

> =IF(AND(K4 = "Large", G4 >= 25000), "Critical", " ")

Note that we use open quotes to return a blank cell if the order is not critical. After copying these formulas down the columns, Figure 2.16 shows a portion of the results.

	A	B	C	D	E	F	G	H	I	J	K	L
1	Purchase Orders											
2												
3	Supplier	Order No.	Item No.	Item Description	Item Cost	Quantity	Cost per order	A/P Terms (Months)	Order Date	Arrival Date	Order Size	Type
4	Hulkey Fasteners	Aug11001	1122	Airframe fasteners	$ 4.25	19,500	$ 82,875.00	30	08/05/11	08/13/11	Large	Critical
5	Alum Sheeting	Aug11002	1243	Airframe fasteners	$ 4.25	10,000	$ 42,500.00	30	08/08/11	08/14/11	Large	Critical
6	Fast-Tie Aerospace	Aug11003	5462	Shielded Cable/ft.	$ 1.05	23,000	$ 24,150.00	30	08/10/11	08/15/11	Large	
7	Fast-Tie Aerospace	Aug11004	5462	Shielded Cable/ft.	$ 1.05	21,500	$ 22,575.00	30	08/15/11	08/22/11	Large	
8	Steelpin Inc.	Aug11005	5319	Shielded Cable/ft.	$ 1.10	17,500	$ 19,250.00	30	08/20/11	08/31/11	Large	
9	Fast-Tie Aerospace	Aug11006	5462	Shielded Cable/ft.	$ 1.05	22,500	$ 23,625.00	30	08/20/11	08/26/11	Large	
10	Steelpin Inc.	Aug11007	4312	Bolt-nut package	$ 3.75	4,250	$ 15,937.50	30	08/25/11	09/01/11	Small	
11	Durrable Products	Aug11008	7258	Pressure Gauge	$ 90.00	100	$ 9,000.00	45	08/25/11	08/28/11	Small	
12	Fast-Tie Aerospace	Aug11009	6321	O-Ring	$ 2.45	1,300	$ 3,185.00	30	08/25/11	09/04/11	Small	
13	Fast-Tie Aerospace	Aug11010	5462	Shielded Cable/ft.	$ 1.05	22,500	$ 23,625.00	30	08/25/11	09/02/11	Large	
14	Steelpin Inc.	Aug11011	5319	Shielded Cable/ft.	$ 1.10	18,100	$ 19,910.00	30	08/25/11	09/05/11	Large	
15	Hulkey Fasteners	Aug11012	3166	Electrical Connector	$ 1.25	5,600	$ 7,000.00	30	08/25/11	08/29/11	Small	

▶ **Figure 2.16**

Classifying Order Sizes Using the IF Function

You may "nest" up to seven IF functions by replacing *value if true* or *value if false* in an IF function with another IF function:

$$=IF(A8=2,(IF(B3=5, \text{"YES"}, \text{""})),15)$$

This says that if cell A8 equals 2, then check the contents of cell B3. If cell B3 is 5, then the value of the function is the text string YES; if not, it is a blank space. However, if cell A8 is not 2, then the value of the function is 15 no matter what cell B3 is. One tip to do this easily is to write out each IF formula first, then successively embed them into one formula.

EXAMPLE 2.12   Calculating the Price of Quantity Discounts

Suppose that a company offers quantity discounts on purchases. For quantities of 1,000 or less, the unit price is $10; for quantities of 1,001 to 5,000, the unit price is $9.00; and for quantities that exceed 5,000, the unit price is $7.50. We may use nested IF functions to compute the total cost for any purchase quantity. Here is the logic. Let Q represent the purchase quantity. If Q <= 1,000, the price is $10 × Q; if this is not true, then we need to check if Q <= 5,000. If so, the price will be $9.00 × Q. Finally, if this is false, then the quantity must exceed 5,000, and the price will be $7.50 × Q. Write the first IF statement as

IF(Q <= 1,000, Q*10, *value if false*)

If this is false, we would have the following IF statement:

IF(Q <= 5,000, Q*9, *value if false*)

If this is false, the value must be Q times 7.5.
Now substitute:

IF(Q <= 1,000, Q*10, *value if false*)
IF(Q <= 1,000, Q*10, IF(Q<= 5,000, Q*9, *value if false*))
IF(Q <= 1,000, Q*10, IF(Q<= 5,000, Q*9, Q*7.5))

Figure 2.17 shows the formula used in the spreadsheet implementation.

▶ **Figure 2.17**

Spreadsheet Implementation of Quantity Discount Calculation

B7		fx	=IF(B6<=1000,B6*B2,IF(B6<=5000,B6*B3,B6*B4))

	A	B
1	Quantity purchased	Price
2	1-1000	$10.00
3	1001-5000	$9.00
4	> 5000	$7.50
5		
6	Quantity	10000
7	Total Cost	$75,000

CHECK YOUR UNDERSTANDING

1. Write an IF function that returns the word "Valid" if the value of cell B16 is greater than or equal to 0 and the word "Invalid" if not.

2. Explain the difference between the AND and OR functions.

3. Write an IF function that performs the following and explain how it should be implemented: If the contents of cell D10 is the text string "Invoice received" and the contents of cell E10 is "Paid," then set the contents of cell F10 to "Order Completed"; otherwise, set the contents of cell F10 to "Order Open."

Lookup Functions for Database Queries

Excel provides some useful **lookup functions** for finding specific data in a spreadsheet. These are the following:

VLOOKUP(*lookup_value, table_array, col_index_num, [range lookup]*) looks up a value in the leftmost column of a table (specified by the *table_array*) and returns a value in the same row from a column you specify (*col_index_num*).

HLOOKUP(*lookup_value, table_array, row_index_num, [range lookup]*) looks up a value in the top row of a table and returns a value in the same column from a row you specify.

INDEX(*array, row_num, col_num*) returns a value or reference of the cell at the intersection of a particular row and column in a given range.

MATCH(*lookup_value, lookup_array, match_type*) returns the relative position of an item in an array that matches a specified value in a specified order.

CHOOSE(*index_num, value1, value2, . . .*) returns a value from a list based on the position in the list, specified by *index_num*.

In the VLOOKUP and HLOOKUP functions, *range lookup* is optional. If this is omitted or set as *True*, then the first column of the table must be sorted in ascending numerical

EXAMPLE 2.13   **Using the VLOOKUP Function**

A database of sales transactions for a firm that sells instructional fitness books and DVDs is provided in the Excel file *Sales Transactions*. The database is sorted by customer ID, and a portion of it is shown in Figure 2.18. Suppose that a customer calls a representative about a payment issue. The representative finds the customer ID—for example, 10007—and needs to look up the type of payment and transaction code. We may use the VLOOKUP function to do this. In the function =VLOOKUP(*lookup_value, table_array, col_index_num*), *lookup_value* represents the customer ID. The *table_array* is the range of the data in the spreadsheet; in this case, it is the range A4:H475. The value for *col_index_num* represents the column in the table range we wish to retrieve. For the type of payment, this is column 3; for the transaction code, this is column 4. Note that the first column is already sorted in ascending numerical order, so we can either omit the *range lookup* argument or set it as

True. Thus, if we enter the formula below in any blank cell of the spreadsheet (we recommend using absolute references for arrays)

=VLOOKUP(10007,$A$4:$H$475,3)

it returns the payment type *Credit*. If we use the following formula:

=VLOOKUP(10007,$A$4:$H$475,4)

the function returns the transaction code 80103311.

Now suppose the database was sorted by transaction code so that the customer ID column is no longer in ascending numerical order, as shown in Figure 2.19. If we use the function =VLOOKUP(10007,$A$4:$H$475,4, True), Excel returns #N/A. However, if we change the *range lookup* argument to False, then the function returns the correct value of the transaction code.

	A	B	C	D	E	F	G	H
1	Sales Transactions: July 14							
2								
3	Cust ID	Region	Payment	Transaction Code	Source	Amount	Product	Time Of Day
4	10001	East	Paypal	93816545	Web	$20.19	DVD	22:19
5	10002	West	Credit	74083490	Web	$17.85	DVD	13:27
6	10003	North	Credit	64942368	Web	$23.98	DVD	14:27
7	10004	West	Paypal	70560957	Email	$23.51	Book	15:38
8	10005	South	Credit	35208817	Web	$15.33	Book	15:21
9	10006	West	Paypal	20978903	Email	$17.30	DVD	13:11
10	10007	East	Credit	80103311	Web	$177.72	Book	21:59
11	10008	West	Credit	14132683	Web	$21.76	Book	4:04
12	10009	West	Paypal	40128225	Web	$15.92	DVD	19:35
13	10010	South	Paypal	49073721	Web	$23.39	DVD	13:26

	A	B	C	D	E	F	G	H
1	Sales Transactions: July 14							
2								
3	Cust ID	Region	Payment	Transaction Code	Source	Amount	Product	Time Of Day
4	10391	West	Credit	10325805	Web	$22.79	Book	0:00
5	10231	North	Paypal	10400774	Web	$216.20	Book	10:33
6	10267	West	Paypal	10754185	Web	$23.01	DVD	17:44
7	10228	West	Credit	10779898	Web	$15.33	DVD	5:05
8	10037	South	Paypal	11165609	Web	$217	Book	0:00
9	10297	North	Credit	11175481	Web	$22.65	Book	6:06
10	10294	West	Paypal	11427628	Web	$15.40	Book	17:16
11	10081	North	Credit	11673210	Web	$16.14	DVD	4:04
12	10129	West	Credit	11739665	Web	$22.03	DVD	14:49
13	10406	East	Credit	12075708	Web	$22.99	Book	9:09
14	10344	East	Credit	12222505	Web	$15.55	DVD	6:06

order. If an exact match for the *lookup_value* is found in the first column, then Excel will return the value of the *col_index_num* of that row. If an exact match is not found, Excel will choose the row with the largest value in the first column that is less than the *lookup_value*. If *range lookup* is *False*, then Excel seeks an exact match in the first column of the table range. If no exact match is found, Excel will return #N/A (not available). We recommend that you specify the range lookup to avoid errors.

The HLOOKUP function works in a similar fashion. For most spreadsheet databases, we would normally need to use the VLOOKUP function. In some modeling situations, however, the HLOOKUP function can be useful if the data are arranged column by column rather than row by row.

The INDEX function works as a lookup procedure by returning the value in a particular row and column of an array. For example, in the *Sales Transactions* database, =INDEX($A$4:$H$475, 7, 4) would retrieve the transaction code 80103311 that is in the 7th row and 4th column of the data array (see Figure 2.18), as the VLOOKUP function did in Example 2.13. The difference is that it relies on the row number rather than the actual value of the customer ID.

In the MATCH function, *lookup_value* is the value that you want to match in *lookup_array*, which is the range of cells being searched. The *match_type* is either -1, 0, or 1. The default is 1. If *match_type* $= 1$, then the function finds the largest value that is less than or equal to *lookup_value*; however, the values in the *lookup_array* must be placed in ascending order. If *match_type* $= 0$, MATCH finds the first value that is exactly equal to *lookup_value*; in this case the values in the *lookup_array* can be in any order. If *match_type* $= -1$, then the function finds the smallest value that is greater than or equal to *lookup_value*; in this case the values in the *lookup_array* must be placed in descending order. Example 2.14 shows how the INDEX and MATCH functions can be used.

The VLOOKUP function will not work if you want to look up something to the left of a specified range (because it uses the first column of the range to find the lookup value). However, we can use the INDEX and MATCH functions easily to do this, as Example 2.14 shows.

EXAMPLE 2.14	Using INDEX and MATCH Functions for Database Queries

Figure 2.20 shows the data in the Excel file *Monthly Product Sales Queries*. Suppose we wish to design a simple query application to input the month and product name, and retrieve the corresponding sales. The three additional worksheets in the workbook show how to do this in three different ways. The *Query1* worksheet (see Figure 2.21) uses the VLOOKUP function with embedded IF statements. The formula in cell I8 is

= VLOOKUP(I5, A4:F15, IF(I6 ="A", 2, IF(I6 = "B", 3, IF(I6 = "C",4, IF(I6 = "D", 5, IF(I6 = "E", 6)))))),FALSE)

The IF functions are used to determine the column in the lookup table to use and, as you can see, is somewhat complex, especially if the table were much larger.

The *Query2* worksheet (not shown here; see the Excel workbook) uses the VLOOKUP and MATCH functions in cell I8. The formula in cell I8 is

=VLOOKUP(I5,A4:F15,MATCH(I6,B3:F3,0) + 1,FALSE)

In this case, the MATCH function is used to identify the column in the table corresponding to the product name in cell I6. Note the use of the "+1" to shift the relative column number of the product to the correct column number in the lookup table.

Finally, the *Query3* worksheet (also not shown here) uses only INDEX and MATCH functions in cell I8. The formula in cell I8 is

=INDEX(A4:F15,MATCH(I5,A4:A15,0),MATCH(I6,A3:F3,0))

The MATCH functions are used as arguments in the INDEX function to identify the row and column numbers in the table based on the month and product name. The INDEX function then retrieves the value in the corresponding row and column. This is perhaps the cleanest formula of the three. By studying these examples carefully, you will better understand how to use these functions in other applications.

► **Figure 2.20**

Monthly Product Sales Queries *Workbook*

	A	B	C	D	E	F
1	Sales Units					
2				Product		
3	Month	A	B	C	D	E
4	January	7,792	5,554	3,105	3,168	10,350
5	February	7,268	3,024	3,228	3,751	8,965
6	March	7,049	5,543	2,147	3,319	6,827
7	April	7,560	5,232	2,636	4,057	8,544
8	May	8,233	5,450	2,726	3,837	7,535
9	June	8,629	3,943	2,705	4,664	9,070
10	July	8,702	5,991	2,891	5,418	8,389
11	August	9,215	3,920	2,782	4,085	7,367
12	September	8,986	4,753	2,524	5,575	5,377
13	October	8,654	4,746	3,258	5,333	7,645
14	November	8,315	3,566	2,144	4,924	8,173
15	December	7,978	5,670	3,071	6,563	6,088

Data | Query1 | Query2 | Query3

► **Figure 2.21**

Query1 *Worksheet in* Monthly Product Sales Queries *Workbook*

	A	B	C	D	E	F	G	H	I
1	Sales Units								Using VLOOKUP + IF
2				Product					
3	Month	A	B	C	D	E		Sales Lookup	
4	January	7,792	5,554	3,105	3,168	10,350			
5	February	7,268	3,024	3,228	3,751	8,965		Month	April
6	March	7,049	5,543	2,147	3,319	6,827		Product	E
7	April	7,560	5,232	2,636	4,057	8,544			
8	May	8,233	5,450	2,726	3,837	7,535		Sales	8,544
9	June	8,629	3,943	2,705	4,664	9,070			
10	July	8,702	5,991	2,891	5,418	8,389			
11	August	9,215	3,920	2,782	4,085	7,367			
12	September	8,986	4,753	2,524	5,575	5,377			
13	October	8,654	4,746	3,258	5,333	7,645			
14	November	8,315	3,566	2,144	4,924	8,173			
15	December	7,978	5,670	3,071	6,563	6,088			

Data | Query1 | Query2 | Query3

EXAMPLE 2.15   Using INDEX and MATCH for a Left Table Lookup

Suppose that, in the *Sales Transactions* database, we wish to find the customer ID associated with a specific transaction code. Refer back to Figure 2.18 or the Excel workbook. Suppose that we enter the transaction code in cell K2 and want to display the customer ID in cell K4. Use the following formula in cell K4:

=INDEX(A4:A475,MATCH(K2,D4:D475,0),1)

Here, the MATCH function is used to identify the row number in the table range that matches the transaction code exactly, and the INDEX function uses this row number and column 1 to identify the associated customer ID.

To illustrate the CHOOSE function, suppose that in the monthly product sales data (see Figure 2.20), we want to select the January sales of product B. We could use =CHOOSE(2,B4,C4,D4,E4,F4), which would return the value 5,554. We may also specify a list of text names, such as =CHOOSE(2,"A","B","C","D","E"), which would return B.

CHECK YOUR UNDERSTANDING

1. Explain the purpose of the VLOOKUP, INDEX, and MATCH functions.

2. In the *Purchase Orders* database, what will =VLOOKUP(4111,C4:J97, 5, TRUE) and =VLOOKUP(4111,C4:J97, 5, FALSE) find? Explain the difference.

3. Write an Excel function that is entered in cell J2 that will find the amount of an order for any customer ID in the *Sales Transactions* database using (1) only VLOOKUP, and (2) only the INDEX and MATCH functions.

Excel Template Design

Many database queries are repetitive; think of customer service representatives who must look up order information, prices, and so on. We may use the logical and lookup functions that we studied in the previous section to create user-friendly **Excel templates** for repetitive database queries. We did this in Example 2.14 for finding sales for a specific month and product name. In this section, we discuss some design approaches and Excel tools.

An Excel template would typically have input cells, result cells, and possibly intermediate calculations. Templates should be "clean," well organized, and easy to use. Validation tools should be used to ensure that users do not inadvertently make errors in input data. Visualizations, such as charts, which will be discussed in the next chapter, can be included for additional insight.

EXAMPLE 2.16   A Tax Bracket Calculator

The table below shows the 2016 U.S. federal income tax rates for the four different types of filing status.

1. Single	2. Married filing jointly or qualifying widow/widower	3. Married filing seperately	4. Head of household	Tax rate
Up to $9,275	Up to $18,550	Up to $9,275	Up to $13,250	10%
$9,276 to $37,650	$18,551 to $75,300	$9,276 to $37,650	$13,251 to $50,400	15%
$37,651 to $91,150	$75,301 to $151,900	$37,651 to $75,950	$50,401 to $130,150	25%
$91,151 to $190,150	$151,901 to $231,450	$75,951 to $115,725	$130,151 to $210,800	28%
$190,151 to $413,350	$231,451 to $413,350	$115,726 to $206,675	$210,801 to $413,350	33%
$413,351 to $415,050	$413,351 to $466,950	$206,676 to $233,475	$413,351 to $441,000	35%
$415,051 or more	$466,951 or more	$233,476 or more	$441,001 or more	39.60%

We will create a simple template (see the Excel file *Tax Bracket Template*) that allows an individual to enter their filing status and taxable income and then returns the corresponding tax bracket. This is shown in Figure 2.22. We can use a lookup table to identify the tax rate. First, note that the main worksheet does not include the lookup table to keep the template clean. The input and results cells are color-coded to separate inputs from the result. We used a nested IF statement to select the proper lookup range based on the filing status code.

An alternative way of doing this would be to use the CHOOSE function as follows ("Lookup" is the name of the worksheet containing the lookup table):

=CHOOSE($B$9,VLOOKUP($B$10,Lookup!$A$2:$E$8,5), VLOOKUP($B$10,Lookup!$B$2:$E$8,4),VLOOKUP($B$10, Lookup!$C$2:$E$8,3),VLOOKUP($B$10,Lookup!$D$2:$E$8,2))

Data Validation Tools

Excel provides various **data validation** tools to reduce the chances that users will make a mistake. First, select the cell range for which data validation will be applied, and then choose *Data Validation* from the *Data Tools* group on the *Data* tab in Excel 2016, or select *Validation* from the *Data* menu on the a Mac.

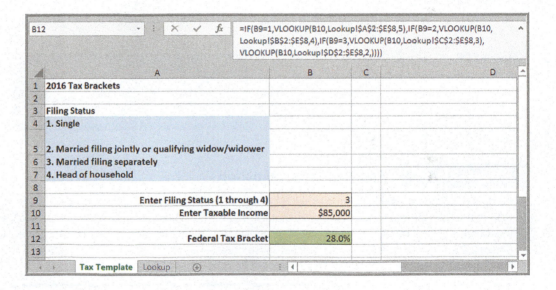

▶ **Figure 2.22**

Excel Template for Identifying Tax Bracket

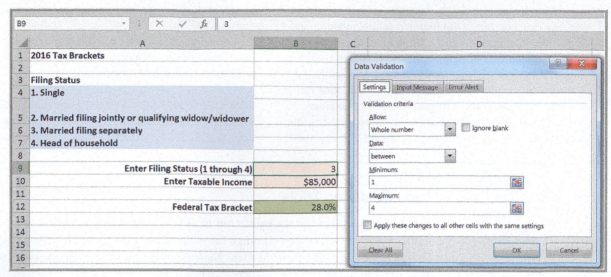

▶ **Figure 2.23**

Data Validation *Dialog*

▶ **Figure 2.24**

Data Validation *Error Alert*

| EXAMPLE 2.17 | **Applying Data Validation** |

For the Tax Bracket template, choose cell B9 (filing status). In the *Data Validation* dialog, specify that the value must be a whole number between 1 and 4, as shown in Figure 2.23. If the input value does not meet these criteria, a default error alert will pop up (see Figure 2.24). This can be customized from the *Error Alert* tab in the *Data Validation* dialog. You may also use the *Input Message* tab to create a prompt to guide the user when an input cell is selected.

Another way of applying data validation for numerical inputs is to use the ISNUMBER function. This function returns TRUE if the cell or range contains a number; otherwise, it returns FALSE. This can help to avoid any typing mistakes resulting from inadvertently pressing a letter or symbol key.

| EXAMPLE 2.18 | **Using the ISNUMBER Function for Data Validation** |

In the Tax template, we will use ISNUMBER to ensure that the taxable income in cell B10 is numeric. In the *Data Validation* dialog, choose *Custom*, and in the *Formula* box, enter the formula =ISNUMBER(B10), as shown in Figure 2.25. An error alert message will pop up if an invalid entry is made.

▶ **Figure 2.25**
ISNUMBER Data Validation

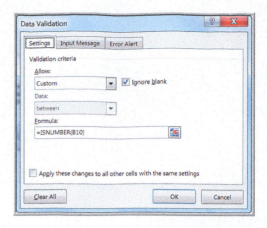

Form Controls

Form controls are buttons, boxes, and other mechanisms for inputting or changing data on spreadsheets easily that can be used to design user-friendly spreadsheets. They allow the user to more easily interface with models to enter or change data without the potential of inadvertently introducing errors in formulas. To use form controls, you must first activate the *Developer* tab on the ribbon. Click the *File* tab, then *Options*, and then *Customize Ribbon*. Under *Customize the Ribbon*, make sure that *Main Tabs* is displayed in the drop-down box, and then click the check box next to *Developer* (which is typically unchecked in a standard Excel installation). You will see the new tab in the Excel ribbon, as shown in Figure 2.26. If you click the *Insert* button in the *Controls* group, you will see the form controls available (*do not* confuse these with the *Active X Controls* in the same menu!). On a Mac, select *Excel* > *Preferences* > *View* and check the box for *Developer* tab. The Mac ribbon is different and shows graphics of the form controls.

The most common form controls are

- *Spin button*—a button used to increase or decrease a numerical value
- *Scroll bar*—a slider used to change a numerical value
- *Check box*—a box used to select or deselect a scenario
- *Option button*—a radio button used to select an option
- *List box*—a box that provides a list of options
- *Combo box*—a box that provides an expandable list of options
- *Group box*—a box that can hold a group of controls

▶ **Figure 2.26**
Excel 2016 Developer *Tab*

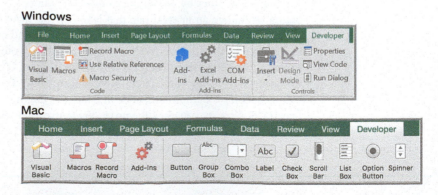

	A	B	C	D	E
1	2016 Tax Brackets				
2					
3	Filing Status				
4	1. Single		Filing Status	Taxable Income	
5	2. Married filing jointly or qualifying widow/widower				
6	3. Married filing separately				
7	4. Head of household				
8					
9	Enter Filing Status (1 through 4)	2		1451	
10	Enter Taxable Income	$145,100			
11					
12	Federal Tax Bracket	25.0%			

▶ **Figure 2.27**

Tax Bracket Template with Form Controls

To insert a form control, click the *Insert* button in the *Controls* tab under the *Developer* menu, click on the control you want to use, and then click within your worksheet. The following example shows how to use both a spin button and scroll bar in the Tax Bracket template.

EXAMPLE 2.19   **Using Form Controls in the Tax Bracket Template**

In the Tax Bracket template, we will use a spin button for the filing status and a scroll bar for the taxable income (in unit increments of $100, from $0 through $500,000). The completed spreadsheet is shown in Figure 2.27.

First, click the *Insert* button in the *Controls* group of the *Developer* tab, select the spin button, click it, and then click somewhere in the worksheet. The spin button (and any form control) can be re-sized by dragging the handles along the edge and moving the button within the worksheet. Move it to a convenient location, and enter the name you wish to use (such as Filing Status) adjacent to it. Next, right click the spin button and select *Format Control*. You will see the dialog box shown in Figure 2.28. Enter the values shown and click *OK*. Now if you click the up or down buttons, the value in cell B9 will change within the specified range.

Next, repeat this process by inserting the scroll bar next to the taxable income cell. Right click the scroll bar and select *Format Control*. For a scroll bar, the maximum value is limited to 30,000, so we need to scale the parameters so that the taxable income can vary up to $500,000. Set the parameters in the dialog for a minimum value of 0 and maximum value of 5,000, with an incremental change of 1. Choose a blank cell for the cell link (say, D9). Then, in cell B10, enter the formula =D9*100. Now when the scroll bar is moved, the income will change from 0 to $500,000. You may also click the arrows on the end of the scroll bar to fine-tune the value. Using these controls, you can easily change the inputs in the template.

Form controls only allow integer increments, so you would also have to scale values to change a number by a fractional value. For example, suppose that we want to use a spin button to change an interest rate that is in cell B8 from 0% to 10% in increments of 0.1% (that is, 0.001). Choose some empty cell, say, C8, and enter a value between 0 and 100 in it. Then enter the formula =C8/1000 in cell B8. Note that if the value in C8 = 40, for example, then the value in cell B8 will be 40/1000 = 0.04, or 4%. Then as the value in cell C8 changes by 1, the value in cell B8 changes by 1/1,000, or 0.1%. In the *Format Control* dialog, specify the minimum value at 0 and the maximum value at 100 and link the button to cell C8. Now as you click the up or down arrows on the spin button, the value in cell C8 changes by 1 and the value in cell B8 changes by 0.1%.

▶ Figure 2.28

Format Control *Dialog*

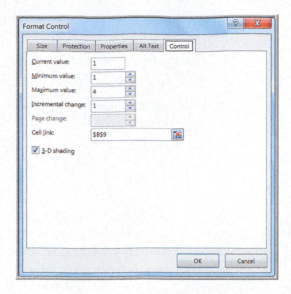

Other form controls can also be used; for instance, in the Tax Bracket template, you might use a list box to select the filing status. We encourage you to experiment and identify creative ways to use them. The next example shows the use of the combo box.

EXAMPLE 2.20   **Using a Combo Box**

Figure 2.29 shows the application of a combo box to compute the total sales for any month in the *Monthly Product Sales* database. We set the cell link to H1; this provides the row number of the month in the data array to use in an INDEX function to compute total sales. The formula in cell H7 is =SUM(INDEX(B4:F15,H1,0)). Using a 0 as the

column number references the entire row, so the function sums all the columns in the row listed in cell H1. In the Format Control dialog, set the Drop Down lines to 12; this will display all 12 months when the arrow button is clicked. A smaller number will require you to scroll down to seem them all.

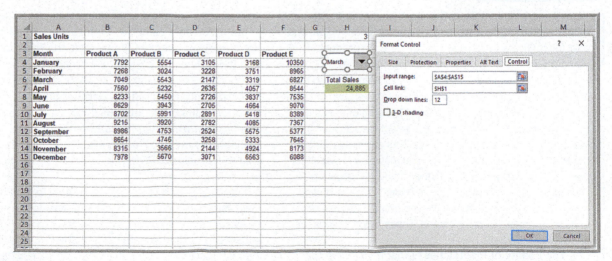

▶ **FIGURE 2.29**

Combo Box Application

1. State the purpose of an Excel template and the key properties that it should have.

2. What is the purpose of *Data Validation* in Excel?

3. What will the function =ISNUMBER(A1) return when the contents of cell A1 is either 10 or the word ten?

4. Explain the value of form controls in Excel templates.

PivotTables

Excel provides a powerful tool for distilling a complex database into meaningful information: **PivotTables** (yes, it is one word!). PivotTables allows you to create custom summaries and charts of key information in the data and to drill down into a large set of data in numerous ways. In this section, we will introduce PivotTables in the context of databases; however, we will also use them in other chapters to facilitate data visualization and statistical analysis of data sets.

To apply PivotTables, first, select any cell in the database and choose *PivotTable* under the *Insert* tab and follow the steps of the wizard. We will use the *Sales Transactions* database (see Figure 2.18). Excel first asks you to select a table or range of data; if you click on any cell within the database before inserting a PivotTable, Excel will default to the complete range of your data. You may either put the PivotTable into a new worksheet or in a blank range of the existing worksheet. Excel then creates a blank PivotTable, as shown in Figure 2.30.

In the *PivotTable Fields* window on the right side of Figure 2.30 is a list of the fields that correspond to the headers in the database. You select which ones you want to include, as row labels, column labels, values, or a *filter*. You should first decide what types of tables you wish to create—that is, what fields you want for the rows, columns, and data values.

EXAMPLE 2.21	Creating a PivotTable

We will use a PivotTable for the *Sales Transactions* database to find the total amount of revenue in each region, and also the total revenue in each region for each product. If you drag the field *Region* from the PivotTable *Fields* list in Figure 2.30 to the *Rows* area and the field *Amount* into the Σ *Values* area, you will create the PivotTable shown in Figure 2.31. The values in column B are the sum of the sales revenues for each region. You may select the range of the value, right click, and format the cells as currency to make the PivotTable more meaningful. You may also replace the names in the headers; for instance, change Row Labels to Region and Sum of Amount to Revenue. The drop-down arrow next to Row Labels allows you to filter the results and display only selected regions. These modifications are shown in Figure 2.32.

The beauty of PivotTables is that if you wish to change the analysis, you can simply drag the field

names out of the field areas (or just uncheck the boxes in the PivotTable *Field* list) or drag them to different areas. For example, if you drag the *Product* field into the *Columns* area, you will create the PivotTable in Figure 2.33, showing the breakdown of revenues by product for each region. Another option is to drag *Product* to the *Rows* area, as shown in Figure 2.34. The order in which the fields are listed will also change the view (try moving *Region* after *Product*). The best way to learn about PivotTables is simply to experiment with them.

Dragging a field into the *Filters* area in the PivotTable *Field* list allows you to add a third dimension to your analysis. For example, moving *Payment* to the *Filters* area allows you to filter revenues by PayPal or credit, as shown in Figure 2.35.

▶ **Figure 2.30**

Blank PivotTable

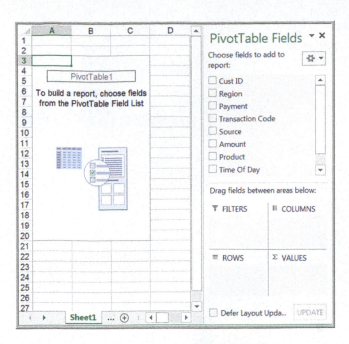

▶ **Figure 2.31**

PivotTable for Total Revenue by Region

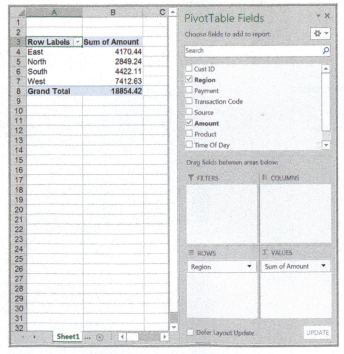

▶ **Figure 2.32**

Re-formatted PivotTable for Total Revenue by Region

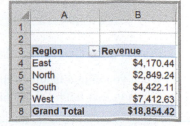

▶ **Figure 2.33**

Revenue Breakdown by Region and Product

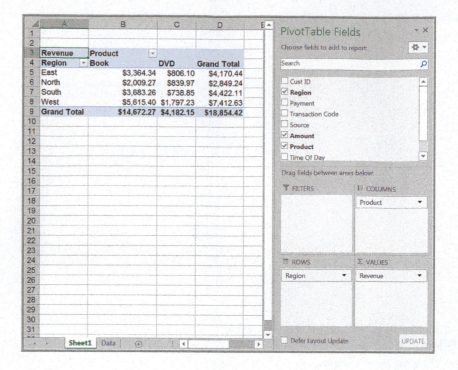

▶ **Figure 2.34**

Alternate PivotTable View of Region and Product Revenue

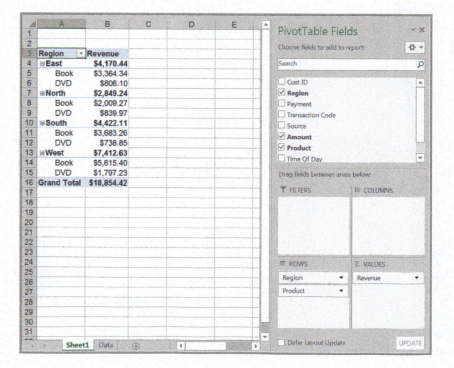

PivotTable Customization

When you drag a field into the Σ *Values* area, the PivotTable defaults to the sum of the values in that field. Often, this is not what you want, but it can be easily changed. For

► Figure 2.35

*PivotTable Filtered by Credit
Payment*

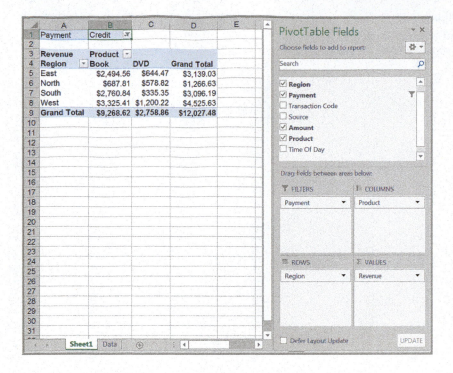

example, suppose that you wanted a count of the number of transactions for each region and product. In Figure 2.31, click the dropdown arrow next to Revenue in the Σ *Values* area and choose *Value Field Settings*. In the dialog shown in Figure 2.36, select *Count* (many other options exist that perform statistical calculations; we will use these in Chapter 4). This results in the PivotTable shown in Figure 2.37 (after reformatting the values to numbers instead of currency). Now we can see how many books and DVDs were sold in each region.

An easier way to reformat data in a PivotTable is to select any cell, right click, and choose *Value Field Settings*. Notice the Number Format button at the bottom left in Figure 2.36. If you click this, you will be able to choose a format for all the data in the PivotTable.

► Figure 2.36

Value Field Settings *Dialog*

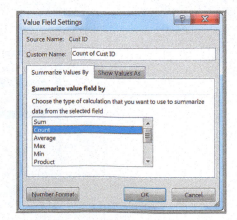

▶ **Figure 2.37**

PivotTable for Count of Regional Sales by Product

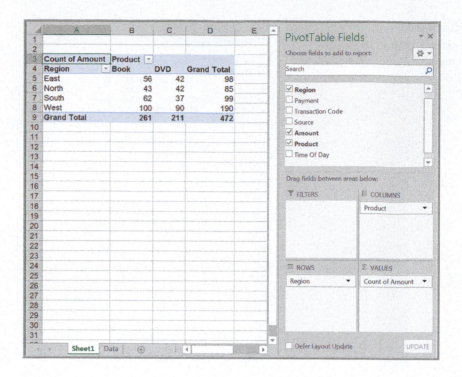

You may express the data in a PivotTable in various percentage views, such as the percent of the grand total, percent of row total, percent of column total, and other options. In the *Value Field Settings* dialog, click the tab *Show Values As*. Select the option in the dropdown box; Figure 2.38 illustrates the selection of *% of Row Total*. Figure 2.39 shows the results; we now can see the percentage breakdown of product sales for each individual region.

Other options for customizing PivotTables can be found in the PivotTable Tools Design tab. Again, we invite you to experiment with them.

▶ **Figure 2.38**

Show Values *Tab in* Value Field Settings

▶ **Figure 2.39**

Percentage of Product Sales by Region

	A	B	C	D
1				
2				
3	Sum of Amount	Product ▾		
4	Region ▾	Book	DVD	Grand Total
5	East	80.67%	19.33%	100.00%
6	North	70.52%	29.48%	100.00%
7	South	83.29%	16.71%	100.00%
8	West	75.75%	24.25%	100.00%
9	Grand Total	77.82%	22.18%	100.00%

Slicers

Excel has a tool called **slicers**—for drilling down to "slice" a PivotTable and display a subset of data. Slicers simplify the process of filtering a PivotTable and are very useful in presentations to quickly show alternate PivotTable views. To create a slicer for any of the columns in the database, click on the PivotTable and choose *Insert Slicer* from the *Analyze* tab in the *PivotTable Tools* ribbon in Windows, or from the *PivotTable Analyze* tab on a Mac.

EXAMPLE 2.22 **Using Slicers**

For the PivotTable we created in Figure 2.33 for the revenues by region and product, let us insert a slicer for the source of the transaction, as shown in Figure 2.40. From the *Insert Slicers* dialog, check the box for *Source*. This results in the slicer window shown in Figure 2.40. If you click on one of the *Source* buttons, Email or Web, the PivotTable reflects only those records corresponding to that source. In Figure 2.40, we are viewing only the total revenues generated through Web sales. If you click on the small red "x" by the filter in the top right of the slicer, you would clear the filter and the PivotTable would show revenues from all sources, as in Figure 2.33.

CHECK YOUR UNDERSTANDING

1. What is the purpose of a PivotTable?

2. Explain how to construct a PivotTable for the *Sales Transactions* database so that it shows the count of the number of transactions using PayPal and credit for each region.

3. What is a slicer? How does it simplify the use of PivotTables?

▶ **Figure 2.40**

Revenues Sliced by Web Sales

	A	B	C	D	E	F	G
1							
2							
3	Sum of Amount	Product ▾			Source		
4	Region ▾	Book	DVD	Grand Total	Email		
5	East	$2,429.41	$684.38	$3,113.79			
6	North	$1,777.94	$587.56	$2,365.50	Web		
7	South	$2,468.54	$533.33	$3,001.87			
8	West	$3,380.14	$1,373.96	$4,754.10			
9	Grand Total	$10,056.03	$3,179.23	$13,235.26			

KEY TERMS

Data set

Data validation

Database

Database functions

Empirical data

Excel table

Excel template

Filtering

Form controls

Logical functions

Lookup functions

Pareto analysis

PivotTables

Slicers

CHAPTER 2 TECHNOLOGY HELP

Useful Excel Functions

IF(*condition, value if true, value if false*) A logical function that returns one value if the condition is true and another if the condition is false.

AND(*condition 1, condition 2, . . .*) A logical function that returns TRUE if all conditions are true and FALSE if not.

OR(*condition 1, condition 2, . . .*) A logical function that returns TRUE if any condition is true and FALSE if not.

VLOOKUP(*lookup_value, table_array, col_index_num, [range lookup]*) Looks up a value in the leftmost column of a table (specified by the *table_array*) and returns a value in the same row from a column you specify (*col_index_num*).

HLOOKUP(*lookup_value, table_array, row_index_num, [range lookup]*) Looks up a value in the top row of a table and returns a value in the same column from a row you specify.

INDEX(*array, row_num, col_num*) Returns a value or reference of the cell at the intersection of a particular row and column in a given range.

MATCH(*lookup_value, lookup_array, match_type*) Returns the relative position of an item in an array that matches a specified value in a specified order.

CHOOSE(*index_num, value1, value2, . . .*) Returns a value from a list based on the position in the list, specified by *index_num*.

DSUM(*database, field, criteria*) Sums values in the *field* column from a specified *database* that includes the column labels, using a *criteria* range that specifies the

records you want to sum using the same format as the *Advanced Filter*. Other useful database functions include DAVERAGE and DCOUNT.

Excel Techniques

Create a range name (Example 2.2):

There are three options.

1. Select a cell; in the *Name* box, type the name and press *Enter*.
2. Select a cell; click *Define Name* on the *Formulas* tab and complete the dialog.
3. Highlight a range with text labels immediately to the right or left of the cell or range, or in the row immediately above or below the cell or range. On the *Formulas* tab, choose *Create from Selection* and complete the dialog.

Create an Excel table (Example 2.3):

Select the range of the data, including headers. Click *Table* from the *Tables* group on the *Insert* tab and make sure that the box for *My Table Has Headers* is checked.

Sort data (Example 2.5):

Select a single cell in the column you want to sort on, and in the *Data* tab, click the "*AZ down arrow*" button to sort from smallest to largest or the "*AZ up arrow*" button to sort from largest to smallest. Click the *Sort* button to specify criteria for more advanced sorting capabilities.

Filter data (Example 2.7):

Select any cell within the database. Then, from the Excel *Data* tab, click on *Filter*. Choose the filtering criteria.

Using the Advanced Filter (Example 2.9):

Copy the headers from the database to an open location in the worksheet. Under the headers, specify the criteria that you want to use to filter the data. Multiple criteria in the same row are logically joined by "and," while criteria in rows are joined by "or." In general, criteria for numerical values can include =, >, <, <> (not equal), >=, or <=. Choose *Advanced* next to the *Filter* in the *Data* tab. In the dialog box, enter the list range of the database and the criteria range.

Apply Data Validation (Example 2.17):

Select the cell range for which data validation will be applied, and then choose *Data Validation* from the *Data Tools* group on the *Data* tab in Excel 2016, or select *Validation* from the *Data* menu on a Mac. Complete the dialog.

Use Form Controls (Examples 2.19 and 2.20):

Activate the *Developer* tab on the ribbon. To insert a form control, click the *Insert* button in the *Controls* tab under the *Developer* menu, click on the control you want to use, and then click within your worksheet. Right click the form control and select *Format Control* and specify the parameters.

Create a PivotTable (Example 2.21):

Select any cell in the database and choose *Pivot-Table* from the *Tables* group under the *Insert* tab and follow the steps of the wizard.

Insert a Slicer in a PivotTable (Example 2.22):

Click on the PivotTable and choose *Insert Slicer* from the *Analyze* tab in the *PivotTable Tools* ribbon in Windows, or from the *PivotTable Analyze* tab on a Mac.

StatCrunch

StatCrunch provides the ability to easily create surveys for collecting data. This is useful for creating your own data sets and databases for class assignments. You can find video tutorials and step-by-step procedures with examples at https://www.statcrunch.com/5.0/example.php. We suggest that you first view the tutorials *Getting started with StatCrunch* and *Working with StatCrunch sessions* in the Basics section. Scroll down to the Surveys section and click on *Creating surveys* and *Administering surveys and analyzing the results.*

PROBLEMS AND EXERCISES

Data Sets and Databases

1. The Excel file *MBA Motivation and Salary Expectations* provides data on students' pre-MBA salary and post-MBA salary expectations. Define range names for each of these ranges and then use the range names in formulas to find the expected salary increase for each of the respondents.

2. The Excel file *Syringe Samples* provides data for 15 samples of an important measurement in the manufacturing process for medical syringes. Define range names for each of the samples and then use these range names in formulas to find the average value for each sample.

3. The *Budget Forecasting* database shows estimated expenses for the last nine months of the coming year. Define range names for each month and type of expense and use the range names in formulas to find the total for each month and type of expense.

Data Queries: Tables, Sorting, and Filtering

4. Convert the *Store and Regional Sales* database to an Excel table. Use the techniques described in Example 2.4 to find:
   a. the total number of units sold.
   b. the total number of units sold in the South region.
   c. the total number of units sold in December.

5. Convert the *Purchase Orders* database to an Excel table. Use the techniques described in Example 2.4 to find:
   a. the total cost of all orders.
   b. the total quantity of airframe fasteners purchased.
   c. the total cost of all orders placed with Manley Valve.

6. The Excel file *Economic Poll* provides some demographic and opinion data on whether the economy is moving in the right direction. Convert these data

into an Excel table and filter the respondents who are homeowners and perceive that the economy is not moving in the right direction. What is the distribution of their political party affiliations?

7. The Excel file *Corporate Default Database* summarizes financial information for 32 companies and their perceived risk of default. Convert these data into an Excel table. Use table-based calculations to find the average credit score, average debt, and average equity for companies with a risk of default, and also for those without a risk of default. Does there appear to be a difference between companies with and without a risk of default?

8. Open the Excel file *Store and Regional Sales Database*. Sort the data by units sold, high to low.

9. In the Excel file *Automobile Quality*, search for and add a new column for the country of origin for each brand. Then sort the data from the lowest to highest by number of problems per 100 vehicles using the sort capability in Excel. What conclusions can you reach?

10. In the *Purchase Orders* database, conduct a Pareto analysis of the cost per order data. What conclusions can you reach?

11. Conduct a Pareto analysis for the number of minutes that flights are late in the *Atlanta Airline Data* Excel file. Interpret the results.

12. Use Excel's filtering capability in the *Purchase Orders* database to extract

   a. all orders for control panels.

   b. all orders for quantities of less than 500 units.

   c. all orders for control panels with quantities of less than 500 units.

13. In the *Sales Transactions* database, use Excel's filtering capability to extract

   a. all orders that used PayPal.

   b. all orders under $100.

   c. all orders that were over $100 and used a credit card.

14. Filter the data in the *Bicycle Inventory* database to obtain only the records for the leisure category. What is the average selling price and total quantity on hand for these bicycles?

15. Use the Excel file *Credit Risk Data* that was introduced in this chapter to perform the following activities:

   a. Compute the combined checking and savings account balance for each record in the *Base Data* worksheet. Then sort the records by the number of months as a customer of the bank. From examining the data, does it appear that customers with a longer association with the bank have more assets?

   b. Apply Pareto analysis to draw conclusions about the combined amount of money in checking and savings accounts.

   c. Use Excel's filtering capability to extract all records for new-car loans. How many individuals with new-car loans are single, married, and divorced?

   d. Use Excel's filtering capability to extract all records for individuals employed less than 12 months. Can you draw any conclusions about the credit risk associated with these individuals?

16. Apply the *Advanced Filter* to the *Credit Risk Data* (*Base Data* worksheet) to find the following:

   a. All new car loans obtained for single females.

   b. All furniture loans obtained for single individuals who rent.

   c. All education loans obtained by unskilled workers.

   d. All used car loans obtained by individuals who are employed 12 months or less.

   e. All car loans (new or used) obtained by males who have been employed at least 36 months.

17. In the *Credit Risk Data* file, use database functions to find the average amount of savings for each of the situations listed in Problem 16.

18. For the *Bicycle Inventory* database, use database functions to find the following:

   a. The total number of leisure bicycles on hand.

   b. The average purchase cost and selling price of road bicycles.

Logical Functions

19. Modify the data in the *Base Data* worksheet in the Excel file *Credit Risk Data* using IF functions to include new columns, classifying the checking and savings account balances as low if the balance is less than $250, medium if at least $250 but less than $2000, and high otherwise.

20. The Excel file *President's Inn Guest Database* provides a list of customers, rooms they occupied, arrival and departure dates, number of occupants, and daily rate for a small bed-and-breakfast inn during one month.[7] Room rates include breakfast and are the same for one or two guests; however, any additional guests must pay an extra $20 per person per day for breakfast. Parties staying for seven days or more receive a 10% discount on the room rate as well as any additional breakfast fees. Modify the spreadsheet to calculate the number of days that each party stayed at the inn and the total revenue for the length of stay.

21. Figure 2.41 shows the 2016 U.S. federal income tax rates for the four different types of filing status. Suppose an individual is single. Write one logical IF function that will determine the correct tax rate for any income that is entered in cell B3 of a spreadsheet. Implement this on a spreadsheet and verify its correctness.

22. For the *Bicycle Inventory* database, write a logical IF function that enters "Markdown" in column H if the purchase cost is less than $100 and the selling price is at least $125; if not, then enter a blank.

23. A manager has a list of items that have been sorted according to an item ID. Some of them are duplicates.

She wants to add a code to the database that assigns a 1 to the item if it is unique, and if there are duplicates, assigns the number of the duplicate. An example is shown below. The first two items are unique, so the repeat code is 1. However, item ID 37692 is listed six times, so the codes are assigned from 1 to 6, and so on. Explain how to assign the correct code using an IF statement.

	A Item ID	B Repeat Code
1	Item ID	Repeat Code
2	35078	1
3	35088	1
4	37692	1
5	37692	2
6	37692	3
7	37692	4
8	37692	5
9	37692	6
10	37712	1
11	37713	1
12	37737	1
13	37737	2

▼ Figure 2.41

2016 Federal Income Tax Rates for Problem 21

1. Single	2. Married filing jointly or qualifying widow/widower	3. Married filing separately	4. Head of household	Tax rate
Up to $9,275	Up to $18,550	Up to $9,275	Up to $13,250	10%
$9,276 to $37,650	$18,551 to $75,300	$9,276 to $37,650	$13,251 to $50,400	15%
$37,651 to $91,150	$75,301 to $151,900	$37,651 to $75,950	$50,401 to $130,150	25%
$91,151 to $190,150	$151,901 to $231,450	$75,951 to $115,725	$130,151 to $210,800	28%
$190,151 to $413,350	$231,451 to $413,350	$115,726 to $206,675	$210,801 to $413,350	33%
$413,351 to $415,050	$413,351 to $466,950	$206,676 to $233,475	$413,351 to $441,000	35%
$415,051 or more	$466,951 or more	$233,476 or more	$441,001 or more	39.60%

---

[7] Based on Kenneth C. Laudon and Jane P. Laudon, *Essentials of Management Information Systems,* 9th ed. (Upper Saddle River, NJ: Prentice Hall, 2011).

Lookup Functions for Database Queries

24. A manager needs to identify some information from the *Purchase Orders* database but has only the order number. Modify the Excel file to use the VLOOKUP function to find the item description and cost per order for the following order numbers: Aug11008, Sep11023, and Oct11020.

25. The Excel file *S&P 500* provides open, high, low, and close values for the S&P 500 index over a period of time.

  a. Enter any date (using the format month/day/year) within the range of the data in cell G2. Use a MATCH function in cell G3 to find the row in the database that corresponds to this date.

  b. Write an INDEX function in cell G4 that finds the closing value on this date using your answer to part a.

  c. Combine the MATCH and INDEX functions in parts a and b into one function that displays the closing value in cell G5.

26. Enhance the *Sales Transactions* database file to perform a database query that finds the transaction code and amount associated with any customer ID input. Apply your results to customer ID 10029.

27. Enhance the *Treasury Yield Rates* database file to perform a database query that finds the rate associated with any date and term. (This is a challenging question because you will not be able to use the term headers in a MATCH function. Hint: Add a row that assigns a code from 1 to 11 for each term. Then use the code for your input value. You might also use a lookup function to convert the code back to the actual term as part of the output.)

28. For this exercise, use the *Purchase Orders* database. Use MATCH and/or INDEX functions to find the following:

  a. The row numbers corresponding to the first and last instance of item number 1369 in column C (be sure column C is sorted by item number).

  b. The order cost associated with the first instance of item 1369 that you identified in part a.

  c. The total cost of all orders for item 1369. Use the answers to parts a. and b. along with the SUM function to do this. In other words, you should use the appropriate INDEX and MATCH functions within the SUM function to find the answer. Validate your results by applying the SUM function directly to the data in column G.

29. Use INDEX and MATCH functions to fill in a table that extracts the amounts shipped between each pair of cities in the Excel file *General Appliance Corporation*, which shows the solution to an optimization model that finds the minimum cost and amounts shipped from two plants to four distribution centers. Your table should be set up as follows, and the formulas for the Amount Shipped column should reference the names in the From and To columns:

From	To	Amount Shipped
Marietta	Cleveland	
Marietta	Baltimore	
Marietta	Chicago	
Marietta	Phoenix	
Minneapolis	Cleveland	
Minneapolis	Baltimore	
Minneapolis	Chicago	
Minneapolis	Phoenix	

30. In the *Purchase Orders* database, we need to find the order number and the supplier that made a purchase totaling $9,045. Write functions for a left table lookup to answer these queries.

31. An auditor found an expense receipt for $179.24 without a name. Write a left table lookup function that would identify the sales rep in the *Travel Expenses* database.

32. Use the CHOOSE function to develop a database query for the *Monthly Product Sales* data that will complete the following table for any month specified in the top left corner:

Month	Product A	Product B	Product C	Product D	Product E
Sales					

Each empty cell should contain only one CHOOSE function. Hint: Use other appropriate functions to determine the row associated with the specified month.

Excel Template Design

33. Suppose that a company offers quantity discounts. If up to 1,000 units are purchased, the unit price is $10; if more than 1,000 and up to 5,000 units are purchased, the unit price is $9; and if more than 5,000 units are purchased, the unit price is $7.50. Develop an Excel template using the VLOOKUP function to find the unit price associated with any order quantity and compute the total cost of the order.

34. Develop an Excel template for computing the number of calories that a person needs each day to maintain his or her weight. Use the following formulas. For a male, the number of calories needed is

$$66.47 + 13.75 \times \text{weight in lb}/2.2 + 5 \\ \times \text{height in inches} \times 2.54 - 6.75 \times \text{age}$$

For a female, the number of calories needed is

$$665.09 + 9.56 \times \text{weight in lb}/2.2 + 1.84 \\ \times \text{height in inches} \times 2.54 - 6.75 \times \text{age}$$

Use form controls to input the height, age, weight, and gender.

35. A college graduate who is living in Cincinnati expects to earn an annual salary of $55,000. The Excel file *Cost of Living Adjustments* shows the comparative salaries in other cities and percentage adjustments for living expenses. Develop an Excel template that allows the user to enter current annual expenses (in Cincinnati) for groceries, housing, utilities, transportation, and health care and compute the corresponding expenses for any selected city and the net salary surplus for both Cincinnati and the selected city. An example of what the output should look like is provided next:

	Cincinnati	Atlanta
Salary	$55,000.00	$60,482.00
Annual Expenses		
Groceries	$7,800.00	$8,970.00
Housing	$14,400.00	$18,000.00
Utilities	$3,000.00	$2,700.00
Transportation	$1,020.00	$1,081.20
Healthcare	$4,200.00	$4,410.00
Total	$30,420.00	$35,161.20
Salary Surplus	$24,580.00	$25,320.80

Use a list box form control to select the city and sliders to input current annual expenses.

36. The Excel file *Payroll Data* provides hourly salaries for a group of employees. Create an Excel template that allows the user to select an employee by employee ID, enter the number of regular hours and overtime hours worked, and display a payroll summary with the employee name, gross pay, federal tax, state tax, Social Security, Medicare withholding deductions, and net pay. Assume that the federal tax rate is 11%, the state tax rate is 2.385%, Social Security withholding is 6.2%, and Medicare withholding is 1.45%. Use a form control to select the employee ID. (Hint: Use the CONCATENATE function to join the first and last name in your template. See Chapter 1 Appendix A1 for a discussion of this function.)

PivotTables

37. Construct PivotTables showing the counts of gender versus carrier and type versus usage in the Excel file *Cell Phone Survey*. What might you conclude from this analysis?

38. Use PivotTables to find the number of loans by different purposes, marital status, and credit risk in the Excel file *Credit Risk Data (Base Data* worksheet*).

39. Use PivotTables to find the number of sales transactions by product and region, total amount of revenue by region, and total amount of revenue by region and product in the *Sales Transactions* database.

40. The Excel file *Retail Survey* provides data about customers' preferences for denim jeans. Use PivotTables and slicers to draw conclusions about how preferences differ by gender and age group and summarize your results in a brief memo.

CASE: PEOPLE'S CHOICE BANK

The People's Choice Bank is a small community bank that has three local branches, in Blue Ash, Delhi, and Anderson Hills. The Excel file *Peoples Choice Bank* is a database of major account transactions for the month of August.

1. Note that it is difficult to determine whether each transaction in the Amount column represents a positive cash inflow or a negative cash outflow without information in other columns. Modify the database as appropriate to more easily analyze the data.

2. Suppose that you have been asked to prepare a summary of these transactions that would be clear and meaningful to the bank's president. Use PivotTables (that are well-designed and properly formatted) to provide a summary of the key information that the president and his direct reports would want to know. Justify and explain your reasoning.

CASE: DROUT ADVERTISING RESEARCH PROJECT[8]

Jamie Drout is interested in perceptions of gender stereotypes within beauty product advertising, which includes soap, deodorant, shampoo, conditioner, lotion, perfume, cologne, makeup, chemical hair color, razors, skin care, feminine care, and salon services, as well as the perceived benefits of empowerment advertising. Gender stereotypes specifically use cultural perceptions of what constitutes an attractive, acceptable, and desirable man or woman, frequently exploiting specific gender roles, and are commonly employed in advertisements for beauty products. Women are represented as delicately feminine, strikingly beautiful, and physically flawless, occupying small amounts of physical space that generally exploit their sexuality; men are represented as strong and masculine, with chiseled physical bodies, occupying large amounts of physical space to maintain their masculinity and power. In contrast, empowerment advertising strategies negate gender stereotypes and visually communicate the unique differences in each individual. In empowerment advertising, men and women are to represent the diversity in beauty, body type, and levels of perceived femininity and masculinity. Jamie's project is focused on understanding consumer perceptions of these advertising strategies.

Jamie conducted a survey using the following questionnaire:

1. What is your gender?

   Male

   Female

2. What is your age?

3. What is the highest level of education you have completed?

   Some High School Classes

   High School Diploma

   Some Undergraduate Courses

   Associate Degree

   Bachelor Degree

   Master Degree

   J.D.

   M.D.

   Doctorate Degree

4. What is your annual income?

   $0 to <$10,000

   $10,000 to <$20,000

   $20,000 to <$30,000

   $30,000 to <$40,000

   $40,000 to <$50,000

   $50,000 to <$60,000

   $60,000 to <$70,000

   $70,000 to <$80,000

   $80,000 to <$90,000

   $90,000 to <$110,000

   $110,000 to <$130,000

   $130,000 to <$150,000

   $150,000 or More

5. On average, how much do you pay for beauty and hygiene products or services per year? Include references to the following products: soap, deodorant, shampoo, conditioner, lotion, perfume, cologne, makeup, chemical hair color, razors, skin care, feminine care, and salon services.

6. On average, how many beauty and hygiene advertisements, if at all, do you think you view or hear per day? Include references to the following advertisements: television, billboard, Internet, radio, newspaper, magazine, and direct mail.

7. On average, how many of those advertisements, if any, specifically subscribe to gender roles and stereotypes?

8. On the following scale, what role, if any, do these advertisements have in reinforcing specific gender stereotypes?

   Drastic

   Influential

   Limited

   Trivial

   None

9. To what extent do you agree that empowerment advertising, which explicitly communicates the unique differences in each individual, would help transform cultural gender stereotypes?

   Strongly agree

   Agree

   Somewhat agree

   Neutral

---

[8] The author expresses appreciation to Jamie Drout for providing this original material from her class project as the basis for this case.

Somewhat disagree

Disagree

Strongly disagree

10. On average, what percentage of advertisements that you view or hear per day currently utilize empowerment advertising?

Assignment: Jamie received 105 responses, which are organized as a database in the Excel file *Drout Advertising Survey*.

1. Explain how the data and subsequent analysis using business analytics might lead to a better understanding of stereotype versus empowerment advertising.

Specifically, state some of the key insights that you would hope to answer by analyzing the data.

2. Create some PivotTables to draw some initial insights from the data.

An important aspect of business analytics is good communication. Write up your answers to this case formally in a well-written report as if you were a consultant to Ms. Drout. This case will continue in Chapters 3, 4, 6, and 7, and you will be asked to use a variety of descriptive analytics tools to analyze the data and interpret the results. Your instructor may ask you to add your insights to the report, culminating in a complete project report that fully analyzes the data and draws appropriate conclusions.

Data Visualization

ESB Professional/Shutterstock

LEARNING OBJECTIVES   After studying this chapter, you will be able to:

- Create Microsoft Excel charts.
- Determine the appropriate chart to visualize different types of data.

- Apply data bars, color scales, icon sets, and sparklines to create other types of visualizations.
- Develop useful dashboards for communicating data and information.

Making sense of large quantities of disparate data is necessary not only for gaining competitive advantage in today's business environment, but also for surviving in it. Converting data into information to understand past and current performance is the core of descriptive analytics and is vital to making good business decisions. The old adage "A picture is worth 1,000 words" is probably truer in today's information–rich environment than ever before. **Data visualization** is the process of displaying data (often in large quantities) in a meaningful fashion to provide insights that will support better decisions. Researchers have observed that data visualization improves decision making, provides managers with better analysis capabilities that reduce reliance on IT professionals, and improves collaboration and information sharing. In your career, you will most likely use data visualization extensively as an analysis tool and to communicate data and information to others. For example, if you work in finance, you can use data visualization to track revenues, costs, and profits over time; to compare performance between years or among different departments; and to track budget performance. In marketing, you can use data visualization to show trends in customer satisfaction, compare sales among different regions, and show the impact of advertising strategies. In operations, you might illustrate the performance of different facilities, product quality, call volumes in a technical support department, or supply chain metrics such as late deliveries.

In this chapter, we will illustrate how to construct and use Excel charts and other Excel visualization tools and how to build dashboards that summarize and communicate key information visually.

The Value of Data Visualization

Raw data are important, particularly when one needs to identify accurate values or compare individual numbers. However, it is quite difficult to identify trends and patterns, find exceptions, or compare groups of data in tabular form. The human brain does a surprisingly good job of processing visual information—if presented in an effective way. Visualizing data provides a way of communicating data at all levels of a business and can reveal surprising patterns and relationships, thereby providing important insights for making decisions. It also helps users to more quickly understand and interpret data, and helps analysts choose the most appropriate data analysis tool. For many unique and intriguing examples of data visualization, visit the Data Visualization Gallery at the U.S. Census Bureau Web site, www.census.gov/dataviz/.

In addition to descriptive analytics, data visualization is important for predictive and prescriptive analytics as well. For example, recall the chart predicting new product sales over time, shown in Figure 1.2 in Chapter 1. This graph conveys the concept much more easily than would either a verbal description or mathematical model. Visualizing a pattern also helps analysts select the most appropriate mathematical function to model the phenomenon. Complex prescriptive models often yield complex results. Visualizing the results often helps in understanding and gaining insight about model output and solutions.

EXAMPLE 3.1   **Tabular Versus Visual Data Analysis**

Figure 3.1 shows the data in the Excel file *Monthly Product Sales*. We can use the data to determine exactly how many units of a certain product were sold in a particular month, or to compare one month to another. For example, we see that sales of product A dropped in February, specifically by 6.7% (computed by the Excel formula =1 − B3/B2). Beyond such calculations, however, it is difficult to draw big–picture conclusions. Figure 3.2 displays a chart of monthly sales for each product. We can easily compare overall sales of different products (product C sells the least, for example) and identify trends (sales of product D are increasing), other patterns (sales of product C are relatively stable, while sales of product B fluctuate more over time), and exceptions (product E's sales fell considerably are September).

▶ **Figure 3.1**

Monthly Product Sales *Data*

	A	Product A	Product B	Product C	Product D	Product E
1	Month	Product A	Product B	Product C	Product D	Product E
2	January	7792	5554	3105	3168	10350
3	February	7268	3024	3228	3751	8965
4	March	7049	5543	2147	3319	6827
5	April	7560	5232	2636	4057	8544
6	May	8233	5450	2726	3837	7535
7	June	8629	3943	2705	4664	9070
8	July	8702	5991	2891	5418	8389
9	August	9215	3920	2782	4085	7367
10	September	8986	4753	2524	5575	5377
11	October	8654	4746	3258	5333	7645
12	November	8315	3566	2144	4924	8173
13	December	7978	5670	3071	6563	6088

▶ **Figure 3.2**

Visualization of Monthly Product Sales *Data*

ANALYTICS IN PRACTICE: Data Visualization for the New York City Police Department's Domain Awareness System[1]

The New York City Police Department (NYPD), the largest state or local police force in the United States, is charged with securing New York City from crime and terrorism. The department had accumulated a tremendous amount of information but had limited means of sharing it among its officers. Much of the information was available only to officers in the precinct house with permission to access stand-alone software applications, with little analytics or data visualization techniques to give the officers any insight.

In 2008, the NYPD began developing a new system, the Domain Awareness System (DAS), which is a citywide network of sensors, databases, devices, software, and infrastructure that informs decision making by delivering analytics and tailored information to officers' smartphones and precinct desktops. The NYPD has used the system to employ a unique combination of analytics and information technology.

A key feature of the system is data visualization. Prior to the DAS, reports simply listed data in numerical tables. The only analysis presented might be a year-to-year percentage change. Using it to pick out geographic clusters or potential patterns was virtually impossible. Today, information is presented in an interactive form. If a user clicks on a number, DAS brings up all the records included in that number and marks them on the map. The NYPD also constructed a data visualization engine to enable the user to explore trends and patterns in the statistics. Bar and pie charts of categorical data and line charts of temporal data are available with the press of a button.

Through improving the efficiency of the NYPD's staff, DAS has generated estimated savings of $50 million per year. Most importantly, the NYPD has used it to combat terrorism and improve its crime-fighting effectiveness. Since DAS was deployed department-wide in 2013, the overall crime index in the city has fallen by six percent.

Tools and Software for Data Visualization

Data visualization ranges from simple Excel charts to more advanced interactive tools and software that allow users to easily view and manipulate data with a few clicks, not only on computers, but also on iPads and other devices. In this chapter, we discuss basic tools available in Excel. Commercial software packages such as Tableau, QlikView, and SAS Visual Analytics offer more powerful tools, especially for applications involving big data. In particular, we suggest that you look at the capabilities of Tableau (www.tableau.com), which we describe in the Appendix to this chapter. Tableau is easy to use and offers a free trial.

CHECK YOUR UNDERSTANDING

1. Explain the pros and cons of tabular versus visual data analysis.

2. How is data visualization used in descriptive, predictive, and prescriptive analytics?

 ## Creating Charts in Microsoft Excel

Microsoft Excel provides a comprehensive charting capability with many features. With a little experimentation, you can create professional charts for business analyses and presentations. These include vertical and horizontal bar charts, line charts, pie charts, area charts,

[1]E. S. Levine, Jessica Tisch, Anthony Tasso, and Michael Joya, "The New York City Police Department's Domain Awareness System," *Interfaces*, Vol. 47, No. 1, January–February 2017, pp. 70–84, © 2017 INFORMS.

scatter plots, and many other special types of charts. We will not guide you through every application but do provide some guidance for new procedures as appropriate.

Certain charts work better for certain types of data, and using the wrong chart can make the data difficult for the user to interpret and understand. While Excel offers many ways to make charts unique and fancy, naive users often focus more on the attention-grabbing aspects of charts rather than their effectiveness in displaying information. So we recommend that you keep charts simple, and avoid such bells and whistles as 3-D bars, cylinders, cones, and so on. We highly recommend books written by Stephen Few, such as *Show Me the Numbers* (Oakland, CA: Analytics Press, 2004), for additional guidance in developing effective data visualizations.

To create a chart in Excel, it is best to first highlight the range of the data you wish to chart. The Excel Help files provide guidance on formatting your data for a particular type of chart. Click the *Insert* tab in the Excel ribbon (Figure 3.3; the Mac ribbon is similar). Click the chart type, and then click a chart subtype that you want to use. Once a basic chart is created, you may use the options in the *Design* (*Chart Design* in Mac) and *Format* tabs to customize your chart (Figure 3.4). In the *Design* tab, you can change the type of chart, data included in the chart, chart layout, and styles. The *Format* tab provides various formatting options. You may also customize charts easily by right clicking on elements of the chart or by using the *Quick Layout* options within the *Design* tab.

You should realize that up to 10% of the male population are affected by color blindness, making it difficult to distinguish between different color variations. Although we generally display charts using Excel's default colors, which often, unfortunately, use red, experts suggest using blue-orange palettes. We suggest that you be aware of this for professional and commercial applications.

Column and Bar Charts

Excel distinguishes between vertical and horizontal bar charts, calling the former **column charts** and the latter **bar charts**. A *clustered column chart* compares values across categories using vertical rectangles; a *stacked column chart* displays the contribution of each value to the total by stacking the rectangles; and a *100% stacked column chart* compares the percentage that each value contributes to a total. Column and bar charts are useful for comparing categorical or ordinal data, for illustrating differences between sets of values, and for showing proportions or percentages of a whole.

▲ **Figure 3.3**

Excel Insert *Ribbon for Windows*

▲ **Figure 3.4**

Excel Chart Design *Ribbon for Windows*

EXAMPLE 3.2   **Creating Column Charts**

The Excel file *EEO Employment Report* provides data on the number of employees in different categories broken down by racial/ethnic group and gender (Figure 3.5). We will construct a simple column chart for the various employment categories for all employees. First, highlight the range C3:K6, which includes the headings and data for each category. Click on the *Column Chart* button found in the *Insert* tab, and then on the first chart type in the list (a clustered column chart). To add a title, click on the *Add Chart Elements* button in the *Design* tab ribbon. Click on "Chart Title" in the chart and change it to "Alabama Employment." The names of the data series can be changed by clicking on the *Select Data* button in the *Data* group of the *Design* tab. In the *Select Data Source* dialog (see Figure 3.6), click on "Series1" and then the *Edit* button. Enter the name of the data series, in this case "All Employees." Change the names of the other data series to "Men" and "Women" in a similar fashion. You can also change the order in which the data series are displayed on the chart using the up and down buttons. The final chart is shown in Figure 3.7.

Be cautious when changing the scale of the numerical axis. The heights or lengths of the bars only accurately reflect the data values if the axis starts at zero. If not, the relative sizes can paint a misleading picture of the true relative values of the data.

Data Label and Data Table Chart Options

Excel provides options for including the numerical data on which charts are based within the charts. Data labels can be added to chart elements to show the actual value of bars,

	A	B	C	D	E	F	G	H	I	J	K
1	Equal Employment Opportunity Commission Report - Number Employed in State of Alabama, 2006										
2											
3	Racial/Ethnic Group and Gender	Total Employment	Officials &	Professionals	Technicians	Sales Workers	Office & Clerical	Craft Workers	Operatives	Laborers	Service Workers
4	ALL EMPLOYEES	632,329	60,258	80,733	39,868	62,019	67,014	61,322	120,810	68,752	71,553
5	Men	349,353	41,777	39,792	19,848	23,727	11,293	55,853	84,724	44,736	27,603
6	Women	282,976	18,481	40,941	20,020	38,292	55,721	5,469	36,086	24,016	43,950
7											
8	WHITE	407,545	51,252	67,622	28,830	41,091	44,565	45,742	67,555	26,712	34,176
9	Men	237,516	36,536	34,842	16,004	17,756	7,656	42,699	50,537	17,802	13,684
10	Women	170,029	14,716	32,780	12,826	23,335	36,909	3,043	17,018	8,910	20,492
11											
12	MINORITY	224,784	9,006	13,111	11,038	20,928	22,449	15,580	53,255	42,040	37,377
13	Men	111,837	5,241	4,950	3,844	5,971	3,637	13,154	34,187	26,934	13,919
14	Women	112,947	3,765	8,161	7,194	14,957	18,812	2,426	19,068	15,106	23,458

▲ **Figure 3.5**

Portion of EEO Employment Report *Data*

▶ **Figure 3.6**

Select Data Source *Dialog*

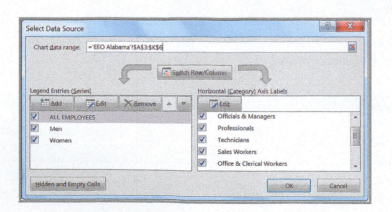

► Figure 3.7

*Column Chart for Alabama
Employment Data*

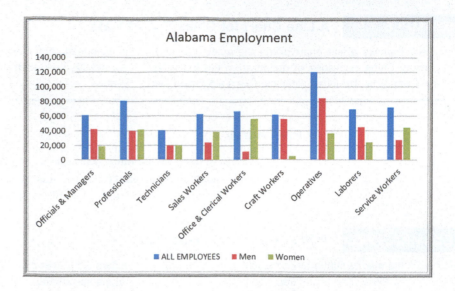

for example. Data tables can also be added; these are usually better than data labels, which can get quite messy. Both can be added from the *Add Chart Element* button in the *Chart Tools Design* tab, or from the *Quick Layout* button, which provides standard design options. Figure 3.8 shows a data table added to the Alabama Employment chart. You can see that the data table provides useful additional information to improve the visualization.

Line Charts

Line charts provide a useful means for displaying data over time, as Example 3.3 illustrates. You may plot multiple data series in line charts; however, they can be difficult to interpret if the magnitude of the data values differs greatly. In that case, it would be advisable to create separate charts for each data series.

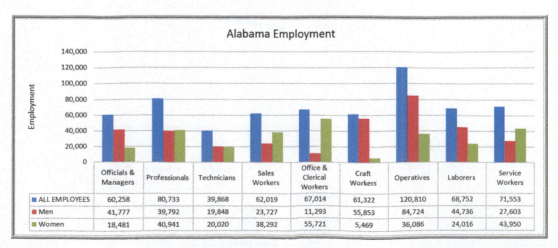

▲ Figure 3.8

Alternate Column Chart Format with a Data Table

| EXAMPLE 3.3 | A Line Chart for China Export Data |

Figure 3.9 shows a line chart giving the amount of U.S. exports to China in billions of dollars from the Excel file *China Trade Data.* The chart clearly shows a significant rise in exports starting in the year 2000, which began to level off around 2008 and then show a sharp increase in subsequent years.

Pie Charts

For many types of data, we are interested in understanding the relative proportion of each data source to the total. A **pie chart** displays this by partitioning a circle into pie-shaped areas showing the relative proportions. Example 3.4 provides one application.

| EXAMPLE 3.4 | A Pie Chart for Census Data |

Consider the marital status of individuals in the U.S. population in the Excel file *Census Education Data*, a portion of which is shown in Figure 3.10. To show the relative proportion in each category, we can use a pie chart, as shown in Figure 3.11. This chart uses a layout option that shows the labels associated with the data as well as the actual proportions as percentages. A different layout that shows the values and/or proportions can also be chosen.

▶ **Figure 3.9**

Line Chart for China Trade Data

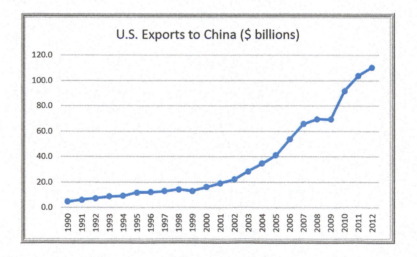

▶ **Figure 3.10**

Portion of Census Education Data

	A	B	C	D	E	F	G
1	Census Education Data						
2		Not a High School Grad	High School Graduate	Some College No Degree	Associate's Degree	Bachelor's Degree	Advanced Degree
18	**Marital Status**						
19	Never Married	4,120,320	7,777,104	4,789,872	1,828,392	5,124,648	2,137,416
20	Married, spouse present	15,516,160	36,382,720	18,084,352	8,346,624	19,154,432	9,523,712
21	Married, spouse absent	1,847,880	2,368,024	1,184,012	465,392	670,712	301,136
22	Separated	1,188,090	1,667,010	842,715	336,165	405,240	165,780
23	Widowed	5,145,683	4,670,488	1,765,010	556,657	977,544	475,195
24	Divorced	2,968,680	7,003,040	3,806,000	1,674,640	2,340,690	1,217,920

▶ **Figure 3.11**

Pie Chart for Marital Status: Not a High School Grad

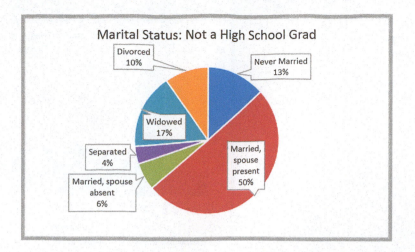

Data visualization professionals don't recommend using pie charts. For example, contrast the pie chart in Figure 3.11 with the column chart in Figure 3.12 for the same data. In the pie chart, it is difficult to compare the relative sizes of areas; however, the bars in the column chart can easily be compared to determine relative ratios of the data. If you do use pie charts, restrict them to small numbers of categories, always ensure that the numbers add to 100%, and use labels to display the group names and actual percentages. Avoid three-dimensional (3-D) pie charts—especially those that are rotated—and keep them simple.

Area Charts

An **area chart** combines the features of a pie chart with those of line charts. Area charts present more information than pie or line charts alone but may clutter the observer's mind with too many details if too many data series are used; thus, they should be used with care.

| EXAMPLE 3.5 | **An Area Chart for Energy Consumption** |

Figure 3.13 displays total energy consumption (billions Btu) and consumption of fossil fuels from the Excel file *Energy Production & Consumption.* This chart shows that although total energy consumption has grown since 1949, the relative proportion of fossil fuel consumption has remained generally consistent at about half of the total, indicating that alternative energy sources have not replaced a significant portion of fossil fuel consumption.

▶ **Figure 3.12**

Alternative Column Chart for Marital Status: Not a High School Grad

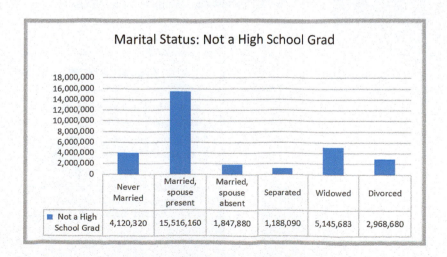

▶ **Figure 3.13**

*Area Chart for Energy
Consumption*

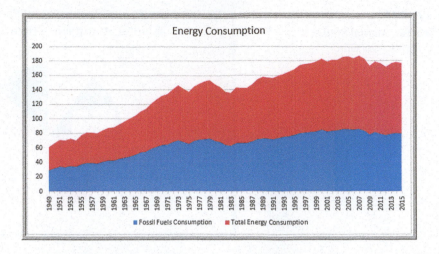

Scatter Charts and Orbit Charts

Scatter charts show the relationship between two variables. To construct a scatter chart, we need observations that consist of pairs of variables. For example, students in a class might have grades for both a midterm and a final exam. A scatter chart would show whether high or low grades on the midterm correspond strongly to high or low grades on the final exam, or whether the relationship is weak or nonexistent.

EXAMPLE 3.6	**A Scatter Chart for Real Estate Data**

Figure 3.14 shows a scatter chart of house size (in square feet) versus the home market value from the Excel file *Home* *Market Value.* The data clearly suggest that higher market values are associated with larger homes.

An **orbit chart** is a scatter chart in which the points are connected in sequence, such as over time. Orbit charts show the "path" that the data take over time, often showing some unusual patterns that can provide unique insights. You can construct an orbit chart by creating a scatter chart with smooth lines and markers from the scatter chart options. Figure 3.15 shows an example from the *Gasoline Sales* Excel file using the first 10 weeks of data.

▶ **Figure 3.14**

*Scatter Chart of House Size
Versus Market Value*

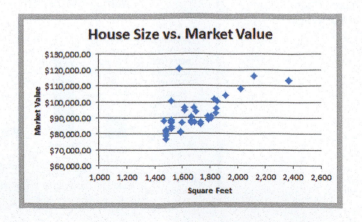

▶ **Figure 3.15**

Orbit Chart

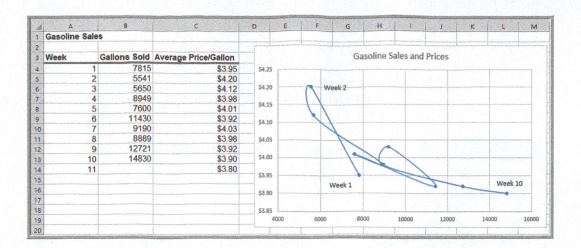

Bubble Charts

A **bubble chart** is a type of scatter chart in which the size of the data marker corresponds to the value of a third variable; consequently, it is a way to plot three variables in two dimensions.

EXAMPLE 3.7   A Bubble Chart for Comparing Stock Characteristics

Figure 3.16 shows a bubble chart for displaying price, P/E (price/earnings) ratio, and market capitalization for five different stocks on one particular day from the Excel file *Stock* *Comparisons*. The position on the chart shows the price and P/E; the size of the bubble represents the market cap in billions of dollars.

▶ **Figure 3.16**

Bubble Chart for Stock Comparisons

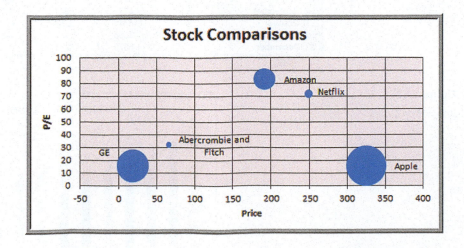

Combination Charts

Often, we wish to display multiple data series on the same chart using different chart types. Excel 2016 for Windows provides a *Combo Chart* option for constructing such a **combination chart**; in Excel 2016 for Mac, it must be done manually. We can also plot a second data series on a secondary axis; this is particularly useful when the scales differ greatly.

EXAMPLE 3.8   **Creating a Combination Chart and Secondary Axis**

Figure 3.17 shows data that have been added to the *Monthly Product Sales* data–a sales goal for product E and the percent of the goal that was actually achieved. We will first construct a chart that shows the sales of product E as compared to the monthly goals. In Excel 2016 for Windows, select the data in columns G and H, and from the *Charts* options in the *Insert* ribbon, select *Insert Combo Chart*. Figure 3.18 shows the result. On a Mac, create a standard column chart using both data series. Then right click the product E goal data series, select *Change Chart Type*, and select *Line*.

Next, to plot product E sales and the percent of the goal achieved, select the data in columns F and H (first select the data in column F, then hold the Ctrl key and select the data in column H). Choose a *Combo Chart* again. Because the percent of goal data series is very small in comparison to product sales, the line chart will look like it is on the *x*-axis. Carefully right click on this data series and select *Format Data Series > Axis > Plot series on secondary axis*. The scale for % Goal will be added on the secondary axis on the right of the chart, as in Figure 3.19.

▶ **Figure 3.17**

Monthly Product Sales with Additional Data

	A	B	C	D	E	F	G	H
1	Sales Units							
2								
3	Month	Product A	Product B	Product C	Product D	Product E	Product E Goal	% Goal
4	January	7792	5554	3105	3168	10350	10000	104%
5	February	7268	3024	3228	3751	8965	9000	100%
6	March	7049	5543	2147	3319	6827	8000	85%
7	April	7560	5232	2636	4057	8544	7000	122%
8	May	8233	5450	2726	3837	7535	7000	108%
9	June	8629	3943	2705	4664	9070	8000	113%
10	July	8702	5991	2891	5418	8389	8000	105%
11	August	9215	3920	2782	4085	7367	7000	105%
12	September	8986	4753	2524	5575	5377	6000	90%
13	October	8654	4746	3258	5333	7645	7000	109%
14	November	8315	3566	2144	4924	8173	8000	102%
15	December	7978	5670	3071	6563	6088	9000	68%

▶ **Figure 3.18**

Combination Chart

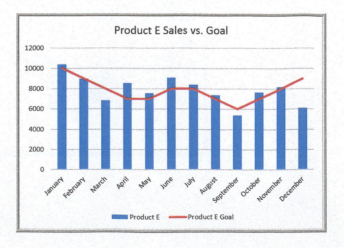

► **Figure 3.19**

Combination Chart with Secondary Axis

Radar Charts

Radar charts show multiple metrics on a spider web. This is a useful chart to compare survey data from one time period to another or to compare performance of different entities such as factories, companies, and so on using the same criteria.

| EXAMPLE 3.9 | A Radar Chart for Survey Responses |

Figure 3.20 shows average survey responses for six questions on a customer satisfaction survey. The radar chart compares the average responses between the first and second quarters. You can easily see which responses increased or decreased.

Stock Charts

A **stock chart** allows you to plot stock prices, such as daily high, low, and close values. It may also be used for scientific data such as temperature changes. We will explain how to create stock charts in Chapter 6 to visualize some statistical results, and again in Chapter 15 to visualize optimization results.

Charts from PivotTables

If you click inside a PivotTable, you can easily insert a chart that visualizes the data in the PivotTable from the *Insert* tab using any of the recommended charts, or by choosing your

► **Figure 3.20**

Radar Chart

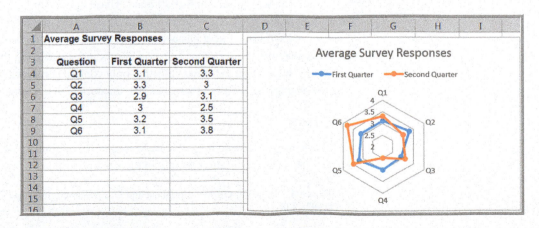

own. In Excel for Windows, this is called a PivotChart (see the Appendix to this chapter for further information about PivotCharts). You can do this on a Mac also; however, charts created from PivotTables will not have the *Filter* buttons as described in the Appendix.

Geographic Data

Many applications of business analytics involve geographic data. For example, finding the best location for production and distribution facilities, analyzing regional sales performance, transporting raw materials and finished goods, and routing vehicles such as delivery trucks involve geographic data. In such problems, data mapping can help in a variety of ways. Visualizing geographic data can highlight key data relationships, identify trends, and uncover business opportunities. In addition, it can often help to spot data errors and help end users understand solutions, thus increasing the likelihood of acceptance of decision models. Companies like Nike use geographic data and information systems for visualizing where products are being distributed and how that relates to demographic and sales information. This information is vital to marketing strategies. The use of prescriptive analytic models in combination with data mapping was instrumental in the success of Procter & Gamble's North American supply chain study, which saved the company in excess of $200 million dollars per year.[2] We discuss this application in Chapter 14.

Excel 2016 for Windows includes a geographic visualization tool called *3D Maps*. We encourage you to explore this. Another excellent option is *Tableau*, which is described in the Appendix to this chapter, in which we will illustrate its geographic visualization capabilities.

CHECK YOUR UNDERSTANDING

1. Summarize the most useful charts available in Excel and the types of applications for which they should be used.

2. What is the difference between a bar and a column chart?

3. Why don't data visualization professionals recommend the use of pie charts?

4. How do you create a combination chart in Excel?

Other Excel Data Visualization Tools

Microsoft Excel offers numerous other tools to help visualize data. These include data bars, color scales, icon sets, and sparklines. These options are part of Excel's *Conditional Formatting* rules, which allow you to visualize different numerical values through the use of colors and symbols. Excel has a variety of standard templates to use, but you may also customize the rules to meet your own conditions and styles using the *New Formatting Rule* option. This allows you to format cells that only contain certain values, those that are above or below average, as well as other rules. We encourage you to experiment with these tools.

Data Bars

Data bars display colored bars that are scaled to the magnitude of the data values (similar to a bar chart) but placed directly within the cells of a range.

---

[2]J. Camm et al., "Blending OR/MS, Judgment and GIS: Restructuring P&G's Supply Chain," *Interfaces*, 27, 1 (1997): 128–142.

▶ **Figure 3.21**

Example of Data Bars

	A	B	C	D	E	F
1	Sales Units					
2						
3	Month	Product A	Product B	Product C	Product D	Product E
4	January	7792	5554	3105	3168	10350
5	February	7268	3024	3228	3751	8965
6	March	7049	5543	2147	3319	6827
7	April	7560	5232	2636	4057	8544
8	May	8233	5450	2726	3837	7535
9	June	8629	3943	2705	4664	9070
10	July	8702	5991	2891	5418	8389
11	August	9215	3920	2782	4085	7367
12	September	8986	4753	2524	5575	5377
13	October	8654	4746	3258	5333	7645
14	November	8315	3566	2144	4924	8173
15	December	7978	5670	3071	6563	6088

EXAMPLE 3.10   **Data Visualization with Data Bars**

Figure 3.21 shows data bars applied to the data in the *Monthly Product Sales* worksheet. Highlight the data in each column, click the *Conditional Formatting* button in the *Styles* group within the *Home* tab, select *Data Bars*, and choose the fill option and color.

You may also display data bars without the data in the cells. A useful tip is to copy the data next to the original data to display the data bars along with the original data.

Figure 3.22 illustrates this for the total monthly sales. We first summed the monthly sales in column G, then copied these to column H. Then highlight the range of data in column H, click *Conditional Formatting*, choose *Data Bars*, and select *More Rules*. In the *Edit Formatting Rule* dialog, check the box *Show Bar Only*. If some data are negative, data bars will display them on the left side of a vertical axis, allowing you to clearly visualize both positive and negative values.

Color Scales

Color scales shade cells based on their numerical value using a color palette. This is another option in the *Conditional Formatting* menu.

EXAMPLE 3.11   **Data Visualization with Color Scales**

Figure 3.23 shows the use of a green-yellow-red color scale for the monthly product sales, which highlights cells containing large values in green, small values in red, and middle values in yellow. The darker the green, the larger the value; the darker the red, the smaller the value. For intermediate values, you can see that the colors blend together. This provides a quick way of identifying the largest and smallest product-month sales values.

▶ **Figure 3.22**

Displaying Data Bars Outside of Data Cells

	A	B	C	D	E	F	G	H
1	Sales Units							
2								
3	Month	Product A	Product B	Product C	Product D	Product E	Total	
4	January	7792	5554	3105	3168	10350	29969	
5	February	7268	3024	3228	3751	8965	26236	
6	March	7049	5543	2147	3319	6827	24885	
7	April	7560	5232	2636	4057	8544	28029	
8	May	8233	5450	2726	3837	7535	27781	
9	June	8629	3943	2705	4664	9070	29011	
10	July	8702	5991	2891	5418	8389	31391	
11	August	9215	3920	2782	4085	7367	27369	
12	September	8986	4753	2524	5575	5377	27215	
13	October	8654	4746	3258	5333	7645	29636	
14	November	8315	3566	2144	4924	8173	27122	
15	December	7978	5670	3071	6563	6088	29370	

► **Figure 3.23**

Example of Color Scales

⊿	A	B	C	D	E	F
1	Sales Units					
2						
3	Month	Product A	Product B	Product C	Product D	Product E
4	January	7792	5554	3105	3168	10350
5	February	7268	3024	3228	3751	8965
6	March	7049	5543	2147	3319	6827
7	April	7560	5232	2636	4057	8544
8	May	8233	5450	2726	3837	7535
9	June	8629	3943	2705	4664	9070
10	July	8702	5991	2891	5418	8389
11	August	9215	3920	2782	4085	7367
12	September	8986	4753	2524	5575	5377
13	October	8654	4746	3258	5333	7645
14	November	8315	3566	2144	4924	8173
15	December	7978	5670	3071	6563	6088

Color-coding of quantitative data is commonly called a **heatmap**. Heatmaps are often used to display geographic data, such as population densities, or other socioeconomic metrics by nation, state, county, and so on.

Icon Sets

Icon sets provide similar information as color scales using various symbols such as arrows (see Figure 3.24) or red/yellow/green stoplights. Many companies use green, yellow, and red stoplight symbols to show good, marginal, and poor performance, respectively, in business dashboards. The next example illustrates this, and also shows how to customize a conditional formatting rule.

EXAMPLE 3.12   **Data Visualization with Customized Icon Sets**

Figure 3.25 shows a set of stoplight icons that code the monthly product sales for each product as green if they are in the top 20% of the data range, red if in the bottom 20%, and yellow if in between. Note that because of the relative differences in the magnitude of sales among products, we created a new rule for each column of data. Highlight the data in a column, click *Conditional Formatting*, select *Icon Sets*, and then select *More Rules*. In the *Edit Formatting Rule* dialog, change the default values of 67 and 33% to 80 and 20%, as shown in the figure. To understand how this

works, look at product B. The minimum value is 3,024 and the maximum is 5,991. So 80% is 3024 + 0.8 × (5991 − 3024) = 5397.6. Thus, any cell value above this is coded as green. Similarly, 20% is 3024 + 0.2 × (5991 − 3024) = 3617.4; any cell value below this is coded as red. In the *Edit Formatting Rule* dialog, you may also change the rule to code the cells based on their actual values by changing *Percent* to *Number* from the drop-down box. For example, you could code all values for product B greater than or equal to 5,000 as green and all values below 4,000 as red.

► **Figure 3.24**

Example of Icon Sets

⊿	A	B	C	D	E	F
1	Sales Units					
2						
3	Month	Product A	Product B	Product C	Product D	Product E
4	January	⬆ 7792	⮕ 5554	⬇ 3105	⬇ 3168	⬆ 10350
5	February	⮕ 7268	⬇ 3024	⬇ 3228	⬇ 3751	⬆ 8965
6	March	⮕ 7049	⮕ 5543	⬇ 2147	⬇ 3319	⮕ 6827
7	April	⮕ 7560	⮕ 5232	⬇ 2636	⬇ 4057	⬆ 8544
8	May	⬆ 8233	⮕ 5450	⬇ 2726	⬇ 3837	⮕ 7535
9	June	⬆ 8629	⬇ 3943	⬇ 2705	⬇ 4664	⬆ 9070
10	July	⬆ 8702	⮕ 5991	⬇ 2891	⮕ 5418	⬆ 8389
11	August	⬆ 9215	⬇ 3920	⬇ 2782	⬇ 4085	⮕ 7367
12	September	⬆ 8986	⮕ 4753	⬇ 2524	⮕ 5575	⬇ 5377
13	October	⬆ 8654	⮕ 4746	⬇ 3258	⮕ 5333	⬆ 7645
14	November	⬆ 8315	⬇ 3566	⬇ 2144	⮕ 4924	⬆ 8173
15	December	⬆ 7978	⮕ 5670	⬇ 3071	⮕ 6563	⮕ 6088

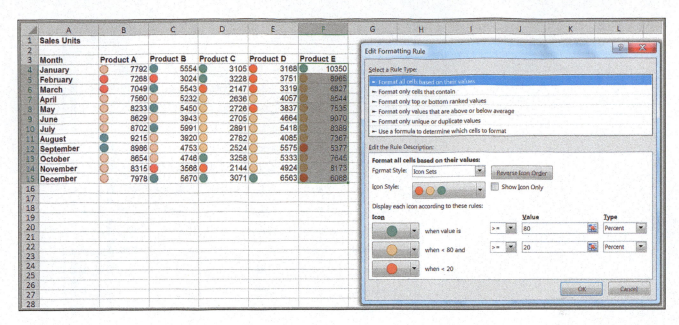

▲ **Figure 3.25**

Creating Customized Icon Sets

Sparklines

Sparklines are graphics that summarize a row or column of data in a single cell. Sparklines were introduced by Edward Tufte, a famous expert on visual presentation of data. He described sparklines as "data-intense, design-simple, word-sized graphics." Excel has three types of sparklines: line, column, and win/loss. Line sparklines are useful for time-series data, while column sparklines are more appropriate for categorical data. Win/loss sparklines are useful for data that move up or down over time. They are found in the *Sparklines* group within the *Insert* tab on the ribbon.

EXAMPLE 3.13   Examples of Sparklines

We will again use the *Monthly Product Sales* data. Figure 3.26 shows line sparklines in row 16 for each product. In column G, we display column sparklines, which are essentially small column charts. Generally, you need to expand the row or column widths to display them effectively. Notice, however, that the lengths of the bars are not scaled properly to the data; for example, in the first one, products D and E are roughly one-third the value of product E, yet the bars are not scaled correctly. Excel defaults to automatic scaling of the minimum and maximum values of the vertical axis. You can change this

and scale the bars properly by choosing *Axis* from *Group* in the *Sparkline Tools Design* tab.

Figure 3.27 shows a modified worksheet in which we computed the percentage change from one month to the next for products A and B. The win loss sparklines in row 16 show the patterns of sales increases and decreases, suggesting that product A has a cyclical pattern while product B changed in a more random fashion. If you click on any cell containing a sparkline, the *Sparkline Tools Design* tab appears, allowing you to customize colors and other options.

▶ Figure 3.26

Line and Column Sparklines

	A	B	C	D	E	F	G
1	Sales Units						
2							
3	Month	Product A	Product B	Product C	Product D	Product E	
4	January	7792	5554	3105	3168	10350	
5	February	7268	3024	3228	3751	8965	
6	March	7049	5543	2147	3319	6827	
7	April	7560	5232	2636	4057	8544	
8	May	8233	5450	2726	3837	7535	
9	June	8629	3943	2705	4664	9070	
10	July	8702	5991	2891	5418	8389	
11	August	9215	3920	2782	4085	7367	
12	September	8986	4753	2524	5575	5377	
13	October	8654	4746	3258	5333	7645	
14	November	8315	3566	2144	4924	8173	
15	December	7978	5670	3071	6563	6088	
16							

▶ Figure 3.27

Win/Loss Sparklines

	A	B	C	D	E
1	Sales Units				
2					
3	Month	Product A	Percent Change	Product B	Percent Change
4	January	7792		5554	
5	February	7268	-6.72%	3024	-45.55%
6	March	7049	-3.01%	5543	83.30%
7	April	7560	7.25%	5232	-5.61%
8	May	8233	8.90%	5450	4.17%
9	June	8629	4.81%	3943	-27.65%
10	July	8702	0.85%	5991	51.94%
11	August	9215	5.90%	3920	-34.57%
12	September	8986	-2.49%	4753	21.25%
13	October	8654	-3.69%	4746	-0.15%
14	November	8315	-3.92%	3566	-24.86%
15	December	7978	-4.05%	5670	59.00%
16					

CHECK YOUR UNDERSTANDING

1. Explain the purpose of data bars, color scales, and icon sets.

2. What is a heatmap, and what are some typical applications?

3. How are sparklines different from standard Excel charts?

Dashboards

Making data visible and accessible to employees at all levels is a hallmark of effective modern organizations. A **dashboard** is a visual representation of a set of key business measures. It is derived from the analogy of an automobile's control panel, which displays speed, gasoline level, temperature, and so on. Dashboards provide important summaries of key business information to help manage a business process or function. For example, the Cincinnati Zoo (see the introduction in Chapter 1) uses hourly, daily, and yearly dashboards that show such metrics as attendance and types of admission, cities where visitors come from, and revenue at different food and retail locations.

Dashboards are particularly useful for senior managers who don't have the time to sift through large amounts of data and need a summary of the state of business performance during monthly or quarterly reviews. Dashboards might include tabular as well as visual data to allow managers to quickly locate key data. Figure 3.28 shows a simple dashboard for the product sales data in Figure 3.1. Dashboards often incorporate color scales or icon sets to quickly pinpoint areas of concern. This one displays the monthly sales for each product individually, sales of all products combined, total annual sales by product, a comparison of the last two months, and monthly percent changes by product.

An effective dashboard should capture all the key information that the users need for making good decisions. Important business metrics are often called **key performance indicators (KPIs)**. People tend to look at data at the top left first, so the most important charts should be positioned there. An important principle in dashboard design is to keep it simple—don't clutter the dashboard with too much information or use formats (such as 3-D charts) that don't clearly convey information.

▲ **Figure 3.28**

Dashboard for Product Sales

CHECK YOUR UNDERSTANDING

1. What is a dashboard?

2. What are the key design principles for dashboards?

3. What is a key performance indicator (KPI)?

ANALYTICS IN PRACTICE: Driving Business Transformation with IBM Business Analytics[3]

Founded in the 1930s and headquartered in Ballinger, Texas, Mueller is a leading retailer and manufacturer of pre-engineered metal buildings and metal roofing products. Today, the company sells its products directly to consumers all over the southwestern United States from 35 locations across Texas, New Mexico, Louisiana, and Oklahoma.

Historically, Mueller saw itself first and foremost as a manufacturer; the retail aspects of the business were a secondary focus. However, in the early 2000s, the company decided to shift the focus of its strategy and become much more retail-centric—getting closer to its end-use customers and driving new business through a better understanding of their needs. To achieve its transformation objective, the company needed to communicate its retail strategy to employees across the organization.

As Mark Lack, Manager of Strategy Analytics and Business Intelligence at Mueller, explains: "The transformation from pure manufacturing to retail-led manufacturing required a more end-customer-focused approach to sales. We wanted a way to track how successfully our sales teams across the country were adapting to this new strategy, and identify where improvements could be made."

To keep track of sales performance, Mueller worked with IBM to deploy IBM Cognos Business Intelligence. The IBM team helped Mueller apply technology to its balanced scorecard process for strategy management in Cognos Metric Studio.

By using a common set of KPIs, Mueller can easily identify the strengths and weaknesses of all of its sales teams through sales performance analytics. "Using Metric Studio in Cognos Business Intelligence, we get a clear picture of each team's strategy performance," says Mark Lack. "Using sales performance insights from Cognos scorecards, we can identify teams that are hitting their targets, and determine the reasons for their success. We can then share this knowledge with underperforming teams, and demonstrate how they can change their way of working to meet their targets."

Instead of just trying to impose or enforce new ways of working, we are able to show sales teams exactly how they are contributing to the business, and explain what they need to do to improve their metrics. It's a much more effective way of driving the changes in behavior that are vital for business transformation."

Recently, IBM Business Analytics Software Services helped Mueller upgrade to IBM Cognos 10. With the new version in place, Mueller has started using a new feature called Business Insight to empower regional sales managers to track and improve the performance of their sales teams by creating their own personalized dashboards.

"Static reports are a good starting point, but people don't enjoy reading through pages of data to find the information they need," comments Mark Lack. "The new version of Cognos gives us the ability to create customized interactive dashboards that give each user immediate insight into their own specific area of the business, and enable them to drill down into the raw data if they need to. It's a much more intuitive and compelling way of using information."

Mueller now uses Cognos to investigate the reasons why some products sell better in certain areas, which of its products have the highest adoption rates, and which have the biggest margins. Using these insights, the company can adapt its strategy to ensure that it markets the right products to the right customers—increasing sales.

By using IBM SPSS Modeler to mine enormous volumes of transactional data, the company aims to reveal patterns and trends that will help to predict future risks and opportunities, as well as uncover unseen problems and anomalies in its current operations. One initial project with IBM SPSS Modeler aims to help Mueller find ways to reduce its fuel costs. Using SPSS Modeler, the company is building a sophisticated statistical model that will automate the process of analyzing fuel transactions for hundreds of vehicles, drivers, and routes.

"With SPSS Modeler, we will be able to determine the average fuel consumption for each vehicle on each route over the course of a week," says Mark Lack. "SPSS will automatically flag up any deviations from the average consumption, and we then drill down to find the root cause. The IBM solution helps us to determine if higher-than-usual fuel transactions are legitimate—for example, a driver covering extra miles—or the result of some other factor, such as fraud."

---

[3]"Mueller builds a customer-focused business," IBM Software, Business Analytics, © IBM Corporation, 2013.

KEY TERMS

Area chart	Icon sets
Bar chart	Key performance indicator (KPI)
Bubble chart	Line chart
Color scales	Orbit chart
Column chart	Pie chart
Combination chart	Radar chart
Dashboard	Scatter chart
Data bars	Sparklines
Data visualization	Stock chart
Heatmap	

CHAPTER 3 TECHNOLOGY HELP

Excel Techniques

Creating a chart (Example 3.2):

Highlight the range of the data to chart. Click the *Insert* tab in the Excel ribbon. Click the chart type, and then click a chart subtype that you want to use. Use the options in the *Design* (*Chart Design* on a Mac) and *Format* tabs within the *Chart Tools* tabs to customize the chart, or use the *Quick Layout* options. To add a title, click on the *Add Chart Elements* button in the *Design* tab ribbon. The names of the data series can be changed by clicking on the *Select Data* button in the *Data* group of the *Design* tab.

Creating combination charts (Example 3.8):

To display multiple data series on the same chart using different chart types in Excel 2016 for Windows, select the data series in two columns, and from the *Charts* options in the *Insert* ribbon, select *Insert Combo Chart*. On a Mac, create a standard column chart using both data series. Then right click the data series you wish to change and select *Change Chart Type*.

Display data bars, color scales, and icon sets (Examples 3.10–3.12):

Highlight the data, click the *Conditional Formatting* button in the *Styles* group within the *Home* tab, and select *Data Bars*, *Color Scales*, or *Icon Sets*. Choose the display you want or select *More Rules* to customize it. You may also select from other predetermined rules from the *Conditional Formatting* menu.

Displaying Sparklines (Example 3.13):

Select *Sparklines* from the *Insert* tab. Choose *Line*, *Column*, or *Win/Loss*, and complete the dialog. Expand the height or width of the cells to improve the visualization.

StatCrunch

StatCrunch provides various ways to chart and visualize data, including charts that are difficult to implement in Excel and are not discussed in this chapter. You can find video tutorials with step-by-step procedures and Study Card examples at https://www.statcrunch.com/5.0/example.php. We suggest that you first view the tutorials *Getting started with StatCrunch* and *Working with StatCrunch sessions*. The following tutorials are listed in the Graphs section on this Web page and explain how to create basic charts:

- Pie charts from raw data
- Pie charts with summary data
- Split and stacked bar plots
- Charting values across multiple columns
- Scatter plots
- Bubble plots

You can also find tutorials for additional charts and methods for customizing colors and styles:

- Dotplots
- Stem and leaf plots
- Boxplots with a group by column
- Painting/annotating graphs

Example: Loading a file

Click the *Data* menu, choose *Load*, and select the option.

Example: Create a Scatter Plot

1. Select the *X column* and *Y column* for the plot.
2. Enter an optional *Where* statement to specify the data rows to be included.
3. Color-code points with an optional *Group by* column.
4. Click *Compute!* to produce the plot.

Example: Create a Pie Chart with Summary Data

1. Select the variable that contains the categories and the variable that contains the counts for the plot.

2. Enter an optional *Where* statement to specify the data rows to be included.

3. Click *Compute!* to produce the plot.

PROBLEMS AND EXERCISES

Creating Charts in Microsoft Excel

1. Create clustered column and stacked column charts for the pre-MBA and post-MBA salary data in the Excel file *MBA Motivation and Salary Expectations*. Discuss which type of chart you feel is better to use to explain the information.

2. Create a line chart showing the growth in the annual CPI in the Excel file *Consumer Price Index*.

3. Create a line chart for the closing prices for all years and a stock chart for the high/low/close prices for August 2013 in the Excel file *S&P 500*.

4. Create a pie chart showing the breakdown of occupations for each year in the *Science and Engineering Jobs* Excel file, and contrast these with simple column charts.

5. Create a stacked area chart contrasting primary energy imports and primary energy exports in the Excel file *Energy Production & Consumption*. What conclusion can you reach?

6. A national homebuilder builds single-family homes and condominium-style townhouses. The Excel file *House Sales* provides information on the selling price, lot cost, type of home, and region of the country (Midwest, South) for closings during one month. Construct a scatter diagram showing the relationship between sales price and lot cost. What conclusion can you reach?

7. The Excel file *Facebook Survey* provides data gathered from a sample of college students. Create a scatter diagram showing the relationship between hours online/week and friends. Friends should be on the *x*-axis and hours online/week on the *y*-axis. What conclusion can you reach?

8. Create a bubble chart for the first five colleges in the Excel file *Colleges and Universities* for which the *x*-axis is the top 10% HS, *y*-axis is acceptance rate, and bubbles represent the expenditures per student.

9. Construct a column chart for the data in the Excel file *State Unemployment Rates* to allow comparison of the January rate with the historical highs and lows. Would any other charts be better for visually conveying this information? Why or why not?

10. The Excel file *Internet Usage* provides data about users of the Internet. Construct stacked bar charts that will allow you to compare any differences due to age or educational attainment and draw any conclusions that you can. Would another type of chart be more appropriate?

11. Construct an appropriate chart to show the relative value of funds in each investment category in the Excel file *Retirement Portfolio*.

12. Construct an appropriate chart or charts to visualize the information in the *Budget Forecasting* Excel file. Explain why you chose the chart(s) you used.

13. A marketing researcher surveyed 92 individuals, asking them if they liked a new product concept or not. The following results are shown:

	Yes	No
Male	30	50
Female	6	6

Convert the data into percentages for each gender class. Then construct a chart of the counts and a chart of the percentages. Discuss what each conveys visually and how the different charts may lead to different interpretations of the data.

Other Excel Data Visualization Tools

14. In the Excel file *Banking Data*, apply the following data visualization tools:

    a. Use data bars to visualize the relative values of median home value.

    b. Use color scales to visualize the relative values of median household wealth.

    c. Use an icon set to show high, medium, and low average bank balances, where high is above $30,000, low is below $10,000, and medium is anywhere in between.

15. Apply three different colors of data bars to lunch, dinner, and delivery sales in the Excel file *Restaurant Sales* to visualize the relative amounts of sales. Then sort the data by the day of the week beginning on Sunday. Compare the nonsorted data with the sorted data and comment on the information content of the visualizations.

16. For the *Store and Regional Sales* database, apply a four-traffic-light icon set to visualize the distribution of the number of units sold for each store, where green corresponds to at least 30 units sold, yellow to at least 20 but less than 30, red to at least 10 but less than 20, and black to below 10.

17. For the Excel file *Closing Stock Prices*,

    a. apply both column and line sparklines to visualize the trends in the prices for each of the four stocks in the file.

    b. compute the daily change in the Dow Jones index and apply a win/loss sparkline to visualize the daily up or down movement in the index.

Dashboards

18. Create a useful dashboard for the data in the Excel file *President's Inn Guest Database*. Use the additional information stated in Problem 20 of Chapter 2: Room rates are the same for one or two guests; however, additional guests must pay an additional $20 per person per day for meals. Guests staying for seven days or more receive a 10% discount. Modify the spreadsheet to calculate the number of days that each party stayed at the inn and the total revenue for the length of stay. Use appropriate charts and layouts and other visualization tools that help to convey the information. Explain why you chose the elements of the dashboard and how a manager might use them.

19. Create a useful dashboard for the data in the Excel file *Restaurant Sales*. Use appropriate charts and layouts and other visualization tools that help to convey the information. Explain why you chose the elements of the dashboard and how a manager might use them.

20. Create a useful dashboard for the data in the Excel file *Store and Regional Sales Database*. Use appropriate charts and layouts and other visualization tools that help to convey the information. Explain why you chose the elements of the dashboard and how a manager might use them.

21. Create a useful dashboard for the data in the Excel file *Corporate Default Database*. Use appropriate charts and layouts and other visualization tools that help to convey the information. Explain why you chose the elements of the dashboard and how a manager might use them.

CASE: PERFORMANCE LAWN EQUIPMENT

Part 1: PLE originally produced lawn mowers, but a significant portion of sales volume over recent years has come from the growing small-tractor market. As we noted in the case in Chapter 1, PLE sells their products worldwide, with sales regions including North America, South America, Europe, and the Pacific Rim. Three years ago, a new region was opened to serve China, where a booming market for small tractors has been established. PLE has always emphasized quality and considers the quality it builds into its products as its primary selling point. In the past two years, PLE has also emphasized the ease of use of their products.

Before digging into the details of operations, Elizabeth Burke wants to gain an overview of PLE's overall business performance and market position by examining the information provided in the database for this company. Specifically, she is asking you to construct appropriate charts for the data in the following worksheets in the *Performance Lawn Equipment Database* and summarize your conclusions from analysis of these charts.

    a. *Dealer Satisfaction*
    b. *End-User Satisfaction*
    c. *Complaints*
    d. *Mower Unit Sales*
    e. *Tractor Unit Sales*
    f. *On-Time Delivery*
    g. *Defects After Delivery*
    h. *Response Time*

Part 2: Propose a monthly dashboard of the most important business information that Ms. Burke can use on a routine basis as data are updated. Create one using the most recent data. Your dashboard should not consist of more than six to eight charts, which should fit comfortably on one screen.

Write a formal report summarizing your results for both parts of this case.

Additional Tools for Data Visualization

In this appendix, we describe some additional charts that are only available in the Windows version of Excel 2016 and also illustrate the capabilities of Tableau, a powerful software package for data visualization.

Hierarchy Charts

Hierarchical data are organized in a tree-like structure. A simple example is the *Purchase Orders* database. At the top level we have suppliers; at the next level we have the items purchased; then the order quantity or cost per order, and so on. A **treemap chart** divides the chart area into rectangles that represent different levels and relative sizes of hierarchical data. Each rectangle is subdivided into smaller rectangles representing the next level in the hierarchy. Rectangles at the top level of the hierarchy are arranged with the largest in the upper left corner of the chart to the smallest in the lower right corner. Within a rectangle, the next level of the hierarchy is also arranged with rectangles from the upper left to the lower right.

EXAMPLE A3.1 **A Treemap for Purchase Order Data**

For the *Purchase Orders* database, we used a PivotTable to create a new worksheet showing the items purchased from each supplier and the total cost of the orders. Highlight the data and select *Insert Hierarchy Chart* from the *Charts* group on the *Insert* tab. Figure A3.1 shows the result. We see that the largest order costs come from Hulkey Fasteners, followed by Durrable Products, Steelpin Inc., and so on. Within each colored area you can also see the relative sizes of the orders from each supplier.

An alternative to a treemap chart is a **sunburst chart**. In this chart, the hierarchy is represented by a series of circles, with the highest level of the hierarchy in the center and lower levels of the hierarchy as rings displayed outside the center. The lowest level of the hierarchy is the outside ring. Figure A3.2 shows a sunburst chart for the *Purchase Orders* data in the previous example.

Waterfall Charts

A **waterfall chart** shows a running total as values are added or subtracted. A common application of waterfall charts is to show the impact of positive or negative cash flows on net income or profit.

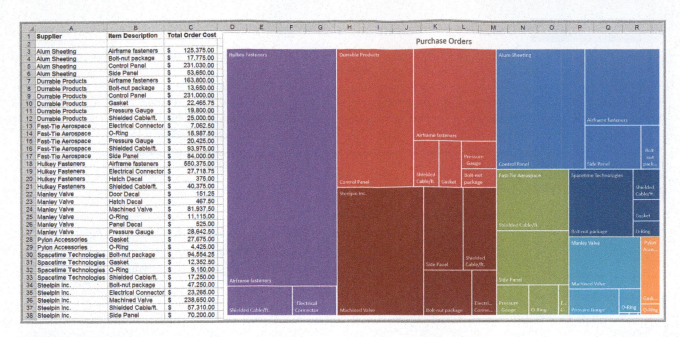

	A	B	C
1	Supplier	Item Description	Total Order Cost
2			
3	Alum Sheeting	Airframe fasteners	$ 125,375.00
4	Alum Sheeting	Bolt-nut package	$ 17,775.00
5	Alum Sheeting	Control Panel	$ 231,030.00
6	Alum Sheeting	Side Panel	$ 53,650.00
7	Durrable Products	Airframe fasteners	$ 163,800.00
8	Durrable Products	Bolt-nut package	$ 13,650.00
9	Durrable Products	Control Panel	$ 231,000.00
10	Durrable Products	Gasket	$ 22,465.75
11	Durrable Products	Pressure Gauge	$ 19,800.00
12	Durrable Products	Shielded Cable/ft.	$ 25,000.00
13	Fast-Tie Aerospace	Electrical Connector	$ 7,062.50
14	Fast-Tie Aerospace	O-Ring	$ 18,987.50
15	Fast-Tie Aerospace	Pressure Gauge	$ 20,425.00
16	Fast-Tie Aerospace	Shielded Cable/ft.	$ 93,975.00
17	Fast-Tie Aerospace	Side Panel	$ 84,000.00
18	Hulkey Fasteners	Airframe fasteners	$ 550,375.00
19	Hulkey Fasteners	Electrical Connector	$ 27,718.75
20	Hulkey Fasteners	Hatch Decal	$ 375.00
21	Hulkey Fasteners	Shielded Cable/ft.	$ 40,375.00
22	Manley Valve	Door Decal	$ 151.25
23	Manley Valve	Hatch Decal	$ 467.50
24	Manley Valve	Machined Valve	$ 81,937.50
25	Manley Valve	O-Ring	$ 11,115.00
26	Manley Valve	Panel Decal	$ 525.00
27	Manley Valve	Pressure Gauge	$ 28,642.50
28	Pylon Accessories	Gasket	$ 27,675.00
29	Pylon Accessories	O-Ring	$ 4,425.00
30	Spacetime Technologies	Bolt-nut package	$ 94,554.25
31	Spacetime Technologies	Gasket	$ 12,352.50
32	Spacetime Technologies	O-Ring	$ 9,150.00
33	Spacetime Technologies	Shielded Cable/ft.	$ 17,250.00
34	Steelpin Inc.	Bolt-nut package	$ 47,250.00
35	Steelpin Inc.	Electrical Connector	$ 23,265.00
36	Steelpin Inc.	Machined Valve	$ 238,650.00
37	Steelpin Inc.	Shielded Cable/ft.	$ 57,310.00
38	Steelpin Inc.	Side Panel	$ 70,200.00

▲ **Figure A3.1**

Purchase Orders Treemap

▶ **Figure A3.2**

Sunburst Chart

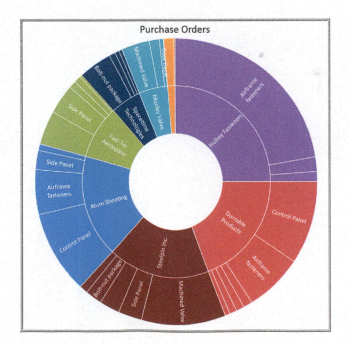

EXAMPLE A3.2   Creating a Waterfall Chart

Figure A3.3 shows the calculation of net income as sales less cost of goods sold, administrative expenses, selling expenses, depreciation expenses, interest expense, and taxes (these data are found in a slightly different format in the Excel file *Net Income Models*, which we will use in Chapter 11). Simply highlight the range A3:B10, choose *Insert Waterfall or Stock Chart* from the *Charts* group on the *Insert* tab, and select the waterfall chart. In the default chart, double click on the net income data point. In the *Format Data Point* window, check the box *Set as total*. This positions the net income bar on the *x*-axis. The chart shows how each component adds to or decreases the net income.

PivotCharts

Microsoft Excel for Windows provides a simple one-click way of creating charts to visualize data in PivotTables called **PivotCharts**. To display a PivotChart for a PivotTable, first click inside the PivotTable. From the *Analyze* tab, click on *PivotChart*. Excel will display an *Insert Chart* dialog that allows you to choose the type of chart you wish to display.

EXAMPLE A3.3   A PivotChart for *Sales Transactions* Data

Figure 2.34 showed a PivotTable for revenues by region and product for the *Sales Transactions* database. To display a column PivotChart, choose *Clustered Column* from the *PivotChart* menu. Figure A3.4 shows the chart generated by Excel. By clicking on the drop-down filter buttons, you can easily change the display.

PivotCharts make it easy to create dashboards. However, you should consider what message you are trying to convey in choosing the design of PivotCharts. In Figure A3.4, for instance, it is easier to compare product sales within a region, while in Figure A3.5, which was created by reversing the order of region and product in the *Rows* area, it is easier to compare regional sales for a given product. PivotCharts are also highly effective in oral presentations since the *Filter* buttons allow you to drill down into the data to respond to questions from the audience.

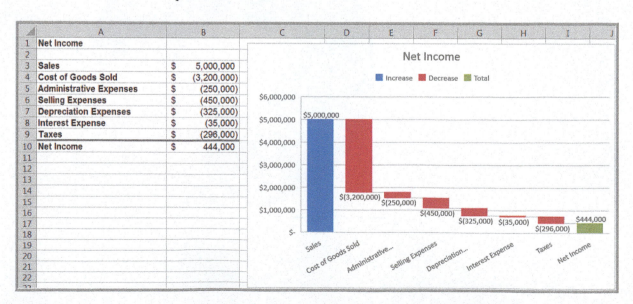

▲ **Figure A3.3**

Waterfall Chart

▶ **Figure A3.4**

*PivotChart for Revenues by
Region and Product*

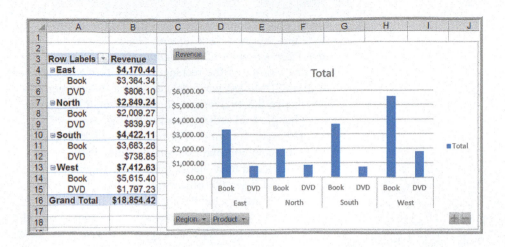

▶ **Figure A3.5**

*Alternate PivotChart for
Revenues by Region and
Product*

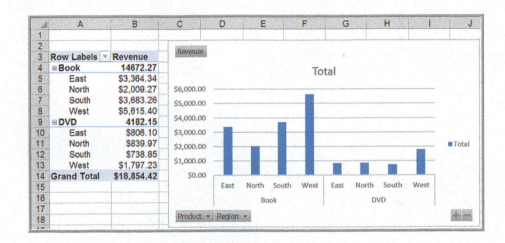

 **Tableau**

Tableau (www.tableau.com) is professional business intelligence software that facilitates data visualization using intuitive drag-and-drop capabilities. It can be used for spreadsheets as well as other types of database formats such as geographic data and has the power to handle big data easily. It also applies best practices in data visualization and facilitates the development of dashboards, which can easily be published and shared on the web and on mobile devices. We briefly illustrate some of its basic capabilities.

Figure A3.6 shows the window for the *Purchase Orders* database after it has been loaded into Tableau. Tableau automatically partitions the data in the left pane as "dimensions," namely, the columns that correspond to names or dates, and "measures," those that correspond to numerical data in the database. Similar to PivotTables, you simply drag these into the areas for *Columns, Rows, Filters*, and *Pages*. In Figure A3.6, we first drag *Item Description* into the *Columns* area and *Supplier* into the *Rows* area; then drag *Cost per order* inside of the rectangular area formed by the rows and columns. Note that this displays SUM(Cost per order) as shown in the green oval. From a drop-down menu, you can change this to average, count, minimum, maximum, and other measures, similar to changing the field settings in a PivotTable.

The *Show Me* window provides one-click visualizations for a wide variety of charts, including common bar, column, and pie charts, along with more advanced charts such as treemaps and heat maps. Tableau intelligently shows only those charts that apply to the data and hides the others (grayed out). Figure A3.7 shows a stacked bar chart, and Figure A3.8 shows a bubble chart view of the data.

Tableau provides free licenses for classroom use and a host of training materials and instructional videos. We encourage you to visit www.tableau.com/products/desktop to see more of the capabilities and features of Tableau.

▶ **Figure A3.6**

Tableau Window

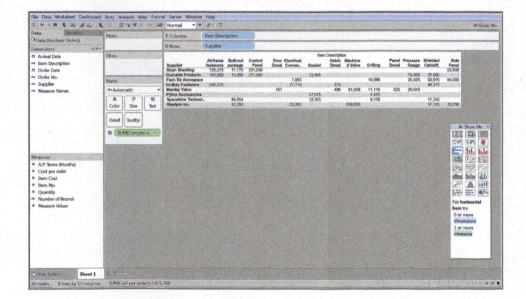

▶ **Figure A3.7**

Tableau Stacked Bar Chart

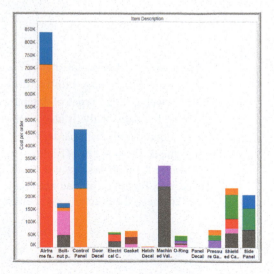

▶ **Figure A3.8**

Tableau Bubble Chart

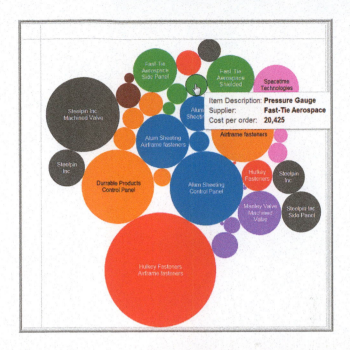

PROBLEMS AND EXERCISES

Hierarchy Charts

1. Create a treemap for the regions, items, and units sold in the Excel file *Store and Regional Sales Database*.

2. Create a sunburst chart for the regions, items, and units sold in the Excel file *Store and Regional Sales Database*.

3. Compute the total expenses for each month in the *Budget Forecasting* Excel file. Then develop waterfall charts for each month to visualize the data.

PivotCharts

4. Use PivotTables to find the number of loans by different purposes, marital status, and credit risk in the Excel file *Credit Risk Data*. Illustrate the results in a PivotChart.

5. Use PivotCharts to create a useful dashboard for the data in the Excel file *President's Inn Guest Database*. Explain why you chose the elements of the dashboard and how a manager might use them.

6. Use PivotCharts to create a useful dashboard for the data in the Excel file *Restaurant Sales*. Explain why you chose the elements of the dashboard and how a manager might use them.

7. Use PivotCharts to create a useful dashboard for the data in the Excel file *Store and Regional Sales*. Explain why you chose the elements of the dashboard and how a manager might use them.

8. Use PivotCharts to create a useful dashboard for the data in the Excel file *Corporate Default Database*. Explain why you chose the elements of the dashboard and how a manager might use them.

9. Use PivotCharts to create a useful dashboard for the data in the Excel file *Peoples Choice Bank*. Explain why you chose the elements of the dashboard and how a manager might use them.

Descriptive Statistics

Nataliiap/Shutterstock

Nataliiap/Shutterstock

LEARNING OBJECTIVES
After studying this chapter, you will be able to:

- Define a metric and explain the concepts of measurement and measures.
- Explain the difference between a discrete metric and continuous metric, and provide examples of each.
- Describe the four groups of data classification: categorical, ordinal, interval, and ratio, and provide examples of each.
- Explain the science of statistics and define the term statistic.
- Construct a frequency distribution for categorical, numerical, and grouped data.
- Construct a relative frequency distribution and histogram.

- Compute cumulative relative frequencies.
- Find percentiles and quartiles for a data set.
- Construct a cross-tabulation (contingency table).
- Explain the difference between a population and a sample.
- Understand statistical notation.
- List different measures of location.
- Compute the mean, median, mode, and midrange of a set of data.
- Use measures of location to make practical business decisions.
- List different measures of dispersion.

- Compute the range, interquartile range, variance, and standard deviation of a set of data.
- Explain Chebyshev's theorem.
- State the empirical rules and apply them to practical data.
- Compute a standardized value (z-score) for observations in a data set.
- Define and compute the coefficient of variation.
- Explain the nature of skewness and kurtosis in a distribution.
- Interpret the coefficients of skewness and kurtosis.
- Use the Excel *Descriptive Statistics* tool to summarize data.
- Calculate the mean, variance, and standard deviation for grouped data.

- Calculate a proportion.
- Use PivotTables to compute the mean, variance, and standard deviation of summarized data.
- Explain the importance of understanding relationships between two variables. Explain the difference between covariance and correlation.
- Calculate measures of covariance and correlation.
- Use the Excel *Correlation* tool.
- Identify outliers in data.
- State the principles of statistical thinking.
- Interpret variation in data from a logical and practical perspective.
- Explain the nature of variation in sample data.

Statistics, as defined by David Hand, past president of the Royal Statistical Society in the UK, is both the science of uncertainty and the technology of extracting information from data.[1] Statistics involves collecting, organizing, analyzing, interpreting, and presenting data. A **statistic** is a summary measure of data. You are undoubtedly familiar with the concept of statistics in daily life as reported in the media: baseball batting averages, airline on-time arrival performance, and economic statistics such as the Consumer Price Index are just a few examples. Statistics helps to understand and quantify uncertainty in data and to incorporate uncertainty for predicting the future.

You have undoubtedly been using statistics informally for many years. Numerical measures, such as your GPA, average incomes, and housing prices, are examples of statistics. While we see averages all the time in sports, finance, and marketing, these are only the tip of the iceberg. Statistics is much more than just computing averages. Statistics provides the means of gaining insight—both numerically and visually—into large quantities of data, understanding uncertainty and risk, and drawing conclusions from sample data that come from very large populations. For example, marketing analysts employ statistics extensively to analyze survey data to understand brand loyalty and satisfaction with goods and services, to segment customers into groups for targeted ads, and to identify factors that drive consumer demand; finance personnel use statistics to evaluate stock and mutual fund performance in order to identify good investment opportunities and to evaluate changes in foreign currency rates; and operations managers use statistics to gauge production and quality performance to determine

---

[1]David Hand, "Statistics: An Overview," in Miodrag Lovric, Ed., *International Encyclopedia of Statistical Science*, Springer Major Reference; http://www.springer.com/statistics/book/978-3-642-04897-5, p. 1504.

process and design improvements. You will find yourself routinely using many of the statistical concepts in this chapter in your daily work.

Statistical methods are essential to business analytics and are used throughout this book. Microsoft Excel supports statistical analysis in two ways:

1. with statistical functions that are entered in worksheet cells directly or embedded in formulas
2. with the Excel *Analysis Toolpak* add-in to perform more complex statistical computations. While previous versions of Excel for Mac did not support the *Analysis Toolpak*, Excel 2016 now includes it.

We use both statistical functions and the *Analysis Toolpak* in many examples.

Descriptive statistics refers to methods of describing and summarizing data using tabular, visual, and quantitative techniques. In this chapter, we focus on both tabular and visual methods, as well as quantitative measures for statistical analysis of data. We begin with a discussion of different types of metrics and ways to classify data.

ANALYTICS IN PRACTICE: Applications of Statistics in Health Care[2]

The science of statistics is vitally important to health care decision makers. Descriptive statistics summarize the utility, efficacy, and costs of medical goods and services. Increasingly, health care organizations employ statistical analysis to measure their performance outcomes. Some examples include the following:

- Descriptive statistics summarize the utility, efficacy, and costs of medical goods and services. For example, government health and human service agencies gauge the overall health and well-being of populations with statistical information.
- Hospitals and other large provider service organizations implement data-driven, continuous quality improvement programs to maximize efficiency. Statistics are important to health care companies in measuring performance success or failure. By establishing benchmarks, or standards of service excellence, quality improvement managers can measure future outcomes. Analysts map the overall growth and viability of a health care company by using statistical data gathered over time.

- Researchers gather data on human population samples. The health care industry benefits from knowing consumer market characteristics such as age, sex, race, income, and disabilities. These demographic statistics can predict the types of services that people are using and the level of care that is affordable to them.
- Statistical information is invaluable in determining what combination of goods and services to produce, which resources to allocate in producing them, and to which populations to offer them.
- Public and private health care administrators, charged with providing continuums of care to diverse populations, compare existing services to community needs. Statistical analysis is a critical component in a needs assessment. Statistics is equally important to pharmaceutical and technology companies in developing product lines that meet the needs of the populations they serve.
- Innovative medicine begins and, sometimes, ends with statistical analysis. Data are collected and carefully reported in clinical trials of new technologies and treatments to weigh products' benefits against their risks.

---

[2]Adapted from Rae Casto, "Why Are Statistics Important in the Health Care Field?" https://www.livestrong.com/article/186334-why-are-statistics-important-in-the-health-care-field/.

Metrics and Data Classification

A **metric** is a unit of measurement that provides a way to objectively quantify performance. For example, senior managers might assess overall business performance using such metrics as net profit, return on investment, market share, and customer satisfaction. A plant manager might monitor such metrics as the proportion of defective parts produced or the number of inventory turns each month. For a Web-based retailer, some useful metrics are the percentage of orders filled accurately and the time taken to fill a customer's order. **Measurement** is the act of obtaining data associated with a metric. **Measures** are numerical values associated with a metric.

Metrics can be either discrete or continuous. A **discrete metric** is one that is derived from counting something. For example, a delivery is either on time or not; an order is complete or incomplete; or an invoice can have one, two, three, or any number of errors. Some discrete metrics associated with these examples would be the number of on-time deliveries, the number of incomplete orders each day, and the number of errors per invoice. **Continuous metrics** are based on a continuous scale of measurement. Any metrics involving dollars, length, time, volume, or weight, for example, are continuous.

Another classification of data is by the type of measurement scale. Data may be classified into four groups:

1. **Categorical (nominal) data**, which are sorted into categories according to specified characteristics. For example, a firm's customers might be classified by their geographical region (e.g., North America, South America, Europe, and Pacific); employees might be classified as managers, supervisors, and associates. The categories bear no quantitative relationship to one another, but we usually assign an arbitrary number to each category to ease the process of managing the data and computing statistics. Categorical data are usually counted or expressed as proportions or percentages.

2. **Ordinal data**, which can be ordered or ranked according to some relationship to one another. College football or basketball rankings are ordinal; a higher ranking signifies a stronger team but does not specify any numerical measure of strength. Ordinal data are more meaningful than categorical data because data can be compared. A common example in business is data from survey scales—for example, rating a service as poor, average, good, very good, or excellent. Such data are categorical but also have a natural order (excellent is better than very good) and, consequently, are ordinal. However, ordinal data have no fixed units of measurement, so we cannot make meaningful numerical statements about differences between categories. Thus, we cannot say that the difference between excellent and very good is the same as between good and average, for example. Similarly, a team ranked number 1 may be far superior to the number 2 team, whereas there may be little difference between teams ranked 9th and 10th.

3. **Interval data**, which are ordinal but have constant differences between observations and have arbitrary zero points. Common examples are time and temperature. Time is relative to global location, and calendars have arbitrary starting dates (compare, for example, the standard Gregorian calendar with the Chinese calendar). Both the Fahrenheit and Celsius scales represent a specified measure of distance—degrees—but have arbitrary zero points. Thus we cannot compute meaningful ratios; for example, we cannot say that 50 degrees is twice as hot as 25 degrees. However, we can compare differences. Another example is SAT or GMAT scores. The scores can be used to rank students, but only differences

between scores provide information on how much better one student performed over another; ratios make little sense. In contrast to ordinal data, interval data allow meaningful comparison of ranges, averages, and other statistics.

In business, data from survey scales, while technically ordinal, are often treated as interval data when numerical scales are associated with categories (for instance, 1 = poor, 2 = average, 3 = good, 4 = very good, 5 = excellent). Strictly speaking, this is not correct because the "distance" between categories may not be perceived as the same (respondents might perceive a larger gap between poor and average than between good and very good, for example). Nevertheless, many users of survey data treat them as interval when analyzing the data, particularly when only a numerical scale is used without descriptive labels.

4. **Ratio data**, which are continuous and have a natural zero point. Most business and economic data, such as dollars and time, fall into this category. For example, the measure dollars has an absolute zero. Ratios of dollar figures are meaningful. For example, knowing that the Seattle region sold $12 million in March whereas the Tampa region sold $6 million means that Seattle sold twice as much as Tampa.

This classification is hierarchical in that each level includes all the information content of the one preceding it. For example, ordinal data are also categorical, and ratio information can be converted to any of the other types of data. Interval information can be converted to ordinal or categorical data but cannot be converted to ratio data without the knowledge of the absolute zero point. Thus, a ratio scale is the strongest form of measurement.

EXAMPLE 4.1   Classifying Data Elements in a Purchasing Database[3]

Figure 4.1 shows a portion of a data set containing all items that an aircraft component manufacturing company has purchased over the past three months. The data provide the supplier; order number; item number, description, and cost; quantity ordered; cost per order; the suppliers' accounts payable (A/P) terms; and the order and arrival dates. We may classify each of these types of data as follows:

- Supplier—categorical
- Order number—ordinal
- Item number—categorical

- Item description—categorical
- Item cost—ratio
- Quantity—ratio
- Cost per order—ratio
- A/P terms—ratio
- Order date—interval
- Arrival date—interval

We might use these data to evaluate the average speed of delivery and rank the suppliers (thus creating ordinal data) by this metric.

	A	B	C	D	E	F	G	H	I	J
1	Purchase Orders									
2										
3	Supplier	Order No.	Item No.	Item Description	Item Cost	Quantity	Cost per order	A/P Terms (Months)	Order Date	Arrival Date
4	Hulkey Fasteners	Aug11001	1122	Airframe fasteners	$ 4.25	19,500	$ 82,875.00	30	08/05/11	08/13/11
5	Alum Sheeting	Aug11002	1243	Airframe fasteners	$ 4.25	10,000	$ 42,500.00	30	08/08/11	08/14/11
6	Fast-Tie Aerospace	Aug11003	5462	Shielded Cable/ft.	$ 1.05	23,000	$ 24,150.00	30	08/10/11	08/15/11
7	Fast-Tie Aerospace	Aug11004	5462	Shielded Cable/ft.	$ 1.05	21,500	$ 22,575.00	30	08/15/11	08/22/11
8	Steelpin Inc.	Aug11005	5319	Shielded Cable/ft.	$ 1.10	17,500	$ 19,250.00	30	08/20/11	08/31/11
9	Fast-Tie Aerospace	Aug11006	5462	Shielded Cable/ft.	$ 1.05	22,500	$ 23,625.00	30	08/20/11	08/26/11
10	Steelpin Inc.	Aug11007	4312	Bolt-nut package	$ 3.75	4,250	$ 15,937.50	30	08/25/11	09/01/11
11	Durrable Products	Aug11008	7258	Pressure Gauge	$ 90.00	100	$ 9,000.00	45	08/25/11	08/28/11
12	Fast-Tie Aerospace	Aug11009	6321	O-Ring	$ 2.45	1,300	$ 3,185.00	30	08/25/11	09/04/11

▲ **Figure 4.1**

Portion of Purchase Orders *Database*

---

[3]Based on Kenneth C. Laudon and Jane P. Laudon, *Essentials of Management Information Systems.* 9th ed. *(Upper Saddle River, NJ: Prentice Hall, 2011).*

CHECK YOUR UNDERSTANDING

1. Explain the science of statistics.

2. What is a metric? How does it differ from a measure?

3. Explain the difference between a discrete and a continuous metric.

4. Describe the four types of measurement scales and give an example of each.

Frequency Distributions and Histograms

A **frequency distribution** is a table that shows the number of observations in each of several nonoverlapping groups. A graphical depiction of a frequency distribution in the form of a column chart is called a **histogram**. Frequency distributions and histograms summarize basic characteristics of data, such as where the data are centered and how broadly data are dispersed. This is usually the first step in using descriptive statistics. In this section, we discuss how to create them for both categorical and numerical data.

Frequency Distributions for Categorical Data

Categorical variables naturally define the groups in a frequency distribution. For example, in the *Purchase Orders* database (see Figure 4.1), orders were placed for the following items:

Airframe fasteners	Machined valve
Bolt-nut package	O-ring
Control panel	Panel decal
Door decal	Pressure gauge
Electrical connector	Shielded cable/ft.
Gasket	Side panel
Hatch decal	

To construct a frequency distribution, we need only count the number of observations that appear in each category. This can be done using the Excel COUNTIF function.

EXAMPLE 4.2   **Constructing a Frequency Distribution for Items in the *Purchase Orders* Database**

First, list the item names in a column on the spreadsheet. We used column A, starting in cell A100, below the existing data array. It is important to use the exact names as used in the data file. To count the number of orders placed for each item, use the function =COUNTIF($D$4:$D$97, *cell_reference*), where *cell_reference* is the cell containing the item name, in this case, cell A101. This is shown in Figure 4.2.

The resulting frequency distribution for the items is shown in Figure 4.3. Thus, the company placed 14 orders for airframe fasteners and 11 orders for the bolt-nut package. We may also construct a column chart to visualize these frequencies, as shown in Figure 4.4. We might wish to sort these using Pareto analysis to gain more insight into the order frequency.

▶ **Figure 4.2**

*Using the COUNTIF Function
to Construct a Frequency
Distribution*

	A	B
100	**Item Description**	**Frequency**
101	Airframe fasteners	=COUNTIF($D$4:$D$97,A101)
102	Bolt-nut package	=COUNTIF($D$4:$D$97,A102)
103	Control Panel	=COUNTIF($D$4:$D$97,A103)
104	Door Decal	=COUNTIF($D$4:$D$97,A104)
105	Electrical Connector	=COUNTIF($D$4:$D$97,A105)
106	Gasket	=COUNTIF($D$4:$D$97,A106)
107	Hatch Decal	=COUNTIF($D$4:$D$97,A107)
108	Machined Valve	=COUNTIF($D$4:$D$97,A108)
109	O-Ring	=COUNTIF($D$4:$D$97,A109)
110	Panel Decal	=COUNTIF($D$4:$D$97,A110)
111	Pressure Gauge	=COUNTIF($D$4:$D$97,A111)
112	Shielded Cable/ft.	=COUNTIF($D$4:$D$97,A112)
113	Side Panel	=COUNTIF($D$4:$D$97,A113)

▶ **Figure 4.3**

*Frequency Distribution for
Items Purchased*

	A	B
100	**Item Description**	**Frequency**
101	Airframe fasteners	14
102	Bolt-nut package	11
103	Control Panel	4
104	Door Decal	2
105	Electrical Connector	8
106	Gasket	10
107	Hatch Decal	2
108	Machined Valve	4
109	O-Ring	12
110	Panel Decal	1
111	Pressure Gauge	7
112	Shielded Cable/ft.	11
113	Side Panel	8

▶ **Figure 4.4**

*Column Chart for Frequency
Distribution of Items
Purchased*

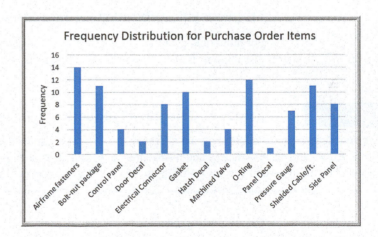

Relative Frequency Distributions

We may express the frequencies as a fraction, or proportion, of the total; this is called the
relative frequency. If a data set has n observations, the relative frequency of category i is
computed as

$$\text{Relative Frequency of Category } i = \frac{\text{Frequency of Category } i}{n} \qquad \textbf{(4.1)}$$

We often multiply the relative frequencies by 100 to express them as percentages. A
relative frequency distribution is a tabular summary of the relative frequencies of all
categories.

	A	B	C
100	Item Description	Frequency	Relative Frequency
101	Airframe fasteners	14	0.1489
102	Bolt-nut package	11	0.1170
103	Control Panel	4	0.0426
104	Door Decal	2	0.0213
105	Electrical Connector	8	0.0851
106	Gasket	10	0.1064
107	Hatch Decal	2	0.0213
108	Machined Valve	4	0.0426
109	O-Ring	12	0.1277
110	Panel Decal	1	0.0106
111	Pressure Gauge	7	0.0745
112	Shielded Cable/ft.	11	0.1170
113	Side Panel	8	0.0851
114	Total	94	1.0000

EXAMPLE 4.3   **Constructing a Relative Frequency Distribution for Items in the *Purchase Orders* Database**

The calculations for relative frequencies are simple. First, sum the frequencies to find the total number (note that the sum of the frequencies must be the same as the total number of observations, *n*). Then divide the frequency of each category by this value. Figure 4.5 shows the relative frequency distribution for the purchase order items. The formula in cell C101, for example, is =B101/$B$114. You then copy this formula down the column to compute the other relative frequencies. Note that the sum of the relative frequencies must equal 1.0. A pie chart of the frequencies is sometimes used to show these proportions visually, although it is more appealing for a smaller number of categories. For a large number of categories, a column or bar chart would work better, as we noted in Chapter 3.

Frequency Distributions for Numerical Data

For numerical data that consist of a small number of values, we may construct a frequency distribution similar to the way we did for categorical data; that is, we simply use COUNTIF to count the frequencies of each value.

EXAMPLE 4.4   **Frequency and Relative Frequency Distribution for A/P Terms**

In the *Purchase Orders* data, the A/P terms are all whole numbers: 15, 25, 30, and 45. A frequency and relative frequency distribution for these data are shown in Figure 4.6. A bar chart showing the proportions, or relative frequencies, in Figure 4.7, clearly shows that the majority of orders had accounts payable terms of 30 months.

Excel Histogram Tool

Frequency distributions and histograms can be created using the *Analysis Toolpak* in Excel. To do this, click the *Data Analysis* tools button in the *Analysis* group under the *Data* tab in the Excel menu bar and select *Histogram* from the list. In the dialog box (see Figure 4.8), specify the *Input Range* corresponding to the data. If you include the column header, then also check the *Labels* box so Excel knows that the range contains a label. The

	A	B	C
117	A/P Terms	Frequency	Relative Frequency
118	15	5	0.0532
119	25	12	0.1277
120	30	64	0.6809
121	45	13	0.1383
122	Total	94	1.0000

► **Figure 4.7**

Bar Chart of Relative Frequencies of A/P Terms

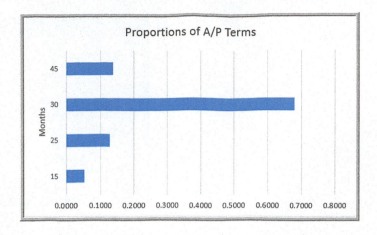

Bin Range defines the groups (Excel calls these "bins") used for the frequency distribution. If you do not specify a *Bin Range*, Excel will automatically determine bin values for the frequency distribution and histogram, which often results in a rather poor choice. If you have discrete values, set up a column of these values in your spreadsheet for the bin range and specify this range in the *Bin Range* field. We describe how to handle continuous data shortly. Check the *Chart Output* box to display a histogram in addition to the frequency distribution. You may also sort the values as a Pareto chart and display the cumulative frequencies by checking the additional boxes.

| EXAMPLE 4.5 | **Using the *Histogram* Tool** |

We will create a frequency distribution and histogram for the A/P Terms variable in the *Purchase Orders* database. Figure 4.9 shows the completed *Histogram* dialog. The input range includes the column header as well as the data in column H. We define the bin range below and enter it in cells H99:H103 (including the header "Months"):

Months
15
25
30
45

If you check the *Labels* box, it is important that both the *Input Range* and the *Bin Range* have labels included in the first row. Figure 4.10 shows the results from this tool.

Grouped Frequency Distributions

For numerical data that have many different discrete values with little repetition or are continuous, we usually group the data into "bins." A grouped frequency distribution requires that we specify

1. the number of groups,
2. the width of each group, and
3. the upper and lower limits of each group.

▶ **Figure 4.8**

Histogram *Tool Dialog*

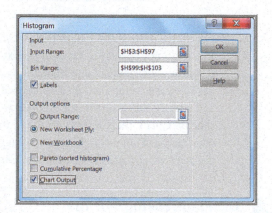

▶ **Figure 4.9**

Histogram *Dialog for A/P Terms Data*

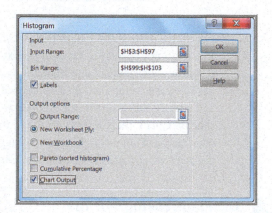

▶ **Figure 4.10**

Excel Frequency Distribution and Histogram for A/P Terms

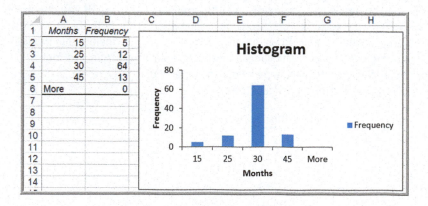

It is important to remember that the groups may not overlap, so that each value is counted in exactly one group.

You should define the groups after examining the range of the data. Generally, you should choose between 5 and 15 groups, and the range of each should be equal. The more data you have, the more groups you should generally use. Note that with fewer groups, the group widths will be wider. Wider group widths provide a "coarse" histogram. Sometimes you need to experiment to find the best number of groups to provide a useful visualization of the data. Choose the lower limit of the first group (LL) as a whole number smaller than the minimum data value and the upper limit of the last group (UL) as a whole number larger than the maximum data value. Generally, it makes sense to choose nice, round whole numbers. Then you may calculate the group width as

$$\text{Group Width} = \frac{\text{UL} - \text{LL}}{\text{Number of Groups}} \qquad (4.2)$$

| EXAMPLE 4.6 | **Constructing a Frequency Distribution and Histogram for Cost per Order** |

In this example, we apply the Excel *Histogram* tool to the cost per order data in column G of the *Purchase Orders* database. The data range from a minimum of $68.75 to a maximum of $127,500. You can find this either by using the MIN and MAX functions or simply by sorting the data. To ensure that all the data will be included in some group, it makes sense to set the lower limit of the first group to $0 and the upper limit of the last group to $130,000. Thus, if we select five groups, using equation (4.2), the width of each group is ($130,000 − 0)/5 = $26,000; if we choose ten groups, the width is ($130,000 − 0)/10 = $13,000. We select five groups. Doing so, the bin range is specified as

Upper Group Limit
$ 0.00
$ 26,000.00
$ 52,000.00
$ 78,000.00
$104,000.00
$130,000.00

This means that the first group includes all values less than or equal to $0; the second group includes all values greater than $0 but less than or equal to $26,000, and so on. Note that the groups do not overlap because the lower limit of one group is strictly greater than the upper limit of the previous group. We suggest using the header "Upper Group Limit" for the bin range to make this clear. In the spreadsheet, this bin range is entered in cells G99:G105. The *Input Range* in the *Histogram* dialog is G4:G97. Figure 4.11 shows the results. These results show that the vast majority of orders were for $26,000 or less and fall rapidly beyond this value. Selecting a larger number of groups might help to better understand the nature of the data. Figure 4.12 shows results using ten groups. This shows that there were a higher percentage of orders for $13,000 or less than for between $13,000 and $26,000.

▶ **Figure 4.11**

Frequency Distribution and Histogram for Cost per Order (Five Groups)

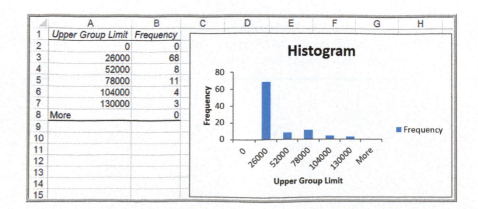

▶ **Figure 4.12**

Frequency Distribution and Histogram for Cost per Order (Ten Groups)

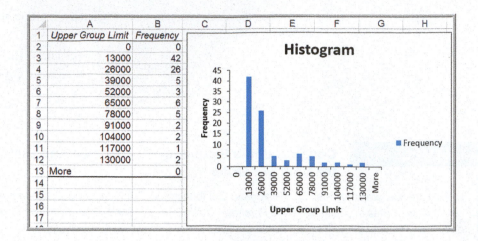

One limitation of the Excel *Histogram* tool is that the frequency distribution and histogram are not linked to the data; thus, if you change any of the data, you must repeat the entire procedure to construct a new frequency distribution and histogram.

Cumulative Relative Frequency Distributions

For numerical data, we may also compute the relative frequency of observations in each group. By summing all the relative frequencies at or below each upper limit, we obtain the cumulative relative frequency. The **cumulative relative frequency** represents the proportion of the total number of observations that fall at or below the upper limit of each group. A tabular summary of cumulative relative frequencies is called a **cumulative relative frequency distribution**.

EXAMPLE 4.7   **Computing Cumulative Relative Frequencies**

Figure 4.13 shows the relative frequency and cumulative relative frequency distributions for the cost per order data in the *Purchase Orders* database using ten groups. The relative frequencies are computed using the same approach as in Example 4.3—namely, by dividing the frequency by the total number of observations (94). In column D, we set the cumulative relative frequency of the first group equal to its relative frequency. Then we add the relative frequency of the next group to the cumulative relative frequency. For example, the cumulative relative frequency in cell D3 is computed as =D2+C3 = 0.000+0.4468 = 0.4468; the cumulative relative frequency in cell D4 is computed as =D3+C4 = 0.4468+0.2766 = 0.7234, and so on. Because relative frequencies must be between 0 and 1 and must add up to 1, the cumulative frequency for the last group must equal 1.

Figure 4.14 shows a chart for the cumulative relative frequency, which is called an **ogive**. From this chart, you can easily estimate the proportion of observations that fall below a certain value. For example, you can see that slightly more than 70% of the data fall at or below $26,000, about 90% of the data fall at or below $78,000, and so on. Note that cumulative frequencies can also be displayed using the *Histogram* tool.

▶ **Figure 4.13**

Cumulative Relative Frequency Distribution for Cost per Order Data

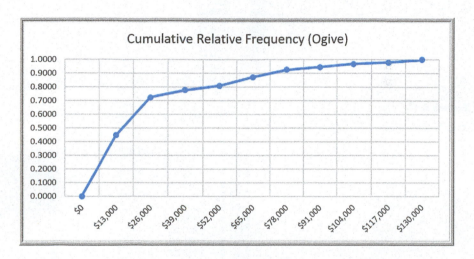

	A	B	C	D
				Cumulative
			Relative	Relative
1	Upper Group Limit	Frequency	Frequency	Frequency
2	0	0	0.0000	0.0000
3	13000	42	0.4468	0.4468
4	26000	26	0.2766	0.7234
5	39000	5	0.0532	0.7766
6	52000	3	0.0319	0.8085
7	65000	6	0.0638	0.8723
8	78000	5	0.0532	0.9255
9	91000	2	0.0213	0.9468
10	104000	2	0.0213	0.9681
11	117000	1	0.0106	0.9787
12	130000	2	0.0213	1.0000
13	More	0	0.0000	1.0000
14	Total	94		

▶ **Figure 4.14**

Ogive for Cost per Order

Constructing Frequency Distributions Using PivotTables

PivotTables make it quite easy to construct frequency distributions. For example, in the *Purchase Orders* data, we can simply build a PivotTable to find a count of the number of orders for each item, resulting in the same summary we saw in Figure 4.5. With Excel for Windows, a PivotChart (see Appendix A3 in Chapter 3) will display the histogram. For continuous numerical data, we can also use PivotTables to construct a grouped frequency distribution.

EXAMPLE 4.8   **Constructing a Grouped Frequency Distribution Using PivotTables**

Using the *Purchase Orders* database, create a PivotTable as shown in Figure 4.15. Note that this simply shows the frequencies of the cost per order values, each of which is unique. It doesn't matter what field is used in the Sum Values area; just ensure that the field settings specify "count." Next, click on any value in the Row Labels column, and from the *Analyze* tab for *PivotTable Tools*, select *Group Field*. Edit the dialog to start at 0 and end at 130000, and use 26000 as the group range, as shown in Figure 4.16. The result is shown in Figure 4.17; this groups the data in the same bins as we saw in Figure 4.11.

▶ **Figure 4.15**

*Count of Cost per Order
PivotTable*

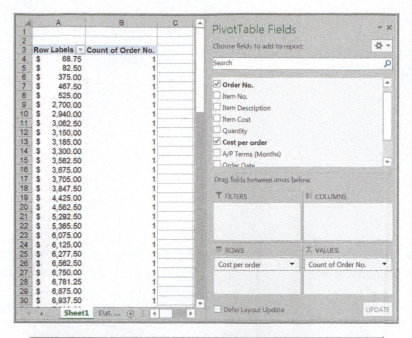

▶ **Figure 4.16**

Group Field Dialog

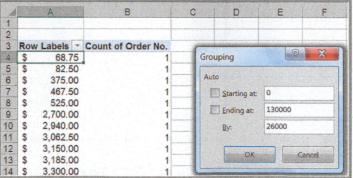

▶ **Figure 4.17**

*Grouped Frequency
Distribution in a PivotTable*

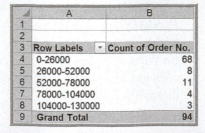

CHECK YOUR UNDERSTANDING

1. What is a frequency distribution?

2. Explain how to calculate a relative frequency.

3. State how to use Excel for constructing frequency distributions for categorical and numerical data.

4. State the features and limitations of the Excel *Histogram* tool.

5. What does an ogive display?

6. Use a PivotTable to construct the frequency distribution in Figure 4.12.

Percentiles and Quartiles

Data are often expressed as *percentiles* and *quartiles*. You are no doubt familiar with percentiles from standardized tests used for college or graduate school entrance examinations (SAT, ACT, GMAT, GRE, etc.). Percentiles specify the percent of other test takers who scored at or below the score of a particular individual. Generally speaking, the **kth percentile** is a value at or below which at least k percent of the observations lie. However, the way by which percentiles are calculated is not standardized. The most common way to compute the kth percentile is to order the data values from smallest to largest and calculate the rank of the kth percentile using the formula

$$\frac{nk}{100} + 0.5 \qquad\qquad (4.3)$$

where n is the number of observations. Round this to the nearest integer, and take the value corresponding to this rank as the kth percentile.

EXAMPLE 4.9   Computing Percentiles

In the *Purchase Orders* data, we have $n = 94$ observations. The rank of the 90th percentile ($k = 90$) for the cost per order data is computed as 94(90)/100 + 0.5 = 85.1, or, rounded, 85. The 85th ordered value is $74,375 and is the 90th percentile. This means that 90% of the costs per order are less than or equal to $74,375, and 10% are higher.

Statistical software use different methods that often involve interpolating between ranks instead of rounding, thus producing different results. The Excel function PERCENTILE.INC(*array, k*) computes the $100 \times k$th percentile of data in the range specified in the *array* field, where k is in the range 0 to 1, inclusive.

EXAMPLE 4.10   Computing Percentiles in Excel

To find the 90th percentile for the cost per order data in the *Purchase Orders* data, use the Excel function = PERCENTILE.INC(G4:G97,0.9). This calculates the 90th percentile as $73,737.50, which is different from using formula (4.3).

Excel also has a tool for sorting data from high to low and computing percentiles associated with each value. Select *Rank and Percentile* from the *Data Analysis* menu in the *Data tab* and specify the range of the data in the dialog. Be sure to check the *Labels in First Row* box if your range includes a header in the spreadsheet.

EXAMPLE 4.11   Excel *Rank and Percentile* Tool

A portion of the results from the *Rank and Percentile* tool for the cost per order data is shown in Figure 4.18. You can see that the Excel value of the 90th percentile that we computed in Example 4.9 as $74,375 is the 90.3rd percentile value.

Quartiles break the data into four parts. The 25th percentile is called the *first quartile*, Q_1; the 50th percentile is called the *second quartile*, Q_2; the 75th percentile is called the *third quartile*, Q_3; and the 100th percentile is the *fourth quartile*, Q_4. One-fourth of the data fall below the first quartile, one-half are below the second quartile, and three-fourths are below the third quartile. We may compute quartiles using the Excel function QUARTILE.INC(*array, quart*), where *array* specifies the range of the data and *quart* is a whole number between 1 and 4, designating the desired quartile.

► **Figure 4.18**

Portion of Rank and Percentile *Tool Results*

	A	B	C	D
1	Point	Cost per order	Rank	Percent
2	74	$127,500.00	1	100.00%
3	62	$121,000.00	2	98.90%
4	71	$110,000.00	3	97.80%
5	16	$103,530.00	4	96.70%
6	73	$ 96,750.00	5	95.60%
7	1	$ 82,875.00	6	94.60%
8	67	$ 81,937.50	7	93.50%
9	82	$ 77,400.00	8	92.40%
10	54	$ 76,500.00	9	91.30%
11	80	$ 74,375.00	10	90.30%
12	68	$ 72,250.00	11	89.20%
13	20	$ 65,875.00	12	88.10%
14	65	$ 64,500.00	13	87.00%
15	28	$ 63,750.00	14	86.00%

EXAMPLE 4.12   **Computing Quartiles in Excel**

For the cost per order data in the *Purchase Orders* database, we may use the Excel function =QUARTILE.INC(G4:G97,k), where k ranges from 1 to 4, to compute the quartiles. The results are as follows:

$k = 1$	First quartile	$6,757.81
$k = 2$	Second quartile	$15,656.25
$k = 3$	Third quartile	$27,593.75
$k = 4$	Fourth quartile	$127,500.00

We may conclude that 25% of the order costs fall at or below $6,757.81; 50% fall at or below $15,656.25; 75% fall at or below $27,593.75, and 100% fall at or below the maximum value of $127,500.

We can extend these ideas to other divisions of the data. For example, *deciles* divide the data into ten sets: the 10th percentile, 20th percentile, and so on. All these types of measures are called **data profiles**, or **fractiles**.

CHECK YOUR UNDERSTANDING

1. Explain how to interpret the 75th percentile.

2. How do quartiles relate to percentiles?

3. What Excel functions can you use to find percentiles and quartiles?

 Cross-Tabulations

One of the most basic statistical tools used to summarize categorical data and examine the relationship between two categorical variables is cross-tabulation. A **cross-tabulation** is a tabular method that displays the number of observations in a data set for different subcategories of two categorical variables. A cross-tabulation table is often called a **contingency table**. The subcategories of the variables must be mutually exclusive and exhaustive, meaning that each observation can be classified into only one subcategory, and, taken together over all subcategories, they must constitute the complete data set. Cross-tabulations are commonly used in marketing research to provide insight into characteristics of different market segments using categorical variables such as gender, educational level, and marital status.

EXAMPLE 4.13 Constructing a Cross-Tabulation

Let us examine the *Sales Transactions* database, a portion of which is shown in Figure 4.19. Suppose we wish to identify the number of books and DVDs ordered by region. A cross-tabulation will have rows corresponding to the different regions and columns corresponding to the products. Within the table we list the count of the number in each pair of categories. A cross-tabulation of these data is shown in Table 4.1. PivotTables make it easy to construct cross-tabulations. Visualizing the data as a chart is a good way of communicating the results. Figure 4.20 shows the differences between product and regional sales.

Expressing the results as percentages of a row or column makes it easier to interpret differences between regions or products, particularly as the totals for each category differ. Table 4.2 shows the percentage of book and DVD sales within each region; this is computed by dividing the counts by the row totals and multiplying by 100 (in Excel, simply divide the count by the total and format the result as a percentage by clicking the % button in the *Number* group within the *Home* tab in the ribbon). In Chapter 2, we saw how to do this easily in a PivotTable. For example, we see that although more books and DVDs are sold in the West region than in the North, the relative percentages of each product are similar, particularly when compared to the East and South regions.

▶ **Figure 4.19**

Portion of Sales Transactions *Database*

	A	B	C	D	E	F	G	H
1	Sales Transactions: July 14							
2								
3	Cust ID	Region	Payment	Transaction Code	Source	Amount	Product	Time Of Day
4	10001	East	Paypal	93816545	Web	$20.19	DVD	22:19
5	10002	West	Credit	74083490	Web	$17.85	DVD	13:27
6	10003	North	Credit	64942368	Web	$23.98	DVD	14:27
7	10004	West	Paypal	70560957	Email	$23.51	Book	15:38
8	10005	South	Credit	35208817	Web	$15.33	Book	15:21
9	10006	West	Paypal	20978903	Email	$17.30	DVD	13:11
10	10007	East	Credit	80103311	Web	$177.72	Book	21:59
11	10008	West	Credit	14132683	Web	$21.76	Book	4:04
12	10009	West	Paypal	40128225	Web	$15.92	DVD	19:35
13	10010	South	Paypal	49073721	Web	$23.39	DVD	13:26

▶ **Table 4.1**

Cross-Tabulation of Sales Transaction Data

Region	Book	DVD	Total
East	56	42	98
North	43	42	85
South	62	37	99
West	100	90	190
Total	261	211	472

▶ **Table 4.2**

Percentage Sales of Products Within Each Region

Region	Book	DVD	Total
East	57.1%	42.9%	100.0%
North	50.6%	49.4%	100.0%
South	62.6%	37.4%	100.0%
West	52.6%	47.4%	100.0%

► **Figure 4.20**

Chart of Regional Sales by Product

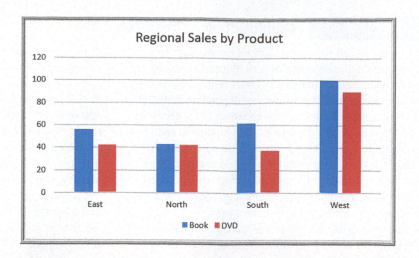

 CHECK YOUR UNDERSTANDING

1. What is a cross-tabulation?

2. Use a PivotTable to construct the cross-tabulation shown in Table 4.1.

Descriptive Statistical Measures

In this section, we introduce numerical measures that provide an effective and efficient way of obtaining meaningful information from data. Before discussing these measures, however, we need to understand the differences between populations and samples.

Populations and Samples

A **population** consists of all items of interest for a particular decision or investigation—for example, *all* individuals in the United States who do not own cell phones, *all* subscribers to Netflix, or *all* stockholders of Google. A company like Netflix keeps extensive records on its customers, making it easy to retrieve data about the entire population of customers. However, it would probably be impossible to identify all individuals who do not own cell phones.

A **sample** is a subset of a population. For example, a list of individuals who rented a comedy from Netflix in the past year would be a sample from the population of all customers. Whether this sample is representative of the population of customers—which depends on how the sample data are intended to be used—may be debatable; nevertheless, it is a sample. Most populations, even if they are finite, are generally too large to deal with effectively or practically. For instance, it would be impractical as well as too expensive to survey the entire population of TV viewers in the United States. Sampling is also clearly necessary when data must be obtained from destructive testing or from a continuous production process. Thus, the purpose of sampling is to obtain sufficient information to draw a valid inference about a population. Market researchers, for example, use sampling to gauge consumer perceptions on new or existing goods and services; auditors use sampling to verify the accuracy of financial statements; and quality-control analysts sample production output to verify quality levels and identify opportunities for improvement.

Most data with which businesses deal are samples. For instance, the *Purchase Orders* and *Sales Transactions* databases that we used in previous chapters represent samples because the purchase order data include only orders placed within a three-month time period, and the sales transactions represent orders placed on only one day, July 14. Therefore, unless it is obvious or noted otherwise, we will assume that any data set is a sample.

Statistical Notation

We typically label the elements of a data set using subscripted variables, x_1, x_2, and so on. In general, x_i represents the ith observation. It is a common practice in statistics to use Greek letters, such as μ (mu), σ (sigma), and π (pi), to represent population measures and italic letters such as $\bar{x}$ (x-bar), s, and p to represent sample statistics. We will use N to represent the number of items in a population and n to represent the number of observations in a sample. Statistical formulas often contain a summation operator, Σ (Greek capital sigma), which means that the terms that follow it are added together. Thus, $\sum_{i=1}^{n} x_i = x_1 + x_2 + \cdots + x_n$. Understanding these conventions and mathematical notation will help you to interpret and apply statistical formulas.

Measures of Location: Mean, Median, Mode, and Midrange

Measures of location provide estimates of a single value that in some fashion represents the "centering" of a set of data. The most common is the *average*. We all use averages routinely in our lives, for example, to measure student accomplishment in college (e.g., grade point average), to measure the performance of athletes (e.g., batting average), and to measure performance in business (e.g., average delivery time).

Arithmetic Mean

The average is formally called the **arithmetic mean** (or simply the **mean**), which is the sum of the observations divided by the number of observations. Mathematically, the mean of a population is denoted by the Greek letter μ, and the mean of a sample is denoted by $\bar{x}$. If a population consists of N observations x_1, x_2,..., x_N, the population mean, μ, is calculated as

$$\mu = \frac{\sum_{i=1}^{N} x_i}{N} \tag{4.4}$$

The mean of a sample of n observations, x_1, x_2,..., x_n, denoted by $\bar{x}$, is calculated as

$$\bar{x} = \frac{\sum_{i=1}^{n} x_i}{n} \tag{4.5}$$

Note that the calculations for the mean are the same whether we are dealing with a population or a sample; only the notation differs. We may also calculate the mean in Excel using the function AVERAGE(*data range*).

One property of the mean is that the sum of the deviations of each observation from the mean is zero:

$$\sum_{i} (x_i - \bar{x}) = 0 \tag{4.6}$$

This simply means that the sum of the deviations above the mean is the same as the sum of the deviations below the mean; essentially, the mean "balances" the values on either side of it. However, it does not suggest that half the data lie above or below the mean—a common misconception among those who don't understand statistics.

In addition, the mean is unique for every set of data and is meaningful for both interval and ratio data. However, it can be affected by **outliers**—observations that are radically different from the rest—which pull the value of the mean toward these values. We discuss more about outliers later in this chapter.

EXAMPLE 4.14   **Computing the Mean Cost per Order**

In the *Purchase Orders* database, suppose that we are interested in finding the mean cost per order. We calculate the mean cost per order by summing the values in column G and then dividing by the number of observations. Using equation (4.5), note that $x_1 = \$2,700$, $x_2 = \$19,250$, and so on, and $n = 94$. The sum of these order costs is $2,471,760. Therefore, the mean cost per order is

$2,471,760/94 = \$26,295.32$. We show these calculations in a separate worksheet, *Mean*, in the *Purchase Orders* Excel workbook. A portion of this worksheet in split-screen mode is shown in Figure 4.21. Alternatively, we used the Excel function =AVERAGE(B2:B95) in this worksheet to arrive at the same value. We encourage you to study the calculations and formulas used.

▶ **Figure 4.21**

Excel Calculations of Mean Cost per Order

	A	B
1	Observation	Cost per order
2	x1	$82,875.00
3	x2	$42,500.00
4	x3	$24,150.00
5	x4	$22,575.00
6	x5	$19,250.00
91	x90	$467.50
92	x91	$9,975.00
93	x92	$30,625.00
94	x93	$21,450.00
95	x94	$68.75
96	**Sum of Cost per Order**	$2,471,760.00
97	**Number of observations**	94
98		
99	**Mean Cost per Order (=B96/B97)**	$26,295.32
100		
101	**Excel AVERAGE function**	$26,295.32

Median

The measure of location that specifies the middle value when the data are arranged from least to greatest is the **median**. Half the data are below the median, and half the data are above it. For an odd number of observations, the median is the middle of the sorted numbers. For an even number of observations, the median is the mean of the two middle numbers. We could use the *Sort* option in Excel to rank-order the data and then determine the median. The Excel function MEDIAN(*data range*) could also be used. The median is meaningful for ratio, interval, and ordinal data. As opposed to the mean, the median is *not* affected by outliers.

EXAMPLE 4.15   **Finding the Median Cost per Order**

In the *Purchase Orders* database, sort the data in column G from smallest to largest. Since we have 94 observations, the median is the average of the 47th and 48th observations. You should verify that the 47th sorted observation is $15,562.50 and the 48th observation is $15,750. Taking the average of these two values results in the median value

of ($15,562.5 + $15,750)/2 = $15,656.25. Thus, we may conclude that the costs of half the orders were less than $15,656.25 and half were above this amount. In this case, the median is not very close in value to the mean. These calculations are shown in the worksheet *Median* in the *Purchase Orders* Excel workbook, as shown in Figure 4.22.

▶ **Figure 4.22**

*Excel Calculations for
Median Cost per Order*

	A	B	C	D
1	Rank	Cost per order		
2	1	$68.75		
3	2	$82.50		
4	3	$375.00		
5	4	$467.50		
45	44	$14,910.00		
46	45	$14,910.00		
47	46	$15,087.50		
48	47	$15,562.50		$15,562.50
49	48	$15,750.00		$15,750.00
50	49	$15,937.50	Average	$15,656.25
51	50	$16,276.75		
52	51	$16,330.00		

Mode

A third measure of location is the **mode**. The mode is the observation that occurs most frequently. The mode is most useful for data sets that contain a relatively small number of unique values. For data sets that have few repeating values, the mode does not provide much practical value. You can easily identify the mode from a frequency distribution by identifying the value having the largest frequency or from a histogram by identifying the highest bar. You may also use the Excel function MODE.SNGL(*data range*). For frequency distributions and histograms of grouped data, the mode is the group with the greatest frequency. Some data sets have multiple modes; to identify these, you can use the Excel function MODE.MULT(*data range*), which returns an array of modal values.

EXAMPLE 4.16   **Finding the Mode**

In the *Purchase Orders* database, from the frequency distribution and histogram for A/P terms in Figure 4.10, we see that the greatest frequency corresponds to a value of 30 months; this is also the highest bar in the histogram.

Therefore, the mode is 30 months. For the grouped frequency distribution and histogram of the cost per order variable in Figure 4.12, we see that the mode corresponds to the group between $0 and $13,000.

Midrange

A fourth measure of location that is used occasionally is the **midrange**. This is simply the average of the largest and smallest values in the data set.

EXAMPLE 4.17   **Computing the Midrange**

We may identify the minimum and maximum values using the Excel functions MIN and MAX or sort the data and find them easily. For the cost per order data, the minimum value

is $68.78 and the maximum value is $127,500. Thus, the midrange is ($127,500 + $68.78)/2 = $63,784.39.

Caution must be exercised when using the midrange because extreme values easily distort the result, as this example illustrates. This is because the midrange uses only two pieces of data, whereas the mean uses *all* the data; thus, it is usually a much rougher estimate than the mean and is often used for only small sample sizes.

Using Measures of Location in Business Decisions

Because everyone is so familiar with the concept of the average in daily life, managers often use the mean inappropriately in business when other statistical information should be considered. The following hypothetical example, which was based on a real situation, illustrates this.

EXAMPLE 4.18   Quoting Computer Repair Times

The Excel file *Computer Repair Times* provides a sample of the times it took to repair and return 250 computers to customers who used the repair services of a national electronics retailer. Computers are shipped to a central facility, where they are repaired and then shipped back to the stores for customer pickup. The mean, median, and mode are all very close and show that the typical repair time is about two weeks (see Figure 4.23). So you might think that if a customer brought in a computer for repair, it would be reasonable to quote a repair time of two weeks. What would happen if the stores quoted all customers a time of two weeks? Clearly about half the customers would be upset because their computers would not be completed by this time.

Figure 4.24 shows a portion of the frequency distribution and histogram for these repair times (see the

Histogram tab in the Excel file). We see that the longest repair time took almost six weeks. So should the company give customers a guaranteed repair time of six weeks? They probably wouldn't have many customers because few would want to wait that long. Instead, the frequency distribution and histogram provide insight into making a more rational decision. You may verify that 90% of the time, repairs are completed within 21 days; on the rare occasions that it takes longer, it generally means that technicians had to order and wait for a part. So it would make sense to tell customers that they could probably expect their computers back within two to three weeks and inform them that it might take longer if a special part was needed.

From this example, we see that using frequency distributions, histograms, and percentiles can provide more useful information than simple measures of location. This leads us to introduce ways of quantifying variability in data, which we call *measures of dispersion*.

▶ **Figure 4.23**

Measures of Location for Computer Repair Times

	A	B
1	**Computer Repair Times**	
2		
3	**Sample**	**Repair Time (Days)**
4	1	18
5	2	15
6	3	17
250	247	31
251	248	6
252	249	17
253	250	13
254		
255	Mean	14.912
256	Median	14
257	Mode	15

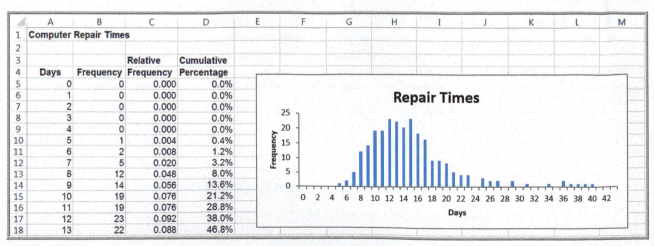

	A	B	C	D	E	F	G	H	I	J	K	L	M
1	**Computer Repair Times**												
2													
3			Relative	Cumulative									
4	**Days**	**Frequency**	**Frequency**	**Percentage**									
5	0	0	0.000	0.0%									
6	1	0	0.000	0.0%									
7	2	0	0.000	0.0%									
8	3	0	0.000	0.0%									
9	4	0	0.000	0.0%									
10	5	1	0.004	0.4%									
11	6	2	0.008	1.2%									
12	7	5	0.020	3.2%									
13	8	12	0.048	8.0%									
14	9	14	0.056	13.6%									
15	10	19	0.076	21.2%									
16	11	19	0.076	28.8%									
17	12	23	0.092	38.0%									
18	13	22	0.088	46.8%									

▲ **Figure 4.24**

Frequency Distribution and Histogram for Computer Repair Times

Measures of Dispersion: Range, Interquartile Range, Variance, and Standard Deviation

Dispersion refers to the degree of variation in the data, that is, the numerical spread (or compactness) of the data. Several statistical measures characterize dispersion: the *range*, *interquartile range*, *variance*, and *standard deviation*.

Range

The **range** is the easiest to compute, and is the difference between the maximum value and the minimum value in the data set. Although Excel does not provide a function for the range, it can be computed easily by the formula =MAX(*data range*) − MIN(*data range*). Like the midrange, the range is affected by outliers and, thus, is often only used for very small data sets.

EXAMPLE 4.19   **Computing the Range**

For the cost per order data in the *Purchase Orders* database, the minimum value is $68.78 and the maximum value is $127,500. Thus, the range is $127,500 − $68.78 = $127,431.22.

Interquartile Range

The difference between the first and third quartiles, $Q_3 - Q_1$, is often called the **interquartile range (IQR)**, or the **midspread**. This includes only the middle 50% of the data and, therefore, is not influenced by extreme values. Thus, it is sometimes used as an alternative measure of dispersion.

EXAMPLE 4.20   **Computing the Interquartile Range**

For the cost per order data in the *Purchase Orders* database, we identified the first and third quartiles as $Q_1 = \$6,757.81$ and $Q_3 = \$27,593.75$ in Example 4.12. Thus, IQR = $27,593.75 − $6,757.81 = $20,835.94. Therefore, the middle 50% of the data are concentrated over a relatively small range of $20,835.94. Note that the upper 25% of the data span the range from $27,593.75 to $127,500, indicating that high costs per order are spread out over a large range of $99,906.25.

Variance

A more commonly used measure of dispersion is the **variance**, whose computation depends on *all* the data. The larger the variance, the more the data are spread out from the mean and the more variability one can expect in the observations. The formula used for calculating the variance is different for populations and samples.

The formula for the variance of a population is

$$\sigma^2 = \frac{\sum_{i=1}^{N}(x_i - \mu)^2}{N} \tag{4.7}$$

where x_i is the value of the *i*th item, N is the number of items in the population, and μ is the population mean. Essentially, the variance is the average of the squared deviations of the observations from the mean.

A significant difference exists between the formulas for computing the variance of a population and that of a sample. The variance of a sample is calculated using the formula

$$s^2 = \frac{\sum_{i=1}^{n}(x_i - \bar{x})^2}{n - 1} \tag{4.8}$$

where n is the number of items in the sample and $\bar{x}$ is the sample mean. It may seem peculiar to use a different denominator to "average" the squared deviations from the mean for populations and samples, but statisticians have shown that the formula for the sample variance provides a more accurate representation of the true population variance. We discuss this more formally in Chapter 6. For now, simply understand that the proper calculations of the population and sample variance use different denominators based on the number of observations in the data.

The Excel function VAR.S(*data range*) may be used to compute the sample variance, s^2, whereas the Excel function VAR.P(*data range*) is used to compute the variance of a population, σ^2.

EXAMPLE 4.21   **Computing the Variance**

Figure 4.25 shows a portion of the Excel worksheet *Variance* in the *Purchase Orders* workbook. To find the variance of the cost per order using equation (4.8), we first need to calculate the mean, as was done in Example 4.14. Then for each observation, calculate the difference between the observation and the mean, as shown in column C. Next, square these differences, as shown in column D. Finally, add these square deviations (cell D96) and divide by $n - 1 = 93$. This results in the variance 890,594,573.82. Alternatively, the Excel function =VAR.S(B2:B95) yields the same result.

▶ **Figure 4.25**

Excel Calculations for Variance of Cost per Order

	A	B	C	D
1	Observation	Cost per order	(xi - mean)	(xi - mean)^2
2	x1	$82,875.00	$56,579.68	$3,201,260,285.21
3	x2	$42,500.00	$16,204.68	$262,591,681.48
4	x3	$24,150.00	-$2,145.32	$4,602,394.25
5	x4	$22,575.00	-$3,720.32	$13,840,774.57
6	x5	$19,250.00	-$7,045.32	$49,636,521.91
91	x90	$467.50	-$25,827.82	$667,076,241.99
92	x91	$9,975.00	-$16,320.32	$266,352,817.12
93	x92	$30,625.00	$4,329.68	$18,746,136.27
94	x93	$21,450.00	-$4,845.32	$23,477,117.66
95	x94	$68.75	-$26,226.57	$687,832,929.32
96	Sum of Cost per Order	$2,471,760.00	Sum of squared deviations	$82,825,295,365.68
97	Number of observations	94		
98				
99	Mean Cost per Order (=B96/B97)	$26,295.32	Variance (=D96/(B97-1))	$890,594,573.82
100				
101			Excel VAR.S function	$890,594,573.82

Note that the dimension of the variance is the square of the dimension of the observations. So for example, the variance of the cost per order is not expressed in dollars, but rather in dollars squared. This makes it difficult to use the variance in practical applications. However, a measure closely related to the variance that can be used in practical applications is the standard deviation.

Standard Deviation

The **standard deviation** is the square root of the variance. For a population, the standard deviation is computed as

$$\sigma = \sqrt{\frac{\sum_{i=1}^{N}(x_i - \mu)^2}{N}}$$

(4.9)

and for samples, it is

$$s = \sqrt{\frac{\sum_{i=1}^{n}(x_i - \bar{x})^2}{n - 1}}$$

(4.10)

The Excel function STDEV.P(*data range*) calculates the standard deviation for a population (σ); the function STDEV.S(*data range*) calculates it for a sample (*s*).

EXAMPLE 4.22   **Computing the Standard Deviation**

We may use the same worksheet calculations as in Example 4.21. All we need to do is to take the square root of the computed variance to find the standard deviation. Thus, the standard deviation of the cost per order is

$\sqrt{890{,}594{,}573.82} = \$29{,}842.8312$. Alternatively, we could use the Excel function =STDEV.S(B2:B95) to find the same value.

The standard deviation is generally easier to interpret than the variance because its units of measure are the same as the units of the data. Thus, it can be more easily related to the mean or other statistics measured in the same units.

The standard deviation is a popular measure of risk, particularly in financial analysis, because many people associate risk with volatility in stock prices. The standard deviation measures the tendency of a fund's monthly returns to vary from their long-term average (as *Fortune* stated in one of its issues, ". . . standard deviation tells you what to expect in the way of dips and rolls. It tells you how scared you'll be.").[4] For example, a mutual fund's return might have averaged 11% with a standard deviation of 10%. Thus, about two-thirds of the time, the annualized monthly return was between 1% and 21%. By contrast, another fund's average return might be 14% but have a standard deviation of 20%. Its returns would have fallen in a range of −6% to 34% and, therefore, it is more risky. Many financial Web sites, such as IFA.com and Morningstar.com, provide standard deviations for market indexes and mutual funds.

To illustrate risk, the Excel file *Closing Stock Prices* (see Figure 4.26) lists daily closing prices for four stocks and the Dow Jones Industrial Average index over a one-month period. The average closing prices for Intel (INTC) and General Electric (GE) are quite

▶ **Figure 4.26**

Excel File Closing Stock Prices

	A	B	C	D	E	F
1	**Closing Stock Prices**					
2						
3	**Date**	**IBM**	**INTC**	**CSCO**	**GE**	**DJ Industrials Index**
4	9/3/2010	$127.58	$18.43	$21.04	$15.39	10447.93
5	9/7/2010	$125.95	$18.12	$20.58	$15.44	10340.69
6	9/8/2010	$126.08	$17.90	$20.64	$15.70	10387.01
7	9/9/2010	$126.36	$18.00	$20.61	$15.91	10415.24
8	9/10/2010	$127.99	$17.97	$20.62	$15.98	10462.77
9	9/13/2010	$129.61	$18.56	$21.26	$16.25	10544.13
10	9/14/2010	$128.85	$18.74	$21.45	$16.16	10526.49
11	9/15/2010	$129.43	$18.72	$21.59	$16.34	10572.73
12	9/16/2010	$129.67	$18.97	$21.93	$16.23	10594.83
13	9/17/2010	$130.19	$18.81	$21.86	$16.29	10607.85
14	9/20/2010	$131.79	$18.93	$21.75	$16.55	10753.62
15	9/21/2010	$131.98	$19.14	$21.64	$16.52	10761.03
16	9/22/2010	$132.57	$19.01	$21.67	$16.50	10739.31
17	9/23/2010	$131.67	$18.98	$21.53	$16.14	10662.42
18	9/24/2010	$134.11	$19.42	$22.09	$16.66	10860.26
19	9/27/2010	$134.65	$19.24	$22.11	$16.43	10812.04
20	9/28/2010	$134.89	$19.51	$21.86	$16.44	10858.14
21	9/29/2010	$135.48	$19.24	$21.87	$16.36	10835.28
22	9/30/2010	$134.14	$19.20	$21.90	$16.25	10788.05
23	10/1/2010	$135.64	$19.32	$21.91	$16.36	10829.68

---

[4]*Fortune,* 1999 Investor's Guide (December 21, 1998).

similar, $18.81 and $16.19, respectively. However, the standard deviation of Intel's price over this time frame was $0.50, whereas GE's was $0.35. GE had less variability and, therefore, less risk. A larger standard deviation implies that while a greater potential of a higher return exists, there is also greater risk of realizing a lower return. Many investment publications and Web sites provide standard deviations of stocks and mutual funds to help investors assess risk in this fashion. We learn more about risk in other chapters.

Chebyshev's Theorem and the Empirical Rules

One of the more important results in statistics is **Chebyshev's theorem**, which states that for *any set of data*, the proportion of values that lie within k standard deviations $(k > 1)$ of the mean is at least $1 - 1/k^2$. Thus, for $k = 2$, at least 3/4, or 75%, of the data lie within two standard deviations of the mean; for $k = 3$, at least 8/9, or 89%, of the data lie within three standard deviations of the mean. We can use these values to provide a basic understanding of the variation in a set of data using only the computed mean and standard deviation.

EXAMPLE 4.23   **Applying Chebyshev's Theorem**

For the cost per order data in the *Purchase Orders* database, a two-standard-deviation interval around the mean is [−$33,390.34, $85,980.98]. If we count the number of observations within this interval, we find that 89 of 94, or 94.7%, fall within two standard deviations

of the mean. A three-standard-deviation interval is [−$63,233.17, $115,823.81], and we see that 92 of 94, or 97.9%, fall in this interval. Both percentages are above the thresholds of 75% and 89% in Chebyshev's theorem.

For many data sets encountered in practice, such as the cost per order data, the percentages are generally much higher than what Chebyshev's theorem specifies. These are reflected in what are called the **empirical rules**:

1. Approximately 68% of the observations will fall within one standard deviation of the mean, or between $\bar{x} - s$ and $\bar{x} + s$.
2. Approximately 95% of the observations will fall within two standard deviations of the mean, or within $\bar{x} \pm 2s$.
3. Approximately 99.7% of the observations will fall within three standard deviations of the mean, or within $\bar{x} \pm 3s$.

We see that the cost per order data reflect these empirical rules rather closely. Depending on the data and the shape of the frequency distribution, the actual percentages may be higher or lower.

To illustrate the empirical rules, suppose that a retailer knows that on average, an order is delivered by standard ground transportation in eight days with a standard deviation of one day. With the first empirical rule, about 68% of the observations will fall between seven and nine days. Using the second empirical rule, 95% of the observations will fall between six and ten days. Finally, using the third rule, almost 100% of the observations will fall between 5 and 11 days. These rules can be used to predict future observations, assuming that the sample data are representative of the future. Generally, rules 2 and 3 are commonly used. Note, however, that using rule 2, there is a 5% chance that an observation will fall outside the predicted range. Since the third empirical rule generally covers nearly all of the observations, it provides a better prediction and reduces the risk that a future observation will fall outside of the predicted range.

As another example, it is important to ensure that the output from a manufacturing process meets the specifications that engineers and designers require. The dimensions for a typical manufactured part are usually specified by a target, or ideal, value as well as a tolerance, or "fudge factor," that recognizes that variation will exist in most manufacturing processes due to factors such as materials, machines, work methods, human performance, and environmental conditions. For example, a part dimension might be specified as 5.00 ± 0.2 cm. This simply means that a part having a dimension between 4.80 and 5.20 cm will be acceptable; anything outside of this range would be classified as defective. To measure how well a manufacturing process can achieve the specifications, we usually take a sample of output, measure the dimension, compute the total variation using the third empirical rule (that is, estimate the total variation by six standard deviations), and then compare the result to the specifications by dividing the specification range by the total variation. The result is called the **process capability index**, denoted as C_p:

$$C_p = \frac{\text{Upper Specification} - \text{Lower Specification}}{\text{Total Variation}} \tag{4.11}$$

Manufacturers use this index to evaluate the quality of their products and determine when they need to make improvements in their processes.

EXAMPLE 4.24 Using Empirical Rules to Measure the Capability of a Manufacturing Process

Figure 4.27 shows a portion of the data collected from a manufacturing process for a part whose dimensions are specified as 5.00 ± 0.2 centimeters. These are provided in the Excel workbook *Manufacturing Measurements*. The mean and standard deviation are first computed in cells J3 and J4 using the Excel AVERAGE and STDEV.S functions (these functions work correctly whether the data are arranged in a single column or in a matrix form). Using the third empirical rule, the total variation is then calculated as the mean plus or minus three standard deviations. In cell J14, C_p is calculated using equation (4.11). A C_p value less than 1.0 is not good; it means that the variation in the process is wider than the specification limits, signifying that some of the parts will not meet the specifications. In practice, many manufacturers want to have C_p values of at least 1.5.

Figure 4.28 shows a frequency distribution and histogram of these data (worksheet *Histogram* in the *Manufacturing Measurements* workbook). Note that the bin values represent the upper limits of the groupings in the histogram; thus, three observations fell at or below 4.8, the lower specification limit. In addition, five observations exceeded the upper specification limit of 5.2. Therefore, 8 of the 200 observations, or 4%, were actually defective, and 96% were acceptable. Although this doesn't meet the empirical rule exactly, you must remember that we are dealing with sample data. Other samples from the same process would have different characteristics, but overall, the empirical rule provides a good estimate of the total variation in the data that we can expect from any sample.

	A	B	C	D	E	F	G	H	I	J
1	**Manufacturing Measurements**									
2										
3	5.21	5.87	4.85	4.95	5.07	4.96	4.96	5.11	Mean	4.99
4	5.02	5.33	4.82	4.86	4.82	4.96	5.06	5.11	Standard deviation	0.117
5	4.90	5.11	5.02	5.13	5.03	4.94	4.86	5.08		
6	5.00	5.07	4.90	4.95	4.85	5.19	4.96	5.03	Mean - 3*Stdev	4.640
7	5.16	4.93	4.73	5.22	4.89	4.91	4.99	4.94	Mean + 3*Stdev	5.340
8	5.03	4.99	5.04	4.81	4.82	5.01	4.94	4.88	Total variaton	0.700
9	4.96	5.04	5.07	4.91	5.18	4.93	5.06	4.91		
10	5.04	5.14	4.81	4.95	5.02	5.05	4.95	4.86	Lower Specification	4.8
11	4.98	5.09	5.04	4.94	5.05	4.96	5.02	4.89	Upper Specification	5.2
12	5.07	5.06	5.03	4.81	4.88	4.92	5.01	4.91	Specification range	0.4
13	5.02	4.85	5.01	5.11	5.08	4.95	5.04	4.87		
14	5.08	4.93	5.14	4.81	4.98	5.08	5.01	4.93	Cp	0.57

▲ Figure 4.27

Calculation of C_p Index

▶ **Figure 4.28**

Frequency Distribution and Histogram of Manufacturing Measurements

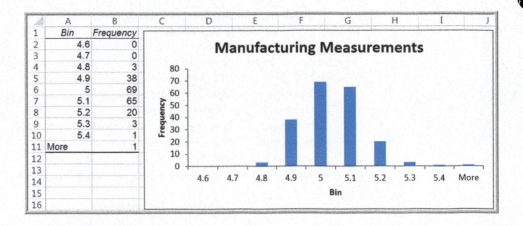

Standardized Values (*Z*-Scores)

A **standardized value**, commonly called a ***z*-score**, provides a relative measure of the distance an observation is from the mean, which is independent of the units of measurement. The *z*-score for the *i*th observation in a data set is calculated as follows:

$$z_i = \frac{x_i - \bar{x}}{s} \qquad (4.12)$$

We subtract the sample mean from the *i*th observation, x_i, and divide the result by the sample standard deviation. In formula (4.12), the numerator represents the distance that x_i is from the sample mean; a negative value indicates that x_i lies to the left of the mean, and a positive value indicates that it lies to the right of the mean. By dividing by the standard deviation, s, we scale the distance from the mean to express it in units of standard deviations. Thus, a *z*-score of 1.0 means that the observation is one standard deviation to the right of the mean; a *z*-score of -1.5 means that the observation is 1.5 standard deviations to the left of the mean. Thus, even though two data sets may have different means and standard deviations, the same *z*-score means that the observations have the same relative distance from their respective means.

Z-scores can be computed easily on a spreadsheet; however, Excel has a function that calculates it directly, STANDARDIZE(*x, mean, standard_dev*).

EXAMPLE 4.25   **Computing *Z*-Scores**

Figure 4.29 shows the calculations of *z*-scores for a portion of the cost per order data. This worksheet may be found in the *Purchase Orders* workbook as *z-scores*. In cells B97 and B98, we compute the mean and standard deviation using the Excel AVERAGE and STDEV.S functions. In column C, we could use either formula (4.12) or the Excel STANDARDIZE function. For example, the formula in cell C2 is =(B2−$B$97)/$B$98, but it could also be calculated as =STANDARDIZE(B2,$B$97,$B$98). Thus, the first observation, $82,875, is 1.90 standard deviations above the mean, whereas observation 94 ($68.75) is 0.88 standard deviations below the mean. Only two observations are more than three standard deviations above the mean. We saw this in Example 4.23, when we applied Chebyshev's theorem to the data.

▶ **Figure 4.29**

Computing Z-Scores for Cost per Order Data

	A	B	C	D
1	Observation	Cost per order	z-score	
2	x1	$82,875.00	1.90	=(B2-$B$96)/$B$97
3	x2	$42,500.00	0.54	
4	x3	$24,150.00	-0.07	
5	x4	$22,575.00	-0.12	
6	x5	$19,250.00	-0.24	
91	x90	$467.50	-0.87	
92	x91	$9,975.00	-0.55	
93	x92	$30,625.00	0.15	
94	x93	$21,450.00	-0.16	
95	x94	$68.75	-0.88	
96	Mean	$26,295.32		
97	Standard Deviation	$29,842.83		

Coefficient of Variation

The **coefficient of variation (CV)** provides a relative measure of the dispersion in data relative to the mean and is defined as

$$CV = \frac{\text{Standard Deviation}}{\text{Mean}} \tag{4.13}$$

Sometimes the coefficient of variation is multiplied by 100 to express it as a percent. This statistic is useful when comparing the variability of two or more data sets when their scales differ.

The coefficient of variation provides a relative measure of risk to return. The smaller the coefficient of variation, the smaller the relative risk is for the return provided. The reciprocal of the coefficient of variation, called **return to risk**, is often used because it is easier to interpret. That is, if the objective is to maximize return, a higher return-to-risk ratio is often considered better. A related measure in finance is the *Sharpe ratio*, which is the ratio of a fund's excess returns (annualized total returns minus Treasury bill returns) to its standard deviation. If several investment opportunities have the same mean but different variances, a rational (risk-averse) investor will select the one that has the smallest variance.[5] This approach to formalizing risk is the basis for modern portfolio theory, which seeks to construct minimum-variance portfolios. As *Fortune* magazine once observed, "It's not that risk is always bad. . . . It's just that when you take chances with your money, you want to be paid for it."[6] One practical application of the coefficient of variation is in comparing stock prices.

EXAMPLE 4.26   **Applying the Coefficient of Variation**

If examining only the standard deviations in the *Closing Stock Prices* worksheet, we might conclude that IBM is more risky than the other stocks. However, the mean stock price of IBM is much greater than the other stocks. Thus, comparing standard deviations directly provides little information. The coefficient of variation provides a more comparable measure. Figure 4.30 shows the calculations of the coefficients of variation for these variables. For IBM, the CV is 0.025; for Intel, 0.027; for Cisco, 0.024; for GE, 0.022; and for the DJIA, 0.016. We see that the coefficients of variation of the stocks are not very different; in fact, Intel is just slightly more risky than IBM relative to its average price. However, an index fund based on the Dow would be less risky than any of the individual stocks.

---

[5]David G. Luenberger, *Investment Science* (New York: Oxford University Press, 1998).
[6]*Fortune*, 1999 Investor's Guide (December 21, 1998).

▶ **Figure 4.30**

Calculating Coefficients of Variation for Closing Stock Prices

	A	B	C	D	E	F
1	Closing Stock Prices					
2						
3	Date	IBM	INTC	CSCO	GE	DJ Industrials Index
4	9/3/2010	$127.58	$18.43	$21.04	$15.39	10447.93
5	9/7/2010	$125.95	$18.12	$20.58	$15.44	10340.69
6	9/8/2010	$126.08	$17.90	$20.64	$15.70	10387.01
22	9/30/2010	$134.14	$19.20	$21.90	$16.25	10788.05
23	10/1/2010	$135.64	$19.32	$21.91	$16.36	10829.68
24	Mean	$130.93	$18.81	$21.50	$16.20	$10,639.98
25	Standard Deviation	$3.22	$0.50	$0.52	$0.35	$171.94
26	Coefficient of Variation	0.025	0.027	0.024	0.022	0.016

Measures of Shape

Histograms of sample data can take on a variety of different shapes. Figure 4.31 shows the histograms for cost per order and A/P terms that we created for the *Purchase Orders* data. The histogram for A/P terms is relatively symmetric, having its modal value in the middle and falling away from the center in roughly the same fashion on either side. However, the cost per order histogram is asymmetrical, or *skewed*; that is, more of the mass is concentrated on one side, and the distribution of values "tails off" to the other. Those that have more mass on the left and tail off to the right, like this example, are called *positively skewed*; those that have more mass on the right and tail off to the left are said to be *negatively skewed*. **Skewness** describes the lack of symmetry of data.

The **coefficient of skewness (CS)** measures the degree of asymmetry of observations around the mean. The coefficient of skewness for a population is computed as

$$CS = \frac{\frac{1}{N}\sum_{i=1}^{N}(x_i - \mu)^3}{\sigma^3} \tag{4.14}$$

It can be computed using the Excel function SKEW.P(*data range*). For sample data, the formula is different, but you may calculate it using the Excel function SKEW(*data range*). If CS is positive, the distribution of values is positively skewed; if negative, it is negatively skewed. The closer CS is to zero, the less the degree of skewness. A coefficient of skewness greater than 1 or less than −1 suggests a high degree of skewness. A value between 0.5 and 1 or between −0.5 and −1 represents moderate skewness. Coefficients between 0.5 and −0.5 indicate relative symmetry.

EXAMPLE 4.27   **Measuring Skewness**

Using the Excel function SKEW in the *Purchase Orders* database, the coefficients of skewness for the cost per order and A/P terms data are calculated as

CS (cost per order) = 1.66
CS (A/P terms) = 0.60

This tells us that the cost per order data are highly positively skewed, whereas the A/P terms data are more symmetric. These are evident from the histograms in Figure 4.31.

Histograms that have only one "peak" are called **unimodal**. (If a histogram has exactly two peaks, we call it **bimodal**. This often signifies a mixture of samples from different populations.) For unimodal histograms that are relatively symmetric, the mode is a fairly good estimate of the mean. For example, the mode for the A/P terms data is clearly 30 months; the mean is 30.638 months. On the other hand, for the cost per order data,

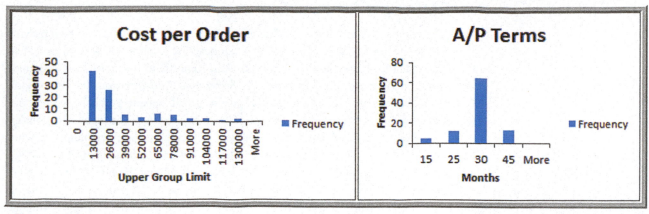

▲ **Figure 4.31**

Histograms of Cost per Order and A/P Terms

▶ **Figure 4.32**

Characteristics of Skewed Distributions

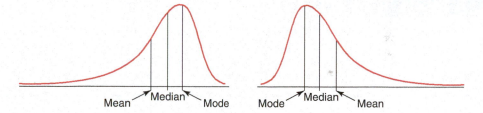

the mode occurs in the group (0, 13,000). The midpoint of the group, $6,500, which can be used as a numerical estimate of the mode, is not very close at all to the true mean of $26,295.32. The high level of skewness pulls the mean away from the mode.

Comparing measures of location can sometimes reveal information about the shape of the distribution of observations. For example, if the distribution was perfectly symmetrical and unimodal, the mean, median, and mode would all be the same. If it was negatively skewed, we would generally find that mean < median < mode, whereas a positive skewness would suggest that mode < median < mean (see Figure 4.32).

Kurtosis refers to the peakedness (that is, high, narrow) or flatness (that is, short, flat-topped) of a histogram. The **coefficient of kurtosis (CK)** measures the degree of kurtosis of a population and is computed as

$$CK = \frac{\frac{1}{N}\sum_{i=1}^{N}(x_i - \mu)^4}{\sigma^4} \tag{4.15}$$

Distributions with values of CK less than 3 are more flat, with a wide degree of dispersion; those with values of CK greater than 3 are more peaked, with less dispersion. Excel computes kurtosis differently; the function KURT (*data range*) computes "excess kurtosis" for sample data (Excel does not have a corresponding function for a population), which is CK − 3. Thus, to interpret kurtosis values in Excel, distributions with values less than 0 are more flat, while those with values greater than 0 are more peaked.

Skewness and kurtosis can help provide more information to evaluate risk than just using the standard deviation. For example, both a negatively and positively skewed distribution may have the same standard deviation, but clearly if the objective is to achieve a high return, the negatively skewed distribution will have higher probabilities of larger returns. The higher the kurtosis, the more area the histogram has in the tails rather than in the middle. This can indicate a greater potential for extreme and possibly catastrophic outcomes.

Excel *Descriptive Statistics* Tool

Excel provides a useful tool for basic data analysis, *Descriptive Statistics*, which provides a summary of numerical statistical measures that describe location, dispersion, and shape for sample data (not a population). Click on *Data Analysis* in the *Analysis* group under the *Data* tab in the Excel menu bar. Select *Descriptive Statistics* from the list of tools. The *Descriptive Statistics* dialog shown in Figure 4.33 will appear. You need to enter only the range of the data, which must be in a *single row* or *column*. If the data are in multiple columns, the tool treats each row or column as a separate data set, depending on which you specify. This means that if you have a single data set arranged in a matrix format, you would have to stack the data in a single column before applying the *Descriptive Statistics* tool. Check the box *Labels in First Row* if labels are included in the input range. You may choose to save the results in the current worksheet or in a new one. For basic summary statistics, check the box *Summary statistics*; you need not check any others.

EXAMPLE 4.28	Using the Descriptive Statistics Tool

We will apply the *Descriptive Statistics* tool to the cost per order and A/P terms data in columns G and H of the *Purchase Orders* database. The results are provided in the *Descriptive Statistics* worksheet in the *Purchase Orders* workbook and are shown in Figure 4.34. The tool provides all the measures we have discussed as well as the standard error, which we discuss in Chapter 6, along with the minimum, maximum, sum, and count.

▶ **Figure 4.33**

Descriptive Statistics *Dialog*

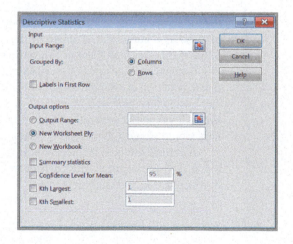

▶ **Figure 4.34**

Purchase Orders *Data Descriptive Statistics Summary*

	A	B	C	D
1	*Cost per order*		*A/P Terms (Months)*	
2				
3	Mean	26295.31915	Mean	30.63829787
4	Standard Error	3078.053014	Standard Error	0.702294026
5	Median	15656.25	Median	30
6	Mode	14910	Mode	30
7	Standard Deviation	29842.8312	Standard Deviation	6.808993205
8	Sample Variance	890594573.8	Sample Variance	46.36238847
9	Kurtosis	2.079637302	Kurtosis	1.512188562
10	Skewness	1.664271519	Skewness	0.599265003
11	Range	127431.25	Range	30
12	Minimum	68.75	Minimum	15
13	Maximum	127500	Maximum	45
14	Sum	2471760	Sum	2880
15	Count	94	Count	94

One important point to note about the use of the tools in the *Analysis Toolpak* versus Excel functions is that while Excel functions dynamically change as the data in the spreadsheet are changed, the results of the *Analysis Toolpak* tools do not. For example, if you compute the average value of a range of numbers directly using the function AVERAGE(*range*), then changing the data in the range will automatically update the result. However, you would have to rerun the *Descriptive Statistics* tool after changing the data.

CHECK YOUR UNDERSTANDING

1. Explain the difference between a population and a sample.

2. List and explain the common measures of location used in statistics.

3. List and explain the common measures of dispersion used in statistics.

4. How are Chebyshev's theorem and the empirical rules useful in business?

5. What is a standardized value (*z*-score)?

6. Explain the value of the coefficient of variation.

7. Define the common measures of shape used in statistics.

8. What information does the Excel *Descriptive Statistics* tool provide?

Computing Descriptive Statistics for Frequency Distributions

When data are summarized in a frequency distribution, we can use the frequencies to compute the mean and variance. The following formulas provide a way to calculate the mean and variance of data expressed as a frequency distribution.

The mean of a population expressed as a frequency distribution may be computed using the formula

$$\mu = \frac{\sum_{i=1}^{N} f_i x_i}{N} \tag{4.16}$$

For samples, the formula is similar:

$$\bar{x} = \frac{\sum_{i=1}^{n} f_i x_i}{n} \tag{4.17}$$

where f_i is the frequency of observation i. Essentially, we multiply the frequency by the value of observation i, add them up, and divide by the number of observations.

We may use similar formulas to compute the population variance

$$\sigma^2 = \frac{\sum_{i=1}^{N} f_i(x_i - \mu)^2}{N} \tag{4.18}$$

and sample variance

$$s^2 = \frac{\sum_{i=1}^{n} f_i(x_i - \bar{x})^2}{n - 1} \qquad (4.19)$$

To find the standard deviation, take the square root of the variance, as we did earlier.

Note the similarities between these formulas and formulas (4.16) and (4.17). In multiplying the values by the frequency, we are essentially adding the same values f_i times. So they really are the same formulas, just expressed differently.

EXAMPLE 4.29   Computing Statistical Measures from Frequency Distributions

In Example 4.4, we constructed a frequency distribution for A/P terms in the *Purchase Orders* database. Figure 4.35 shows the calculations of the mean and sample variance using formulas (4.17) and (4.19) for the frequency distribution. In column C, we multiply the frequency by the value of the observations [the numerator in formula (4.17)] and then divide by n, the sum of the frequencies in column B, to find the mean. Columns D, E, and F provide the calculations needed to find the variance. We divide the sum of the data in column F by $n - 1 = 93$ to find the variance.

In some situations, data may already be grouped in a frequency distribution, and we may not have access to the raw data. This is often the case when extracting information from government databases such as the Census Bureau or Bureau of Labor Statistics. In these situations, we cannot compute the mean or variance using the standard formulas.

If the data are grouped into k cells in a frequency distribution, we can use modified versions of these formulas to estimate the mean and variance by replacing x_i with a representative value (such as the midpoint, M) for all the observations in each cell group and summing over all groups. The corresponding formulas follow. In these formulas, k is the number of groups and M_i is the midpoint of group i.

Population mean estimate for grouped data:

$$\mu = \frac{\sum_{i=1}^{k} f_i M_i}{N} \qquad (4.20)$$

Sample mean estimate for grouped data:

$$\bar{x} = \frac{\sum_{i=1}^{k} f_i M_i}{n} \qquad (4.21)$$

▶ **Figure 4.35**

Calculations of Mean and Variance Using a Frequency Distribution

	A	B	C	D	E	F
1	Frequency Distribution Calculations					
2						
3	A/P Terms (x)	Frequency (f)	f*x	x - Mean	(x - Mean)^2	f*(x - Mean)^2
4	15	5	75	-15.6383	244.5563603	1222.781802
5	25	12	300	-5.6383	31.7904029	381.4848348
6	30	64	1920	-0.6383	0.407424174	26.07514713
7	45	13	585	14.3617	206.258488	2681.360344
8	Sum	94	2880			4311.702128
9						
10		Mean	30.6383		Sample variance	46.36238847

▶ **FIGURE 4.36**

Grouped Frequency Calculations for Cost per Order

	A	B	C	D	E	F	G
1	Grouped Frequency Distribution Calculations						
2							
3	Cost/Order Group	Midpoint (x)	Frequency (f)	f*x	x - Mean	(x - Mean)^2	f*(x-Mean)^2
4	0 to 26000	13000	68	884000	-14936.17	223089180.6	15170064282
5	26000 to 52000	39000	8	312000	11063.8298	122408329.6	979266636.5
6	52000 to 78000	65000	11	715000	37063.8298	1373727478	15111002263
7	78000 to 104000	91000	4	364000	63063.8298	3977046627	15908186510
8	104000 to 130000	117000	3	351000	89063.8298	7932365776	23797097329
9		Sum	94	2626000			70965617021
10							
11			Mean	27936.2		Sample variance	763071150.8

Population variance estimate for grouped data:

$$\sigma^2 = \frac{\sum_{i=1}^{k} f_i (M_i - \mu)^2}{N} \tag{4.22}$$

Sample variance estimate for grouped data:

$$s^2 = \frac{\sum_{i=1}^{k} f_i (M_i - \bar{x})^2}{n - 1} \tag{4.23}$$

EXAMPLE 4.30 **Computing Descriptive Statistics for a Grouped Frequency Distribution**

In Figure 4.11, we illustrated a grouped frequency distribution for cost per order in the *Purchase Orders* database. Figure 4.36 shows the calculations of the mean and sample variance using formulas (4.21) and (4.23). It is important to understand that because we have not used all the original data in computing these statistics, they are only estimates of the true values.

CHECK YOUR UNDERSTANDING

1. Explain the process of computing the mean and variance from a frequency distribution and from a grouped frequency distribution.

2. Use Excel to find the mean and variance for the grouped frequency distribution in Figure 4.12.

Descriptive Statistics for Categorical Data: The Proportion

Statistics such as means and variances are not appropriate for categorical data. Instead, we are generally interested in the fraction of data that have a certain characteristic. The formal statistical measure is called the **proportion**, usually denoted by p. The proportion is computed using the formula

$$p = x/n \tag{4.24}$$

where x is the number of observations having a certain characteristic and n is the sample size. Note that proportions are analogous to relative frequencies for categorical data.

Proportions are key descriptive statistics for categorical data, such as demographic data from surveys, defects or errors in quality-control applications, or consumer preferences in market research.

EXAMPLE 4.31   **Computing a Proportion**

In the *Purchase Orders* database, column A lists the name of the supplier for each order. We may use the Excel function =COUNTIF(*data range, criteria*) to count the number of observations meeting specified characteristics. For instance, to find the number of orders placed with Spacetime Technologies, we used the function =COUNTIF(A4:A97, "Spacetime Technologies"). This returns a value of 12. Because 94 orders were placed, the proportion of orders placed with Spacetime Technologies is $p = 12/94 = 0.128$.

It is important to realize that proportions are numbers between 0 and 1. Although we often convert these to percentages—for example, 12.8% of orders were placed with Spacetime Technologies in the last example—we must be careful to use the decimal expression of a proportion when statistical formulas require it.

CHECK YOUR UNDERSTANDING

1. What is a proportion, and how is it computed?

2. Find the proportion of orders placed for bolt-nut packages in the *Purchase Orders* database.

Statistics in PivotTables

We introduced PivotTables in Chapter 2. PivotTables also have the functionality to calculate many basic statistical measures from the data summaries. If you look at the *Value Field Settings* dialog shown in Figure 4.37, you can see that you can calculate the average, standard deviation, and variance of a value field.

▶ **Figure 4.37**

Value Field Settings *Dialog*

Value Field Settings
Source Name: Checking
Custom Name:
Summarize Values By · Show Values As
Summarize value field by
Choose the type of calculation that you want to use to summarize data from the selected field
Sum
Count
Average
Max
Min
Product
Count Numbers
StdDev
StdDevp
Var
Varp
Number Format · OK · Cancel

▶ **Figure 4.38**

PivotTable for Average Checking and Savings Account Balances by Job

Row Labels	Average of Checking	Average of Savings
Management	$606.94	$1,616.83
Skilled	$1,079.24	$1,836.43
Unemployed	$1,697.64	$2,760.91
Unskilled	$1,140.27	$1,741.44
Grand Total	$1,048.01	$1,812.56

EXAMPLE 4.32   **Statistical Measures in PivotTables**

In the *Credit Risk Data* Excel file, suppose that we want to find the average amount of money in checking and savings accounts by job classification. Create a Pivot-Table, and in the *PivotTable Field List*, move Job to the *Row Labels field* and Checking and Savings to the *Values* field. Then change the field settings from "Sum of Checking" and "Sum of Savings" to the averages. The result is shown in Figure 4.38; we have also formatted the values as currency using the *Number Format* button in the dialog. In a similar fashion, you could find the standard deviation or variance of each group by selecting the appropriate field settings.

CHECK YOUR UNDERSTANDING

1. What statistical information can be displayed in a PivotTable?

2. Use a PivotTable to find the standard deviation of the amount of money in checking and savings accounts for each job classification in the *Credit Risk Data* Excel file.

Measures of Association

Two variables have a strong statistical relationship with one another if they appear to move together. We see many examples on a daily basis; for instance, attendance at baseball games is often closely related to the win percentage of the team, and ice cream sales likely have a strong relationship with daily temperature. We can examine relationships between two variables visually using scatter charts, which we introduced in Chapter 3.

When two variables appear to be related, you might suspect a cause-and-effect relationship. Sometimes, however, statistical relationships exist even though a change in one variable is not *caused* by a change in the other. For example, the *New York Times* reported a strong statistical relationship between the golf handicaps of corporate CEOs and their companies' stock market performance over three years. CEOs who were better-than-average golfers were likely to deliver above-average returns to shareholders.[7] Clearly, the ability to golf would not cause better business performance. Therefore, you must be cautious in drawing inferences about causal relationships based solely on statistical relationships. (On the other hand, you might want to spend more time out on the practice range!)

Understanding the relationships between variables is extremely important in making good business decisions, particularly when cause-and-effect relationships can be justified. When a company understands how internal factors such as product quality, employee training, and pricing affect such external measures as profitability and customer satisfaction, it can make better decisions. Thus, it is helpful to have statistical tools for measuring these relationships.

---

[7]Adam Bryant, "CEOs' Golf Games Linked to Companies' Performance," *Cincinnati Enquirer*, June 7, 1998, El.

	A	B	C	D	E	F	G
1	**Colleges and Universities**						
2							
3	School	Type	Median SAT	Acceptance Rate	Expenditures/Student	Top 10% HS	Graduation %
4	Amherst	Lib Arts	1315	22%	$ 26,636	85	93
5	Barnard	Lib Arts	1220	53%	$ 17,653	69	80
6	Bates	Lib Arts	1240	36%	$ 17,554	58	88
7	Berkeley	University	1176	37%	$ 23,665	95	68
8	Bowdoin	Lib Arts	1300	24%	$ 25,703	78	90
9	Brown	University	1281	24%	$ 24,201	80	90
10	Bryn Mawr	Lib Arts	1255	56%	$ 18,847	70	84

▲ **Figure 4.39**

Portion of Excel File Colleges and Universities

The Excel file *Colleges and Universities*, a portion of which is shown in Figure 4.39, contains data from 49 top liberal arts and research universities across the United States. Several questions might be raised about statistical relationships among these variables. For instance, does a higher percentage of students in the top 10% of their high school class suggest a higher graduation rate? Is acceptance rate related to the amount spent per student? Do schools with lower acceptance rates tend to accept students with higher SAT scores? Questions such as these can be addressed by computing statistical measures of association between the variables.

Covariance

Covariance is a measure of the linear association between two variables, X and Y. Like the variance, different formulas are used for populations and samples. Computationally, covariance of a population is the average of the products of deviations of each observation from its respective mean:

$$\text{cov}(X, Y) = \frac{\sum_{i=1}^{N}(x_i - \mu_x)(y_i - \mu_y)}{N} \tag{4.25}$$

To better understand the covariance, let us examine formula (4.25). The covariance between X and Y is the average of the product of the deviations of each pair of observations from their respective means. Suppose that large (small) values of X are generally associated with large (small) values of Y. Then, in most cases, both x_i and y_i are either above or below their respective means. If so, the product of the deviations from the means will be a positive number and when added together and averaged will give a positive value for the covariance. On the other hand, if small (large) values of X are associated with large (small) values of Y, then one of the deviations from the mean will generally be negative while the other is positive. When multiplied together, a negative value results, and the value of the covariance will be negative. Thus, the larger the absolute value of the covariance, the higher the degree of linear association between the two variables. The sign of the covariance tells us whether there is a direct relationship (that is, one variable increases as the other increases) or an inverse relationship (that is, one variable increases while the other decreases, or vice versa). We can generally identify the strength of any linear association between two variables and the sign of the covariance by constructing a scatter diagram. The Excel function COVARIANCE.P(*array1, array2*) computes the covariance of a population.

The sample covariance is computed as

$$\text{cov}(X, Y) = \frac{\sum_{i=1}^{n}(x_i - \bar{x})(y_i - \bar{y})}{n - 1} \tag{4.26}$$

Similar to the sample variance, note the use of $n - 1$ in the denominator. The Excel function COVARIANCE.S(*array1, array2*) computes the covariance of a sample.

EXAMPLE 4.33 **Computing the Covariance**

Figure 4.40 shows a scatter chart of graduation rates (*Y* variable) versus median SAT scores (*X* variable) for the *Colleges and Universities* data. It appears that as the median SAT scores increase, the graduation rate also increases; thus, we would expect to see a positive covariance. Figure

4.41 shows the calculations using formula (4.26); these are provided in the worksheet *Covariance Calculations* in the *Colleges and Universities* Excel workbook. The Excel function =COVARIANCE.S(B2:B50,C2:C50) in cell F55 verifies the calculations.

Correlation

The numerical value of the covariance is generally difficult to interpret because it depends on the units of measurement of the variables. For example, if we expressed the graduation rate as a true proportion rather than as a percentage in the previous example, the numerical value of the covariance would be smaller, although the linear association between the variables would be the same.

Correlation is a measure of the linear relationship between two variables, *X* and *Y*, which does not depend on the units of measurement. Correlation is measured by the **correlation coefficient**, also known as the **Pearson product moment correlation coefficient**. The correlation coefficient for a population is computed as

$$\rho_{XY} = \frac{\text{cov}(X, Y)}{\sigma_X \sigma_Y} \tag{4.27}$$

▶ **Figure 4.40**

Scatter Chart of Graduation Rate Versus Median SAT

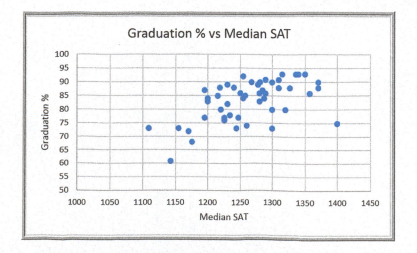

▶ Figure 4.41

Covariance Calculations for Graduation Rate and Median SAT

	A	B	C	D	E	F
1		Graduation % (X)	Median SAT (Y)	X - Mean(X)	Y - Mean(Y)	(X - Mean(X))(Y-Mean(Y))
2		93	1315	9.755	51.898	506.2698875
3		80	1220	-3.245	-43.102	139.8617243
4		88	1240	4.755	-23.102	-109.8525614
47		86	1250	2.755	-13.102	-36.09745939
48		91	1290	7.755	26.898	208.5964182
49		93	1336	9.755	72.898	711.1270304
50		93	1350	9.755	86.898	847.698459
51	Mean	83.245	1263.102		Sum	12641.77551
52					Count	49
53					Covariance	263.3703231
54						
55					COVARIANCE.S	263.3703231

where σ_X is the standard deviation of X, and σ_Y is the standard deviation of Y. By dividing the covariance by the product of the standard deviations, we are essentially scaling the numerical value of the covariance to a number between -1 and 1.

In a similar fashion, the **sample correlation coefficient** is computed as

$$r_{XY} = \frac{\text{cov}(X, Y)}{s_X s_Y} \tag{4.28}$$

Excel's CORREL(*array1, array2*) function computes the correlation coefficient of two data arrays.

A correlation of 0 indicates that the two variables have no linear relationship to each other. Thus, if one changes, we cannot reasonably predict what the other variable might do. A positive correlation coefficient indicates a linear relationship when one variable increases while the other also increases. A negative correlation coefficient indicates a linear relationship when one variable that increases while the other decreases. In economics, for instance, a price-elastic product has a negative correlation between price and sales; as price increases, sales decrease, and vice versa. Visualizations of correlation are illustrated in Figure 4.42. Note that although Figure 4.42(d) shows a clear relationship between the variables, the relationship is not linear and the correlation is zero.

▶ Figure 4.42

Examples of Correlation

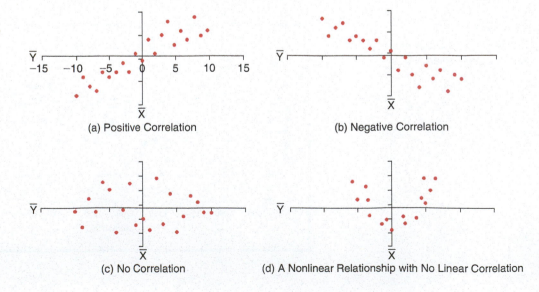

(a) Positive Correlation

(b) Negative Correlation

(c) No Correlation

(d) A Nonlinear Relationship with No Linear Correlation

	A	B	C	D	E	F
1		Graduation % (X)	Median SAT (Y)	X - Mean(X)	Y - Mean(Y)	(X - Mean(X))(Y-Mean(Y))
2		93	1315	9.755	51.898	506.2698875
3		80	1220	-3.245	-43.102	139.8617243
4		88	1240	4.755	-23.102	-109.8525614
47		86	1250	2.755	-13.102	-36.09745939
48		91	1290	7.755	26.898	208.5964182
49		93	1336	9.755	72.898	711.1270304
50		93	1350	9.755	86.898	847.698459
51	Mean	83.245	1263.102		Sum	12641.77551
52	Standard Deviation	7.449	62.676		Count	49
53					Covariance	263.3703231
54					Correlation	0.564146827
55						
56					CORREL Function	0.564146827

▲ **Figure 4.43**

Correlation Calculations for Graduation Rate and Median SAT

EXAMPLE 4.34   **Computing the Correlation Coefficient**

Figure 4.43 shows the calculations for computing the sample correlation coefficient for the graduation rate and median SAT variables in the *Colleges and Universities* data file. We first compute the standard deviation of each variable in cells B52 and C52 and then divide the covariance by the product of these standard deviations in cell F54. Cell F56 shows the same result using the Excel function =CORREL(B2:B50,C2:C50).

When using the CORREL function, it does not matter if the data represent samples or populations. In other words,

$$CORREL(array1, array2) = \frac{COVARIANCE.P(array1, array2)}{STDEV.P(array1) \times STDEV.P(array2)}$$

and

$$CORREL(array1, array2) = \frac{COVARIANCE.S(array1, array2)}{STDEV.S(array1) \times STDEV.S(array2)}$$

For instance, in Example 4.34, if we assume that the data are populations, we find that the population standard deviation for X is 7.372 and the population standard deviation for Y is 62.034 (using the function STDEV.P). By dividing the population covariance, 257.995 (using the function COVARIANCE.P), by the product of these standard deviations, we find that the correlation coefficient is still 0.564 as computed by the CORREL function.

Excel Correlation Tool

The *Data Analysis Correlation* tool computes correlation coefficients for more than two arrays. Select *Correlation* from the *Data Analysis* tool list. The dialog is shown in Figure 4.44. You need to input only the range of the data (which must be in contiguous columns; if not, you must move them in your worksheet), specify whether the data are grouped by rows or columns (most applications will be grouped by columns), and indicate whether the first row contains data labels. The output of this tool is a matrix giving the correlation between each pair of variables. This tool provides the same output as the CORREL function for each pair of variables.

▶ **Figure 4.44**

Excel Correlation *Tool Dialog*

▶ **Figure 4.45**

Correlation Results for Colleges and Universities Data

	A	B	C	D	E	F
1		*Median SAT*	*Acceptance Rate*	*Expenditures/Student*	*Top 10% HS*	*Graduation %*
2	Median SAT	1				
3	Acceptance Rate	-0.601901959	1			
4	Expenditures/Student	0.572741729	-0.284254415	1		
5	Top 10% HS	0.503467995	-0.609720972	0.505782049	1	
6	Graduation %	0.564146827	-0.55037751	0.042503514	0.138612667	1

EXAMPLE 4.35   **Using the *Correlation* Tool**

The correlation matrix among all the variables in the *Colleges and Universities* data file is shown in Figure 4.45. None of the correlations are very strong. The moderate positive correlation between the graduation rate and SAT scores indicates that schools with higher median SATs have higher graduation rates. We see a moderate negative correlation between acceptance rate and graduation rate,

indicating that schools with lower acceptance rates have higher graduation rates. We also see that the acceptance rate is also negatively correlated with the median SAT and top 10% HS, suggesting that schools with lower acceptance rates have higher student profiles. The correlations with expenditures/student also suggest that schools with higher student profiles spend more money per student.

CHECK YOUR UNDERSTANDING

1. Explain the difference between covariance and correlation.

2. What Excel functions and tools can you use to find the covariance and correlation of populations and samples?

3. Explain how to interpret the correlation coefficient.

Outliers

Earlier we noted that the mean and range are sensitive to outliers—unusually large or small values in the data. Outliers can make a significant difference in the results we obtain from statistical analyses. An important statistical question is how to identify them. The first thing to do from a practical perspective is to check the data for possible errors, such as a misplaced decimal point or an incorrect transcription to a computer file. Histograms can help to identify possible outliers visually. We might use the empirical rule and z-scores to identify an outlier as one that is more than three standard deviations from the mean. We can also identify outliers based on the interquartile range. "Mild" outliers are often defined as being between $1.5 \times IQR$ and $3 \times IQR$ to the left of Q_1 or to the right of Q_3, and "extreme" outliers as more than $3 \times IQR$ away from these quartiles. Basically, there is no standard definition of what constitutes an outlier other than an unusual observation as compared with the rest. However, it is important to try to identify outliers and determine their significance when conducting business analytic studies.

EXAMPLE 4.36   Investigating Outliers

The Excel data file *Home Market Value* provides a sample of data for homes in a neighborhood (Figure 4.46). Figure 4.47 shows *z*-score calculations for the square feet and market value variables. None of the *z*-scores for either of these variables exceed 3 (these calculations can be found in the worksheet *Outliers* in the Excel *Home Market Value* workbook). However, while individual variables might not exhibit outliers, combinations of them might. We see this in the scatter diagram in Figure 4.48. The last observation has a high market value ($120,700) but a relatively small house size

(1,581 square feet). The point on the scatter diagram does not seem to coincide with the rest of the data.

The question is what to do with possible outliers. They should not be blindly eliminated unless there is a legitimate reason for doing so—for instance, if the last home in the Home Market Value example has an outdoor pool that makes it significantly different from the rest of the neighborhood. Statisticians often suggest that analyses should be run with and without the outliers so that the results can be compared and examined critically.

▶ **Figure 4.46**

Portion of Home Market Value *Excel File*

	A	B	C
1	Home Market Value		
2			
3	House Age	Square Feet	Market Value
4	33	1,812	$90,000.00
5	32	1,914	$104,400.00
6	32	1,842	$93,300.00
7	33	1,812	$91,000.00
8	32	1,836	$101,900.00
9	33	2,028	$108,500.00
10	32	1,732	$87,600.00
11	33	1,850	$96,000.00

▶ **Figure 4.47**

Computing Z-Scores for Examining Outliers

	A	B	C	D	E
1	Home Market Value				
2					
3	House Age	Square Feet	z-score	Market Value	z-score
4	33	1,812	0.5300	$90,000.00	-0.196
5	32	1,914	0.9931	$104,400.00	1.168
6	32	1,842	0.6662	$93,300.00	0.117
7	33	1,812	0.5300	$91,000.00	-0.101
41	27	1,484	-0.9592	$81,300.00	-1.020
42	27	1,520	-0.7957	$100,700.00	0.818
43	28	1,520	-0.7957	$87,200.00	-0.461
44	27	1,684	-0.0511	$96,700.00	0.439
45	27	1,581	-0.5188	$120,700.00	2.713
46	Mean	1,695		92,069	
47	Standard Deviation	220.257		10553.083	

▶ **Figure 4.48**

Scatter Diagram of House Size Versus Market Value

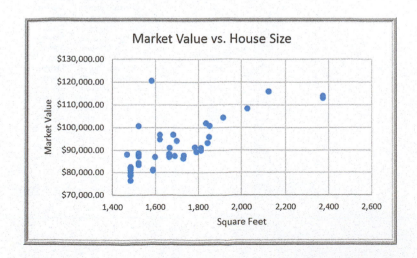

CHECK YOUR UNDERSTANDING

1. What rules are typically used to identify outliers?

2. How should you handle outliers in your analysis?

Using Descriptive Statistics to Analyze Survey Data

Many business applications deal with survey data. Descriptive statistics tools are extremely valuable for summarizing and analyzing survey data. Consider, for example, the data in the Excel file *Insurance Survey*, which is shown in Figure 4.49. We may compute various statistical measures and apply statistical tools to understand these data, such as the following:

- Frequency distributions and histograms for the ratio variables (age, years employed, and satisfaction)
- Descriptive statistical measures for the ratio variables using the *Descriptive Statistics* tool
- Proportions for various attributes of the categorical variables in the sample, such as the proportion of males and females, different educational levels, and marital status
- PivotTables that break down the averages of ratio variables by gender, education, and marital status
- Cross-tabulations by gender and education, gender and marital status, and education and marital status
- Z-scores for examination of potential outliers

► **Figure 4.49**

Insurance Survey *Data*

	Age	Gender	Education	Marital Status	Years Employed	Satisfaction*	Premium/Deductible**
1	Insurance Survey						
2							
3	Age	Gender	Education	Marital Status	Years Employed	Satisfaction*	Premium/Deductible**
4	36	F	Some college	Divorced	4	4	N
5	55	F	Some college	Divorced	2	1	N
6	61	M	Graduate degree	Widowed	26	3	N
7	65	F	Some college	Married	9	4	N
8	53	F	Graduate degree	Married	6	4	N
9	50	F	Graduate degree	Married	10	5	N
10	28	F	College graduate	Married	4	5	N
11	62	F	College graduate	Divorced	9	3	N
12	48	M	Graduate degree	Married	6	5	N
13	31	M	Graduate degree	Married	1	5	N
14	57	F	College graduate	Married	4	5	N
15	44	M	College graduate	Married	2	3	N
16	38	M	Some college	Married	3	2	N
17	27	M	Some college	Married	2	3	N
18	56	M	Graduate degree	Married	4	4	Y
19	43	F	College graduate	Married	5	3	Y
20	45	M	College graduate	Married	15	3	Y
21	42	F	College graduate	Married	12	3	Y
22	29	M	Graduate degree	Single	10	5	N
23	28	F	Some college	Married	3	4	Y
24	36	M	Some college	Divorced	15	4	Y
25	49	F	Graduate degree	Married	2	5	N
26	46	F	College graduate	Divorced	20	4	N
27	52	F	College graduate	Married	18	2	N
28							
29	*Measured from 1-5 with 5 being highly satisfied.						
30	**Would you be willing to pay a lower premium for a higher deductible?						

Problem 69 at the end of this chapter asks you to use these approaches to analyze these data. These tools can be supplemented by various data visualization tools that we discussed in the previous chapter. Such analysis can provide the basis for formal reports and presentations that one might use to communicate and explain results to managers.

CHECK YOUR UNDERSTANDING

1. Summarize statistical tools that are typically used to analyze survey data.

2. Explain the importance of using statistics for communicating survey data.

Statistical Thinking in Business Decisions

The importance of applying statistical concepts to make good business decisions and improve performance cannot be overemphasized. **Statistical thinking** is a philosophy of learning and action for improvement that is based on the principles that

- all work occurs in a system of interconnected processes,
- variation exists in all processes, and
- better performance results from understanding and reducing variation.[8]

Work gets done in any organization through *processes*—systematic ways of doing things that achieve desired results. Understanding business processes provides the context for determining the effects of variation and the proper type of action to be taken. Any process contains many sources of variation. In manufacturing, for example, different batches of material vary in strength, thickness, or moisture content. During manufacturing, tools experience wear, vibrations cause changes in machine settings, and electrical fluctuations cause variations in power. Workers may not position parts on fixtures consistently, and physical and emotional stress may affect workers' consistency. In addition, measurement gauges and human inspection capabilities are not uniform, resulting in measurement error. Similar phenomena occur in service processes because of variation in employee and customer behavior, application of technology, and so on. Reducing variation results in more consistency in manufacturing and service processes, fewer errors, happier customers, and better accuracy of such things as delivery time quotes.

Although variation exists everywhere, many managers often do not recognize it or consider it in their decisions. How often do managers make decisions based on one or two data points without looking at the pattern of variation, see trends in data that aren't justified, or try to manipulate measures they cannot truly control? Unfortunately, the answer is quite often. For example, if sales in some region fell from the previous quarter, a regional manager might quickly blame her sales staff for not working hard enough, even though the drop in sales may simply be the result of uncontrollable variation. Usually, it is simply a matter of ignorance of how to deal with variation in data. This is where business analytics can play a significant role. Statistical analysis can provide better insight into the facts and nature of relationships among the many factors that may have contributed to an event and enable managers to make better decisions.

[8]Galen Britz, Don Emerling, Lynne Hare, Roger Hoerl, and Janice Shade, "How to Teach Others to Apply Statistical Thinking," *Quality Progress* (June 1997): 67–79.

EXAMPLE 4.37   Applying Statistical Thinking

Figure 4.50 shows a portion of the data in the Excel file *Surgery Infections* that document the number of infections that occurred after surgeries over 36 months at one hospital, along with a line chart of the number of infections. (We will assume that the number of surgeries performed each month was the same.) The number of infections tripled in months 2 and 3 as compared to the first month. Is this indicative of a trend caused by failure of some health care protocol or simply random variation? Should action be taken to determine a cause? From a statistical perspective, three points are insufficient to conclude that a trend exists. It is more appropriate to look at a larger sample of data and study the pattern of variation.

Over the 36 months, the data clearly indicate that variation exists in the monthly infection rates. The number of infections seems to fluctuate between 0 and 3 with the exception of month 12. However, a visual analysis of the chart cannot necessarily lead to a valid conclusion. So let's apply some statistical thinking. The average number of infections is 1.583, and the standard deviation is 1.180.

If we apply the empirical rule that most observations should fall within three standard deviations of the mean, we would expect them to fall between −1.957 (clearly, the number of infections cannot be negative, so let's set this value to zero) and 5.124. This means that, from a statistical perspective, we can expect almost all the observations to fall within these limits. Figure 4.51 shows the chart displaying these ranges. The number of infections for month 12 clearly exceeds the upper range value and suggests that the number of infections for this month is statistically different from the rest. The hospital administrator should seek to investigate what may have happened that month and try to prevent similar occurrences.

Similar analyses are used routinely in quality control and other business applications to monitor performance statistically. The proper analytical calculations depend on the type of measure and other factors and are explained fully in books dedicated to quality control and quality management.

Variability in Samples

Because we usually deal with sample data in business analytics applications, it is extremely important to understand that different samples from any population will vary; that is, they will have different means, standard deviations, and other statistical measures and will have differences in the shapes of histograms. In particular, samples are extremely sensitive to the sample size—the number of observations included in the samples.

▶ **Figure 4.50**

Portion of Excel File
Surgery Infections

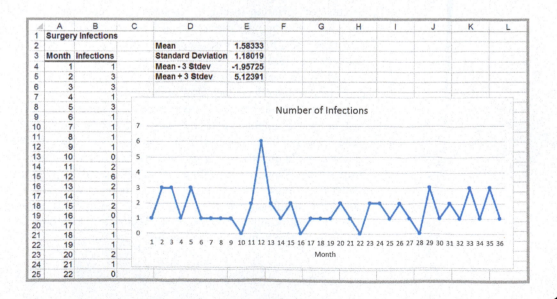

▶ **Figure 4.51**

Infections with Empirical Rule Ranges

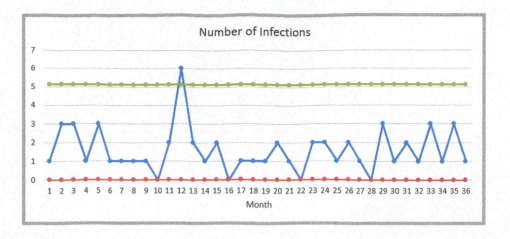

EXAMPLE 4.38   Variation in Sample Data

For the 250 observations in the *Computer Repair Times* database, we can determine that the average repair time is 14.9 days, and the variance of the repair times is 35.50. Suppose we selected some smaller samples from these data. Figure 4.52 shows two samples of size 50 randomly selected from the 250 repair times. Observe that the means and variances

differ from each other as well as from the mean and variance of the entire sample shown in Figure 4.24. In addition, the histograms show a slightly different profile. In Figure 4.53, we show the results for two smaller samples of size 25. Here we actually see *more* variability in both the statistical measures and the histograms as compared with the entire data set.

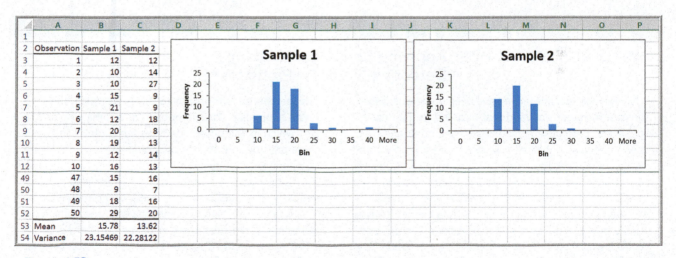

▲ **Figure 4.52**

Two Samples of Size 50 from Computer Repair Times

> This example demonstrates that it is important to understand the variability in sample data and that statistical information drawn from a sample may not accurately represent the population from which it comes. This is one of the most important concepts in applying business analytics. We explore this topic more in Chapter 6.

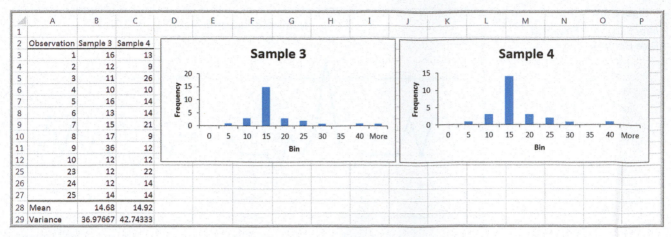

▲ Figure 4.53

Two Samples of Size 25 from Computer Repair Times

CHECK YOUR UNDERSTANDING		

CHECK YOUR UNDERSTANDING

1. What are the principles of statistical thinking?

2. What factors cause variation in sample data?

3. Why is understanding variation in sample data important?

ANALYTICS IN PRACTICE: Applying Statistical Thinking to Detecting Financial Problems[9]

Over the past decade, there have been numerous discoveries of management fraud that have led to the downfall of several prominent companies. These companies had been effective in hiding their financial difficulties, and investors and creditors are now seeking ways to identify financial problems before scandals occur. Even with the passage of the Sarbanes-Oxley Act in July 2002, which helped to improve the quality of the data being disclosed to the public, it is still possible to misjudge an organization's financial strength without analytical evaluation. Several warning signs exist, but there is no systematic and objective way to determine whether a given financial metric, such as a write-off or insider-trading pattern, is high or unusual.

Researchers have proposed using statistical thinking to detect anomalies. They propose an "anomaly detection score," which is the difference between a target financial measure and the company's own past performance or its competitors' current performance using standard deviations. This technique is a variation of a standardized z-score. Specifically, their approach involves comparing performance to past performance (within analysis) and comparing performance to the performance of the company's peers over the same period (between analyses). They created two types of exceptional anomaly scores: z-between (Z_b) to address the variation between companies and z-within (Z_w) to address the variation within the company. These measures quantify the number of standard deviations a

(continued)

[9]Based on Deniz Senturk, Christina LaComb, Radu Neagu, and Murat Doganaksoy, "Detect Financial Problems With Six Sigma," *Quality Progress* (April 2006): 41–47.

company's financial measure deviates from the average. Using these measures, the researchers applied the technique to 25 case studies. These included several high-profile companies that had been charged with financial statement fraud by the SEC or had admitted accounting errors, causing a restatement of their financials. The method was able to identify anomalies for critical metrics known by experts to be warning signs for financial statement fraud. These warning signs were consistent when compared with expert postmortem commentary on the high-profile fraud cases. More importantly, they signaled anomalous behavior at least six quarters before an SEC investigation announcement with fewer than 5% false negatives and 40% false positives.

langstrup/123RF

KEY TERMS

Arithmetic mean (mean)
Bimodal
Categorical (nominal) data
Chebyshev's theorem
Coefficient of kurtosis (CK)
Coefficient of skewness (CS)
Coefficient of variation (CV)
Contingency table
Continuous metric
Correlation
Correlation coefficient (Pearson product moment correlation coefficient)
Covariance
Cross-tabulation
Cumulative relative frequency
Cumulative relative frequency distribution
Data profile (fractile)
Descriptive statistics
Discrete metric
Dispersion
Empirical rules
Frequency distribution
Histogram
Interquartile range (IQR or midspread)
Interval data
kth percentile
Kurtosis

Measure
Measurement
Median
Metric
Midrange
Mode
Ogive
Ordinal data
Outlier
Population
Process capability index
Proportion
Quartile
Range
Ratio data
Relative frequency
Relative frequency distribution
Return to risk
Sample
Sample correlation coefficient
Skewness
Standard deviation
Standardized value (z-score)
Statistic
Statistical thinking
Statistics
Unimodal
Variance

Useful Excel Functions

COUNTIF(*range, cell_reference*) Used to count frequencies to construct a frequency distribution for non-grouped data or to find a proportion.

PERCENTILE.INC(*array, k*) Computes the $100 \times k$th percentile of data in the range specified in the array field, where *k* is between 0 and 1, inclusive.

QUARTILE.INC(*array, quart*) Computes quartiles, where *array* specifies the range of the data and *quart* is a whole number between 1 and 4, designating the desired quartile.

AVERAGE(*data range*) Finds the average, or arithmetic mean.

MEDIAN(*data range*) Finds the median.

MODE.SNGL(*data range*) Finds the mode.

MODE.MULT(*data range*) Finds an array of modes.

VAR.S(*data range*) Computes the sample variance.

VAR.P(*data range*) Computes the population variance.

STDEV.S(*data range*) Computes the sample standard deviation.

STDEV.P(*data range*) Computes the population standard deviation.

STANDARDIZE(*x, mean, standard_dev*) Calculates a standardized value, or *z*-score, for the value *x* given the mean and standard deviation.

SKEW.P(*data range*) Finds the coefficient of skewness for population data.

SKEW(*data range*) Finds the coefficient of skewness for sample data.

KURT(*data range*) Computes the "excess kurtosis" for sample data.

COVARIANCE.P(*array1, array2*) Computes the covariance between two sets of data for a population.

COVARIANCE.S(*array1, array2*) Computes the covariance between two sets of data for a sample.

CORREL(*array1, array2*) Computes the correlation coefficient between two sets of data for either a sample or a population.

Excel Techniques

Histogram tool (Examples 4.5 and 4.6):
Click the *Data Analysis* tools button in the *Analysis* group under the *Data* tab in the Excel menu bar and select *Histogram* from the list. In the dialog box, specify the *Input Range* corresponding to the data and check the *Labels* box if appropriate. Specify a *Bin Range* (suggested). Check the *Chart Output* box to display a histogram in addition to the frequency distribution. Check the other optional boxes to display a Pareto distribution or cumulative frequencies.

Descriptive Statistics tool (Example 4.28):
Each data set to be analyzed must be in a single row or column. Click on *Data Analysis* in the *Analysis* group under the *Data* tab in the Excel menu bar. Select *Descriptive Statistics* from the list. Check the box *Labels in First Row* if labels are included in the input range. Choose whether to save the results in the current worksheet or in a new one. For basic summary statistics, check the box *Summary statistics*.

Displaying statistical information in PivotTables (Example 4.32):
Create a PivotTable. In the *Value Field Settings* dialog, choose the average, standard deviation, or variance of a value field.

Correlation tool (Example 4.35):
Select *Correlation* from the *Data Analysis* tool list. Input the range of the data (which must be in contiguous columns), specify whether the data are grouped by rows or columns, and indicate whether the first row contains data labels. The output provides correlations between every pair of variables.

StatCrunch

StatCrunch provides many tools for computing and visualizing statistical data. You can find video tutorials and step-by-step procedures with examples at https://www.statcrunch.com/5.0/example.php. We suggest that you first view the tutorials *Getting started with StatCrunch* and *Working with StatCrunch sessions*. The following tutorials, which are located under the Graphs, Summary Statistics and Tables, and Regression and Correlation groups on this Web page, explain how to create frequency distributions, histograms, summary statistics, and correlation tables:

■ Frequency tables
■ Histograms
■ Simple bar plots with raw data

- Simple bar plots with summary data
- Summary statistics for columns
- Summary statistics for rows
- Correlations between columns

Example: Create a Bar Plot

1. Choose the *With data* option to use data consisting of individual outcomes in the data table.
   a. Select the column(s) to be displayed.

   b. Enter an optional *Where* statement to specify the data rows to be included.
   c. Select an optional *Group by* column to do a side-by-side bar plot.
2. Choose the *With summary* option to use summary information consisting of categories and counts.
   a. Select the column containing the categories.
   b. Select the column containing the counts.
3. Click *Compute!* to construct the bar plot(s).

PROBLEMS AND EXERCISES

Metrics and Data Classification

1. A survey handed out to individuals at a major shopping mall in a small Florida city in July asked the following:

   - Gender
   - Age
   - Ethnicity
   - Length of residency
   - Overall satisfaction with city services (using a scale of 1–5, ranging from poor to excellent)
   - Quality of schools (using a scale of 1–5, ranging from poor to excellent)

   What types of data (categorical, ordinal, interval, or ratio) do each of the survey items represent and why?

2. Classify each of the data elements in the *Sales Transactions* database as categorical, ordinal, interval, or ratio data and explain why.

3. Identify each of the variables in the Excel file *Credit Approval Decisions* as categorical, ordinal, interval, or ratio and explain why.

4. Identify each of the variables in the Excel file *Corporate Default Database* as categorical, ordinal, interval, or ratio and explain why.

5. Classify each of the variables in the Excel file *Weddings* as categorical, ordinal, interval, or ratio and explain why.

Frequency Distributions and Histograms

6. Use the COUNTIF function to construct a frequency distribution of the types of loans in the Excel file *Credit Risk Data*, develop a column chart to express the results visually, and compute the relative frequencies.

7. Use the COUNTIF function to construct frequency distributions for gender, preferred style, and purchase influence in the Excel file *Retail Survey*, develop column charts to express the results visually, and compute the relative frequencies.

8. A community health status survey obtained the following demographic information from the respondents:

Age	Frequency
18 to 29	250
30 to 45	740
46 to 64	560
65 to 80	370

   Compute the relative frequencies and cumulative relative frequencies of the age groups.

9. Use the *Histogram* tool to construct frequency distributions and histograms for weekly usage and waiting time in the Excel file *Car Sharing Survey*. Do not group the data into bins.

10. Construct frequency distributions and histograms using the Excel *Histogram* tool for the gross sales and gross profit data in the Excel file *Sales Data*. First let Excel automatically determine the number of bins and bin ranges. Then determine a more appropriate set of bins and rerun the *Histogram* tool.

11. Use the *Histogram* tool to develop a frequency distribution and histogram for the number of months as a customer of the bank in the Excel file *Credit Risk Data*. Use your judgment for determining the number of bins to use. Compute the relative and cumulative relative frequencies and use a line chart to construct an ogive.

12. Use the *Histogram* tool to construct frequency distributions and histograms for the numerical data in the Excel file *Cell Phone Survey*. Compute the relative frequencies and cumulative relative frequencies and create charts for ogives.

13. Use a PivotTable to construct a frequency distribution and histogram of lunch sales amounts in the *Restaurant Sales* database.

14. Use a PivotTable to develop a frequency distribution with six bins for the age of individuals in the *Base Data* worksheet in the *Credit Risk Data* file. Compute and chart the relative and cumulative relative frequencies.

15. Use a PivotTable to construct a frequency distribution and histogram for GPA in the Excel file *Grade Point Averages*.

Percentiles and Quartiles

16. Find the 20th and 80th percentiles of home prices in the Excel file *Home Market Value*. Use formula (4.3), the Excel PERCENTILE.INC function, and the *Rank and Percentile* tool, and compare the results.

17. Find the 10th and 90th percentiles and the first and third quartiles for the time difference between the scheduled and actual arrival times in the *Atlanta Airline Data* Excel file. Use formula (4.3), the Excel PERCENTILE.INC function, and the *Rank and Percentile* tool, and compare the results.

18. Find the first, second, and third quartiles for the combined amounts of checking and savings accounts in the Excel file *Credit Risk Data* and interpret the results.

19. Find the first, second, and third quartiles for the sales amounts in the *Sales Transactions* database and interpret the results.

Cross-Tabulations

20. Use a PivotTable to construct a cross-tabulation for loan purpose and credit risk for the *Base Data* worksheet in the Excel file *Credit Risk Data*.

21. Use a PivotTable to construct a cross-tabulation for marital status and housing type for the *Base Data* worksheet in the Excel file *Credit Risk Data*.

22. Use PivotTables to construct cross-tabulations for (1) gender and preferred style and (2) gender and purchase influence in the Excel file *Retail Survey*.

23. Use PivotTables to construct cross-tabulations between each pair of variables in the Excel file *Soda Preferences*.

Descriptive Statistical Measures

24. Find the mean, median, and midrange for the data in the Excel file *Automobile Quality* using the appropriate Excel functions or formulas.

25. In the Excel file *Facebook Survey*, find the mean, median, and midrange for hours online/week and number of friends in the sample using the appropriate Excel function or formulas. Compare these measures of location.

26. Compute the mean, median, midrange, and mode for each of the importance factors in the Excel file *Coffee Shop Preferences* using the appropriate Excel functions.

27. Considering the data in the Excel file *Home Market Value* as a population of homeowners on this street, compute the mean, variance, and standard deviation for each of the variables using a spreadsheet and formulas (4.4), (4.7), and (4.9). Verify your calculations using the appropriate Excel function.

28. Considering the data in the Excel file *Home Market Value* as a sample of homeowners on this street, compute the mean, variance, and standard deviation for each of the variables using formulas (4.5), (4.8), and (4.10). Verify your calculations using the appropriate Excel function.

29. In the Excel file *Facebook Survey*, find the range, variance, standard deviation, and interquartile range for hours online/week and number of friends in the sample using the appropriate Excel functions. Compare these measures of dispersion.

30. Using the appropriate Excel functions, compute the range, variance, standard deviation, and interquartile range for each of the importance factors in the Excel file *Coffee Shop Preferences*.

31. For the Excel file *Tablet Computer Sales*, find the mean, standard deviation, and interquartile range of units sold per week. Show that Chebyshev's theorem holds for the data and determine how accurate the empirical rules are.

32. The Excel file *Atlanta Airline Data* provides arrival and taxi-in time statistics for one day at Atlanta Hartsfield International Airport. Find the mean and standard

deviation of the difference between the scheduled and actual arrival times and the taxi-in time to the gate. Compute the z-scores for each of these variables.

33. Compute the mean and standard deviation of the data in the *Cost of Living Adjustments* Excel file. Then compute the z-scores for the comparative salaries and housing adjustments and interpret your results.

34. Compute the coefficient of variation for each variable in the Excel file *Home Market Value*. Which has the least and greatest relative dispersion?

35. Find 30 days of stock prices for three companies in different industries. The average stock prices should have a wide range of values. Using the data, compute and interpret the coefficient of variation.

36. Apply the *Descriptive Statistics* tool for subsets of liberal arts colleges and research universities in the Excel file *Colleges and Universities*. Compare the two types of colleges. What can you conclude?

37. Use the *Descriptive Statistics* tool to summarize the percent gross profit, gross sales, and gross profit in the Excel file *Sales Data*.

38. Use the *Descriptive Statistics* tool to summarize the responses in the Excel file *Job Satisfaction*. What information can you conclude from this analysis?

39. The *Data* worksheet in the Excel file *Airport Service Times* lists a large sample of the time in seconds to process customers at a ticket counter. The second worksheet shows a frequency distribution and histogram of the data.
    a. Summarize the data using the *Descriptive Statistics* tool. What can you say about the shape of the distribution of times?
    b. Find the 90th percentile.
    c. How might the airline use these results to manage its ticketing counter operations?

Computing Descriptive Statistics for Frequency Distributions

40. Construct a frequency distribution for overall satisfaction in the Excel file *Helpdesk Survey*. Develop a spreadsheet to estimate the sample mean and sample variance using formulas (4.17) and (4.19). Check your results using Excel functions with the original data.

41. In Problem 9, we asked you to use the *Histogram* tool to construct frequency distributions and histograms for weekly usage and waiting time in the

Excel file *Car Sharing Survey* without grouping the data into bins. Use your results and formulas (4.17) and (4.19) to find the mean and sample variance. Check your results using Excel functions with the original data.

42. A community health status survey obtained the following demographic information from the respondents:

Age	Frequency
18 to 29	250
30 to 45	740
46 to 64	560
65 to 80	370

Develop a spreadsheet to estimate the sample mean and sample standard deviation of age using formulas (4.17) and (4.19).

43. A marketing study of 800 adults in the 18–34 age group reported the following information:
    - Spent less than $100 on children's clothing per year: 50 responses
    - Spent $100–$499.99 on children's clothing per year: 275 responses
    - Spent $500–$999.99 on children's clothing per year: 175 responses
    - Spent nothing: the remainder

    Develop a spreadsheet to estimate the sample mean and sample standard deviation of spending on children's clothing for this age group using formulas (4.21) and (4.23).

44. The data in the Excel file *Church Contributions* were reported on annual giving for a church. Estimate the mean and standard deviation of the annual contributions by implementing formulas (4.20) and (4.22) on a spreadsheet, assuming these data represent the entire population of parishioners.

Descriptive Statistics for Categorical Data: The Proportion

45. The Excel file *EEO Employment Report* shows the number of people employed in different professions for various racial and ethnic groups. For all employees and each racial/ethnic group, find the proportions of men and women in each profession.

46. In the Excel file *Bicycle Inventory*, find the proportion of bicycle models that sell for less than $200.

47. In the Excel file *Laptop Survey*, find the proportion of respondents who own Dell, Apple, and HP computers.

48. In the *Sales Transactions* database, find the proportion of customers who used PayPal, the proportion of customers who used credit cards, the proportion that purchased a book, and the proportion that purchased a DVD.

49. In the Excel file *Economic Poll*, find the proportions of each categorical variable.

Statistics in PivotTables

50. Create a PivotTable to find the mean and standard deviation of the amount of travel expenses for each sales representative in the Excel file *Travel Expenses*.

51. Create a PivotTable for the data in the Excel file *Weddings* to analyze the average wedding cost by type of payor and value rating. What conclusions do you reach?

52. The Excel File *Rin's Gym* provides sample data on member body characteristics and gym activity. Create PivotTables to find the following:
    a. Cross-tabulations of gender versus body type.
    b. Mean running times, run distance, weight-lifting days, lifting session times, and time spent in the gym by gender. Summarize your conclusions.

53. Use a PivotTable to find the mean annual income by level of education for the data in the Excel file *Education and Income*.

54. In the Excel file *Debt and Retirement Savings*, use a PivotTable to find the mean and standard deviation of income, long-term debt, and retirement savings for both single and married individuals.

55. Using PivotTables, find the mean and standard deviation of sales by region in the *Sales Transactions* database.

56. The Excel file *Freshman College Data* shows data for four years at a large urban university. Use PivotTables to examine differences in high school GPA performance and first-year retention among different colleges at this university. What conclusions do you reach?

57. Call centers have high turnover rates because of the stressful environment. The national average is approximately 50%. The director of human resources for a large bank has compiled data from about 70 former employees at one of the bank's call centers (see the Excel file *Call Center Data*). Use PivotTables to find these statistics:
    a. The average length of service for males and females in the sample
    b. The average length of service for individuals with and without a college degree
    c. The average length of service for males and females with and without prior call center experience

58. For the *Peoples Choice Bank* database, use PivotTables to find the average transaction amount for each account and branch.

59. A national homebuilder builds single-family homes and condominium-style townhouses. The Excel file *House Sales* provides information on the selling price, lot cost, type of home, and region of the country for closings during one month. Use PivotTables to find the average selling price and lot cost for each type of home in each region of the market. What conclusions might you reach from this information?

Measures of Association

60. In the Excel file *Weddings*, determine the correlation between the wedding costs and attendance.

61. For the data in the Excel file *Rin's Gym*, find the sample covariances and correlations among height, weight, and calculated BMI.

62. The Excel file *Beverage Sales* lists a sample of weekday sales at a convenience store, along with the daily high temperature. Compute the covariance and correlation between temperature and sales.

63. For the *President's Inn Guest Database*, find the mean length of stay and mean number of guests per party. Is there any correlation between the size of the party and the length of stay?

64. For the Excel file *TV Viewing Survey*, is there a significant correlation between (1) the number of TVs in a home and the hours of viewing per week, and (2) age and hours of viewing per week?

65. For the Excel file *Credit Risk Data*, compute the correlation between age and months employed, age and combined checking and savings account balance, and the number of months as a customer and amount of money in the bank. Interpret your results.

Outliers

66. Compute the z-scores for the data in the Excel file *Airport Service Times*. How many observations fall farther than three standard deviations from either side of the mean? Would you consider these as outliers? Why or why not?

67. Examine the z-scores you computed in Problem 32 for the *Atlanta Airline Data*. Do they suggest any outliers in the data?

68. In the Excel file *Weddings*, find the mean, median, and standard deviation of the wedding costs. What would you tell a newly engaged couple about what cost to expect? Consider the effect of possible outliers in the data.

Using Descriptive Statistics to Analyze Survey Data

69. Use the following approaches to analyze the survey data in the Excel file *Insurance Survey*.

    - Frequency distributions and histograms for the ratio variables (age, years employed, and satisfaction)
    - Descriptive statistical measures for the ratio variables using the *Descriptive Statistics* tool
    - Proportions for various attributes of the categorical variables in the sample, such as the proportion of males and females, different educational levels, and marital status
    - PivotTables that break down the averages of ratio variables by gender, education, and marital status
    - Cross-tabulations by gender and education, gender and marital status, and education and marital status
    - Z-scores for examination of potential outliers

70. The Excel file *Auto Survey* contains a sample of data about vehicles owned, whether they were purchased new or used, and other types of data. Use the appropriate statistical tools to analyze these data. Summarize the observations that you can make from these results.

71. A producer of computer-aided design software for the aerospace industry receives numerous calls for technical support. Tracking software is used to monitor response and resolution times. In addition, the company surveys customers who request support using the following scale:

    0—Did not meet expectations

    1—Marginally met expectations

    2—Met expectations

    3—Exceeded expectations

    4—Greatly exceeded expectations

    The questions are as follows:

    Q1: Did the support representative explain the process for resolving your problem?

    Q2: Did the support representative keep you informed about the status of progress in resolving your problem?

    Q3: Was the support representative courteous and professional?

    Q4: Was your problem resolved?

    Q5: Was your problem resolved in an acceptable amount of time?

    Q6: Overall, how did you find the service provided by our technical support department?

    A final question asks the customer to rate the overall quality of the product using a scale: 0—very poor, 1—poor, 2—good, 3—very good, 4—excellent. A sample of survey responses and associated resolution and response data are provided in the Excel file *Customer Support Survey*. Use whatever Excel tools you deem appropriate to analyze these sample data and write a report to the manager explaining your findings and conclusions.

72. The Excel file *News Preferences* provides data about how individuals prefer to get news and what types of news they typically read. Use PivotTables and slicers to draw conclusions about these data from a statistical perspective, and summarize your results in a brief memo.

Statistical Thinking in Business Decisions

73. Use the *Manufacturing Measurements* data to compute sample means, assuming that each row in the data file represents a sample from the manufacturing process. Plot the sample means on a line chart, compute the standard deviation of the sample means (not the individual observations!), add empirical rule ranges for the sample means, and interpret your results.

74. A Midwest pharmaceutical company manufactures individual syringes with a self-contained, single dose of an injectable drug.[10] In the manufacturing process, sterile liquid drug is poured into glass syringes and sealed with a rubber stopper. The remaining stage involves insertion of the cartridge into plastic syringes and the electrical "tacking" of the containment cap at a precisely determined length of the syringe. A cap that

---

[10]Based on LeRoy A. Franklin and Samar N. Mukherjee, "An SPC Case Study on Stabilizing Syringe Lengths," *Quality Engineering* 12, 1 (1999–2000): 65–71.

is tacked at a shorter-than-desired length (less than 4.920 inches) leads to pressure on the cartridge stopper and, hence, partial or complete activation of the syringe. Such syringes must then be scrapped. If the cap is tacked at a longer-than-desired length (4.980 inches or longer), the tacking is incomplete or inadequate, which can lead to cap loss and a potential cartridge loss in shipment and handling. Such syringes can be reworked manually to attach the cap at a lower position. However, this process requires a 100% inspection of the tacked syringes and results in increased cost for the items. This final production step seemed to be producing more and more scrap and reworked syringes over successive weeks.

The Excel file *Syringe Samples* provides samples taken every 15 minutes from the manufacturing process. Use statistical analysis and statistical-thinking concepts to draw conclusions.

75. Find the mean and variance of a deck of 52 cards, where an ace is counted as 11 and a picture card as 10. Construct a frequency distribution and histogram of the card values. Shuffle the deck and deal two samples of 20 cards (starting with a full deck each time); compute the mean and variance and construct a histogram. How does the sample data differ from the population data? Repeat this experiment for two samples of 5 cards and summarize your conclusions.

CASE: DROUT ADVERTISING RESEARCH PROJECT

The background for this case was introduced in Chapter 2. For this part of the case, summarize the numerical data using frequency distributions and histograms, cross-tabulations, PivotTables, and descriptive statistics measures; find proportions for categorical variables; examine correlations; and so on. Write up your findings in a formal document, or add your findings to the report you completed for the case in Chapter 2 at the discretion of your instructor.

CASE: PERFORMANCE LAWN EQUIPMENT

Elizabeth Burke has received several questions from other PLE managers regarding quality, customer satisfaction, and operational performance. She would like you to summarize some data from the *Performance Lawn Equipment Database* using statistical tools and analysis:

1. Frequency distributions and histograms for the data in the *Customer Survey* worksheet
2. Descriptive statistical measures for engine production time in the worksheet *Engines*
3. A frequency distribution and histogram for the blade weight samples in the worksheet *Blade Weight*

4. The proportion of samples that failed the functional performance test in the worksheet *Mower Test*
5. PivotTables that summarize the data and provide useful insights in the worksheet *Employee Retention*
6. Correlations among the satisfaction survey variables in the worksheet *Purchasing Survey*

Write a report that clearly shows these results and explain the key insights.

Additional Charts for Descriptive Statistics in Excel for Windows

Excel 2016 for Windows provides two additional charts for descriptive statistics that are not available on the Mac platform. The first is the *Histogram* chart, which can be chosen from the *Insert > Chart* option. You may customize the number of bins and bin widths by right clicking the *x*-axis and choosing *Format Axis* from the pop-up menu. Figure A4.1 shows an example for the cost per order in the *Purchase Orders* database. The histogram is patterned after Figure 4.11. Select the bin width to be 26,000, and choose the underflow bin (which is the first group) to be the upper group limit of the first bin (26,000), and the overflow bin to be the upper group limit of the last bin (130,000). One useful feature of this chart is that you can easily change the number of bins to create the most visually appealing histogram.

The second chart is called a **box-and-whisker** chart. A box-and-whisker chart displays five key statistics of the data set: the minimum value, first quartile, median, third

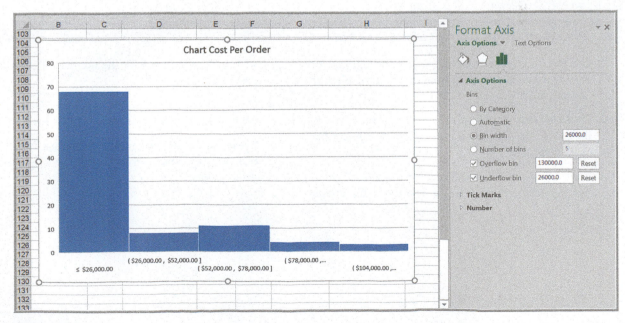

▶ **Figure A4.1**

Histogram of Cost per Order for Purchase Orders *Database*

quartile, and maximum value. Figure A4.2 shows a box-and-whisker chart for the sales data in the *Monthly Product Sales* database. The "whiskers" extend on either side of the box to show the minimum and maximum values in the data set. The box encloses the first and third quartiles (that is, the interquartile range, IQR), and the line inside the box shows the median value. Very long whiskers suggest possible outliers in the data. Box-and-whiskers charts show the relative dispersion in the data and the general shape of the distribution (see the discussion of skewness in the chapter). For example, the distribution of product E is relatively symmetric, while that of product D shows more skewness.

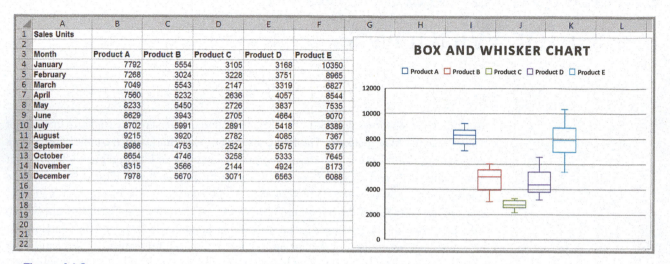

▲ **Figure A4.2**

Box-and-Whisker Chart

PROBLEMS AND EXERCISES

1. Use the *Histogram* chart to construct histograms for weekly usage and waiting time in the Excel file *Car Sharing Survey*.

2. Construct histograms using the *Histogram* chart for the gross sales and gross profit data in the Excel file *Sales Data*.

3. For the *Base Data* in the Excel file *Credit Risk Data*, use the *Histogram* chart to develop a histogram with a bin width of 5 for the number of months as a customer of the bank.

4. Create box-and-whisker charts for each type of expense in the Excel file *Budget Forecasting*.

5. Create box-and-whisker charts for each type of cost of living adjustment in the Excel file *Cost of Living Adjustments*.

Probability Distributions and Data Modeling

PeterVrabel/Shutterstock

LEARNING OBJECTIVES   After studying this chapter, you will be able to:

- Explain the concept of probability and provide examples of the three definitional perspectives of probability.
- Use probability rules and formulas to perform probability calculations.
- Explain conditional probability and how it can be applied in a business context.
- Compute conditional probabilities from cross-tabulation data.

- Determine if two events are independent using probability arguments.
- Apply the multiplication law of probability.
- Explain the difference between a discrete and a continuous random variable.
- Define a probability distribution.
- Verify the properties of a probability mass function.
- Use the cumulative distribution function to compute probabilities over intervals.
- Compute the expected value and variance of a discrete random variable.
- Use expected values to support simple business decisions.
- Calculate probabilities for the Bernoulli, binomial, and Poisson distributions, using the probability mass function and Excel functions.

- Explain how a probability density function differs from a probability mass function.
- List the key properties of probability density functions.
- Use the probability density and cumulative distribution functions to calculate probabilities for a uniform distribution.
- Describe the normal and standard normal distributions and use Excel functions to calculate probabilities.
- Use the standard normal distribution table and z-values to compute normal probabilities.
- Describe properties of the exponential and triangular distributions and compute probabilities.
- Explain the concepts of distribution fitting and data modeling.
- Apply the Chi Square Goodness of Fit test.

Most business decisions involve some elements of uncertainty and randomness. For example, the times to repair computers in the *Computer Repair Times* Excel file that we discussed in Chapter 4 showed quite a bit of uncertainty that we needed to understand to provide information to customers about their computer repairs. We also saw that different samples of repair times result in different means, variances, and frequency distributions. Therefore, it would be beneficial to be able to identify some general characteristics of repair times that would apply to the entire population—even those repairs that have not yet taken place. In other situations, we may not have any data for analysis and simply need to make some judgmental assumptions about future uncertainties. For example, to develop a model to predict the profitability of a new and innovative product, we would need to make reliable assumptions about sales and consumer behavior without any prior data on which to base them. Characterizing the nature of distributions of data and specifying uncertain assumptions in decision models relies on fundamental knowledge of probability concepts and probability distributions—the subject of this chapter.

Understanding probability and probability distributions is important in all areas of business. For example, marketing analysts who wish to predict future sales might use probability to assess the likelihood that consumers will purchase their products. In finance, probability is useful in assessing risk of capital investments or product development initiatives. Operations managers apply it routinely in quality control, inventory management, design for product reliability, and customer service policies. Company executives use probability to make competitive decisions and analyze long-term strategies. Coaches and managers of sports teams use probability to make tactical decisions, such as the best football play to make when faced with third down and 4 yards to go on the opponent's 17-yard line. You use probability concepts in daily life more than you probably realize, such as when deciding whether or not to play golf or go to the beach based on the weather forecast. To experience using probability to make decisions in a game setting, check out one of the author's favorite games, Qwixx!

Basic Concepts of Probability

The notion of probability is used everywhere, both in business and in our daily lives, from market research and stock market predictions to the World Series of Poker and weather forecasts. In business, managers need to know such things as the likelihood that a new product will be profitable or the chances that a project will be completed on time. Probability quantifies the uncertainty that we encounter all around us and is an important building block for business analytics applications. **Probability** is the likelihood that an outcome—such as whether a new product will be profitable or not or whether a project will be completed within 15 weeks—occurs. Probabilities are expressed as values between 0 and 1, although many people convert them to percentages. The statement that there is a 10% chance that oil prices will rise next quarter is another way of stating that the probability of a rise in oil prices is 0.1. The closer the probability is to 1, the more likely it is that the outcome will occur.

Experiments and Sample Spaces

To formally discuss probability, we need some new terminology. An **experiment** is a process that results in an outcome. An experiment might be as simple as rolling two dice, observing and recording weather conditions, conducting a market research study, or watching the stock market. The **outcome** of an experiment is a result that we observe; it might be the sum of two dice, a description of the weather, the proportion of consumers who favor a new product, or the change in the Dow Jones Industrial Average (DJIA) at the end of a week. The collection of all possible outcomes of an experiment is called the **sample space**. For instance, if we roll two fair dice, the possible outcomes for the sum are the numbers 2 through 12; if we observe the weather, the outcome might be clear, partly cloudy, or cloudy; the outcomes for customer reaction to a new product in a market research study would be favorable or unfavorable, and the weekly change in the DJIA can theoretically be any real number (positive, negative, or zero). Note that a sample space may consist of a small number of discrete outcomes or an infinite number of outcomes.

Combinations and Permutations

Enumerating and counting the outcomes for an experiment can sometimes be difficult, particularly when the experiment consists of multiple steps. A bit of logic and visualization often helps.

EXAMPLE 5.1   Rolling Two Dice

Suppose we roll two dice. The first roll can be 1, 2, 3, 4, 5, or 6. For each of these outcomes, the second roll can also be 1, 2, 3, 4, 5, or 6. Thus, the outcomes of the experiment are (1, 1), (1, 2), (1, 3), . . . , (6, 4), (6, 5), (6, 6). We can visualize this as a tree diagram in Figure 5.1. A **tree diagram** is a visualization of a multistep experiment. Counting the outcomes, we find there are 36.

In general, for an experiment with k steps, the number of outcomes is

$$n_1 \times n_2 \times \ldots \times n_k \qquad (5.1)$$

where n_i is the number of possible outcomes in step i. Applying this rule to the dice rolls, we have $n_1 = 6$ and $n_2 = 6$; therefore, the total number of outcomes is $6 \times 6 = 36$.

In some experiments, we want to select n objects from a set of N objects.

EXAMPLE 5.2   Selecting *n* Objects from *N*

In a group of five students, an instructor might wish to select three of them to make presentations. How many different ways can this be done? Note that the same student cannot be selected more than once. You might use the analogy of drawing three cards from a deck of five without replacing

the cards that are drawn. The first student selected can be student 1, 2, 3, 4, or 5. However, if student 1 is selected first, then the second student can be 2, 3, 4, or 5. Then, if the second student selected is student 4, the third student can be 2, 3, or 5. (Try to draw a tree diagram for this!)

▶ **Figure 5.1**

Tree Diagram for Rolling Two Dice

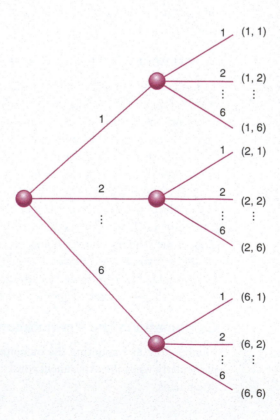

First Roll   Second Roll   Outcome

Counting the number of outcomes is not as easy as you might think. Your first thought might be to use formula (5.1) for Example 5.2: We can choose five outcomes in the first step, four in the second, and three in the third, which would result in $5 \times 4 \times 3 = 60$. However, many of these are duplicates, for instance, (1, 2, 3), (1, 3, 2), (2, 1, 3), (2, 3, 1), (3, 1, 2), and (3, 2, 1). Since the order does not matter, we only want to count unique outcomes, which we call **combinations**. The number of combinations for selecting n objects from a set of N is

$$C(n, N) = \binom{N}{n} = \frac{N!}{n!(N - n)!} \tag{5.2}$$

The notation ! means factorial, and any number $x!$ is computed as $x \times (x - 1) \times (x - 2)\ldots \times 2 \times 1$, where x is a nonnegative integer. For instance, $4! = 4 \times 3 \times 2 \times 1 = 24$. Zero factorial, $0!$, is defined to be 1.

EXAMPLE 5.3   **Applying the Combinations Formula**

In Example 5.2, the number of ways of selecting three students from a group of five is

$$C(3, 5) = \binom{5}{3} = \frac{5!}{3!(5 - 3)!} = \frac{(5)(4)(3)(2)(1)}{3(2)(1) \times (2)(1)} = 10$$

If we want to select n objects from N *and* the order is important, then we call the outcomes **permutations**. In general, the number of permutations of n objects selected from N is

$$P(n, N) = n!\binom{N}{n} = \frac{N!}{(N - n)!} \tag{5.3}$$

EXAMPLE 5.4   **Applying the Permutations Formula**

In Example 5.2, suppose we want to count the number of ways of selecting three students from a group of five where the order is important (for instance, knowing which student presents first, second, and third). Applying formula (5.3), we have

The permutations formula is easier to apply because we don't have to think of how many outcomes can occur in each step of the process, particularly when the number of steps is large.

$$P(3, 5) = 3!\binom{5}{3} = \frac{5!}{(5 - 3)!} = \frac{(5)(4)(3)(2)(1)}{(2)(1)} = 60$$

Probability Definitions

Probability may be defined from one of three perspectives. First, if the process that generates the outcomes is known, probabilities can be deduced from theoretical arguments; this is the *classical definition* of probability.

EXAMPLE 5.5   Classical Definition of Probability

Suppose we roll two dice. If we examine the outcomes described in Example 5.1 and add the values of the dice, we may find the probabilities of rolling each value between 2 and 12. The probability of rolling any number is the ratio of the number of ways of rolling that number to the total number of possible outcomes. For instance, the probability of rolling a 2 is 1/36, the probability of rolling a 3 is 2/36 = 1/18, and the probability of rolling a 7 is 6/36 = 1/6. Similarly, if two consumers are asked whether

or not they like a new product, there could be four possible outcomes:

1. (like, like)
2. (like, dislike)
3. (dislike, like)
4. (dislike, dislike)

If these are assumed to be equally likely, the probability that *at least* one consumer would respond unfavorably is 3/4.

The second approach to probability, called the *relative frequency definition*, is based on empirical data. The probability that an outcome will occur is simply the relative frequency associated with that outcome.

EXAMPLE 5.6   Relative Frequency Definition of Probability

Using the sample of computer repair times in the Excel file *Computer Repair Times*, we developed the relative frequency distribution in Chapter 4, shown again in Figure 5.2. We could state that the probability that a computer would be repaired in as little as four days is 0, the probability that

it would be repaired in exactly ten days is 0.076, and so on. In using the relative frequency definition, it is important to understand that as more data become available, the distribution of outcomes and, hence, the probabilities may change.

Finally, the *subjective definition* of probability is based on judgment and experience, as financial analysts might use in predicting a 75% chance that the DJIA will increase 10% over the next year, or as sports experts might predict at the start of the football season, a 1-in-5 chance (0.20 probability) of a certain team making it to the Super Bowl.

Which definition to use depends on the specific application and the information we have available. We will see various examples that draw upon each of these perspectives.

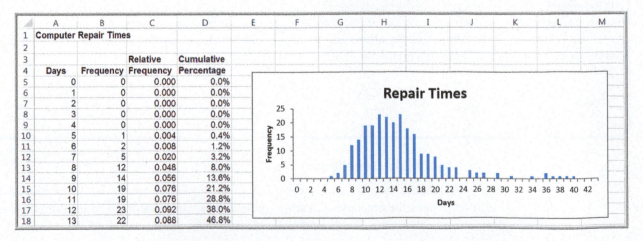

▶ **Figure 5.2**

Distribution of Computer Repair Times

Probability Rules and Formulas

Suppose we label the n outcomes in a sample space as $O_1, O_2, \ldots, O_n$, where O_i represents the ith outcome in the sample space. Let $P(O_i)$ be the probability associated with the outcome O_i. Two basic facts govern probability:

- The probability associated with any outcome must be between 0 and 1, inclusive, or

$$0 \leq P(O_i) \leq 1 \text{ for each outcome } O_i \tag{5.4}$$

- The sum of the probabilities over all possible outcomes must be 1, or

$$P(O_1) + P(O_2) + \cdots + P(O_n) = 1 \tag{5.5}$$

An **event** is a collection of one or more outcomes from a sample space. Examples of events are rolling a sum of 7 or 11 with two dice, completing a computer repair in between 7 and 14 days, and obtaining a positive weekly change in the DJIA. This leads to the following rule:

Rule 1. The probability of any event is the sum of the probabilities of the outcomes that comprise that event.

| EXAMPLE 5.7 | Computing the Probability of an Event |

Consider the event of rolling a 7 or 11 on two dice. The probability of rolling a 7 is $\frac{6}{36}$ and the probability of rolling an 11 is $\frac{2}{36}$; thus, the probability of rolling a 7 or 11 is $\frac{6}{36} + \frac{2}{36} = \frac{8}{36}$. Similarly, the probability of repairing a computer in seven days or less is the sum of the probabilities of the outcomes

$O_1 = 0, O_2 = 1, O_3 = 2, O_4 = 3, O_5 = 4, O_6 = 5, O_7 = 6,$ and $O_8 = 7$ days, or $0.004 + 0.008 + 0.020 = 0.032$ (note that the probabilities $P(O_1) = P(O_2) = P(O_3) = P(O_4) = P(O_5) = 0$; see Figure 5.2).

If A is any event, the **complement** of A, denoted A^c, consists of all outcomes in the sample space not in A.

Rule 2. The probability of the complement of any event A is $P(A^c) = 1 - P(A)$.

| EXAMPLE 5.8 | Computing the Probability of the Complement of an Event |

If $A = \{7, 11\}$ in the dice example, then $A^c = \{2, 3, 4, 5, 6, 8, 9, 10, 12\}$. Thus, the probability of rolling any sum other than a 7 or 11 is $P(A^c) = 1 - \frac{8}{36} = \frac{28}{36}$. If $A = \{0, 1, 2, 3, 4, 5, 6, 7\}$ in the computer repair example,

$A^c = \{8, 9, \ldots, 42\}$ and $P(A^c) = 1 - 0.032 = 0.968$. This is the probability of completing the repair in more than a week.

The **union** of two events contains all outcomes that belong to either of the two events. To illustrate this with rolling the sum of dice, let A be the event $\{7, 11\}$ and B be the event $\{2, 3, 12\}$. The union of A and B is the event $\{2, 3, 7, 11, 12\}$. If A and B are two events, the probability that some outcome in either A or B (that is, the union of A and B) occurs is denoted as $P(A \text{ or } B)$. Finding this probability depends on whether the events are mutually exclusive or not.

Two events are **mutually exclusive** if they have no outcomes in common. The events A and B in the dice example are mutually exclusive. When events are mutually exclusive, the following rule applies:

Rule 3. If events A and B are mutually exclusive, then $P(A \text{ or } B) = P(A) + P(B)$.

EXAMPLE 5.9   Computing the Probability of Mutually Exclusive Events

For the dice example, the probability of event $A = \{7, 11\}$ is $P(A) = \frac{8}{36}$, and the probability of event $B = \{2, 3, 12\}$ is $P(B) = \frac{4}{36}$. Therefore, the probability that either event A or B occurs, that is, the sum of the dice is 2, 3, 7, 11, or 12, is $\frac{8}{36} + \frac{4}{36} = \frac{12}{36}$.

If two events are *not* mutually exclusive, then adding their probabilities would result in double-counting some outcomes, so an adjustment is necessary. This leads to the following rule:

> **Rule 4.** If two events A and B are not mutually exclusive, then $P(A \text{ or } B) = P(A) + P(B) - P(A \text{ and } B)$.

Here, $(A \text{ and } B)$ represents the **intersection** of events A and B—that is, all outcomes belonging to both A and B.

EXAMPLE 5.10   Computing the Probability of Non mutually Exclusive Events

In the dice example, let us define the events $A = \{2, 3, 12\}$ and $B = \{\text{even number}\}$. Then A and B are not mutually exclusive because the events have the numbers 2 and 12 in common. Thus, the intersection $(A \text{ and } B) = \{2, 12\}$. Therefore, $P(A \text{ or } B) = P(\{2, 3, 12\}) + P(\text{even number}) - P(A \text{ and } B) = \frac{4}{36} + \frac{18}{36} - \frac{2}{36} = \frac{20}{36}$.

Joint and Marginal Probability

In many applications, more than one event occurs simultaneously, or in statistical terminology, *jointly*. We will only discuss the simple case of two events. For instance, suppose that a sample of 100 individuals were asked to evaluate their preference for three new proposed energy drinks in a blind taste test. The sample space consists of two types of outcomes corresponding to each individual: gender (F = female or M = male) and brand preference (B_1, B_2, or B_3). We may define a new sample space consisting of the outcomes that reflect the different combinations of outcomes from these two sample spaces. Thus, for any respondent in the blind taste test, we have six possible (mutually exclusive) combinations of outcomes:

1. O_1 = the respondent is female and prefers brand 1
2. O_2 = the respondent is female and prefers brand 2
3. O_3 = the respondent is female and prefers brand 3
4. O_4 = the respondent is male and prefers brand 1
5. O_5 = the respondent is male and prefers brand 2
6. O_6 = the respondent is male and prefers brand 3

Here, the probability of each of these events is the intersection of the gender and brand preference event. For example, $P(O_1) = P(F \text{ and } B_1)$, $P(O_2) = P(F \text{ and } B_2)$, and so on. The probability of the intersection of two events is called a **joint probability**. The probability of an event, irrespective of the outcome of the other joint event, is called a **marginal probability**. Thus, $P(F)$, $P(M)$, $P(B_1)$, $P(B_2)$, and $P(B_3)$ would be marginal probabilities.

EXAMPLE 5.11 Applying Probability Rules to Joint Events

Figure 5.3 shows a portion of the data file *Energy Drink Survey*, along with a cross-tabulation constructed from a Pivot-Table. The joint probabilities of gender and brand preference are easily calculated by dividing the number of respondents corresponding to each of the six outcomes listed above O sub 1 through O sub 6 by the total number of respondents, 100. Thus, $P(F \text{ and } B_1) = P(O_1) = 9/100 = 0.09$, $P(F \text{ and } B_2) = P(O_2) = 6/100 = 0.06$, and so on. Note that the sum of the probabilities of all these outcomes is 1.

We see that the event F (respondent is female) is composed of the outcomes O_1, O_2, and O_3, and therefore $P(F) = P(O_1) + P(O_2) + P(O_3) = 0.37$ using rule 1. The complement of this event is M; that is, the respondent is male. Note that $P(M) = 0.63 = 1 - P(F)$, as reflected by rule 2. The event B_1 is composed of the outcomes O_1 and O_4, and thus, $P(B_1) = P(O_1) + P(O_4) = 0.34$. Similarly, we find that $P(B_2) = 0.23$ and $P(B_3) = 0.43$.

Events F and M are mutually exclusive, as are events B_1, B_2, and B_3, since a respondent may be only male or

female and prefer exactly one of the three brands. We can use rule 3 to find, for example, $P(B_1 \text{ or } B_2) = 0.34 + 0.23 = 0.57$. Events F and B_1, however, are not mutually exclusive because a respondent can both be female and prefer brand 1. Therefore, using rule 4, we have $P(F \text{ or } B_1) = P(F) + P(B_1) - P(F \text{ and } B_1) = 0.37 + 0.34 - 0.09 = 0.62$.

The joint probabilities can easily be computed, as we have seen, by dividing the values in the cross-tabulation by the total, 100. Below the PivotTable in Figure 5.3 is a **joint probability table**, which summarizes these joint probabilities.

The marginal probabilities are given in the margins of the joint probability table by summing the rows and columns. Note, for example, that $P(F) = P(F \text{ and } B_1) + P(F \text{ and } B_2) + P(F \text{ and } B_3) = 0.09 + 0.06 + 0.22 = 0.37$. Similarly, $P(B_1) = P(F \text{ and } B_1) + P(M \text{ and } B_1) = 0.09 + 0.25 = 0.34$.

This discussion of joint probabilities leads to the following probability rule:

Rule 5. If event A is comprised of the outcomes $\{A_1, A_2, \ldots, A_n\}$ and event B is comprised of the outcomes $\{B_1, B_2, \ldots, B_n\}$, then

$$P(A_i) = P(A_i \text{ and } B_1) + P(A_i \text{ and } B_2) + \cdots + P(A_i \text{ and } B_n)$$
$$P(B_i) = P(A_1 \text{ and } B_i) + P(A_2 \text{ and } B_i) + \cdots + P(A_n \text{ and } B_i)$$

	A	B	C	D	E	F	G	H	I
1	Energy Drink Survey								
2									
3	Respondent	Gender	Brand Preference						
4	1	Male	Brand 3		Count of Respondent	Column Labels			
5	2	Female	Brand 3		Row Labels	Brand 1	Brand 2	Brand 3	Grand Total
6	3	Male	Brand 3		Female	9	6	22	37
7	4	Male	Brand 1		Male	25	17	21	63
8	5	Male	Brand 1		Grand Total	34	23	43	100
9	6	Female	Brand 2						
10	7	Male	Brand 2						
11	8	Female	Brand 2		Joint Probability Table	Brand 1	Brand 2	Brand 3	Grand Total
12	9	Male	Brand 1		Female	0.09	0.06	0.22	0.37
13	10	Female	Brand 3		Male	0.25	0.17	0.21	0.63
14	11	Male	Brand 3		Grand Total	0.34	0.23	0.43	1
15	12	Male	Brand 2						
16	13	Female	Brand 3						

▲ **Figure 5.3**

Portion of Excel File Energy Drink Survey

Conditional Probability

Conditional probability is the probability of occurrence of one event A, given that another event B is known to be true or has already occurred.

EXAMPLE 5.12   Computing a Conditional Probability in a Cross-Tabulation

We will use the information shown in the energy drink survey example in Figure 5.3 to illustrate how to compute conditional probabilities from a cross-tabulation or joint probability table.

Suppose that we know that a respondent is male. What is the probability that he prefers brand 1? From the Pivot-Table, note that there are only 63 males in the group and

of these, 25 prefer brand 1. Therefore, the probability that a male respondent prefers brand 1 is $\frac{25}{63}$. We could have obtained the same result from the joint probability table by dividing the joint probability 0.25 (the probability that the respondent is male and prefers brand 1) by the marginal probability 0.63 (the probability that the respondent is male).

Conditional probabilities are useful in analyzing data in cross-tabulations, as well as in other types of applications. Many companies save purchase histories of customers to predict future sales. Conditional probabilities can help to predict future purchases based on past purchases.

EXAMPLE 5.13   Conditional Probability in Marketing

The Excel file *Apple Purchase History* presents a hypothetical history of consumer purchases of Apple products, showing the first and second purchase for a sample of 200 customers who have made repeat purchases (see Figure 5.4). The PivotTable in Figure 5.5 shows the count of the type of second purchase given that each product was purchased first. For example, 13 customers purchased iMacs as their first Apple product. Then the conditional

probability of purchasing an iPad given that the customer first purchased an iMac is $\frac{2}{13} = 0.15$. Similarly, 74 customers purchased a MacBook as their first purchase; the conditional probability of purchasing an iPhone if a customer first purchased a MacBook is $\frac{26}{74} = 0.35$. By understanding which products are more likely to be purchased by customers who already own other products, companies can better target advertising strategies.

▶ **Figure 5.4**

Portion of Excel File Apple Purchase History

	A	B
1	Apple Products Purchase History	
2		
3	**First Purchase**	**Second Purchase**
4	iPod	iMac
5	iPhone	MacBook
6	iMac	iPhone
7	iPhone	iPod
8	iPod	iPhone
9	MacBook	iPod
10	iPhone	MacBook
11	MacBook	iPhone
12	iPod	MacBook

▶ **Figure 5.5**

PivotTable of Purchase Behavior

	A	B	C	D	E	F	G
1							
2							
3	Count of Second Purchase	Column Labels					
4	Row Labels	iMac	iPad	iPhone	iPod	MacBook	Grand Total
5	iMac			2	3	2	6	13
6	iPad	1			1	2	10	14
7	iPhone	3	4			14	21	42
8	iPod	3	12	12			30	57
9	MacBook	8	16	26	24			74
10	Grand Total	15	34	42	42	67	200

In general, the conditional probability of an event A given that event B is known to have occurred is

$$P(A|B) = \frac{P(A \text{ and } B)}{P(B)} \qquad (5.6)$$

EXAMPLE 5.14   Using the Conditional Probability Formula

Using the data from the energy drink survey example, substitute B_1 for A and M for B in formula (5.6). This results in the conditional probability of B_1 given M:

$$P(B_1|M) = \frac{P(B_1 \text{ and } M)}{P(M)} = \frac{0.25}{0.63} = 0.397$$

Similarly, the probability of preferring brand 1 if the respondent is female is

$$P(B_1|F) = \frac{P(B_1 \text{ and } F)}{P(F)} = \frac{0.09}{0.37} = 0.243$$

The following table summarizes the conditional probabilities of brand preference given gender.

P(Brand|Gender)	Brand 1	Brand 2	Brand 3
Male	0.397	0.270	0.333
Female	0.243	0.162	0.595

Such information can be important in marketing efforts. Knowing that there is a difference in preference by gender can help focus advertising. For example, we see that about 40% of males prefer brand 1, whereas only about 24% of females do, and a higher proportion of females prefer brand 3. This suggests that it would make more sense to focus on advertising brand 1 more in male-oriented media and brand 3 in female-oriented media.

We read the notation $P(A|B)$ as "the probability of A given B."

The conditional probability formula may be used in other ways. For example, multiplying both sides of formula (5.6) by $P(B)$, we obtain $P(A \text{ and } B) = P(A|B) P(B)$. Note that we may switch the roles of A and B and write $P(B \text{ and } A) = P(B|A) P(A)$. But $P(B \text{ and } A)$ is the same as $P(A \text{ and } B)$; thus we can express $P(A \text{ and } B)$ in two ways:

$$P(A \text{ and } B) = P(A|B) P(B) = P(B|A) P(A) \qquad (5.7)$$

This is often called the **multiplication law of probability**.

We may use this concept to express the probability of an event in a joint probability table in a different way. Using the energy drink survey in Figure 5.3 again, note that

$$P(F) = P(F \text{ and Brand 1}) + P(F \text{ and Brand 2}) + P(F \text{ and Brand 3})$$

Using formula (5.7), we can express the joint probabilities $P(A \text{ and } B)$ by $P(A|B) P(B)$. Therefore,

$$P(F) = P(F|\text{Brand 1}) P(\text{Brand 1}) + P(F|\text{Brand 2}) P(\text{Brand 2}) + P(F|\text{Brand 3})$$
$$P(\text{Brand 3}) = (0.265)(0.34) + (0.261)(0.23) + (0.512)(0.43) = 0.37 (\text{within rounding precision}).$$

We can express this calculation using the following extension of the multiplication law of probability. Suppose $B_1, B_2, \ldots, B_n$ are mutually exclusive events whose union comprises the entire sample space. Then

$$P(A) = P(A|B_1)P(B_1) + P(A|B_2)P(B_2) + \cdots + P(A|B_n)P(B_n) \qquad (5.8)$$

EXAMPLE 5.15   **Using the Multiplication Law of Probability**

Texas Holdem has become a popular game because of the publicity surrounding the World Series of Poker. At the beginning of a game, players each receive two cards face down (we won't worry about how the rest of the game is played). Suppose that a player receives an ace on her first card. The probability that she will end up with "pocket aces" (two aces in the hand) is P(ace on first card and ace on second card) $= P$(ace on second card | ace on first card) $\times$

P(ace on first card). Since the probability of an ace on the first card is 4/52 and the probability of an ace on the second card if she has already drawn an ace is 3/51, we have

P(ace on first card and ace on second card)
$= P$(ace on second card | ace on first card)
$\quad \times P$ (ace on first card)
$= \left(\dfrac{3}{51}\right) \times \left(\dfrac{4}{52}\right) = 0.004525$

In Example 5.14, we see that the probability of preferring a brand depends on gender. We may say that brand preference and gender are not independent. We may formalize this concept by defining the notion of **independent events**: *Two events A and B are independent if $P(A|B) = P(A)$.*

EXAMPLE 5.16   **Determining if Two Events Are Independent**

We use this definition in the energy drink survey example. Recall that the conditional probabilities of brand preference given gender are

P(Brand|Gender)	Brand 1	Brand 2	Brand 3
Male	0.397	0.270	0.333
Female	0.243	0.162	0.595

We see that whereas $P(B_1|M) = 0.397$, $P(B_1)$ was shown to be 0.34 in Example 5.11; thus, these two events are not independent.

Finally, we see that if two events are independent, then we can simplify the multiplication law of probability in equation (5.7) by substituting $P(A)$ for $P(A|B)$:

$$P(A \text{ and } B) = P(A)P(B) = P(B)\,P(A) \tag{5.9}$$

EXAMPLE 5.17   **Using the Multiplication Law for Independent Events**

Suppose A is the event that a sum of 6 is first rolled on a pair of dice and B is the event of rolling a sum of 2, 3, or 12 on the next roll. These events are

independent because the roll of a pair of dice does not depend on the previous roll. Then we may compute $P(A \text{ and } B) = P(A)P(B) = \left(\frac{5}{36}\right)\left(\frac{4}{36}\right) = \frac{20}{1296}$.

CHECK YOUR UNDERSTANDING

1. Define the terms experiment, outcome, and sample space.
2. Explain the difference between a permutation and a combination.
3. Give an example of each of the three definitions of probability.
4. What are the two key facts that govern probability?
5. What is an event? Explain how to compute $P(A \text{ or } B)$ for two events A and B.
6. Explain the concepts of joint, marginal, and conditional probability, and independent events.

Random Variables and Probability Distributions

Some experiments naturally have numerical outcomes, such as a sum of dice, the time it takes to repair computers, or the weekly change in a stock market index. For other experiments, such as obtaining consumer response to a new product, the sample space is categorical. To have a consistent mathematical basis for dealing with probability, we would like the outcomes of all experiments to be numerical. A **random variable** is a numerical description of the outcome of an experiment. Formally, a random variable is a function that assigns a real number to each element of a sample space. If we have categorical outcomes, we can associate an arbitrary numerical value to them. For example, if a consumer likes a product in a market research study, we might assign this outcome a value of 1; if the consumer dislikes the product, we might assign this outcome a value of 0. Random variables are usually denoted by capital italic letters, such as X or Y.

Random variables may be discrete or continuous. A **discrete random variable** is one for which the number of possible outcomes can be counted. A **continuous random variable** has outcomes over one or more continuous intervals of real numbers.

EXAMPLE 5.18   Discrete and Continuous Random Variables

The outcomes of the sum of rolling two dice (the numbers 2 through 12) and customer reactions to a product (like or dislike) are discrete random variables. The number of outcomes may be finite or theoretically infinite, such as the number of hits on a Web site link during some period of time—we cannot place a guaranteed upper limit on this number; nevertheless, the number of hits can be counted. Examples of continuous random variables are the weekly change in the DJIA, the daily temperature, the time to complete a task, the time between failures of a machine, and the return on an investment.

A **probability distribution** is the characterization of the possible values that a random variable may assume along with the probability of assuming these values. A probability distribution can be either discrete or continuous, depending on the nature of the random variable it models. Discrete distributions are easier to understand and work with, and we deal with them first.

We may develop a probability distribution using any one of the three perspectives of probability. First, if we can quantify the probabilities associated with the values of a random variable from theoretical arguments, then we can easily define the probability distribution.

EXAMPLE 5.19   Probability Distribution of Dice Rolls

The probabilities of the sum of outcomes for rolling two dice are calculated by counting the number of ways to roll each sum divided by the total number of possible outcomes. These, along with an Excel column chart depicting the probability distribution, are shown from the Excel file *Dice Rolls* in Figure 5.6.

Second, we can calculate the relative frequencies from a sample of empirical data to develop a probability distribution. Thus, the relative frequency distribution of computer repair times (Figure 5.2) is an example. Because this is based on sample data, we usually call this an **empirical probability distribution**. An empirical probability distribution is an approximation of the probability distribution of the associated random variable, whereas the probability distribution of a random variable, such as the one derived from counting arguments, is a theoretical model of the random variable.

▶ **Figure 5.6**

Probability Distribution of Rolls of Two Dice

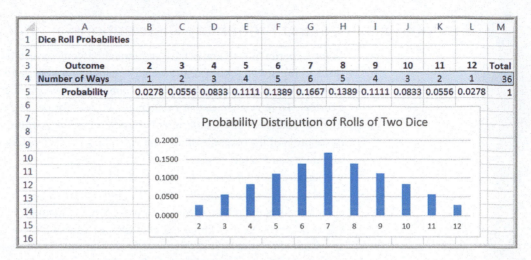

Finally, we could simply specify a probability distribution using subjective values and expert judgment. This is often done in creating decision models for phenomena for which we have no historical data.

EXAMPLE 5.20 **A Subjective Probability Distribution**

Figure 5.7 shows a hypothetical example of the distribution of one expert's assessment of how the DJIA might change in the next year. This might have been created purely by intuition and expert judgment, but we hope it would be supported by some extensive analysis of past and current data using business analytics tools.

Researchers have identified many common types of probability distributions that are useful in a variety of applications of business analytics. A working knowledge of common families of probability distributions is important for several reasons. First, it can help you to understand the underlying process that generates sample data. We investigate the relationship between distributions and samples later. Second, many phenomena in business and nature follow some theoretical distribution and, therefore, are useful in building decision models. Finally, working with distributions is essential in computing probabilities of occurrence of outcomes to assess risk and make decisions.

▶ **Figure 5.7**

Subjective Probability Distribution of DJIA Change

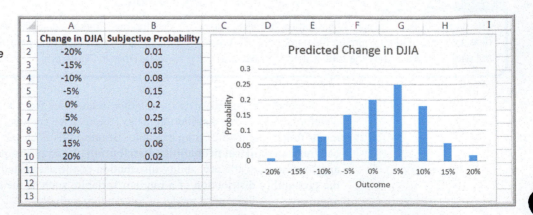

CHECK YOUR UNDERSTANDING

1. Explain the difference between a discrete and a continuous random variable, and give an example of each.

2. What is a probability distribution?

3. What is an empirical probability distribution?

4. Why is it important to understand the common types of probability distributions?

Discrete Probability Distributions

For a discrete random variable X, the probability distribution of the discrete outcomes is called a **probability mass function** and is denoted by a mathematical function, $f(x)$. The symbol x_i represents the ith value of the random variable X, and $f(x_i)$ is the probability associated with x_i.

EXAMPLE 5.21   **Probability Mass Function for Rolling Two Dice**

For instance, in Figure 5.6 for the dice example, the values of the random variable X, which represents the sum of the rolls of two dice, are $x_1 = 2$, $x_2 = 3$, $x_3 = 4$, $x_4 = 5$, $x_5 = 6$, $x_6 = 7$, $x_7 = 8$, $x_8 = 9$, $x_9 = 10$, $x_{10} = 11$, and $x_{11} = 12$. The probability mass function for X is

$$f(x_1) = \frac{1}{36} = 0.0278$$

$$f(x_2) = \frac{2}{36} = 0.0556$$

$$f(x_3) = \frac{3}{36} = 0.0833$$

$$f(x_4) = \frac{4}{36} = 0.1111$$

$$f(x_5) = \frac{5}{36} = 0.1389$$

$$f(x_6) = \frac{6}{36} = 0.1667$$

$$f(x_7) = \frac{5}{36} = 0.1389$$

$$f(x_8) = \frac{4}{36} = 0.1111$$

$$f(x_9) = \frac{3}{36} = 0.0833$$

$$f(x_{10}) = \frac{2}{36} = 0.0556$$

$$f(x_{11}) = \frac{1}{36} = 0.0278$$

A probability mass function has the properties that (1) the probability of each outcome must be between 0 and 1, inclusive, and (2) the sum of all probabilities must add to 1; that is,

$$0 \leq f(x_i) \leq 1 \quad \text{for all } i \tag{5.10}$$

$$\sum_i f(x_i) = 1 \tag{5.11}$$

You can easily verify that this holds in each of the examples we have described.

A **cumulative distribution function**, $F(x)$, specifies the probability that the random variable X assumes a value *less than or equal to* a specified value, x. This is also denoted as $P(X \leq x)$ and reads as "the probability that the random variable X is less than or equal to x."

EXAMPLE 5.22   Using the Cumulative Distribution Function

The cumulative distribution function for the sum of rolling two dice is shown in Figure 5.8, along with an Excel line chart that describes it visually from the worksheet *CumDist* in the *Dice Rolls* Excel file. To use this, suppose we want to know the probability of rolling a 6 or less. We simply look up the cumulative probability for 6, which is 0.4167. Alternatively, we could locate the point for $x = 6$ in the chart and estimate the probability from the graph. Also note that since the probability of rolling a 6 or less is 0.4167, then the probability of the complementary event (rolling a 7 or more) is $1 - 0.4167 = 0.5833$. We can also use the cumulative distribution function to find probabilities over intervals. For example, to find the probability of rolling a number between 4 and 8, $P(4 \leq X \leq 8)$, we can find $P(X \leq 8)$ and subtract $P(X \leq 3)$; that is,

$$P(4 \leq X \leq 8) = P(X \leq 8) - P(X \leq 3)$$
$$= 0.7222 - 0.0833 = 0.6389$$

A word of caution. Be careful with the endpoints when computing probabilities over intervals for discrete distributions; because 4 is included in the interval we wish to compute, we need to subtract $P(X \leq 3)$, not $P(X \leq 4)$.

Expected Value of a Discrete Random Variable

The **expected value** of a random variable corresponds to the notion of the mean, or average, for a sample. For a discrete random variable X, the expected value, denoted $E[X]$, is the weighted average of all possible outcomes, where the weights are the probabilities:

$$E[X] = \sum_{i=1}^{\infty} x_i f(x_i) \tag{5.12}$$

Note the similarity to computing the population mean using formula (4.16) in Chapter 4:

$$\mu = \frac{\sum_{i=1}^{N} f_i x_i}{N}$$

If we write this as the sum of x_i multiplied by (f_i/N), then we can think of f_i/N as the probability of x_i. Then this expression for the mean has the same basic form as the expected value formula.

▶ **Figure 5.8**

Cumulative Distribution Function for Rolling Two Dice

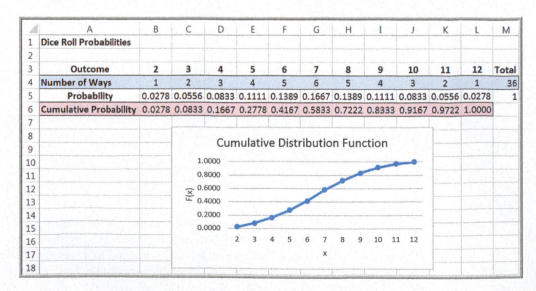

	A	B	C	D	E	F	G	H	I	J	K	L	M
1	**Dice Roll Probabilities**												
2													
3	**Outcome**	2	3	4	5	6	7	8	9	10	11	12	**Total**
4	**Number of Ways**	1	2	3	4	5	6	5	4	3	2	1	36
5	**Probability**	0.0278	0.0556	0.0833	0.1111	0.1389	0.1667	0.1389	0.1111	0.0833	0.0556	0.0278	1
6	**Cumulative Probability**	0.0278	0.0833	0.1667	0.2778	0.4167	0.5833	0.7222	0.8333	0.9167	0.9722	1.0000	
7													

EXAMPLE 5.23   Computing the Expected Value

We may apply formula (5.12) to the probability distribution for the sum of rolling two dice. We multiply the outcome 2 by its probability 1/36, add this to the product of the outcome 3 and its probability, and so on. Continuing in this fashion, the expected value is

$$E[X] = 2(0.0278) + 3(0.0556) + 4(0.0833) + 5(0.0111)$$
$$+ 6(0.1389) + 7(0.1667) + 8(0.1389) + 9(0.111)$$
$$+ 10(0.0833) + 11(0.0556) + 12(0.0278) = 7$$

Figure 5.9 shows these calculations in an Excel spreadsheet (worksheet *Expected Value* in the *Dice Rolls* Excel file). Note that you can use the SUMPRODUCT function to easily calculate the expected value. For this example, we would use =SUMPRODUCT(A4:A14, B4:B14). As expected (no pun intended), the average value of the sum of the roll of two dice is 7.

Using Expected Value in Making Decisions

Expected value can be helpful in making a variety of decisions, even those we see in daily life.

EXAMPLE 5.24   Expected Value on Television

One of the author's favorite examples stemmed from a task in season 1 of the former TV show, *The Apprentice*. Teams were required to select an artist and sell his or her art for the highest total amount of money. One team selected a mainstream artist who specialized in abstract art that sold for between $1,000 and $2,000; the second team chose an avant-garde artist whose surrealist and rather controversial art was priced much higher. Guess who won? The first team did, because the probability of selling a piece of mainstream art was much higher than the avant-garde artist whose bizarre art (the team members themselves didn't even like it!) had a very low probability of a sale. A simple expected value calculation would have easily predicted the winner.

A popular game show that took TV audiences by storm several years ago was called *Deal or No Deal*. The game involved a set of numbered briefcases that contain amounts of money from 1 cent to $1,000,000. Contestants begin choosing cases to be opened and removed, and their amounts are shown. After each set of cases is opened,

the banker offers the contestant an amount of money to quit the game, which the contestant may either choose or reject. Early in the game, the banker's offers are usually less than the expected value of the remaining cases, providing an incentive to continue. However, as the number of remaining cases becomes small, the banker's offers approach or may even exceed the average of the remaining cases. Most people press on until the bitter end and often walk away with a smaller amount than they could have had they been able to estimate the expected value of the remaining cases and make a more rational decision. In one case, a contestant had five briefcases left with $100, $400, $1,000, $50,000, and $300,000. Because the choice of each case is equally likely, the expected value was 0.2($100 + $400 + $1000 + $50,000 + $300,000) = $70,300, and the banker offered $80,000 to quit. Instead, she said "No deal" and proceeded to open the $300,000 suitcase, eliminating it from the game. She took the next banker's offer of $21,000, which was more than 60% larger than the expected value of the remaining cases.[1]

It is important to understand that the expected value is a "long-run average" and is appropriate for decisions that occur on a repeated basis. For one-time decisions, however, you need to consider the downside risk and the upside potential of the decision. The following example illustrates this.

---

[1]"Deal or No Deal: A Statistical Deal." www.pearsonified.com/2006/03/deal_or_no_deal_the_real_deal.php

► **Figure 5.9**

Expected Value Calculations for Rolling Two Dice

	A	B	C
1	Expected Value Calculations		
2			
3	Outcome, x	Probability, f(x)	x*f(x)
4	2	0.0278	0.0556
5	3	0.0556	0.1667
6	4	0.0833	0.3333
7	5	0.1111	0.5556
8	6	0.1389	0.8333
9	7	0.1667	1.1667
10	8	0.1389	1.1111
11	9	0.1111	1.0000
12	10	0.0833	0.8333
13	11	0.0556	0.6111
14	12	0.0278	0.3333
15		Expected value	7.0000

EXAMPLE 5.25   Expected Value of a Charitable Raffle

Suppose that you are offered the chance to buy one of 1,000 tickets sold in a charity raffle for $50, with the prize being $25,000. Clearly, the probability of winning is $\frac{1}{1,000}$, or 0.001, whereas the probability of losing is $1 - 0.001 = 0.999$. The random variable X is your net winnings, and its probability distribution is

x	f(x)
− $50	0.999
$24,950	0.001

The expected value, $E[X]$, is $-\$50(0.999) + \$24,950(0.001) = -\$25.00$. This means that if you played this game

repeatedly over the long run, you would lose an average of $25.00 *each time* you play. Of course, for any *one* game, you would either lose $50 or win $24,950. So the question becomes, Is the risk of losing $50 worth the potential of winning $24,950? Although the expected value is negative, you might take the chance because the upside potential is large relative to what you might lose, and, after all, it is for charity. However, if your potential loss is large, you might not take the chance, even if the expected value were positive.

Decisions based on expected values are common in real estate development, day trading, and pharmaceutical research projects. Drug development is a good example. The cost of research and development projects in the pharmaceutical industry is generally in the hundreds of millions of dollars and often approaches $1 billion. Many projects never make it to clinical trials or might not get approved by the Food and Drug Administration. Statistics indicate that 7 of 10 products fail to return the cost of the company's capital. However, large firms can absorb such losses because the return from one or two blockbuster drugs can easily offset these losses. On an average basis, drug companies make a net profit from these decisions.

EXAMPLE 5.26   Airline Revenue Management

Let us consider a simplified version of the typical revenue management process that airlines use. At any date prior to a scheduled flight, airlines must make a decision as to whether to reduce ticket prices to stimulate demand for unfilled seats. If the airline does not discount the fare, empty seats might not be sold and the airline will lose revenue. If the airline discounts the remaining seats too early

(and could have sold them at the higher fare), they will lose profit. The decision depends on the probability p of selling a full-fare ticket if they choose not to discount the price. Because an airline makes hundreds or thousands of such decisions each day, the expected value approach is appropriate.

Assume that only two fares are available: full and discount. Suppose that a full-fare ticket is $560, the discount fare is $400, and $p = 0.75$. For simplification, assume that if the price is reduced, then any remaining seats would be sold at that price. The expected value of not discounting the price is $0.25(0) + 0.75(\$560) = \$420$. Because this is higher than the discounted price, the airline should not discount at this time. In reality, airlines constantly update the probability p based on the information they collect and analyze in a database. When the value of p drops below the break-even point: $\$400 = p(\$560)$, or $p = 0.714$, then it is beneficial to discount. It can also work in reverse; if demand is such that the probability that a higher-fare ticket would be sold, then the price may be adjusted upward. This is why published fares constantly change and why you may receive last-minute discount offers or may pay higher prices if you wait too long to book a reservation. Other industries such as hotels and cruise lines use similar decision strategies.

Variance of a Discrete Random Variable

We may compute the variance, Var[X], of a discrete random variable X as a weighted average of the squared deviations from the expected value:

$$\text{Var}\,[X] = \sum_{i=1}^{\infty} (x_i - E[X])^2 f(x_i) \tag{5.13}$$

EXAMPLE 5.27   Computing the Variance of a Random Variable

We may apply formula (5.13) to calculate the variance of the probability distribution for the sum of rolling two dice.

Figure 5.10 shows these calculations in an Excel spreadsheet (worksheet *Variance* in *Random Variable Calculations* Excel file).

Similar to our discussion in Chapter 4, the variance measures the uncertainty of the random variable; the higher the variance, the higher the uncertainty of the outcome. Although variances are easier to work with mathematically, we usually measure the variability of a random variable by its standard deviation, which is simply the square root of the variance.

Bernoulli Distribution

The **Bernoulli distribution** characterizes a random variable having two possible outcomes, each with a constant probability of occurrence. Typically, these outcomes represent "success" ($x = 1$), having probability p, and "failure" ($x = 0$), having probability

▶ **Figure 5.10**

Variance Calculations for Rolling Two Dice

	A	B	C	D	E	F
1	Variance Calculations					
2						
3	Outcome, x	Probability, f(x)	x*f(x)	(x - E[X])	(x - E[X])^2	(x - E[X])^2*f(x)
4	2	0.0278	0.0556	-5.0000	25.0000	0.6944
5	3	0.0556	0.1667	-4.0000	16.0000	0.8889
6	4	0.0833	0.3333	-3.0000	9.0000	0.7500
7	5	0.1111	0.5556	-2.0000	4.0000	0.4444
8	6	0.1389	0.8333	-1.0000	1.0000	0.1389
9	7	0.1667	1.1667	0.0000	0.0000	0.0000
10	8	0.1389	1.1111	1.0000	1.0000	0.1389
11	9	0.1111	1.0000	2.0000	4.0000	0.4444
12	10	0.0833	0.8333	3.0000	9.0000	0.7500
13	11	0.0556	0.6111	4.0000	16.0000	0.8889
14	12	0.0278	0.3333	5.0000	25.0000	0.6944
15		Expected value	7.0000		Variance	5.8333

$1 - p$. A success can be any outcome you define. For example, in attempting to boot a new computer just off the assembly line, we might define a success as "does not boot up" in defining a Bernoulli random variable to characterize the probability distribution of a defective product. Thus, success need not be a favorable result in the traditional sense.

The probability mass function of the Bernoulli distribution is

$$f(x) = \begin{cases} p & \text{if } x = 1 \\ 1 - p & \text{if } x = 0 \end{cases} \tag{5.14}$$

where p represents the probability of success. The expected value is p, and the variance is $p(1 - p)$.

EXAMPLE 5.28 Using the Bernoulli Distribution

A Bernoulli distribution might be used to model whether an individual responds positively ($x = 1$) or negatively ($x = 0$) to a telemarketing promotion. For example, if you estimate that 20% of customers contacted will make a purchase, the probability distribution that describes whether or not a particular individual makes a purchase is Bernoulli with $p = 0.2$. Think of the following experiment. Suppose that you have a box with 100 marbles, 20 red and 80 white. For each customer, select one marble at random (and then replace it). The outcome will have a Bernoulli distribution. If a red marble is chosen, then that customer makes a purchase; if it is white, the customer does not make a purchase.

Binomial Distribution

The **binomial distribution** models n independent repetitions of a Bernoulli experiment, each with a probability p of success. The random variable X represents the number of successes in these n experiments. In the telemarketing example, suppose that we call $n = 10$ customers, each of which has a probability $p = 0.2$ of making a purchase. Then the probability distribution of the number of positive responses obtained from ten customers is binomial. Using the binomial distribution, we can calculate the probability that exactly x customers out of the ten will make a purchase for any value of x between 0 and 10, inclusive. A binomial distribution might also be used to model the results of sampling inspection in a production operation or the effects of drug research on a sample of patients.

The probability mass function for the binomial distribution is

$$f(x) = \begin{cases} \binom{n}{x} p^x (1 - p)^{n-x}, & \text{for } x = 0, 1, 2, \ldots, n \\ 0, & \text{otherwise} \end{cases} \tag{5.15}$$

We saw the notation $\binom{n}{x}$ earlier in this chapter when discussing combinations; it represents the number of ways of choosing x distinct items from a group of n items and is computed using formula (5.2).

EXAMPLE 5.29 Computing Binomial Probabilities

We may use formula (5.15) to compute binomial probabilities. For example, if the probability that any individual will make a purchase from a telemarketing solicitation is 0.2, then the probability distribution that x individuals out of ten calls will make a purchase is

$$f(x) = \begin{cases} \binom{10}{x}(0.2)^x(0.8)^{10-x}, & \text{for } x = 0, 1, 2, \ldots, n \\ 0, & \text{otherwise} \end{cases}$$

Thus, to find the probability that three people will make a purchase among the ten calls, we compute

$$f(3) = \binom{10}{3}(0.2)^3(0.8)^{10-3}$$

$$= (10!/(3!7!))(0.008)(0.2097152)$$

$$= 120(0.008)(0.2097152) = 0.20133$$

The formula for the probability mass function for the binomial distribution is rather complex, and binomial probabilities are tedious to compute by hand; however, they can easily be computed in Excel using the function

$$\text{BINOM.DIST}(\textit{number\_s, trials, probability\_s, cumulative})$$

In this function, *number_s* plays the role of *x*, and *probability_s* is the same as *p*. If *cumulative* is set to TRUE, then this function will provide cumulative probabilities; otherwise the default is FALSE, and it provides values of the probability mass function, $f(x)$.

EXAMPLE 5.30   Using Excel's Binomial Distribution Function

Figure 5.11 shows the results of using this function to compute the distribution for the previous example (Excel file *Binomial Probabilities*). For instance, the probability that exactly three individuals will make a purchase is BINOM.DIST(A10, $B$3, $B$4, FALSE) = 0.20133 = $f(3)$.

The probability that three or fewer individuals will make a purchase is BINOM.DIST(A10, $B$3, $B$4, TRUE) = 0.87913 = $F(3)$. Correspondingly, the probability that more than three out of ten individuals will make a purchase is $1 - F(3) = 1 - 0.87913 = 0.12087$.

The expected value of the binomial distribution is np, and the variance is $np(1 - p)$. The binomial distribution can assume different shapes and amounts of skewness, depending on the parameters. Figure 5.12 shows an example when $p = 0.8$. For larger values of p, the binomial distribution is negatively skewed; for smaller values, it is positively skewed. When $p = 0.5$, the distribution is symmetric.

Poisson Distribution

The **Poisson distribution** is a discrete distribution used to model the number of occurrences in some unit of measure—for example, the number of customers arriving at a Subway store during a weekday lunch hour, the number of failures of a machine during a month, the number of visits to a Web page during 1 minute, or the number of errors per line of software code.

► **Figure 5.11**

Computing Binomial Probabilities in Excel

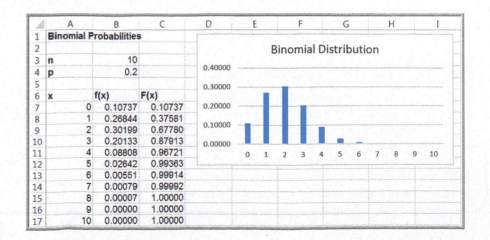

▶ **Figure 5.12**

Example of the Binomial Distribution with p = 0.8

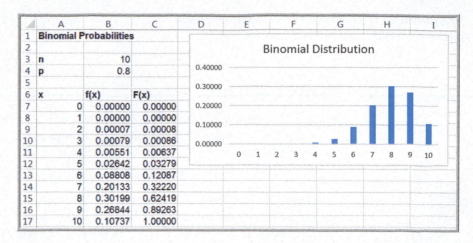

The Poisson distribution assumes no limit on the number of occurrences (meaning that the random variable X may assume any nonnegative integer value), that occurrences are independent, and that the average number of occurrences per unit is a constant, λ (Greek lowercase lambda). The expected value of the Poisson distribution is λ, and the variance also is equal to λ.

The probability mass function for the Poisson distribution is

$$f(x) = \begin{cases} \dfrac{e^{-\lambda}\lambda^x}{x!}, & \text{for } x = 0, 1, 2, \ldots \\ 0, & \text{otherwise} \end{cases} \tag{5.16}$$

EXAMPLE 5.31   Computing Poisson Probabilities

Suppose that, on average, the number of customers arriving at Subway during lunch hour is 12 customers per hour. The probability that exactly x customers will arrive during the hour is given by a Poisson distribution with a mean of 12. The probability that exactly x customers will arrive during the hour would be calculated using formula (5.16):

$$f(x) = \begin{cases} \dfrac{e^{-12}12^x}{x!}, & \text{for } x = 0, 1, 2, \ldots \\ 0, & \text{otherwise} \end{cases}$$

Substituting $x = 5$ in this formula, the probability that exactly five customers will arrive is $f(5) = 0.01274$.

Like the binomial, Poisson probabilities are cumbersome to compute by hand. Probabilities can easily be computed in Excel using the function POISSON.DIST(*x, mean, cumulative*).

EXAMPLE 5.32   Using Excel's Poisson Distribution Function

Figure 5.13 shows the results of using this function to compute the distribution for Example 5.31 with $\lambda = 12$ (see the Excel file *Poisson Probabilities*). Thus, the probability of exactly one arrival during the lunch hour is calculated by the Excel function =POISSON.DIST(A7, $B$3, FALSE) = 0.00007 = $f(1)$; the probability of four arrivals or fewer is calculated by =POISSON.DIST(A10,

$B$3, TRUE) = 0.00760 = $F(4)$, and so on. Because the possible values of a Poisson random variable are infinite, we have not shown the complete distribution. As x gets large, the probabilities become quite small. Like the binomial, the specific shape of the distribution depends on the value of the parameter λ; the distribution is more skewed for smaller values.

▶ **Figure 5.13**

*Computing Poisson
Probabilities in Excel*

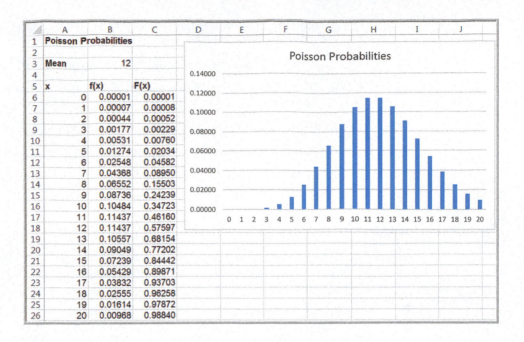

	A	B	C
1	Poisson Probabilities		
2			
3	Mean	12	
4			
5	x	f(x)	F(x)
6	0	0.00001	0.00001
7	1	0.00007	0.00008
8	2	0.00044	0.00052
9	3	0.00177	0.00229
10	4	0.00531	0.00760
11	5	0.01274	0.02034
12	6	0.02548	0.04582
13	7	0.04368	0.08950
14	8	0.06552	0.15503
15	9	0.08736	0.24239
16	10	0.10484	0.34723
17	11	0.11437	0.46160
18	12	0.11437	0.57597
19	13	0.10557	0.68154
20	14	0.09049	0.77202
21	15	0.07239	0.84442
22	16	0.05429	0.89871
23	17	0.03832	0.93703
24	18	0.02555	0.96258
25	19	0.01614	0.97872
26	20	0.00968	0.98840

ANALYTICS IN PRACTICE: Using the Poisson Distribution for Modeling Bids on Priceline[2]

Priceline is well known for allowing customers to name their own prices (but not the service providers) in bidding for services such as airline flights or hotel stays. Some hotels take advantage of Priceline's approach to fill empty rooms for leisure travelers while not diluting the business market by offering discount rates through traditional channels. In one study using business analytics to develop a model to optimize pricing strategies for Kimpton Hotels, which develops, owns, or manages more than 40 independent boutique lifestyle hotels in the United States and Canada, the distribution of the number of bids for a given number of days before arrival was modeled as a Poisson distribution because it corresponded well with data that were observed. For example, the average number of bids placed per day three days before arrival on a weekend (the random variable X) was 6.3. Therefore, the distribution used in the model was $f(x) = e^{-6.3}6.3^x/x!$, where x is the number of bids placed. The analytic model helped to determine the prices to post on Priceline and the inventory allocation for each price. After using the model, rooms sold via Priceline increased 11% in one year, and the average rate for these rooms increased 3.7%.

Fantasista/Fotolia

[2]Based on Chris K. Anderson, "Setting Prices on Priceline," *Interfaces*, 39, 4 (July–August 2009): 307–315.

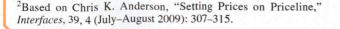

CHECK YOUR UNDERSTANDING

1. What properties must a probability mass function have?

2. What is the difference between a probability mass function and a cumulative distribution function?

3. Explain how to compute the expected value and variance of a discrete random variable.

4. How can expected value concepts be used in business decisions?

5. Provide examples of situations for which the Bernoulli, binomial, and Poisson distributions can be used.

Continuous Probability Distributions

As we noted earlier, a continuous random variable is defined over one or more intervals of real numbers and, therefore, has an infinite number of possible outcomes. Suppose that the expert who predicted the probabilities associated with next year's change in the DJIA in Figure 5.7 kept refining the estimates over larger and larger ranges of values. Figure 5.14 shows what such a probability distribution might look like using 2.5% increments rather than 5%. Notice that the distribution is similar in shape to the one in Figure 5.7 but simply has more outcomes. If this refinement process continues, then the distribution will approach the shape of a smooth curve, as shown in the figure. Such a curve that characterizes outcomes of a continuous random variable is called a **probability density function** and is described by a mathematical function $f(x)$.

Properties of Probability Density Functions

A probability density function has the following properties:

1. $f(x) \geq 0$ *for all values of x.* This means that a graph of the density function must lie at or above the *x*-axis.
2. *The total area under the density function above the x-axis is 1.* This is analogous to the property that the sum of all probabilities of a discrete random variable must add to 1.

▶ **Figure 5.14**

Refined Probability Distribution of DJIA Change

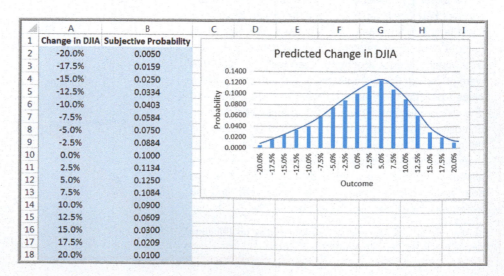

3. $P(X = x) = 0$. For continuous random variables, it does not make mathematical sense to attempt to define a probability for a specific value of x because there are an infinite number of values.
4. *Probabilities of continuous random variables are only defined over intervals.* Thus, we may calculate probabilities between two numbers a and b, $P(a \leq X \leq b)$, or to the left or right of a number c—for example, $P(X < c)$ and $P(X > c)$.
5. $P(a \leq X \leq b)$ *is the area under the density function between a and b.*

The cumulative distribution function for a continuous random variable is denoted the same way as for discrete random variables, $F(x)$, and represents the probability that the random variable X is less than or equal to x, $P(X \leq x)$. Intuitively, $F(x)$ represents the area under the density function to the left of x. $F(x)$ can often be derived mathematically from $f(x)$.

Knowing $F(x)$ makes it easy to compute probabilities over intervals for continuous distributions. The probability that X is between a and b is equal to the difference of the cumulative distribution function evaluated at these two points; that is,

$$P(a \leq X \leq b) = P(X \leq b) - P(X \leq a) = F(b) - F(a) \tag{5.17}$$

For continuous distributions, we need not be concerned about the endpoints, as we were with discrete distributions, because $P(a \leq X \leq b)$ is the same as $P(a < X < b)$ as a result of property 3 above.

The formal definitions of expected value and variance for a continuous random variable are similar to those for a discrete random variable; however, to understand them, we must rely on notions of calculus, so we do not discuss them in this book. We simply state them when appropriate.

Uniform Distribution

The **uniform distribution** characterizes a continuous random variable for which all outcomes between some minimum and maximum value are equally likely. The uniform distribution is often assumed in business analytics applications when little is known about a random variable other than reasonable estimates for minimum and maximum values. The parameters a and b are chosen judgmentally to reflect a modeler's best guess about the range of the random variable.

For a uniform distribution with a minimum value a and a maximum value b, the density function is

$$f(x) = \begin{cases} \dfrac{1}{b-a}, & \text{for } a \leq x \leq b \\ 0, & \text{otherwise} \end{cases} \tag{5.18}$$

and the cumulative distribution function is

$$F(x) = \begin{cases} 0, & \text{if } x < a \\ \dfrac{x-a}{b-a}, & \text{if } a \leq x \leq b \\ 1, & \text{if } b < x \end{cases} \tag{5.19}$$

Although Excel does not provide a function to compute uniform probabilities, the formulas are simple enough to incorporate into a spreadsheet. Probabilities are also easy to compute for the uniform distribution because of the simple geometric shape of the density function, as Example 5.33 illustrates.

EXAMPLE 5.33 Computing Uniform Probabilities

Suppose that sales revenue, X, for a product varies uniformly each week between $a = \$1,000$ and $b = \$2,000$. The density function is $f(x) = 1/(2,000 - 1,000) = 1/1,000$ and is shown in Figure 5.15. Note that the area under the density function is 1, which you can easily verify by multiplying height by the width of the rectangle.

Suppose we wish to find the probability that sales revenue will be less than $x = \$1,300$. We could do this in two ways. First, compute the area under the density function using geometry, as shown in Figure 5.16. The area is $(1/1,000)(300) = 0.30$. Alternatively, we could use formula (5.19) to compute $f(1,300)$:

$$F(1,300) = (1,300 - 1,000)/(2,000 - 1,000) = 0.30$$

In either case, the probability is 0.30.

Now suppose we wish to find the probability that revenue will be between \$1,500 and \$1,700. Again, using geometrical arguments (see Figure 5.17), the area of the rectangle between \$1,500 and \$1,700 is $(1/1,000)(200) = 0.2$. We may also use formula (5.17) and compute it as follows:

$$P(1,500 \le X \le 1,700) = P(X \le 1,700) - P(X \le 1,500)$$
$$= F(1,700) - F(1,500)$$
$$= \frac{(1,700 - 1,000)}{(2,000 - 1,000)} - \frac{(1,500 - 1,000)}{(2,000 - 1,000)}$$
$$= 0.7 - 0.5 = 0.2$$

The expected value and variance of a uniform random variable X are computed as follows:

$$E[X] = \frac{a + b}{2} \tag{5.20}$$

$$\text{Var}[X] = \frac{(b - a)^2}{12} \tag{5.21}$$

A variation of the uniform distribution is one for which the random variable is restricted to integer values between a and b (also integers); this is called a **discrete uniform distribution**. An example of a discrete uniform distribution is the roll of a single die. Each of the numbers 1 through 6 has a $\frac{1}{6}$ probability of occurrence.

▶ **Figure 5.15**

Uniform Probability Density Function

▶ **Figure 5.16**

Probability that X < $1,300

▶ **Figure 5.17**

P($1,500 < X < $1,700)

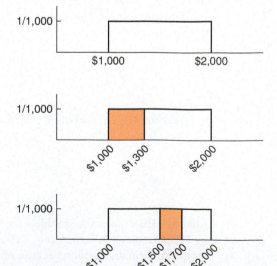

Normal Distribution

The **normal distribution** is a continuous distribution that is described by the familiar bell-shaped curve and is perhaps the most important distribution used in statistics. The normal distribution is observed in many natural phenomena. Test scores such as the SAT, deviations from specifications of machined items, human height and weight, and many other measurements are often normally distributed.

The normal distribution is characterized by two parameters: the mean, μ, and the standard deviation, σ. Thus, as μ changes, the location of the distribution on the x-axis also changes, and as σ is decreased or increased, the distribution becomes narrower or wider, respectively. Figure 5.18 shows some examples.

The normal distribution has the following properties:

1. The distribution is symmetric, so its measure of skewness is zero.
2. The mean, median, and mode are all equal. Thus, half the area falls above the mean and half falls below it.
3. The range of X is unbounded, meaning that the tails of the distribution extend to negative and positive infinity.
4. The empirical rules apply exactly for the normal distribution; the area under the density function within ± 1 standard deviation is 68.3%, the area under the density function within ± 2 standard deviations is 95.4%, and the area under the density function within ± 3 standard deviations is 99.7%.

Normal probabilities cannot be computed using a mathematical formula. Instead, we may use the Excel function NORM.DIST(x, *mean*, *standard_deviation*, *cumulative*). NORM.DIST(x, *mean*, *standard_deviation*, *TRUE*) calculates the cumulative probability $F(x) = P(X \le x)$ for a specified mean and standard deviation. (If *cumulative* is set to *FALSE*, the function simply calculates the value of the density function $f(x)$, which has little practical application other than tabulating values of the density function. This was used to draw the distributions in Figure 5.18 using Excel.)

▶ **Figure 5.18**

Examples of Normal Distributions

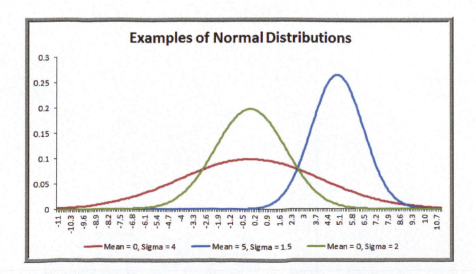

EXAMPLE 5.34   Using the NORM.DIST Function to Compute Normal Probabilities

Suppose that a company has determined that the distribution of customer demand (X) is normal with a mean of 750 units/month and a standard deviation of 100 units/month. Figure 5.19 shows some cumulative probabilities calculated with the NORM.DIST function (see the Excel file *Normal Probabilities*). The company would like to know the following:

1. What is the probability that demand will be at most 900 units?
2. What is the probability that demand will exceed 700 units?
3. What is the probability that demand will be between 700 and 900 units?

To answer the questions, first draw a picture. This helps to ensure that you know what area you are trying to calculate and how to use the formulas for working with a cumulative distribution correctly.

- **Question 1.** Figure 5.20(a) shows the probability that demand will be at most 900 units, or $P(X \leq 900)$.

This is simply the cumulative probability for $x = 900$, which can be calculated using the Excel function =NORM.DIST(900, 750, 100, TRUE) = 0.9332.

- **Question 2.** Figure 5.20(b) shows the probability that demand will exceed 700 units, $P(X > 700)$. Using the principles we have previously discussed, this can be found by subtracting $P(X < 700)$ from 1:

$$P(X > 700) = 1 - P(X < 700) = 1 - F(700)$$
$$= 1 - 0.3085 = 0.6915$$

This can be computed in Excel using the formula =1 − NORM.DIST(700, 750, 100, TRUE).

- **Question 3.** The probability that demand will be between 700 and 900, $P(700 \leq X \leq 900)$, is illustrated in Figure 5.20(c). This is calculated by

$$P(700 \leq X \leq 900) = P(X \leq 900) - P(X \leq 700)$$
$$= F(900) - F(700) = 0.9332 - 0.3085 = 0.6247$$

In Excel, we would use the formula =NORM.DIST (900, 750, 100, TRUE) − NORM.DIST(700, 750, 100, TRUE).

The NORM.INV Function

With the NORM.DIST function, we are given a value, x, of the random variable X and can find the cumulative probability to the left of x. Now let's reverse the problem. Suppose that we know the cumulative probability but don't know the value of x. How can we find it? We are often faced with such a question in many applications. The Excel function NORM.INV(*probability*, *mean*, *standard_dev*) can be used to do this. In this function, *probability* is the cumulative probability value corresponding to the value of x we seek, and "INV" stands for inverse.

▶ **Figure 5.19**

Normal Probability Calculations in Excel

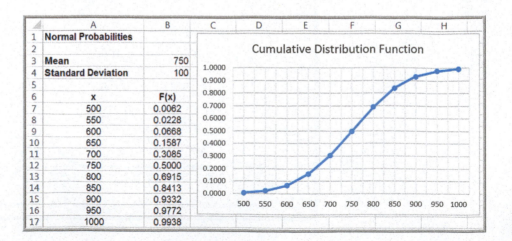

	A	B
1	**Normal Probabilities**	
2		
3	**Mean**	750
4	**Standard Deviation**	100
5		
6	x	F(x)
7	500	0.0062
8	550	0.0228
9	600	0.0668
10	650	0.1587
11	700	0.3085
12	750	0.5000
13	800	0.6915
14	850	0.8413
15	900	0.9332
16	950	0.9772
17	1000	0.9938

▶ **Figure 5.20**

Computing Normal Probabilities

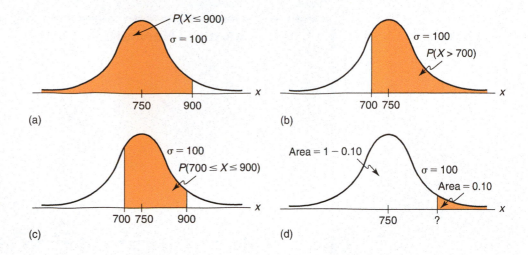

EXAMPLE 5.35   **Using the NORM.INV Function**

In the previous example, what level of demand would be exceeded at most 10% of the time? Here, we need to find the value of x so that $P(X > x) = 0.10$. This is illustrated in Figure 5.20(d). Because the area in the upper tail of the normal distribution is 0.10, the cumulative probability must be $1 - 0.10 = 0.90$. From Figure 5.19,

we can see that the correct value must be somewhere between 850 and 900 because $F(850) = 0.8413$ and $F(900) = 0.9332$. We can find the exact value using the Excel function =NORM.INV(0.90, 750, 100) = 878.155. Therefore, a demand of approximately 878 will satisfy the criterion.

Standard Normal Distribution

Figure 5.21 provides a sketch of a special case of the normal distribution called the **standard normal distribution**—the normal distribution with $\mu = 0$ and $\sigma = 1$. This distribution is important in performing many probability calculations. A standard normal random variable is usually denoted by Z, and its density function by $f(z)$. The scale along the z-axis represents the number of standard deviations from the mean of zero. The Excel function NORM.S.DIST(z) finds probabilities for the standard normal distribution.

EXAMPLE 5.36   **Computing Probabilities with the Standard Normal Distribution**

We have previously noted that the empirical rules apply to any normal distribution. Let us find the areas under the standard normal distribution within 1, 2, and 3 standard deviations of the mean. These can be found by using the function NORM.S.DIST(z). Figure 5.22 shows a tabulation of the cumulative probabilities for z ranging from −3 to +3 and calculations of the areas within 1, 2, and 3 standard deviations of the mean. We apply formula (5.17) to find the difference between the cumulative

probabilities, $F(b) - F(a)$. For example, the area within 1 standard deviation of the mean is found by calculating $P(-1 < Z < 1) = F(1) - F(-1) =$ NORM.S.DIST(1) −NORM.S.DIST(−1) = $0.84134 - 0.15866 = 0.6827$ (the difference due to decimal rounding). As the empirical rules stated, about 68% of the area falls within 1 standard deviation; 95% within 2 standard deviations; and more than 99% within 3 standard deviations of the mean.

▶ **Figure 5.21**

Standard Normal Distribution

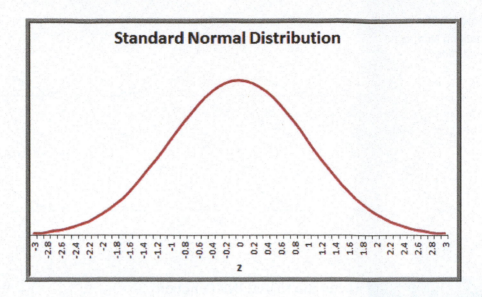

Standard Normal Distribution

▶ **Figure 5.22**

Computing Standard Normal Probabilities

	A	B	C	D	E	F	G	H
1	**Standard Normal Probabilities**							
2								
3	z	F(z)		a	b	F(a)	F(b)	F(b) - F(a)
4	-3	0.00135		-1	1	0.15866	0.84134	0.6827
5	-2	0.02275		-2	2	0.02275	0.97725	0.9545
6	-1	0.15866		-3	3	0.00135	0.99865	0.9973
7	0	0.50000						
8	1	0.84134						
9	2	0.97725						
10	3	0.99865						

Using Standard Normal Distribution Tables

Although it is quite easy to use Excel to compute normal probabilities, tables of the standard normal distribution are commonly found in textbooks and professional references when a computer is not available. Such a table is provided in Table A.1 of Appendix A at the end of this book. The table allows you to look up the cumulative probability for any value of z between −3.00 and +3.00.

One of the advantages of the standard normal distribution is that we may compute probabilities for any normal random variable X having a mean μ and standard deviation σ by converting it to a standard normal random variable Z. We introduced the concept of standardized values (z-scores) for sample data in Chapter 4. Here, we use a similar formula to convert a value x from an arbitrary normal distribution into an equivalent standard normal value, z:

$$z = \frac{x - \mu}{\sigma}$$

(5.22)

EXAMPLE 5.37   Computing Probabilities with Standard Normal Tables

We will answer the first question posed in Example 5.34: What is the probability that demand will be at most $x = 900$ units if the distribution of customer demand (X) is normal with a mean of 750 units/month and a standard deviation of 100 units/month? Using formula (5.22), convert x to a standard normal value:

$$z = \frac{900-750}{100} = 1.5$$

Note that 900 is 150 units higher than the mean of 750; since the standard deviation is 100, this simply means that 900 is 1.5 standard deviations above the mean, which is the value of z. Using Table A.1 in Appendix A, we see that the cumulative probability for $z = 1.5$ is 0.9332, which is the same answer we found for Example 5.34.

Exponential Distribution

The **exponential distribution** is a continuous distribution that models the time between randomly occurring events. Thus, it is often used in such applications as modeling the time between customer arrivals to a service system or the time to or between failures of machines, lightbulbs, hard drives, and other mechanical or electrical components.

Similar to the Poisson distribution, the exponential distribution has one parameter, λ. In fact, the exponential distribution is closely related to the Poisson distribution; if the number of events occurring *during* an interval of time has a Poisson distribution, then the time *between* events is exponentially distributed. For instance, if the number of arrivals at a bank is Poisson-distributed, say with mean $\lambda = 12/\text{hour}$, then the time between arrivals is exponential, with mean $\mu = 1/12$ hour, or 5 minutes.

The exponential distribution has the density function

$$f(x) = \lambda e^{-\lambda x}, \quad \text{for } x \geq 0 \tag{5.23}$$

and its cumulative distribution function is

$$F(x) = 1 - e^{-\lambda x}, \quad \text{for } x \geq 0 \tag{5.24}$$

Sometimes, the exponential distribution is expressed in terms of the mean μ rather than the rate λ. To do this, simply substitute $1/\mu$ for λ in the preceding formulas.

The expected value of the exponential distribution is $1/\lambda$ and the variance is $(1/\lambda)^2$. Figure 5.23 provides a sketch of the exponential distribution. The exponential distribution has the properties that it is bounded below by 0, it has its greatest density at 0, and the density declines as x increases. The Excel function EXPON.DIST(x, *lambda*, *cumulative*) can be used to compute exponential probabilities. As with other Excel probability distribution functions, *cumulative* is either TRUE or FALSE, with TRUE providing the cumulative distribution function.

EXAMPLE 5.38   Using the Exponential Distribution

Suppose that the mean time to failure of a critical component of an engine is $\mu = 8{,}000$ hours. Therefore, $\lambda = 1/\mu = 1/8{,}000$ failures/hour. The probability that the component will fail before x hours is given by the cumulative distribution function $F(x)$. Figure 5.24 shows a portion of the cumulative distribution function, which may be found in the Excel file *Exponential Probabilities*. For example, the probability of failing before 5,000 hours is $F(5{,}000) = 0.4647$.

▶ **Figure 5.23**

Example of an Exponential Distribution ($\lambda = 1$)

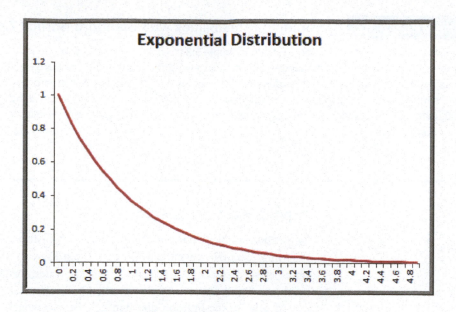

▶ **Figure 5.24**

Computing Exponential Probabilities in Excel

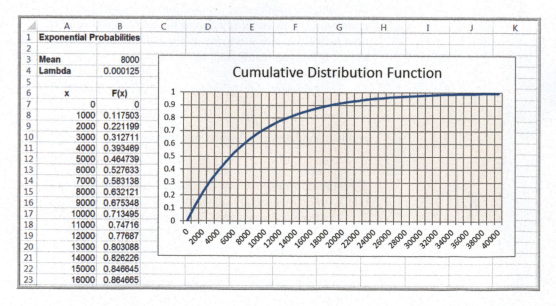

Triangular Distribution

The **triangular distribution** is defined by three parameters: the minimum, a; maximum, b; and most likely, c. Outcomes near the most likely value have a higher chance of occurring than those at the extremes. By varying the most likely value, the triangular distribution can be symmetric or skewed in either direction, as shown in Figure 5.25. Because the triangular distribution can assume different shapes, it is useful to model a wide variety of phenomena. For example, triangular distribution is often used when no data are available to characterize an uncertain variable and the distribution must be estimated judgmentally.

The mean of the triangular distribution is calculated as

$$\text{Mean} = (a + c + b)/3 \tag{5.25}$$

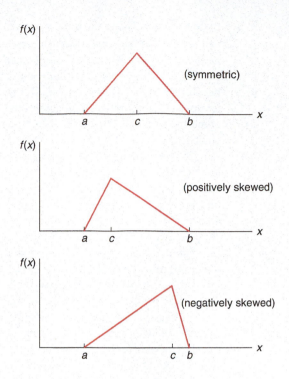

► **Figure 5.25**

Examples of Triangular Distributions

The variance is

$$\text{Variance} = (a^2 + c^2 + b^2 - a \times c - a \times b - c \times b)/18 \qquad \textbf{(5.26)}$$

CHECK YOUR UNDERSTANDING

1. How does a continuous probability distribution differ from a discrete probability distribution? What properties must it have?

2. State how to compute probabilities over intervals for continuous distributions.

3. Explain the uniform distribution and how it might be used.

4. What properties does a normal distribution have?

5. Explain how to use Excel functions NORM.DIST and NORM.INV to compute normal probabilities.

6. What is a standard normal distribution? How can you convert a random variable X having an arbitrary normal distribution into a standard normal random variable Z?

7. What typical situations do the exponential and triangular distributions model?

Data Modeling and Distribution Fitting

In many applications of business analytics, we need to collect sample data of important variables such as customer demand, purchase behavior, machine failure times, and service activity times, to name just a few, to gain an understanding of the distributions of these variables. Using the tools we have studied, we may construct frequency distributions and histograms and compute basic descriptive statistical measures to better understand the nature of the data. However, sample data are just that—samples.

Using sample data may limit our ability to predict uncertain events that may occur because potential values *outside* the range of the sample data are not included. A better approach is to identify the underlying probability distribution from which sample data come by "fitting" a theoretical distribution to the data and verifying the goodness of fit statistically.

To select an appropriate theoretical distribution that fits sample data, we might begin by examining a histogram of the data to look for the distinctive shapes of particular distributions. For example, normal data are symmetric, with a peak in the middle. Exponential data are very positively skewed, with no negative values. This approach is not, of course, always accurate or valid, and sometimes it can be difficult to apply, especially if sample sizes are small. However, it may narrow the search down to a few potential distributions.

Summary statistics, such as the mean, median, standard deviation, and coefficient of variation, often provide information about the nature of the distribution. For instance, normally distributed data tend to have a fairly low coefficient of variation (however, this may not be true if the mean is small). For normally distributed data, we would also expect the median and mean to be approximately the same. For exponentially distributed data, however, the median will be less than the mean. Also, we would expect the mean to be about equal to the standard deviation, or, equivalently, the coefficient of variation would be close to 1. We could also look at the skewness index. For instance, normal data are not skewed. The following examples illustrate some of these ideas.

The examination of histograms and summary statistics might provide some idea of the appropriate distribution; however, a better approach is to analytically fit the data to the best type of probability distribution.

EXAMPLE 5.39   Analyzing Airline Passenger Data

An airline operates a daily route between two medium-sized cities using a 70-seat regional jet. The flight is rarely booked to capacity but often accommodates business travelers who book at the last minute at a high price. Figure 5.26 shows the number of passengers for a sample of 25 flights (Excel file *Airline Passengers*). The histogram shows a relatively symmetric distribution. The mean, median, and mode are all similar, although there is some degree of positive skewness. From our discussion in Chapter 4 about the variability of samples, it is important to recognize that although the histogram in Figure 5.26 does not look perfectly normal, this is a relatively small sample that can exhibit a lot of variability compared with the population from which it is drawn. Thus, based on these characteristics, it would not be unreasonable to assume a normal distribution for the purpose of developing a predictive or prescriptive analytics model.

EXAMPLE 5.40   Analyzing Airport Service Times

Figure 5.27 shows a portion of the data and statistical analysis of 812 samples of service times at an airport's ticketing counter (Excel file *Airport Service Times*). The data certainly do not appear to be normal or uniform. The histogram appears to look like an exponential distribution, and this might be a reasonable choice. However, there is a difference between the mean and standard deviation, which we would not expect to see for an exponential distribution. Some other exotic distribution that we have not introduced might be more appropriate. Identifying the best fitting distribution can be done using sophisticated software.

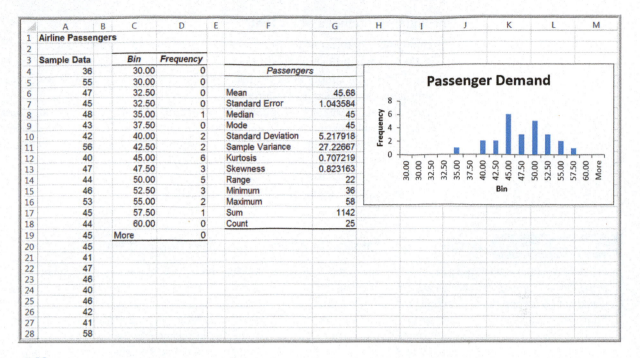

▶ Figure 5.26

Data and Statistics for Passenger Demand

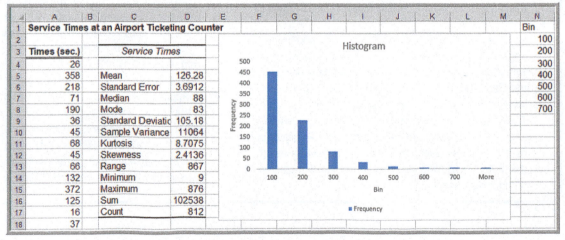

▶ Figure 5.27

Airport Service Times Statistics

Goodness of Fit: Testing for Normality of an Empirical Distribution

The basis for fitting data to a probability distribution is a statistical procedure called **goodness of fit**. Goodness of fit attempts to draw a conclusion about the *nature* of the distribution. For instance, in Example 5.39 we suggested that it might be reasonable to assume that the distribution of passenger demand is normal. Goodness of fit would provide objective, analytical support for this assumption. We can do this using a procedure called the chi-square goodness of fit test.

Chi-Square Goodness of Fit Test

The **chi-square goodness of fit test** can be used to determine whether sample data are representative of some probability distribution. We will explain the theory behind this test in Chapter 7; here, you simply need to understand the calculations. Example 5.41 illustrates its application for a normal distribution.

The chi-square test can be applied to many distributions; the only modification to the procedure that must be made is to use the proper function to find the cumulative probabilities of the observed frequency distribution. For example, if we suspect that data follow an exponential distribution, then use =EXPON.DIST(*upper limit, lambda, TRUE*), where *lambda* is 1/sample mean. For a uniform distribution, the probabilities can be computed easily using simple geometry, like we did in Example 5.33.

EXAMPLE 5.41   Determining Normality for Airline Passenger Demand Data

In Example 5.39, we analyzed passenger demand data in the *Airline Passengers* Excel file and concluded that it would not be unreasonable to assume that the data stem from a normal distribution. To test for this statistically, we do the following (this is shown in Figure 5.28 and in the Excel file *Chi Square Goodness of Fit Test*):

1. Start with the frequency distribution of the data; these are the observed frequencies. In Figure 5.28 we show this, as well as the descriptive statistics, in columns H through N.
2. Find the cumulative probability corresponding to the upper limit for each bin, assuming a normal distribution with the sample mean and sample standard deviation of the data. This is found using the function =NORM.DIST (*upper limit, sample mean, sample standard deviation, TRUE*) and is shown in column C of the spreadsheet.
3. Find the probability associated with each bin using formula (5.17). See question 3 in Example 5.34 for using the NORM.DIST function. This is computed in column D of the spreadsheet.

4. Multiply the bin probabilities by the number of observations (in this case, 25). These values are the expected frequencies that you would see if the normal distribution were the correct data model. These are computed in column E of the spreadsheet.
5. Compute the difference between the observed and expected frequencies for each bin, square the result, and divide by the expected frequency, $(O - E)^2/E$, for each bin (see column F of the spreadsheet).
6. Sum the chi-square calculations for all these bins (cell F17). This sum is called the **chi-square statistic**.
7. Compute the **chi-square critical value** using the function =CHISQ.INV.RT(0.05, number of bins − 3). This is shown in cell F18 in the spreadsheet.
8. Compare the chi-square statistic with the critical value. If the chi-square statistic is less than or equal to the critical value, then the data can be reasonably assumed to come from a normal distribution having the sample mean and standard deviation. If not, then conclude that the normal distribution is not appropriate to model the data.

	A	B	C	D	E	F	G	H	I	J	K	L	M	N
1	Chi Square Goodness of Fit Test (Normality)							Airline Passengers						
2														
3								Sample Data		Bin	Frequency			
4	Observed Data		Normal Assumption			Chi-Square Calculations		36		30.00	0		Passengers	
5	Bin	Frequency	Cumulative Probability	Bin Probability	Expected Frequency	(Observed - Expected)^2/Expected		55		30.00	0			
6	35.00	1	0.020	0.020	0.508	0.4751		47		32.50	0		Mean	45.68
7	37.50	0	0.058	0.038	0.953	0.9535		45		32.50	0		Standard Error	1.04358357
8	40.00	2	0.138	0.080	1.992	0.0000		48		35.00	1		Median	45
9	42.50	2	0.271	0.133	3.324	0.5271		43		37.50	0		Mode	45
10	45.00	6	0.448	0.177	4.426	0.5598		42		40.00	2		Standard Deviation	5.217917848
11	47.50	3	0.636	0.188	4.706	0.6182		56		42.50	2		Sample Variance	27.22666667
12	50.00	5	0.796	0.160	3.994	0.2534		40		45.00	6		Kurtosis	0.707219447
13	52.50	3	0.904	0.108	2.706	0.0318		47		47.50	3		Skewness	0.82316313
14	55.00	2	0.963	0.059	1.464	0.1962		44		50.00	5		Range	22
15	57.50	1	0.988	0.025	0.632	0.2139		46		52.50	3		Minimum	36
16								53		55.00	2		Maximum	58
17					Chi Squared Statistic	3.8290		45		57.50	1		Sum	1142
18					Critical value	14.067		44		60.00	0		Count	25
19								45		More	0			

► **Figure 5.28**

Calculations for Chi-Square Test for Normality

The number of degrees of freedom used in the CHISQ.INV.RT function depends on the type of distribution you are fitting. In general, degrees of freedom is the number of bins minus one minus the number of parameters of the distribution that are estimated from the data. For the normal distribution, we need to estimate two parameters — the mean and standard deviation; thus, degrees of freedom equals the number of bins minus 3 (see step 7 in Example 5.41). For the exponential distribution, we need only estimate one parameter, lambda, so the degrees of freedom are the number of bins minus 2.

ANALYTICS IN PRACTICE: The Value of Good Data Modeling in Advertising

To illustrate the importance of identifying the correct distribution in decision modeling, we discuss an example in advertising.[3] The amount that companies spend on the creative component of advertising (that is, making better ads) is traditionally quite small relative to the overall media budget. One expert noted that the expenditure on creative development was about 5% of that spent on the media delivery campaign.

Whatever money is spent on creative development is usually directed through a single advertising agency. However, one theory that has been proposed is that more should be spent on creative ad development, and the expenditures should be spread across a number of competitive advertising agencies. In research studies of this theory, the distribution of advertising effectiveness was assumed to be normal. In reality, data collected on the response to consumer product ads show that this distribution is actually quite skewed and, therefore, not normally distributed. Using the wrong assumption in any model or application can produce erroneous results. In this situation, the skewness actually provides an advantage for advertisers, making it more effective to obtain ideas from a variety of advertising agencies.

A mathematical model (called Gross's model) relates the relative contributions of creative and media dollars to total advertising effectiveness and is often used to identify the best number of draft ads to purchase. This model includes factors of ad development cost, total media spending budget, the distribution of effectiveness across ads (assumed to be normal), and the unreliability of identifying the most effective ad from a set of independently generated alternatives. Gross's model concluded that large gains were possible if

Victor Correia/Shutterstock.com

multiple ads were obtained from independent sources, and the best ad is selected.

Since the data observed on ad effectiveness was clearly skewed, other researchers examined ad effectiveness by studying standard industry data on ad recall without requiring the assumption of normally distributed effects. This analysis found that the best of a number of ads was more effective than any single ad. Further analysis revealed that the optimal number of ads to commission can vary significantly, depending on the shape of the distribution of effectiveness for a single ad.

The researchers developed an alternative to Gross's model. From their analyses, they found that as the number of draft ads was increased, the effectiveness of the best ad also increased. Both the optimal number of draft ads and the payoff from creating multiple independent drafts were higher *when the correct distribution was used* than the results reported in Gross's original study.

[3]Based on G. C. O'Connor, T. R. Willemain, and J. MacLachlan, "The Value of Competition Among Agencies in Developing Ad Campaigns: Revisiting Gross's Model," *Journal of Advertising*, 25, 1 (1996): 51–62.

CHECK YOUR UNDERSTANDING

1. Explain the concept of "fitting" a probability distribution to data.

2. What are some simple approaches to help you fit a distribution?

3. Explain how to use the chi-square goodness of fit test to determine whether sample data can be adequately represented by a particular probability distribution.

KEY TERMS

Bernoulli distribution	Joint probability
Binomial distribution	Joint probability table
Chi-square critical value	Marginal probability
Chi-square goodness of fit test	Multiplication law of probability
Chi-square statistic	Mutually exclusive
Combination	Normal distribution
Complement	Outcome
Conditional probability	Permutation
Continuous random variable	Poisson distribution
Cumulative distribution function	Probability
Discrete random variable	Probability density function
Discrete uniform distribution	Probability distribution
Empirical probability distribution	Probability mass function
Event	Random variable
Expected value	Sample space
Experiment	Standard normal distribution
Exponential distribution	Tree diagram
Goodness of fit	Triangular distribution
Independent events	Uniform distribution
Intersection	Union

CHAPTER 5 TECHNOLOGY HELP

Useful Excel Functions

BINOM.DIST(*number _s, trials, probability _s, cumulative*) Computes probabilities for the binomial distribution.

POISSON.DIST(*x, mean, cumulative*) Computes probabilities for the Poisson distribution.

NORM.DIST(*x, mean, standard_deviation, cumulative*) Computes probabilities for the normal distribution.

NORM.INV(*probability, mean, standard_dev*) Finds the value of *x* for a normal distribution having a specified cumulative *probability*.

NORM.S.DIST(*z*) Finds probabilities for the standard normal distribution.

EXPON.DIST(*x, lambda, cumulative*) Computes probabilities for the exponential distribution.

CHISQ.INV.RT(*probability, deg_freedom*) Returns the value of chi-square that has a right-tail area equal to *probability* for a specified degree of freedom. Used for the chi-square test

StatCrunch

StatCrunch provides graphical calculations for all the distributions we studied in this chapter (binomial, discrete uniform, exponential, normal, Poisson, uniform, and custom, which is a user-defined discrete distribution) as well as many other advanced distributions. You can find video tutorials and step-by-step procedures with examples at https://www.statcrunch.com/5.0/example.php. We suggest that you first view the tutorials *Getting started with StatCrunch* and *Working with StatCrunch sessions*.

The following tutorials are listed under the heading Graphical Calculators:

- Continuous distributions
- Discrete distributions

Example: Using Graphical Calculators

1. Select the name of the desired distribution from the menu listing (for example, Binomial, Normal).
2. In the first line below the plot in the calculator window, specify the distribution parameters. As examples, with the normal distribution, specify the mean and standard deviation; with the binomial distribution, specify n and p.

3. In the second line below the plot, specify the direction of the desired probability.
   a. To compute a probability, enter a value to the right of the direction selector and leave the remaining field empty (for example, $P(X < 3) = $ ____).
   b. To determine the point that will provide a specified probability, enter the probability to the right of the equal sign and leave the other field empty (for example, $P(X < $ ____$) = 0.25$). This option is available only for continuous distributions.
4. Click *Compute!* to fill in the empty fields and to update the graph of the distribution.

PROBLEMS AND EXERCISES

Basic Concepts of Probability

1. Lauren drinks a variety of soft drinks. Over the past month, she has had 17 diet colas, 3 cans of lemonade, and 5 cans of root beer in no particular order or pattern.
   a. Given this history, what is the probability that her next drink will be a diet cola? Lemonade? Root beer?
   b. What definition of probability did you use to answer this question?

2. A home improvement company in Cincinnati ran a promotion: If a customer bought new windows before December 10, then if Cincinnati had a white Christmas, the purchase would be free. In the fine print, a "white Christmas" was defined as four inches of snowfall on December 25. The National Weather Service summarized Christmas Day snowfall from 1893 to 2014 as follows:

Inches of snow	Frequency
0	68
Trace	28
0.1–0.4	11
0.5–0.9	7
1.0–2.9	6
3.0 or more	1

What is the probability that the company would have to reimburse customers who took this offer?

3. Consider the experiment of drawing two cards without replacement from a deck consisting of only the ace through 10 of a single suit (for example, only hearts).

   a. Describe the outcomes of this experiment. List the elements of the sample space.
   b. Define the event A_i to be the set of outcomes for which the sum of the values of the cards is i (with an ace $= 1$). List the outcomes associated with A_i for $i = 3$ to 19.
   c. What is the probability of obtaining a sum of the two cards equaling from 3 to 19?

4. Three coins are dropped on a table.
   a. List all possible outcomes in the sample space.
   b. Find the probability associated with each outcome.

5. A market research company surveyed consumers to determine their ranked preferences of energy drinks among the brands Monster, Red Bull, and Rockstar.
   a. What are the outcomes of this experiment for one respondent?
   b. What is the probability that one respondent will rank Red Bull first?
   c. What is the probability that two respondents will both rank Red Bull first?

6. Refer to the card scenario described in Problem 3.
   a. Let A be the event "total card value is odd." Find $P(A)$ and $P(A^c)$.
   b. What is the probability that the sum of the two cards will be more than 14?
   c. What is the probably that the sum of the two cards will not exceed 12?

7. Refer to the coin scenario described in Problem 4.

   a. Let *A* be the event "exactly two heads." Find *P*(*A*).

   b. Let *B* be the event "at most one head." Find *P*(*B*).

   c. Let *C* be the event "at least two heads." Find *P*(*C*).

   d. Are the events *A* and *B* mutually exclusive? Find *P*(*A* or *B*).

   e. Are the events *A* and *C* mutually exclusive? Find *P*(*A* or *C*).

8. Roulette is played at a table similar to the one in Figure 5.29. A wheel with the numbers 1 through 36 (evenly distributed with the colors red and black) and two green numbers, 0 and 00, rotate in a shallow bowl with a curved wall. A small ball is spun on the inside of the wall and drops into a pocket corresponding to one of the numbers. Players may make 11 different types of bets by placing chips on different areas of the table. These include bets on a single number, two adjacent numbers, a row of three numbers, a block of four numbers, two adjacent rows of six numbers, and the five number combinations of 0, 00, 1, 2, and 3; bets on the numbers 1–18 or 19–36; the first, second, or third group of 12 numbers; a column of 12 numbers; even or odd; and red or black. Payoffs differ by bet. For instance, a single-number bet pays 35 to 1 if it wins; a three-number bet pays 11 to 1; a column bet pays 2 to 1; and a color bet pays even money.

   Define the following events: *C1* = column 1 number, *C2* = column 2 number, *C3* = column 3 number, *O* = odd number, *E* = even number, *G* = green number, *F12* = first 12 numbers, *S12* = second 12 numbers, and *T12* = third 12 numbers.

   a. Find the probability of each of these events.

   b. Find *P*(*G* or *O*), *P*(*O* or *F12*), *P*(*C1* or *C3*), *P*(*E* and *F12*), *P*(*E* or *F12*), *P*(*S12* and *T12*), and *P*(*O* or *C2*).

9. Students in the new MBA class at a state university have the following specialization profile:

Finance—83

Marketing—36

Operations and Supply Chain Management—72

Information Systems—59

Find the probability that a student is either a finance or a marketing major. Are the events finance specialization and marketing specialization mutually exclusive? If so, what assumption must be made?

10. An airline tracks data on its flight arrivals. Over the past six months, 70 flights on one route arrived early, 150 arrived on time, 15 were late, and 25 were canceled.

   a. What is the probability that a flight is early? On time? Late? Canceled?

   b. Are these outcomes mutually exclusive?

   c. What is the probability that a flight is either early or on time?

11. A survey of 200 college graduates who have been working for at least three years found that 75 owned only mutual funds, 35 owned only stocks, and 90 owned both.

   a. What is the probability that an individual owns a stock? A mutual fund?

   b. What is the probability that an individual owns neither any stocks nor mutual funds?

   c. What is the probability that an individual owns either a stock or a mutual fund?

▶ **Figure 5.29**

Layout of a Typical Roulette Table

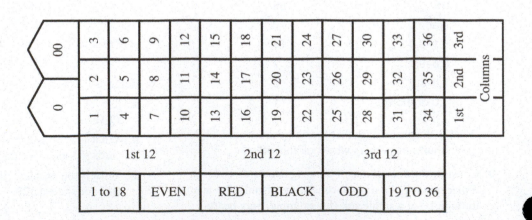

12. Row 26 of the Excel file *Census Education Data* gives the number of employed persons in the civilian labor force having a specific educational level.

  a. Find the probability that an employed person has attained each of the educational levels listed in the data.

  b. Suppose that *A* is the event "An employed person has some type of college degree" and *B* is the event "An employed person has at least some college." Find the probabilities of these events. Are they mutually exclusive? Why or why not?

  c. Find the probability *P*(*A* or *B*). Explain what this means.

13. A survey of shopping habits found the percentage of respondents that use technology for shopping as shown in Figure 5.30. For example, 17.39% only use online coupons; 21.74% use online coupons and check prices online before shopping, and so on.

  a. What is the probability that a shopper will check prices online before shopping?

  b. What is the probability that a shopper will use a smart phone to save money?

  c. What is the probability that a shopper will use online coupons?

  d. What is the probability that a shopper will not use any of these technologies?

  e. What is the probability that a shopper will check prices online and use online coupons but not use a smart phone?

  f. If a shopper checks prices online, what is the probability that he or she will use a smart phone?

  g. What is the probability that a shopper will check prices online but not use online coupons or a smart phone?

14. A Canadian business school summarized the gender and residency of its incoming class as follows:

Gender	Residency				
	Canada	United States	Europe	Asia	Other
Male	125	18	17	50	8
Female	103	8	10	92	4

  a. Construct a joint probability table.

  b. Calculate the marginal probabilities.

  c. What is the probability that a female student is from outside Canada or the United States?

15. In Example 4.13, we developed the following cross-tabulation of sales transaction data:

Region	Book	DVD	Total
East	56	42	98
North	43	42	85
South	62	37	99
West	100	90	190
Total	261	211	472

  a. Find the marginal probabilities that a sale originated in each of the four regions and the marginal probability of each type of sale (book or DVD).

  b. Find the conditional probabilities of selling a book given that the customer resides in each region.

16. Use the Civilian Labor Force data in the Excel file *Census Education Data* to find the following:

  a. *P*(unemployed and advanced degree)

  b. *P*(unemployed | advanced degree)

  c. *P*(not a high school grad | unemployed)

  d. Are the events "unemployed" and "at least a high school graduate" independent?

▶ **Figure 5.30**

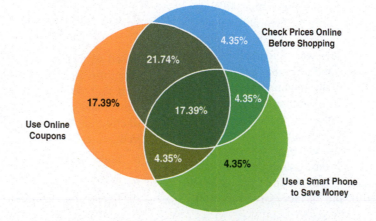

17. Using the data in the Excel file *Consumer Transportation Survey*, develop a cross-tabulation for Gender and Vehicle Driven; then convert this table into probabilities.

    a. What is the probability that a respondent is female?

    b. What is the probability that a respondent drives an SUV?

    c. What is the probability that a respondent is male and drives a minivan?

    d. What is the probability that a female respondent drives either a truck or an SUV?

    e. If it is known that an individual drives a car, what is the probability that the individual is female?

    f. If it is known that an individual is male, what is the probability that he drives an SUV?

    g. Determine whether the random variable "gender" and the event "vehicle driven" are statistically independent. What would this mean for advertisers?

18. A home pregnancy test is not always accurate. Suppose the probability is 0.015 that the test indicates that a woman is pregnant when she actually is not, and the probability is 0.025 that the test indicates that a woman is not pregnant when she really is. Assume that the probability that a woman who takes the test is actually pregnant is 0.7. What is the probability that a woman is pregnant if the test yields a not-pregnant result?

19. In the scenario in Problem 3, what is the probability of drawing an ace first, followed by a 2? How does this differ if the first card is replaced in the deck? Clearly explain what formulas you use and why.

20. In the roulette example described in Problem 8, what is the probability that the outcome will be green twice in a row? What is the probability that the outcome will be black twice in a row?

21. A consumer products company found that 48% of successful products also received favorable results from test market research, whereas 12% had unfavorable results but nevertheless were successful. That is, P(successful product and favorable test market) = 0.48, P(successful product and unfavorable test market) = 0.12. They also found that 28% of unsuccessful products had unfavorable research results, whereas 12% of them had favorable research results; that is, P(unsuccessful product and favorable test market) = 0.12, and P(unsuccessful product and unfavorable test market) = 0.28. Find the probabilities of successful and unsuccessful products given known test market results, that is, P(successful product | favorable test market), P(successful product | unfavorable test market), P(unsuccessful product | favorable test market), and P(unsuccessful product | unfavorable test market).

Discrete Probability Distributions

22. An investor estimates that there is a 1 in 10 chance that a stock purchase will lose 20% of its value, a 2 in 10 chance that it will break even, a 3 in 10 chance that it will gain 15%, and a 4 in 10 chance that it will gain 30%. What is the expected return as a percentage based on these estimates?

23. The weekly demand of a slow-moving product has the following probability mass function:

Demand, x	Probability, f(x)
0	0.2
1	0.4
2	0.3
3	0.1
4 or more	0

Find the expected value, variance, and standard deviation of weekly demand.

24. Construct the probability distribution for the value of a two-card hand dealt from a standard deck of 52 cards (all face cards have a value of 10 and an ace has a value of 11).

    a. What is the probability of being dealt 21?

    b. What is the probability of being dealt 16?

    c. Construct a chart for the cumulative distribution function. What is the probability of being dealt a 16 or less? Between 12 and 16? Between 17 and 20?

    d. Find the expected value and standard deviation of a two-card hand.

25. Based on the data in the Excel file *Consumer Transportation Survey*, develop a probability mass function and cumulative distribution function (both tabular and as charts) for the random variable number of children. What is the probability that an individual in this survey has

    a. fewer than three children?

    b. at least one child?

    c. five or more children?

26. A major application of analytics in marketing is determining customer retention. Suppose that the probability of a long-distance carrier's customer leaving for another carrier from one month to the next is 0.12. What distribution models the retention of an individual customer? What is the expected value and standard deviation?

27. The Excel file *Call Center Data* shows that in a sample of 70 individuals, 27 had prior call center experience. If we assume that the probability that any potential hire will also have experience with a probability of 27/70, what is the probability that among ten potential hires, more than half of them will have experience? Define the parameter(s) for this distribution based on the data.

28. If a cell phone company conducted a telemarketing campaign to generate new clients and the probability of successfully gaining a new customer was 0.07, what is the probability that contacting 50 potential customers would result in at least 5 new customers?

29. During one year, a particular mutual fund outperformed the S&P 500 index 32 out of 52 weeks. Find the probability that it would perform as well or better again.

30. A popular resort hotel has 300 rooms and is usually fully booked. About 7% of the time a reservation is canceled before the 6:00 p.m. deadline with no penalty. What is the probability that at least 285 rooms will be occupied? Use the binomial distribution to find the exact value.

31. A telephone call center where people place marketing calls to customers has a probability of success of 0.06. Find the number of calls needed to ensure that there is a probability of approximately 0.90 of obtaining five or more successful calls.

32. A financial consultant has an average of eight customers who contact him each day for consultations for which he receives a fee; assume a Poisson distribution. The consultant's overhead requires that he consults with at least six customers each day in order to cover his expenses. Find the probabilities of 0–10 customer contacts in a given day. What is the probability that he will consult with at least six customers?

33. The number and frequency of Atlantic hurricanes annually from 1940 through 2015 is shown here. This means, for instance, that no hurricanes occurred during 5 of these years, only one hurricane occurred in 16 of these years, and so on.

Number	Frequency
0	5
1	16
2	20
3	14
4	4
5	5
6	5
7	3
8	2
10	1
12	1

  a. Find the probabilities of 0–12 hurricanes each season using these data.

  b. Find the mean number of hurricanes.

  c. Assuming a Poisson distribution and using the mean number of hurricanes per season from part b, compute the probabilities of experiencing 0–12 hurricanes in a season. Compare these to your answer to part a. How accurately does a Poisson distribution model this phenomenon? Construct a chart to visualize these results.

Continuous Probability Distributions

34. Verify that the function corresponding to the following figure is a valid probability density function. Then find the following probabilities:

  a. $P(x < 8)$

  b. $P(x > 7)$

  c. $P(6 < x < 10)$

  d. $P(8 < x < 11)$

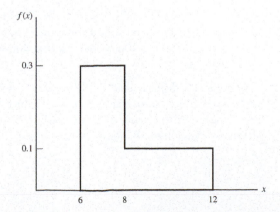

35. The time required to play a game of Battleship™ is uniformly distributed between 20 and 60 minutes.

   a. Find the expected value and variance of the time to complete the game.

   b. What is the probability of finishing within 30 minutes?

   c. What is the probability that the game would take longer than 40 minutes?

36. A contractor has estimated that the minimum number of days to remodel a bathroom for a client is ten. He also estimates that 80% of similar jobs are completed within 18 days. If the remodeling time is uniformly distributed, what should be the parameters of the uniform distribution?

37. In determining automobile mileage ratings, it was found that the mpg (X) for a certain model is normally distributed, with a mean of 31 mpg and a standard deviation of 1.8 mpg. Find the following:

   a. $P(X < 30)$

   b. $P(28 < X < 32)$

   c. $P(X > 34)$

   d. $P(X > 30)$

   e. The mileage rating that the upper 10% of cars achieve.

38. The distribution of SAT scores in math for an incoming class of business students has a mean of 610 and standard deviation of 20. Assume that the scores are normally distributed.

   a. Find the probability that an individual's SAT score is less than 600.

   b. Find the probability that an individual's SAT score is between 590 and 620.

   c. Find the probability that an individual's SAT score is greater than 650.

   d. What scores will the top 5% of students have?

   e. Find the standardized values for students scoring 540, 600, 650, and 700 on the test. Explain what these mean.

39. A popular soft drink is sold in 2-liter (2,000-milliliter) bottles. Because of variation in the filling process, bottles have a mean of 2,000 milliliters and a standard deviation of 18, normally distributed.

   a. If the process fills the bottle by more than 30 milliliters, the overflow will cause a machine malfunction. What is the probability of this occurring?

   b. What is the probability of underfilling the bottles by at least 10 milliliters?

40. A supplier contract calls for a key dimension of a part to be between 1.96 and 2.04 centimeters. The supplier has determined that the standard deviation of its process, which is normally distributed, is 0.03 centimeter.

   a. If the actual mean of the process is 1.98, what fraction of parts will meet specifications?

   b. If the mean is adjusted to 2.00, what fraction of parts will meet specifications?

   c. How small must the standard deviation be to ensure that no more than 2% of parts are nonconforming, assuming the mean is 2.00?

41. Historical data show that customers who download music from a popular Web service spend approximately $22 per month, with a standard deviation of $3. Find the probability that a customer will spend at least $20 per month. How much (or more) do the top 10% of customers spend?

42. A lightbulb is warranted to last for 5,000 hours. If the time to failure is exponentially distributed with a true mean of 4,800 hours, what is the probability that it will last at least 5,000 hours?

43. The actual delivery time from Giodanni's Pizza is exponentially distributed with a mean of 25 minutes.

   a. What is the probability that the delivery time will exceed 30 minutes?

   b. What proportion of deliveries will be completed within 20 minutes?

Data Modeling and Distribution Fitting

44. Apply the chi-square goodness of fit test to the data in the *Airport Service Times* Excel file to determine if a normal distribution models the data. Use bins of width 100. Note that for a normal distribution, the number of degrees of freedom for the CHISQ.INV. RT function should be the number of bins minus 3 as discussed in the chapter.

45. Apply the chi-square goodness of fit test to the data in the *Airport Service Times* Excel file to determine if an exponential distribution models the data. Use bins of width 100. Note that for an exponential distribution, the number of degrees of freedom for the CHISQ.INV.RT function should be the number of bins minus 2 as discussed in the chapter.

46. Compute the daily change of the closing price for the data in the Excel file *S&P 500*. Compute descriptive statistics, a frequency distribution, and histogram for the closing prices (using a bin width of 25). What probability distribution would you propose as a good fit for the data? Verify your choice using the chi-square goodness of fit test.

CASE: PERFORMANCE LAWN EQUIPMENT

PLE collects a variety of data from special studies, many of which are related to the quality of its products. The company collects data about functional test performance of its mowers after assembly; results from the past 30 days are given in the worksheet *Mower Test* in the *Performance Lawn Equipment Database*. In addition, many in-process measurements are taken to ensure that manufacturing processes remain in control and can produce according to design specifications. The worksheet *Blade Weight* shows 350 measurements of blade weights taken from the manufacturing process that produces mower blades during the most recent shift. Elizabeth Burke has asked you to study these data from an analytics perspective. Drawing upon your experience, you have developed a number of questions.

1. For the mower test data, what distribution might be appropriate to model the failure of an individual mower?
2. What fraction of mowers fails the functional performance test using all the mower test data?
3. What is the probability of having x failures in the next 100 mowers tested, for x from 0 to 20?

4. What is the average blade weight and how much variability is occurring in the measurements of blade weights?
5. Assuming that the data are normal, what is the probability that blade weights from this process will exceed 5.20?
6. What is the probability that blade weights will be less than 4.80?
7. What is the actual percent of blade weights that exceed 5.20 or are less than 4.80 from the data in the worksheet?
8. Is the process that makes the blades stable over time? That is, are there any apparent changes in the pattern of the blade weights?
9. Could any of the blade weights be considered outliers, which might indicate a problem with the manufacturing process or materials?
10. Is the assumption that blade weights are normally distributed justified?

Summarize all your findings to these questions in a well-written report.

Sampling and Estimation

Robert Brown Stock/Shutterstock

LEARNING OBJECTIVES   After studying this chapter, you will be able to:

- Describe the elements of a sampling plan.
- Explain the difference between subjective and probabilistic sampling.
- State two types of subjective sampling.
- Explain how to conduct simple random sampling and use Excel to find a simple random sample from an Excel database.
- Explain systematic, stratified, and cluster sampling and sampling from a continuous process.
- Explain the importance of unbiased estimators.

- Describe the difference between sampling error and nonsampling error.
- Explain how the average, standard deviation, and distribution of means of samples change as the sample size increases.
- Define the sampling distribution of the mean.
- Calculate the standard error of the mean.
- Explain the practical importance of the central limit theorem.
- Use the standard error in probability calculations.

- Explain how an interval estimate differs from a point estimate.
- Define and give examples of confidence intervals.
- Calculate confidence intervals for population means and proportions using the formulas in the chapter and the appropriate Excel functions.
- Explain how confidence intervals change as the level of confidence increases or decreases.

- Describe the difference between the *t*-distribution and the normal distribution.
- Use and visualize confidence intervals to draw conclusions about population parameters.
- Compute a prediction interval and explain how it differs from a confidence interval.
- Compute sample sizes needed to ensure a confidence interval for means and proportions with a specified margin of error.

We discussed the difference between population and samples in Chapter 4. Sampling is the foundation of statistical analysis. We use sample data in business analytics applications for many purposes. For example, we might wish to estimate the mean, variance, or proportion of a very large or unknown population; provide values for inputs in decision models; understand customer satisfaction; reach a conclusion as to which of several sales strategies is most effective; or understand if a change in a process resulted in an improvement. In this chapter, we discuss sampling methods, how they are used to estimate population parameters, and how we can assess the error inherent in sampling.

In your career, you will discover many opportunities to apply statistical sampling and estimation. For example, marketing analysts routinely use surveys to understand customer demographics, brand and product attribute preferences, and customer satisfaction. Accounting professionals use sampling techniques to audit invoices, accounts receivable, and freight bills. Operations managers use sampling to verify the quality of incoming materials and production output. Human resource managers use surveys and sampling techniques to evaluate employee satisfaction. Companies in the entertainment industry routinely use sampling to rate TV shows, movies, and video games. So, every time you are asked to complete a survey, you are seeing sampling and estimation at work!

Statistical Sampling

The first step in sampling is to design an effective sampling plan that will yield representative samples of the populations under study. A **sampling plan** is a description of the approach that is used to obtain samples from a population prior to any data collection activity. A sampling plan states

- the objectives of the sampling activity,
- the target population,
- the **population frame** (the list from which the sample is selected),

- the method of sampling,
- the operational procedures for collecting the data, and
- the statistical tools that will be used to analyze the data.

EXAMPLE 6.1   A Sampling Plan for a Market Research Study

Suppose that a company wants to understand how golfers might respond to a membership program that provides discounts at golf courses in the golfers' locality as well as across the country. The *objective* of a sampling study might be to estimate the proportion of golfers who would likely subscribe to this program. The *target population* might be all golfers over 25 years old. However, identifying all golfers in America might be impossible. A practical *population frame* might be a list of golfers who have purchased equipment from national golf or sporting goods companies through which the discount card will be sold. The *operational procedures* for collecting the data might be an e-mail link to a survey site or direct-mail questionnaire. The data might be stored in an Excel database; *statistical tools* such as PivotTables and simple descriptive statistics would be used to segment the respondents into different demographic groups and estimate their likelihood of responding positively.

Sampling Methods

Many types of sampling methods exist. Sampling methods can be *subjective* or *probabilistic*. Subjective methods include **judgment sampling**, in which expert judgment is used to select the sample (survey the "best" customers), and **convenience sampling**, in which samples are selected based on the ease with which the data can be collected (survey all customers who happen to visit this month). Probabilistic sampling involves selecting the items in the sample using some random procedure. Probabilistic sampling is necessary to draw valid statistical conclusions.

The most common probabilistic sampling approach is simple random sampling. **Simple random sampling** involves selecting items from a population so that every subset of a given size has an equal chance of being selected. If the population data are stored in a database, simple random samples can generally be easily obtained.

EXAMPLE 6.2   Simple Random Sampling with Excel

Suppose that we wish to sample from the Excel database *Sales Transactions*. Excel provides a tool to generate a random set of values from a given population size. Click on *Data Analysis* in the *Analysis* group of the *Data* tab and select *Sampling*. This brings up the dialog shown in Figure 6.1. In the *Input Range* box, we specify the data range from which the sample will be taken. This tool requires that the data sampled be numeric, so in this example we sample from the first column of the data set, which corresponds to the customer ID number. There are two options for sampling:

1. Sampling can be *Periodic*, and we will be prompted for the *Period*, which is the interval between sample observations from the beginning of the data set. For instance, if a period of 5 is used, observations 5, 10, 15, and so on will be selected as samples.

2. Sampling can also be *Random*, and we will be prompted for the *Number of Samples*. Excel will then randomly select this number of samples from the specified data set. However, this tool generates random samples *with replacement*, so we must be careful to check for duplicate observations in the sample created.

Figure 6.2 shows 20 random samples generated by the tool. We sorted them in ascending order to make it easier to identify duplicates. As you can see, two of the customers were duplicated by the tool.

▶ **Figure 6.1**

Excel Sampling *Tool Dialog*

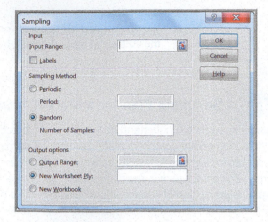

▶ **Figure 6.2**

*Samples Generated Using
the Excel* Sampling *Tool*

	A
1	**Sample of Customer IDs**
2	10009
3	10092
4	10102
5	10118
6	10167
7	10176
8	10256
9	10261
10	10266
11	10293
12	10320
13	10336
14	10355
15	10355
16	10377
17	10393
18	10413
19	10438
20	10438
21	10455

Other methods of sampling include the following:

- *Systematic (Periodic) Sampling.* **Systematic**, or **periodic**, **sampling** is a sampling plan (one of the options in the Excel *Sampling* tool) that selects every *n*th item from the population. For example, to sample 250 names from a list of 400,000, the first name could be selected at random from the first 1,600, and then every 1,600th name could be selected. This approach can be used for telephone sampling when supported by an automatic dialer that is programmed to dial numbers in a systematic manner. However, systematic sampling is not the same as simple random sampling because for any sample, every possible sample of a given size in the population does not have an equal chance of being selected. In some situations, this approach can induce significant bias if the population has some underlying pattern. For instance, sampling orders received every seven days may not yield a representative sample if customers tend to send orders on certain days every week.
- *Stratified Sampling.* **Stratified sampling** applies to populations that are divided into natural subsets (called *strata*) and allocates the appropriate proportion of

samples to each stratum. For example, a large city may be divided into political districts called wards. Each ward has a different number of citizens. A stratified sample would choose a sample of individuals in each ward proportionate to its size. This approach ensures that each stratum is weighted by its size relative to the population and can provide better results than simple random sampling if the items in each stratum are not homogeneous. However, issues of cost or significance of certain strata might make a disproportionate sample more useful. For example, the ethnic or racial mix of each ward might be significantly different, making it difficult for a stratified sample to obtain the desired information.

- *Cluster Sampling.* **Cluster sampling** is based on dividing a population into subgroups (clusters), sampling a set of clusters, and (usually) conducting a complete census within the clusters sampled. For instance, a company might segment its customers into small geographical regions. A cluster sample would consist of a random sample of the geographical regions, and all customers within these regions would be surveyed (which might be easier because regional lists might be easier to produce and mail).

- *Sampling from a Continuous Process.* Selecting a sample from a continuous manufacturing process can be accomplished in two main ways. First, select a time at random; then select the next *n* items produced after that time. Second, select *n* times at random; then select the next item produced after each of these times. The first approach generally ensures that the observations will come from a homogeneous population; however, the second approach might include items from different populations if the characteristics of the process should change over time, so caution should be used.

CHECK YOUR UNDERSTANDING

1. State the major elements of a sampling plan.
2. What is simple random sampling?
3. Describe and give an example of systematic, stratified, cluster, and continuous process sampling methods.

ANALYTICS IN PRACTICE: Using Sampling Techniques to Improve Distribution[1]

U.S. breweries rely on a three-tier distribution system to deliver product to retail outlets, such as supermarkets and convenience stores, and on-premise accounts, such as bars and restaurants. The three tiers are the manufacturer, wholesaler (distributor), and retailer. A distribution network must be as efficient and cost effective as possible to deliver to the market a fresh product that is damage free and is delivered at the right place at the right time.

To understand distributor performance related to overall effectiveness, MillerCoors brewery defined seven attributes of proper distribution and collected data from 500 of

(continued)

[1]Based on Tony Gojanovic and Ernie Jimenez, "Brewed Awakening: Beer Maker Uses Statistical Methods to Improve How Its Products Are Distributed," *Quality Progress* (April 2010).

its distributors. A field quality specialist (FQS) audits distributors within an assigned region of the country and collects data on these attributes. The FQS uses a handheld device to scan the universal product code on each package to identify the product type and amount. When audits are complete, data are summarized and uploaded from the handheld device into a master database.

This distributor auditing uses stratified random sampling with proportional allocation of samples based on the distributor's market share. In addition to providing a more representative sample and better logistical control of sampling, stratified random sampling enhances statistical precision when data are aggregated by market area served by the distributor. This enhanced precision is a consequence of smaller and typically homogeneous market regions, which are able to provide realistic estimates of variability, especially when compared to another market region that is markedly different.

Randomization of retail accounts is achieved through a specially designed program based on the GPS location of the distributor and serviced retail accounts. The sampling strategy ultimately addresses a specific distributor's performance related to out-of-code product, damaged product, and out-of-rotation product at the retail level. All in all, more than 6,000 of the brewery's national retail

Stephen Finn/Shutterstock.com

accounts are audited during a sampling year. Data collected by the FQSs during the year are used to develop a performance ranking of distributors and identify opportunities for improvement.

Estimating Population Parameters

Sample data provide the basis for many useful analyses to support decision making. **Estimation** involves assessing the value of an unknown population parameter—such as a population mean, population proportion, or population variance—using sample data. **Estimators** are the measures used to estimate population parameters; for example, we use the sample mean $\bar{x}$ to estimate a population mean μ. The sample variance s^2 estimates a population variance σ^2, and the sample proportion p estimates a population proportion π. A **point estimate** is a single number derived from sample data that is used to estimate the value of a population parameter.

Unbiased Estimators

It seems quite intuitive that the sample mean should provide a good point estimate for the population mean. However, it may not be clear why the formula for the sample variance that we introduced in Chapter 4 has a denominator of $n - 1$, particularly because it is different from the formula for the population variance (see formulas (4.7) and (4.8) in Chapter 4). In these formulas, the population variance is computed by

$$\sigma^2 = \frac{\sum_{i=1}^{n}(x_i - \mu)^2}{N}$$

whereas the sample variance is computed by the formula

$$s^2 = \frac{\sum\limits_{i=1}^{n}(x_i - \bar{x})^2}{n - 1}$$

Why is this so? Statisticians develop many types of estimators, and from a theoretical as well as a practical perspective, it is important that they "truly estimate" the population parameters they are supposed to estimate. Suppose that we perform an experiment in which we repeatedly sampled from a population and computed a point estimate for a population parameter. Each individual point estimate will vary from the population parameter; however, we would hope that the long-term average (expected value) of all possible point estimates would equal the population parameter. If the expected value of an estimator equals the population parameter it is intended to estimate, the estimator is said to be *unbiased*. If this is not true, the estimator is called *biased* and will not provide correct results.

Fortunately, all the estimators we have introduced are unbiased and, therefore, are meaningful for making decisions involving the population parameter. In particular, statisticians have shown that the denominator $n - 1$ used in computing s^2 is necessary to provide an unbiased estimator of σ^2. If we simply divided by the number of observations, the estimator would tend to underestimate the true variance.

Errors in Point Estimation

One of the drawbacks of using point estimates is that they do not provide any indication of the magnitude of the potential error in the estimate. A major metropolitan newspaper reported that, based on a Bureau of Labor Statistics survey, college professors were the highest-paid workers in the region, with an average salary of $150,004. Actual averages for two local universities were less than $70,000. What happened? As reported in a follow-up story, the sample size was very small and included a large number of highly paid medical school faculty; as a result, there was a significant error in the point estimate that was used.

When we sample, the estimators we use—such as a sample mean, sample proportion, or sample variance—are actually random variables that are characterized by some distribution. By knowing what this distribution is, we can use probability theory to quantify the uncertainty associated with the estimator. To understand this, we first need to discuss sampling error and sampling distributions.

In Chapter 4, we observed that different samples from the same population have different characteristics—for example, variations in the mean, standard deviation, frequency distribution, and so on. **Sampling (statistical) error** occurs because samples are only a subset of the total population. Sampling error is inherent in any sampling process, and although it can be minimized, it cannot be totally avoided. Another type of error, called **nonsampling error**, occurs when the sample does not represent the target population adequately. This is generally a result of poor sample design, such as using a convenience sample when a simple random sample would have been more appropriate, or choosing the wrong population frame. It may also result from inadequate data reliability, which we discussed in Chapter 1. To draw good conclusions from samples, analysts need to eliminate nonsampling error and understand the nature of sampling error.

Sampling error depends on the size of the sample relative to the population. Thus, determining the number of samples to take is essentially a statistical issue that is based on the accuracy of the estimates needed to draw a useful conclusion. We discuss this later in this chapter. However, from a practical standpoint, one must also consider the cost of sampling and sometimes make a trade-off between cost and the information that is obtained.

Understanding Sampling Error

Suppose that we estimate the mean of a population using the sample mean. How can we determine how accurate we are? In other words, can we make an informed statement about how far the sample mean might be from the true population mean? We could gain some insight into this question by performing a sampling experiment as Example 6.3 illustrates.

If we apply the empirical rules to these results, we can estimate the sampling error associated with one of the sample sizes we have chosen (see Example 6.4).

EXAMPLE 6.3   A Sampling Experiment

Let us choose a population that is uniformly distributed between $a = 0$ and $b = 10$. Formulas (5.20) and (5.21) state that the expected value (population mean) is $(0 + 10)/2 = 5$, and the variance is $(10 - 0)^2/12 = 8.333$. We use the Excel *Random Number Generation* tool described in Chapter 5 to generate 25 samples, each of size ten, from this population. Figure 6.3 shows a portion of a spreadsheet for this experiment, along with a histogram of the data (on the left side) that shows that the 250 observations are approximately uniformly distributed. (This is available in the Excel file *Sampling Experiment*.)

In row 12, we compute the mean of each sample. These statistics vary a lot from the population values because of sampling error. The histogram on the right shows the distribution of the 25 sample means, which vary from less than 4 to more than 6. Now let's compute the average and standard deviation of the sample means in row 12 (cells

AB12 and AB13). Note that the average of all the sample means is quite close to the true population mean of 5.0.

Now let us repeat this experiment for larger sample sizes. Table 6.1 shows some results. Notice that as the sample size gets larger, the averages of the 25 sample means are all still close to the expected value of 5; however, the standard deviation of the 25 sample means becomes smaller for increasing sample sizes, meaning that the means of samples are clustered closer together around the true expected value. Figure 6.4 shows comparative histograms of the sample means for each of these cases. These illustrate the conclusions we just made and, perhaps even more surprisingly, the distribution of the sample means appears to assume the shape of a normal distribution for larger sample sizes. In our experiment, we used only 25 sample means. If we had used a much-larger number, the distributions would have been more well defined.

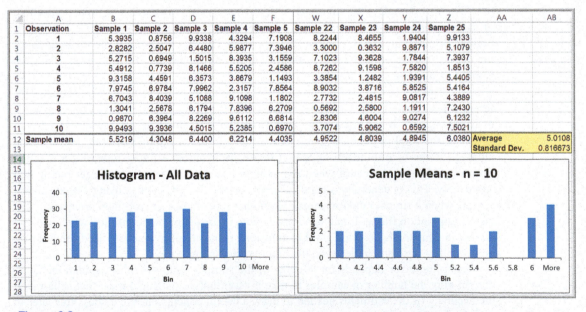

▲ **Figure 6.3**

Portion of Spreadsheet for **Sampling Experiment**

▶ **Table 6.1**

Results from Sampling Experiment

Sample Size	Average of 25 Sample Means	Standard Deviation of 25 Sample Means
10	5.0108	0.816673
25	5.0779	0.451351
100	4.9137	0.301941
500	4.9754	0.078993

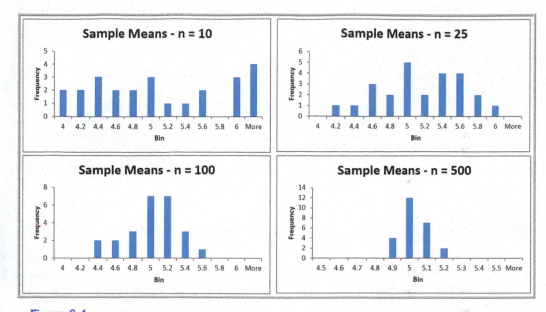

▲ **Figure 6.4**

Histograms of Sample Means for Increasing Sample Sizes

EXAMPLE 6.4   **Estimating Sampling Error Using the Empirical Rules**

Using the standard deviations in Table 6.1 in Example 6.3 and the empirical rule for three standard deviations around the mean, we could state, for example, that using a sample size of ten, the distribution of sample means should fall within three standard deviations of the population mean, 5, or approximately from $5.0 - 3(0.816673) = 2.55$ to $5.0 + 3(0.816673) = 7.45$. Thus, there is considerable error in estimating the mean using a sample of only ten. For a sample of size 25, we would expect the sample means to fall between $5.0 - (0.451351) = 3.65$ and $5.0 + 3(0.451351) = 6.35$. Note that as the sample size increased, the error decreased. For sample sizes of 100 and 500, the intervals are [4.09, 5.91] and [4.76, 5.24].

CHECK YOUR UNDERSTANDING

1. What is an unbiased estimator, and why is it important in statistics?

2. Explain the difference between sampling and nonsampling error.

3. What happens to the distribution of the sample mean as the sample size increases?

Sampling Distributions

We can quantify the sampling error in estimating the mean for any unknown population. To do this, we need to characterize the sampling distribution of the mean.

Sampling Distribution of the Mean

The means of *all possible* samples of a fixed size n from some population will form a distribution that we call the **sampling distribution of the mean**. The histograms in Figure 6.4 are approximations to the sampling distributions of the mean based on 25 samples. Statisticians have shown two key results about the sampling distribution of the mean. First, the standard deviation of the sampling distribution of the mean, called the **standard error of the mean**, is computed as

$$\text{Standard Error of the Mean} = \sigma/\sqrt{n} \qquad (6.1)$$

where σ is the standard deviation of the population from which the individual observations are drawn and n is the sample size. From this formula, we see that as n increases, the standard error decreases, just as our experiment demonstrated. This suggests that the estimates of the mean that we obtain from larger sample sizes provide greater accuracy in estimating the true population mean. In other words, *larger sample sizes have less sampling error.*

EXAMPLE 6.5	**Computing the Standard Error of the Mean**

For our experiment, we know that the variance of the population is 8.33 (because the values were uniformly distributed). Therefore, the standard deviation of the population is $\sigma = 2.89$. We may compute the standard error of the mean for each of the sample sizes in our experiment using formula (6.1). For example, with $n = 10$, we have

Standard Error of the Mean $= \sigma/\sqrt{n} = 2.89/\sqrt{10} = 0.914$

For the remaining data in Table 6.1 we have the following:

Sample Size, n	Standard Error of the Mean
10	0.914
25	0.5778
100	0.289
500	0.129

The standard deviations shown in Table 6.1 are simply estimates of the standard error of the mean based on using only 25 samples. If we compare these estimates with the theoretical values in the previous example, we see that they are close but not exactly the same. This is because the true standard error is based on *all possible* sample means in the sampling distribution, whereas we used only 25. If you repeat the experiment with a larger number of samples, the observed values of the standard error would be closer to these theoretical values.

In practice, we will never know the true population standard deviation and generally take only a limited sample of n observations. However, we may estimate the standard error of the mean using the sample data by simply dividing the sample standard deviation by the square root of n.

The second result that statisticians have shown is called the **central limit theorem**, one of the most important practical results in statistics that makes systematic inference possible. The central limit theorem states that if the sample size is large enough, the sampling distribution of the mean is approximately normally distributed, *regardless* of the distribution of the population, and that the mean of the sampling distribution will be the same as that of the population. This is exactly what we observed in our experiment. The distribution of the population was uniform, yet the sampling distribution of the mean converges to the shape of a normal distribution as the sample size increases. The central limit theorem also states that if the population is normally distributed, then the sampling distribution of the mean will also be normal for *any* sample size. The central limit theorem allows us to use the theory we learned about calculating probabilities for normal distributions to draw conclusions about sample means.

Applying the Sampling Distribution of the Mean

The key to applying sampling distribution of the mean correctly is to understand whether the probability that you wish to compute relates to an individual observation or to the mean of a sample. If it relates to the mean of a sample, then you must use the sampling distribution of the mean, whose standard deviation is the standard error, $\sigma/\sqrt{n}$.

EXAMPLE 6.6   Using the Standard Error in Probability Calculations

Suppose that the size of individual customer orders (in dollars), X, from a major discount book publisher Web site is normally distributed with a mean of \$36 and standard deviation of \$8. The probability that the next individual who places an order at the Web site will make a purchase of more than \$40 can be found by calculating

$1 - \text{NORM.DIST}(40, 36, 8, \text{TRUE}) = 1 - 0.6915 = 0.3085$

Now suppose that a sample of 16 customers is chosen. What is the probability that the *mean purchase* for these 16 customers will exceed \$40? To find this, we must realize that we must use the sampling distribution of the mean to carry out the appropriate calculations. The sampling distribution

of the mean will have a mean of \$36 but a standard error of $\$8/\sqrt{16} = \$2$. Then the probability that the mean purchase exceeds \$40 for a sample size of n = 16 is

$1 - \text{NORM.DIST}(40, 36, 2, \text{TRUE}) = 1 - 0.9772 = 0.0228$

Although about 30% of individuals will make purchases exceeding \$40, the chance that 16 customers will collectively average more than \$40 is much smaller. It would be very unlikely for all 16 customers to make high-volume purchases, because some individual purchases would as likely be less than \$36 as more, making the variability of the mean purchase amount for the sample of 16 much smaller than that for individuals.

CHECK YOUR UNDERSTANDING

1. What is the sampling distribution of the mean?
2. Define the standard error of the mean, and explain how to compute it.
3. What does the central limit theorem state, and of what value is it?
4. Explain when to use the standard error in probability calculations versus using the standard deviation of the population.

 Interval Estimates

An **interval estimate** provides a range for a population characteristic based on a sample. Intervals are quite useful in statistics because they provide more information than a point estimate. Intervals specify a range of plausible values for the characteristic of interest and a way of assessing "how plausible" they are. In general, a $100(1 - \alpha)\%$ **probability interval** is any interval $[A, B]$ such that the probability of falling between A and B is $1 - \alpha$. Probability intervals are often centered on the mean or median. For instance, in a normal distribution, the mean plus or minus 1 standard deviation describes an approximate 68% probability interval around the mean. As another example, the 5th and 95th percentiles in a data set constitute a 90% probability interval.

EXAMPLE 6.7   Interval Estimates in the News

We see interval estimates in the news all the time when trying to estimate the mean or proportion of a population. Interval estimates are often constructed by taking a point estimate and adding and subtracting a margin of error that is based on the sample size. For example, a Gallup poll might report that 56% of voters support a certain candidate with a margin of error of $\pm 3\%$. We would conclude that the true percentage of voters that support the candidate is most likely between 53% and 59%. Therefore, we would have a lot of confidence in predicting that the candidate would win a forthcoming election. If, however, the poll showed a 52% level of support with a margin of error of $\pm 4\%$, we might not be as confident in predicting a win because the true percentage of supportive voters is likely to be somewhere between 48% and 56%.

The question you might be asking at this point is how to calculate the error associated with a point estimate. In national surveys and political polls, such margins of error are usually stated, but they are never properly explained. To understand them, we need to introduce the concept of confidence intervals.

Confidence Intervals

Confidence interval estimates provide a way of assessing the accuracy of a point estimate. A **confidence interval** is a range of values between which the value of the population parameter is believed to be, along with a probability that the interval correctly estimates the true (unknown) population parameter. This probability is called the **level of confidence**, denoted by $1 - \alpha$, where α is a number between 0 and 1. The level of confidence is usually expressed as a percent; common values are 90%, 95%, or 99%. (Note that if the level of confidence is 90%, then $\alpha = 0.1$.) The margin of error depends on the level of confidence and the sample size. For example, suppose that the margin of error for some sample size and a level of confidence of 95% is calculated to be 2.0. One sample might yield a point estimate of 10. Then, a 95% confidence interval would be [8, 12]. However, this interval may or may not include the true population mean. If we take a different sample, we will most likely have a different point estimate, say, 10.4, which, given the same margin of error, would yield the interval estimate [8.4, 12.4]. Again, this may or may not include the true population mean. If we chose 100 different samples, leading to 100 different interval estimates, we would expect that 95% of them—the level of confidence—would contain the true population mean. We would say we are "95% confident" that the interval we obtain from sample data contains the true population mean. The higher the confidence level, the more assurance we have that the interval contains the true population parameter. As the confidence level increases, the confidence interval becomes wider to provide higher levels of assurance. You can view α as the risk of incorrectly concluding that the confidence interval contains the true mean.

When national surveys or political polls report an interval estimate, they are actually confidence intervals. However, the level of confidence is generally not stated because the average person would probably not understand the concept or terminology. While not stated, you can probably assume that the level of confidence is 95%, as this is the most common value used in practice (however, the Bureau of Labor Statistics tends to use 90% quite often).

Many different types of confidence intervals may be developed. The formulas used depend on the population parameter we are trying to estimate and possibly other characteristics or assumptions about the population. We illustrate a few types of confidence intervals.

Confidence Interval for the Mean with Known Population Standard Deviation

The simplest type of confidence interval is for the mean of a population where the standard deviation is assumed to be known. You should realize, however, that in nearly all practical sampling applications, the population standard deviation will *not* be known. However, in some applications, such as measurements of parts from an automated machine, a process might have a very stable variance that has been established over a long history, and it can reasonably be assumed that the standard deviation is known.

A $100(1 - \alpha)\%$ confidence interval for the population mean μ based on a sample of size n with a sample mean $\bar{x}$ and a known population standard deviation σ is given by

$$\bar{x} \pm z_{\alpha/2}(\sigma/\sqrt{n}) \tag{6.2}$$

Note that this formula is simply the sample mean (point estimate) plus or minus a margin of error.

The margin of error is a number $z_{\alpha/2}$ multiplied by the standard error of the sampling distribution of the mean, $\sigma/\sqrt{n}$. The value $z_{\alpha/2}$ represents the value of a standard normal random variable that has an upper tail probability of $\alpha/2$ or, equivalently, a cumulative probability of $1 - \alpha/2$. It may be found from the standard normal table (see Table A.1 in Appendix A at the end of the book) or may be computed in Excel using the value of the function NORM.S.INV($1 - \alpha/2$). For example, if $\alpha = 0.05$ (for a 95% confidence interval), then =NORM.S.INV(0.975) = 1.96; if $\alpha = 0.10$ (for a 90% confidence interval), then =NORM.S.INV(0.95) = 1.645, and so on.

Although formula (6.2) can easily be implemented in a spreadsheet, the Excel function CONFIDENCE.NORM(*alpha, standard_deviation, size*) can be used to compute the margin of error term, $z_{\alpha/2}\,\sigma/\sqrt{n}$; thus, the confidence interval is the sample mean $\pm$ CONFIDENCE.NORM(*alpha, standard_deviation, size*).

EXAMPLE 6.8   Computing a Confidence Interval with a Known Standard Deviation

In a production process for filling bottles of liquid detergent, historical data have shown that the variance in the volume is constant; however, clogs in the filling machine often affect the average volume. The historical standard deviation is 15 milliliters. In filling 800-milliliter bottles, a sample of 25 found an average volume of 796 milliliters. Using formula (6.2), a 95% confidence interval for the population mean is

$$\bar{x} \pm z_{\alpha/2}\,(\sigma/\sqrt{n})$$

$$= 796 \pm 1.96(15/\sqrt{25}) = 796 \pm 5.88, \text{ or } [790.12, 801.88]$$

The worksheet *Population Mean Sigma Known* in the Excel workbook *Confidence Intervals* computes this interval using the CONFIDENCE.NORM function to compute the margin of error in cell B9, as shown in Figure 6.5.

▶ **Figure 6.5**

Confidence Interval for Mean Liquid Detergent Filling Volume

	A	B	C	D	E	F
1	Confidence Interval for Population Mean, Standard Deviation Known					
2						
3	Alpha	0.05				
4	Standard deviation	15				
5	Sample size	25				
6	Sample average	796				
7						
8	Confidence Interval	95%				
9	Error	5.879892				
10	Lower	790.1201				
11	Upper	801.8799				

As the level of confidence, $1 - \alpha$, decreases, $z_{\alpha/2}$ decreases, and the confidence interval becomes narrower. For example, a 90% confidence interval will be narrower than a 95% confidence interval. Similarly, a 99% confidence interval will be wider than a 95% confidence interval. Essentially, you must trade off a higher level of accuracy with the risk that the confidence interval does not contain the true mean. Smaller risk will result in a wider confidence interval. However, you can also see that as the sample size increases, the standard error decreases, making the confidence interval narrower and providing a more accurate interval estimate for the same level of risk. So if you wish to reduce the risk, you should consider increasing the sample size.

The *t*-Distribution

In most practical applications, the standard deviation of the population is unknown, and we need to calculate the confidence interval differently. Before we can discuss how to compute this type of confidence interval, we need to introduce a new probability distribution called the *t*-distribution. The *t*-distribution is actually a family of probability distributions with a shape similar to the standard normal distribution. Different *t*-distributions are distinguished by an additional parameter, **degrees of freedom** (*df*). The *t*-distribution has a larger variance than the standard normal, thus making confidence intervals wider than those obtained from the standard normal distribution, in essence, correcting for the uncertainty about the true standard deviation, which is not known. As the number of degrees of freedom increases, the *t*-distribution converges to the standard normal distribution (Figure 6.6). When sample sizes get to be as large as 120, the distributions are virtually identical; even for sample sizes as low as 30 to 35, it becomes difficult to distinguish between the two. Thus, for large sample sizes, many people use *z*-values as in formula (6.2) to establish confidence intervals even when the population standard deviation is unknown but estimated from the sample data. We must point out, however, that for any sample size, the *true* sampling distribution of the mean is the *t*-distribution; so when in doubt, use the *t*-distribution as we will illustrate in the next section.

The concept of degrees of freedom can be puzzling. It can best be explained by examining the formula for the sample variance:

$$s^2 = \frac{\sum_{i=1}^{n}(x_i - \bar{x})^2}{n - 1}$$

Note that to compute s^2, we first need to compute the sample mean, $\bar{x}$. If we know the value of the mean, then we need know only $n - 1$ distinct observations; the *n*th is completely determined. (For instance, if the mean of three numbers is 4 and you know that two of the numbers are 2 and 4, you can easily determine that the third number must be 6.) The

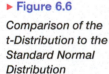

▶ **Figure 6.6**

Comparison of the t-Distribution to the Standard Normal Distribution

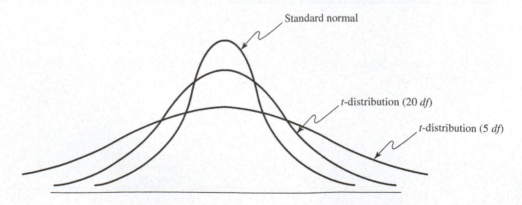

number of sample values that are free to vary defines the number of degrees of freedom; in general, *df* equals the number of sample values minus the number of estimated parameters. Because the sample variance uses one estimated parameter, the mean, the *t*-distribution used in confidence interval calculations has $n - 1$ degrees of freedom. Because the *t*-distribution explicitly accounts for the effect of the sample size in estimating the population variance, it is the proper one to use for any sample size. However, for large samples, the difference between *t*- and *z*-values is very small, as we noted earlier.

Confidence Interval for the Mean with Unknown Population Standard Deviation

The formula for a $100(1 - \alpha)\%$ confidence interval for the mean μ when the population standard deviation is unknown is

$$\bar{x} \pm t_{\alpha/2, n-1}(s/\sqrt{n}) \tag{6.3}$$

where $t_{\alpha/2, n-1}$ is the value from the *t*-distribution with $n - 1$ degrees of freedom, giving an upper-tail probability of $\alpha/2$. We may find *t*-values in Table A.2 in Appendix A at the end of the book or by using the Excel function T.INV$(1 - \alpha/2, n - 1)$ or the function T.INV.2T$(\alpha, n - 1)$. The Excel function CONFIDENCE.T(*alpha, standard_deviation, size*) can be used to compute the margin of error term, $t_{\alpha/2, n-1}(s/\sqrt{n})$; thus, the confidence interval is the sample mean $\pm$ CONFIDENCE.T(*alpha, standard_deviation, size*).

EXAMPLE 6.9 **Computing a Confidence Interval with Unknown Standard Deviation**

In the Excel file *Credit Approval Decisions*, a large bank has sample data used in making credit approval decisions (see Figure 6.7). Suppose that we want to find a 95% confidence interval for the mean revolving balance for the population of applicants who own a home. First, sort the data by homeowner and compute the mean and standard deviation of the revolving balance for the sample of homeowners. This results in $\bar{x} = \$12,630.37$ and $s = \$5,393.38$. The sample

size is $n = 27$, so the standard error $s/\sqrt{n} = \$1,037.96$. The *t*-distribution has 26 degrees of freedom; therefore, $t_{.025,26} = 2.056$. Using formula (6.3), the confidence interval is $\$12,630.37 \pm 2.056(\$1,037.96)$ or [$10,497, $14,764]. The worksheet *Population Mean Sigma Unknown* in the Excel workbook *Confidence Intervals* computes this interval using the CONFIDENCE.T function to compute the margin of error in cell B10, as shown in Figure 6.8.

Confidence Interval for a Proportion

For categorical variables such as gender (male or female), education (high school, college, post-graduate), and so on, we are usually interested in the *proportion* of observations in a sample that has a certain characteristic. An unbiased estimator of a population

	A	B	C	D	E	F
1	**Credit Approval Decisions**					
2						
3	**Homeowner**	**Credit Score**	**Years of Credit History**	**Revolving Balance**	**Revolving Utilization**	**Decision**
4	Y	725	20	$ 11,320	25%	Approve
5	Y	573	9	$ 7,200	70%	Reject
6	Y	677	11	$ 20,000	55%	Approve
7	N	625	15	$ 12,800	65%	Reject
8	N	527	12	$ 5,700	75%	Reject
9	Y	795	22	$ 9,000	12%	Approve
10	N	733	7	$ 35,200	20%	Approve

▲ **Figure 6.7**

Portion of Excel File Credit Approval Decisions

▶ **Figure 6.8**

*Confidence Interval for
Mean Revolving Balance of
Homeowners*

	A	B	C	D	E
1	Confidence Interval for Population Mean, Standard Deviation Unknown				
2					
3	Alpha	0.05			
4	Sample standard deviation	5393.38			
5	Sample size	27			
6	Sample average	12630.37			
7					
8	Confidence Interval	95%			
9	t-value	2.056			
10	Error	2133.55			
11	Lower	10496.82			
12	Upper	14763.92			

proportion π (this is not the *number* pi $= 3.14159 \ldots$) is the statistic $\hat{p} = x/n$ (the **sample proportion**), where x is the number in the sample having the desired characteristic and n is the sample size.

A $100(1 - \alpha)\%$ confidence interval for the proportion is

$$\hat{p} \pm z_{\alpha/2}\sqrt{\frac{\hat{p}(1-\hat{p})}{n}} \qquad (6.4)$$

Notice that as with the mean, the confidence interval is the point estimate plus or minus some margin of error. In this case, $\sqrt{\hat{p}(1-\hat{p})/n}$ is the standard error for the sampling distribution of the proportion. Excel does not have a function for computing the margin of error, but it can easily be implemented on a spreadsheet. The value of $z_{\alpha/2}$ is found using the Excel function NORM.S.INV$(1 - \alpha/2)$.

EXAMPLE 6.10   **Computing a Confidence Interval for a Proportion**

The last column in the Excel file *Insurance Survey* (see Figure 6.9) describes whether a sample of employees would be willing to pay a lower premium for a higher deductible for their health insurance. Suppose we are interested in the proportion of individuals who answered yes. We may easily confirm that 6 out of the 24 employees, or 25%, answered yes. Thus, a point estimate for the proportion answering yes is $\hat{p} = 0.25$. Using formula (6.4), we find that a 95% confidence interval for the proportion of employees answering yes is

$$0.25 \pm 1.96\sqrt{\frac{0.25(0.75)}{24}} = 0.25 \pm 0.173, \text{ or } [0.077, 0.423]$$

The worksheet *Proportion* in the Excel workbook *Confidence Intervals* computes this interval, as shown in Figure 6.10. Notice that this is a fairly wide confidence interval, suggesting that we have quite a bit of uncertainty as to the true value of the population proportion. This is because of the relatively small sample size.

	A	B	C	D	E	F	G
1	Insurance Survey						
2							
3	Age	Gender	Education	Marital Status	Years Employed	Satisfaction*	Premium/Deductible**
4	36	F	Some college	Divorced	4	4	N
5	55	F	Some college	Divorced	2	1	N
6	61	M	Graduate degree	Widowed	26	3	N
7	65	F	Some college	Married	9	4	N
8	53	F	Graduate degree	Married	6	4	N
9	50	F	Graduate degree	Married	10	5	N
10	28	F	College graduate	Married	4	5	N
11	62	F	College graduate	Divorced	9	3	N
12	48	M	Graduate degree	Married	6	5	N

▲ **Figure 6.9**

Portion of Excel File Insurance Survey

▶ **Figure 6.10**

Confidence Interval for the Proportion

	A	B
1	Confidence Interval for a Proportion	
2		
3	Alpha	0.05
4	Sample proportion	0.25
5	Sample size	24
6		
7	Confidence Interval	95%
8	z-value	1.96
9	Standard error	0.088388
10	Lower	0.076762
11	Upper	0.423238

Additional Types of Confidence Intervals

Confidence intervals may be calculated for other population parameters such as a variance or standard deviation and also for differences in the means or proportions of two populations. The concepts are similar to the types of confidence intervals we have discussed, but many of the formulas are rather complex and more difficult to implement on a spreadsheet. Some advanced software packages and spreadsheet add-ins provide additional support. Therefore, we do not discuss them in this book, but we do suggest that you consult other books and statistical references should you need to use them, now that you understand the basic concepts underlying them.

CHECK YOUR UNDERSTANDING

1. What is an interval estimate?

2. Explain the difference between a probability interval and a confidence interval.

3. What does the level of confidence specify?

4. Explain the differences in the formulas used to compute confidence intervals for the mean with known and unknown standard deviations.

5. How does the *t*-distribution compare with the normal distribution?

Using Confidence Intervals for Decision Making

Confidence intervals can be used in many ways to support business decisions.

EXAMPLE 6.11 **Drawing a Conclusion About a Population Mean Using a Confidence Interval**

In packaging a commodity product such as laundry detergent, the manufacturer must ensure that the packages contain the stated amount to meet government regulations. In Example 6.8, we saw an example where the required volume is 800 milliliters, yet the sample average was only 796 milliliters. Does this indicate a serious problem? Not necessarily. The 95% confidence interval for the mean we computed in Figure 6.5 was [790.12, 801.88]. Although the sample mean is less than 800, the sample does not provide sufficient evidence to draw the conclusion that the population mean is less than 800 because 800 is contained within

the confidence interval. In fact, it is just as plausible that the population mean is 801. We cannot tell definitively because of the sampling error. However, suppose that the sample average is 792. Using the Excel worksheet *Population Mean Sigma Known* in the workbook *Confidence Intervals*, we find that the confidence interval for the mean would be [786.12, 797.88]. In this case, we would conclude that it is highly unlikely that the population mean is 800 milliliters because the confidence interval falls completely below 800; the manufacturer should check and adjust the equipment to meet the standard.

The next example shows how to interpret a confidence interval for a proportion.

EXAMPLE 6.12 Using a Confidence Interval to Predict Election Returns

Suppose that an exit poll of 1,300 voters found that 692 voted for a particular candidate in a two-person race. This represents a proportion of 53.23% of the sample. Could we conclude that the candidate will likely win the election? A 95% confidence interval for the proportion is [0.505, 0.559]. This suggests that the population proportion of voters who favor this candidate is highly likely to exceed 50%, so it is safe to predict the winner. On the other hand,

suppose that only 670 of the 1,300 voters voted for the candidate, a sample proportion of 0.515. The confidence interval for the population proportion is [0.488, 0.543]. Even though the sample proportion is larger than 50%, the sampling error is large, and the confidence interval suggests that it is reasonably likely that the true population proportion could be less than 50%, so it would not be wise to predict the winner based on this information.

Data Visualization for Confidence Interval Comparison

We may use an Excel stock chart to visualize confidence intervals easily (the stock chart requires that you have at least three variables). This is particularly useful for comparing confidence intervals for different groups.

EXAMPLE 6.13 Creating a Stock Chart for Confidence Intervals

Suppose that we constructed confidence intervals for monthly credit card debt for a sample of banking customers in four age groups. The data are shown below.

Age Group	Upper CI	Lower CI	Mean
25–34	$2,103	$711	$1,407.14
35–44	$1,617	$872	$1,244.44
45–54	$1,114	$468	$ 791.30
55–64	$1,931	$309	$1,120.00

The columns should correspond to the variable name, upper CI limit, lower CI limit, and mean, in that order. Next, highlight the range of this table and insert an Excel High-Low-Close Stock Chart. Right-click one of the confidence intervals in the chart and choose *Format High-Low Lines* . . . . In the drop-down menu for *High-Low Line Options*, choose one of the series (Upper CI, Lower CI, or Mean) to format the markers.

In the *Format Data Series* pane, click the paint icon and then *Marker*, making sure to expand the *Marker Options* menu. Choose the type of marker you wish and

increase the width of the markers to make them more visible. We chose a "+" for the upper CI limit, a dash for the Lower CI limit, and a diamond for the mean. This results in the chart shown in Figure 6.11.

We may use confidence intervals and their visualizations to determine whether the means of two populations are significantly different from each other. The rule is simple: If the confidence intervals overlap, then you cannot conclude statistically that the means differ. However, if the confidence intervals do not overlap, then you can conclude that there is a statistical difference in the means. In Example 6.13, for instance, suppose we wish to determine whether the mean of the 35–44 age group is significantly different from that of the 45–54 age group. In Figure 6.11, we see that the confidence intervals overlap, so we cannot conclude that the means are significantly different. In the next chapter, we will discuss these concepts in a more formalized manner using a statistical approach called hypothesis testing.

▶ **Figure 6.11**

Visualization of Confidence Intervals Using an Excel Stock Chart

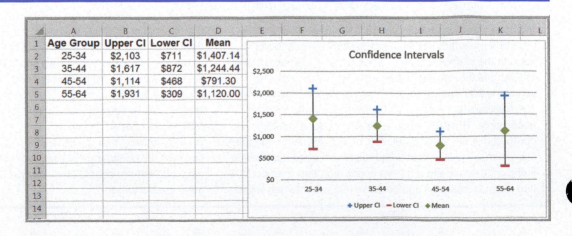

CHECK YOUR UNDERSTANDING

1. Explain how to use confidence intervals to draw conclusions about the population mean or proportion.

2. How can confidence interval visualization be used to draw conclusions about a population?

Prediction Intervals

Another type of interval used in estimation is a prediction interval. A **prediction interval** is one that provides a range for predicting the value of a new observation from the same population. This is different from a confidence interval, which provides an interval estimate of a population parameter, such as the mean or proportion. A confidence interval is associated with the *sampling distribution* of a statistic, but a prediction interval is associated with the distribution of the random variable itself.

When the population standard deviation is unknown, a $100(1 - \alpha)\%$ prediction interval for a new observation is

$$\bar{x} \pm t_{\alpha/2, n-1}\left(s\sqrt{1 + \frac{1}{n}}\right) \tag{6.5}$$

Note that this interval is wider than the confidence interval in formula (6.3) because of the additional value of 1 under the square root. This is because, in addition to estimating the population mean, we must also account for the variability of the new observation around the mean.

One important thing to realize also is that in formula (6.3) for a confidence interval, as n gets large, the error term tends to zero so the confidence interval converges on the mean. However, in the prediction interval formula (6.5), as n gets large, the error term converges to $t_{\alpha/2, n-1}(s)$, which is simply a $100(1 - \alpha)\%$ probability interval. Because we are trying to predict a new observation from the population, there will always be uncertainty.

EXAMPLE 6.14 **Computing a Prediction Interval**

In estimating the revolving balance in the Excel file *Credit Approval Decisions* in Example 6.9, we may use formula (6.5) to compute a 95% prediction interval for the revolving balance of a new homeowner as

$$\$12,630.37 \pm 2.056(\$5,393.38)\sqrt{1 + \frac{1}{27}}, \text{ or }$$
$$[\$1,338.10, \$23,922.64]$$

Note that compared with Example 6.9, the size of the prediction interval is considerably wider than that of the confidence interval.

CHECK YOUR UNDERSTANDING

1. How does a prediction interval differ from a confidence interval?

2. Why are prediction intervals wider than confidence intervals?

Confidence Intervals and Sample Size

An important question in sampling is the size of the sample to take. Note that in all the formulas for confidence intervals, the sample size plays a critical role in determining the width of the confidence interval. As the sample size increases, the width of the confidence interval decreases, providing a more accurate estimate of the true population parameter. In many applications, we would like to control the margin of error in a confidence interval. For example, in reporting voter preferences, we might wish to ensure that the margin of error is $\pm 2\%$. Fortunately, it is relatively easy to determine the appropriate sample size needed to estimate the population parameter within a specified level of precision.

The formulas for determining sample sizes to achieve a given margin of error are based on the confidence interval half-widths. For example, consider the confidence interval for the mean with a known population standard deviation we introduced in formula (6.2):

$$\bar{x} \pm z_{\alpha/2}\left(\frac{\sigma}{\sqrt{n}}\right)$$

Suppose we want the width of the confidence interval on either side of the mean (that is, the margin of error) to be at most E. In other words,

$$E \geq z_{\alpha/2}\left(\frac{\sigma}{\sqrt{n}}\right)$$

Solving for n, we find:

$$n \geq (z_{\alpha/2})^2\frac{\sigma^2}{E^2} \tag{6.6}$$

Of course, we generally do not know the population standard deviation prior to finding the sample size. A commonsense approach would be to take an initial sample to estimate the population standard deviation using the sample standard deviation s and determine the required sample size, collecting additional data if needed. If the half-width of the resulting confidence interval is within the required margin of error, then we clearly have achieved our goal. If not, we can use the new sample standard deviation s to determine a new sample size and collect additional data as needed. Note that if s changes significantly, we still might not have achieved the desired precision and might have to repeat the process. Usually, however, this will be unnecessary.

In a similar fashion, we can compute the sample size required to achieve a desired confidence interval half-width for a proportion by solving the following equation (based on formula (6.4) using the population proportion π in the margin of error term) for n:

$$E \geq z_{\alpha/2}\sqrt{\pi(1-\pi)/n}$$

This yields

$$n \geq (z_{\alpha/2})^2\frac{\pi(1-\pi)}{E^2} \tag{6.7}$$

In practice, the value of π will not be known. You could use the sample proportion from a preliminary sample as an estimate of π to plan the sample size, but this might require several iterations and additional samples to find the sample size that yields the required precision. When no information is available, the most conservative estimate is to set $\pi = 0.5$. This maximizes the quantity $\pi(1 - \pi)$ in the formula, resulting in the sample size that will guarantee the required precision no matter what the true proportion is.

EXAMPLE 6.15   Sample Size Determination for the Mean

In the liquid detergent example (Example 6.8), the confidence interval we computed in Figure 6.5 was [790.12, 801.88]. The width of the confidence interval is ± 5.88 milliliters, which represents the sampling error. Suppose the manufacturer would like the sampling error to be at most 3 milliliters. Using formula (6.6), we may compute the required sample size as follows:

$$n \geq (z_{\alpha/2})^2 \frac{(\sigma^2)}{E^2}$$

$$= (1.96)^2 \frac{(15^2)}{3^2} = 96.04$$

Rounding up, we find that 97 samples would be needed. To verify this, Figure 6.12 shows that if a sample of 97 is used along with the same sample mean and standard deviation, the confidence interval does indeed have a sampling error less than 3 milliliters.

EXAMPLE 6.16   Sample Size Determination for a Proportion

For the voting example we discussed, suppose that we wish to determine the number of voters to poll to ensure a sampling error of at most $\pm 2\%$. As we stated, when no information is available, the most conservative approach is to use 0.5 for the estimate of the true proportion. Using formula (6.7) with $\pi = 0.5$, the number of voters to poll to obtain a 95% confidence interval on the proportion of voters that choose a particular candidate with a precision of ± 0.02 or less is

$$n \geq (z_{\alpha/2})^2 \frac{\pi(1 - \pi)}{E^2}$$

$$= (1.96)^2 \frac{(0.5)(1 - 0.5)}{0.02^2} = 2,401$$

CHECK YOUR UNDERSTANDING

1. Explain why it is important to determine the proper sample size to use in a survey.

2. How do you find the sample size for a proportion if no good estimate of the population proportion is available?

► **Figure 6.12**

Confidence Interval for the Mean Using Sample Size = 97

	A	B	C	D	E	F
1	Confidence Interval for Population Mean, Standard Deviation Known					
2						
3	Alpha	0.05				
4	Standard deviation	15				
5	Sample size	97				
6	Sample average	796				
7						
8	Confidence Interval	95%				
9	Error	2.985063				
10	Lower	793.0149				
11	Upper	798.9851				

KEY TERMS

Central limit theorem	Population frame
Cluster sampling	Prediction interval
Confidence interval	Probability interval
Convenience sampling	Sample proportion
Degrees of freedom (*df*)	Sampling distribution of the mean
Estimation	Sampling plan
Estimators	Sampling (statistical) error
Interval estimate	Simple random sampling
Judgment sampling	Standard error of the mean
Level of confidence	Stratified sampling
Nonsampling error	Systematic (or periodic) sampling
Point estimate	*t*-Distribution

CHAPTER 6 TECHNOLOGY HELP

Useful Excel Functions

NORM.S.INV($1 - \alpha/2$) Finds the value of $z_{\alpha/2}$, a standard normal random variable that has an upper tail probability of $\alpha/2$ or, equivalently, a cumulative probability of $1 - \alpha/2$. Used for confidence intervals for the mean with a known population standard deviation.

CONFIDENCE.NORM(*alpha, standard_deviation, size*) Computes the margin of error for a confidence interval for the mean with a known population standard deviation.

T.INV($1 - \alpha/2, n - 1$) or T.INV.2T($\alpha, n - 1$) Finds the value from the *t*-distribution with $n - 1$ degrees of freedom, giving an upper-tail probability of $\alpha/2$. Used for confidence intervals for the mean with an unknown population standard deviation.

CONFIDENCE.T(*alpha, standard_deviation, size*) Computes the margin of error for a confidence interval for the mean with an unknown population standard deviation.

Excel Templates

Confidence Intervals (Examples 6.8–6.10):

Open the Excel file *Confidence Intervals*. Select the worksheet tab corresponding to the type of confidence interval to calculate and enter the appropriate data.

StatCrunch

StatCrunch provides calculations of confidence intervals. You can find video tutorials and step-by-step procedures with examples at https://www.statcrunch.com/5.0/example.php. We suggest that you first view the tutorials *Getting started with StatCrunch* and *Working with StatCrunch sessions*. The following tutorials, listed under Confidence

Intervals For on this Web page, explain how to calculate confidence intervals in StatCrunch:

- A mean with raw data
- A mean with summary data
- A proportion with raw data
- A proportion with summary data
- The difference between two means with raw data
- The difference between two means with summary data
- The difference between means with paired data
- The difference between two proportions with raw data
- The difference between two proportions with summary data

Example: Calculating a Confidence Interval for the Mean with Raw Data

1. Choose the *Stat > T Stats > One Sample > With Data* menu option.
2. Select the column containing the data.
3. Under *Perform*, choose *Confidence interval for* μ.
4. Change the confidence level if needed and click *Compute!*.

Example: Calculating a Confidence Interval for the Difference in Proportions with Summary Data

1. Choose the *Stat > Proportion Stats > Two Sample > With Summary* menu option.
2. Under *Sample 1*, enter the *# of successes* and the *# of observations*. Under *Sample 2*, enter the *# of successes* and the *# of observations*.
3. Under *Perform*, choose *Confidence interval for* $p_1 - p_2$.
4. Change the confidence level if needed and click *Compute!*.

PROBLEMS AND EXERCISES

Statistical Sampling

1. Your college or university wishes to obtain reliable information about student perceptions of administrative communication. Describe how to design a sampling plan for this situation based on your knowledge of the structure and organization of your college or university. How would you implement simple random sampling, stratified sampling, and cluster sampling for this study? What would be the pros and cons of using each of these methods?

2. Number the rows in the Excel file *Credit Risk Data* to identify each record. The bank wants to sample from this database to conduct a more detailed audit. Use the Excel *Sampling* tool to find a simple random sample of 20 unique records.

3. Describe how to apply stratified sampling to sample from the *Credit Risk Data* file based on the different types of loans. Implement your process in Excel to choose a random sample consisting of 10% of the records for each type of loan.

4. Find the current 30 stocks that comprise the Dow Jones Industrial Average. Set up an Excel spreadsheet for their names, market capitalization, and one or two other key financial statistics (search Yahoo! Finance or a similar Web source). Using the Excel *Sampling* tool, obtain a random sample of five stocks, compute point estimates for the mean and standard deviation, and compare them to the population parameters.

Estimating Population Parameters

5. Repeat the sampling experiment in Example 6.3 for sample sizes 50, 100, 250, and 500. Compare your results to the example and use the empirical rules to analyze the sampling error. For each sample, also find the standard error of the mean using formula (6.1).

6. Find point estimates for the mean and standard deviation of the Months Customer data in the *Credit Risk Data* file. Draw five random samples of sizes 50 and 250 from the data using the *Sampling* tool. Use the empirical rules to analyze the sampling error and state your conclusions.

Sampling Distributions

7. Based upon extensive data from a national high school educational testing program, the standard deviation of national test scores for mathematics was found to be 120 points. If a sample of 225 students are given the test, what would be the standard error of the mean?

8. Based upon extensive data from a national high school educational testing program, the standard deviation of national test scores for critical reading was found to be 115 points. If a sample of 500 students are given the test, what would be the standard error of the mean?

9. Suppose that the mean score for the mathematics test cited in Problem 7 is 610. What is the probability that a random sample of 225 students will have a mean score of more than 625? Less than 600?

10. Suppose that the mean score for the critical reading test cited in Problem 8 is 580. What is the probability that a random sample of 500 students will have a mean score of more than 590? Less than 575?

11. In determining automobile mileage ratings, it was found that the mpg in the city for a certain model is normally distributed, with a mean of 30 mpg and a standard deviation of 1.7 mpg. Suppose that the car manufacturer samples five cars from its assembly line and tests them for mileage ratings.
    a. What is the distribution of the mean mpg for the sample?
    b. What is the probability that the mean mpg of the sample will be greater than 31 mpg?
    c. What is the probability that the mean mpg of the sample will be less than 29.5 mpg?

12. A popular soft drink is sold in 2-liter (2,000-milliliter) bottles. Because of variation in the filling process, bottles have a mean of 2,000 milliliters and a standard deviation of 16, normally distributed.
    a. If the manufacturer samples 100 bottles, what is the probability that the mean fill of the sample is less than 1,995 milliliters?
    b. What mean fill will be exceeded only 10% of the time for the sample of 100 bottles?

Interval Estimates

13. From a sample of 22 graduate students, the mean number of months of work experience prior to entering an MBA program was 34.86. The national standard deviation is known to be 19 months. What is a 95% confidence interval for the population mean?

Compute the confidence interval using the appropriate formula and verify your results using the Excel *Confidence Intervals* template.

14. A sample of 33 airline passengers found that the average check-in time is 2.167 minutes. Based on long-term data, the population standard deviation is known to be 0.48 minutes. Find a 95% confidence interval for the mean check-in time. Use the appropriate formula and verify your result using the *Confidence Intervals* workbook.

15. A survey of 26 college freshmen found that they average 6.85 hours of sleep each night. A 90% confidence interval had a margin of error of 0.497.

    a. What are the lower and upper limits of the confidence interval?

    b. What was the standard deviation, assuming that the population standard deviation is known?

16. A sample of 20 international students attending an urban U.S. university found that the average amount budgeted for expenses per month was $1,612.50 with a standard deviation of $1,179.64. Find a 95% confidence interval for the mean monthly expense budget of the population of international students. Use the appropriate formula and verify your result using the *Confidence Intervals* workbook.

17. A sample of 40 individuals at a shopping mall found that the mean number of visits to a restaurant per week was 2.88 with a standard deviation of 1.59. Find a 99% confidence interval for the mean number of restaurant visits. Use the appropriate formula and verify your result using the *Confidence Intervals* workbook.

18. A survey of 23 individuals found that they spent an average of $39.48 on headphones to use for exercising. The margin of error for a 95% confidence interval was found to be 21.2.

    a. What are the lower and upper limits of the confidence interval?

    b. What was the standard deviation of the sample?

19. For the data in the Excel file *Grade Point Averages*, find 90%, 95%, and 99% confidence intervals for the mean GPA. Compute the confidence intervals using the appropriate formulas and verify your results using the Excel *Confidence Intervals* template.

20. For the data in the Excel file *Debt and Retirement Savings*, find 95% confidence intervals for the mean income, long-term debt, and retirement savings. Use the appropriate formulas and Excel functions.

21. Find the standard deviation of the total assets held by the bank in the Excel file *Credit Risk Data*.

    a. Treating the records in the database as a population, use your sample in Problem 2 and compute 90%, 95%, and 99% confidence intervals for the total assets held in the bank by loan applicants using formula (6.2) and any appropriate Excel functions. Explain the differences as the level of confidence increases.

    b. How do your confidence intervals differ if you assume that the population standard deviation is not known but estimated using your sample data?

22. A marketing study found that the mean spending in 15 categories of consumer items for 297 respondents in the 18–34 age group was $91.86 with a standard deviation of $50.90. For 536 respondents in the 35+ age group, the mean and standard deviation were $81.53 and $45.29, respectively. Develop 95% confidence intervals for the mean spending amounts for each age group. Interpret your results. Use the appropriate formulas and Excel functions.

23. Using the data in the Excel file *Accounting Professionals*, find and interpret 95% confidence intervals for the following:

    a. mean years of service

    b. proportion of employees who have a graduate degree

    Use the appropriate formulas and Excel functions.

24. The Excel file *Restaurant Sales* provides sample information on lunch, dinner, and delivery sales for a local Italian restaurant. Develop 95% confidence intervals for the mean of each of these variables, as well as total sales for weekdays and weekends (Saturdays and Sundays). Use the appropriate formulas and Excel functions.

25. A bank estimated that the standard error for a 95% confidence interval for the proportion of a certain demographic group that may default on a loan is 0.31. The lower confidence limit was calculated as 0.15. What is the upper confidence limit?

26. A bank sampled its customers to determine the proportion of customers who use their debit card at least once each month. A sample of 50 customers found that only 12 use their debit card monthly. Find 95%

and 99% confidence intervals for the proportion of customers who use their debit card monthly. Use the appropriate formula and verify your result using the *Confidence Intervals* workbook.

27. If, based on a sample size of 850, a political candidate finds that 458 people would vote for him in a two-person race, what is the 95% confidence interval for his expected proportion of the vote? Would he be confident of winning based on this poll? Use the appropriate formula and verify your result using the *Confidence Intervals* workbook.

28. If, based on a sample size of 200, a political candidate found that 125 people would vote for her in a two-person race, what is the 99% confidence interval for her expected proportion of the vote? Would she be confident of winning based on this poll? Use the appropriate formula and verify your result using the *Confidence Intervals* workbook.

29. Using the data in the worksheet *Consumer Transportation Survey*, develop 95% confidence intervals for the following:

    a. the proportion of individuals who are satisfied with their vehicle

    b. the proportion of individuals who have at least one child

    Use the appropriate formula and verify your result using the *Confidence Intervals* workbook.

Using Confidence Intervals for Decision Making

30. A survey of 50 young professionals found that they spent an average of $19.31 when dining out, with a standard deviation of $12.11. Can you conclude statistically that the population mean is greater than $18?

31. A survey of 240 individuals found that one-third of them use their cell phones primarily for e-mail. Can you conclude statistically that the population proportion who use cell phones primarily for e-mail is less than 0.40?

32. A manufacturer conducted a survey among 500 randomly selected target market households in the test market for its new disposable diapers. The objective of the survey was to determine the market share for its new brand. If the sample point estimate for market share is 16%, develop a 95% confidence interval. Can the company reasonably conclude that it has a 20% market share? How about an 18% market share?

33. Using data in the Excel file *Colleges and Universities*, find 95% confidence intervals for the median SAT for each of the two groups, liberal arts colleges and research universities. Based on these confidence intervals, does there appear to be a difference in the median SAT scores between the two groups?

34. The Excel file *Baseball Attendance* shows the attendance in thousands at San Francisco Giants' baseball games for the ten years before the Oakland Athletics moved to the Bay Area in 1968, as well as the combined attendance for both teams for the next 11 years. Develop 95% confidence intervals for the mean attendance of each of the two groups. Based on these confidence intervals, would you conclude that attendance changed after the move?

35. A study of nonfatal occupational injuries in the United States found that about 31% of all injuries in the service sector involved the back. The National Institute for Occupational Safety and Health (NIOSH) recommended conducting a comprehensive ergonomics assessment of jobs and workstations. In response to this information, Mark Glassmeyer developed a unique ergonomic handcart to help field service engineers be more productive and also to reduce back injuries from lifting parts and equipment during service calls. Using a sample of 382 field service engineers who were provided with these carts, Mr. Glassmeyer collected the following data:

	Year 1 (without Cart)	Year 2 (with Cart)
Average call time	8.27 hours	7.98 hours
Standard deviation call time	1.15 hours	1.21 hours
Proportion of back injuries	0.018	0.010

Find 95% confidence intervals for the average call times and proportion of back injuries in each year. What conclusions would you reach based on your results?

36. For the data in the Excel file *Education and Income*, find 95% confidence intervals for the mean annual income of males and the mean annual income of females. Can you conclude that the mean income of one group is larger than the other?

37. For the data in the Excel file *Debt and Retirement Savings*, find 95% confidence intervals for the mean income, long-term debt, and retirement savings for

individuals who are single and individuals who are married. What conclusions can you reach by comparing the two groups?

38. A survey of 21 patients at a hospital found 95% confidence intervals for the following:

	Satisfaction with Quality of Care from Nurses	Satisfaction with Quality of Care from Specialists
Sample Proportion	16/21	13/21
Upper	1.019	0.797
Lower	0.505	0.441

Construct a stock chart-based visualization of these confidence intervals. What conclusions can you draw from the chart?

Prediction Intervals

39. Using the data in the worksheet *Consumer Transportation Survey*, develop 95% and 99% prediction intervals for the following:
   a. the hours per week that an individual will spend in his or her vehicle
   b. the number of miles driven per week

40. The Excel file *Restaurant Sales* provides sample information on lunch, dinner, and delivery sales for a local Italian restaurant. Develop 95% prediction intervals for the daily dollar sales of each of these variables and also for the total sales dollars on a weekend day.

41. For the Excel file *Credit Approval Decisions*, find 95% prediction intervals for the credit

scores and revolving balance of homeowners and non-homeowners. How do they compare?

Confidence Intervals and Sample Size

42. Trade associations, such as the United Dairy Industry Association, frequently conduct surveys to identify characteristics of their membership. If this organization conducted a survey to estimate the annual per capita consumption of milk and wanted to be 95% confident that the estimate was no more than 0.75 gallon away from the actual average, what sample size is needed? Past data have indicated that the standard deviation of consumption is approximately 6 gallons.

43. If a manufacturer conducted a survey among randomly selected target market households and wanted to be 95% confident that the difference between the sample estimate and the actual market share for its new product was no more than 3%, what sample size would be needed?

44. An Oregon wine association wants to determine the proportion of West Coast consumers who would spend at least $30 on Willamette Valley pinot noir at a 99% confidence level. If they wish to have an error of no more than 4%, how large a sample must they take? Based on surveys of cruise passengers who have visited the wineries, the association estimates that the proportion is around 0.12.

45. A community hospital wants to estimate the body mass index (BMI) of its local population. To estimate the BMI with an error of at most 1.0 at a 99% confidence level, what sample size should they use? The standard deviation based on available hospital patient data is 5.0.

CASE: DROUT ADVERTISING RESEARCH PROJECT

The background for this case was introduced in Chapter 2. This is a continuation of the case in Chapter 4. For this part of the case, compute confidence intervals for means and proportions and analyze the sampling errors, possibly

suggesting larger sample sizes to obtain more precise estimates. Write up your findings in a formal report or add your findings to the report you completed for the case in Chapter 4, depending on your instructor's requirements.

CASE: PERFORMANCE LAWN EQUIPMENT

In reviewing your previous reports, several questions came to Elizabeth Burke's mind. Use point and interval estimates to help answer these questions.

1. What proportion of customers rate the company with "top box" survey responses (which is defined as scale levels 4 and 5) on quality, ease of use, price, and service in the *Customer Survey* worksheet? How do these proportions differ by geographic region?
2. What estimates, with reasonable assurance, can PLE give customers for response times to customer service calls?
3. Engineering has collected data on alternative process costs for building transmissions in the worksheet *Transmission Costs*. Can you determine whether one of the proposed processes is better than the current process?
4. What would be a confidence interval for the proportion of failures of mower test performance as in the worksheet *Mower Test*?
5. For the data in the worksheet *Blade Weight*, what is the sampling distribution of the mean, the overall mean, and the standard error of the mean? Is a normal distribution an appropriate assumption for the sampling distribution of the mean?
6. How many blade weights must be measured to find a 95% confidence interval for the mean blade weight with a sampling error of at most 0.05? What if the sampling error is specified as 0.02?

Answer these questions and summarize your results in a formal report to Ms. Burke.

Statistical Inference

Jirsak/Shutterstock

LEARNING OBJECTIVES   After studying this chapter, you will be able to:

- Explain the purpose of hypothesis testing.
- Explain the difference between the null and alternative hypotheses.
- List the steps in the hypothesis-testing procedure.
- State the proper forms of hypotheses for one-sample hypothesis tests.
- Correctly formulate hypotheses.
- List the four possible outcome results from a hypothesis test.

- Explain the difference between Type I and Type II errors.
- State how to increase the power of a test.
- Choose the proper test statistic for hypothesis tests involving means and proportions.
- Explain how to draw a conclusion for one- and two-tailed hypothesis tests.
- Use *p*-values to draw conclusions about hypothesis tests.

- State the proper forms of hypotheses for two-sample hypothesis tests.
- Select and use Excel *Analysis Toolpak* procedures for two-sample hypothesis tests.
- Explain the purpose of analysis of variance.

- Use the Excel *ANOVA* tool to conduct an analysis of variance test.
- List the assumptions of ANOVA.
- Conduct and interpret the results of a chi-square test for independence.

Managers need to know if the decisions they have made or are planning to make are effective. For example, they might want to answer questions like the following: Did an advertising campaign increase sales? Will product placement in a grocery store make a difference? Did a new assembly method improve productivity or quality in a factory? Many applications of business analytics involve seeking statistical evidence that decisions or process changes have met their objectives. **Statistical inference** focuses on drawing conclusions about populations from samples. Statistical inference includes estimation of population parameters and hypothesis testing, which is a technique that allows you to draw valid statistical conclusions about the value of population parameters or differences among them.

Hypothesis Testing

Hypothesis testing involves drawing inferences about two contrasting propositions (each called a **hypothesis**) relating to the value of one or more population parameters, such as the mean, proportion, standard deviation, or variance. One of these propositions (called the **null hypothesis**) describes the existing theory or a belief that is accepted as valid unless strong statistical evidence exists to the contrary. The second proposition (called the **alternative hypothesis**) is the complement of the null hypothesis; it must be true if the null hypothesis is false. The null hypothesis is denoted by H_0, and the alternative hypothesis is denoted by H_1. Using sample data, we either

1. *reject* the null hypothesis and conclude that the sample data provide sufficient statistical evidence to support the alternative hypothesis, or
2. *fail to reject* the null hypothesis and conclude that the sample data do not support the alternative hypothesis.

If we fail to reject the null hypothesis, then we can only accept as valid the existing theory or belief, but we can never prove it.

Hypothesis-Testing Procedure

Conducting a hypothesis test involves several steps:

1. Identifying the population parameter of interest and formulating the hypotheses to test

| EXAMPLE 7.1 | A Legal Analogy for Hypothesis Testing |

A good analogy for hypothesis testing is the U.S. legal system. In our system of justice, a defendant is innocent until proven guilty. The null hypothesis—our belief in the absence of any contradictory evidence—is not guilty, whereas the alternative hypothesis is guilty. If the evidence (sample data) strongly indicates that the defendant is guilty, then we reject the assumption of innocence. If the evidence is not sufficient to indicate guilt, then we cannot reject the not guilty hypothesis; however, we haven't *proven* that the defendant is innocent. In reality, you can only conclude that a defendant is guilty from the evidence; you still have not proven it!

2. Selecting a *level of significance*, which defines the risk of drawing an incorrect conclusion when the assumed hypothesis is actually true
3. Determining a decision rule on which to base a conclusion
4. Collecting data and calculating a test statistic
5. Applying the decision rule to the test statistic and drawing a conclusion

We will apply this procedure to two different types of hypothesis tests: the first involving a single population (called one-sample tests) and, later, tests involving more than one population (multiple-sample tests).

CHECK YOUR UNDERSTANDING

1. Explain the difference between the null and alternative hypotheses.

2. List the steps in the general hypothesis-testing procedure.

One-Sample Hypothesis Tests

A **one-sample hypothesis test** is one that involves a single population parameter, such as the mean, proportion, standard deviation, and so on. To conduct the test, we use a single sample of data from the population. We may conduct three types of one-sample hypothesis tests:

H_0: population parameter $\geq$ constant vs. H_1: population parameter $<$ constant

H_0: population parameter $\leq$ constant vs. H_1: population parameter $>$ constant

H_0: population parameter $=$ constant vs. H_1: population parameter $\neq$ constant

Notice that one-sample tests always compare a population parameter to some constant. For one-sample tests, the statements of the null hypotheses are expressed as $\geq$, $\leq$, or $=$. It is *not correct* to formulate a null hypothesis using $>$, $<$, or $\neq$.

How do we determine the proper form of the null and alternative hypotheses? Hypothesis testing always *assumes* that H_0 is true and uses sample data to determine whether H_1 is more likely to be true. Statistically, we cannot "prove" that H_0 is true; we can only *fail to reject* it. Thus, if we cannot reject the null hypothesis, we have shown only that there is insufficient evidence to conclude that the alternative hypothesis is true. However, rejecting the null hypothesis provides strong evidence (in a statistical sense) that the null hypothesis is not true and that the alternative hypothesis is true. Therefore, what we wish to provide evidence for statistically should be identified as the alternative hypothesis.

EXAMPLE 7.2 Formulating a One-Sample Test of Hypothesis

CadSoft, a producer of computer-aided design software for the aerospace industry, receives numerous calls for technical support. In the past, the average response time has been at least 25 minutes. The company has upgraded its information systems and believes that this will help reduce response time. As a result, it believes that the average response time can be reduced to less than 25 minutes. The company collected a sample of 44 response times in the Excel file *CadSoft Technical Support Response Times* (see Figure 7.1).

If the new information system makes a difference, then data should be able to confirm that the mean response time is less than 25 minutes; this defines the alternative hypothesis, H_1.

Therefore, the proper statements of the null and alternative hypotheses are

H_0: population mean response time $\geq$ 25 minutes

H_1: population mean response time $<$ 25 minutes

We would typically write this using the proper symbol for the population parameter. In this case, letting μ be the mean response time, we would write:

$$H_0: \mu \geq 25$$
$$H_1: \mu < 25$$

▶ **Figure 7.1**

Portion of Excel file Technical Support Response Times

▲	A	B	C	D	E
1	CadSoft Technical Support Response Times				
2					
3	Customer	Time (min)			
4	1	20			
5	2	12			
6	3	15			
7	4	11			
8	5	22			
9	6	6			
10	7	39			

Understanding Potential Errors in Hypothesis Testing

We already know that sample data can show considerable variation; therefore, conclusions based on sample data may be wrong. Hypothesis testing can result in one of four different outcomes:

1. The null hypothesis is actually *true*, and the test *correctly fails to reject it.*
2. The null hypothesis is actually *false*, and the hypothesis test *correctly reaches this conclusion.*
3. The null hypothesis is actually *true*, but the hypothesis test *incorrectly rejects it* (called **Type I error**).
4. The null hypothesis is actually *false*, but the hypothesis test *incorrectly fails to reject it* (called **Type II error**).

The probability of making a Type I error, that is, $P(\text{rejecting } H_0 | H_0 \text{ is true})$, is denoted by α and is called the **level of significance**. This defines the likelihood that you will make the incorrect conclusion that the alternative hypothesis is true when, in fact, the null hypothesis is true. The value of α can be controlled by the decision maker and is selected before the test is conducted. Commonly used levels for α are 0.10, 0.05, and 0.01.

The probability of *correctly failing to reject* the null hypothesis, or $P(\text{not rejecting } H_0 | H_0 \text{ is true})$, is called the **confidence coefficient** and is calculated as $1 - \alpha$. A confidence coefficient of 0.95 means that we expect 95 out of 100 samples to support the null hypothesis rather than the alternate hypothesis when H_0 is actually true.

Unfortunately, we cannot control the probability of a Type II error, P(not rejecting $H_0 | H_0$ is false), which is denoted by β. Unlike α, β cannot be specified in advance but depends on the true value of the (unknown) population parameter (see Example 7.3).

The value $1 - \beta$ is called the **power of the test** and represents the probability of *correctly rejecting* the null hypothesis when it is indeed false, or P(rejecting $H_0 | H_0$ is false). We would like the power of the test to be high (equivalently, we would like the probability of a Type II error to be low) to allow us to make a valid conclusion. The power of the test is sensitive to the sample size; small sample sizes generally result in a low value of $1 - \beta$. The power of the test can be increased by taking larger samples, which enable us to detect small differences between the sample statistics and population parameters with more accuracy. However, a larger sample size incurs higher costs, giving new meaning to the adage there is no such thing as a free lunch. This suggests that if you choose a small level of significance, you should try to compensate by having a large sample size when you conduct the test.

EXAMPLE 7.3	**How β Depends on the True Population Mean**

Consider the hypotheses in the CadSoft example:

H_0: mean response time $\geq$ 25 minutes

H_1: mean response time $<$ 25 minutes

If the true mean response from which the sample is drawn is, say, 15 minutes, we would expect to have a much smaller probability of incorrectly concluding that the null hypothesis is true than when the true mean response is 24 minutes, for example. If the true mean were 15 minutes, the sample mean would very likely be much less than 25, leading us to reject H_0. If the true mean were 24 minutes, even though it is less than 25, we would have a much higher probability of failing to reject H_0 because a higher likelihood exists that the sample mean would be greater than 25 due to sampling error. Thus, the farther away the true mean response time is from the hypothesized value, the smaller is β. Generally, as α decreases, β increases, so the decision maker must consider the trade-offs of these risks. So if you choose a level of significance of 0.01 instead of 0.05 and keep the sample size constant, you would reduce the probability of a Type I error but increase the probability of a Type II error.

Selecting the Test Statistic

The next step is to collect sample data and use the data to draw a conclusion. The decision to reject or fail to reject a null hypothesis is based on computing a *test statistic* from the sample data. The test statistic used depends on the type of hypothesis test. Different types of hypothesis tests use different test statistics, and it is important to use the correct one. The proper test statistic often depends on certain assumptions about the population—for example, whether or not the standard deviation is known. In the vast majority of practical applications, the population standard deviation is unknown. The following formulas show two types of one-sample hypothesis tests for means and their associated test statistics. The value of μ_0 is the hypothesized value of the population mean, that is, the "constant" in the hypothesis formulation.

Type of Test	**Test Statistic**	
One-sample test for mean, σ known	$z = \dfrac{\bar{x} - \mu_0}{\sigma / \sqrt{n}}$	(7.1)
One-sample test for mean, σ unknown	$t = \dfrac{\bar{x} - \mu_0}{s / \sqrt{n}}$	(7.2)

With sample data, we generally do not know the population standard deviation, and therefore we only will illustrate an example where σ is unknown. We will summarize the procedure when σ is known later in this chapter.

| EXAMPLE 7.4 | **Computing the Test Statistic** |

For the CadSoft example, the average response time for the sample of 44 customers is $\bar{x} = 21.91$ minutes and the sample standard deviation is $s = 19.49$. The hypothesized mean is $\mu_0 = 25$. You might wonder why we have to test the hypothesis statistically when the sample average of 21.91 is clearly less than 25. We do because of sampling error. It is quite possible that the population mean truly is 25 or more and that we were lucky to draw a sample whose mean was smaller. Because of potential sampling error, it would be dangerous to conclude that the company was meeting its goal just by looking at the sample mean without better statistical evidence.

Because we don't know the value of the population standard deviation, the proper test statistic to use is formula (7.2):

$$t = \frac{\bar{x} - \mu_0}{s/\sqrt{n}}$$

Therefore, the value of the test statistic is

$$t = \frac{\bar{x} - \mu_0}{s/\sqrt{n}} = \frac{21.91 - 25}{19.49/\sqrt{44}} = \frac{-3.09}{2.938} = -1.05$$

Observe that the numerator is the distance between the sample mean (21.91) and the hypothesized value (25). By dividing by the standard error, the value of t represents the number of standard errors the sample mean is from the hypothesized value. In this case, the sample mean is 1.05 standard errors below the hypothesized value of 25. This notion provides the fundamental basis for the hypothesis test—if the sample mean is "too far" away from the hypothesized value, then the null hypothesis should be rejected.

Finding Critical Values and Drawing a Conclusion

The conclusion to reject or fail to reject H_0 is based on comparing the value of the test statistic to a "critical value" from the sampling distribution of the test statistic when the null hypothesis is true and the chosen level of significance, α. The sampling distribution of the test statistic is usually the normal distribution, t-distribution, or some other well-known distribution. For example, the sampling distribution of the z-test statistic in formula (7.1) is a standard normal distribution; the t-test statistic in formula (7.2) has a t-distribution with $n - 1$ degrees of freedom. For a one-tailed test, the critical value is the number of standard errors away from the hypothesized value for which the probability of exceeding the critical value is α. If $\alpha = 0.05$, for example, then we are saying that there is only a 5% chance that a sample mean will be that far away from the hypothesized value purely because of sampling error. Should this occur, it suggests that the true population mean is different from what was hypothesized.

The critical value divides the sampling distribution into two parts, a *rejection region* and a *nonrejection region*. If the null hypothesis is false, it is more likely that the test statistic will fall into the rejection region. If it does, we reject the null hypothesis; otherwise, we fail to reject it. The rejection region is chosen so that the probability of the test statistic falling into it if H_0 is true is the probability of a Type I error, α.

The rejection region occurs in the tails of the sampling distribution of the test statistic and depends on the structure of the hypothesis test, as shown in Figure 7.2. If the null hypothesis is structured as $=$ and the alternative hypothesis as $\neq$, then we would reject H_0 if the test statistic is *either* significantly high or low. In this case, the rejection region will occur in *both* the upper and lower tail of the distribution [see Figure 7.2(a)]. This is called a **two-tailed test of hypothesis**. Because the probability that the test statistic falls into the rejection region, given that H_0 is true, is α, each tail has an area of $\alpha/2$.

▶ **Figure 7.2**

Illustration of Rejection Regions in Hypothesis Testing

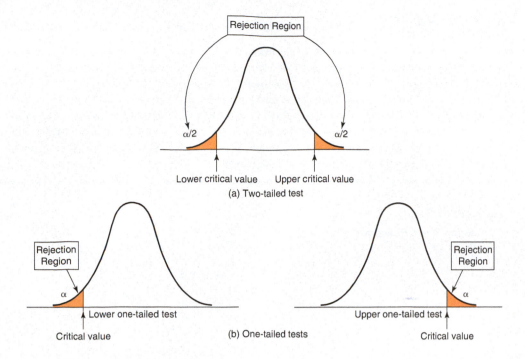

(a) Two-tailed test

(b) One-tailed tests

The other types of hypothesis tests, which specify a direction of relationship (where H_0 is either $\geq$ or $\leq$), are called **one-tailed tests of hypothesis**. In this case, the rejection region occurs only in one tail of the distribution [see Figure 7.2(b)]. Determining the correct tail of the distribution to use as the rejection region for a one-tailed test is easy. If H_1 is stated as $<$, the rejection region is in the lower tail; if H_1 is stated as $>$, the rejection region is in the upper tail (just think of the inequality associated with the alternative hypothesis as an arrow pointing to the proper tail direction).

Two-tailed tests have both upper and lower critical values, whereas one-tailed tests have either a lower or an upper critical value. For standard normal and *t*-distributions, which have a mean of zero, lower-tail critical values are negative; upper-tail critical values are positive.

Critical values make it easy to determine whether or not the test statistic falls in the rejection region of the proper sampling distribution. For example, for an upper one-tailed test, if the test statistic is greater than the critical value, the decision would be to reject the null hypothesis. Similarly, for a lower one-tailed test, if the test statistic is less than the critical value, we would reject the null hypothesis. For a two-tailed test, if the test statistic is *either* greater than the upper critical value or less than the lower critical value, the decision would be to reject the null hypothesis.

The critical value for a one-sample, one-tailed test when the standard deviation is unknown is the value of the *t*-distribution with $n - 1$ degrees of freedom that provides a tail area of alpha, that is, $t_{\alpha, n-1}$. We may find *t*-values in Table A.2 in Appendix A at the end of the book or by using the Excel function T.INV$(1 - \alpha, n - 1)$. Again, for a lower-tailed test, the critical value is negative, and for a two-tailed test, we would use $\alpha/2$.

EXAMPLE 7.5   Finding the Critical Value and Drawing a Conclusion

For the CadSoft example, if the level of significance is 0.05, then the critical value is $t_{0.05,43}$ is found in Excel using $=$T.INV$(0.95, 43) = 1.68$. Because the t-distribution is symmetric with a mean of 0 and this is a lower-tailed test, we use the negative of this number (-1.68) as the critical value.

By comparing the value of the t-test statistic with this critical value, we see that the test statistic computed in Example 7.4 does not fall below the critical value (that is,

$-1.05 > -1.68$) and is not in the rejection region. Therefore, we cannot reject H_0 and cannot conclude that the mean response time has improved to less than 25 minutes. Figure 7.3 illustrates the conclusion we reached. Even though the sample mean is less than 25, we cannot conclude that the population mean response time is less than 25 because of the large amount of sampling error.

▶ **Figure 7.3**

t-*Test for Mean Response Time*

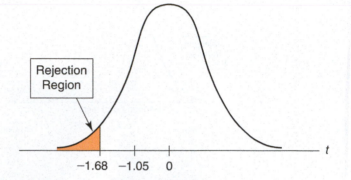

Two-Tailed Test of Hypothesis for the Mean

Basically, all hypothesis tests are similar; you just have to ensure that you select the correct test statistic, critical value, and rejection region, depending on the type of hypothesis. The following example illustrates a two-tailed test of hypothesis for the mean. For a two-tailed test using the t-distribution, we use the Excel function T.INV$(1 - \alpha/2, n - 1)$ or T.INV.2T$(\alpha, n - 1)$ to find the critical value, $t_{\alpha/2,n-1}$. Be careful when using T.INV.2T! Use α and not $\alpha/2$ in this function.

EXAMPLE 7.6   Conducting a Two-Tailed Hypothesis Test for the Mean

Figure 7.4 shows a portion of data collected in a survey of 34 respondents by a travel agency (provided in the Excel file *Vacation Survey*). Suppose that the travel agency wanted to target individuals who were approximately 35 years old. Thus, we wish to test whether the average age of respondents is equal to 35. The hypothesis to test is

$$H_0: \text{mean age} = 35$$
$$H_1: \text{mean age} \neq 35$$

The sample mean is computed to be 38.676, and the sample standard deviation is 7.857.

We use the t-test statistic:

$$t = \frac{\bar{x} - \mu_0}{s/\sqrt{n}} = \frac{38.676 - 35}{.857/\sqrt{34}} = 2.73$$

In this case, the sample mean is 2.73 standard errors above the hypothesized mean of 35. However, because this is a two-tailed test, the rejection region and decision rule are different. For a level of significance α, we reject H_0 if the t-test statistic falls either below the negative critical value, $-t_{\alpha/2,n-1}$, or above the positive critical value, $t_{\alpha/2,n-1}$. Using either Table A.2 in Appendix A at the back of this book or the Excel function $=$T.INV.2T$(.05, 33)$ to calculate $t_{0.025,33}$, we obtain 2.0345. Thus, the critical values are ± 2.0345. Because the t-test statistic does *not* fall between these values, we must reject the null hypothesis that the average age is 35 (see Figure 7.5).

► **Figure 7.4**

Portion of Vacation
Survey *Data*

	A	B	C	D	E
1	Vacation Survey				
2					
3	Age	Gender	Relationship Status	Vacations per Year	Number of Children
4	24	Male	Married	2	0
5	26	Female	Married	4	0
6	28	Male	Married	2	2
7	33	Male	Married	4	0
8	45	Male	Married	2	0
9	49	Male	Married	1	2
10	29	Male	Married	4	0

► **Figure 7.5**

*Illustration of a Two-Tailed
Test for Example 7.6*

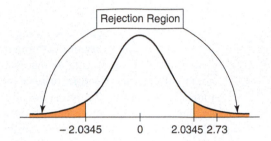

Rejection Region

−2.0345   0   2.0345 2.73

Summary of One-Sample Hypothesis Tests for the Mean

Case 1: (σ Unknown)

1. Determine whether the proper hypotheses represent a lower-tailed, upper-tailed, or two-tailed test.
2. Calculate the test statistic using formula (7.2).
3. Find the critical value.
   a. If it is a lower-tailed test, the critical value is found using the Excel function T.INV($1 - \alpha, n - 1$). Note the minus sign!
   b. If it is an upper-tailed test, the critical value is found using the Excel function T.INV($1 - \alpha, n - 1$).
   c. If you have a two-tailed test, use T.INV.2T($\alpha, n - 1$); the critical values will be both positive and negative.
4. Compare the test statistic to the critical value(s) and draw a conclusion to either reject the null hypothesis or fail to reject it.

When the population standard deviation is known, the process is exactly the same; only the formulas and Excel functions differ.

Case 2: (σ Known)

1. Determine whether the proper hypotheses represent a lower-tailed, upper-tailed, or two-tailed test.
2. Calculate the test statistic using formula (7.1).
3. Find the critical value. The critical value for a one-sample, one-tailed test when the standard deviation is known is the value of the normal distribution that has a tail area of alpha. This may be found by using Table A.1 in Appendix A to find the z-value corresponding to an area of $1 - \alpha$ or using the Excel function

NORM.S.INV$(1 - \alpha)$. Remember that for a lower-tailed test, the critical value is negative. For a two-tailed test, use $\alpha/2$.

    a. If it is a lower-tailed test, the critical value is found using the Excel function NORM.S.INV$(1 - \alpha)$. Note the minus sign!

    b. If it is an upper-tailed test, the critical value is found using the Excel function NORM.S.INV$(1 - \alpha)$.

    c. If you have a two-tailed test, use NORM.S.INV$(1 - \alpha/2)$; the critical values will be both positive and negative.

  4. Compare the test statistic to the critical value(s) and draw a conclusion to either reject the null hypothesis or fail to reject it.

p-Values

An alternative approach to comparing a test statistic to a critical value in hypothesis testing is to find the probability of obtaining a test statistic value equal to or more extreme than that obtained from the sample data when the null hypothesis is true. This probability is commonly called a ***p*-value**, or **observed significance level**. The magnitude of a *p*-value indicates the compatibility or incompatibility of the data with the null hypothesis. The smaller the *p*-value, the greater the statistical incompatibility with the null hypothesis. The *p*-value does *not* measure the probability that the null hypothesis is true!

It is common practice to compare *p*-values with the level of significance chosen for a hypothesis test. To draw a conclusion, compare the *p*-value to the chosen level of significance α; whenever $p < \alpha$, reject the null hypothesis and otherwise fail to reject it. Using *p*-values makes it easy to draw conclusions about hypothesis tests. If the population standard deviation is known, the *p*-value for a lower one-tailed test is the probability to the left of the test statistic z and is found by =NORM.S.DIST(z, TRUE). For an upper one-tailed test, the *p*-value is the probability to the right of the test statistic z and is found by =1− NORM.S.DIST(z, TRUE). For a two-tailed test, the *p*-value is the probability to the left of the negative z-value plus the probability to the right of the positive z-value and can be computed in Excel as =2*(1 − NORM.S.DIST(ABS(z), TRUE). When the population standard deviation is unknown, the *p*-value for a lower one-tailed test is the probability to the left of the test statistic t in the t-distribution and is found by =T.DIST($t, n - 1$, TRUE). For an upper one-tailed test, the *p*-value is the probability to the right of the test statistic t and is found by =1− T.DIST($t, n - 1$, TRUE). For a two-tailed test, the *p*-value can be found using the Excel function =T.DIST.2T(ABS(t), $n - 1$).

The statistical community has questioned this practice. The American Statistical Association has noted[1]

> *Practices that reduce data analysis or scientific inference to mechanical "bright-line" rules (such as "p < 0.05") for justifying scientific claims or conclusions can lead to erroneous beliefs and poor decision-making. A conclusion does not immediately become "true" on one side of the divide and "false" on the other. Researchers should bring many contextual factors into play to derive scientific inferences, including the design of a study, the quality of the measurements, the external evidence for the phenomenon under study, and the validity of assumptions that underlie the data analysis. Pragmatic considerations often require binary, "yes-no" decisions, but this does not mean that p-values alone can ensure that a decision is correct or incorrect. The widespread use of "statistical significance" (generally interpreted as "p ≤ 0.05") as a license for making a claim of a scientific finding (or implied truth) leads to considerable distortion of the scientific process.*

As a result, *p*-values should be interpreted cautiously.

---

[1]Ronald L. Wasserstein and Nicole A. Lazar (2016), "The ASA's statement on *p*-values: context, process, and purpose," The American Statistician, DOI: 10.1080/00031305.2016.1154108

EXAMPLE 7.7   Using *p*-Values

For the CadSoft example (see Example 7.4), the *t*-test statistic for the hypothesis test in the response-time example is −1.05. If the true mean is really 25, then the *p*-value is the probability of obtaining a test statistic of −1.05 or less (the area to the left of −1.05 in Figure 7.3). We can calculate the *p*-value using the Excel function =T.DIST(−1.05, 43, TRUE) = 0.15 . Because p = 0.15 is not less than $\alpha = 0.05$, we do not reject H_0. In other words, there is about a 15% chance that the test statistic would

be −1.05 or smaller if the null hypothesis were true. This is a fairly high probability, so it would be difficult to conclude that the true mean is less than 25 and we could attribute the fact that the test statistic is less than the hypothesized value to sampling error alone and not reject the null hypothesis.

For the *Vacation Survey* two-tailed hypothesis test in Example 7.6, the *p*-value for this test is 0.010, which can also be computed by the Excel function =T.DIST.2T(2.73, 33); therefore, since 0.010 < 0.05, we reject H_0.

One-Sample Tests for Proportions

Many important business measures, such as market share or the fraction of deliveries received on time, are expressed as proportions. We may conduct a test of hypothesis about a population proportion in a similar fashion as we did for means. The test statistic for a one-sample test for proportions is

$$z = \frac{\hat{p} - \pi_0}{\sqrt{\pi_0(1 - \pi_0)/n}} \tag{7.3}$$

where π_0 is the hypothesized value and $\hat{p}$ is the sample proportion. Similar to the test statistic for means, the *z*-test statistic shows the number of standard errors that the sample proportion is from the hypothesized value. The sampling distribution of this test statistic has a standard normal distribution.

For a lower-tailed test, the *p*-value would be computed by the area to the left of the test statistic; that is, =NORM.S.DIST(*z*, TRUE). For an upper-tailed test, the *p*-value would be the area to the right of the test statistic, or =1− NORM.S.DIST(*z*, TRUE). If we had a two-tailed test, the *p*-value is computed by the Excel formula =2*(1−NORM.S.DIST(ABS(*z*), TRUE)).

EXAMPLE 7.8   A One-Sample Test for the Proportion

CadSoft also sampled 44 customers and asked them to rate the overall quality of the company's software product using the following scale:

   0—very poor
   1—poor
   2—good
   3—very good
   4—excellent

These data can be found in the Excel File *CadSoft Product Satisfaction Survey*. The firm tracks customer satisfaction of quality by measuring the proportion of responses in the top two categories. In the past, this proportion has averaged about 75%. For these data, 35 of the 44 responses, or 79.5%, are in the top two categories. Is there sufficient evidence to conclude that this satisfaction measure has significantly exceeded 75%

using a significance level of 0.05? Answering this question involves testing the hypotheses about the population proportion π:

$$H_0: \pi \le 0.75$$

$$H_1: \pi > 0.75$$

This is an upper-tailed, one-tailed test. The test statistic is computed using formula (7.3):

$$z = \frac{0.795 - 0.75}{\sqrt{0.75(1 - 0.75)/44}} = 0.69$$

In this case, the sample proportion of 0.795 is 0.69 standard error above the hypothesized value of 0.75. Because this is an upper-tailed test, we reject H_0 if the value of the test statistic is larger than the critical value. Because the sampling distribution of *z* is a standard normal, the critical

(continued)

value of z for a level of significance of 0.05 is found by the Excel function NORM.S. INV$(0.95) = 1.645$. Because the test statistic does not exceed the critical value, we cannot reject the null hypothesis that the proportion is no greater than 0.75. Thus, even though the sample proportion exceeds 0.75, we cannot conclude statistically that the customer satisfaction ratings have significantly improved. We could attribute this to sampling error and the relatively small sample size. The p-value can be found by computing the area to the right of the test statistic in the standard normal distribution using the Excel formula $= 1 -$ NORM.S.DIST$(0.69, $ TRUE$) = 0.25$. Note that the p-value is greater than the significance level of 0.05, leading to the same conclusion of not rejecting the null hypothesis.

Confidence Intervals and Hypothesis Tests

A close relationship exists between confidence intervals and hypothesis tests. For example, suppose we construct a 95% confidence interval for the mean. If we wish to test the hypotheses

$$H_0: \mu = \mu_0$$
$$H_1: \mu \neq \mu_0$$

at a 5% level of significance, we simply check whether the hypothesized value μ_0 falls within the confidence interval. If it does not, then we reject H_0; if it does, then we cannot reject H_0.

For one-tailed tests, we need to examine on which side of the hypothesized value the confidence interval falls. For a lower-tailed test, if the confidence interval falls entirely below the hypothesized value, we reject the null hypothesis. For an upper-tailed test, if the confidence interval falls entirely above the hypothesized value, we also reject the null hypothesis.

An Excel Template for One-Sample Hypothesis Tests

The Excel file *One Sample Hypothesis Tests* provides template worksheets for conducting hypothesis tests for means (with σ known and unknown) and proportions based on the formulas and Excel functions that we have introduced in this section. The templates provide results for upper and lower one-tailed tests, and also for a two-tailed test of hypothesis. Figure 7.6 shows the templates for the CadSoft examples. We strongly suggest that you examine the Excel formulas and compare them to the examples and rules that we have presented to better understand the templates.

CHECK YOUR UNDERSTANDING

1. State the three types of one-sample hypothesis tests.

2. Why is it important to identify the alternative hypothesis based on what we wish to provide statistical evidence for?

3. Explain the concepts of Type I and Type II errors.

4. Explain the difference between a one-tailed and two-tailed hypothesis test.

5. How do you use a test statistic to draw a conclusion about a hypothesis test?

6. What is a p-value, and how is it used?

▶ **Figure 7.6**

Excel Templates for One-Sample Hypothesis Tests

	A	B	C	D	E
1	One Sample Test for the Mean			Hypothesis test results	
2	Population standard deviation unknown				
3				Lower one-tailed test	
4	Sample Size	44		Critical t-value	-1.681070703
5	Sample Mean	21.91		p-value	0.149416269
6	Sample Standard Deviation	19.49		Conclusion	Do Not Reject Null Hypothesis
7	Hypothesized value	25			
8	Level of significance	0.05		Upper one-tailed test	
9				Critical t-value	1.681070703
10	t-statistic	-1.0517		p-value	0.850583731
11	One-tailed t-value	1.6811		Conclusion	Do Not Reject Null Hypothesis
12	Two-tailed t-value (+ and -)	2.0167			
13				Two-tailed test	
14				Critical t-value (+ and -)	2.016692199
15				p-value	0.298832538
16				Conclusion	Do Not Reject Null Hypothesis

	A	B	C	D	E
1	One Sample Test for the Mean			Hypothesis test results	
2	Population Standard Deviation Known				
3				Lower one-tailed test	
4	Sample Size	44		Critical z-value	-1.644853627
5	Sample Mean	21.91		p-value	0.146479106
6	Population Standard Deviation	19.49		Conclusion	Do Not Reject Null Hypothesis
7	Hypothesized value	25			
8	Level of significance	0.05		Upper one-tailed test	
9				Critical z-value	1.644853627
10	z-statistic	-1.0517		p-value	0.853520894
11	One-tailed z-value	1.6449		Conclusion	Do Not Reject Null Hypothesis
12	Two-tailed z-value (+ and -)	1.96			
13				Two-tailed test	
14				Critical z-value (+ and -)	1.959963985
15				p-value	0.292958211
16				Conclusion	Do Not Reject Null Hypothesis

	A	B	C	D	E
1	One Sample Test for the Proportion			Hypothesis test results	
2					
3	Sample Size	44		Lower one-tailed test	
4	Sample Proportion	0.795		Critical z-value	-1.644853627
5	Hypothesized value	0.75		p-value	0.754697699
6	Level of significance	0.05		Conclusion	Do Not Reject Null Hypothesis
7					
8	z-statistic	0.6893		Upper one-tailed test	
9	One-tailed z-value	1.6449		Critical z-value	1.644853627
10	Two-tailed z-value (+ and -)	1.96		p-value	0.245302301
11				Conclusion	Do Not Reject Null Hypothesis
12					
13				Two-tailed test	
14				Critical z-value (+ and -)	1.959963985
15				p-value	0.490604602
16				Conclusion	Do Not Reject Null Hypothesis

Two-Sample Hypothesis Tests

Many practical applications of hypothesis testing involve comparing two populations for differences in means, proportions, or other population parameters. Such tests can confirm differences between suppliers, performance at two factory locations, new and old work methods or reward and recognition programs, and many other situations. Similar to one-sample tests, two-sample hypothesis tests for differences in population parameters have one of the following forms:

1. *Lower-tailed test H_0*: population parameter (1) − population parameter (2) $\geq D_0$ vs. H_1: population parameter (1) − population parameter (2) $< D_0$. This test seeks evidence that the difference between population parameter (1) and population parameter (2) is less than some value, D_0. When $D_0 = 0$, the test simply seeks to conclude whether population parameter (1) is smaller than population parameter (2).
2. *Upper-tailed test H_0*: population parameter (1) − population parameter (2) $\leq D_0$ vs. H_1: population parameter (1) − population parameter (2) $> D_0$. This test seeks evidence that the difference between population parameter (1) and population parameter (2) is greater than some value, D_0. When $D_0 = 0$, the test simply seeks to conclude whether population parameter (1) is larger than population parameter (2).
3. *Two-tailed test H_0*: population parameter (1) − population parameter (2) $= D_0$ vs. H_1: population parameter (1) − population parameter (2) $\neq D_0$. This test seeks evidence that the difference between the population parameters is equal to D_0. When $D_0 = 0$, we are seeking evidence that population parameter (1) differs from parameter (2).

In most applications, $D_0 = 0$, and we are simply seeking to compare the population parameters. However, there are situations when we might want to determine if the parameters differ by some non-zero amount; for example, "job classification A makes at least $5,000 more than job classification B."

These hypothesis-testing procedures are similar to those previously discussed in the sense of computing a test statistic and comparing it to a critical value. However, the test statistics for two-sample tests are more complicated than those for one-sample tests and we will not delve into the mathematical details. Fortunately, Excel provides several tools for conducting two-sample tests, and we will use these in our examples. Table 7.1 summarizes the Excel *Analysis Toolpak* procedures that we will use.

Two-Sample Tests for Differences in Means

In a two-sample test for differences in means, we always test hypotheses of the form

$$H_0: \mu_1 - \mu_2 \{\geq, \leq, \text{ or } =\} 0$$
$$H_1: \mu_1 - \mu_2 \{<, >, \text{ or } \neq\} 0 \tag{7.4}$$

▶ **Table 7.1**

Excel Analysis Toolpak Procedures for Two-Sample Hypothesis Tests

Type of Test	Excel Procedure
Two-sample test for means, σ^2 known	Excel z-test: Two-sample for means
Two-sample test for means, σ^2 unknown and assumed unequal	Excel t-test: Two-sample assuming unequal variances
Two-sample test for means, σ^2 unknown and assumed equal	Excel t-test: Two-sample assuming equal variances
Paired two-sample test for means	Excel t-test: Paired two-sample for means
Two-sample test for equality of variances	Excel F-test: Two-sample for variances

EXAMPLE 7.9	**Comparing Supplier Performance**

The last two columns in the *Purchase Orders* data file provide the order date and arrival date of all orders placed with each supplier. The time between placement of an order and its arrival is commonly called the lead time. We may compute the lead time by subtracting the Excel date function values from each other (Arrival Date − Order Date), as shown in Figure 7.7.

Figure 7.8 shows a PivotTable for the average lead time for each supplier. Purchasing managers have noted that they order many of the same types of items from Alum Sheeting and Durable Products and are considering dropping Alum Sheeting from their supplier base if its lead time is significantly longer than that of Durable Products. Therefore, they would like to test the hypothesis

$$H_0: \mu_1 - \mu_2 \leq 0$$

$$H_1: \mu_1 - \mu_2 > 0$$

where μ_1 = mean lead time for Alum Sheeting and μ_2 = mean lead time for Durable Products.

Rejecting the null hypothesis suggests that the average lead time for Alum Sheeting is statistically longer than Durable Products. However, if we cannot reject the null hypothesis, then even though the mean lead time for Alum Sheeting is longer, the difference would most likely be due to sampling error, and we could not conclude that there is a statistically significant difference.

Selection of the proper test statistic and Excel procedure for a two-sample test for means depends on whether the population variances are known, and if not, whether they are assumed to be equal.

1. *Population variance is known.* In Excel, choose *z-Test: Two-Sample for Means* from the *Data Analysis* menu. This test uses a test statistic that is based on the standard normal distribution.
2. *Population variance is unknown and assumed unequal.* From the *Data Analysis* menu, choose *t-Test: Two-Sample Assuming Unequal Variances*. The test statistic for this case has a *t*-distribution.
3. *Population variance is unknown but assumed equal.* In Excel, choose *t-Test: Two-Sample Assuming Equal Variances.* The test statistic also has a *t*-distribution, but it is different from the unequal variance case.

These tools calculate the test statistic, the *p*-value for both a one-tailed and two-tailed test, and the critical values for one-tailed and two-tailed tests. For the *z*-test with known population variances, these are called z, $P(Z \leq z)$ *one-tail* or $P(Z \leq z)$ *two-tail*, and z *Critical one-tail* or z *Critical two-tail*, respectively. For the *t*-tests, these are called t Stat, $P(T \leq t)$ *one-tail* or $P(T \leq t)$ *two-tail*, and t *Critical one-tail* or t *Critical two-tail*, respectively.

Caution: You must be *very careful* in interpreting the output information from these Excel tools and apply the following rules:

1. If the test statistic is negative, the one-tailed *p*-value is the correct *p*-value for a lower-tailed test; however, for an upper-tailed test, you must subtract this number from 1.0 to get the correct *p*-value.

▼ **Figure 7.7**

Portion of Purchase Orders *Database with Lead Time Calculations*

	A	B	C	D	E	F	G	H	I	J	K
1	Purchase Orders										
2											
3	Supplier	Order No.	Item No.	Item Description	Item Cost	Quantity	Cost per order	A/P Terms (Months)	Order Date	Arrival Date	Lead Time
4	Hulkey Fasteners	Aug11001	1122	Airframe fasteners	$ 4.25	19,500	$ 82,875.00	30	08/05/11	08/13/11	8
5	Alum Sheeting	Aug11002	1243	Airframe fasteners	$ 4.25	10,000	$ 42,500.00	30	08/08/11	08/14/11	6
6	Fast-Tie Aerospace	Aug11003	5462	Shielded Cable/ft.	$ 1.05	23,000	$ 24,150.00	30	08/10/11	08/15/11	5
7	Fast-Tie Aerospace	Aug11004	5462	Shielded Cable/ft.	$ 1.05	21,500	$ 22,575.00	30	08/15/11	08/22/11	7
8	Steelpin Inc.	Aug11005	5319	Shielded Cable/ft.	$ 1.10	17,500	$ 19,250.00	30	08/20/11	08/31/11	11
9	Fast-Tie Aerospace	Aug11006	5462	Shielded Cable/ft.	$ 1.05	22,500	$ 23,625.00	30	08/20/11	08/26/11	6
10	Steelpin Inc.	Aug11007	4312	Bolt-nut package	$ 3.75	4,250	$ 15,937.50	30	08/25/11	09/01/11	7

▶ **Figure 7.8**

Pivot Table for Average Supplier Lead Time

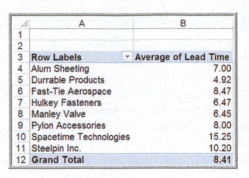

Row Labels	Average of Lead Time
Alum Sheeting	7.00
Durrable Products	4.92
Fast-Tie Aerospace	8.47
Hulkey Fasteners	6.47
Manley Valve	6.45
Pylon Accessories	8.00
Spacetime Technologies	15.25
Steelpin Inc.	10.20
Grand Total	**8.41**

2. If the test statistic is nonnegative (positive or zero), then the *p*-value in the output is the correct *p*-value for an upper-tailed test; but for a lower-tailed test, you must subtract this number from 1.0 to get the correct *p*-value.

3. For a lower-tailed test, you must change the sign of the one-tailed critical value.

Only rarely are the population variances known; also, it is often difficult to justify the assumption that the variances of each population are equal. Therefore, in most practical situations, we use the *t-Test: Two-Sample Assuming Unequal Variances*. This procedure also works well with small sample sizes if the populations are approximately normal. It is recommended that the size of each sample be approximately the same and total 20 or more. If the populations are highly skewed, then larger sample sizes are recommended.

EXAMPLE 7.10   **Testing the Hypotheses for Supplier Lead-Time Performance**

To conduct the hypothesis test for comparing the lead times for Alum Sheeting and Durrable Products, first sort the data by supplier and then select *t-Test: Two-Sample Assuming Unequal Variances* from the *Data Analysis* menu. The dialog is shown in Figure 7.9. The dialog prompts you for the range of the data for each variable, hypothesized mean difference, whether the ranges have labels, and the level of significance α. If you leave the box *Hypothesized Mean Difference* blank or enter zero, the test is for equality of means. However, the tool allows you to specify a value D_0 to test the hypothesis $H_0: \mu_1 - \mu_2 = D_0$ if you want to test whether the population means have a certain distance between them. In this example, the *Variable 1* range defines the lead times for Alum Sheeting and the *Variable 2* range for Durrable Products.

Figure 7.10 shows the results from the tool. The tool provides information for both one-tailed and two-tailed tests. Because this is a one-tailed test, we use the highlighted information in Figure 7.10 to draw our conclusions. For this example, *t Stat* is positive and we have an upper-tailed test; therefore, using the rules stated earlier, the *p*-value is 0.00166. Based on this alone, we reject the null hypothesis and must conclude that Alum Sheeting has a statistically longer average lead time than Durrable Products. We may draw the same conclusion by comparing the value of *t Stat* with the critical value *t Critical one-tail*. Being an upper-tailed test, the value of *t Critical one-tail* is 1.812. Comparing this with the value of *t Stat*, we would reject H_0 only if *t Stat* > *t Critical one-tail*. Since *t Stat* is greater than *t Critical one-tail*, we reject the null hypothesis.

Two-Sample Test for Means with Paired Samples

In the previous example for testing differences in the mean supplier lead times, we used independent samples; that is, the orders in each supplier's sample were not related to each other. In many situations, data from two samples are naturally paired or matched. For example, suppose that a sample of assembly line workers perform a task using two different types of work methods, and the plant manager wants to determine if any differences exist between the two methods. In collecting the data, each worker will have performed

▶ **Figure 7.9**

*Dialog for Two-Sample
t-Test, Sigma Unknown*

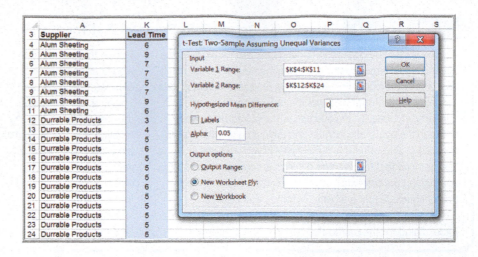

▶ **Figure 7.10**

*Results for Two-Sample Test
for Lead-Time Performance*

	A	B	C
1	t-Test: Two-Sample Assuming Unequal Variances		
2		Alum Sheeting	Durrable Products
3		Variable 1	Variable 2
4	Mean	7	4.923076923
5	Variance	2	0.576923077
6	Observations	8	13
7	Hypothesized Mean Difference	0	
8	df	10	
9	t Stat	3.827958507	
10	P(T<=t) one-tail	0.001664976	
11	t Critical one-tail	1.812461123	
12	P(T<=t) two-tail	0.003329952	
13	t Critical two-tail	2.228138852	

the task using each method. Had we used independent samples, we would have randomly selected two different groups of employees and assigned one work method to one group and the alternative method to the second group. Each worker would have performed the task using only one of the methods. As another example, suppose that we wish to compare retail prices of grocery items between two competing grocery stores. It makes little sense to compare different samples of items from each store. Instead, we would select a sample of grocery items and find the price charged for the same items by each store. In this case, the samples are paired because each item would have a price from each of the two stores.

When paired samples are used, a paired *t*-test is more accurate than assuming that the data come from independent populations. The null hypothesis we test revolves around the mean difference (μ_D) between the paired samples; that is,

$$H_0: \mu_D \ \{\geq, \leq, \text{ or} =\} \ 0$$

$$H_1: \mu_D \ \{<, >, \text{ or} \neq\} \ 0$$

The test uses the average difference between the paired data and the standard deviation of the differences similar to a one-sample test.

Excel has a *Data Analysis* tool, *t-Test: Paired Two-Sample for Means*, for conducting this type of test. In the dialog, you need to enter only the variable ranges and hypothesized mean difference.

EXAMPLE 7.11   Using the Paired Two-Sample Test for Means

The Excel file *Pile Foundation* contains the estimates used in a bid and actual auger-cast pile lengths that engineers ultimately had to use for a foundation-engineering project. The contractor's past experience suggested that the bid information was generally accurate, so the average of the paired differences between the actual pile lengths and estimated lengths should be close to zero. After this project was completed, the contractor found that the average difference between the actual lengths and the estimated lengths was 6.38. Could the contractor conclude that the bid information was poor?

Figure 7.11 shows a portion of the data and the Excel dialog for the paired two-sample test. Figure 7.12

shows the output from the Excel tool using a significance level of 0.05, where *Variable 1* is the estimated lengths, and *Variable 2* is the actual lengths. This is a two-tailed test, so in Figure 7.12 we interpret the results using only the two-tail information that is highlighted. The critical values are ± 1.968, and because *t Stat* is much smaller than the lower critical value, we must reject the null hypothesis and conclude that the mean of the differences between the estimates and the actual pile lengths is statistically significant. Note that the *p*-value is essentially zero, verifying this conclusion.

Two-Sample Test for Equality of Variances

Understanding variation in business processes is very important, as we have stated before. For instance, does one location or group of employees show higher variability than others? We can test for equality of variances between two samples using a new type of test, the *F*-test. To use this test, we must assume that both samples are drawn from normal populations. The hypotheses we test are

▶ **Figure 7.11**

Portion of Excel File Pile Foundation

▶ **Figure 7.12**

Excel Output for Paired Two-Sample Test for Means

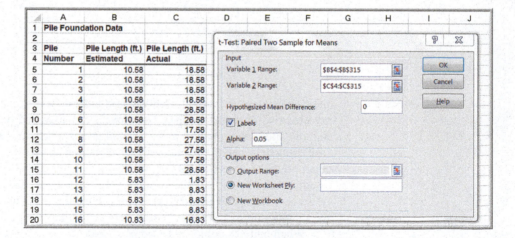

$$H_0: \sigma_1^2 - \sigma_2^2 = 0$$
$$H_1: \sigma_1^2 - \sigma_2^2 \neq 0 \tag{7.5}$$

To test these hypotheses, we collect samples of n_1 observations from population 1 and n_2 observations from population 2. The test uses an F-test statistic, which is the ratio of the variances of the two samples:

$$F = \frac{s_1^2}{s_2^2} \tag{7.6}$$

Because the test statistic is a ratio of variances, an alternate way of expressing the hypotheses that some statisticians prefer is

$$H_0: \frac{\sigma_1^2}{\sigma_2^2} = 1$$
$$H_1: \frac{\sigma_1^2}{\sigma_2^2} \neq 1$$

However, this form is equivalent to (7.5).

The sampling distribution of this statistic is called the F-distribution. Similar to the t-distribution, it is characterized by degrees of freedom; however, the F-distribution has *two* degrees of freedom, one associated with the numerator of the F-statistic, $n_1 - 1$, and one associated with the denominator of the F-statistic, $n_2 - 1$. Table A.4 in Appendix A at the end of the book provides only upper-tail critical values, and the distribution is not symmetric, as is the standard normal distribution or the t-distribution. Therefore, although the hypothesis test is really a two-tailed test, we will simplify it as a one-tailed test to make it easy to use tables of the F-distribution and interpret the results of the Excel tool that we will use. We do this by ensuring that when we compute F, we take the ratio of the larger sample variance to the smaller sample variance.

EXAMPLE 7.12   Applying the *F*-Test for Equality of Variances

To illustrate the F-test, suppose that we wish to determine whether the variance of lead times is the same for Alum Sheeting and Durable Products in the *Purchase Orders* data. The F-test can be applied using the Excel *Data Analysis* tool *F-test Two-Sample for Variances*. The dialog prompts you to enter the range of the sample data for each variable. As we noted, you should ensure that the first variable has the larger variance; this might require you to calculate the variances *before* you use the tool. In this case, the variance of the lead times for Alum Sheeting is larger than the variance for Durable Products (see Figure 7.10),

so this is assigned to *Variable 1*. Note also that if we choose $\alpha = 0.05$, we must enter 0.025 for the level of significance in the Excel dialog. The results are shown in Figure 7.13.

The value of the F-statistic, F, is 3.467. We compare this with the upper-tailed critical value, *F Critical one-tail*, which is 3.607. Because $F < F$ *Critical one-tail*, we cannot reject the null hypothesis and conclude that the variances are not significantly different from each other. Note that the p-value is $P(F<=f)$ *one tail* $= 0.0286$. Although the level of significance is 0.05, remember that we must compare this to $\alpha/2 = 0.025$ because we are using only upper-tailed information.

▶ **Figure 7.13**

Results for Two-Sample F-Test for Equality of Variances

	A	B	C
1	F-Test Two-Sample for Variances		
2		Alum Sheeting	Durable Products
3		Variable 1	Variable 2
4	Mean	7	4.923076923
5	Variance	2	0.576923077
6	Observations	8	13
7	df	7	12
8	F	3.466666667	
9	P(F<=f) one-tail	0.028595441	
10	F Critical one-tail	3.606514642	

If the variances differ significantly from each other, we would expect F to be much larger than 1; the closer F is to 1, the more likely it is that the variances are the same. Therefore, we need only to compare F to the upper-tailed critical value. Hence, for a level of significance α, we find the critical value $F_{\alpha/2, df1, df2}$ of the F-distribution, and then we reject the null hypothesis if the F-test statistic exceeds the critical value. Note that we are using $\alpha/2$ to find the critical value, not α. This is because we are using only the upper-tailed information on which to base our conclusion.

The F-test for equality of variances is often used before testing for the difference in means so that the proper test (population variance is unknown and assumed unequal or population variance is unknown and assumed equal, which we discussed earlier in this chapter) is selected.

CHECK YOUR UNDERSTANDING

1. State the only correct forms of two-sample hypothesis tests.
2. What Excel procedures are available for conducting two-sample hypothesis tests?
3. How does a paired sample test for means differ from other two-sample hypothesis tests for means?
4. Explain how to conduct a test for equality of variances.

Analysis of Variance (ANOVA)

To this point, we have discussed hypothesis tests that compare a population parameter to a constant value or that compare the means of two different populations. Often, we would like to compare the means of more than two groups to determine if all are equal or if any are significantly different from the rest.

EXAMPLE 7.13 **Differences in *Insurance Survey* Data**

In the Excel data file *Insurance Survey*, we might be interested in whether any significant differences exist in satisfaction among individuals with different levels of education. We could sort the data by educational level and then create a table similar to the one shown below.

Although the average satisfaction for each group is somewhat different and it appears that the mean satisfaction of individuals with a graduate degree is higher, we cannot tell conclusively whether or not these differences are significant because of sampling error.

	College Graduate	Graduate Degree	Some College
	5	3	4
	3	4	1
	5	5	4
	3	5	2
	3	5	3
	3	4	4
	3	5	4
	4	5	
	2		
Average	3.444	4.500	3.143
Count	9	8	7

In statistical terminology, the variable of interest is called a **factor**. In this example, the factor is the educational level, and we have three categorical levels of this factor, college graduate, graduate degree, and some college. Thus, it would appear that we will have to perform three different pairwise tests to establish whether any significant differences exist among them. As the number of factor levels increases, you can easily see that the number of pairwise tests grows large very quickly.

Fortunately, other statistical tools exist that eliminate the need for such a tedious approach. **Analysis of variance (ANOVA)** is one of them. Suppose we have m groups. The null hypothesis for ANOVA is that the population means of all m groups are equal; the alternative hypothesis is that at least one mean differs from the rest:

$$H_0: \mu_1 = \mu_2 = \cdots = \mu_m$$

H_1: at least one mean is different from the others

ANOVA derives its name from the fact that we are analyzing variances in the data; essentially, ANOVA computes a measure of the variance between the means of each group and a measure of the variance within the groups and examines a test statistic that is the ratio of these measures. This test statistic can be shown to have an F-distribution (similar to the test for equality of variances). If the F-statistic is large enough based on the level of significance chosen and exceeds a critical value, we would reject the null hypothesis. Excel provides a *Data Analysis* tool, *ANOVA: Single Factor*, to conduct analysis of variance.

| EXAMPLE 7.14 | **Applying the Excel ANOVA Tool** |

To test the null hypothesis that the mean satisfaction for all educational levels in the Excel file *Insurance Survey* are equal against the alternative hypothesis that at least one mean is different, select *ANOVA: Single Factor* from the *Data Analysis* options. First, you must set up the worksheet so that the data you wish to use are displayed in contiguous columns, as shown in Example 7.13. In the dialog shown in Figure 7.14, specify the input range of the data (which must be in contiguous columns) and whether it is stored in rows or columns (that is, whether each factor level or group is a row or column in the range). The sample size for each factor level need not be the same, but the input range must be a rectangular region that contains all data. You must also specify the level of significance (α).

The results for this example are given in Figure 7.15. The output report begins with a summary report of basic statistics for each group. The ANOVA section reports the details of the hypothesis test. You needn't worry about all the mathematical details. The important information to interpret the test is given in the columns labeled F (the F-test statistic), *P-value* (the p-value for the test), and *F crit* (the critical value from the F-distribution). In this example, $F = 3.92$, and the critical value from the F-distribution is 3.4668. Here $F > F\ crit$; therefore, we must reject the null hypothesis and conclude that there are significant differences in the means of the groups—that is, the mean satisfaction is not the same among the three educational levels. Alternatively, we see that the p-value is smaller than the chosen level of significance, 0.05, leading to the same conclusion.

► **Figure 7.14**

ANOVA: Single Factor
Dialog

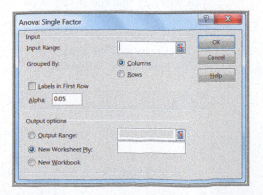

▶ **Figure 7.15**

ANOVA Results for Insurance Survey Data

	A	B	C	D	E	F	G
1	Anova: Single Factor						
2							
3	SUMMARY						
4	*Groups*	*Count*	*Sum*	*Average*	*Variance*		
5	College graduate	9	31	3.444444444	1.027777778		
6	Graduate degree	8	36	4.5	0.571428571		
7	Some college	7	22	3.142857143	1.476190476		
8							
9							
10	ANOVA						
11	*Source of Variation*	*SS*	*df*	*MS*	*F*	*P-value*	*F crit*
12	Between Groups	7.878968254	2	3.939484127	3.924651732	0.035635398	3.466800112
13	Within Groups	21.07936508	21	1.003779289			
14							
15	Total	28.95833333	23				

Although ANOVA can identify a difference among the means of multiple populations, it cannot determine which means are different from the rest. To do this, we may use the Tukey-Kramer multiple comparison procedure. Unfortunately, Excel does not provide this tool, but it may be found in other statistical software.

Assumptions of ANOVA

ANOVA requires assumptions that the m groups or factor levels being studied represent populations whose outcome measures

1. are randomly and independently obtained,
2. are normally distributed, and
3. have equal variances.

If these assumptions are violated, then the level of significance and the power of the test can be affected. Usually, the first assumption is easily validated when random samples are chosen for the data. ANOVA is fairly robust to departures from normality, so in most cases this isn't a serious issue. If sample sizes are equal, violation of the third assumption does not have serious effects on the statistical conclusions; however, with unequal sample sizes, it can.

When the assumptions underlying ANOVA are violated, you may use a *nonparametric test* that does not require these assumptions; we refer you to more comprehensive texts on statistics for further information and examples.

Finally, we wish to point out that students often use ANOVA to compare the equality of means of exactly two populations. It is important to realize that by doing this, you are making the assumption that the populations *have equal variances* (assumption 3). Thus, you will find that the p-values for both ANOVA and the *t-Test: Two-Sample Assuming Equal Variances* will be the same and lead to the same conclusion. However, if the variances are unequal, as is generally the case with sample data, ANOVA may lead to an erroneous conclusion. We recommend that you do not use ANOVA for comparing the means of two populations, but instead use the appropriate *t*-test that assumes unequal variances.

CHECK YOUR UNDERSTANDING

1. What hypotheses does ANOVA test?

2. How can you use the *F, P-value*, and *F crit* values from the Excel *ANOVA: Single Factor* procedure to draw a conclusion?

3. What assumptions does ANOVA rely upon?

Chi-Square Test for Independence

A common problem in business is to determine whether two categorical variables are independent. We introduced the concept of independent events in Chapter 5. In the energy drink survey example (Example 5.11), we used conditional probabilities to determine whether brand preference was independent of gender. However, with sample data, sampling error can make it difficult to properly assess the independence of categorical variables. We would never expect the joint probabilities to be exactly the same as the product of the marginal probabilities because of sampling error even if the two variables are statistically independent. Testing for independence is important in marketing applications.

EXAMPLE 7.15   Independence and Marketing Strategy

Figure 7.16 shows a portion of the sample data used in Chapter 5 for brand preferences of energy drinks (Excel file *Energy Drink Survey*) and the cross-tabulation of the results. A key marketing question is whether the proportion of males who prefer a particular brand is no different from the proportion of females. For instance, of the 63 males, 25 (40%) prefer brand 1. If gender and brand preference are indeed independent, we would expect that about the same proportion of the sample of females would also prefer brand 1. In actuality, only 9 of 37 (24%) prefer brand 1. Similarly,

27% of males prefer brand 2 versus 16% of females; and 33% of males prefer brand 3 versus 59% of females. However, we do not know whether these differences are simply due to sampling error or represent a significant difference. Knowing whether gender and brand preference are independent can help marketing personnel better target advertising campaigns. If they are not independent, then advertising should be targeted differently to males and females, whereas if they are independent, it would not matter.

We can test for independence by using a hypothesis test called the *chi-square test for independence.* The chi-square test for independence tests the following hypotheses:

H_0: the two categorical variables are independent

H_1: the two categorical variables are dependent

The chi-square test is an example of a *nonparametric test*, that is, one that does not depend on restrictive statistical assumptions, as ANOVA does. This makes it a widely applicable and popular tool for understanding relationships among categorical data. The first step in the procedure is to compute the expected frequency in each cell of

	A	B	C	D	E	F	G	H	I
1	Energy Drink Survey								
2									
3	Respondent	Gender	Brand Preference						
4	1	Male	Brand 3		Count of Respondent	Column Labels			
5	2	Female	Brand 3		Row Labels	Brand 1		Brand 2	Brand 3	Grand Total
6	3	Male	Brand 3		Female		9	6	22	37
7	4	Male	Brand 1		Male		25	17	21	63
8	5	Male	Brand 1		Grand Total		34	23	43	100
9	6	Female	Brand 2						
10	7	Male	Brand 2						

▲ Figure 7.16

Portion of Energy Drink Survey *and Cross-Tabulation*

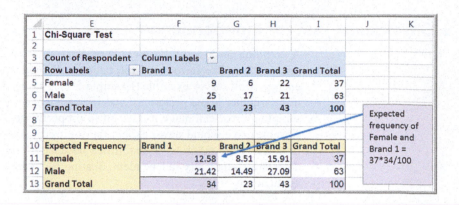

the cross-tabulation if the two variables are independent. This is easily done using the following:

$$\text{Expected Frequency in Row } i \text{ and Column } j = \frac{(\text{Grand Total Row } i)(\text{Grand Total Column } j)}{\text{Total Number of Observations}}$$

$$(7.7)$$

EXAMPLE 7.16   Computing Expected Frequencies

For the *Energy Drink Survey* data, we may compute the expected frequencies using the data from the cross-tabulation and formula (7.7). For example, the expected frequency of females who prefer brand 1 is (37)(34)/100 = 12.58. This

can easily be implemented in Excel. Figure 7.17 shows the results (see the Excel file *Chi-Square Test*). The formula in cell F11, for example, is =$I5*F$7/$I$7, which can be copied to the other cells to complete the calculations.

Next, we compute a test statistic, called a **chi-square statistic**, which is the sum of the squares of the differences between observed frequency, f_o, and expected frequency, f_e, divided by the expected frequency in each cell:

$$\chi^2 = \sum \frac{(f_o - f_e)^2}{f_e}$$

$$(7.8)$$

The closer the observed frequencies are to the expected frequencies, the smaller the value of the chi-square statistic will be. The sampling distribution of χ^2 is a special distribution called the **chi-square** (χ^2) **distribution**. The chi-square distribution is characterized by degrees of freedom, similar to the *t*-distribution. Table A.3 in Appendix A in the back of this book provides critical values of the chi-square distribution for selected values of α. We compare the chi-square statistic for a specified level of significance α to the critical value from a chi-square distribution with $(r - 1)(c - 1)$ degrees of freedom, where r and c are the number of rows and columns in the cross-tabulation table, respectively. The Excel function CHISQ.INV.RT(*probability, deg_freedom*) returns the value of χ^2 that has a right-tail area equal to *probability* for a specified degree of freedom. By setting *probability* equal to the level of significance, we can obtain the critical value for the hypothesis test. If the test statistic exceeds the critical value for a specified level of significance, we reject H_0. The Excel function CHISQ.TEST(*actual_range, expected_range*) computes the *p*-value for the chi-square test.

▶ **Figure 7.18**

Excel Implementation of Chi-Square Test

	E	F	G	H	I
1	Chi-Square Test				
2					
3	Count of Respondent	Column Labels			
4	Row Labels	Brand 1	Brand 2	Brand 3	Grand Total
5	Female	9	6	22	37
6	Male	25	17	21	63
7	Grand Total	34	23	43	100
8					
9					
10	Expected Frequency	Brand 1	Brand 2	Brand 3	Grand Total
11	Female	12.58	8.51	15.91	37
12	Male	21.42	14.49	27.09	63
13	Grand Total	34	23	43	100
14					
15					
16	Chi Square Statistic	Brand 1	Brand 2	Brand 3	Grand Total
17	Female	1.02	0.74	2.33	4.09
18	Male	0.60	0.43	1.37	2.40
19	Grand Total	1.62	1.18	3.70	6.49
20					
21		Chi-square critical value			5.99146455
22		p-value			0.03892134

EXAMPLE 7.17	**Conducting the Chi-Square Test for Independence**

For the *Energy Drink Survey* data, Figure 7.18 shows the calculations of the chi-square statistic using formula (7.8). For example, the formula in cell F17 is =(F5 − F11)^2/F11, which can be copied to the other cells. The grand total in the lower right cell is the value of x^2. In this case, the chi-square test statistic is 6.4924. Since the cross-tabulation has $r = 2$ rows and $c = 3$ columns, we have $(2 - 1)(3 - 1) = 2$ degrees of freedom for the chi-square distribution. Using $\alpha = 0.05$, the Excel function CHISQ.INV.RT(0.05, 2) returns the critical value 5.99146. Because the test statistic exceeds the critical value, we reject the null hypothesis that the two categorical variables are independent.

Alternatively, we could simply use the CHISQ.TEST function to find the *p*-value for the test and base our conclusion on that without computing the chi-square statistic. For this example, the function CHISQ.TEST(F5:H6, F11:H12) returns the *p*-value of 0.0389, which is less than $\alpha = 0.05$; therefore, we reject the null hypothesis that brand and gender are independent. Because a higher proportion of females prefer brand 3 while a higher proportion of males prefer brand 1, marketers might advertise brand 3 in female-related media, while brand 1 should receive higher emphasis in male-related media.

Cautions in Using the Chi-Square Test

First, when using PivotTables to construct a cross-tabulation and implement the chi-square test in Excel similar to Figure 7.17, be extremely cautious of blank cells in the PivotTable. Blank cells will not be counted in the chi-square calculations and will lead to errors. If you have blank cells in the PivotTable, simply replace them with zeros, or right click in the PivotTable, choose *PivotTable Options*, and enter 0 in the field for the checkbox *For empty cells show*.

Second, the chi-square test assumes adequate expected cell frequencies. A rule of thumb is that there be no more than 20% of cells with expected frequencies smaller than 5, and no expected frequencies of zero. More advanced statistical procedures exist to handle this, but you might consider aggregating some of the rows or columns in a logical fashion to enforce this assumption, as the Example 7.18 illustrates. This, of course, results in fewer rows or columns, which changes the number of degrees of freedom.

EXAMPLE 7.18   Violations of Chi-Square Assumptions

A survey of 100 students at a university queried their beverage preferences at a local coffee shop. The results are shown in the table below.

	Brewed coffee	Iced coffee	Espresso	Cappuccino	Latte	Mocha	Iced blended	Tea	Total
Female	6	10	2	4	7	9	10	8	56
Male	16	5	8	2	3	2	7	1	44
Total	22	15	10	6	10	11	17	9	100

The expected frequencies are shown next.

Expected frequencies									
	Brewed coffee	Iced coffee	Espresso	Cappuccino	Latte	Mocha	Iced blended	Tea	Total
Female	12.32	8.4	5.6	3.36	5.6	6.16	9.52	5.04	56
Male	9.68	6.6	4.4	2.64	4.4	4.84	7.48	3.96	44
Total	22	15	10	6	10	11	17	9	100

If we were to conduct a chi-square test of independence, we would see that of the 16 cells, five, or over 30%, have frequencies smaller than 5. Four of them are in the Cappuccino, Latte, and Mocha columns; these can be aggregated into one column called Hot Specialty beverages.

	Brewed coffee	Iced coffee	Espresso	Hot Specialty	Iced blended	Tea	Total
Female	6	10	2	20	10	8	56
Male	16	5	8	7	7	1	44
	22	15	10	27	17	9	100

Now only 2 of 12 cells have an expected frequency less than 5; this now meets the assumptions of the chi-square test.

Expected frequencies							
	Brewed coffee	Iced coffee	Espresso	Hot Specialty	Iced blended	Tea	Total
Female	12.32	8.4	5.6	15.12	9.52	5.04	56
Male	9.68	6.6	4.4	11.88	7.48	3.96	44
	22	15	10	27	17	9	100

Chi-Square Goodness of Fit Test

In Chapter 5, we introduced the chi-square goodness of fit test to determine whether data can be reasonably assumed to come from a specified distribution like the normal or exponential distribution. What we actually did was to conduct a hypothesis test:

H_0: data are sampled from a specified distribution

H_1: data are not sampled from the specified distribution

You can see the similarity to the chi-square test for independence in how the actual and expected frequencies are used to calculate the chi-square statistic. If the chi-square statistic is larger than the critical value, reject the null hypothesis; otherwise, fail to reject it and conclude that the specified distribution is a reasonable model for the sample data. The key difference is how we used the hypothesized probability distribution to compute the expected frequencies.

CHECK YOUR UNDERSTANDING

1. What hypothesis does the chi-square test for independence test?

2. Why is it important to understand independence of categorical variables in business applications?

3. Explain the steps required to conduct chi-square tests for independence and normality in Excel.

4. What cautions should you be aware of in using the chi-square test for independence?

ANALYTICS IN PRACTICE: Using Hypothesis Tests and Business Analytics in a Help Desk Service Improvement Project[2]

Schlumberger is an international oilfield-services provider headquartered in Houston, Texas. Through an outsourcing contract, they supply help desk services for a global telecom company that offers wireline communications and integrated telecom services to more than 2 million cellular subscribers. The help desk, located in Ecuador, faced increasing customer complaints and losses in dollars and cycle times. The company drew upon the analytics capability of one of the help desk managers to investigate and solve the problem. The data showed that the average solution time for issues reported to the help desk was 9.75 hours. The company set a goal to reduce the average solution time by 50%. In addition, the number of issues reported to the help desk had reached an average of 30,000 per month. Reducing the total number of issues reported to the help desk would allow the company to address those issues that hadn't been resolved because of a lack of time, and to reduce the number of abandoned calls. They set a goal to identify preventable issues so that customers would not have to contact the help desk in the first place, and set a target of 15,000 issues.

As part of their analysis, they observed that the average solution time for help desk technicians working at the call center seemed to be lower than the average for technicians working on site with clients. They conducted a hypothesis test structured around the question: Is there a difference between having help desk employees working at an off-site facility rather than on site within the client's main office? The null hypothesis was that there was no significant difference; the alternative hypothesis was

Hurst Photo/Shutterstock

that there was a significant difference. Using a two-sample t-test to assess whether the call center and the help desk are statistically different from each other, they found no statistically significant advantage in keeping help desk employees working at the call center. As a result, they moved help desk agents to the client's main office area. Using a variety of other analytical techniques, they were able to make changes to their process, resulting in the following:

- a decrease in the number of help desk issues of 32%
- improved capability to meet the target of 15,000 total issues
- a reduction in the average desktop solution time from 9.75 hours to 1 hour, an improvement of 89.7%
- a reduction in the call-abandonment rate from 44% to 26%
- a reduction of 69% in help desk operating costs

[2]Based on Francisco, Endara M. "Help Desk Improves Service and Saves Money with Six Sigma," American Society for Quality, http://asq.org/economic-case/markets/pdf/help-desk-24490.pdf, accessed 8/19/11.

KEY TERMS

Alternative hypothesis
Analysis of variance (ANOVA)
Chi-square distribution
Chi-square statistic
Confidence coefficient
Factor
Hypothesis
Hypothesis testing
Level of significance

Null hypothesis
One-sample hypothesis test
One-tailed test of hypothesis
p-Value (observed significance level)
Power of the test
Statistical inference
Two-tailed test of hypothesis
Type I error
Type II error

CHAPTER 7 TECHNOLOGY HELP

Useful Excel Functions and Formulas

NORM.S.INV$(1 - \alpha)$ Finds the critical value for a one-sample, one-tailed test when the standard deviation is known. For a lower-tailed test, the critical value is negative; for a two-tailed test, use $\alpha/2$.

T.INV$(1 - \alpha, n - 1)$ Finds the critical value for a one-sample, one-tailed test for the mean when the population standard deviation is unknown. For a lower-tailed test, the critical value is negative; for a two-tailed test, use $\alpha/2$.

T.INV.2T$(\alpha, n - 1)$ Finds the critical value for a one-sample, two-tailed test of hypothesis when the population standard deviation is unknown. The critical value for the lower tail is negative. Note that this function is the same as T.INV$(1 - \alpha/2, n - 1)$.

T.DIST$(t, n - 1, \text{TRUE})$ Finds the *p*-value for a lower-tailed hypothesis test for the mean when the population standard deviation is unknown.

$1 -$ T.DIST$(t, n - 1, \text{TRUE})$ Finds the *p*-value for an upper-tailed hypothesis test for the mean when the population standard deviation is unknown using the test statistic *t*.

T.DIST.2T$(\text{ABS}(t), n - 1)$ Finds the *p*-value for a two-tailed hypothesis test for the mean when the population standard deviation is unknown for the test statistic *t*.

NORM.S.DIST(z, TRUE) Finds the *p*-value for a lower-tailed test for a proportion using the test statistic *z*.

$1 -$ NORM.S.DIST(z, TRUE) Finds the *p*-value for an upper-tailed test for a proportion using the test statistic *z*.

$2*$NORM.S.DIST$(\text{ABS}(z), \text{TRUE})$ Finds the *p*-value for a two-tailed test for a proportion using the test statistic *z*.

CHISQ.INV.RT(*probability*, *deg_ freedom*) Returns the value of chi-square that has a right-tail area equal to *probability* for a specified degree of freedom. By setting *probability* equal to the level of significance, we obtain the critical value for the chi-square hypothesis test for independence.

CHISQ.TEST(*actual_range*, *expected_range*) Computes the *p*-value for the chi-square test for independence.

Excel Templates

One-Sample Hypothesis Tests (Figure 7.6):
> Open the Excel file *One Sample Hypothesis Tests*. Select the worksheet tab corresponding to the type of hypothesis test to perform and enter the appropriate data.

Excel Techniques

Excel z-Test: Two-Sample for Means:
> Performs a two-sample hypothesis test for means when σ^2 for each population is known. Select *z-Test: Two-Sample for Means* from the *Data Analysis* menu and enter the data ranges, hypothesized mean difference, and value of α. Check *Labels* if headers are included.

Excel t-Test: Two-Sample Assuming Unequal Variances (Example 7.10):
> Performs a two-sample test for means when σ^2 for each population is unknown and assumed unequal. Select *t-Test: Two-Sample Assuming Unequal Variances* from the *Data Analysis* menu and enter

the data ranges, hypothesized mean difference, and value of α. Check *Labels* if headers are included.

Excel t-Test: Two-Sample Assuming Equal Variances: Performs a two-sample test for means when σ^2 for each population is unknown and assumed equal. Select *t-Test: Two-Sample Assuming Equal Variances* from the *Data Analysis* menu and enter the data ranges, hypothesized mean difference, and value of α. Check *Labels* if headers are included.

For each of the above tests, apply the following rules:
1. If the test statistic is negative, the one-tailed *p*-value is the correct *p*-value for a lower-tailed test; however, for an upper-tailed test, you must subtract this number from 1.0 to get the correct *p*-value.
2. If the test statistic is nonnegative (positive or zero), then the *p*-value in the output is the correct *p*-value for an upper-tailed test, but for a lower-tailed test, you must subtract this number from 1.0 to get the correct *p*-value.
3. For a lower-tailed test, you must change the sign of the one-tailed critical value.

Excel t-Test: Paired Two-Sample for Means (Example 7.11): Performs a two-sample test for means using paired samples. Select *t-Test: Paired Two-Sample for Means* from the *Data Analysis* menu and enter the data ranges, hypothesized mean difference, and value of α. Check *Labels* if headers are included.

Excel F-Test: Two-Sample for Variances (Example 7.12): Performs a two-sample test for equality of variances. Select *F-Test: Two-Sample for Variances* from the *Data Analysis* menu and enter the data ranges, hypothesized mean difference, and value of α. Make sure that the first variable has the larger variance and enter $\alpha/2$, not α, for the level of significance in the dialog. Check *Labels* if headers are included.

ANOVA (Example 7.14): Set up the worksheet so that the data you wish to use are displayed in contiguous columns. Select *ANOVA: Single Factor* from the *Data Analysis* menu. Enter the input range as the rectangular array that contains all the data; check *Labels in the First Row* if headers are included.

StatCrunch

StatCrunch provides the ability to conduct the hypothesis tests, analysis of variance, and chi-square tests that we learned about in this chapter. You can find video tutorials and step-by-step procedures with examples at https://www.statcrunch.com/5.0/example.php. We suggest that you first view the tutorials *Getting started with Stat-Crunch* and *Working with StatCrunch sessions*. The following tutorials explain how to conduct hypothesis tests in StatCrunch. These are located under the headings Traditional Hypothesis Tests For, ANOVA, and Summary Statistics and Tables.

- A mean with raw data
- A mean with summary data
- A proportion with raw data
- A proportion with summary data
- The difference between two means with raw data
- The difference between two means with summary data
- The difference between means with paired data
- The difference between two proportions with raw data
- The difference between two proportions with summary data
- One way ANOVA
- Contingency tables from raw data
- Contingency tables from summary data

Example: Testing a Hypothesis for a Mean with Raw Data

1. Choose the *Stat > T Stats > One Sample > With Data* menu option.
2. Select the column containing the sample data values.
3. Under *Perform*, select *Hypothesis test for u* .
4. Enter the null mean and choose $\neq, <$, or $>$ for the alternative.
5. Click *Compute!* to view the results.

Example: Conducting a Two-Sample Hypothesis Test for the Mean with Summary Data

1. Choose the *Stat > T Stats > One Sample > With Data* menu option.
2. Choose the *With Summary* option to enter the sample mean, sample standard deviation, and sample size for both samples.
3. Uncheck the *Pool variances* option if desired.
4. Select the *Hypothesis test* option, enter the difference in means for the null hypothesis, and choose $\neq, <$, or $>$ for the alternative.
5. Click *Compute!* to view the results.

For all hypothesis tests, assume that the level of significance is 0.05 unless otherwise stated.

Hypothesis Testing

1. When President Donald Trump took office, he believed that the reason he did not win the popular vote was because 3 to 5 million people voted illegally. Explain how hypothesis testing might be used in a similar fashion as the legal analogy example.

One-Sample Hypothesis Tests

2. According to an article in a business publication, the average tenure of a U.S. worker is 4.6 years. Formulate an appropriate one-sample test of hypothesis to test this belief.

3. A national magazine stated that at most, 16% of millennials have a gym membership. Formulate an appropriate one-sample test of hypothesis to test this.

4. Looking at the data and information in the Excel file *Car Sharing Survey*, state some examples of interesting hypothesis tests by proposing null and alternative hypotheses similar to those in Example 7.2.

5. Looking at the data and information in the Excel file *Retail Survey*, state some examples of interesting hypothesis tests by proposing null and alternative hypotheses similar to those in Example 7.2.

6. Looking at the data and information in the Excel file *TV Viewing Survey*, state some examples of interesting hypothesis tests by proposing null and alternative hypotheses similar to those in Example 7.2.

7. The price of a certain combo meal at different franchises of a national fast food company varies from $5.00 to $7.36 and has a known standard deviation of $2.11. A sample of 23 students in an online course that includes students across the country stated that their average price is $5.75. The students have also stated that they are generally unwilling to pay more than $6.50 for this meal. Formulate and conduct a hypothesis test to determine if you can conclude that the population mean is less than $6.50.

8. A business school has a goal that the average number of years of work experience of its MBA applicants is more than three years. Historical data suggest that the variance has been constant at around six months, and thus, the population variance can be assumed to be known. Based on last year's applicants, it was found that among a sample of 47, the average number of years of work experience is 3.1. Can the school state emphatically that it is meeting its goal? Formulate the appropriate hypothesis test and conduct the test.

9. According to a magazine, people read an average of more than two books in a month. A survey of 25 random individuals found that the mean number of books they read was 2.1 with a standard deviation of 1.24.
   a. To test the magazine's claim, what should the appropriate hypotheses be?
   b. Compute the test statistic.
   c. Using a level of significance of 0.05, what is the critical value?
   d. Find the *p*-value for the test.
   e. What is your conclusion?

10. A national poll stated that most Americans exercise more than two days a week. A random sample of 105 Americans found that the average number of days they exercised each week was 2.17, with a standard deviation of 2.14. Is there sufficient evidence to support the poll results? Formulate the appropriate hypothesis test and draw a conclusion.

11. A bank found that in recent years, the average monthly charge on its credit card was $1,350. With an improving economy, they suspect that this amount has increased. A sample of 42 customers resulted in an average monthly charge of $1,376.54 with a standard deviation of $183.89. Do these data provide statistical evidence that the average monthly charge has increased? Formulate the appropriate hypothesis test and draw a conclusion.

12. Using the data in the Excel file *Consumer Transportation Survey*, test the following null hypotheses:
   a. Individuals spend at least eight hours per week in their vehicles.
   b. Individuals drive an average of 600 miles per week.
   c. The average age of SUV drivers is no greater than 35.
   d. At least 80% of individuals are satisfied with their vehicles.

Perform the calculations using the correct formulas and Excel functions, and compare your results with the *One Sample Hypothesis Test* Excel template to verify them.

13. A retailer believes that its new advertising strategy will increase sales. Previously, the mean spending in 15 categories of consumer items in both the 18–34 and 35+ age groups was $70.00.

    a. Formulate a hypothesis test to determine if the mean spending in these categories has statistically increased.

    b. After the new advertising campaign was launched, a marketing study found that the mean spending for 300 respondents in the 18–34 age group was $75.86, with a standard deviation of $50.90. Is there sufficient evidence to conclude that the advertising strategy significantly increased sales in this age group?

    c. The marketing study also determined that for 700 respondents in the 35+ age group, the mean and standard deviation were $68.53 and $45.29, respectively. Is there sufficient evidence to conclude that the advertising strategy significantly increased sales in this age group?

14. A university believes that the average retirement age among their faculty is now 70 instead of the historical value of 65. A sample of 85 faculty found that the average of their expected retirement age is 68.4 with a standard deviation of 3.6. Is the university's belief valid?

15. Call centers typically have high turnover. The director of human resources for a large bank has compiled data on about 70 former employees at one of the bank's call centers in the Excel file *Call Center Data*. In writing an article about call center working conditions, a reporter has claimed that the average tenure is no more than two years. Formulate and test a hypothesis using these data to determine if this claim can be disputed.

16. Using the data in the Excel file *Airport Service Times*, determine if the airline can claim that its average service time is less than 2.5 minutes.

17. A computer repair firm believes that its average repair time is less than two weeks. Using the data in the Excel file *Computer Repair Times*, determine if the company can continue to support this claim.

18. The State of Ohio Department of Education has a mandated ninth-grade proficiency test that covers writing, reading, mathematics, citizenship (social studies), and science. The Excel file *Ohio Education Performance* provides data on success rates (defined as the percent of students passing) in school districts in the greater Cincinnati metropolitan area along with state averages. Test null hypotheses that the average scores in the Cincinnati area for each test and also for the composite score are equal to the state averages.

19. Formulate and test hypotheses to determine if statistical evidence suggests that the graduation rate for (1) top liberal arts colleges or (2) research universities in the Excel file *Colleges and Universities* exceeds 90%. Do the data support a conclusion that the graduation rates exceed 85%? Would your conclusions change if the level of significance was 0.01 instead of 0.05?

20. An industry trade publication stated that the average profit per customer for this industry was greater than $4,500. The Excel file *Sales Data* provides data on a sample of customers. Using a test of hypothesis, do the data support this claim or not?

21. A financial advisor believes that the proportion of investors who are risk-averse (that is, try to avoid risk in their investment decisions) is at least 0.7. A survey of 32 investors found that 20 of them were risk-averse. Formulate a one-sample hypothesis test for a proportion to test this belief.

22. The U.S. Department of Labor stated that the national unemployment rate was 4.9%. Of a sample of 90 people in one city, 5.2% were unemployed. Can the city conclude that their unemployment rate is the same as the national average?

23. The Excel file *Room Inspection* provides data for 100 room inspections at each of 25 hotels in a major chain. Management would like the proportion of nonconforming rooms to be less than 2%. Formulate a one-sample hypothesis test for a proportion and perform the calculations using the correct formulas and Excel functions, and compare your results with the *One Sample Hypothesis Test* Excel template to verify them.

24. An online bookseller is considering selling an e-reader but will do so only if they have evidence that the proportion of customers who will likely purchase one is more than 0.4. Of a survey of 25 customers, 8 of them stated that they would likely purchase an e-reader. What should the bookseller do?

25. An employer is considering negotiating its pricing structure for health insurance with its provider if there is sufficient evidence that customers will be willing to pay a lower premium for a higher deductible. Specifically, they want at least 30% of their employees to be willing to do this. Using the sample data in the Excel file *Insurance Survey*, determine what decision they should make.

Two-Sample Hypothesis Tests

26. A two-sample test for means was conducted to test whether the mean number of movies watched each month differed between males and females. The Excel *Data Analysis* tool results are shown below.

t-Test: Two-Sample Assuming Unequal Variances

	Female	Male
Mean	5.6	7.5
Variance	6.267	21.833
Observations	10	10
Hypothesized Mean Difference	0	
df	14	
t Stat	−1.133	
P(T<=t) one-tail	0.138	
t Critical one-tail	1.76	
P(T<=t) two-tail	0.276	
t Critical two-tail	2.144	

    a. Explain how to use this information to draw a conclusion if the null hypothesis is $H_0: \mu_F - \mu_M \leq 0$. Clearly state the correct critical value and *p*-value and your conclusion.

    b. Explain how to use this information to draw a conclusion if the null hypothesis is $H_0: \mu_F - \mu_M \geq 0$. Clearly state the correct critical value and *p*-value and your conclusion.

    c. Explain how to use this information to draw a conclusion if the null hypothesis is $H_0: \mu_F - \mu_M = 0$. Clearly state the correct critical value and *p*-value and your conclusion.

27. A two-sample test for means was conducted to determine if the completion time for continuing education programs for nurses differed when costs are paid by employers (Yes) versus when individuals paid out of his or her own funds (No). The Excel *Data Analysis* tool results are shown next.

t-Test: Two-Sample Assuming Unequal Variances

	Yes	No
Mean	33.846	28.125
Variance	446.474	349.554
Observations	13	8
Hypothesized Mean Difference	0	
df	16	
t Stat	0.648	
P(T<=t) one-tail	0.263	
t Critical one-tail	1.746	
P(T<=t) two-tail	0.526	
t Critical two-tail	2.12	

    a. Explain how to use this information to draw a conclusion if the null hypothesis is $H_0: \mu_Y - \mu_N \leq 0$. Clearly state the correct critical value and *p*-value and your conclusion.

    b. Explain how to use this information to draw a conclusion if the null hypothesis is $H_0: \mu_Y - \mu_N \geq 0$. Clearly state the correct critical value and *p*-value and your conclusion.

    c. Explain how to use this information to draw a conclusion if the null hypothesis is $H_0: \mu_Y - \mu_N = 0$. Clearly state the correct critical value and *p*-value and your conclusion.

28. The director of human resources for a large bank has compiled data on about 70 former employees at one of the bank's call centers (see the Excel file *Call Center Data*). For each of the following, assume equal variances of the two populations.

    a. Test the null hypothesis that the average length of service for males is the same as for females.

    b. Test the null hypothesis that the average length of service for individuals without prior call center experience is the same as those with experience.

    c. Test the null hypothesis that the average length of service for individuals with a college degree is the same as for individuals without a college degree.

    d. Now conduct tests of hypotheses for equality of variances. Were your assumptions of equal variances valid? If not, repeat the test(s) for means using the unequal variance test.

29. Using the Excel file *Facebook Survey*, determine if the mean number of hours spent online per week is the same for males as it is for females.

30. Determine if there is evidence to conclude that the mean number of vacations taken by married individuals is less than the number taken by single/divorced individuals using the data in the Excel file *Vacation Survey*. Use a level of significance of 0.05. Would your conclusion change if the level of significance were 0.01?

31. The Excel file *Accounting Professionals* provides the results of a survey of 27 employees in a tax division of a Fortune 100 company.

  a. Test the null hypothesis that the average number of years of service is the same for males and females.

  b. Test the null hypothesis that the average years of undergraduate study is the same for males and females.

32. For the data in the Excel file *Coffee Shop Preferences*, conduct a hypothesis test to determine if price and taste ratings are the same for large/chain stores versus small/independent coffee shops.

33. In the Excel file *Cell Phone Survey*, test the hypothesis that the mean responses for value for the dollar and customer service do not differ by gender.

34. In the Excel file *Credit Risk Data*, test the hypotheses that the number of months employed is the same for applicants with low credit risk as for those with high credit risk evaluations. Use a level of significance of 0.01.

35. For the data in the Excel file *TV Viewing Survey*, conduct a hypothesis test to determine if married individuals watch less TV than single individuals.

36. Determine if there is evidence to conclude that the mean GPA of males who plan to attend graduate school is higher than that of females who plan to attend graduate school using the data in the Excel file *Graduate School Survey*.

37. A producer of computer-aided design software for the aerospace industry receives numerous calls for technical support. Tracking software is used to monitor response and resolution times. In addition, the company surveys customers who request support using the following scale: 0—did not meet expectations; 1—marginally met expectations; 2—met expectations; 3—exceeded expectations; 4—greatly exceeded expectations. The questions are as follows:

Q1: Did the support representative explain the process for resolving your problem?

Q2: Did the support representative keep you informed about the status of progress in resolving your problem?

Q3: Was the support representative courteous and professional?

Q4: Was your problem resolved?

Q5: Was your problem resolved in an acceptable amount of time?

Q6: Overall, how did you find the service provided by our technical support department?

A final question asks the customer to rate the overall quality of the product using a scale of 0—very poor; 1—poor; 2—good; 3—very good; 4—excellent. A sample of survey responses and associated resolution and response data are provided in the Excel file *Customer Support Survey*.

  a. The company has set a service standard of one day for the mean resolution time. Does evidence exist that the resolution time is more than one day? How do the outliers in the data affect your result? What should you do about them?

  b. Test the hypothesis that the average service index is equal to the average engineer index.

38. Using the data in the Excel file *Ohio Education Performance*, test the hypotheses that the mean difference in writing and reading scores is zero and that the mean difference in math and science scores is zero. Use the paired-sample procedure.

39. The Excel file *Unions and Labor Law Data* reports the percent of public and private-sector employees in unions in 1982 for each state, along with indicators of whether the states had a bargaining law that covered public employees or right-to-work laws.

  a. Test the hypothesis that the mean percent of employees in unions for both the public sector and private sector is the same for states having bargaining laws as for those who do not.

  b. Test the hypothesis that the mean percent of employees in unions for both the public sector and private sector is the same for states having right-to-work laws as for those who do not.

40. Using the data in the Excel file *Student Grades*, which represent exam scores in one section of a large statistics course, test the hypothesis that the variance in grades is the same for both tests.

41. In the Excel file *Restaurant Sales*, determine if the variance of weekday sales is the same as that of weekend sales for each of the three variables (lunch, dinner, and delivery).

Analysis of Variance (ANOVA)

42. For the Excel file *Job Satisfaction*, use ANOVA to determine if the mean overall job satisfaction ratings differ by department.

43. A college is trying to determine if there is a significant difference in the mean GMAT score of students from different undergraduate backgrounds who apply to the MBA program. The Excel file *GMAT Scores* contains data from a sample of students. What conclusion can be reached using ANOVA?

44. For the data in the Excel file *Helpdesk Survey*, determine if the mean overall satisfaction ratings differ by the ratings given for response time.

45. Using the data in the Excel file *Cell Phone Survey*, apply ANOVA to determine if the mean response for value for the dollar is the same for different types of cell phones.

46. For the data in the Excel file *Freshman College Data*,

   a. use ANOVA to determine whether significant differences exist in the mean retention rate for the different colleges over the four-year period.

   b. use ANOVA to determine if significant differences exist in the mean ACT scores among the different colleges.

   c. use ANOVA to determine if significant differences exist in the mean SAT scores among the different colleges.

Chi-Square Tests for Independence and Normality

47. The cross-tabulation data given below represent the number of males and females in a work group who feel overstressed and those who don't.

Overstressed?	Female	Male
No	8	4
Yes	4	6

   a. Write the hypotheses for the chi-square test for independence.

   b. Find the expected frequencies.

   c. Compute the chi-square statistic using a level of significance of 0.05.

   d. Find the chi-square critical value and p-value and draw a conclusion.

48. The cross-tabulation data given below represent the number of males and females in a survey who have or have not visited an urgent care facility in the last month.

   a. Write the hypotheses for the chi-square test for independence.

   b. Find the expected frequencies.

   c. Compute the chi-square statistic using a level of significance of 0.05.

   d. Find the chi-square critical value and p-value and draw a conclusion.

Visited Urgent Care?	Female	Male
Yes	14	6
No	2	7

49. The following cross-tabulation shows the number of people who rated a customer service representative as friendly and polite based on whether the representative greeted them first.

   a. Write the hypotheses for the chi-square test for independence.

   b. Find the expected frequencies.

   c. Compute the chi-square statistic using a level of significance of 0.01.

   d. Find the chi-square critical value and p-value and draw a conclusion.

Staff Greeting	Friendliness/Politeness	
	No	Yes
No	11	5
Yes	13	20

50. Conduct the chi-square test for independence using the aggregated results for Example 7.18 using a level of significance of 0.05.

51. For the cross-tabulation data shown on the next page, which represent a count of the type of books that individuals prefer, verify that the assumptions of the chi-square test are not met, and aggregate the data in an appropriate way so that the assumptions are met. Then conduct a chi-square test for independence.

Gender	Fiction	Nonfiction	Romance	Autobiography
Male	8	2	1	6
Female	7	2	2	0

52. An online bookseller is trying to determine whether customers' gender is independent of the genre of books they most often purchase. A sample of customers revealed the following:

Gender	Literature	Genre Magazine	Nonfiction	Popular Fiction
Female	12	28	9	37
Male	8	15	22	29

What can the bookseller conclude?

53. A survey of college students determined their preference for cell phone providers. The following data were obtained:

Gender	Provider T-Mobile	AT&T	Verizon	Other
Male	12	39	27	16
Female	8	22	24	12

Can we conclude that gender and cell phone provider are independent? If not, what implications does this have for marketing?

54. For the data in the Excel file *Graduate School Survey*, perform a chi-square test for independence to determine if plans to attend graduate school are independent of gender.

55. For the data in the Excel file *New Account Processing*, perform chi-square tests for independence to determine if certification is independent of gender and if certification is independent of having prior industry background.

56. Conduct a chi-square test to determine if a normal distribution can reasonably model the data in the Excel file *Airport Service Times*.

57. Conduct a chi-square test to determine if a normal distribution can reasonably model the expense data in the Excel file *Travel Expenses*.

58. Conduct a chi-square test to determine if a normal distribution can reasonably model the sum of the midterm and final exam grades in the Excel file *Student Grades*.

CASE: DROUT ADVERTISING RESEARCH PROJECT

The background for this case was introduced in Chapter 2. This is a continuation of the case in Chapter 6. For this part of the case, propose and test some meaningful hypotheses that will help Ms. Drout understand and explain the results. Include two-sample tests, ANOVA, and/or chi-square tests for independence as appropriate. Write up your conclusions in a formal report, or add your findings to the report you completed for the case in Chapter 6 as per your instructor's requirements. If you have accumulated all sections of this case into one report, polish it up so that it is as professional as possible, drawing final conclusions about the perceptions of the role of advertising in the reinforcement of gender stereotypes and the impact of empowerment advertising.

CASE: PERFORMANCE LAWN EQUIPMENT

Elizabeth Burke has identified some additional questions she would like you to answer using the *Performance Lawn Equipment Database*.

1. Are there significant differences in ratings of specific product/service attributes in the *Customer Survey* worksheet?

2. In the worksheet *On-Time Delivery*, has the proportion of on-time deliveries in 2018 significantly improved since 2014?

3. Have the data in the worksheet *Defects After Delivery* changed significantly over the past five years?

4. Although engineering has collected data on alternative process costs for building transmissions in the

worksheet *Transmission Costs*, why didn't they reach a conclusion as to whether one of the proposed processes is better than the current process?

5. Are there differences in employee retention due to gender, college graduation status, or whether the employee is from the local area in the data in the worksheet *Employee Retention*?

Conduct appropriate statistical analyses and hypothesis tests to answer these questions and summarize your results in a formal report to Ms. Burke.

Trendlines and Regression Analysis

Luca Bertolli/123RF

LEARNING OBJECTIVES After studying this chapter, you will be able to:

- Explain the purpose of regression analysis and provide examples in business.
- Use a scatter chart to identify the type of relationship between two variables.
- List the common types of mathematical functions used in predictive modeling.

- Use the Excel *Trendline* tool to fit models to data.
- Explain how least-squares regression finds the best-fitting regression model.
- Use Excel functions to find least-squares regression coefficients.

- Use the Excel *Regression* tool for both single and multiple linear regressions.
- Interpret the regression statistics of the Excel *Regression* tool.
- Interpret significance of regression from the Excel *Regression* tool output.
- Draw conclusions for tests of hypotheses about regression coefficients.
- Interpret confidence intervals for regression coefficients.
- Calculate standard residuals.
- List the assumptions of regression analysis and describe methods to verify them.

- Explain the differences in the Excel *Regression* tool output for simple and multiple linear regression models.
- Apply a systematic approach to build good regression models.
- Explain the importance of understanding multicollinearity in regression models.
- Build regression models for categorical data using dummy variables.
- Test for interactions in regression models with categorical variables.
- Identify when curvilinear regression models are more appropriate than linear models.

The late management and quality guru Dr. W. Edwards Deming once stated that all management is prediction. He was implying that when managers make decisions, they must consider what will happen in the future. For example, financial analysts cannot make good investment decisions without predicting the future movement of stock prices, and airlines cannot determine when to purchase jet fuel—and how much—without predicting how oil prices will change. *Trendlines* show the movement of such attributes over time.

Many other applications of predictive analytics involve modeling relationships between one or more independent variables and some dependent variable (called *regression analysis*). For example, we might wish to predict the level of sales based on the price we set, or to predict sales based on the U.S. GDP (gross domestic product) and the ten-year treasury bond rate to capture the influence of the business cycle.[1] A marketing researcher might want to predict the intent of buying a particular automobile model based on a survey that measures consumer attitudes toward the brand, negative word of mouth, and income level.[2] An insurance company may need to predict the number and amount of claims based on the demographics of its customers. Human resource managers might need to predict the need for different workforce skills in order to develop hiring and training plans. Fantasy sports hobbyists may want to predict the value of players based on various performance attributes. Thus, you will likely find numerous opportunities to apply the tools we learn in this chapter to your career and personal interests.

[1]James R. Morris and John P. Daley, *Introduction to Financial Models for Management and Planning* (Boca Raton, FL: Chapman & Hall/CRC, 2009): 257.

[2]Alvin C. Burns and Ronald F. Bush, *Basic Marketing Research Using Microsoft Excel Data Analysis*, 2nd ed. (Upper Saddle River, NJ: Prentice Hall, 2008): 450.

Trendlines and regression analysis are tools for building predictive models. Our principal focus is to gain a basic understanding of how to use and interpret trendlines and regression models, statistical issues associated with interpreting regression analysis results, and practical issues in using trendlines and regression as tools for making and evaluating decisions.

Modeling Relationships and Trends in Data

Understanding both the mathematics and the descriptive properties of different functional relationships is important in building predictive analytical models. We often begin by creating a chart of the data to understand them and choose the appropriate type of functional relationship to incorporate into an analytical model. For cross-sectional data, we use a scatter chart; for time-series data, we use a line chart.

Common types of mathematical functions used in predictive analytical models include the following:

- **Linear function:** $y = a + bx$. Linear functions show steady increases or decreases over the range of x. This is the simplest type of function used in predictive models. It is easy to understand and, over small ranges of values, can approximate behavior rather well.
- **Logarithmic function:** $y = \ln(x)$. Logarithmic functions are used when the rate of change in a variable increases or decreases quickly and then levels out, such as with diminishing returns to scale. Logarithmic functions are often used in marketing models where constant percentage increases in advertising, for instance, result in constant, absolute increases in sales.
- **Polynomial function:** $y = ax^2 + bx + c$ (second order—quadratic function), $y = ax^3 + bx^2 + cx + d$ (third order—cubic function), and so on. A second-order polynomial is parabolic in nature and has only one hill or valley; a third-order polynomial has one or two hills or valleys. Revenue models that incorporate price elasticity are often polynomial functions.
- **Power function:** $y = ax^b$. Power functions define phenomena that increase at a specific rate. Learning curves that express improving times in performing a task are often modeled with power functions having $a > 0$ and $b < 0$.
- **Exponential function:** $y = ab^x$. Exponential functions have the property that y rises or falls at constantly increasing rates. For example, the perceived brightness of a lightbulb grows at a decreasing rate as the wattage increases. In this case, a would be a positive number and b would be between 0 and 1. The exponential function is often defined as $y = ae^x$, where $b = e$, the base of natural logarithms (approximately 2.71828).

The Excel *Trendline* tool provides a convenient method for determining the best-fitting functional relationship among these alternatives for a set of data. First, click the chart to which you wish to add a trendline; this will display the *Chart Tools* menu. Select the *Chart Tools Design* tab, and then click *Add Chart Element* from the *Chart Layouts* group. From the *Trendline* submenu, you can select one of the options (*Linear* is the most common) or *More Trendline Options* . . . . If you select *More Trendline Options*, you will get the *Format Trendline* pane in the worksheet (see Figure 8.1). A simpler way of doing all this is to right click on the data series in the chart and choose *Add trendline* from the pop-up menu—try it! Select the radio button for the type of functional relationship you wish

▶ **Figure 8.1**

Excel Format Trendline *Pane*

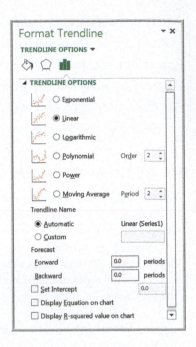

to fit to the data. Check the boxes for *Display Equation on chart* and *Display R-squared value on chart*. You may then close the *Format Trendline* pane. Excel will display the results on the chart you have selected; you may move the equation and R-squared value for better readability by dragging them to a different location. To clear a trendline, right click on it and select *Delete*.

R^2 (**R-squared**) is a measure of the "fit" of the line to the data. The value of R^2 will be between 0 and 1. The larger the value of R^2, the better the fit. We will discuss this further in the context of regression analysis.

Trendlines can be used to model relationships between variables and understand how the dependent variable behaves as the independent variable changes. For example, the demand-prediction models that we introduced in Chapter 1 (Examples 1.7 and 1.8) would generally be developed by analyzing data.

EXAMPLE 8.1   **Modeling a Price-Demand Function**

A market research study has collected data on sales volumes for different levels of pricing of a particular product. The data and a scatter diagram are shown in Figure 8.2 (Excel file *Price-Sales Data*). The relationship between price and sales clearly appears to be linear, so a linear trendline was fit to the data. The resulting model is

Sales = 20,512 − 9.5116 × Price

If the price is $125, we can estimate the the level of sales as

Sales = 20,512 − 9.5116 × 125 = 19,323

This model can be used as the demand function in other marketing or financial analyses.

Trendlines are also used extensively in modeling trends over time—that is, when the variable x in the functional relationships represents time. For example, an analyst for an airline needs to predict where fuel prices are going, and an investment analyst would want to predict the price of stocks or key economic indicators.

EXAMPLE 8.2   **Predicting Crude Oil Prices**

Figure 8.3 shows a chart of historical data on crude oil prices on the first Friday of each month from January 2006 through June 2008 (data are in the Excel file *Crude Oil Prices*). Using the *Trendline* tool, we can try to fit the various functions to these data (here *x* represents the number of months starting with January 2006). The results are as follows:

Exponential: $y = 50.49e^{0.021x}$ $R^2 = 0.664$

Logarithmetic: $y = 13.02\ln(x) + 39.60$ $R^2 = 0.382$

Polynomial (second order):
$y = 0.130x^2 - 2.40x + 68.01$ $R^2 = 0.905$

Polynomial (third order):
$y = 0.005x^3 - 0.111x^2 + 0.648x + 59.497$
$R^2 = 0.928$

Power: $y = 45.96x^{0.0169}$ $R^2 = 0.397$

The best-fitting model among these, which has the largest R^2, is the third-order polynomial, shown in Figure 8.4.

▶ **Figure 8.2**

Scatter Chart with Fitted Linear Function

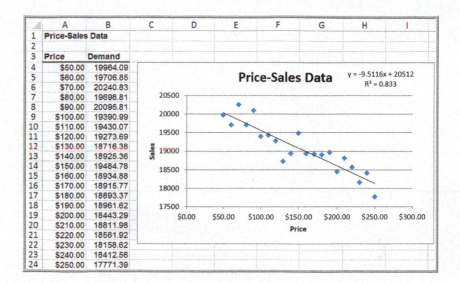

▶ **Figure 8.3**

Chart of Crude Oil Prices Data

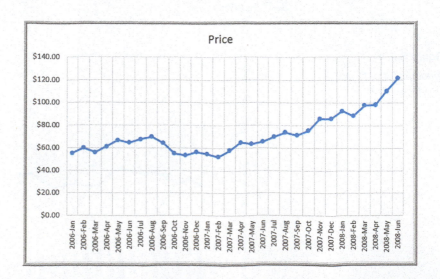

Be cautious when using polynomial functions. The R^2 value will continue to increase as the order of the polynomial increases; that is, a third-order polynomial will provide a better fit than a second-order polynomial, and so on. Higher-order polynomials will generally not be very smooth and will be difficult to interpret visually. Thus, we don't recommend going beyond a third-order polynomial when fitting data. Use your eye to make a good judgment!

Of course, the proper model to use depends on the scope of the data. As the chart in Figure 8.3 shows, crude oil prices were relatively stable until early 2007 and then began to increase rapidly. If the early data are included, the long-term functional relationship might not adequately express the short-term trend. For example, fitting a model to only the data beginning with January 2007 yields these models:

Exponential:	$y = 50.56\,e^{0.044x}$	$R^2 = 0.969$
Polynomial (second order):	$y = 0.121x^2 + 1.23x + 53.48$	$R^2 = 0.968$
Linear:	$y = 3.55x + 45.76$	$R^2 = 0.944$

The difference in prediction can be significant. For example, predicting the price six months after the last data point ($x = 36$) yields \$172.25 for the third-order polynomial fit with all the data and \$246.45 for the exponential model with only the recent data. Thus, you must be careful to select the proper amount of data for the analysis. The question then becomes one of choosing the best assumptions for the model. Is it reasonable to assume that prices would increase exponentially, or perhaps at a slower rate, such as with the linear model fit? Or would they level off and start falling? Clearly, factors other than historical trends would enter into this choice. In the latter half of 2008, oil prices plunged; thus, all predictive models are risky.

CHECK YOUR UNDERSTANDING

1. State the common types of mathematical functions used in predictive analytics and their properties.
2. Explain how to use the *Trendline* tool in Excel.
3. What does R^2 measure?

▶ **Figure 8.4**

Polynomial Fit of Crude Oil Prices *Data*

ANALYTICS IN PRACTICE: Using Predictive Trendline Models at Procter & Gamble[3]

Procter & Gamble (P&G) laundry products are global household brands that include Tide, Dash, and Gain and are offered in several physical product forms, including powders, liquids, pods, tablets, and bars. These products are manufactured at more than 30 sites and sold in more than 150 countries worldwide. The design of laundry product formulations (that is, ingredient composition of chemical mixtures) has become more complex over the years because of challenges such as product-portfolio expansion, rapidly changing ingredient costs and availability, and increasing competitive activity.

P&G's research and development organization is at the forefront of the development and adoption of modeling tools that enable the company to make better decisions on product formulation, processing, and manufacturing. These include empirical models that predict chemical reactions during manufacturing, in-use physical properties of the product, technical performance of the product, and even consumer acceptance rates. These tools enable researchers to instantly predict a product's physical properties and performance, integrate models, and balance production trade-offs using a variety of predictive and prescriptive capabilities. Predictive models the company has used include third-order polynomial functions that capture the two performance qualities of a mixture: stain removal and whiteness. For example, stain removal performance is predicted by the stain removal index (SRI) response function, which has the following form: $SRI = C_0 + C_1 v_1 + C_2 v_2 + C_3 v_1 v_2 + \ldots$, where C_i represent coefficients and v_i represent design variables: such as wash concentrations (milligrams per liter) of chemical ingredients and wash conditions (for example, temperature).

Simple Linear Regression

Regression analysis is a tool for building mathematical and statistical models that characterize relationships between a dependent variable (which must be a ratio variable and not categorical) and one or more independent, or explanatory, variables, all of which are numerical (but may be either ratio or categorical).

Two broad categories of regression models are often used in business settings: (1) regression models of cross-sectional data and (2) regression models of time-series data, in which the independent variables are time or some function of time and the focus is on predicting the future. Time-series regression is an important tool in *forecasting*, which is the subject of Chapter 9.

A regression model that involves a single independent variable is called *simple regression*. A regression model that involves two or more independent variables is called *multiple regression*. In the remainder of this chapter, we describe how to develop and analyze both simple and multiple regression models.

Simple linear regression involves finding a linear relationship between one independent variable, *X*, and one dependent variable, *Y*. The relationship between two variables can assume many forms, as illustrated in Figure 8.5. The relationship may be linear or nonlinear, or there may be no relationship at all. Because we are focusing our discussion on linear regression models, the first thing to do is to verify that the relationship is linear, as in Figure 8.5(a). We would not expect to see the data line up perfectly along a straight line; we simply want to verify that the general relationship is linear. If the relationship is clearly nonlinear, as in Figure 8.5(b), then alternative approaches must be used, and if no relationship is evident, as in Figure 8.5(c), then it is pointless to even consider developing a linear regression model.

[3]Adapted from Nats Esquejo, Kevin Miller, Kevin Norwood, Ivan Oliveira, Rob Pratt, and Ming Zhao, "Statistical and Optimization Techniques for Laundry Portfolio Optimization at Procter & Gamble," *Interfaces*, Vol. 45, No. 5, September–October 2015, pp. 444–461.

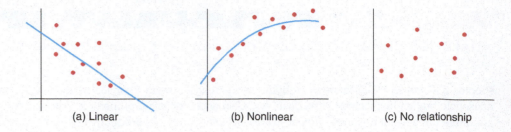

(a) Linear   (b) Nonlinear   (c) No relationship

To determine if a linear relationship exists between the variables, we recommend that you create a scatter chart that can show the relationship between variables visually.

EXAMPLE 8.3 *Home Market Value* **Data**

The market value of a house is typically related to its size. In the Excel file *Home Market Value* (see Figure 8.6), data obtained from a county auditor provide information about the age, square footage, and current market value of houses in a particular subdivision. We might wish to investigate the relationship between the market value and the size of the home. The independent variable, X, is the number of square feet, and the dependent variable, Y, is the market value.

Figure 8.7 shows a scatter chart of the market value in relation to the size of the home. In general, we see that higher market values are associated with larger house sizes, and the relationship is approximately linear. Therefore, we could conclude that simple linear regression would be an appropriate technique for predicting market value based on house size.

▶ **Figure 8.6**

Portion of Home Market
Value

	A	B	C
1	Home Market Value		
2			
3	House Age	Square Feet	Market Value
4	33	1,812	$90,000.00
5	32	1,914	$104,400.00
6	32	1,842	$93,300.00
7	33	1,812	$91,000.00
8	32	1,836	$101,900.00
9	33	2,028	$108,500.00
10	32	1,732	$87,600.00

▶ **Figure 8.7**

*Scatter Chart of Market
Value Versus Home Size*

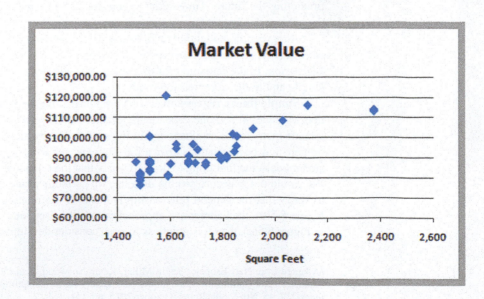

Finding the Best-Fitting Regression Line

The idea behind simple linear regression is to express the relationship between the dependent variable Y and the independent variable X by a simple linear equation,

$$Y = a + bX$$

where a is the y-intercept and b is the slope of the line. For the home market value example, this would be

$$\text{Market Value} = a + b \times \text{Square Feet}$$

If we draw a straight line through the data, some of the points will fall above the line, some will fall below it, and a few might fall on the line itself. Figure 8.8 shows two possible straight lines that pass through the data. Clearly, you would choose A as the better-fitting line over B because all the points are closer to the line and the line appears to be in the middle of the data. The only difference between the lines is the value of the slope and intercept; thus, we seek to determine the values of the slope and intercept that provide the best-fitting line. We can find the best-fitting line using the Excel *Trendline* tool (with the *Linear* option chosen), as described earlier in this chapter.

EXAMPLE 8.4   Using Excel to Find the Best Regression Line

When using the *Trendline* tool for simple linear regression in the *Home Market Value* example, be sure the Linear option is selected (it is the default option when you use the tool). Figure 8.9 shows the best-fitting regression line. The equation is

Market Value = $32,673 + $35.036 × Square Feet

The value of the regression line can be explained as follows. Suppose we wanted to estimate the home market value for any home in the population from which the sample data were gathered. If all we knew were the market values, then the best estimate of the market value for any home would simply be the sample mean, which is $92,069. Thus, no matter if the house has 1,500 square feet or 2,200 square feet, the best estimate of market value would still be $92,069. Because the market values vary from about $75,000 to more than $120,000, there is quite a bit of uncertainty in using the

mean as the estimate. However, from the scatter chart, we see that larger homes tend to have higher market values. Therefore, if we know that a home has 2,200 square feet, we would expect the market value estimate to be higher than that for one that has only 1,500 square feet. For example, the estimated market value of a home with 2,200 square feet would be

Market Value = $32,673 + $35.036 × 2,200 = $109,752

whereas the estimated value for a home with 1,500 square feet would be

Market Value = $32,673 + $35.036 × 1,500 = $85,227

The regression model explains the differences in market value as a function of the house size and provides better estimates than simply using the average of the sample data.

Using Regression Models for Prediction

As we saw in Example 8.4, once we have determined the slope and intercept of the best-fitting line, we can simply substitute a value for the independent variable X in order to predict the dependent variable Y. However, there is one important caution: It is dangerous to extrapolate a regression model outside the ranges covered by the observations. For instance, if you want to predict the market value of a house that has 3,000 square feet, the results may or may not be accurate because the regression model estimates did not use any observations greater than 2,400 square feet. We cannot be sure that a linear extrapolation will hold and should not use the model to make such predictions.

▶ **Figure 8.8**

Two Possible Regression Lines

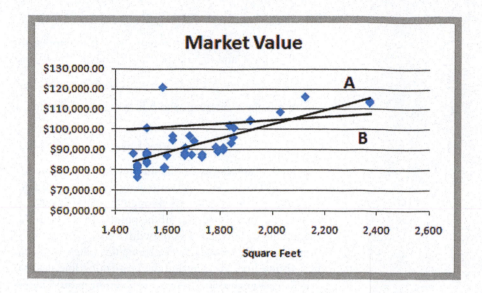

▶ **Figure 8.9**

Best-Fitting Simple Linear Regression Line

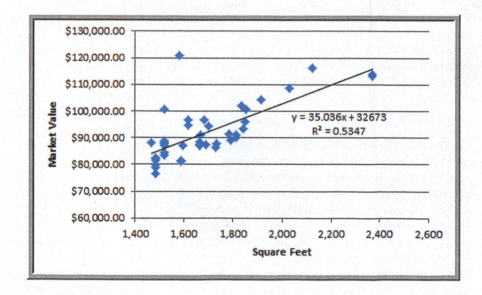

Least-Squares Regression

The mathematical basis for the best-fitting regression line is called **least-squares regression**. In regression analysis, we assume that the values of the dependent variable, *Y*, in the sample data are drawn from some unknown population for each value of the independent variable, *X*. For example, in the *Home Market Value* data, the first and fourth observations come from a population of homes having 1,812 square feet; the second observation comes from a population of homes having 1,914 square feet; and so on.

Because we are assuming that a linear relationship exists, the expected value of *Y* is $\beta_0 + \beta_1 X$ for each value of *X*. The coefficients β_0 and β_1 are population parameters that represent the intercept and slope, respectively, of the population from which a sample of observations is taken. The intercept is the mean value of Y when $X = 0$, and the slope is the change in the mean value of Y as X changes by one unit.

Thus, for a specific value of X, we have many possible values of Y that vary around the mean. To account for this, we add an error term, ε (the Greek letter epsilon), to the mean. This defines a simple linear regression model:

$$Y = \beta_0 + \beta_1 X + \varepsilon \qquad (8.1)$$

However, because we don't know the entire population, we don't know the true values of β_0 and β_1. In practice, we must estimate these as best we can from the sample data. Define b_0 and b_1 to be estimates of β_0 and β_1. Thus, the estimated simple linear regression equation is

$$\hat{Y} = b_0 + b_1 X \qquad (8.2)$$

Let X_i be the value of the independent variable of the ith observation. When the value of the independent variable is X_i, then $\hat{Y}_i = b_0 + b_1 X_i$ is the estimated value of Y for X_i.

One way to quantify the relationship between each point and the estimated regression equation is to measure the vertical distance between them, as illustrated in Figure 8.10. We can think of these differences, e_i, as the observed errors (often called **residuals**) associated with estimating the value of the dependent variable using the regression line. Thus, the error associated with the ith observation is

$$e_i = Y_i - \hat{Y}_i \qquad (8.3)$$

The best-fitting line should minimize some measure of these errors. Because some errors will be negative and others positive, we might take their absolute value or simply square them. Mathematically, it is easier to work with the squares of the errors.

Adding the squares of the errors, we obtain the following function:

$$\sum_{i=1}^{n} e_i^2 = \sum_{i=1}^{n} (Y_i - \hat{Y}_i)^2 = \sum_{i=1}^{n} (Y_i - [b_0 + b_1 X_i])^2 \qquad (8.4)$$

If we can find the best values of the slope and intercept that minimize the sum of squares (hence the name "least squares") of the observed errors e_i, we will have found the best-fitting regression line. Note that X_i and Y_i are the values of the sample data and that b_0 and b_1 are unknowns in equation (8.4). Using calculus, we can show that the solution that minimizes the sum of squares of the observed errors is

$$b_1 = \frac{\sum_{i=1}^{n} X_i Y_i - n \bar{X}\, \bar{Y}}{\sum_{i=1}^{n} X_i^2 - n \bar{X}^2} \qquad (8.5)$$

$$b_0 = \bar{Y} - b_1 \bar{X} \qquad (8.6)$$

▶ **Figure 8.10**

Measuring the Errors in a Regression Model

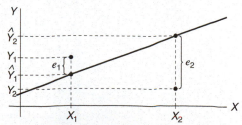

Errors associated with individual observations

Although the calculations for the least-squares coefficients appear to be somewhat complicated, they can easily be performed on an Excel spreadsheet. Even better, Excel has built-in capabilities for doing this. For example, you may use the functions INTERCEPT(*known_y's, known_x's*) and SLOPE(*known_y's, known_x's*) to find the least-squares coefficients b_0 and b_1. We may use the Excel function TREND(*known_y's, known_x's, new_x's*) to estimate Y for any value of X.

EXAMPLE 8.5   Using Excel Functions to Find Least-Squares Coefficients

For the *Home Market Value* Excel file, the range of the dependent variable Y (market value) is C4:C45; the range of the independent variable X (square feet) is B4:B45. The Excel formula =INTERCEPT(C4:C45, B4:B45) yields $b_0 = 32,673$ and =SLOPE(C4:C45, B4:B45) yields $b_1 = 35.036$, as we saw in Example 8.4. The slope tells us that for every additional square foot, the market value increases by $35.036.

Using the TREND function, we may estimate the market value for a house with 1,750 square feet with the formula =*TREND(C4:C45, B4:B45, 1750)* = $93,987.

We could stop at this point, because we have found the best-fitting line for the observed data. However, there is a lot more to regression analysis from a statistical perspective, because we are working with sample data—and usually rather small samples—which we know have a lot of variation as compared with the full population. Therefore, it is important to understand some of the statistical properties associated with regression analysis.

Simple Linear Regression with Excel

Regression analysis software tools available in Excel provide a variety of information about the statistical properties of regression analysis. The Excel *Regression* tool can be used for both simple and multiple linear regressions. For now, we focus on using the tool just for simple linear regression.

From the *Data Analysis* menu in the *Analysis* group under the *Data* tab, select the *Regression* tool. The dialog box shown in Figure 8.11 is displayed. In the box for the *Input Y Range*, specify the range of the dependent variable values. In the box for the *Input X Range*, specify the range for the independent variable values. Check *Labels* if your data range contains a descriptive label (we highly recommend using this). You have the option of forcing the

▶ **Figure 8.11**

Excel Regression *Tool Dialog*

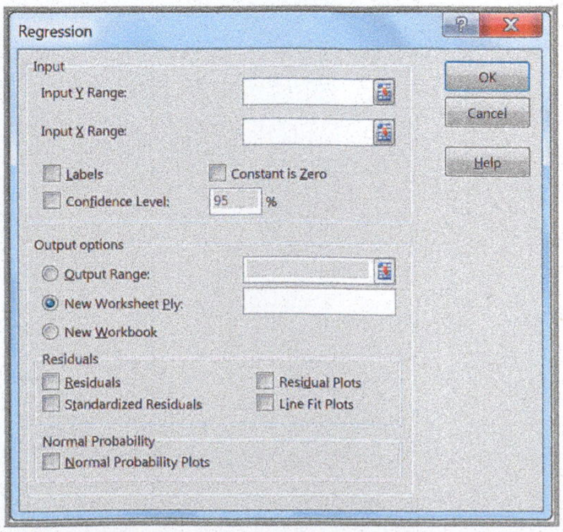

intercept to zero by checking *Constant is Zero*; however, you will usually not check this box because adding an intercept term allows a better fit to the data. You also can set a *Confidence Level* (the default of 95% is commonly used) to provide confidence intervals for the intercept and slope parameters. In the *Residuals* section, you have the option of including a residuals output table by checking the boxes for *Residuals, Standardized Residuals, Residual Plots*, and *Line Fit Plots. Residual Plots* generates a chart for each independent variable versus the residual, and *Line Fit Plots* generates a scatter chart with the values predicted by the regression model included (however, creating a scatter chart with an added trendline is visually superior to what this tool provides). Finally, you may also choose to have Excel construct a normal probability plot for the dependent variable, which transforms the cumulative probability scale (vertical axis) so that the graph of the cumulative normal distribution is a straight line. The closer the points are to a straight line, the better the fit to a normal distribution.

Figure 8.12 shows the basic regression analysis output provided by the Excel *Regression* tool for the *Home Market Value* data. The output consists of three sections: *Regression Statistics* (rows 3–8), *ANOVA* (rows 10–14), and an unlabeled section at the bottom (rows 16–18) with other statistical information. The least-squares estimates of the slope and intercept are found in the *Coefficients* column in the bottom section of the output.

In the *Regression Statistics* section, *Multiple R* is another name for the sample correlation coefficient, *r*, which was introduced in Chapter 4. Values of *r* range from −1 to 1, where the sign is determined by the sign of the slope of the regression line. A *Multiple R* value greater than 0 indicates positive correlation, that is, as the independent variable increases, the dependent variable does also; a value less than 0 indicates negative correlation—as *X* increases, *Y* decreases. A value of 0 indicates no correlation.

R Square (R^2) is called the **coefficient of determination**. Earlier we noted that R^2 is a measure of how well the regression line fits the data; this value is also provided by the *Trendline* tool. Specifically, R^2 gives the proportion of variation in the dependent variable that is explained by the independent variable of the regression model. The value of R^2 is between 0 and 1. A value of 1.0 indicates a perfect fit, and all data points lie on the regression line, whereas a value of 0 indicates that no relationship exists. Although we would like high values of R^2, it is difficult to specify a "good" value that signifies a strong relationship because this depends on the application. For example, in scientific applications such as calibrating physical measurement equipment, R^2 values close to 1 would be expected. In marketing research studies, an R^2 of 0.6 or more is considered very good; however, in many social science applications, values in the neighborhood of 0.3 might be considered acceptable.

▶ **Figure 8.12**

Basic Regression Analysis Output for Home Market Value *Example*

	A	B	C	D	E	F	G
1	Regression Analysis						
2							
3	*Regression Statistics*						
4	Multiple R	0.731255223					
5	R Square	0.534734202					
6	Adjusted R Square	0.523102557					
7	Standard Error	7287.722712					
8	Observations	42					
9							
10	ANOVA						
11		*df*	*SS*	*MS*	*F*	*Significance F*	
12	Regression	1	2441633669	2441633669	45.97236277	3.79802E-08	
13	Residual	40	2124436093	53110902.32			
14	Total	41	4566069762				
15							
16		*Coefficients*	*Standard Error*	*t Stat*	*P-value*	*Lower 95%*	*Upper 95%*
17	Intercept	32673.2199	8831.950745	3.699434116	0.000649604	14823.18178	50523.25802
18	Square Feet	35.03637258	5.16738385	6.780292234	3.79802E-08	24.59270036	45.48004481

Adjusted R Square is a statistic that modifies the value of R^2 by incorporating the sample size and the number of explanatory variables in the model. Although it does not give the actual percent of variation explained by the model as R^2 does, it is useful when comparing this model with other models that include additional explanatory variables. We discuss it more fully in the context of multiple linear regression later in this chapter.

Standard Error in the Excel output is the variability of the observed Y-values from the predicted values $(\hat{Y})$. This is formally called the **standard error of the estimate**, S_{YX}. If the data are clustered close to the regression line, then the standard error will be small; the more scattered the data, the larger the standard error.

EXAMPLE 8.6 **Interpreting Regression Statistics for Simple Linear Regression**

After running the Excel *Regression* tool, the first things to look for are the values of the slope and intercept, namely, the estimates b_1 and b_0 in the regression model. In the *Home Market Value* example, we see that the intercept is 32,673, and the slope (coefficient of the independent variable, square feet) is 35.036, just as we had computed earlier. In the *Regression Statistics* section, $R^2 = 0.5347$. This means that approximately 53% of the variation in market value is explained by square feet. The remaining variation is due to other factors that were not included in the model. The standard error of the estimate is $7,287.72. If we compare this to the standard deviation of the market value, which is $10,553, we see that the variation around the regression line ($7,287.72) is *less* than the variation around the sample mean ($10,553). This is because the independent variable in the regression model explains some of the variation.

Regression as Analysis of Variance

In Chapter 7, we introduced analysis of variance (ANOVA), which conducts an F-test to determine whether variation due to a particular factor, such as the differences in sample means, is significantly greater than that due to error. ANOVA is commonly applied to regression to test for *significance of regression*. For a simple linear regression model, **significance of regression** is simply a hypothesis test of whether the regression coefficient β_1 (slope of the independent variable) is zero:

$$H_0: \beta_1 = 0$$
$$H_1: \beta_1 \neq 0 \tag{8.7}$$

If we reject the null hypothesis, then we may conclude that the slope of the independent variable is not zero and, therefore, is statistically significant in the sense that it explains some of the variation of the dependent variable around the mean. Similar to our discussion in Chapter 7, you needn't worry about the mathematical details of how F is computed, or even its value, especially since the tool does not provide the critical value for the test. What is important is the value of *Significance F*, which is the p-value for the F-test. If *Significance F* is less than the level of significance (typically 0.05), we would reject the null hypothesis.

EXAMPLE 8.7 **Interpreting Significance of Regression**

For the *Home Market Value* example, the ANOVA test is shown in rows 10–14 in Figure 8.12. *Significance F*, that is, the p-value associated with the hypothesis test

$$H_0: \beta_1 = 0$$
$$H_1: \beta_1 \neq 0$$

is essentially zero (3.798×10^{-8}). Therefore, assuming a level of significance of 0.05, we must reject the null hypothesis and conclude that the slope—the coefficient for square feet—is not zero. This means that home size is a statistically significant variable in explaining the variation in market value.

Testing Hypotheses for Regression Coefficients

Rows 17–18 of the Excel output, in addition to specifying the least-squares coefficients, provide additional information for testing hypotheses associated with the intercept and slope. Specifically, we may test the null hypothesis that β_0 or β_1 equals zero. Usually, it makes little sense to test or interpret the hypothesis that $\beta_0 = 0$ unless the intercept has a significant physical meaning in the context of the application. For simple linear regression, testing the null hypothesis $H_0: \beta_1 = 0$ is the same as the significance of regression test that we described earlier.

The t-test for the slope is similar to the one-sample test for the mean that we described in Chapter 7. The test statistic is

$$ t = \frac{b_1 - 0}{\text{Standard Error}} \tag{8.8} $$

and is given in the column labeled t Stat in the Excel output. Although the critical value of the t-distribution is not provided, the output does provide the p-value for the test.

EXAMPLE 8.8	**Interpreting Hypothesis Tests for Regression Coefficients**

For the *Home Market Value* example, note that the value of t Stat is computed by dividing the coefficient by the standard error using formula (8.8). For instance, t Stat for the slope is $35.03637258/5.16738385 = 6.780292232$. Because Excel does not provide the critical value with which to compare the t Stat value, we may use the p-value to draw a conclusion. Because the p-values for both coefficients are essentially zero, we would conclude that neither coefficient is statistically equal to zero. Note that the p-value associated with the test for the slope coefficient, square feet, is equal to the *Significance F* value. This will always be true for a regression model with one independent variable because it is the only explanatory variable. However, as we shall see, this will not be the case for multiple regression models.

Confidence Intervals for Regression Coefficients

Confidence intervals (*Lower 95%* and *Upper 95%* values in the output) provide information about the unknown values of the true regression coefficients, accounting for sampling error. They tell us what we can reasonably expect to be the ranges for the population intercept and slope at a 95% confidence level. By using the confidence interval limits for the model parameters, we can determine how estimates using the model might vary.

We may also use confidence intervals to test hypotheses about the regression coefficients. For example, in Figure 8.12, we see that neither confidence interval includes zero; therefore, we can conclude that β_0 and β_1 are statistically different from zero. Similarly, we can use them to test the hypotheses that the regression coefficients equal some value other than zero. For example, to test the hypotheses

$$ H_0: \beta_1 = B_1 $$
$$ H_1: \beta_1 \neq B_1 $$

we need only check whether B_1 falls within the confidence interval for the slope. If it does not, then we reject the null hypothesis; otherwise, we fail to reject it.

EXAMPLE 8.9 **Interpreting Confidence Intervals for Regression Coefficients**

For the *Home Market Value* data, a 95% confidence interval for the intercept is [14,823, 50,523]. Similarly, a 95% confidence interval for the slope is [24.59, 45.48]. Although the regression model is $\hat{Y} = 32,673 + 35.036X$, the confidence intervals suggest a bit of uncertainty about predictions using the model. Thus, although we estimated that a house with 1,750 square feet has a

market value of 32,673 + 35.036(1,750) = \$93,986, if the true population parameters are at the extremes of the confidence intervals, the estimate might be as low as 14,823 + 24.59(1,750) = \$57,856 or as high as 50,523 + 45.48(1,750) = \$130,113. Narrower confidence intervals provide more accuracy in our predictions.

CHECK YOUR UNDERSTANDING

1. What is regression analysis? What is the difference between simple and multiple regression?
2. How can you determine whether simple linear regression would be an appropriate technique to use?
3. Why does simple linear regression provide better predictions than simply using the sample mean of the dependent variable?
4. Explain the concepts underlying least-squares regression.
5. Explain how to interpret the results from the Excel *Regression* tool.

Residual Analysis and Regression Assumptions

Recall that residuals are the observed errors, which are the differences between the actual values and the estimated values of the dependent variable using the regression equation. Figure 8.13 shows a portion of the residual table generated by the Excel *Regression* tool. The residual output includes, for each observation, the predicted value using the estimated regression equation, the residual, and the standard residual. The residual is simply the difference between the actual value of the dependent variable and the predicted value, or $Y_i - \hat{Y}_i$. Figure 8.14 shows the residual plot generated by the Excel tool. This chart is actually a scatter chart of the residuals with the values of the independent variable on the x-axis.

Standard residuals are residuals divided by their standard deviation. Standard residuals describe how far each residual is from its mean in units of standard deviations (similar to a z-value for a standard normal distribution). Standard residuals are useful in checking assumptions underlying regression analysis, which we will address shortly, and to detect outliers that may bias the results. Recall that an outlier is an extreme value that is different from the rest of the data. A single outlier can make a significant difference in the regression

▶ **Figure 8.13**

Portion of Residual Output

	A	B	C	D
22	RESIDUAL OUTPUT			
23				
24	*Observation*	*Predicted Market Value*	*Residuals*	*Standard Residuals*
25	1	96159.12702	-6159.127018	-0.855636403
26	2	99732.83702	4667.162978	0.64837022
27	3	97210.2182	-3910.218196	-0.543214164
28	4	96159.12702	-5159.127018	-0.716714702
29	5	96999.99996	4900.00004	0.680716341

▶ **Figure 8.14**

Residual Plot for Square Feet

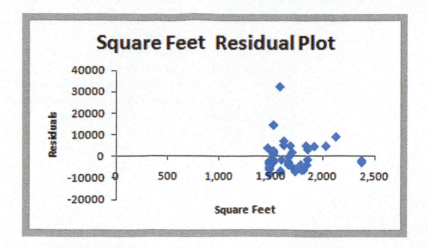

equation, changing the slope and intercept and, hence, how they would be interpreted and used in practice. Some consider a standardized residual outside of ± 2 standard deviations as an outlier. A more conservative rule of thumb would be to consider outliers outside of a ± 3 standard deviation range. (Commercial software packages have more sophisticated techniques for identifying outliers.)

EXAMPLE 8.10   **Interpreting Residual Output**

For the *Home Market Value* data, the first observation has a market value of $90,000 and the regression model predicts $96,159.13. Thus, the residual is $90,000 - 96,159.13 = -\$6,159.13$. The standard deviation of the residuals can be computed as 7,198.299. By dividing the residual by this value, we have the standardized residual for the first observation. The value of -0.8556 tells us that the first observation is about 0.85 standard deviation below the regression line. If we check the values of all the standardized residuals, you will find that the value of the last data point is 4.53, meaning that the market value of this home, having only 1,581 square

feet, is more than 4 standard deviations above the predicted value and would clearly be identified as an outlier. (If you look back at Figure 8.7, you may have noticed that this point appears to be quite different from the rest of the data.) You might question whether this observation belongs in the data, because the house has a large value despite a relatively small size. The explanation might be an outdoor pool or an unusually large plot of land. Because this value will influence the regression results and may not be representative of the other homes in the neighborhood, you might consider dropping this observation and recomputing the regression model.

Checking Assumptions

The statistical hypothesis tests associated with regression analysis are predicated on some key assumptions about the data.

1. *Linearity.* This is usually checked by examining a scatter diagram of the data or examining the residual plot. If the model is appropriate, then the residuals should appear to be randomly scattered about zero, with no apparent pattern. If the residuals exhibit some well-defined pattern, such as a linear trend or a parabolic shape, then there is good evidence that some other functional form might better fit the data.

2. *Normality of errors.* Regression analysis assumes that the errors for each individual value of X are normally distributed, with a mean of zero. This can be verified either by examining a histogram of the standard residuals and inspecting for a bell-shaped distribution or by using more formal goodness of fit tests. It is usually difficult to evaluate normality with small sample sizes. However, regression analysis is fairly robust against departures from normality, so in most cases, this is not a serious issue.

3. *Homoscedasticity.* The third assumption is **homoscedasticity**, which means that the variation about the regression line is constant for all values of the independent variable. This can also be evaluated by examining the residual plot and looking for large differences in the variances at different values of the independent variable. Caution should be exercised when looking at residual plots. In many applications, the model is derived from limited data, and multiple observations for different values of X are not available, making it difficult to draw definitive conclusions about homoscedasticity. If this assumption is seriously violated, then techniques other than least squares should be used for estimating the regression model.

4. *Independence of errors.* Finally, residuals should be independent for each value of the independent variable. For cross-sectional data, this assumption is usually not a problem. However, when time is the independent variable, this is an important assumption. If successive observations appear to be correlated—for example, by becoming larger over time or exhibiting a cyclical type of pattern—then this assumption is violated. Correlation among successive observations over time is called **autocorrelation** and can be identified by residual plots having clusters of residuals with the same sign. Autocorrelation can be evaluated more formally using a statistical test based on a measure called the Durbin–Watson statistic. The Durbin–Watson statistic is

$$D = \frac{\sum_{i=2}^{n} (e_i - e_{i-1})^2}{\sum_{i=1}^{n} e_i^2} \tag{8.9}$$

This is a ratio of the squared differences in successive residuals to the sum of the squares of all residuals. D will range from 0 to 4. When successive residuals are positively autocorrelated, D will approach 0. Critical values of the statistic have been tabulated based on the sample size and number of independent variables that allow you to conclude that there is either evidence of autocorrelation or no evidence of autocorrelation or the test is inconclusive. For most practical purposes, values below 1 suggest autocorrelation; values above 1.5 and below 2.5 suggest no autocorrelation; and values above 2.5 suggest negative autocorrelation. This can become an issue when using regression in forecasting, which we discuss in the next chapter. Some software packages compute this statistic; however, Excel does not.

When assumptions of regression are violated, statistical inferences drawn from the hypothesis tests may not be valid. Thus, before drawing inferences about regression models and performing hypothesis tests, these assumptions should be checked. However, other than linearity, these assumptions are not needed solely for model fitting and estimation purposes.

► **Figure 8.15**

Histogram of Standard Residuals

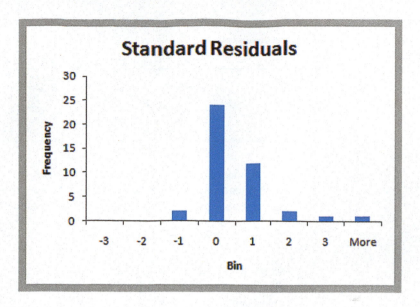

EXAMPLE 8.11 **Checking Regression Assumptions for the *Home Market Value* Data**

Linearity: The scatter diagram of the market value data appears to be linear; looking at the residual plot in Figure 8.14 also confirms no pattern in the residuals.

Normality of errors: Figure 8.15 shows a histogram of the standard residuals for the market value data. The distribution appears to be somewhat positively skewed (particularly with the outlier) but does not appear to be a serious

departure from normality, particularly as the sample size is small.

Homoscedasticity: In the residual plot in Figure 8.14, we see no serious differences in the spread of the data for different values of *X*, particularly if the outlier is eliminated.

Independence of errors: Because the data are cross-sectional, we can assume that this assumption holds.

CHECK YOUR UNDERSTANDING

1. What are standard residuals, and how can they be used?
2. Explain the assumptions behind regression analysis.
3. How can the assumptions of regression analysis be checked?

Multiple Linear Regression

Many colleges try to predict student performance as a function of several characteristics. In the Excel file *Colleges and Universities* (see Figure 8.16), suppose that we wish to predict the graduation rate as a function of the other variables—median SAT, acceptance rate, expenditures/student, and percent in the top 10% of their high school class. It is logical to propose that schools with students who have higher SAT scores, a lower acceptance rate, a larger budget, and a higher percentage of students in the top 10% of their high school classes will tend to retain and graduate more students.

▶ **Figure 8.16**

Portion of Excel File Colleges and Universities

	A	B	C	D	E	F	G
1	**Colleges and Universities**						
2							
3	**School**	**Type**	**Median SAT**	**Acceptance Rate**	**Expenditures/Student**	**Top 10% HS**	**Graduation %**
4	Amherst	Lib Arts	1315	22%	$ 26,636	85	93
5	Barnard	Lib Arts	1220	53%	$ 17,653	69	80
6	Bates	Lib Arts	1240	36%	$ 17,554	58	88
7	Berkeley	University	1176	37%	$ 23,665	95	68
8	Bowdoin	Lib Arts	1300	24%	$ 25,703	78	90
9	Brown	University	1281	24%	$ 24,201	80	90

A linear regression model with more than one independent variable is called a **multiple linear regression** model. Simple linear regression is just a special case of multiple linear regression. A multiple linear regression model has the form

$$Y = \beta_0 + \beta_1 X_1 + \beta_2 X_2 + \cdots + \beta_k X_k + \varepsilon \tag{8.10}$$

where Y is the dependent variable, $X_1, \ldots, X_k$ are the independent (explanatory) variables, β_0 is the intercept term, $\beta_1, \ldots, \beta_k$ are the regression coefficients for the independent variables, and ε is the error term.

Similar to simple linear regression, we estimate the regression coefficients—called **partial regression coefficients**—$b_0, b_1, b_2, \ldots, b_k$, then use the model

$$\hat{Y} = b_0 + b_1 X_1 + b_2 X_2 + \cdots + b_k X_k \tag{8.11}$$

to predict the value of the dependent variable. The partial regression coefficients represent the expected change in the dependent variable when the associated independent variable is increased by one unit *while the values of all other independent variables are held constant.*

For the college and university data, the proposed model would be

Graduation % = $b_0 + b_1$ SAT + b_2 ACCEPTANCE + b_3 EXPENDITURES + b_4 TOP10%HS

Thus, b_2 would represent an estimate of the change in the graduation rate for a unit increase in the acceptance rate while holding all other variables constant.

As with simple linear regression, multiple linear regression uses least squares to estimate the intercept and slope coefficients that minimize the sum of squared error terms over all observations. The principal assumptions discussed for simple linear regression also hold here. The Excel *Regression* tool can easily perform multiple linear regression; you need to specify only the full range for the independent variable data in the dialog. One caution when using the tool: *The independent variables in the spreadsheet must be in contiguous columns.* So you may have to manually move the columns of data around before applying the tool.

The results from the *Regression* tool are in the same format as we saw for simple linear regression. However, some key differences exist. *Multiple R* and *R Square* (or R^2) are called the **multiple correlation coefficient** and the **coefficient of multiple determination**, respectively, in the context of multiple regression. They indicate the strength of association between the dependent and independent variables. Similar to simple linear regression, R^2 explains the percentage of variation in the dependent variable that is explained by the set of independent variables in the model.

The interpretation of the ANOVA section is quite different from that in simple linear regression. For multiple linear regression, ANOVA tests for significance of the *entire model*. That is, it computes an *F*-statistic for testing the hypotheses

$$H_0: \beta_1 = \beta_2 = \cdots = \beta_k = 0$$

$$H_1: \text{at least one } \beta_j \text{ is not } 0$$

The null hypothesis states that no linear relationship exists between the dependent and *any* of the independent variables, whereas the alternative hypothesis states that the dependent variable has a linear relationship with *at least* one independent variable. If the null hypothesis is rejected, we cannot conclude that a relationship exists with every independent variable individually.

The multiple linear regression output also provides information to test hypotheses about *each* of the individual regression coefficients. Specifically, we may test the null hypothesis that β_0 (the intercept) or any β_i equals zero. If we reject the null hypothesis that the slope associated with independent variable *i* is zero, $H_0: \beta_i = 0$, then we may state that independent variable *i* is *significant* in the regression model; that is, it contributes to reducing the variation in the dependent variable and improves the ability of the model to better predict the dependent variable. However, if we cannot reject H_0, then that independent variable is not significant and probably should not be included in the model. We see how to use this information to identify the best model in the next section.

Finally, for multiple regression models, a residual plot is generated for each independent variable. This allows you to assess the linearity and homoscedasticity assumptions of regression.

EXAMPLE 8.12   **Interpreting Regression Results for the *Colleges and Universities* Data**

The multiple regression results for the college and university data are shown in Figure 8.17.

From the *Coefficients* section, we see that the model is

Graduation % =
17.92 + 0.072 SAT − 24.859 ACCEPTANCE
− 0.000136 EXPENDITURES − 0.163 TOP10% HS

The signs of some coefficients make sense; higher SAT scores and lower acceptance rates suggest higher graduation rates. However, we might expect that larger student expenditures and a higher percentage of top high school students would also positively influence the graduation rate. Perhaps the problem occurred because some of the best students are more demanding and change schools if their needs are not being met, some entrepreneurial students might pursue other interests before graduation, or there is sampling error. As with simple linear regression, the model should be used only for values of the independent variables within the range of the data.

The value of R^2 (0.53) indicates that 53% of the variation in the dependent variable is explained by these independent variables. This suggests that other factors not included in the model, perhaps campus living conditions, social opportunities, and so on, might also influence the graduation rate.

From the *ANOVA* section, we may test for significance of regression. At a 5% significance level, we reject the null hypothesis because *Significance F* is essentially zero. Therefore, we may conclude that at least one slope is statistically different from zero.

Looking at the *p*-values for the independent variables in the last section, we see that all are less than 0.05; therefore, we reject the null hypothesis that each partial regression coefficient is zero and conclude that each of them is statistically significant.

Figure 8.18 shows one of the residual plots from the Excel output. The assumptions appear to be met, and the other residual plots (not shown) also validate these assumptions. The normal probability plot (also not shown) does not suggest any serious departures from normality.

▶ **Figure 8.17**

Multiple Regression Results for Colleges and Universities Data

	A	B	C	D	E	F	G
1	SUMMARY OUTPUT						
2							
3	Regression Statistics						
4	Multiple R	0.731044486					
5	R Square	0.534426041					
6	Adjusted R Square	0.492101135					
7	Standard Error	5.30833812					
8	Observations	49					
9							
10	ANOVA						
11		df	SS	MS	F	Significance F	
12	Regression	4	1423.209266	355.8023166	12.62675098	6.33158E-07	
13	Residual	44	1239.851958	28.1784536			
14	Total	48	2663.061224				
15							
16		Coefficients	Standard Error	t Stat	P-value	Lower 95%	Upper 95%
17	Intercept	17.92095587	24.55722367	0.729763108	0.469402466	-31.57087643	67.41278818
18	Median SAT	0.072006285	0.017983915	4.003927007	0.000236106	0.035762085	0.108250485
19	Acceptance Rate	-24.8592318	8.315184822	-2.989618672	0.004559569	-41.61738567	-8.101077939
20	Expenditures/Student	-0.00013565	6.59314E-05	-2.057438385	0.045600178	-0.000268526	-2.77379E-06
21	Top 10% HS	-0.162764489	0.079344518	-2.051364015	0.046213848	-0.322672857	-0.00285612

▶ **Figure 8.18**

Residual Plot for Top 10% HS Variable

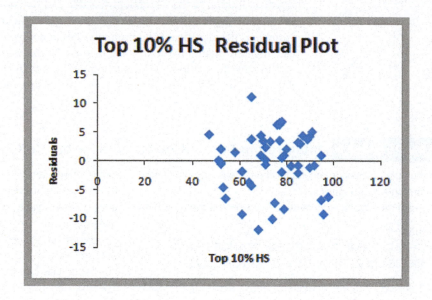

ANALYTICS IN PRACTICE: Using Linear Regression and Interactive Risk Simulators to Predict Performance at Aramark[4]

Aramark is a leader in professional services, providing award-winning food services, facilities management, and uniform and career apparel to health care institutions, universities and school districts, stadiums and arenas, and businesses around the world. Headquartered in Philadelphia, Aramark has approximately 255,000 employees serving clients in 22 countries.

Aramark's Global Risk Management Department (GRM) needed a way to determine the statistical relationships between key business metrics (for example, employee tenure, employee engagement, a trained workforce, account tenure, service offerings) and risk metrics (for example, OSHA rate, workers' compensation rate, customer injuries) to understand the impact of these risks on the business.

---

[4]The author expresses his appreciation to John Toczek, Manager of Decision Support and Analytics at Aramark Corporation.

GRM also needed a simple tool that field operators and the risk management team could use to predict the impact of business decisions on risk metrics before those decisions were implemented. Typical questions they would want to ask were, What would happen to our OSHA rate if we increased the percentage of part time labor? and How could we impact turnover if operations improved safety performance?

Aramark maintains extensive historical data. For example, the GRM group keeps track of data such as OSHA rates, slip/trip/fall rates, injury costs, and level of compliance with safety standards; the Human Resources department monitors turnover and percentage of part-time labor; the Payroll Department keeps data on average wages; and the Training and Organizational Development Department collects data on employee engagement. Excel-based linear regression was used to determine the relationships between the dependent variables (such as OSHA rate, slip/trip/fall rate, claim cost, and turnover) and the independent variables (such as the percentage of part-time labor, average wage, employee engagement, and safety compliance).

Although the regression models provided the basic analytical support that Aramark needed, the GRM team used a novel approach to implement the models for use by their clients. They developed "interactive risk simulators," which are simple online tools that allowed users to manipulate the values of the independent variables in the regression models using interactive sliders that correspond to the business metrics and instantaneously view the values of the dependent variables (the risk metrics) on gauges similar to those found on the dashboard of a car.

Figure 8.19 illustrates the structure of the simulators. The gauges are updated instantly as the user adjusts the sliders, showing how changes in the business environment affect the risk metrics. This visual representation made the models easy to use and understand, particularly for non-technical employees.

GRM sent out more than 200 surveys to multiple levels of the organization to assess the usefulness of the interactive risk simulators. One hundred percent of respondents answered "Yes" to "Were the simulators easy to use?" and 78% of respondents answered "Yes" to "Would these simulators be useful in running your business and helping you make decisions?" The deployment of interactive risk simulators to the field has been met with overwhelming positive response and recognition from leadership within all lines of business, including frontline managers, food-service directors, district managers, and general managers.

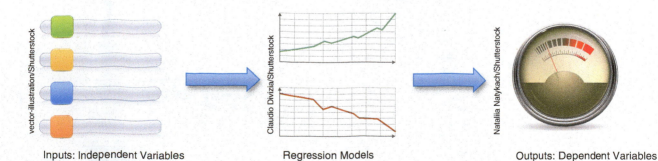

Inputs: Independent Variables   Regression Models   Outputs: Dependent Variables

▲ **Figure 8.19**

Structure of an Interactive
Risk Simulator

CHECK YOUR UNDERSTANDING

1. What is a multiple linear regression model?
2. How do you interpret a partial regression coefficient?
3. What hypothesis is tested with ANOVA for a multiple linear regression model?
4. How are ANOVA results in multiple linear regression used to test hypotheses for individual regression coefficients?

Building Good Regression Models

In the colleges and universities regression example, all the independent variables were found to be significant by evaluating the *p*-values of the regression analysis. This will not always be the case and leads to the question of how to build good regression models that include the "best" set of variables.

Figure 8.20 shows a portion of the Excel file *Banking Data*, which provides data acquired from banking and census records for different ZIP codes in the bank's current market. Such information can be useful in targeting advertising for new customers or for choosing locations for branch offices. The data show the median age of the population, median years of education, median income, median home value, median household wealth, and average bank balance.

Figure 8.21 shows the results of regression analysis used to predict the average bank balance as a function of the other variables. Although the independent variables explain

▶ **Figure 8.20**

Portion of Banking Data

	A	B	C	D	E	F
1	Banking Data					
2						
3	Median	Median Years	Median	Median	Median Household	Average Bank
4	Age	Education	Income	Home Value	Wealth	Balance
5	35.9	14.8	$91,033	$183,104	$220,741	$38,517
6	37.7	13.8	$86,748	$163,843	$223,152	$40,618
7	36.8	13.8	$72,245	$142,732	$176,926	$35,206
8	35.3	13.2	$70,639	$145,024	$166,260	$33,434
9	35.3	13.2	$64,879	$135,951	$148,868	$28,162
10	34.8	13.7	$75,591	$155,334	$188,310	$36,708

▶ **Figure 8.21**

Regression Analysis Results for Banking Data

	A	B	C	D	E	F	G
1	SUMMARY OUTPUT						
2							
3	Regression Statistics						
4	Multiple R	0.97309221					
5	R Square	0.946908448					
6	Adjusted R Square	0.944143263					
7	Standard Error	2055.64333					
8	Observations	102					
9							
10	ANOVA						
11		df	SS	MS	F	Significance F	
12	Regression	5	7235179873	1447035975	342.4394584	1.5184E-59	
13	Residual	96	405664271.9	4225669.499			
14	Total	101	7640844145				
15							
16		Coefficients	Standard Error	t Stat	P-value	Lower 95%	Upper 95%
17	Intercept	-10710.64278	4260.976308	-2.513659314	0.013613179	-19168.61391	-2252.671659
18	Age	318.6649626	60.98611242	5.225205378	1.01152E-06	197.6084862	439.721439
19	Education	621.8603472	318.9595184	1.949652891	0.054135377	-11.26929279	1254.989987
20	Income	0.146323453	0.040781001	3.588029937	0.000526666	0.065373806	0.227273101
21	Home Value	0.009183067	0.011038075	0.831944635	0.407504891	-0.012727338	0.031093473
22	Wealth	0.074331533	0.011189265	6.643111131	1.84838E-09	0.052121017	0.096542049

more than 94% of the variation in the average bank balance, you can see that at a 0.05 significance level, the *p*-values indicate that both education and home value do not appear to be significant. A good regression model should include only significant independent variables. However, it is not always clear exactly what will happen when we add or remove variables from a model; variables that are (or are not) significant in one model may (or may not) be significant in another. Therefore, you should *not* consider dropping all insignificant variables at one time, but rather take a more structured approach.

This suggests a systematic approach to building good regression models:

1. Construct a model with all available independent variables. Check for significance of the independent variables by examining the *p*-values.
2. Identify the independent variable having the largest *p*-value that exceeds the chosen level of significance.
3. Remove the variable identified in step 2 from the model. (Don't remove all variables with *p*-values that exceed α at the same time, but remove only one at a time.)
4. Continue until all variables are significant.

In essence, this approach seeks to find a model that has only significant independent variables.

EXAMPLE 8.13   Identifying the Best Regression Model

We will apply the preceding approach to the *Banking Data* example. The first step is to identify the variable with the largest *p*-value exceeding 0.05; in this case, it is home value, and we remove it from the model and rerun the *Regression* tool. Figure 8.22 shows the results after removing home value. All the *p*-values are now less than 0.05, so this now appears to be the best model. Notice that the

p-value for education, which was larger than 0.05 in the first regression analysis, dropped below 0.05 after home value was removed. This phenomenon often occurs when multicollinearity (discussed in the next section) is present and emphasizes the importance of not removing all variables with large *p*-values from the original model at the same time.

▶ **Figure 8.22**

Regression Results Without Home Value

Adding an independent variable to a regression model will *always* result in R^2 equal to or greater than the R^2 of the original model. This is true even when the new independent variable has little true relationship with the dependent variable. Thus, trying to maximize R^2 is not a useful criterion. A better way of evaluating the relative fit of different models is to use adjusted R^2. Adjusted R^2 reflects both the number of independent variables and the sample size and may either increase or decrease when an independent variable is added or dropped, thus providing an indication of the value of adding or removing independent variables in the model. An increase in adjusted R^2 indicates that the model has improved.

A criterion used to determine if a variable should be removed based on the adjusted R^2 is the t-statistic. If $|t| < 1$, then the standard error will decrease and adjusted R^2 will increase if the variable is removed. If $|t| > 1$, then the opposite will occur. In the banking regression results, we see that the t-statistic for home value is less than 1; therefore, we expect the adjusted R^2 to increase if we remove this variable. You can follow the same iterative approach outlined before, except using t-values instead of p-values. One might, for example, choose the smallest $|t|$ at each step. This approach will improve the adjusted R^2 but may result in a model with insignificant independent variables.

These approaches using the p-values or t-statistics may involve considerable experimentation to identify the best set of variables that result in the largest adjusted R^2. For large numbers of independent variables, the number of potential models can be overwhelming. For example, there are $2^{10} = 1,024$ possible models that can be developed from a set of ten independent variables. This can make it difficult to effectively screen out insignificant variables. Fortunately, automated methods—stepwise regression and best subsets—exist that facilitate this process.

Correlation and Multicollinearity

As we have learned previously, correlation, a numerical value between -1 and $+1$, measures the linear relationship between pairs of variables. The higher the absolute value of the correlation, the greater the strength of the relationship. The sign simply indicates whether variables tend to increase together (positive) or not (negative). Therefore, examining correlations between the dependent and independent variables, which can be done using the Excel *Correlation* tool, can be useful in selecting variables to include in a multiple regression model because a strong correlation indicates a strong linear relationship. However, strong correlations *among the independent variables* can be problematic. This can potentially signify a phenomenon called **multicollinearity**, a condition occurring when two or more independent variables in the same regression model contain high levels of the same information and, consequently, are strongly correlated with one another and can predict each other better than the dependent variable. When significant multicollinearity is present, it becomes difficult to isolate the effect of one independent variable on the dependent variable, and the signs of coefficients may be the opposite of what they should be, making it difficult to interpret regression coefficients. Also, p-values can be inflated, resulting in the conclusion not to reject the null hypothesis for significance of regression when it should be rejected.

Some experts suggest that correlations between independent variables exceeding an absolute value of 0.7 may indicate multicollinearity. However, multicollinearity is best measured using a statistic called the *variance inflation factor (VIF)* for each independent variable. Sophisticated software packages usually compute these; unfortunately, Excel does not.

EXAMPLE 8.14   Identifying Potential Multicollinearity

Figure 8.23 shows the correlation matrix for the variables in the *Colleges and Universities* data. You can see that SAT and acceptance rate have moderate correlations with the dependent variable, graduation %, but the correlation between expenditures/student and top 10% HS with graduation % are relatively low. The strongest correlation, however, is between two independent variables: top 10% HS and acceptance rate. However, the value of −0.6097 does not exceed the recommended threshold of 0.7, so we can likely assume that multicollinearity is not a problem here (a more advanced analysis using VIF calculations does indeed confirm that multicollinearity does not exist).

In contrast, Figure 8.24 shows the correlation matrix for all the data in the banking example. Note that large correlations exist between education and home value and also between wealth and income (in fact, the variance inflation factors do indicate significant multicollinearity). If we remove wealth from the model, the adjusted R^2 drops to 0.9201, but we discover that education is no longer significant. Dropping education and leaving only age and income in the model results in an adjusted R^2 of 0.9202. However, if we remove income from the model instead of wealth, the adjusted R^2 drops to only 0.9345, and all remaining variables (age, education, and wealth) are significant (see Figure 8.25). The R^2 value for the model with these three variables is 0.9365.

▶ **Figure 8.23**

Correlation Matrix for Colleges and Universities Data

	A	B	C	D	E	F
1		Median SAT	Acceptance Rate	Expenditures/Student	Top 10% HS	Graduation %
2	Median SAT	1				
3	Acceptance Rate	-0.601901959	1			
4	Expenditures/Student	0.572741729	-0.284254415	1		
5	Top 10% HS	0.503467995	-0.609720972	0.505782049	1	
6	Graduation %	0.564146827	-0.55037751	0.042503514	0.138612667	1

▶ **Figure 8.24**

Correlation Matrix for Banking Data

	A	B	C	D	E	F	G
1		Age	Education	Income	Home Value	Wealth	Balance
2	Age	1					
3	Education	0.173407147	1				
4	Income	0.4771474	0.57539402	1			
5	Home Value	0.386493114	0.753521067	0.795355158	1		
6	Wealth	0.468091791	0.469413035	0.946665447	0.698477789	1	
7	Balance	0.565466834	0.55488066	0.951684494	0.766387128	0.948711734	1

▶ **Figure 8.25**

Regression Results for Age, Education, and Wealth as Independent Variables

	A	B	C	D	E	F	G
1	SUMMARY OUTPUT						
2							
3	*Regression Statistics*						
4	Multiple R	0.967710981					
5	R Square	0.936464543					
6	Adjusted R Square	0.93451958					
7	Standard Error	2225.695322					
8	Observations	102					
9							
10	ANOVA						
11		df	SS	MS	F	Significance F	
12	Regression	3	7155379617	2385126539	481.4819367	1.71667E-58	
13	Residual	98	485464527.3	4953719.667			
14	Total	101	7640844145				
15							
16		Coefficients	Standard Error	t Stat	P-value	Lower 95%	Upper 95%
17	Intercept	-17732.45142	3801.662822	-4.664393517	9.79978E-06	-25276.72757	-10188.17528
18	Age	367.8214086	64.59823831	5.693985134	1.2977E-07	239.6283071	496.0145102
19	Education	1300.308712	249.9731413	5.201793703	1.08292E-06	804.2451489	1796.372276
20	Wealth	0.116467903	0.004679827	24.88722652	3.75813E-44	0.107180939	0.125754866

Practical Issues in Trendline and Regression Modeling

Finding a good regression model often requires some experimentation and trial and error. From a practical perspective, the independent variables selected should make some sense in attempting to explain the dependent variable (that is, you should have some reason to believe that changes in the independent variable will cause changes in the dependent variable even though causation cannot be proven statistically). Logic should guide your model development. In many applications, behavioral, economic, or physical theory might suggest that certain variables should belong in a model. Remember that additional variables do contribute to a higher R^2 and, therefore, help to explain a larger proportion of the variation. Even though a variable with a large p-value is not statistically significant, it could simply be the result of sampling error and a modeler might wish to keep it.

Good modelers also try to have as simple a model as possible—an age-old principle known as **parsimony**—with the fewest number of explanatory variables that will provide an adequate interpretation of the dependent variable. In the physical and management sciences, some of the most powerful theories are the simplest. Thus, a model for the banking data that includes only age, education, and wealth is simpler than one with four variables; because of the multicollinearity issue, there would be little to gain by including income in the model. Whether the model explains 93% or 94% of the variation in bank deposits would probably make little difference. Therefore, building good regression models relies as much on experience and judgment as it does on technical analysis.

An issue that one often faces in using trendlines and regression is **overfitting** the model. It is important to realize that sample data may have unusual variability that is different from the population; if we fit a model too closely to the sample data, we risk not fitting it well to the population in which we are interested. For instance, in fitting the crude oil prices in Example 8.2, we noted that the R^2 value will increase if we fit higher-order polynomial functions to the data. While this might provide a better mathematical fit to the sample data, doing so can make it difficult to explain the phenomena rationally. The same thing can happen with multiple regression. If we add too many terms to the model, then the model may not adequately predict other values from the population. Overfitting can be mitigated by using good logic, intuition, physical or behavioral theory, and parsimony, as we have discussed.

CHECK YOUR UNDERSTANDING

1. How should you properly use p-values from multiple linear regression to build a good model?
2. Explain how to use the t-statistic to build the best multiple linear regression model.
3. What is multicollinearity, and why is it important to identify in multiple linear regression models?
4. Explain the concepts of parsimony and overfitting in building regression models.

Regression with Categorical Independent Variables

Some data of interest in a regression study may be ordinal or nominal. This is common when including demographic data in marketing studies, for example. Because regression analysis requires numerical data, we could include categorical variables by *coding* the variables. For example, if one variable represents whether an individual is a college graduate or not, we might code no as 0 and yes as 1. Such variables are often called **dummy variables**.

EXAMPLE 8.15 A Model with Categorical Variables

The Excel file *Employee Salaries*, shown in Figure 8.26, provides salary and age data for 35 employees, along with an indicator of whether or not the employees have an MBA (yes or no). The MBA indicator variable is categorical; thus, we code it by replacing no with 0 and yes with 1.

If we are interested in predicting salary as a function of the other variables, we would propose the model

$$Y = \beta_0 + \beta_1 X_1 + \beta_2 X_2 + \varepsilon$$

where Y = salary, X_1 = age, and X_2 = MBA indicator (0 or 1).

After coding the MBA indicator column in the data file, we begin by running a regression on the entire data set, yielding the output shown in Figure 8.27. Note that the model explains about 95% of the variation, and the p-values of both variables are significant. The model is

Salary = 893.59 + 1,044.15 × Age + 14,767.23 × MBA

Thus, a 30-year-old with an MBA would have an estimated salary of

Salary = 893.59 + 1,044.15 × 30 + 14,767.23 × 1
= $ 46,985.32

This model suggests that having an MBA increases the salary of this group of employees by almost $15,000. Note that by substituting either 0 or 1 for MBA, we obtain two models:

No MBA: Salary = 893.59 + 1,044.15 × Age
MBA: Salary = 15,660.82 + 1,044.15 × Age

The only difference between them is the intercept. The models suggest that the rate of salary increase for age is the same for both groups. Of course, this may not be true. Individuals with MBAs might earn relatively higher salaries as they get older. In other words, the slope of age may *depend* on the value of MBA.

▶ **Figure 8.26**

Portion of Excel File
Employee Salaries

	A	B	C	D
1	Employee Salary Data			
2				
3	Employee	Salary	Age	MBA
4	1	$ 28,260	25	No
5	2	$ 43,392	28	Yes
6	3	$ 56,322	37	Yes
7	4	$ 26,086	23	No
8	5	$ 36,807	32	No

▶ **Figure 8.27**

Initial Regression Model for
Employee Salaries

	A	B	C	D	E	F	G
1	SUMMARY OUTPUT						
2							
3	*Regression Statistics*						
4	Multiple R	0.976118476					
5	R Square	0.952807278					
6	Adjusted R Square	0.949857733					
7	Standard Error	2941.914352					
8	Observations	35					
9							
10	ANOVA						
11		df	SS	MS	F	Significance F	
12	Regression	2	5591651177	2795825589	323.0353318	6.05341E-22	
13	Residual	32	276955521.7	8654860.054			
14	Total	34	5868606699				
15							
16		Coefficients	Standard Error	t Stat	P-value	Lower 95%	Upper 95%
17	Intercept	893.5875971	1824.575283	0.489751015	0.627650922	-2822.950634	4610.125828
18	Age	1044.146043	42.14128238	24.77727265	1.8878E-22	958.3070599	1129.985026
19	MBA	14767.23159	1351.801764	10.92411031	2.49752E-12	12013.7015	17520.76168

An **interaction** occurs when the effect of one variable (for example, the slope) is dependent on another variable. We can test for interactions by defining a new variable as the product of the two variables, $X_3 = X_1 \times X_2$, and testing whether this variable is significant, leading to an alternative model.

EXAMPLE 8.16   **Incorporating Interaction Terms in a Regression Model**

For the *Employee Salaries* example, we define an interaction term as the product of age (X_1) and MBA (X_2) by defining $X_3 = X_1 \times X_2$. The new model is

$$Y = \beta_0 + \beta_1 X_1 + \beta_2 X_2 + \beta_3 X_3 + \varepsilon$$

In the worksheet, we need to create a new column (called Interaction) by multiplying MBA by age for each observation (see Figure 8.28). The regression results are shown in Figure 8.29.

From Figure 8.29, we see that the adjusted R^2 increases; however, the *p*-value for the MBA indicator variable is 0.33, indicating that this variable is not significant. We would typically drop this variable and run a regression using only age and the interaction term (Figure 8.30), and obtain the model:

$$\text{Salary} = 3{,}323.11 + 984.25 \times \text{Age} + 425.58 \times \text{MBA} \times \text{Age}$$

We see that salary depends not only on whether an employee holds an MBA, but also on age, and is more realistic than the original model.

However, statisticians recommend that if interactions are significant, first-order terms should be kept in the model regardless of their p-values. Thus, using the output in Figure 8.29, we have:

$$\text{Salary} = 3902.51 + 971.31 \times \text{Age} - 2971.08 \times \text{MBA} + 501.85 \times \text{MBA} \times \text{Age}$$

Actually, there is little difference in the predictive ability of either model.

► Figure 8.28

Portion of Employee Salaries *Modified for Interaction Term*

	A	B	C	D	E
1	Employee Salary Data				
2					
3	Employee	Salary	Age	MBA	Interaction
4	1	$ 28,260	25	0	0
5	2	$ 43,392	28	1	28
6	3	$ 56,322	37	1	37
7	4	$ 26,086	23	0	0

► Figure 8.29

Regression Results with Interaction Term

	A	B	C	D	E	F	G
1	SUMMARY OUTPUT						
2							
3	*Regression Statistics*						
4	Multiple R	0.989321416					
5	R Square	0.978756863					
6	Adjusted R Square	0.976701076					
7	Standard Error	2005.37675					
8	Observations	35					
9							
10	ANOVA						
11		*df*	*SS*	*MS*	*F*	*Significance F*	
12	Regression	3	5743939086	1914646362	476.098288	5.31397E-26	
13	Residual	31	124667613.2	4021535.91			
14	Total	34	5868606699				
15							
16		Coefficients	Standard Error	t Stat	P-value	Lower 95%	Upper 95%
17	Intercept	3902.509386	1336.39766	2.920170772	0.006467654	1176.908389	6628.110383
18	Age	971.3090382	31.06887722	31.26308786	5.23658E-25	907.9436454	1034.674431
19	MBA	-2971.080074	3026.24236	-0.98177202	0.333812767	-9143.142058	3200.981911
20	Interaction	501.8483604	81.55221742	6.153705887	7.9295E-07	335.5215164	668.1752044

▶ **Figure 8.30**

Regression Model for Salary Data After Dropping MBA

	A	B	C	D	E	F	G
1	SUMMARY OUTPUT						
2							
3	*Regression Statistics*						
4	Multiple R	0.98898754					
5	R Square	0.978096355					
6	Adjusted R Square	0.976727377					
7	Standard Error	2004.24453					
8	Observations	35					
9							
10	ANOVA						
11		*df*	*SS*	*MS*	*F*	*Significance F*	
12	Regression	2	5740062823	2870031411	714.4720368	2.80713E-27	
13	Residual	32	128543876.4	4016996.136			
14	Total	34	5868606699				
15							
16		*Coefficients*	*Standard Error*	*t Stat*	*P-value*	*Lower 95%*	*Upper 95%*
17	Intercept	3323.109564	1198.353141	2.773063675	0.009184278	882.1440943	5764.075033
18	Age	984.2455409	28.12039088	35.00113299	4.40388E-27	926.9661791	1041.524903
19	Interaction	425.5845915	24.81794165	17.14826304	1.08793E-17	375.0320986	476.1370843

Categorical Variables with More Than Two Levels

When a categorical variable has only two levels, as in the previous example, we coded the levels as 0 and 1 and added a new variable to the model. However, when a categorical variable has $k > 2$ levels, we need to add $k - 1$ additional variables to the model.

EXAMPLE 8.17 **A Regression Model with Multiple Levels of Categorical Variables**

The Excel file *Surface Finish* provides measurements of the surface finish of 35 parts produced on a lathe, along with the revolutions per minute (RPM) of the spindle and one of four types of cutting tools used (see Figure 8.31). The engineer who collected the data is interested in predicting the surface finish as a function of RPM and type of tool.

Intuition might suggest defining a dummy variable for each tool type; however, doing so will cause numerical instability in the data and cause the regression tool to crash. Instead, we will need $k - 1 = 3$ dummy variables corresponding to three of the levels of the categorical variable. The level left out will correspond to a reference, or baseline, value. Therefore, because we have $k = 4$ levels of tool type, we will define a regression model of the form

$$Y = \beta_0 + \beta_1 X_1 + \beta_2 X_2 + \beta_3 X_3 + \beta_4 X_4 + \varepsilon$$

where $Y = $ *surface finish*, $X_1 = $ RPM, $X_2 = 1$ if tool type is B and 0 if not, $X_3 = 1$ if tool type is C and 0 if not, and $X_4 = 1$ if tool type is D and 0 if not.

Note that when $X_2 = X_3 = X_4 = 0$, then, by default, the tool type is A. Substituting these values for each tool type into the model, we obtain

$$\text{Tool type A: } Y = \beta_0 + \beta_1 X_1 + \varepsilon$$
$$\text{Tool type B: } Y = \beta_0 + \beta_1 X_1 + \beta_2 + \varepsilon$$

$$\text{Tool type C: } Y = \beta_0 + \beta_1 X_1 + \beta_3 + \varepsilon$$
$$\text{Tool type D: } Y = \beta_0 + \beta_1 X_1 + \beta_4 + \varepsilon$$

For a fixed value of RPM (X_1), the slopes corresponding to the dummy variables represent the difference between the surface finish using that tool type and the baseline using tool type A.

To incorporate these dummy variables into the regression model, we add three columns to the data, as shown in Figure 8.32. Using these data, we obtain the regression results shown in Figure 8.33. The resulting model is

$$\text{Surface Finish} = 24.49 + 0.098 \times \text{RPM} - 13.31 \times \text{Type B}$$
$$- 20.49 \times \text{Type C} - 26.04 \times \text{Type D}$$

Almost 99% of the variation in surface finish is explained by the model, and all variables are significant. The models for each individual tool are

$$\text{Tool A: Surface Finish} = 24.49 + 0.098 \times \text{RPM} - 13.31 \times 0$$
$$- 20.49 \times 0 - 26.04 \times 0$$
$$= 24.49 + 0.098 \times \text{RPM}$$
$$\text{Tool B: Surface Finish} = 24.49 + 0.098 \times \text{RPM} - 13.31 \times 1$$
$$- 20.49 \times 0 - 26.04 \times 0$$
$$= 11.18 + 0.098 \times \text{RPM}$$

(continued)

Tool C: Surface Finish $= 24.49 + 0.098 \times RPM - 13.31 \times 0$
$$- 20.49 \times 1 - 26.04 \times 0$$
$$= 4.00 + 0.098 \times RPM$$
Tool D: Surface Finish $= 24.49 + 0.098 \times RPM - 13.31 \times 0$
$$- 20.49 \times 0 - 26.04 \times 1$$
$$= -1.55 + 0.098 \times RPM$$

Note that the only differences among these models are the intercepts; the slopes associated with RPM are the same. This suggests that we might wish to test for interactions between the type of cutting tool and RPM; we leave this to you as an exercise.

▶ **Figure 8.31**

Portion of Excel File Surface Finish

	A	B	C	D
1	Surface Finish Data			
2				
3	Part	Surface Finish	RPM	Cutting Tool
4	1	45.44	225	A
5	2	42.03	200	A
6	3	50.10	250	A
7	4	48.75	245	A
8	5	47.92	235	A
9	6	47.79	237	A
10	7	52.26	265	A
11	8	50.52	259	A
12	9	45.58	221	A
13	10	44.78	218	A
14	11	33.50	224	B
15	12	31.23	212	B
16	13	37.52	248	B
17	14	37.13	260	B
18	15	34.70	243	B

▶ **Figure 8.32**

Data Matrix for Surface Finish *with Dummy Variables*

	A	B	C	D	E	F
1	Surface Finish Data					
2						
3	Part	Surface Finish	RPM	Type B	Type C	Type D
4	1	45.44	225	0	0	0
5	2	42.03	200	0	0	0
6	3	50.10	250	0	0	0
7	4	48.75	245	0	0	0
8	5	47.92	235	0	0	0
9	6	47.79	237	0	0	0
10	7	52.26	265	0	0	0
11	8	50.52	259	0	0	0
12	9	45.58	221	0	0	0
13	10	44.78	218	0	0	0
14	11	33.50	224	1	0	0
15	12	31.23	212	1	0	0
16	13	37.52	248	1	0	0
17	14	37.13	260	1	0	0
18	15	34.70	243	1	0	0
19	16	33.92	238	1	0	0
20	17	32.13	224	1	0	0
21	18	35.47	251	1	0	0
22	19	33.49	232	1	0	0
23	20	32.29	216	1	0	0
24	21	27.44	225	0	1	0
25	22	24.03	200	0	1	0
26	23	27.33	250	0	1	0
27	24	27.20	245	0	1	0
28	25	27.10	235	0	1	0
29	26	27.30	237	0	1	0
30	27	28.30	265	0	1	0
31	28	28.40	259	0	1	0
32	29	26.80	221	0	1	0
33	30	26.40	218	0	1	0
34	31	21.40	224	0	0	1
35	32	20.50	212	0	0	1
36	33	21.90	248	0	0	1
37	34	22.13	260	0	0	1
38	35	22.40	243	0	0	1

► **Figure 8.33**

Surface Finish *Regression Model Results*

	A	B	C	D	E	F	G
1	SUMMARY OUTPUT						
2							
3	*Regression Statistics*						
4	Multiple R	0.994447053					
5	R Square	0.988924942					
6	Adjusted R Square	0.987448267					
7	Standard Error	1.089163115					
8	Observations	35					
9							
10	ANOVA						
11		*df*	*SS*	*MS*	*F*	*Significance F*	
12	Regression	4	3177.784271	794.4460678	669.6973322	7.32449E-29	
13	Residual	30	35.58828875	1.186276292			
14	Total	34	3213.37256				
15							
16		*Coefficients*	*Standard Error*	*t Stat*	*P-value*	*Lower 95%*	*Upper 95%*
17	Intercept	24.49437244	2.473298088	9.903526211	5.73134E-11	19.44322388	29.54552101
18	RPM	0.097760627	0.010399996	9.400064035	1.89415E-10	0.076521002	0.119000252
19	Type B	-13.31056756	0.487142953	-27.32374035	9.37003E-23	-14.3054462	-12.31568893
20	Type C	-20.487	0.487088553	-42.06011387	3.12134E-28	-21.48176754	-19.49223246
21	Type D	-26.03674519	0.596886375	-43.62094073	1.06415E-28	-27.25574979	-24.81774059

CHECK YOUR UNDERSTANDING

1. Why is it necessary to code categorical independent variables in regression modeling?

2. What is an interaction? How can you test for interactions with regression?

3. Explain how to build a regression model if a categorical variable has more than two levels.

Regression Models with Nonlinear Terms

Linear regression models are not appropriate for every situation. A scatter chart of the data might show a nonlinear relationship, or the residuals for a linear fit might result in a nonlinear pattern. In such cases, we might propose a nonlinear model to explain the relationship. For instance, a second-order polynomial model would be

$$Y = \beta_0 + \beta_1 X + \beta_2 X^2 + \varepsilon$$

Sometimes, this is called a **curvilinear regression model**. In this model, β_1 represents the linear effect of X on Y, and β_2 represents the curvilinear effect. However, although this model appears to be quite different from ordinary linear regression models, it is still *linear in the parameters* (the betas, which are the unknowns that we are trying to estimate). In other words, all terms are a product of a beta coefficient and some function of the data, which are simply numerical values. In such cases, we can still apply least squares to estimate the regression coefficients.

Curvilinear regression models are also often used in forecasting when the independent variable is time. This and other applications of regression in forecasting are discussed in the next chapter.

EXAMPLE 8.18 Modeling Beverage Sales Using Curvilinear Regression

The Excel file *Beverage Sales* provides data on the sales of cold beverages at a small restaurant with a large outdoor patio during the summer months (see Figure 8.34). The owner has observed that sales tend to increase on hotter days. Figure 8.35 shows linear regression results for these data. The U-shape of the residual plot (a second-order polynomial trendline was fit to the residual data) suggests that a linear relationship is not appropriate. To apply a curvilinear regression model, add a column to the data matrix by squaring

the temperatures. Now, both temperature and temperature squared are the independent variables. Figure 8.36 shows the results for the curvilinear regression model. The model is

$$\text{Sales} = 142{,}850 - 3{,}643.17 \times \text{Temperature} + 23.3 \times \text{Temperature}^2$$

Note that the adjusted R^2 has increased significantly from the linear model and that the residual plots now show more random patterns.

▶ **Figure 8.34**

Portion of Excel File Beverage Sales

	A	B
1	Beverage Sales	
2		
3	Temperature	Sales
4	85	$ 1,810
5	90	$ 4,825
6	79	$ 438
7	82	$ 775
8	84	$ 1,213
9	96	$ 8,692

▶ **Figure 8.35**

Linear Regression Results for Beverage Sales

	A	B	C	D	E	F	G
1	SUMMARY OUTPUT						
2							
3	*Regression Statistics*						
4	Multiple R	0.922351218					
5	R Square	0.850731769					
6	Adjusted R Square	0.842875547					
7	Standard Error	1041.057399					
8	Observations	21					
9							
10	ANOVA						
11		*df*	*SS*	*MS*	*F*	*Significance F*	
12	Regression	1	117362193.6	117362193.6	108.2876347	2.7611E-09	
13	Residual	19	20592209.67	1083800.509			
14	Total	20	137954403.2				
15							
16		*Coefficients*	*Standard Error*	*t Stat*	*P-value*	*Lower 95%*	*Upper 95%*
17	Intercept	-32511.24671	3408.723477	-9.53766034	1.12197E-08	-39645.78695	-25376.70648
18	Temperature	408.6026284	39.26555335	10.40613447	2.7611E-09	326.4188807	490.786376

Temperature Residual Plot (residuals vs. temperature, ranging 70–100).

▶ **Figure 8.36**

Curvilinear Regression Results for Beverage Sales

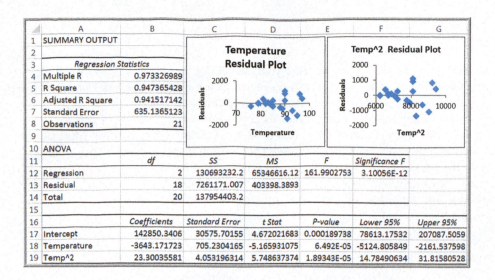

	A	B	C	D	E	F	G
1	SUMMARY OUTPUT						
2							
3	*Regression Statistics*						
4	Multiple R	0.973326989					
5	R Square	0.947365428					
6	Adjusted R Square	0.941517142					
7	Standard Error	635.1365123					
8	Observations	21					
9							
10	ANOVA						
11		*df*	*SS*	*MS*	*F*	*Significance F*	
12	Regression	2	130693232.2	65346616.12	161.9902753	3.10056E-12	
13	Residual	18	7261171.007	403398.3893			
14	Total	20	137954403.2				
15							
16		*Coefficients*	*Standard Error*	*t Stat*	*P-value*	*Lower 95%*	*Upper 95%*
17	Intercept	142850.3406	30575.70155	4.672021683	0.000189738	78613.17532	207087.5059
18	Temperature	-3643.171723	705.2304165	-5.165931075	6.492E-05	-5124.805849	-2161.537598
19	Temp^2	23.30035581	4.053196314	5.748637374	1.89343E-05	14.78490634	31.81580528

CHECK YOUR UNDERSTANDING

1. What is a curvilinear regression model? How are these models often used?
2. How can you tell if a curvilinear regression model should be used?

KEY TERMS

Autocorrelation
Coefficient of determination (R^2)
Coefficient of multiple determination
Curvilinear regression model
Dummy variables
Exponential function
Homoscedasticity
Interaction
Least-squares regression
Linear function
Logarithmic function
Multicollinearity
Multiple correlation coefficient

Multiple linear regression
Overfitting
Parsimony
Partial regression coefficient
Polynomial function
Power function
R^2 (R-squared)
Regression analysis
Residuals
Significance of regression
Simple linear regression
Standard error of the estimate, S_{YX}
Standard residuals

CHAPTER 8 TECHNOLOGY HELP

Useful Excel Functions

INTERCEPT(*known_y's, known_x's*) Finds the intercept for a simple linear regression model.

SLOPE(*known_y's, known_x's*) Finds the slope for a simple linear regression model.

TREND(*known_y's, known_x's, new_x's*) Estimates *Y* for any value of *X*.

Excel Techniques

Trendline tool (Examples 8.1 and 8.2):

Create a line chart or scatter chart for your data. Click the chart; this will display the *Chart Tools* menu. Select the *Chart Tools Design* tab, and then click *Add Chart Element* from the *Chart Layouts* group. From the *Trendline* sub-menu, you can select one of the options or *More Trendline Options* . . . . Selecting *More Trendline Options* will display the *Format Trendline* pane in the worksheet. An alternative is to right click on the data series in the chart and choose *Add trendline* from the pop-up menu. Select the radio button for the trendline. Check the boxes for *Display Equation on chart* and *Display R-squared value on chart*.

Regression tool:

From the *Data Analysis* menu in the *Analysis* group under the *Data* tab, select the *Regression* tool. In the dialog box for the *Input Y Range*, specify the range of the dependent variable values. In the box for the *Input X Range*, specify the range for the independent variable values. Check *Labels* as appropriate. Check the optional boxes to compute and display *Residuals*, *Standardized Residuals*, *Residual Plots*, *Line Fit Plots*, and *Normal Probability Plots*.

StatCrunch

StatCrunch provides a tool for simple and multiple linear regression. You can find video tutorials and step-by-step procedures with examples at https://www.statcrunch.com/5.0/example.php. We suggest that you first view the tutorials *Getting started with StatCrunch* and *Working with StatCrunch sessions*. The following tutorial is located under the Regression and Correlation group on this Web page and explains how to perform simple linear regression.

Example: Simple Linear Regression

1. Choose *Stat* > *Regression* > *Simple Linear*.
2. Select the *X variable* (independent variable) and *Y variable* (dependent variable) for the regression.

3. Enter an optional *Where* statement to specify the data rows to be included.
4. Compare results across groups by selecting an optional *Group* by column.
5. Click *Compute!* to view the results.

Example: Multiple Linear Regression

1. Choose *Stat* > *Regression* > *Multiple Linear*.
2. Select the *Y variable* (dependent variable) for the regression.
3. Select the *X variables* (independent variables) for the regression.
4. Create interaction terms by selecting two or more variables and clicking the *Add 1* or *Add All* button. Interaction terms will then be shown in the area to the right. To delete an interaction term, select it and click the *Delete* button.
5. To center the variables in each interaction term, check the *Center interaction terms* option.
6. Enter an optional *Where* statement to specify the data rows to be included.
7. Select an optional *Group by* column to group results. A separate regression analysis will be constructed for each distinct value of the *Group by* column.
8. Click *Compute!* to view the regression results.

Analytic Solver

Analytic Solver provides a set of advanced techniques for regression modeling. See the online supplement *Using Best Subsets for Regression*. We suggest that you first read the online supplement *Getting Started with Analytic Solver Basic*. This provides information for both instructors and students on how to register for and access Analytic Solver.

PROBLEMS AND EXERCISES

Modeling Relationships and Trends in Data

1. Using the data in the Excel file *Student Grades*, construct a scatter chart for midterm versus final exam grades and add a linear trendline. What is the model? If a student scores 75 on the midterm, what would you predict her grade on the final exam to be?

2. A consumer products company has collected some data relating monthly demand to the price of one of its products:

Price	Demand
$11	2,180
$13	2,020
$17	1,980
$19	1,900

Construct a scatter chart and identify the type of model that would best represent these data. Use the *Trendline* tool to find the model equation.

3. Each worksheet in the Excel file *LineFit Data* contains a set of data that describes a functional relationship between the dependent variable y and the independent variable x. Construct a line chart of each data set and use the *Trendline* tool to determine the best-fitting functions to model these data sets. Do not consider polynomials beyond the third degree.

4. Using the data in the Excel file *Demographics*, determine if a linear relationship exists between unemployment rates and cost of living indexes by constructing a scatter chart. Visually, do there appear to be any outliers? If so, delete them and then find the best-fitting function using the *Trendline* tool. What would you conclude about the strength of any relationship?

5. Use the data in the Excel file *MBA Motivation and Salary Expectations* to determine how well a linear function models the relationship between pre- and post-MBA salaries. Compare it to a third-order polynomial. Would the polynomial model have a logical explanation?

6. Using the data in the Excel file *Weddings*, construct scatter charts to determine whether any linear relationship appears to exist between (1) the wedding cost (X) and attendance (Y), (2) the wedding cost (X) and the value rating (Y), and (3) the couple's income (X) and wedding cost (Y) only for the weddings paid for by the bride and groom. Then find the best-fitting functions using the *Trendline* tool for each of these charts.

Simple Linear Regression

7. Using the results of fitting the *Home Market Value* regression line in Example 8.4, compute the errors associated with each observation using formula (8.3) and construct a frequency distribution and histogram.

8. Set up an Excel worksheet to apply formulas (8.5) and (8.6) to compute the values of b_0 and b_1 for the data in the Excel file *Home Market Value* and verify

that you obtain the same values as in Examples 8.4 and 8.5.

9. The managing director of a consulting group has the following monthly data on total overhead costs and professional labor hours to bill to clients:[5]

Overhead Costs	Billable Hours
$355,000	3,000
$400,000	4,000
$425,000	5,000
$477,000	6,000
$560,000	7,000
$580,000	8,000

a. Develop a simple linear regression model between billable hours and overhead costs.

b. Interpret the coefficients of your regression model. Specifically, what does the fixed component of the model mean to the consulting firm?

c. If a special job was available requiring 4,500 billable hours that would contribute a margin of $240,000 before overhead, would the job be attractive?

10. Using the data in the Excel file *Demographics*, apply the Excel *Regression* tool using unemployment rate as the dependent variable and cost of living index as the independent variable. Interpret all key regression results, hypothesis tests, and confidence intervals in the output.

11. The Excel file *National Football League* provides various data on professional football for one season.

a. Construct a scatter diagram for points/game and yards/game in the Excel file. Does there appear to be a linear relationship?

b. Use the *Regression* tool to develop a model for predicting points/game as a function of yards/game. Explain the statistical significance of the model and the R^2 value.

12. Using the data in the Excel file *Student Grades*, apply the Excel *Regression* tool using the midterm grade as the independent variable and the final exam grade as the dependent variable. Interpret all key regression results, hypothesis tests, and confidence intervals in the output.

[5]Modified from Charles T. Horngren, George Foster, and Srikant M. Datar, *Cost Accounting: A Managerial Emphasis*, 9th ed. (Englewood Cliffs, NJ: Prentice Hall, 1997): 371.

13. Using the Excel file *Weddings*, apply the *Regression* tool using the wedding cost as the dependent variable and attendance as the independent variable.

    a. What is the regression model?

    b. Interpret all key regression results, hypothesis tests, and confidence intervals in the output.

    c. If a couple is planning a wedding for 175 guests, how much should they budget?

14. Using the Excel file *Weddings*, apply the Excel *Regression* tool using the wedding cost as the dependent variable and the couple's income as the independent variable, only for those weddings paid for by the bride and groom. Interpret all key regression results, hypothesis tests, and confidence intervals in the output.

15. A deep-foundation engineering contractor has bid on a foundation system for a new building housing the world headquarters for a Fortune 500 company. A part of the project consists of installing 311 auger cast piles. The contractor was given bid information for cost-estimating purposes, which consisted of the estimated depth of each pile; however, actual drill footage of each pile could not be determined exactly until construction was performed. The Excel file *Pile Foundation* contains the estimates and actual pile lengths after the project was completed.

    a. Construct a scatter chart and interpret it.

    b. Develop a linear regression model to estimate the actual pile length as a function of the estimated pile lengths. What do you conclude?

Residual Analysis and Regression Assumptions

16. Use the results for Problem 10 (*Demographics*) to analyze the residuals to determine if the assumptions underlying the regression analysis are valid. In addition, use the standard residuals to determine if any possible outliers exist.

17. Use the results for Problem 11 (*National Football League*) to analyze the residuals to determine if the assumptions underlying the regression analysis are valid. In addition, use the standard residuals to determine if any possible outliers exist.

18. Use the results for Problem 12 (*Student Grades*) to analyze the residuals to determine if the assumptions underlying the regression analysis are valid. In addition, use the standard residuals to determine if any possible outliers exist.

19. Use the results for Problem 13 (*Weddings*) to analyze the residuals to determine if the assumptions underlying the regression analysis are valid. In addition, use the standard residuals to determine if any possible outliers exist.

Multiple Linear Regression

20. The Excel file *Concert Sales* provides data on sales dollars and the number of radio, TV, and newspaper ads promoting the concerts for a group of cities. Develop simple linear regression models for predicting sales as a function of the number of each type of ad. Compare these results to a multiple linear regression model using both independent variables. State each model and explain *R-Square*, *Significance F*, and *p*-values.

21. Using the data in the Excel file *Home Market Value*, develop a multiple linear regression model for estimating the market value as a function of both the age and size of the house. State the model and explain R^2, *Significance F*, and *p*-values.

22. Use the data in the Excel file *Cost of Living Adjustments* to find a multiple regression model to predict the salary as a function of all the adjusted cost of living rates. State the model and explain R^2, *Significance F*, and *p*-values.

23. Use the data in the Excel file *Job Satisfaction* to find a multiple regression model to predict the overall satisfaction as a function of all the other variables. State the model and explain R^2, *Significance F*, and *p*-values.

Building Good Regression Models

24. Using the data in the Excel file *Home Market Value*, find the best multiple linear regression model for estimating the market value as a function of both the age and size of the house. Predict the value of a house that is 30 years old and has 1,800 square feet, and one that is 5 years old and has 2,800 square feet.

25. Use the data in the Excel file *Cost of Living Adjustments* to find the best multiple regression model to predict the salary as a function of the adjusted cost of living rates. What would the comparable salary be for a city with the following adjustments: groceries: 4%; housing: 9%; utilities: 2%; transportation: 1%; and health care: 8%?

26. Use the data in the Excel file *Job Satisfaction* to find the best multiple regression model to predict the

overall satisfaction as a function of the other variables. What managerial implications does your result have?

27. The Excel file *Cereal Data* provides a variety of nutritional information about 67 cereals in a supermarket. Use regression analysis to find the best model that explains the relationship between calories and the other variables. Keep in mind the principle of parsimony.

28. The Excel file *Salary Data* provides information on current salary, beginning salary, previous experience (in months) when hired, and total years of education for a sample of 100 employees in a firm.

    a. Find the multiple regression model for predicting current salary as a function of the other variables.

    b. Find the best model for predicting current salary using the *t*-value criterion.

29. The Excel file *Credit Approval Decisions* provides information on credit history for a sample of banking customers. Use regression analysis to identify the best model for predicting the credit score as a function of the other numerical variables, using both the *p*-value and *t*-statistic criteria. How do the models compare? Which would you choose?

30. In the Excel file *Freshman College Data*, a school wants to predict the first-year retention rate using the high school metrics (ACT and SAT scores, GPA, and the percentage in the top 10% or 20% of their high school classes). Use the *Correlation* tool to find the correlation matrix for all variables. What can you say about potential multicollinearity? Then identify the best regression model for predicting the first-year retention rate using both the *p*-value and *t*-statistic criteria. How do models using the *p*-value and *t*-statistic criteria compare?

31. The Excel file *Major League Baseball* provides data for one season.

    a. Construct and examine the correlation matrix. Is multicollinearity a potential problem?

    b. Suggest an appropriate set of independent variables that predict the number of wins by examining the correlation matrix.

    c. Find the best multiple regression model for predicting the number of wins having only significant independent variables. How good is your model? Does it use the same variables you thought were appropriate in part b?

32. The Excel file *Golfing Statistics* provides data for a portion of the 2010 professional season for the top 25 golfers.

    a. Find the best multiple regression model for predicting earnings/event as a function of the other variables.

    b. Find the best multiple regression model for predicting average score as a function of GIR, putts/round, driving accuracy, and driving distance.

33. Use the *p*-value criterion to find the best model for predicting the number of points scored per game by football teams using the data in the Excel file *National Football League*. Does the model make logical sense?

34. The State of Ohio Department of Education has a mandated ninth-grade proficiency test that covers writing, reading, mathematics, citizenship (social studies), and science. The Excel file *Ohio Education Performance* provides data on success rates (defined as the percent of students passing) in school districts in the greater Cincinnati metropolitan area along with state averages.

    a. Suggest the best regression model to predict math success as a function of success in the other subjects by examining the correlation matrix; then run the *Regression* tool for this set of variables. (Note: "All" is not an academic subject!)

    b. Develop a multiple regression model to predict math success as a function of success in all other subjects using the systematic approach described in this chapter. Is multicollinearity a problem?

    c. Compare the models in parts a and b. Are they the same? Why or why not?

Regression with Categorical Independent Variables

35. A national homebuilder builds single-family homes and condominium-style townhouses. The Excel file *House Sales* provides information on the selling price, lot cost, type of home, and region of the country (M = Midwest, S = South) for closings during one month.

    a. Develop a multiple regression model for sales price as a function of lot cost and type of home without any interaction term.

    b. Determine if an interaction exists between lot cost and type of home and find the best model.

c. What is the predicted price for either a single-family home or a townhouse with a lot cost of $30,000?

36. For the Excel file *Auto Survey*:

a. Find the best regression model to predict miles/gallon as a function of vehicle age and mileage.

b. Using your result from part a, add the categorical variable purchased to the model. Does this change your result?

c. Determine whether any significant interaction exists between vehicle age and purchased variables.

37. For the Excel file *Job Satisfaction*, develop a regression model for overall satisfaction as a function of years of service and department that has the largest R^2. (Note that the categorical variable department has multiple levels and will require the use of multiple dummy variables, similar to Example 8.17.) Which department, if any, has the highest impact on satisfaction?

Regression Models with Nonlinear Terms

38. Cost functions are often nonlinear with volume because production facilities are often able to produce larger quantities at lower rates than smaller quantities.[6] Using the following data, apply simple linear regression and examine the residual plot. What do you conclude? Construct a scatter chart and use the *Trendline* feature to identify the best type of curvilinear trendline that maximizes R^2.

Units Produced	Cost
500	$12,500
1,000	$25,000
1,500	$32,500
2,000	$40,000
2,500	$45,000
3,000	$50,000

39. The Helicopter Division of Aerospatiale is studying assembly costs at its Marseilles plant.[7] Past data indicates the following number of labor hours per helicopter:

Helicopter Number	Labor Hours
1	2,000
2	1,400
3	1,238
4	1,142
5	1,075
6	1,029
7	985
8	957

reduction in labor hours over time is often called a "learning curve" phenomenon. Using these data, apply simple linear regression and examine the residual plot. What do you conclude? Construct a scatter chart and use the Excel *Trendline* feature to identify the best type of curvilinear trendline (but not going beyond a second-order polynomial) that maximizes R^2.

CASE: PERFORMANCE LAWN EQUIPMENT

In reviewing the data in the *Performance Lawn Equipment Database*, Elizabeth Burke noticed that defects received from suppliers have decreased (worksheet *Defects After Delivery*). Upon investigation, she learned that in 2014, PLE experienced some quality problems due to an increasing number of defects in materials received from suppliers. The company instituted an initiative in August 2015 to work with suppliers to reduce these defects, to more closely coordinate deliveries, and to improve materials quality through reengineering supplier production policies. Ms. Burke noted that the program appeared to reverse an

increasing trend in defects; she would like to predict what might have happened had the supplier initiative not been implemented and how the number of defects might further be reduced in the near future.

In meeting with PLE's human resources director, Ms. Burke also discovered a concern about the high rate of turnover in its field service staff. Senior managers have suggested that the department look closer at its recruiting policies, particularly to try to identify the characteristics of individuals that lead to greater retention. However, in a recent staff meeting, HR managers could not agree on these characteristics.

[6]Horngren, Foster, and Datar, *Cost Accounting: A Managerial Emphasis*, 9th ed.: 349.
[7]Horngren, Foster, and Datar, *Cost Accounting: A Managerial Emphasis*, 9th ed.: 349.

Some argued that years of education and grade point averages were good predictors. Others argued that hiring more mature applicants would lead to greater retention. To study these factors, the staff agreed to conduct a statistical study to determine the effect that years of education, college grade point average, and age when hired have on retention. A sample of 40 field service engineers hired ten years ago was selected to determine the influence of these variables on how long each individual stayed with the company. Data are compiled in the *Employee Retention* worksheet.

Finally, as part of its efforts to remain competitive, PLE tries to keep up with the latest in production technology. This is especially important in the highly competitive lawn mower line, where competitors can gain a real advantage if they develop more cost-effective means of production. The lawn mower division therefore spends a great deal of effort in testing new technology. When new

production technology is introduced, firms often experience learning, resulting in a gradual decrease in the time required to produce successive units. Generally, the rate of improvement declines until the production time levels off. One example is the production of a new design for lawn mower engines. To determine the time required to produce these engines, PLE produced 50 units on its production line; test results are given on the worksheet *Engines* in the database. Because PLE is continually developing new technology, understanding the rate of learning can be useful in estimating future production costs without having to run extensive prototype trials, and Ms. Burke would like a better handle on this.

Use trendlines and regression analysis to assist her in evaluating the data in these three worksheets and reaching useful conclusions. Summarize your work in a formal report with all appropriate results and analyses.

Forecasting Techniques

rawpixel/123RF

LEARNING OBJECTIVES   After studying this chapter, you will be able to:

- Explain how judgmental approaches are used for forecasting.
- List different types of statistical forecasting models.
- Apply moving average and exponential smoothing models to stationary time series.
- State three error metrics used for measuring forecast accuracy and explain the differences among them.
- Apply double exponential smoothing models to time series with a linear trend.

- Use Holt-Winters and regression models to forecast time series with seasonality.
- Apply Holt-Winters forecasting models to time series with both trend and seasonality.
- Identify the appropriate choice of forecasting model based on the characteristics of a time series.
- Explain how regression techniques can be used to forecast with explanatory or causal variables.

Managers require good forecasts of future events to make good decisions. For example, forecasts of interest rates, energy prices, and other economic indicators are needed for financial planning; sales forecasts are needed to plan production and workforce capacity; and forecasts of trends in demographics, consumer behavior, and technological innovation are needed for long-term strategic planning. The government also invests significant resources on predicting short-run U.S. business performance using the index of leading economic indicators (LEI). This index focuses on the performance of individual businesses, which often is highly correlated with the performance of the overall economy and is used to forecast economic trends for the nation as a whole. In this chapter, we introduce some common methods and approaches to forecasting, including both qualitative and quantitative techniques.

Business analysts may choose from a wide range of forecasting techniques to support decision making. Selecting the appropriate method depends on the characteristics of the forecasting problem, such as the time horizon of the variable being forecast, as well as available information on which the forecast will be based. Three major categories of forecasting approaches are *qualitative and judgmental techniques*, *statistical time-series models*, and *explanatory/causal methods*. In this chapter, we introduce a variety of forecasting techniques and use basic Excel tools and linear regression to implement them on spreadsheets.

ANALYTICS IN PRACTICE: Forecasting Call-Center Demand at L.L.Bean[1]

Many of you are familiar with L.L.Bean, a retailer of high-quality outdoor gear. A large percentage of the company's sales are generated through orders to its call center (the call center can account for over 70% of the total sales volume). Calls to the L.L.Bean call center are classified into two types: telemarketing (TM) and telephone inquiry (TI). TM calls are those made for placing an order, whereas TI calls involve customer inquiries, such as order status or order problems. TM calls and TI calls differ in duration and volume. The annual call volume for TM calls is much higher than that for TI calls, but the duration of a TI call is generally much longer than the duration of a TM call.

Accurately forecasting the demand of TM and TI calls is very important to L.L.Bean to reduce costs. Accurate forecasts allow for properly planning the number of agents to have on hand at any point in time. Too few agents result in lost sales, loss of customer loyalty, excessive queue times, and increased phone charges. Too many agents obviously result in unnecessary labor costs.

L.L.Bean developed analytical forecasting models for both TM and TI calls. These models took into account historical trends, seasonal factors, and external explanatory variables such as holidays and catalog mailings. The estimated benefit from better precision from the two forecasting models was approximately $300,000 per year.

[1]Andrews, B.H., and S. M. Cunningham, "L.L. Bean Improves Call-Center Forecasting," *Interfaces*, Vol. 25, No. 6, pp.1-13, November-December, 1995.

Qualitative and Judgmental Forecasting

Qualitative and judgmental techniques rely on experience and intuition; they are necessary when historical data are not available or when the decision maker needs to forecast far into the future. For example, a forecast of when the next generation of a microprocessor will be available and what capabilities it might have will depend greatly on the opinions and expertise of individuals who understand the technology. Another use of judgmental methods is to incorporate nonquantitative information, such as the impact of government regulations or competitor behavior, in a quantitative forecast. Judgmental techniques range from such simple methods as a manager's opinion or a group-based jury of executive opinion to more structured approaches such as historical analogy and the Delphi method.

Historical Analogy

One judgmental approach is **historical analogy**, in which a forecast is obtained through a comparative analysis with a previous situation. For example, if a new product is being introduced, the response of consumers to marketing campaigns to similar, previous products can be used as a basis to predict how the new marketing campaign might fare. Of course, temporal changes or other unique factors might not be fully considered in such an approach. However, a great deal of insight can often be gained through an analysis of past experiences.

EXAMPLE 9.1 **Predicting the Price of Oil**

In early 1998, the price of oil was about $22 a barrel. However, in mid-1998, the price of a barrel of oil dropped to around $11. The reasons for this price drop included an oversupply of oil from new production in the Caspian Sea region, high production in non-OPEC regions, and lower-than-normal demand. In similar circumstances in the past, OPEC would meet and take action to raise the price of

oil. Thus, from historical analogy, we might forecast a rise in the price of oil. OPEC members did, in fact, meet in mid-1998 and agreed to cut their production, but nobody believed that they would actually cooperate effectively, and the price continued to drop for a time. Subsequently, in 2000, the price of oil rose dramatically, falling again in late 2001.

Analogies often provide good forecasts, but you need to be careful to recognize new or different circumstances. Another analogy is international conflict relative to the price of oil. Should war break out, the price would be expected to rise, analogous to what it has done in the past.

The Delphi Method

A popular judgmental forecasting approach, called the **Delphi method**, uses a panel of experts, whose identities are typically kept confidential from one another, to respond to a sequence of questionnaires. After each round of responses, individual opinions, edited to ensure anonymity, are shared, allowing each to see what the other experts think. Seeing other experts' opinions helps to reinforce those in agreement and to influence those who did not agree to possibly consider other factors. In the next round, the experts revise their estimates, and the process is repeated, usually for no more than two or three rounds. The Delphi method promotes unbiased exchanges of ideas and discussion and usually results in some convergence of opinion. It is one of the better approaches to forecasting long-range trends and impacts.

Indicators and Indexes

Indicators and indexes generally play an important role in developing judgmental forecasts. **Indicators** are measures that are believed to influence the behavior of a variable we wish to forecast. By monitoring changes in indicators, we expect to gain insight about the future behavior of the variable to help forecast the future.

EXAMPLE 9.2   Economic Indicators

One variable that is important to the nation's economy is the gross domestic product (GDP), which is a measure of the value of all goods and services produced in the United States. Despite its shortcomings (for instance, unpaid work such as housekeeping and child care is not measured; production of poor-quality output inflates the measure, as does work expended on corrective action), it is a practical and useful measure of economic performance. Like most time series, the GDP rises and falls in a cyclical fashion. Predicting future trends in the GDP is often done by analyzing *leading indicators*—series that tend to rise and fall for some

predictable length of time prior to the peaks and valleys of the GDP. One example of a leading indicator is the formation of business enterprises; if the rate of new businesses grows, we would expect the GDP to increase in the near future. Additional examples of leading indicators are the percent change in the money supply (M1) and net change in business loans. Other indicators, called *lagging indicators*, tend to have peaks and valleys that follow those of the GDP. Some lagging indicators are the Consumer Price Index, prime rate, business investment expenditures, and inventories on hand. The GDP can be used to predict future trends in these indicators.

Indicators are often combined quantitatively into an **index**, a single measure that weights multiple indicators, thus providing a measure of overall expectation. For example, financial analysts use the Dow Jones Industrial Average as an index of general stock market performance. Indexes do not provide a complete forecast but rather a better picture of direction of change and thus play an important role in judgmental forecasting.

EXAMPLE 9.3   Leading Economic Indicators

The Department of Commerce initiated an index of leading indicators to help predict future economic performance. Components of the index include the following:

- average weekly hours, manufacturing
- average weekly initial claims, unemployment insurance
- new orders, consumer goods, materials
- vendor performance—slower deliveries
- new orders, nondefense capital goods
- building permits, private housing
- stock prices, 500 common stocks (Standard & Poor)
- money supply

- interest rate spread
- index of consumer expectations (University of Michigan)

Business Conditions Digest included more than 100 time series in seven economic areas. This publication was discontinued in March 1990, but information related to the index of leading indicators was continued in *Survey of Current Business*. In December 1995, the U.S. Department of Commerce sold this data source to the Conference Board, which now markets the information under the title *Business Cycle Indicators*; information can be obtained at its Web site (www.conference-board.org). The site includes excellent current information about the calculation of the index as well as its current components.

CHECK YOUR UNDERSTANDING

1. Explain how historical analogy and the Delphi method are used for judgmental forecasting.

2. Define an indicator and index. How are they used in forecasting?

Statistical Forecasting Models

Statistical time-series models find greater applicability for short-range forecasting problems. A **time series** is a stream of historical data, such as weekly sales. We characterize the values of a time series over T periods as A_t, $t = 1, 2, \ldots, T$. Time-series models assume that whatever forces have influenced sales in the recent past will continue into the near future; thus, forecasts are developed by extrapolating these data into the future. Time series generally have one or more of the following components: random behavior, trends, seasonal effects, and cyclical effects. Time series that do not have trend, seasonal, or cyclical effects but are relatively constant and exhibit only random behavior are called **stationary time series**.

Many forecasts are based on analysis of historical time-series data and are predicated on the assumption that the future is an extrapolation of the past. A **trend** is a gradual upward or downward movement of a time series over time.

EXAMPLE 9.4	**Identifying Trends in a Time Series**

Figure 9.1 shows a chart of total energy consumption from the data in the Excel file *Energy Production & Consumption*. This time series shows an upward trend. However, we see that energy consumption was rising quite rapidly in a linear fashion during the 1960s, then leveled off for a while and began increasing at a slower rate through the 1980s and 1990s. At the end of the time series, we actually see a slight downward trend. This time series, then, is composed of several different short trends.

Time series may also exhibit short-term *seasonal effects* well as longer-term *cyclical effects*, or nonlinear trends. A **seasonal effect** is one that repeats at fixed intervals of time, typically a year, month, week, or day. At a neighborhood grocery store, for instance, short-term seasonal patterns may occur over a week, with the heaviest volume of customers on weekends; seasonal patterns may also be evident during the course of a day, with higher volumes in the mornings and late afternoons. Figure 9.2 shows seasonal changes in natural gas usage for a homeowner over the course of a year (Excel file *Gas & Electric*). **Cyclical effects** describe ups and downs over a much longer time frame, such as several years. Figure 9.3 shows a chart of the data in the Excel file *Federal Funds Rates*. We see some evidence of long-term cycles in the time series driven by economic factors, such as periods of inflation and recession.

Although visual inspection of a time series to identify trend, seasonal, or cyclical effects may work in a naïve fashion, such unscientific approaches may be a bit unsettling to a manager making important decisions. Subtle effects and interactions of seasonal and cyclical factors may not be evident from simple visual extrapolation of data. Statistical methods, which involve more formal analyses of time series, are invaluable in developing good forecasts. A variety of statistically based forecasting methods for time series

► **Figure 9.1**

Total Energy Consumption Time Series

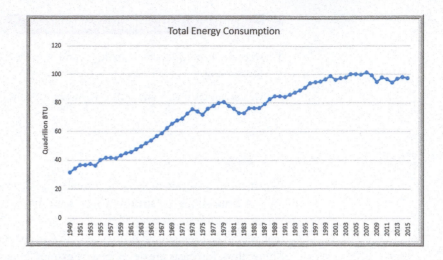

► **Figure 9.2**

Seasonal Effects in Natural Gas Usage

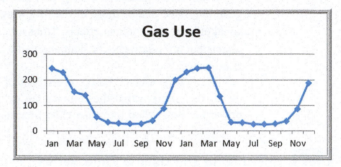

► **Figure 9.3**

Cyclical Effects in Federal Funds Rates

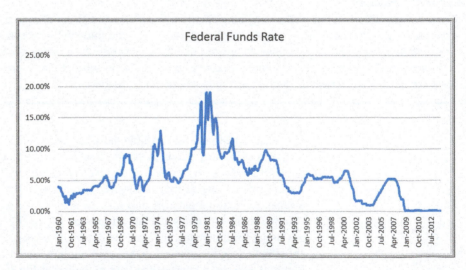

are commonly used. Among the most popular are *moving average methods*, *exponential smoothing*, and *regression analysis*. These can be implemented very easily on a spreadsheet using basic functions and *Data Analysis* tools available in Microsoft Excel, as well as with more powerful software such as *Analytic Solver* (see the online supplement described in the Technology Help section). Moving average and exponential smoothing models work best for time series that do not exhibit trends or seasonal factors. For time series that involve trends and/or seasonal factors, other techniques have been developed. These include double moving average and exponential smoothing models, seasonal additive and multiplicative models, and Holt-Winters additive and multiplicative models.

CHECK YOUR UNDERSTANDING

1. What is a time series?
2. What components often comprise a time series?
3. Explain the difference between seasonal and cyclical effects in a time series.

Forecasting Models for Stationary Time Series

Two simple approaches that are useful over short time periods when trend, seasonal, or cyclical effects are not significant are moving average and exponential smoothing models.

Moving Average Models

The **simple moving average** method is a smoothing method based on the idea of averaging random fluctuations in the time series to identify the underlying direction in which the time series is changing. Because the moving average method assumes that future observations will be similar to the recent past, it is most useful as a short-range forecasting method. Although this method is very simple, it has proven to be quite useful in stable environments, such as inventory management, in which it is necessary to develop forecasts for a large number of items.

Specifically, the simple moving average forecast for the next period is computed as the average of the most recent k observations. Specifically, let A_t represent the observed value in period t, and F_t the forecast for period t. Then the forecast for the period $t + 1$ is computed as

$$F_{t+1} = \frac{A_t + A_{t-1} + \ldots + A_{t-k+1}}{k} \tag{9.1}$$

The value of k is somewhat arbitrary, although its choice affects the accuracy of the forecast. The larger the value of k, the more the current forecast is dependent on older data, and the forecast will not react as quickly to fluctuations in the time series. The smaller the value of k, the quicker the forecast responds to changes in the time series. Also, when k is larger, extreme values have less effect on the forecasts. (In the next section, we discuss how to select k by examining errors associated with different values.)

EXAMPLE 9.5   Moving Average Forecasting

The Excel file *Tablet Computer Sales* contains data for the number of units sold for the past 17 weeks. Figure 9.4 shows a chart of these data. The time series appears to be relatively stable, without trend, seasonal, or cyclical effects; thus, a moving average model would be appropriate. The observed values for weeks 15, 16, and 17 are $A_{15} = 82$, $A_{16} = 71$, and $A_{17} = 50$. Setting $k = 3$, and using equation (9.1), the three-period moving average forecast for week 18 is

$$F_{18} = \frac{(A_{17} + A_{16} + A_{15})}{3} = \frac{82 + 71 + 50}{3} = 67.67$$

Moving average forecasts can be generated easily on a spreadsheet. Figure 9.5 shows the computations for a three-period moving average forecast of tablet computer sales. You should verify these by hand to confirm your understanding. Figure 9.6 shows a chart that contrasts the data with the forecasted values. Although the forecast tracks the actual data fairly well, notice that the forecast overestimates the actual value when the trend is down, while it underestimates the actual value when the trend is up. This is because it uses past data, and thus lags the changes in the data.

▶ **Figure 9.4**

*Chart of Weekly Tablet
Computer Sales*

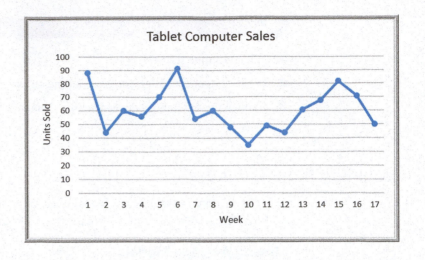

▶ **Figure 9.5**

*Excel Implementation of
Moving Average Forecast*

	A	B	C	D	E	F
1	**Tablet Computer Sales**					
2			**Moving Average**			
3	**Week**	**Units Sold**	**Forecast**			
4	1	88				
5	2	44				
6	3	60				
7	4	56	64.00			
8	5	70	53.33			
9	6	91	62.00			
10	7	54	72.33			
11	8	60	71.67			
12	9	48	68.33			
13	10	35	54.00			
14	11	49	47.67			
15	12	44	44.00			
16	13	61	42.67			
17	14	68	51.33			
18	15	82	57.67			
19	16	71	70.33			
20	17	50	73.67			
21	18		67.67			
22						

Forecast for week 4
=AVERAGE(B4:B6)

Forecast for week 18
=AVERAGE(B18:B20)

▶ **Figure 9.6**

*Chart of Units Sold and
Moving Average Forecast*

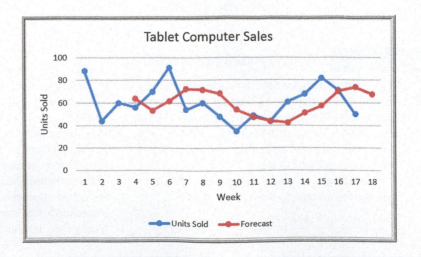

Moving average forecasts can also be obtained from Excel's *Data Analysis* options.

EXAMPLE 9.6 **Using Excel's *Moving Average* Tool**

For the *Tablet Computer Sales* Excel file, select *Data Analysis* and then *Moving Average* from the *Analysis* group. Excel displays the dialog box shown in Figure 9.7. You need to enter the *Input Range* of the data, the *Interval* (the value of k), and the first cell of the *Output Range*. To align the actual data with the forecasted values in the worksheet, select the first cell of the *Output Range* to be one row below the first row of the time series. (See Figure 9.8. The first value of the time series starts in cell B4, so the output range is chosen to start in cell C5.). You may also obtain a chart of the data and the moving averages, as well as a column of standard errors, by checking

the appropriate boxes. However, we *do not recommend* using the chart or error options because the forecasts generated by this tool are not properly aligned with the data (the forecast value aligned with a particular data point represents the forecast for the *next* month) and, thus, can be misleading. Rather, we recommend that you generate your own chart, as we did in Figure 9.6. Figure 9.8 shows the results produced by the *Moving Average* tool (with some customization of the formatting). Note that the forecast for week 18 is aligned with the actual value for week 17 on the chart. Compare this to Figure 9.6 and you can see the difference.

Error Metrics and Forecast Accuracy

The quality of a forecast depends on how accurate it is in predicting future values of a time series. In the simple moving average model, different values for k will produce different forecasts. How do we know which is the best value for k? The error, or residual, in a forecast is the difference between the forecast and the actual value of the time series (once it is known). In Figure 9.6, the forecast error is simply the vertical distance between the forecast and the data for the same time period.

To analyze the effectiveness of different forecasting models, we can define *error metrics*, which compare quantitatively the forecast with the actual observations. Three metrics that are commonly used are the *mean absolute deviation*, *mean square error*, and *mean absolute percentage error*. The **mean absolute deviation (MAD)** is the absolute difference between the actual value and the forecast, averaged over a range of forecasted values:

$$\text{MAD} = \frac{\sum_{t=1}^{n} |A_t - F_t|}{n} \tag{9.2}$$

where A_t is the actual value of the time series at time t, F_t is the forecast value for time t, and n is the number of forecast values (*not* the number of data points since we do not have

▶ Figure 9.7

Excel Moving Average *Tool Dialog*

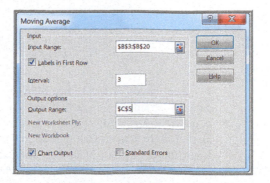

▶ **Figure 9.8**

Results of Excel Moving
Average *Tool (Note misalignment of forecasts with actual
sales in the chart.)*

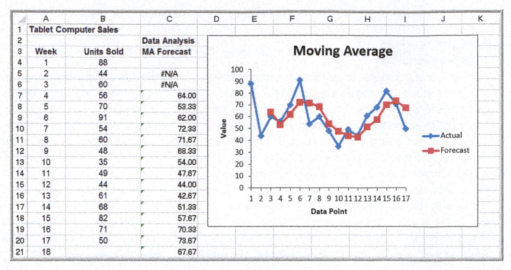

	A	B	C	D	E	F	G	H	I	J	K
1	Tablet Computer Sales										
2			Data Analysis								
3	Week	Units Sold	MA Forecast								
4	1	88									
5	2	44	#N/A								
6	3	60	#N/A								
7	4	56	64.00								
8	5	70	53.33								
9	6	91	62.00								
10	7	54	72.33								
11	8	60	71.67								
12	9	48	68.33								
13	10	35	54.00								
14	11	49	47.67								
15	12	44	44.00								
16	13	61	42.67								
17	14	68	51.33								
18	15	82	57.67								
19	16	71	70.33								
20	17	50	73.67								
21	18		67.67								

a forecast value associated with the first k data points). MAD provides a robust measure of error and is less affected by extreme observations.

Mean square error (MSE) is probably the most commonly used error metric. It penalizes larger errors because squaring larger numbers has a greater impact than squaring smaller numbers. The formula for MSE is

$$\text{MSE} = \frac{\sum_{t=1}^{n}(A_t - F_t)^2}{n} \tag{9.3}$$

Again, n represents the number of forecast values used in computing the average. Sometimes the square root of MSE, called the **root mean square error (RMSE)**, is used:

$$\text{RMSE} = \sqrt{\frac{\sum_{t=1}^{n}(A_t - F_t)^2}{n}} \tag{9.4}$$

Note that unlike MSE, RMSE is expressed in the same units as the data (similar to the difference between a standard deviation and a variance), allowing for more practical comparisons.

A fourth commonly used metric is **mean absolute percentage error (MAPE)**. MAPE is the average of absolute errors divided by actual observation values.

$$\text{MAPE} = \frac{\sum_{t=1}^{n}\left|\dfrac{A_t - F_t}{A_t}\right|}{n} \times 100 \tag{9.5}$$

The values of MAD and MSE depend on the measurement scale of the time-series data. For example, forecasting profit in the range of millions of dollars would result in very large MAD and MSE values, even for very accurate forecasting models. On the other hand, market share is measured in proportions; therefore, even bad forecasting models will have small values of MAD and MSE. Thus, these measures have no meaning except in comparison with other models used to forecast the same data. Generally, MAD is less affected by extreme observations and is preferable to MSE if such extreme observations are considered rare events with no special meaning. MAPE is different in that the measurement scale is eliminated by dividing the absolute error by the time-series data value. This allows a better relative comparison. Although these comments provide some guidelines, there is no universal agreement on which measure is best.

EXAMPLE 9.7   **Using Error Metrics to Compare Moving Average Forecasts**

The metrics we have described can be used to compare different moving average forecasts for the *Tablet Computer Sales* data. A spreadsheet that shows the forecasts as well as the calculations of the error metrics for two-, three-, and four-period moving average models is given in Figure 9.9. The error is the difference between the actual value of the units sold and the forecast. To compute MAD, we first compute the absolute values of the errors and then average them. For MSE, we compute the squared errors and then find the average (we may easily compute RMSE if desired). For MAPE, we find the absolute values of the errors divided by the actual observation multiplied by 100 and then average them. The results suggest that a two-period moving average model provides the best forecast among these alternatives because the error metrics are all smaller than for the other models.

Exponential Smoothing Models

A versatile, yet highly effective, approach for short-range forecasting is **simple exponential smoothing**. The basic simple exponential smoothing model is

$$F_{t+1} = (1 - \alpha)F_t + \alpha A_t$$
$$= F_t + \alpha(A_t - F_t) \tag{9.6}$$

where F_{t+1} is the forecast for time period $t + 1$, F_t is the forecast for period t, A_t is the observed value in period t, and α is a constant between 0 and 1 called the **smoothing constant**. To begin, set F_1 and F_2 equal to the actual observation in period 1, A_1.

Using the two forms of the forecast equation just given, we can interpret the simple exponential smoothing model in two ways. In the first model, the forecast for the next period, F_{t+1}, is a weighted average of the forecast made for period t, F_t, and the actual

	A	B	C	D	E	F	G	H	I	J	K	L	M	N	O	P	Q
1	Tablet Computer Sales																
2			k = 2					k = 3					k = 4				
3	Week	Units Sold	Forecast	Error	Absolute Deviation	Squared Error	Absolute % Error	Forecast	Error	Absolute Deviation	Squared Error	Absolute % Error	Forecast	Error	Absolute Deviation	Squared Error	Absolute % Error
4	1	88															
5	2	44															
6	3	60	66.00	-6.00	6.00	36.00	10.00										
7	4	56	52.00	4.00	4.00	16.00	7.14	64.00	-8.00	8.00	64.00	14.29					
8	5	70	58.00	12.00	12.00	144.00	17.14	53.33	16.67	16.67	277.78	23.81	62.00	8.00	8.00	64.00	11.43
9	6	91	63.00	28.00	28.00	784.00	30.77	62.00	29.00	29.00	841.00	31.87	57.50	33.50	33.50	1122.25	36.81
10	7	54	80.50	-26.50	26.50	702.25	49.07	72.33	-18.33	18.33	336.11	33.95	69.25	-15.25	15.25	232.56	28.24
11	8	60	72.50	-12.50	12.50	156.25	20.83	71.67	-11.67	11.67	136.11	19.44	67.75	-7.75	7.75	60.06	12.92
12	9	48	57.00	-9.00	9.00	81.00	18.75	68.33	-20.33	20.33	413.44	42.36	68.75	-20.75	20.75	430.56	43.23
13	10	35	54.00	-19.00	19.00	361.00	54.29	54.00	-19.00	19.00	361.00	54.29	63.25	-28.25	28.25	798.06	80.71
14	11	49	41.50	7.50	7.50	56.25	15.31	47.67	1.33	1.33	1.78	2.72	49.25	-0.25	0.25	0.06	0.51
15	12	44	42.00	2.00	2.00	4.00	4.55	44.00	0.00	0.00	0.00	0.00	48.00	-4.00	4.00	16.00	9.09
16	13	61	46.50	14.50	14.50	210.25	23.77	42.67	18.33	18.33	336.11	30.05	44.00	17.00	17.00	289.00	27.87
17	14	68	52.50	15.50	15.50	240.25	22.79	51.33	16.67	16.67	277.78	24.51	47.25	20.75	20.75	430.56	30.51
18	15	82	64.50	17.50	17.50	306.25	21.34	57.67	24.33	24.33	592.11	29.67	55.50	26.50	26.50	702.25	32.32
19	16	71	75.00	-4.00	4.00	16.00	5.63	70.33	0.67	0.67	0.44	0.94	63.75	7.25	7.25	52.56	10.21
20	17	50	76.50	-26.50	26.50	702.25	53.00	73.67	-23.67	23.67	560.11	47.33	70.50	-20.50	20.50	420.25	41.00
21	18		60.50		13.63	254.38	23.63	67.67		14.86	299.84	25.37	67.75		16.13	355.25	28.07
22					MAD	MSE	MAPE			MAD	MSE	MAPE			MAD	MSE	MAPE

▲ **Figure 9.9**

Error Metrics Alternative Moving Average Forecasts

observation in period t, A_t. The second form of the model, obtained by simply rearranging terms, states that the forecast for the next period, F_{t+1}, equals the forecast for the last period, F_t, plus a fraction α of the forecast error made in period t, $A_t - F_t$. Thus, to make a forecast once we have selected the smoothing constant, we need to know only the previous forecast and the actual value. By repeated substitution for F_t in the equation, it is easy to demonstrate that F_{t+1} is a decreasingly weighted average of all past time-series data. Thus, the forecast actually reflects *all* the data, provided that α is strictly between 0 and 1.

EXAMPLE 9.8   **Using Exponential Smoothing to Forecast *Tablet Computer Sales***

For the *Tablet Computer Sales* data, the forecast for week 2 is 88, the actual observation for week 1. Suppose we choose $\alpha = 0.7$; then the forecast for week 3 would be

Week 3 Forecast = $(1 - 0.7)(88) + (0.7)(44) = 57.2$

The actual observation for week 3 is 60; thus, the forecast for week 4 would be

Week 4 Forecast = $(1 - 0.7)(57.2) + (0.7)(60) = 59.16$

Because the simple exponential smoothing model requires only the previous forecast and the current time-series value, it is very easy to calculate; thus, it is highly suitable for environments such as inventory systems, where many forecasts must be made. The smoothing constant α is usually chosen by experimentation in the same manner as choosing the number of periods to use in the moving average model. Different values of α affect how quickly the model responds to changes in the time series. For instance, a value of $\alpha = 0$ would simply repeat last period's forecast, whereas $\alpha = 1$ would forecast the last period's actual demand. The closer α is to 1, the quicker the model responds to changes in the time series, because it puts more weight on the actual current observation than on the forecast. Likewise, the closer α is to 0, the more weight is put on the prior forecast, so the model would respond to changes more slowly.

EXAMPLE 9.9   **Finding the Best Exponential Smoothing Model for *Tablet Computer Sales***

An Excel spreadsheet for evaluating exponential smoothing models for the *Tablet Computer Sales* data using values of α between 0.1 and 0.9 is shown in Figure 9.10. Note that in computing the error measures, the first row is not included

because we do not have a forecast for the first period, week 1. A smoothing constant of $\alpha = 0.6$ provides the lowest error for all three metrics.

Excel has a *Data Analysis* tool for exponential smoothing. However, as opposed to the *Moving Average* tool, the chart generated by the *Exponential Smoothing* tool does correctly align the forecasts with the actual data. We will see this in the next example.

EXAMPLE 9.10   **Using Excel's Exponential Smoothing Tool**

In the *Table Computer Sales* example, from the *Analysis* group, select *Data Analysis* and then *Exponential Smoothing*. In the dialog (Figure 9.11), as in the *Moving Average* dialog, you must enter the *Input Range* of the time-series data, the *Damping Factor*, $(1 - \alpha)$—not the smoothing constant as we have defined it—and the first cell of the *Output*

Range, which should be adjacent to the first data point. You also have options for labels, to chart output, and to obtain standard errors. The results are shown in Figure 9.12. You can see that the exponential smoothing model follows the pattern of the data quite closely, although it tends to lag with an increasing trend in the data.

▶ **Figure 9.10**

*Exponential Smooth-
ing Forecasts for* Tablet
Computer Sales

	A	B	C	D	E	F	G	H	I	J	K
1	Tablet Computer Sales										
2							Smoothing Constant				
3	Week	Units Sold	0.10	0.20	0.30	0.40	0.50	0.60	0.70	0.80	0.90
4	1	88	88.00	88.00	88.00	88.00	88.00	88.00	88.00	88.00	88.00
5	2	44	88.00	88.00	88.00	88.00	88.00	88.00	88.00	88.00	88.00
6	3	60	83.60	79.20	74.80	70.40	66.00	61.60	57.20	52.80	48.40
7	4	56	81.24	75.36	70.36	66.24	63.00	60.64	59.16	58.56	58.84
8	5	70	78.72	71.49	66.05	62.14	59.50	57.86	56.95	56.51	56.28
9	6	91	77.84	71.19	67.24	65.29	64.75	65.14	66.08	67.30	68.63
10	7	54	79.16	75.15	74.37	75.57	77.88	80.66	83.53	86.26	88.76
11	8	60	76.64	70.92	68.26	66.94	65.94	64.66	62.86	60.45	57.48
12	9	48	74.98	68.74	65.78	64.17	62.97	61.87	60.86	60.09	59.75
13	10	35	72.28	64.59	60.45	57.70	55.48	53.55	51.86	50.42	49.17
14	11	49	68.55	58.67	52.81	48.62	45.24	42.42	40.06	38.08	36.42
15	12	44	66.60	56.74	51.67	48.77	47.12	46.37	46.32	46.82	47.74
16	13	61	64.34	54.19	49.37	46.86	45.56	44.95	44.70	44.56	44.37
17	14	68	64.00	55.55	52.86	52.52	53.28	54.58	56.11	57.71	59.34
18	15	82	64.40	58.04	57.40	58.71	60.64	62.63	64.43	65.94	67.13
19	16	71	66.16	62.83	64.78	68.03	71.32	74.25	76.73	78.79	80.51
20	17	50	66.65	64.47	66.65	69.22	71.16	72.30	72.72	72.56	71.95
21	18		64.98	61.57	61.65	61.53	60.58	58.92	56.82	54.51	52.20
22		MAD	19.33	17.16	16.15	15.36	14.93	14.71	14.72	14.88	15.36
23		MSE	496.07	390.84	359.18	346.56	340.77	338.41	339.03	343.32	352.38
24		MAPE	38.28%	32.71%	30.12%	28.36%	27.54%	27.09%	27.09%	27.38%	28.23%

▶ **Figure 9.11**

 Exponential Smoothing *Tool
Dialog*

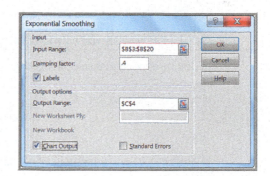

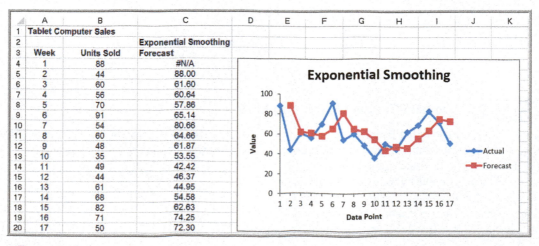

▲ **Figure 9.12**

Excel Exponential Smoothing Forecasts for α = 0.6

CHECK YOUR UNDERSTANDING

1. Explain how the simple moving average method works.

2. What are the three different types of error metrics that can be used to evaluate forecast accuracy?

3. Explain how simple exponential smoothing works.

4. Describe how to use the Excel *Data Analysis* tools for stationary time series.

Forecasting Models for Time Series with a Linear Trend

For time series with a linear trend but no significant seasonal components, **double exponential smoothing** models or regression-based models are more appropriate. These are based on the linear trend equation

$$F_{t+k} = a_t + b_t k \tag{9.7}$$

That is, the forecast for k periods into the future from period t is a function of a base value a_t, also known as the *level*, and a *trend*, or slope, b_t.

Double Exponential Smoothing

In double exponential smoothing, the estimates of a_t and b_t are obtained from the following equations:

$$a_t = \alpha A_t + (1 - \alpha)(a_{t-1} + b_{t-1})$$
$$b_t = \beta(a_t - a_{t-1}) + (1 - \beta)b_{t-1} \tag{9.8}$$

In essence, we are smoothing both parameters of the linear trend model. Initial values are chosen for a_1 as A_1 and b_1 as $A_2 - A_1$. Equations (9.8) must then be used to compute a_t and b_t for the entire time series to be able to generate forecasts into the future. As with simple exponential smoothing, we are free to choose the values of α and β. The forecast for k periods beyond the last period (period T) is

$$F_{T+k} = a_T + b_T(k) \tag{9.9}$$

| EXAMPLE 9.11 | **Double Exponential Smoothing** |

Figure 9.13 shows a portion of the Excel file *Coal Production*, which provides data on total tons produced from 1960 through 2011. The data appear to follow a linear trend. We will apply double exponential smoothing on just the first ten years of the data to illustrate the process.

Figure 9.14 shows a spreadsheet implementation of this using $\alpha = 0.6$ and $\beta = 0.4$ (Excel file *Coal Production*

Double Exponential Smoothing). We first initialize the values of a_1 and b_1:

$$a_1 = A_1 = 434,329,000$$
$$b_1 = A_2 - A_1 = 420,423,000 - 434,329,000$$
$$= -13,906,000$$

If we used equation (9.7) to forecast for period 2, we would obtain

$$F_2 = F_{1+1} = a_1 + b_1(1) = 434{,}329{,}000 + (-13{,}906{,}000)(1)$$
$$= 420{,}423{,}000$$

Note that this is the same as the actual value in period 2 because we used A_2 to calculate b_1, so we cannot make a true forecast until period 3. Continuing, we calculate a_2 and b_2 using formulas (9.8):

$$a_2 = \alpha A_2 + (1 - \alpha)(a_1 + b_1) = 0.6(420{,}423{,}000)$$
$$+ 0.4(434{,}329{,}000 - 13{,}906{,}000) = 420{,}423{,}000$$
$$b_2 = \beta(a_2 - a_1) + (1 - \beta)b_1$$
$$= 0.4(420{,}423{,}000 - 434{,}329{,}000)$$
$$+ 0.6(-13{,}906{,}000) = -13{,}906{,}000$$

Then the forecast for period 3 is

$$F_3 = F_{2+1} = a_2 + b_2(1) = 420{,}423{,}000$$
$$+ (-13{,}906{,}000)(1) = 406{,}517{,}000$$

The other forecasts are calculated in a similar manner and are shown on the spreadsheet. The value of MAD for these forecasts is 23,471,063. We can experiment with different values of α and β to find a better fit.

Using equation (9.9), the forecasting model for k periods beyond period 10 is

$$F_{10+k} = a_{10} + b_{10}(k) = 576{,}753{,}344 + 11{,}962{,}665(k)$$

For example, the forecast for period 11 is $F_{11} = 576{,}753{,}344 + 11{,}962{,}665(1) = 588{,}716{,}009$, and the forecast for period 12 would be

$$F_{12} = a_{10} + b_{10}(2) = 576{,}753{,}344 + 11{,}962{,}665(2)$$
$$= 600{,}678{,}674$$

▶ **Figure 9.13**

Portion of Excel File Coal Production

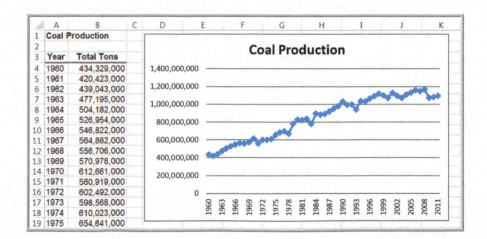

▼ **Figure 9.14**

Excel Implementation of Double Exponential Smoothing

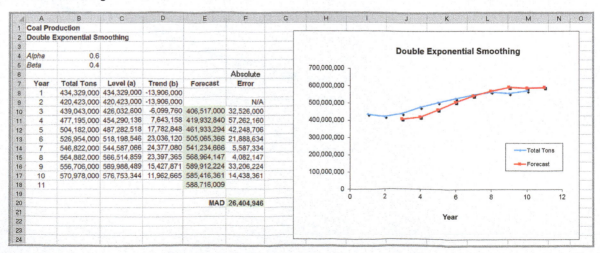

Regression-Based Forecasting for Time Series with a Linear Trend

Equation 9.7 looks similar to the equation for simple linear regression. We introduced regression in the previous chapter as a means of developing relationships between a dependent and independent variables. Simple linear regression can be applied to forecasting using time as the independent variable.

EXAMPLE 9.12 **Forecasting Using Trendlines**

For the data in the Excel file *Coal Production*, a linear trendline, shown in Figure 9.15, gives an R^2 value of 0.95 (the fitted model assumes that the years are numbered 1 through 52, not as actual dates). The model is

Tons = 438,819,885.29 + 15,413,536.97 × Year

Thus, a forecast for 2012 (year 53) would be

Tons = 438,819,885.29 + 15,413,536.97 × (53)
= 1,255,737,345

Note, however, that the linear model does not adequately predict the recent drop in production after 2008.

In Chapter 8, we noted that an important assumption for using regression analysis is the lack of autocorrelation among the data. When autocorrelation is present, successive observations are correlated with one another; for example, large observations tend to follow other large observations, and small observations also tend to follow one another. This can often be seen by examining the residual plot when the data are ordered by time. Figure 9.16 shows the time-ordered residual plot from the Excel *Regression* tool for the coal production example. The residuals do not appear to be random; rather, successive observations seem to be related to one another. This suggests autocorrelation, indicating that other approaches, called *autoregressive models*, are more appropriate. However, these are more advanced than the level of this book and are not discussed here.

▶ **Figure 9.15**

Trendline-Based Forecast for Coal Production Data

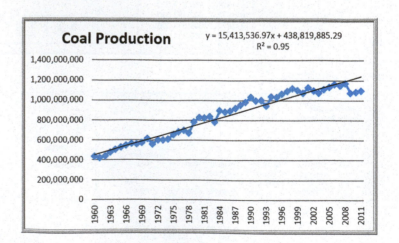

▶ Figure 9.16

Residual Plot for Linear Regression Forecasting Model

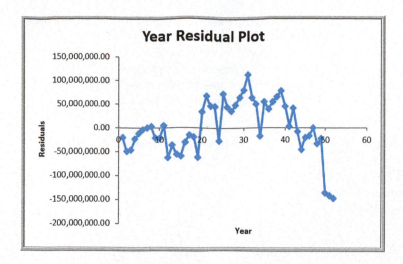

1. What techniques are used to forecast time series with a linear trend?

2. Explain the process for double exponential smoothing.

3. How can trendlines be used in forecasting?

Forecasting Time Series with Seasonality

Quite often, time-series data exhibit seasonality. A "season" can be a year, a quarter, a month, or even a week, depending on the application. Most often, it is a year. We saw an example of this in Figure 9.2. When time series exhibit seasonality, different techniques provide better forecasts than the ones we have described.

Regression-Based Seasonal Forecasting Models

One approach is to use linear regression. Multiple linear regression models with categorical variables can be used for time series with seasonality. To do this, we use dummy categorical variables for the seasonal components.

EXAMPLE 9.13   Regression-Based Forecasting for Natural Gas Usage

With monthly data, as we have for natural gas usage in the *Gas & Electric* Excel file, we have a seasonal categorical variable with $k = 12$ levels. As discussed in Chapter 8, we construct the regression model using $k - 1$ dummy variables. We will use January as the reference month; therefore, this variable does not appear in the model:

$$\text{Gas Usage} = \beta_0 + \beta_1 \text{ Time} + \beta_2 \text{ February} + \beta_3 \text{ March} + \beta_4 \text{ April} + \beta_5 \text{ May} + \beta_6 \text{ June} + \beta_7 \text{ July} + \beta_8 \text{ August} + \beta_9 \text{ September} + \beta_{10} \text{ October} + \beta_{11} \text{ November} + \beta_{12} \text{ December}$$

This coding scheme results in the data matrix shown in Figure 9.17. This model picks up trends from the regression coefficient for time and seasonality from the dummy variables for each month. The forecast for the next January will be $\beta_0 + \beta_1(25)$. The variable coefficients (betas) for each of the other 11 months will show the adjustment relative to January. For example, the forecast for next February will be $\beta_0 + \beta_1(26) + \beta_2(1)$, and so on.

Figure 9.18 shows the results of using the *Regression* tool in Excel after eliminating insignificant variables (time

(continued)

and Feb). Because the data show no clear linear trend, the variable time could not explain any significant variation in the data. The dummy variable for February was probably insignificant because the historical gas usage for both January and February were very close to each other. The R^2 for this model is 0.971, which is very good. The final regression model is

Gas Usage = 236.75 − 36.75 March − 99.25 April
− 192.25 May − 203.25 June − 208.25 July
− 209.75 August − 208.25 September
− 196.75 October − 149.75 November
− 43.25 December

▶ **Figure 9.17**

Data Matrix for Seasonal Regression Model

	A	B	C	D	E	F	G	H	I	J	K	L	M	N
1	Gas and Electric Usage													
2														
3	Month	Gas Use	Time	Feb	Mar	Apr	May	Jun	Jul	Aug	Sep	Oct	Nov	Dec
4	Jan	244	1	0	0	0	0	0	0	0	0	0	0	0
5	Feb	228	2	1	0	0	0	0	0	0	0	0	0	0
6	Mar	153	3	0	1	0	0	0	0	0	0	0	0	0
7	Apr	140	4	0	0	1	0	0	0	0	0	0	0	0
8	May	55	5	0	0	0	1	0	0	0	0	0	0	0
9	Jun	34	6	0	0	0	0	1	0	0	0	0	0	0
10	Jul	30	7	0	0	0	0	0	1	0	0	0	0	0
11	Aug	28	8	0	0	0	0	0	0	1	0	0	0	0
12	Sep	29	9	0	0	0	0	0	0	0	1	0	0	0
13	Oct	41	10	0	0	0	0	0	0	0	0	1	0	0
14	Nov	88	11	0	0	0	0	0	0	0	0	0	1	0
15	Dec	199	12	0	0	0	0	0	0	0	0	0	0	1
16	Jan	230	13	0	0	0	0	0	0	0	0	0	0	0
17	Feb	245	14	1	0	0	0	0	0	0	0	0	0	0
18	Mar	247	15	0	1	0	0	0	0	0	0	0	0	0
19	Apr	135	16	0	0	1	0	0	0	0	0	0	0	0
20	May	34	17	0	0	0	1	0	0	0	0	0	0	0
21	Jun	33	18	0	0	0	0	1	0	0	0	0	0	0
22	Jul	27	19	0	0	0	0	0	1	0	0	0	0	0
23	Aug	26	20	0	0	0	0	0	0	1	0	0	0	0
24	Sep	28	21	0	0	0	0	0	0	0	1	0	0	0
25	Oct	39	22	0	0	0	0	0	0	0	0	1	0	0
26	Nov	86	23	0	0	0	0	0	0	0	0	0	1	0
27	Dec	188	24	0	0	0	0	0	0	0	0	0	0	1

▼ **Figure 9.18**

Final Regression Model for Forecasting Gas Usage

	A	B	C	D	E	F	G	H	I
1	SUMMARY OUTPUT								
2									
3	*Regression Statistics*								
4	Multiple R	0.985480895							
5	R Square	0.971172595							
6	Adjusted R Square	0.948997667							
7	Standard Error	19.54432831							
8	Observations	24							
9									
10	ANOVA								
11		*df*	*SS*	*MS*	*F*	*Significance F*			
12	Regression	10	167292.2083	16729.22083	43.79597661	2.33344E-08			
13	Residual	13	4965.75	381.9807692					
14	Total	23	172257.9583						
15									
16		*Coefficients*	*Standard Error*	*t Stat*	*P-value*	*Lower 95%*	*Upper 95%*	*Lower 95.0%*	*Upper 95.0%*
17	Intercept	236.75	9.772164157	24.22697738	3.33921E-12	215.6385228	257.8614772	215.6385228	257.8614772
18	Mar	-36.75	16.92588482	-2.171230656	0.049016211	-73.31615105	-0.183848953	-73.31615105	-0.183848953
19	Apr	-99.25	16.92588482	-5.863799799	5.55744E-05	-135.816151	-62.68384895	-135.816151	-62.68384895
20	May	-192.25	16.92588482	-11.35834268	4.02824E-08	-228.816151	-155.683849	-228.816151	-155.683849
21	Jun	-203.25	16.92588482	-12.00823485	2.07264E-08	-239.816151	-166.683849	-239.816151	-166.683849
22	Jul	-208.25	16.92588482	-12.30364038	1.54767E-08	-244.816151	-171.683849	-244.816151	-171.683849
23	Aug	-209.75	16.92588482	-12.39226204	1.41949E-08	-246.316151	-173.183849	-246.316151	-173.183849
24	Sep	-208.25	16.92588482	-12.30364038	1.54767E-08	-244.816151	-171.683849	-244.816151	-171.683849
25	Oct	-196.75	16.92588482	-11.62420766	3.05791E-08	-233.316151	-160.183849	-233.316151	-160.183849
26	Nov	-149.75	16.92588482	-8.847395666	7.30451E-07	-186.316151	-113.183849	-186.316151	-113.183849
27	Dec	-43.25	16.92588482	-2.555257847	0.023953114	-79.81615105	-6.683848953	-79.81615105	-6.683848953

Holt-Winters Models for Forecasting Time Series with Seasonality and No Trend

The methods we describe here and in the next section are based on the work of two researchers, C.C. Holt, who developed the basic approach, and P.R. Winters, who extended Holt's work. Hence, these approaches are commonly referred to as **Holt-Winters models**. Holt-Winters models are similar to exponential smoothing models in that smoothing constants are used to smooth out variations in the level and seasonal patterns over time.

For time series with seasonality but no trend, we can use one of two models, the **Holt-Winters additive seasonality model with no trend**,

$$F_{t+k} = a_t + S_{t-s+k} \tag{9.10}$$

or the **Holt-Winters multiplicative seasonality model with no trend**:

$$F_{t+k} = a_t S_{t-s+k} \tag{9.11}$$

The additive model applies to time series with relatively stable seasonality, whereas the multiplicative model apples to time series whose amplitude increases or decreases over time. Therefore, a chart of the time series should be viewed first to identify the type of model to use.

In both models, S_j is the seasonal factor for period j and s is the number of periods in a season. The forecast for period $t + k$ is adjusted up or down from a level (a_t) by the seasonal factor. The multiplicative model is more appropriate when the seasonal factors are increasing or decreasing over time.

Holt-Winters Additive Seasonality Model with No Trend

The level and seasonal factors are estimated in the additive model using the following equations:

$$\text{Level component: } a_t = \alpha(A_t - S_{t-s}) + (1 - \alpha)a_{t-1}$$
$$\text{Seasonal component: } S_t = \gamma(A_t - a_t) + (1 - \gamma)S_{t-s} \tag{9.12}$$

where α and γ are smoothing constants. The first equation estimates the level for period t as a weighted average of the deseasonalized data for period t, ($A_t - S_{t-s}$), and the previous period's level. The seasonal factors are updated as well using the second equation. The seasonal factor is a weighted average of the estimated seasonal component for period t, ($A_t - a_t$), and the seasonal factor for the last period of that season type. Then the forecast for the next period is $F_{t+1} = a_t + S_{t-s+1}$.

To begin, we need to estimate the level and seasonal factors for the first s periods (that is, the length of a season; for an annual season with quarterly data, this would be the first 4 periods; for monthly data, it would be the first 12 periods, and so on) before we can use the smoothing equations. We do this as follows:

$$a_t = \frac{1}{s} \sum_{i=1}^{s} A_i, \text{ for } t = 1, 2, \ldots, s \tag{9.13}$$

and

$$S_t = A_t - a_t, \text{ for } t = 1, 2, \ldots, s \tag{9.14}$$

Then we can use the smoothing equations to update a_t and S_t and calculate forecasts.

| EXAMPLE 9.14 | **Using the Holt-Winters Additive Seasonality Model with No Trend** |

We will use the data for gas usage in the Excel file *Gas & Electric*. Looking at the chart in the Excel file (Figure 9.2), we see that the time series appears to have stable seasonality; thus the additive model would be most appropriate. We arbitrarily select $\alpha = 0.4$ and $\gamma = 0.9$.

First, initialize the values of a_t and S_t for $t = 1$ to 12 using equations (9.13) and (9.14) as $a_t = (244 + 228 + \ldots + 88 + 199)/12 = 105.75$, for $t = 1$ to 12. Then,

$$S_1 = A_1 - a_1 = 244 - 105.75 = 138.25$$
$$S_2 = A_2 - a_2 = 228 - 105.75 = 122.25$$

and so on, up to S_{12}. These are shown in the Excel file *Holt Winters Additive Model for Seasonality* (Figure 9.19).

Using the seasonal factor for the first period, the forecast for period 13 is

$$F_{13} = a_{12} + S_1 = 105.75 + 138.25 = 244$$

We may now update the parameters based on the observed value for period 13, $A_{13} = 230$:

$$a_{13} = \alpha(A_{13} - S_1) + (1 - \alpha)a_{12}$$
$$= 0.4(230 - 138.25) + (1 - 0.4)105.75$$
$$= 100.15$$

$$S_{13} = \gamma(A_{13} - a_{13}) + (1 - \gamma)S_1$$
$$= 0.9(230 - 100.15) + (1 - 0.9)138.25$$
$$= 130.69$$

The forecast for period 14 is then

$$F_{14} = a_{13} + S_2 = 100.15 + 122.25 = 222.40$$

Other forecasted values are calculated in a similar manner.

To forecast beyond the range of the last observed value, use formula (9.10) with $t = T$ and $k \geq 1$:

$$F_{T+k} = a_T + S_{T-s+k}$$

The forecast for the period 25 ($T = 24$ and $k = 1$) is

$$F_{25} = a_{24} + S_{13} = 105.80 + 130.69 = 236.49$$

To find F_{26}, use $T = 24$ and $k = 2$:

$$F_{26} = a_{24} + S_{14} = 105.80 + 134.45 = 240.25$$

We can easily experiment with different values of the smoothing constants by changing the values of α and γ in cells B3 and B4, and compare error measures such as MAD.

Holt-Winters Multiplicative Seasonality Model with No Trend

The multiplicative seasonal model has the same basic smoothing structure as the additive seasonal model with some key differences:

$$\text{Level component: } a_t = \alpha(A_t/S_{t-s}) + (1 - \alpha)a_{t-1}$$
$$\text{Seasonal component: } S_t = \gamma(A_t/a_t) + (1 - \gamma)S_{t-s} \quad (9.15)$$

where α and γ are again the smoothing constants. Here, A_t/S_{t-s} is the deseasonalized estimate for period t. Large values of α put more emphasis on this term in estimating the level for period t. The term A_t/a_t is an estimate of the seasonal factor for period t. Large values of γ put more emphasis on this in the estimate of the seasonal factor. The forecast for the next period is $F_{t+1} = a_t S_{t-s+1}$.

As with the additive model, first initialize the values for the level and seasonal factors. In this case, the formulas are

$$a_t = \frac{1}{s}\sum_{i=1}^{s} A_i, \text{ for } t = 1, 2, \ldots, s$$
$$S_t = A_t/a_t, \text{ for } t = 1, 2, \ldots, s \quad (9.16)$$

This model can be implemented on a spreadsheet in a similar fashion as the additive model.

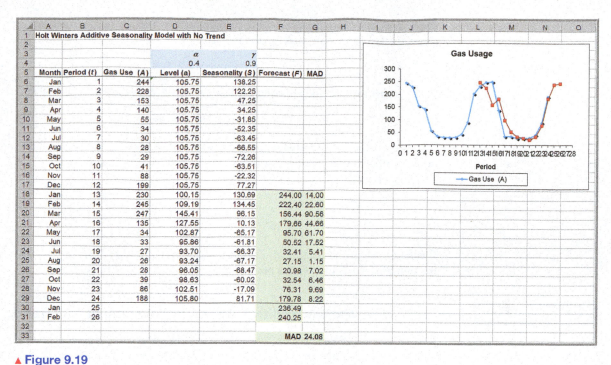

▲ Figure 9.19

Excel Implementation of Holt-Winters Additive Model for Seasonality

Holt-Winters Models for Forecasting Time Series with Seasonality and Trend

Many time series exhibit both trend and seasonality. Such might be the case for growing sales of a seasonal product. These models combine elements of both the trend and seasonal models. Two types of Holt-Winters smoothing models are often used.

The **Holt-Winters additive seasonality model** with trend is based on the equation

$$F_{t+1} = a_t + b_t + S_{t-s+1} \tag{9.17}$$

and the **Holt-Winters multiplicative seasonality model** with trend is

$$F_{t+1} = (a_t + b_t)S_{t-s+1} \tag{9.18}$$

The additive model applies to time series with relatively stable seasonality and a trend, whereas the multiplicative model applies to time series whose amplitude increases or decreases over time along with a trend. As with the no trend case, a chart of the time series should be viewed first to identify the appropriate type of model to use. Three parameters, α, β, and γ, are used to smooth the level, trend, and seasonal factors in the time series.

Holt-Winters Additive Seasonality Model with Trend

This model is similar to the additive model that incorporates only seasonality, but with the addition of a trend component:

$$\begin{aligned}
\text{Level component: } a_t &= \alpha(A_t - S_{t-s}) + (1 - \alpha)(a_{t-1} + b_{t-1}) \\
\text{Trend component: } b_t &= \beta(a_t - a_{t-1}) + (1 - \beta)b_{t-1} \\
\text{Seasonal component: } S_t &= \gamma(A_t - a_t) + (1 - \gamma)S_{t-s}
\end{aligned} \tag{9.19}$$

Here, α, β, and γ are the smoothing parameters for level, trend, and seasonal components, respectively. The forecast for period $t + 1$ is

$$F_{t+1} = a_t + b_t + S_{t-s+1} \tag{9.20}$$

The forecast for k periods beyond the last period of observed data (period T) is

$$F_{T+k} = a_T + b_T k + S_{T-s+k} \tag{9.21}$$

The initial values for level and seasonal factors are the same as in the additive seasonality model without trend, that is, formulas (9.13) and (9.14). The initial values for the trend component are

$$b_t = \frac{1}{s} \sum_{i=1}^{s} \frac{(A_{s+i} - A_i)}{s}, \text{ for } t = 1, 2, \ldots, s \tag{9.22}$$

Basically, we are averaging $A_{s+1} - A_1, A_{s+2} - A_2, \ldots,$ and $A_{s+s} - A_s$. These become the trend factors for the first s periods.

EXAMPLE 9.15   **Using the Holt-Winters Additive Model for Seasonality and Trend**

We will use the data in the Excel file *New Car Sales*, which contains three years of monthly retail sales data. Figure 9.20 shows a chart of the data. Seasonality exists in the time series and appears to be stable, and there is an increasing trend; therefore, the Holt-Winters additive model with seasonality and trend would appear to be the most appropriate model. We arbitrarily select $\alpha = 0.3$, $\beta = 0.2$, and $\gamma = 0.9$. Figure 9.21 shows the Excel implementation (Excel file *Holt Winters Additive Model for Seasonality and Trend*).

The level and seasonal factors are initialized using formulas (9.13) and (9.14). The trend factors are initialized using formula (9.22):

$$b_t = [(42{,}227 - 39{,}810)/12 + (45{,}422 - 40{,}081)/12$$
$$+ \ldots + (49{,}278 - 44{,}186)/12]/12$$
$$= 462.48, \text{ for } t = 1, 2, \ldots, 12$$

Using Equation (9.20), the forecast for period 13 is

$$F_{13} = a_{12} + b_{12} + S_1 = 45{,}427.33$$
$$+ 462.48 + (-5{,}617.33) = 39{,}810.00$$

We may now update the parameters based on the observed value for period 13, $A_{13} = 42{,}227$:

$$a_{13} = \alpha(A_{13} - S_1) + (1 - \alpha)(a_{12} + b_{12})$$
$$= 0.3(42{,}227 - (-5{,}617.33))$$
$$+ 0.7(45{,}427.33 + 462.48) = 46{,}476.17$$

$$b_{13} = \beta(a_{13} - a_{12}) + (1 - \beta)b_{12}$$
$$= 0.2(46{,}476.17 - 45{,}427.33)$$
$$+ 0.8(462.48) = 579.75$$

$$S_{13} = \gamma(A_{13} - a_{13}) + (1 - \gamma)S_1$$
$$= 0.9(42{,}227 - 45427.33)$$
$$+ 0.1(-5{,}617.33) = -4385.99$$

Other calculations are similar. To forecast period 37, we use formula (9.21):

$$F_{36+1} = a_{36} + b_{36}(1) + S_{36-12+1} = 49{,}267 \text{ units}$$

Notice from the chart that the additive model appears to fit the data extremely well. Again, we can easily experiment with different values of the smoothing constants to find a best fit using an error measure such as MAD or MSE.

Holt-Winters Multiplicative Seasonality Model with Trend

The Holt-Winters multiplicative model is similar to the additive model for seasonality, but with a trend component:

$$\begin{aligned} \text{Level component: } a_t &= \alpha(A_t/S_{t-s}) + (1 - \alpha)(a_{t-1} + b_{t-1}) \\ \text{Trend component: } b_t &= \beta(a_t - a_{t-1}) + (1 - \beta)b_{t-1} \\ \text{Seasonal component: } S_t &= \gamma(A_t/a_t) + (1 - \gamma)S_{t-s} \end{aligned} \tag{9.23}$$

▶ **Figure 9.20**

Portion of Excel File New Car Sales

	A	B	C	D
1	New Car Retail Sales			
2				
3	Year	Month	Units	
4	1	Jan	39,810	
5	1	Feb	40,081	
6	1	Mar	47,440	
7	1	Apr	47,297	
8	1	May	49,211	
9	1	Jun	51,479	
10	1	Jul	46,466	
11	1	Aug	45,208	
12	1	Sep	44,800	
13	1	Oct	46,989	
14	1	Nov	42,161	
15	1	Dec	44,186	
16	2	Jan	42,227	
17	2	Feb	45,422	
18	2	Mar	54,075	
19	2	Apr	50,926	
20	2	May	53,572	

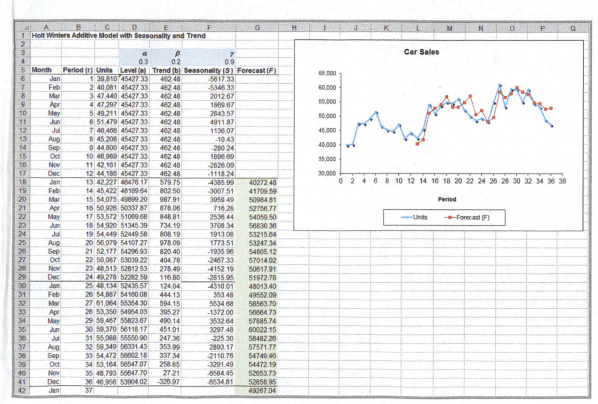

	A Month	B Period (r)	C Units	D Level (a)	E Trend (b)	F Seasonality (S)	G Forecast (F)
1	Holt Winters Additive Model with Seasonality and Trend						
2							
3				α	β		γ
4				0.3	0.2		0.9
5	Month	Period (r)	Units	Level (a)	Trend (b)	Seasonality (S)	Forecast (F)
6	Jan	1	39,810	45427.33	462.48	-5617.33	
7	Feb	2	40,081	45427.33	462.48	-5346.33	
8	Mar	3	47,440	45427.33	462.48	2012.67	
9	Apr	4	47,297	45427.33	462.48	1669.67	
10	May	5	49,211	45427.33	462.48	2843.57	
11	Jun	6	51,479	45427.33	462.48	4911.87	
12	Jul	7	46,466	45427.33	462.48	1136.07	
13	Aug	8	45,208	45427.33	462.48	-10.43	
14	Sep	9	44,800	45427.33	462.48	-280.24	
15	Oct	10	46,989	45427.33	462.48	1896.69	
16	Nov	11	42,161	45427.33	462.48	-2826.09	
17	Dec	12	44,186	45427.33	462.48	-1118.24	
18	Jan	13	42,227	46476.17	579.75	-4385.99	40272.48
19	Feb	14	45,422	48169.64	802.50	-3007.51	41709.59
20	Mar	15	54,075	49899.20	987.91	3959.49	50984.81
21	Apr	16	50,926	50337.87	878.06	716.28	52756.77
22	May	17	53,572	51069.68	848.81	2536.44	54059.50
23	Jun	18	54,920	51345.39	734.19	3708.34	56830.36
24	Jul	19	54,449	52449.58	808.19	1913.08	53215.64
25	Aug	20	56,079	54107.27	978.09	1773.51	53247.34
26	Sep	21	52,177	54296.93	820.40	-1935.96	54805.12
27	Oct	22	50,087	53039.22	404.78	-2467.33	57014.02
28	Nov	23	48,513	52812.53	278.49	-4152.19	50617.91
29	Dec	24	49,278	52282.59	116.80	-2815.95	51972.78
30	Jan	25	48,134	52435.57	124.04	-4310.01	48013.40
31	Feb	26	54,887	54160.08	444.13	353.48	49552.09
32	Mar	27	61,064	55354.30	594.15	5534.68	58563.70
33	Apr	28	53,350	54954.03	395.27	-1372.00	56664.73
34	May	29	59,467	55823.67	490.14	3532.64	57885.74
35	Jun	30	59,370	56118.17	451.01	3297.48	60022.15
36	Jul	31	55,088	55550.90	247.36	-225.30	58482.26
37	Aug	32	59,349	56331.43	353.99	2893.17	57571.77
38	Sep	33	54,472	56602.18	337.34	-2110.76	54749.46
39	Oct	34	53,164	56547.07	258.85	-3291.49	54472.19
40	Nov	35	48,793	55647.70	27.21	-6584.45	52653.73
41	Dec	36	46,956	53904.02	-326.97	-6534.81	52858.95
42	Jan	37					49267.04

▲ **Figure 9.21**

Excel Implementation of Holt-Winters Additive Model for Seasonality and Trend

The forecast for period $t + 1$ is

$$F_{t+1} = (a_t + b_t)S_{t-s+1} \qquad (9.24)$$

The forecast for k periods beyond the last period of observed data (period T) is

$$F_{T+k} = (a_T + b_T k)S_{T-s+k} \qquad (9.25)$$

▼ TABLE 9.1

Forecasting Model Choice

	No Seasonality	**Seasonality**
No trend	Simple moving average or simple exponential smoothing	Holt-Winters additive or multiplicative seasonality models without trend or multiple regression
Trend	Double exponential smoothing	Holt-Winters additive or multiplicative seasonality models with trend

Initialization is performed in the same way as for the multiplicative model without trend. Again, this model can be implemented in a similar fashion on a spreadsheet as the additive model.

Selecting Appropriate Time-Series-Based Forecasting Models

Table 9.1 summarizes the choice of forecasting approaches.

 CHECK YOUR UNDERSTANDING

1. Explain how to use dummy variables and set up worksheets to apply regression analysis for seasonal forecasting models.

2. Explain how to use the Holt-Winters model for seasonality and no trend.

3. Explain how to use the Holt-Winters model for seasonality and trend.

4. State what forecasting techniques are most appropriate for time series that may exhibit seasonality and/or trend.

 ## Regression Forecasting with Causal Variables

In many forecasting applications, other independent variables besides time, such as economic indexes or demographic factors, may influence the time series. For example, a manufacturer of hospital equipment might include such variables as hospital capital spending and changes in the proportion of people over the age of 65 in building models to forecast future sales. Explanatory/causal models, often called **econometric models**, seek to identify factors that explain statistically the patterns observed in the variable being forecast, usually with regression analysis. We will use a simple example of forecasting gasoline sales to illustrate econometric modeling.

EXAMPLE 9.16 **Forecasting Gasoline Sales Using Simple Linear Regression**

Figure 9.22 shows gasoline sales over ten weeks during June through August along with the average price per gallon and a chart of the gasoline sales time series with a fitted trendline (Excel file *Gasoline Sales*). During the summer months, it is not unusual to see an increase in sales as more people go on vacations. The chart shows a linear trend, although R^2 is not very high. The trendline is

$$\text{Sales} = 4,790.1 + 812.99 \text{ Week}$$

Using this model, we would predict sales for week 11 as

$$\text{Sales} = 4,790.1 + 812.99(11) = 13,733 \text{ Gallons}$$

▶ **Figure 9.22**

Gasoline Sales *Data and Trendline*

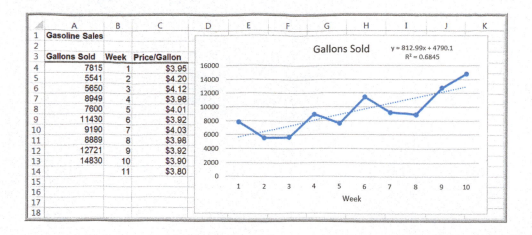

In the *Gasoline Sales* data, we also see that the average price per gallon changes each week, and this may influence consumer sales. Therefore, the sales trend might not simply be a factor of steadily increasing demand, but it might also be influenced by the average price per gallon. The average price per gallon can be considered as a *causal variable*. Multiple linear regression provides a technique for building forecasting models that incorporate not only time, but also other potential causal variables.

EXAMPLE 9.17   **Incorporating Causal Variables in a Regression-Based Forecasting Model**

For the *Gasoline Sales* data, we can incorporate the price/gallon by using two independent variables. This results in the multiple regression model

$$\text{Sales} = \beta_0 + \beta_1 \text{ Week} + \beta_2 \text{ Price/Gallon}$$

The results are shown in Figure 9.23, and the regression model is

$$\text{Sales} = 72{,}333.08 + 508.67 \times \text{Week} - 16{,}463.20 \times \text{Price/Gallon}$$

Notice that the R^2 value is higher when both variables are included, explaining more than 86% of the variation in the data. If the company estimates that the average price for the next week will drop to $3.80, the model would forecast the sales for week 11 as

$$\text{Sales} = 72{,}333.08 + 508.67 \times 11 - 16{,}463.20 \times 3.80$$
$$= 15{,}368 \text{ gallons}$$

CHECK YOUR UNDERSTANDING

1. Explain the purpose of explanatory/causal (econometric) models.

2. Describe how to use multiple linear regression for causal forecasting.

The Practice of Forecasting

Surveys of forecasting practices have shown that both judgmental and quantitative methods are used for forecasting sales of product lines or product families as well as for broad company and industry forecasts. Simple time-series models are used for short- and medium-range forecasts, whereas regression analysis is the most popular method for long-range forecasting. However, many companies rely on judgmental methods far more than

► **Figure 9.23**

*Regression Results for
Gasoline Sales*

	A	B	C	D	E	F	G
1	SUMMARY OUTPUT						
2							
3	*Regression Statistics*						
4	Multiple R	0.930528528					
5	R Square	0.865883342					
6	Adjusted R Square	0.827564297					
7	Standard Error	1235.400329					
8	Observations	10					
9							
10	ANOVA						
11		*df*	*SS*	*MS*	*F*	*Significance F*	
12	Regression	2	68974748.7	34487374.35	22.59668368	0.000883465	
13	Residual	7	10683497.8	1526213.972			
14	Total	9	79658246.5				
15							
16		*Coefficients*	*Standard Error*	*t Stat*	*P-value*	*Lower 95%*	*Upper 95%*
17	Intercept	72333.08447	21969.92267	3.292368642	0.013259225	20382.47252	124283.6964
18	Week	508.6681395	168.1770861	3.024598364	0.019260863	110.9925232	906.3437559
19	Price/Gallon	-16463.19901	5351.082403	-3.076611005	0.017900405	-29116.49823	-3809.899786

quantitative methods, and almost half judgmentally adjust quantitative forecasts. In this chapter, we focus on these three approaches to forecasting.

In practice, managers use a variety of judgmental and quantitative forecasting techniques. Statistical methods alone cannot account for such factors as sales promotions, unusual environmental disturbances, new product introductions, large one-time orders, and so on. Many managers begin with a statistical forecast and adjust it to account for intangible factors. Others may develop independent judgmental and statistical forecasts and then combine them, either objectively by averaging or in a subjective manner. It is important to compare quantitatively generated forecasts to judgmental forecasts to see if the forecasting method is adding value in terms of an improved forecast. It is impossible to provide universal guidance as to which approaches are best, because they depend on a variety of factors, including the presence or absence of trends and seasonality, the number of data points available, the length of the forecast time horizon, and the experience and knowledge of the forecaster. Often, quantitative approaches will miss significant changes in the data, such as reversal of trends, whereas qualitative forecasts may catch them, particularly when using indicators as discussed earlier in this chapter.

ANALYTICS IN PRACTICE: Forecasting at NBCUniversal[2]

NBCUniversal (NBCU), a subsidiary of Comcast, is one of the world's leading media and entertainment companies in the distribution, production, and marketing of entertainment, news, and information. The television broadcast year in the United States starts in the third week of September. The major broadcast networks announce their programming schedules for the new broadcast year in the middle of May. Shortly thereafter, the sale of advertising time, which generates the majority of revenues, begins. The broadcast networks sell 60% to 80% of their airtime inventory during a brief period starting in late May and lasting two to three weeks. This sales period is known as *the upfront market*. Immediately after announcing their program schedules, the networks finalize their ratings forecasts and estimate the market demand. The ratings forecasts are projections of the numbers of people in each of several

---

[2]Based on Srinivas Bollapragada, Salil Gupta, Brett Hurwitz, Paul Miles, and Rajesh Tyagi, "NBC-Universal Uses a Novel Qualitative Forecasting Technique to Predict Advertising Demand," *Interfaces*, 38, 2 (March–April 2008): 103–111.

demographic groups who are expected to watch each airing of the shows in the program schedule for the entire broadcast year. After they finalize their ratings projections and market-demand estimates, the networks set the rate cards that contain the prices for commercials on all their shows and develop pricing strategies.

Forecasting upfront market demand has always been a challenge. NBCU initially relied on historical patterns, expert knowledge, and intuition for estimating demand. Later, it tried time-series forecasting models based on historical demand and leading economic indicator data and implemented the models in a Microsoft Excel–based system. However, these models proved to be unsatisfactory because of the unique nature of NBCU's demand population. The time-series models had fit and prediction errors in the range of 5% to 12% based on the historical data. These errors were considered reasonable, but the sales executives were reluctant to use the models because the models did not consider several qualitative factors that they believed influence the demand. As a result, they did not trust the forecasts that these models generated; therefore, they never used them. Analytics staff at NBCU then decided to develop a qualitative demand forecasting model that captures the knowledge of the sales experts.

Their approach incorporates the Delphi method and "grass-roots forecasting," which is based on the concept of

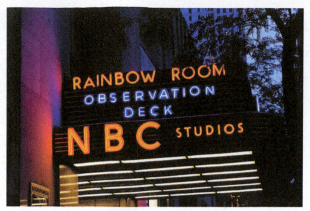
Sean Pavone/Shutterstock

asking those who are close to the end consumer, such as salespeople, about the customers' purchasing plans, along with historical data to develop forecasts. Since 2004, more than 200 sales and finance personnel at NBCU have been using the system to support sales decisions during the upfront market when NBCU signs advertising deals worth more than $4.5 billion. The system enables NBCU to sell and analyze pricing scenarios across all NBCU's television properties with ease and sophistication while predicting demand with a high accuracy. NBCU's sales leaders credit the system with having given them a unique competitive advantage.

CHECK YOUR UNDERSTANDING

1. Discuss practical considerations in forecasting.
2. How are the different types of forecasting methods used together?

KEY TERMS

Cyclical effect
Delphi method
Double exponential smoothing
Econometric model
Historical analogy
Holt-Winters additive seasonality model with no trend
Holt-Winters additive seasonality model with trend
Holt-Winters models
Holt-Winters multiplicative seasonality model with no trend
Holt-Winters multiplicative seasonality model with trend

Index
Indicator
Mean absolute deviation (MAD)
Mean absolute percentage error (MAPE)
Mean square error (MSE)
Root mean square error (RMSE)
Seasonal effect
Simple exponential smoothing
Simple moving average
Smoothing constant
Stationary time series
Time series
Trend

CHAPTER 9 TECHNOLOGY HELP

Excel Techniques

Moving Average tool (Example 9.6):

Select *Data Analysis* and then *Moving Average* from the *Analysis* group. In the dialog box, enter the *Input Range* of the data, the *Interval* (the value of k), and the first cell of the *Output Range*. To align the actual data with the forecasted values in the worksheet, select the first cell of the *Output Range* to be one row below the first value. We do not recommend using the chart or error options because the forecasts generated by this tool are not properly aligned with the data.

Exponential Smoothing tool (Example 9.10):

Select *Data Analysis* and then *Exponential Smoothing* from the *Analysis* group. In the dialog, enter

the *Input Range* of the time-series data, the *Damping Factor*, which is $(1 - \alpha)$—not the smoothing constant α—and the first cell of the *Output Range*, which should be adjacent to the first data point. You also have options for labels, to chart output and to obtain standard errors.

Analytic Solver

Analytic Solver provides a set of tools for forecasting. See the online supplement *Using Forecasting Techniques in Analytic Solver*. We suggest that you first read the online supplement *Getting Started with Analytic Solver Basic*. This provides information for both instructors and students on how to register for and access Analytic Solver.

PROBLEMS AND EXERCISES

Qualitative and Judgmental Forecasting

1. Identify some business applications in which judgmental forecasting techniques such as historical analogy and the Delphi method would be useful.

2. Find the Conference Board's economic forecast for the U.S. economy at https://www.conference-board.org/data/usforecast.cfm. Write a short report about your findings, focusing on the qualitative factors that are used in the forecasts.

Statistical Forecasting Models

3. For each of the time series in the following Excel files, construct a line chart of the data and identify the characteristics of the time series (that is, random, stationary, trend, seasonal, or cyclical).

   a. *Closing Stock Prices*

   b. *Unemployment Rates*

   c. *New Car Sales*

   d. *Housing Starts*

   e. *Prime Rate*

4. The Excel file *Energy Production & Consumption* provides data on production, imports, exports, and consumption.

   a. Develop line charts for each variable and identify the characteristics of the time series (that is, random, stationary, trend, seasonal, or cyclical).

   b. In forecasting the future, discuss whether all or only a portion of the data should be used.

5. Find the Conference Board's annual consumer prices at https://www.conference-board.org/data/consumerdata.cfm. These are available as Excel workbooks on the Web page (click on the Data Tables link). Identify characteristics of the different time series (that is, random, stationary, trend, seasonal, or cyclical) and summarize your findings in a short report.

Forecasting Models for Stationary Time Series

6. For the data in the Excel file *Gasoline Prices*, do the following:

   a. Develop a spreadsheet for forecasting prices using a simple three-period moving average.

   b. Compute MAD, MSE, and MAPE error measures.

7. The Excel file *Unemployment Rates* provides data on monthly rates for four years. Compare 3- and 12-month moving average forecasts using the MAD criterion. Explain why the 3-month model yields better results.

8. The Excel file *Closing Stock Prices* provides data for four stocks and the Dow Jones Industrial Average over a one-month period.

   a. Develop a spreadsheet for forecasting each of the stock prices and the DJIA using a simple two-period and three-period moving average.

   b. Compute MAD, MSE, and MAPE and determine whether two or three moving average periods is better.

9. The Excel file *Closing Stock Prices* provides data for four stocks and the Dow Jones Industrial Average over a one-month period.

   a. Develop a spreadsheet model for forecasting each of the stock prices using simple exponential smoothing with a smoothing constant of 0.3.

   b. Compare your results to the output from Excel's *Data Analysis* tool.

   c. Compute MAD, MSE, and MAPE.

   d. Does a smoothing constant of 0.1 or 0.5 yield better results?

10. For the data in the Excel file *Gasoline Prices*, do the following:

    a. Develop a spreadsheet for forecasting prices using simple exponential smoothing with smoothing constants from 0.1 to 0.9 in increments of 0.1.

    b. Using MAD, MSE, and MAPE as guidance, find the best smoothing constant.

Forecasting Models for Time Series with a Linear Trend

11. For the coal production example using double exponential smoothing (Example 9.11), experiment with the spreadsheet model to find the best values of α and β that minimize MAD.

12. Consider the data in the Excel file *Consumer Price Index*. Use the double exponential smoothing procedure to find forecasts for the next two years.

13. Consider the prices for the Dow Jones Industrials in the Excel file *Closing Stock Prices*. Use simple linear regression to forecast the data. What would be the forecasts for the next three days?

14. Consider the data in the Excel file *Consumer Price Index*. Use simple linear regression to forecast the data. What would be the forecasts for the next two years?

15. Consider the data in the Excel file *Nuclear Power*. Use simple linear regression to forecast the data. What would be the forecasts for the next three years?

Forecasting Time Series with Seasonality

16. Develop a multiple regression model with categorical variables that incorporate seasonality for forecasting the temperature in Washington, DC, using the data for the years 1999 and 2000 in the Excel file *Washington DC Weather*. Use the model to generate forecasts for the next nine months and compare the forecasts to the actual observations in the data for the year 2001.

17. Develop a multiple regression model with categorical variables that incorporate seasonality for forecasting sales using the last three years of data in the Excel file *New Car Sales*.

18. Develop a multiple regression model with categorical variables that incorporate seasonality for forecasting 2010 housing starts using the 2008 and 2009 data in the Excel file *Housing Starts*.

19. Develop a multiple regression model with categorical variables that incorporate seasonality for forecasting for the data in the Excel file *Coal Consumption*.

20. For Example 9.14, experiment with the *Holt-Winters Additive Model for Seasonality* spreadsheet to find the best combination of α and γ to minimize MAD.

21. Modify the Excel file *Holt-Winters Additive Model for Seasonality* (see Example 9.14) to implement the Holt-Winters multiplicative seasonality model with no trend.

22. For Example 9.15, experiment with the *Holt-Winters Additive Model for Seasonality and Trend* spreadsheet model to find the best combination of α, β, and γ to minimize MSE.

23. Modify the Excel file *Holt-Winters Additive Model for Seasonality and Trend* for the *New Car Sales* data (see Example 9.15) to implement the Holt-Winters multiplicative seasonality model with trend. Try to find the best combination of α, β, and γ to minimize MSE, and compare your results with Problem 22.

24. The Excel file *CD Interest Rates* provides annual average interest rates on six-month certificate of deposits. Compare the Holt-Winters additive and multiplicative models using $\alpha = 0.7$, $\beta = 0.3$, $\gamma = 0.1$, and a season of six years. Try to fine-tune the parameters using the MSE criterion.

25. Using Table 9.1, determine the most appropriate forecasting technique for the data in the Excel file *DJIA December Close* and implement the model.

26. Using Table 9.1, determine the most appropriate forecasting technique for the data in the Excel file *Mortgage Rates* and implement the model.

27. Using Table 9.1, determine the most appropriate forecasting technique for the data in the Excel file *Prime Rate* and implement the model.

28. Using Table 9.1, determine the most appropriate forecasting technique for the data in the Excel file *Treasury Yield Rates* and implement the model.

29. The Excel file *Olympic Track and Field Data* provides the gold medal-winning distances for the high jump, discus throw, and long jump for the modern Olympic Games. Develop forecasting models for each of the events.

Regression Forecasting with Causal Variables

30. Data in the Excel File *Microprocessor Data* shows the demand for one type of chip used in industrial equipment from a small manufacturer.

    a. Construct a chart of the data. What appears to happen when a new chip is introduced?

    b. Develop a causal regression model to forecast demand that includes both time and the introduction of a new chip as explanatory variables.

    c. What would the forecast be for the next month if a new chip is introduced? What would it be if a new chip is not introduced?

CASE: PERFORMANCE LAWN EQUIPMENT

An important part of planning manufacturing capacity is having a good forecast of sales. Elizabeth Burke is interested in forecasting sales of mowers and tractors in each marketing region as well as industry sales to assess future changes in market share. She also wants to forecast future increases in production costs. Using the data in the *Performance Lawn Equipment Database*, develop forecasting models for these data and prepare a report of your results with appropriate charts and output from Excel.

Introduction to Data Mining

Laborant/Shutterstock

LEARNING OBJECTIVES   After studying this chapter, you will be able to:

- Define data mining and some common approaches used in data mining.
- Explain how cluster analysis is used to explore and reduce data.
- Explain the purpose of classification methods and how to measure classification performance, and the use of training and validation data.

- Understand *k*-nearest neighbors and discriminant analysis for classification.
- Describe association rule mining and its use in market basket analysis.
- Use correlation analysis for cause-and-effect modeling.

In an article in *Analytics* magazine, Talha Omer observed that using a cell phone to make a voice call leaves behind a significant amount of data. "The cell phone provider knows every person you called, how long you talked, what time you called and whether your call was successful or if was dropped. It also knows where you are, where you make most of your calls from, which promotion you are responding to, how many times you have bought before, and so on."[1] Considering the fact that the vast majority of people today use cell phones, a huge amount of data about consumer behavior is available. Similarly, many stores now use loyalty cards. At supermarkets, drugstores, retail stores, and other outlets, loyalty cards enable consumers to take advantage of sale prices available only to those who use the card. However, when they do, the cards leave behind a digital trail of data about purchasing patterns. How can a business exploit these data? If they can better understand patterns and hidden relationships in the data, they can not only understand buying habits but also customize advertisements, promotions, coupons, and so on for each individual customer and send targeted text messages and e-mail offers (we're not talking spam here, but registered users who opt into such messages).

Data mining is a rapidly growing field of business analytics that is focused on better understanding characteristics and patterns among variables in large databases using a variety of statistical and analytical tools. Many of the tools that we have studied in previous chapters, such as data visualization, data summarization, PivotTables, and correlation and regression analysis, are used extensively in data mining. However, as the amount of data has grown exponentially, many other statistical and analytical methods have been developed to identify relationships among variables in large data sets and understand hidden patterns that they may contain.

Many data-mining procedures require advanced statistical knowledge to understand the underlying theory and special software to implement them. Therefore, our focus is on simple applications and understanding the purpose and application of data-mining techniques rather than their theoretical underpinnings. In an optional online supplement, we describe the use of Analytic Solver for implementing data-mining procedures.

The Scope of Data Mining

Data mining can be considered part descriptive and part prescriptive analytics. In descriptive analytics, data-mining tools help analysts to identify patterns in data. Excel charts and PivotTables, for example, are useful tools for describing patterns and analyzing data sets;

[1]Talha Omer, "From Business Intelligence to Analytics," *Analytics* (January/February 2011): 20. www.analytics-magazine.org.

however, they require manual intervention. Regression analysis and forecasting models help us to predict relationships or future values of variables of interest. As some researchers observe, "the boundaries between prediction and description are not sharp (some of the predictive models can be descriptive, to the degree that they are understandable, and vice versa)."[2] In most business applications, the purpose of descriptive analytics is to help managers predict the future or make better decisions that will impact future performance, so we can generally state that data mining is primarily a predictive analytic approach.

Some common approaches in data mining include the following:

- *Cluster analysis*. Some basic techniques in data mining involve data exploration and "data reduction"—that is, breaking down large sets of data into more-manageable groups or segments that provide better insight. We have seen numerous techniques earlier in this book for data exploration and data reduction. For example, charts, frequency distributions and histograms, and summary statistics provide basic information about the characteristics of data. PivotTables, in particular, are very useful in exploring data from different perspectives and for data reduction. Data mining software provide a variety of tools and techniques for data exploration that complement or extend the concepts and tools we have studied in previous chapters. This involves identifying groups in which the elements of the groups are in some way similar. This approach is often used to understand differences among customers and segment them into homogenous groups. For example, Macy's department stores identified four types of customers defined by their lifestyle: "Katherine," a traditional, classic dresser who doesn't take a lot of risks and likes quality; "Julie," neotraditional and slightly more edgy but still classic; "Erin," a contemporary customer who loves newness and shops by brand; and "Alex," the fashion customer who wants only the latest and greatest (they have male versions also).[3] Such segmentation is useful in design and marketing activities to better target product offerings. These techniques have also been used to identify characteristics of successful employees and improve recruiting and hiring practices.
- *Classification*. Classification is the process of analyzing data to predict how to classify a new data element. An example of classification is spam filtering in an e-mail client. By examining textual characteristics of a message (subject header, key words, and so on), the message is classified as junk or not. Classification methods can help predict whether a credit card transaction may be fraudulent, whether a loan applicant is high risk, or whether a consumer will respond to an advertisement.
- *Association*. Association is the process of analyzing databases to identify natural associations among variables and create rules for target marketing or buying recommendations. For example, Netflix uses association to understand what types of movies a customer likes and provides recommendations based on the data. Amazon.com also makes recommendations based on past purchases. Supermarket loyalty cards collect data on customers' purchasing habits and print coupons at the point of purchase based on what was currently bought.
- *Cause-and-effect modeling*. Cause-and-effect modeling is the process of developing analytic models to describe the relationship between metrics that drive business performance—for instance, profitability, customer satisfaction, or

---

[2]Usama Fayyad, Gregory Piatetsky-Shapiro, and Padhraic Smyth, "From Data Mining to Knowledge Discovery in Databases," *AI Magazine*, American Association for Artificial Intelligence (Fall 1996): 37–54.
[3]"Here's Mr. Macy," *Fortune* (November 28, 2005): 139–142.

employee satisfaction. Understanding the drivers of performance can lead to better decisions to improve performance. For example, the controls group of Johnson Controls, Inc., examined the relationship between satisfaction and contract renewal rates. They found that 91% of contract renewals came from customers who were either satisfied or very satisfied, and customers who were not satisfied had a much higher defection rate. Their model predicted that a one-percentage-point increase in the overall satisfaction score was worth $13 million in service contract renewals annually. As a result, they identified decisions that would improve customer satisfaction.[4] Regression and correlation analysis are key tools for cause-and-effect modeling.

CHECK YOUR UNDERSTANDING

1. What is the purpose of data mining?

2. Explain the basic concepts of cluster analysis, classification, association, and cause-and-effect modeling.

Cluster Analysis

Cluster analysis, also called *data segmentation*, is a set of techniques that seek to group or segment a collection of objects (that is, observations or records) into subsets or clusters, such that those within each cluster are more closely related to one another than objects assigned to different clusters. The objects within clusters should exhibit a high amount of similarity, whereas those in different clusters will be dissimilar.

Cluster analysis is a data-reduction technique in the sense that it can take a large number of observations, such as customer surveys or questionnaires, and reduce the information into smaller, homogenous groups that can be interpreted more easily. The segmentation of customers into smaller groups, for example, can be used to customize advertising or promotions. As opposed to many other data-mining techniques, cluster analysis is primarily descriptive, and we cannot draw statistical inferences about a sample using it. In addition, the clusters identified are not unique and depend on the specific procedure used; therefore, it does not result in a definitive answer but only provides new ways of looking at data. Nevertheless, it is a widely used technique.

There are two major methods of clustering—hierarchical clustering and *k*-means clustering. In **hierarchical clustering**, the data are not partitioned into a particular cluster in a single step. Instead, a series of partitions takes place, which may run from a single cluster containing all objects to *n* clusters, each containing a single object. Hierarchical clustering is subdivided into **agglomerative clustering methods**, which proceed by series of fusions of the *n* objects into groups, and **divisive clustering methods**, which separate *n* objects successively into finer groupings. Figure 10.1 illustrates the differences between these two types of methods. Agglomerative techniques are more commonly used.

An agglomerative hierarchical clustering procedure produces a series of partitions of the data, $P_n, P_{n-1}, \ldots, P_1$. P_n consists of *n* single-object clusters, and P_1 consists of a single group containing all *n* observations. At each particular stage, the method joins together the two clusters that are closest together (most similar). At the first stage, this consists of simply joining together the two objects that are closest together. Different methods use different ways of defining distance (or similarity) between clusters.

[4]Steve Hoisington and Earl Naumann, "The Loyalty Elephant," *Quality Progress* (February 2003): 33–41.

▶ **Figure 10.1**

Agglomerative Versus Divisive Clustering

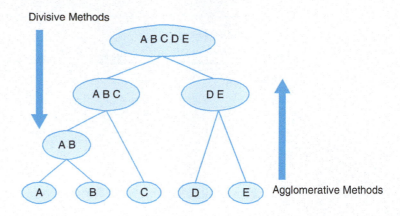

Measuring Distance Between Objects

The most commonly used measure of distance between objects is **Euclidean distance**. This is an extension of the way in which the distance between two points on a plane is computed as the hypotenuse of a right triangle (see Figure 10.2). The Euclidean distance measure between two points $(x_1, x_2, \ldots, x_n)$ and $(y_1, y_2, \ldots, y_n)$ is

$$\sqrt{(x_1 - y_1)^2 + (x_2 - y_2)^2 + \cdots + (x_n - y_n)^2} \qquad \textbf{(10.1)}$$

Some clustering methods use the squared Euclidean distance (that is, without the square root) because it speeds up the calculations.

EXAMPLE 10.1   **Applying the Euclidean Distance Measure**

Figure 10.3 shows a portion of the Excel file *Colleges and Universities*. The characteristics of these institutions differ quite widely. Suppose that we wish to cluster them into more homogeneous groups based on the median SAT, acceptance rate, expenditures/student, percentage of students in the top 10% of their high school, and graduation rate. We can use the Euclidean distance measure in formula (10.1) to measure the distance between them. For example, the distance between Amherst and Barnard is

$$\sqrt{(1315 - 1220)^2 + (22\% - 53\%)^2 + (26{,}636 - 17{,}653)^2 + (85 - 69)^2 + (93 - 80)^2}$$

$$= 8{,}983.53$$

We can implement this easily by using the Excel function SUMXMY2(*array_x*, *array_y*), which sums the squares of the differences in two corresponding ranges or arrays. Therefore, the distance between Amherst and Barnard would be computed by the Excel formula = SQRT(SUMXMY2(C4:G4, C5:G5)).

▶ **Figure 10.2**

Computing the Euclidean Distance Between Two Points

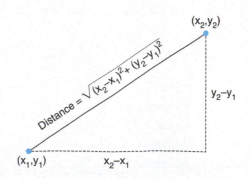

▶ **Figure 10.3**

Portion of the Excel File Colleges and Universities

	A	B	C	D	E	F	G
1	**Colleges and Universities**						
2							
3	School	Type	Median SAT	Acceptance Rate	Expenditures/Student	Top 10% HS	Graduation %
4	Amherst	Lib Arts	1315	22%	$ 26,636	85	93
5	Barnard	Lib Arts	1220	53%	$ 17,653	69	80
6	Bates	Lib Arts	1240	36%	$ 17,554	58	88
7	Berkeley	University	1176	37%	$ 23,665	95	68
8	Bowdoin	Lib Arts	1300	24%	$ 25,703	78	90
9	Brown	University	1281	24%	$ 24,201	80	90
10	Bryn Mawr	Lib Arts	1255	56%	$ 18,847	70	84

Normalizing Distance Measures

When the data have different orders of magnitude, the distance measure can easily be dominated by the large values. Therefore, it is customary to standardize (or **normalize**) the data by converting them into *z*-scores. These are computed in the Excel file *Colleges and Universities Cluster Analysis Worksheet.* Using these, the distance measure between Amherst and Barnard is

$$\sqrt{\begin{aligned}&((0.8280 - (-0.6877))^2 + (-1.2042 - 1.1141)^2 + (-0.2214 - (-0.0824))^2 + \\ &(0.7967 - (-0.3840))^2 + (1.3097 - (-0.4356))^2\end{aligned}}$$

$$= 3.5284$$

A distance matrix between the first five colleges is shown in Table 10.1.

Clustering Methods

One of the simplest agglomerative hierarchical clustering methods is **single linkage clustering**, which is an agglomerative method that keeps forming clusters from the individual objects until only one cluster is left. In the single linkage method, the distance between two clusters *r* and *s*, *D(r, s),* is defined as the minimum distance between any object in cluster *r* and any object in cluster *s*. In other words, the distance between two clusters is given by the value of the shortest link between the clusters. Initially, each cluster simply consists of an individual object. At each stage of clustering, we find the two clusters with the minimum distance between them and merge them together.

Another method that is basically the opposite of single linkage clustering is called **complete linkage clustering**. In this method, the distance between clusters is defined as the distance between the most distant pair of objects, one from each cluster. A third method is **average linkage clustering**. Here the distance between two clusters is defined as the average of distances between all pairs of objects, where each pair is made up of one object from each group. Other methods are **average group linkage clustering**, which uses the mean values for each variable to compute distances between clusters, and **Ward's hierarchical clustering**

▶ **Table 10.1**

Normalized Distance Matrix for First Five Colleges

	Amherst	Barnard	Bates	Berkeley	Bowdoin
Amherst	0	3.5284	2.7007	4.2454	0.7158
Barnard		0	1.8790	2.8901	2.9744
Bates			0	3.9837	2.0615
Berkeley				0	3.8954
Bowdoin					0

method, which uses a sum-of-squares criterion. Different methods generally yield different results, so it is best to experiment and compare the results. In the following example, we illustrate single linkage clustering.

EXAMPLE 10.2   **Single Linkage Clustering**

We will apply single linkage clustering to the first five schools in the Excel file *Colleges and Universities Cluster Analysis Worksheet*. Looking at the distance matrix in Table 10.1, we see that the smallest distance occurs between Amherst and Bowdoin (0.7158). Thus, we join these two into a cluster. Next, recalculate the distance between this cluster and the remaining colleges by finding the minimum distance between any college in the cluster and the others. This results in the distance matrix shown in Table 10.2. Note that the smallest distance between either Amherst or Bowdoin and Barnard, for instance, is MIN(3.5284, 2.9744). This becomes the distance between the Amherst/Bowdoin cluster and Barnard.

In Table 10.2, the smallest distance is between Barnard and Bates (1.879). Therefore, we join these two colleges into a second cluster. This results in the distance matrix shown in Table 10.3.

Next, we join the Amherst/Bowdoin and Barnard/Bates clusters together, as the smallest distance in Table 10.3 is 2.06125. This results in the distance matrix shown in Table 10.4. Only one option remains, that is, to join Berkeley to the cluster of other colleges.

If we examine the original data, we can see that Amherst and Bowdoin, and Barnard and Bates have similar profiles, but that Berkeley is quite different:

School	Type	Median SAT	Acceptance Tare (%)	Expenditures/ Students	Top 10% HS	Graduation %
Amherst	Lib Arts	1315	22	$26,636.00	85	93
Bowdoin	Lib Arts	1300	24	$25,703.00	78	90
Bamard	Lib Arts	1220	53	$17,653.00	69	80
Bates	Lib Arts	1240	36	$17,554.00	58	88
Berkeley	University	1176	37	$23,665.00	95	68

▶ **Table 10.2**

Distance Matrix After First Clustering

	Amherst/Bowdoin	Barnard	Bates	Berkeley
Amherst/Bowdoin	0	2.9744	2.0615	3.8954
Barnard		0	1.879	2.8901
Bates			0	3.8937
Berkeley				0

▶ **Table 10.3**

Distance Matrix After Second Clustering

	Amherst/Bowdoin	Barnard/Bates	Berkeley
Amherst/Bowdoin	0	2.0615	3.8954
Barnard/Bates		0	2.8901
Berkeley			0

▶ **Table 10.4**

Distance Matrix After Third Clustering

	Amherst/Bowdoin/Barnard/Bates	Berkeley
Amherst/Bowdoin/Barnard/Bates	0	2.8901
Berkeley		0

▶ **Figure 10.4**

Dendogram for Colleges *and* Universities *Example*

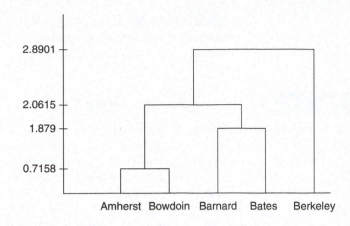

At various stages of the clustering process, there are different numbers of clusters. We can visualize this using a **dendogram**, which is shown in Figure 10.4. The *y*-axis measures the intercluster distance. A dendogram shows the sequence in which clusters are formed as you move up the diagram. At the top, we see that all clusters are merged into a single cluster. If you draw a horizontal line through the dendogram at any value along the *y*-axis, you can identify the number of clusters and the objects in each of them. For example, if you draw a line at the distance value of 2.0, you can see that we have the three clusters: {Amherst, Bowdoin}, {Barnard, Bates}, and {Berkeley}.

CHECK YOUR UNDERSTANDING

1. What is the difference between agglomerative and divisive clustering methods?

2. How are distances between objects measured in cluster analysis?

3. Explain how single linkage clustering works.

Classification

Classification methods seek to classify a categorical outcome into one of two or more categories based on various data attributes. For each record in a database, we have a categorical variable of interest (for example, purchase or not purchase, high risk or no risk), and a number of additional predictor variables (age, income, gender, education, assets, etc.). For a given set of predictor variables, we would like to assign the best value of the categorical variable. We will be illustrating various classification techniques using the Excel database *Credit Approval Decisions*.

A portion of this database is shown in Figure 10.5. In this database, the categorical variable of interest is the decision to approve or reject a credit application. The remaining variables are the predictor variables. Because we are working with numerical data, however, we need to code the Homeowner and Decision fields numerically. We code the Homeowner attribute "Y" as 1 and "N" as 0; similarly, we code the Decision attribute "Approve" as 1 and "Reject" as 0. Figure 10.6 shows a portion of the modified database (Excel file *Credit Approval Decisions Coded*).

▶ **Figure 10.5**

Portion of the Excel File Credit Approval Decisions

	A	B	C	D	E	F
1	Credit Approval Decisions					
2						
3	Homeowner	Credit Score	Years of Credit History	Revolving Balance	Revolving Utilization	Decision
4	Y	725	20	$ 11,320	25%	Approve
5	Y	573	9	$ 7,200	70%	Reject
6	Y	677	11	$ 20,000	55%	Approve
7	N	625	15	$ 12,800	65%	Reject
8	N	527	12	$ 5,700	75%	Reject
9	Y	795	22	$ 9,000	12%	Approve
10	N	733	7	$ 35,200	20%	Approve
11	N	620	5	$ 22,800	62%	Reject
12	Y	591	17	$ 16,500	50%	Reject
13	Y	660	24	$ 9,200	35%	Approve

▶ **Figure 10.6**

Modified Excel File with Numerically Coded Variables

	A	B	C	D	E	F
1	Coded Credit Approval Decisions					
2						
3	Homeowner	Credit Score	Years of Credit History	Revolving Balance	Revolving Utilization	Decision
4	1	725	20	$ 11,320	25%	1
5	1	573	9	$ 7,200	70%	0
6	1	677	11	$ 20,000	55%	1
7	0	625	15	$ 12,800	65%	0
8	0	527	12	$ 5,700	75%	0
9	1	795	22	$ 9,000	12%	1
10	0	733	7	$ 35,200	20%	1
11	0	620	5	$ 22,800	62%	0
12	1	591	17	$ 16,500	50%	0
13	1	660	24	$ 9,200	35%	1

An Intuitive Explanation of Classification

To develop an intuitive understanding of classification, we consider only the credit score and years of credit history as predictor variables.

EXAMPLE 10.3 **Classifying Credit-Approval Decisions Intuitively**

Figure 10.7 shows a chart of the credit scores and years of credit history in the *Credit Approval Decisions* data. The chart plots the credit scores of loan applicants on the *x*-axis and the years of credit history on the *y*-axis. The large bubbles represent the applicants whose credit applications were rejected; the small bubbles represent those that were approved. With a few exceptions (the points at the bottom right corresponding to high credit scores with just a few years of credit history that were rejected), there appears to be a clear separation of the points. When the credit score is greater than 640, the applications were approved, but most applications with credit scores of 640 or less were rejected. Thus, we might propose a simple classification

rule: approve an application with a credit score greater than 640.

Another way of classifying the groups is to use both the credit score and years of credit history by visually drawing a straight line to separate the groups, as shown in Figure 10.8. This line passes through the points (763, 2) and (595, 18). Using a little algebra, we can calculate the equation of the line as

$$Years = -0.095 \times Credit\ Score + 74.66$$

Therefore, we can propose a different classification rule: whenever *Years* + 0.095 × *Credit score* ≤ 74.66, the application is rejected; otherwise, it is approved. Here again, however, we see some misclassification.

Although this is easy to do intuitively for only two predictor variables, it is more difficult to do when we have more predictor variables. Therefore, more-sophisticated procedures are needed, as we will discuss.

► **Figure 10.7**

*Chart of Credit-Approval
Decisions*

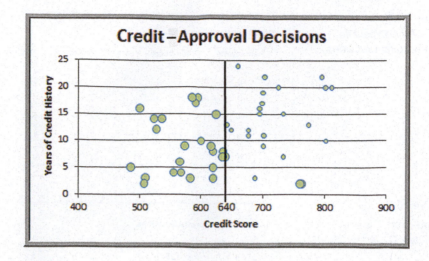

► **Figure 10.8**

*Alternate Credit-Approval
Classification Scheme*

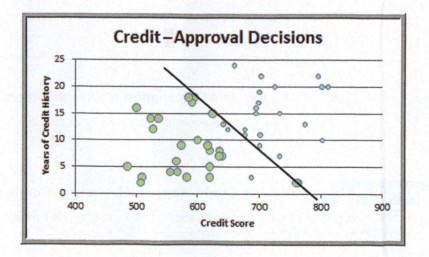

Measuring Classification Performance

As we saw in the previous example, errors may occur with any classification rule, resulting in misclassification. One way to judge the effectiveness of a classification rule is to find the probability of making a misclassification error and summarizing the results in a **classification matrix**, which shows the number of cases that were classified either correctly or incorrectly.

| EXAMPLE 10.4 | **Classification Matrix for Credit-Approval Classification Rules** |

In the credit-approval decision example, using just the credit score to classify the applications, we see that in two cases, applicants with credit scores exceeding 640 were rejected, out of a total of 50 data points. Table 10.5 shows a classification matrix for the credit score rule in Figure 10.7. The off-diagonal elements in the table are the frequencies of misclassification, whereas the diagonal elements are the numbers that were correctly classified. Therefore, the probability of misclassification was $\frac{2}{50}$, or 0.04. We leave it as an exercise for you to develop a classification matrix for the second rule.

► **Table 10.5**

Classification Matrix for Credit Score Rule

	Predicted Classification	
Actual Classification	**Decision = 1**	**Decision = 0**
Decision = 1	23	2
Decision = 0	0	25

The purpose of developing a classification model is to be able to classify new records. After a classification scheme is chosen and the best model is developed based on existing data, we use the predictor variables as inputs to the model to predict the output.

| EXAMPLE 10.5 | **Classifying Records for Credit Decisions Using Credit Scores and Years of Credit History** |

The Excel files *Credit Approval Decisions* and *Credit Approval Decisions Coded* include a small set of new records that we wish to classify in the worksheet *Records to Classify*. These records are shown in Figure 10.9. If we use the simple credit score rule from Example 10.3, that a score of more than 640 is needed to approve an application, then we would classify the decision for the first, third, and sixth records to be 1 and the rest to be 0. If we use the alternate rule developed in Example 10.3, which includes both the credit score and years of credit history—that is, reject the application if *Years* + 0.095 × *Credit Score* ≤ 74.66—then the decisions would be as follows. Only the last record would be approved.

Homeowner	Credit Score	Years of Credit History	Revolving Balance	Revolving Utilization	Years + 0.095* Credit Score	Decision
1	700	8	$21,000.00	15%	74.50	0
0	520	1	$4,000.00	90%	50.40	0
1	650	10	$8,500.00	25%	71.75	0
0	602	7	$16,300.00	70%	64.19	0
0	549	2	$2,500.00	90%	54.16	0
1	742	15	$16,700.00	18%	85.49	1

Classification Techniques

We will describe two different data-mining approaches used for classification: *k*-nearest neighbors and discriminant analysis.

▶ **Figure 10.9**

*Additional Data in the
Excel File* Credit Approval
Decisions Coded

	A	B	C	D	E	F
1						
2	Homeowner	Credit Score	Years of Credit History	Revolving Balance	Revolving Utilization	Decision
3	1	700	8	$21,000	15%	
4	0	520	1	$4,000	90%	
5	1	650	10	$8,500.00	25%	
6	0	602	7	$16,300.00	70%	
7	0	549	2	$2,500.00	90%	
8	1	742	15	$16,700.00	18%	

k-Nearest Neighbors (k-NN)

The **k-nearest neighbors (k-NN) algorithm** is a classification scheme that attempts to find records in a database that are similar to one we wish to classify. Similarity is based on the "closeness" of a record to numerical predictors in the other records. In the *Credit Approval Decisions* database, we have the predictors Homeowner, Credit Score, Years of Credit History, Revolving Balance, and Revolving Utilization. We seek to classify the decision to approve or reject the credit application.

Suppose that the values of the predictors of two records X and Y are labeled $(x_1, x_2, \ldots, x_n)$ and $(y_1, y_2, \ldots, y_n)$. We measure the distance between two records by the Euclidean distance in formula (10.1). Because predictors often have different scales, they are often standardized before computing the distance.

Suppose we have a record X that we want to classify. The nearest neighbor to that record is the one that has the smallest distance from it. The 1-NN rule then classifies record X in the same category as its nearest neighbor. We can extend this idea to a k-NN rule by finding the k-nearest neighbors to each record we want to classify, and then assigning the classification as the classification of a majority of the k-nearest neighbors. The choice of k is somewhat arbitrary. If k is too small, the classification of a record is very sensitive to the classification of the single record to which it is closest. A larger k reduces this variability, but making k too large introduces bias into the classification decisions. For example, if k is the count of the entire data set, all records will be classified the same way. Like the smoothing constants for moving average or exponential smoothing forecasting, some experimentation is needed to find the best value of k to minimize the misclassification rate. Data-mining software usually provides the ability to select a maximum value for k and evaluate the performance of the algorithm on all values of k up to the maximum specified value. Typically, values of k between 1 and 20 are used, depending on the size of the data sets, and odd numbers are often used to avoid ties in computing the majority classification of the nearest neighbors.

EXAMPLE 10.6   Using k-NN for Classifying Credit-Approval Decisions

The Excel file *Credit Approval Decisions Classification Data* provides normalized data for the credit-approval decision records (see Figure 10.10). We would like to classify the new records using the decisions that have already been made.

Consider the first new record, 51. Suppose we set $k = 1$ and find the nearest neighbor to record 51. Using the

Euclidean distance measure in formula (10.1), we find that the record having the minimum distance from record 51 is record 27. Since the credit decision was to approve, we would classify record 51 as an approval.

We can easily implement the search for the nearest neighbor in Excel using the SMALL, MATCH, and

VLOOKUP functions. To find the *k*th smallest value in an array, use the function =SMALL(*array*, *k*). To identify the record associated with this value, use the MATCH function with *match_type* = 0 for an exact match. Since the records are numbered 1 through 50, this will identify the correct record number. Then we can use the VLOOKUP function to identify the decision associated with the record. The formulas used in the example file are shown below.

Nearest Neighbors

k	Distance	Record	Decision
1	=SMALL($O$4:$O$53, 1)	=MATCH(R25, $O$4:$O$53, 0)	=VLOOKUP(S25, $A$4:$G$53, 7)
2	=SMALL($O$4:$O$53, 2)	=MATCH(R26, $O$4:$O$53, 0)	=VLOOKUP(S26, $A$4:$G$53, 7)
3	=SMALL($O$4:$O$53, 3)	=MATCH(R27, $O$4:$O$53, 0)	=VLOOKUP(S27, $A$4:$G$53, 7)
4	=SMALL($O$4:$O$53, 4)	=MATCH(R28, $O$4:$O$53, 0)	=VLOOKUP(S28, $A$4:$G$53, 7)
5	=SMALL($O$4:$O$53, 5)	=MATCH(R29, $O$4:$O$53, 0)	=VLOOKUP(S29, $A$4:$G$53, 7)

Using larger values of *k* helps to smooth the data and mitigate overfitting. Therefore, if *k* = 5, we find the following:

Nearest Neighbors

k	Distance	Record	Decision
1	1.04535	27	Approve
2	1.14457	46	Approve
3	1.17652	26	Approve
4	1.22300	23	Approve
5	1.35578	3	Approve

Because all of these records have an approve decision, we would classify record 51 as approve also. In general, we would use the majority decision, although other rules, which can impact classification error rates, can also be applied.

	I	J	K	L	M	N	O
1			Normalized Data				
2							Record 51
3	Homeowner	Credit Score	Years of Credit History	Revolving Balance	Revolving Utilization	Decision	Distance
4	0.914	0.808	1.442	-0.291	-0.704	Approve	2.28710
5	0.914	-0.881	-0.369	-0.748	0.682	Reject	2.68896
6	0.914	0.275	-0.040	0.673	0.220	Approve	1.35578
7	-1.073	-0.303	0.619	-0.126	0.528	Reject	3.02751
8	-1.073	-1.392	0.125	-0.915	0.836	Reject	3.79052
9	0.914	1.585	1.772	-0.546	-1.104	Approve	2.86543
10	-1.073	0.897	-0.698	2.360	-0.858	Approve	2.57188
11	-1.073	-0.359	-1.028	0.983	0.435	Reject	2.66695
12	0.914	-0.681	0.948	0.284	0.066	Reject	2.25230
13	0.914	0.086	2.101	-0.526	-0.396	Approve	3.03864
14	0.914	0.530	1.278	0.895	-0.919	Approve	1.81707
15	0.914	-1.692	0.784	-0.160	1.082	Reject	3.45596
16	0.914	-0.970	-0.863	-0.693	0.682	Reject	2.72095
17	-1.073	-0.359	-1.357	2.604	1.205	Reject	3.69320
18	0.914	1.352	0.290	-0.870	-1.258	Approve	2.03698
19	0.914	1.663	-0.204	-0.382	-1.320	Approve	1.68678
20	-1.073	-0.136	-0.698	0.373	0.343	Reject	2.53381
21	-1.073	-1.436	0.454	1.450	0.959	Reject	3.62116
22	0.914	1.763	1.442	-0.060	-1.381	Approve	2.50458
23	-1.073	1.230	-1.522	-0.304	0.682	Reject	3.07594
24	-1.073	-1.081	-1.192	-1.270	1.605	Reject	4.24708
25	-1.073	-0.392	-0.369	-0.615	-0.427	Reject	2.66841
26	0.914	-0.114	0.290	0.229	-0.704	Approve	1.22300
27	-1.073	0.397	-1.357	-1.181	-1.135	Approve	2.91816
28	0.914	-0.036	0.125	-0.715	-1.320	Approve	1.75921
29	0.914	0.475	0.619	0.706	-0.796	Approve	1.17652

Records to Classify

Record	Homeowner	Credit Score	Years of Credit History	Revolving Balance	Revolving Utilization	Decision
51	1	700	8	$21,000	15%	
52	0	520	1	$4,000	90%	
53	1	650	10	$8,500.00	25%	
54	0	602	7	$16,300.00	70%	
55	0	549	2	$2,500.00	90%	
56	1	742	15	$16,700.00	18%	

Normalized Records

Record	Homeowner	Credit Score	Years of Credit History	Revolving Balance	Revolving Utilization	Decision
51	0.914	0.530	-0.534	0.784	-1.012	
52	-1.073	-1.469	-1.686	-1.103	1.297	
53	0.914	-0.025	-0.204	-0.604	-0.704	
54	-1.073	-0.559	-0.698	0.262	0.682	
55	-1.073	-1.147	-1.522	-1.270	1.297	
56	0.914	0.997	0.619	0.306	-0.919	

Nearest Neighbors

k	Distance	Record	Decision
1	1.04535	27	Approve
2	1.14457	46	Approve
3	1.17652	26	Approve
4	1.22300	23	Approve
5	1.35578	3	Approve

▲ **Figure 10.10**

Portion of Credit Approval Decisions Classification Data *Excel File*

Discriminant Analysis

Discriminant analysis is a technique for classifying a set of observations into predefined classes. The purpose is to determine the class of an observation based on a set of predictor variables. We will illustrate discriminant analysis using the *Credit Approval Decisions* data. With only two classification groups, we can apply regression analysis. Unfortunately, when there are more than two, linear regression cannot be applied, and special software must be used.

EXAMPLE 10.7   **Classifying Credit Decisions Using Discriminant Analysis**

For the credit-approval data, we want to model the decision (approve or reject) as a function of the other variables. Thus, we use the following regression model, where Y represents the decision (0 or 1):

$Y = b_0 + b_1 \times$ Homeowner $+ b_2 \times$ Credit Score
$\quad + b_3 \times$ Years Credit History $+ b_4 \times$ Revolving Balance
$\quad + b_5 \times$ Revolving Utilization

The estimated value of the decision variable is called a **discriminant score**. The regression results are shown in Figure 10.11. Because Y can assume only two values, it cannot be normally distributed; therefore, the statistical results cannot be interpreted in their usual fashion. The estimated regression function is

$Y = 0.567 + 0149 \times$ Homeowner $+ 0.000465 \times$ Credit Score
$\quad + 0.00420 \times$ Years Credit History $+ 0 \times$ Revolving Balance
$\quad - 1.0986 \times$ Revolving Utilization

For example, the discriminant score for the first record would be calculated as

$Y = 0.567 + 0.149 \times 1 + 0.000465 \times 725 + 0.00420 \times 20$
$\quad + 0 \times 11320 - 1.0986 \times 25\% = 0.862$

The Excel file *Credit Approval Decisions Discriminant Analysis* shows the results (Figure 10.12). Below the data, we calculate the averages for each group of decisions. (Note that the data have been sorted by decision to facilitate computing averages.)

Next, we need a rule for classifying observations using the discriminant scores. This is done by computing a **cut-off value** so that if a discriminant score is less than or equal to it, the observation is assigned to one group; otherwise, it is assigned to the other group. While there are several ways of doing this, one simple way is to use the midpoint of the average discriminant scores:

Cut-Off Value $= (0.9083 + 0.0781)/2 = 0.4932$

We see that all approval decisions have discriminant scores above this cut-off value, while all rejection decisions have scores below it. Data-mining software has more sophisticated ways of performing the classifications. We may use this cut-off value to classify the new records. This is shown in Figure 10.13.

▶ **Figure 10.11**

Regression Results

	J	K	L	M	N	O	P	Q	R
2	SUMMARY OUTPUT								
4	*Regression Statistics*								
5	Multiple R	0.911190975							
6	R Square	0.830268994							
7	Adjusted R Square	0.810981379							
8	Standard Error	0.218884522							
9	Observations	50							
11	ANOVA								
12		*df*	*SS*	*MS*	*F*	*Significance F*			
13	Regression	5	10.3119409	2.062388181	43.04674383	7.33307E-16			
14	Residual	44	2.108059097	0.047910434					
15	Total	49	12.42						
17		*Coefficients*	*Standard Error*	*t Stat*	*P-value*	*Lower 95%*	*Upper 95%*	*Lower 95.0%*	*Upper 95.0%*
18	Intercept	0.567045347	0.478648652	1.184679712	0.242503847	-0.39760763	1.53169832	-0.39760763	1.53169832
19	Homeowner	0.149103522	0.090877595	1.640707181	0.107988621	-0.03404824	0.33225528	-0.03404824	0.33225528
20	Credit Score	0.000464676	0.00059988	0.774615018	0.442710032	-0.0007443	0.00167365	-0.0007443	0.00167365
21	Years of Credit Histor	0.004198118	0.006744824	0.622420643	0.536877785	-0.00939518	0.01779142	-0.00939518	0.01779142
22	Revolving Balance	-8.6449E-07	3.79441E-06	-0.227833217	0.820831342	-8.5116E-06	6.7826E-06	-8.5116E-06	6.7826E-06
23	Revolving Utilization	-1.09861334	0.196984059	-5.57716874	1.40586E-06	-1.49560862	-0.70161805	-1.49560862	-0.70161805

► **Figure 10.12**

Discriminant Calculations

	Homeowner	Credit Score	Years of Credit History	Revolving Balance	Revolving Utilization	Decision	Discriminant Score
Coded Credit Approval Decisions							
	1	725	20	$ 11,320	25%	1	0.8526
	1	677	11	$ 20,000	55%	1	0.4554
	1	795	22	$ 9,000	12%	1	1.0383
	0	733	7	$ 35,200	20%	1	0.6869
	1	660	24	$ 9,200	35%	1	0.7311
	1	700	19	$ 22,000	18%	1	0.9044
	1	774	13	$ 6,100	7%	1	1.0482
	1	802	10	$ 10,500	5%	1	1.0668
	1	811	20	$ 13,400	3%	1	1.1324
	1	642	13	$ 16,000	25%	1	0.7806
	0	688	3	$ 3,300	11%	1	0.7756
	1	649	12	$ 7,500	5%	1	1.0067
	1	695	15	$ 20,300	22%	1	0.8428
	1	701	9	$ 11,700	15%	1	0.9048
	1	677	12	$ 7,600	9%	1	0.9757
	1	699	17	$ 12,800	27%	1	0.8046
	1	703	22	$ 10,000	20%	1	0.9068
	1	695	16	$ 9,700	11%	1	0.9770
	1	774	13	$ 6,100	7%	1	1.0482
	1	802	10	$ 10,500	5%	1	1.0668
	1	801	20	$ 13,400	3%	1	1.1278
	1	702	11	$ 11,700	15%	1	0.9136
	1	733	15	$ 13,000	24%	1	0.8448
	1	573	9	$ 7,200	70%	0	0.2449
	0	625	15	$ 12,800	65%	0	0.1953
	0	527	12	$ 5,700	75%	0	0.0334
	0	620	5	$ 22,800	62%	0	0.1753
	1	591	17	$ 16,500	50%	0	0.4986
	1	500	16	$ 12,500	83%	0	0.0930
	1	565	6	$ 7,700	70%	0	0.2282
	0	620	3	$ 37,400	87%	0	-0.1204
	0	640	7	$ 17,300	59%	0	0.2307
	0	523	14	$ 27,000	79%	0	-0.0224
	0	763	2	$ 11,200	70%	0	0.1513
	0	555	4	$ 2,500	100%	0	-0.2590
	0	617	9	$ 8,400	34%	0	0.5107
	0	635	7	$ 29,100	85%	0	-0.0675
	0	507	2	$ 2,000	100%	0	-0.2893
	0	485	5	$ 1,000	80%	0	-0.0664
	0	582	3	$ 8,500	65%	0	0.1286
	0	585	18	$ 31,000	78%	0	0.0307
	1	620	8	$ 16,200	55%	0	0.4196
	0	640	7	$ 17,300	59%	0	0.2307
	0	536	14	$ 27,000	79%	0	-0.0164
	0	760	2	$ 11,200	70%	0	0.1499
	0	567	4	$ 2,200	95%	0	-0.1983
	0	600	10	$ 12,050	81%	0	-0.0125
	1	636	8	$ 29,100	85%	0	0.0863
	0	509	3	$ 2,000	100%	0	-0.2842
	0	595	18	$ 29,000	78%	0	0.0371
Averages							
Approve	0.9130	723.3913	14.5217	12622.6087	0.1648	1	0.9083
Reject	0.2222	591.7037	8.4444	15061.1111	0.7459	0	0.0781

Records to Classify	Homeowner	Credit Score	Years of Credit History	Revolving Balance	Revolving Utilization	Discriminant Score	Decision
	1	700	8	$21,000	15%	0.8921	Approve
	0	520	1	$4,000	90%	-0.1793	Reject
	1	650	10	$8,500.00	25%	0.7782	Approve
	0	602	7	$16,300.00	70%	0.0930	Reject
	0	549	2	$2,500.00	90%	-0.1604	Reject
	1	742	15	$16,700.00	18%	0.9117	Approve

▲ **Figure 10.13**

Classifying New Records Using Discriminant Scores

CHECK YOUR UNDERSTANDING

1. Explain the purpose of classification.

2. How is classification performance measured?

3. Explain the *k*-nearest neighbors algorithm for classification.

4. Describe when regression can be used for discriminant analysis.

Association

Association rules identify attributes that frequently occur together in a given data set. **Association rule mining**, often called *affinity analysis*, seeks to uncover interesting associations and/or correlation relationships among large sets of data. A typical and widely used example of association rule mining is **market basket analysis**. For example, supermarkets routinely collect data using barcode scanners. Each record lists all items bought by a customer for a single-purchase transaction. Such databases consist of a large number of transaction records. Managers would be interested to know if certain groups of items are consistently purchased together. They could use these data for adjusting store layouts (placing items optimally with respect to each other), for cross-selling, for promotions, for catalog design, and to identify customer segments based on buying patterns. Association rule mining is how companies such as Netflix and Amazon.com make recommendations based on past movie rentals or item purchases, for example.

EXAMPLE 10.8   **Custom Computer Configuration**

Figure 10.14 shows a portion of the Excel file *PC Purchase Data*. The data represent the configurations for a small number of orders of laptops placed over the Web. The main options from which customers can choose are the type of processor, screen size, memory, and hard drive. A "1" signifies that a customer selected a particular option.

If the manufacturer can better understand what types of components are often ordered together, it can speed up final assembly by having partially completed laptops with the most popular combinations of components configured prior to order, thereby reducing delivery time and improving customer satisfaction.

Association rules provide information in the form of if-then statements. These rules are computed from the data but, unlike the if-then rules of logic, association rules are probabilistic in nature. In association analysis, the antecedent (the "if" part) and consequent (the "then" part) are sets of items (called *item sets*) that are disjoint (do not have any items in common).

To measure the strength of association, an association rule has two numbers that express the degree of uncertainty about the rule. The first number is called the **support for the (association) rule**. The support is simply the number of transactions that include all items in the antecedent and consequent parts of the rule. (The support is sometimes

▼ Figure 10.14

Portion of the Excel File PC Purchase Data

	A	B	C	D	E	F	G	H	I	J	K	L
1	PC Purchase Data											
2												
3		Processor			Screen Size			Memory			Hard Drive	
4												
5	Intel Core i3	Intel Core i5	Intel Core i7	10-inch screen	12-inch screen	15-inch screen	2 GB	4 GB	8 GB	320 GB	500 GB	750 GB
6	0	1	0	0	1	0	0	1	0	0	1	0
7	0	1	0	0	0	1	0	0	1	0	0	1
8	0	1	0	0	1	0	0	1	0	1	0	0
9	1	0	0	0	1	0	0	0	1	0	1	0
10	0	0	1	0	0	1	0	0	1	0	0	1
11	0	0	1	0	1	0	0	1	0	0	0	1
12	0	0	1	0	0	1	0	0	1	0	0	1
13	1	0	0	0	1	0	0	1	0	0	1	0
14	0	1	0	1	0	0	1	0	0	0	1	0

expressed as a percentage of the total number of records in the database.) One way to think of support is that it is the probability that a randomly selected transaction from the database will contain all items in the antecedent and the consequent. The second number is the **confidence of the (association) rule**. Confidence is the ratio of the number of transactions that include all items in the consequent as well as the antecedent (namely, the support) to the number of transactions that include all items in the antecedent. The confidence is the conditional probability that a randomly selected transaction will include all the items in the consequent given that the transaction includes all the items in the antecedent:

$$\text{Confidence} = \text{P}(\text{Consequent}\,|\,\text{Antecedent}) = \frac{P(\text{Antecedent and Consequent})}{P(\text{Antecedent})} \quad \textbf{(10.2)}$$

The higher the confidence, the more confident we are that the association rule provides useful information.

Another measure of the strength of an association rule is **lift**, which is defined as the ratio of confidence to expected confidence. Expected confidence is the number of transactions that include the consequent divided by the total number of transactions. Expected confidence assumes independence between the consequent and the antecedent. Lift provides information about the increase in probability of the "then" (consequent) given the "if" (antecedent) part. The higher the lift ratio, the stronger the association rule; a value greater than 1.0 is usually a good minimum.

EXAMPLE 10.9   Measuring Strength of Association

Suppose that a supermarket database has 100,000 point-of-sale transactions, out of which 2,000 include both items A and B and 800 of these include item C. The association rule "If A and B are purchased, then C is also purchased" has a support of 800 transactions (alternatively, 0.8% = 800/100,000) and a confidence of 40% (= 800/2,000). Suppose the number of total transactions for C is 5,000. Then expected confidence is 5,000/100,000 = 5%, and lift = confidence/expected confidence = 40%/5% = 8.

Association rule mining requires special data-mining software to identify good rules. However, we can obtain an intuitive understanding of this technique by examining correlations, as the next example illustrates.

EXAMPLE 10.10   Using Correlations to Explore Associations

Figure 10.15 shows the correlation matrix for the data in the *PC Purchase Data* file. Of course, this only shows the correlation between pairs of variables; however, it can provide some insight for understanding associations. Higher correlations have been highlighted. For example, we see that the highest correlation is between the Intel Core i7 and a 750-GB hard drive. Twelve records have the Core i7, and 17 records have a 750-GB hard drive. If we compute the SUMPRODUCT of these two columns in the data, we find that 8 of the 67 records have both of these components. A simple association rule is *If an Intel Core i7 is chosen* (the antecedent), *then a 750-GB hard drive is purchased* (the consequent). The support for this rule is 8, and the confidence is (8/67)/(12/67) = 8/12 = 67%. The expected confidence is 17/67; thus, the lift is (8/12)/(17/67) = 2.63.

We also see a moderate correlation between the Core i7 and an 8-GB memory. Thus, we might propose the rule *If an Intel Core i7 and 8-GB memory are purchased* (the antecedent), *then a 750-GB hard drive is purchased* (the consequent). In this case, only four records have all three; hence the support is 4. Six records have both components of the antecedent; therefore, the confidence would

(continued)

be (4/67)/(6/67) = 4/6 = 67%. The expected confidence is 17/67; therefore, the lift is (4/6)/(17/67) = 2.63. Neither rule has a stronger association than the other.

Finally, we also see a moderate correlation between a 15-inch screen and the 750-GB hard drive, as well as between a 15-inch screen and 8-GB memory. Nineteen records have a 15-inch screen, 13 have 8-GB memory, and

6 have both. Five records have all three items. Consider the rule *If a 15-inch screen and 8-GB memory are purchased, then a 750-GB hard drive is purchased*. Thus, the support is 5, and the confidence is (5/67)/(6/67) = 5/6 = 83%. The expected confidence is 17/67, so the lift is (5/6)/(17/67) = 3.28. This rule has a higher lift ratio and is therefore a stronger association rule.

	A	Intel Core i3	Intel Core i5	Intel Core i7	10 inch screen	12 inch screen	15 inch screen	2 GB	4 GB	8 GB	320 GB	500 GB	750 GB
2	Intel Core i3	1											
3	Intel Core i5	-0.68884672	1										
4	Intel Core i7	-0.32659863	-0.46017899	1									
5	10 inch screen	0.279261486	-0.06166099	-0.26162798	1								
6	12 inch screen	0.031339159	-0.1052632	0.098863947	-0.535569542	1							
7	15 inch screen	-0.29888445	0.174976377	0.137915917	-0.352396093	-0.601585208	1						
8	2 GB	0.103561074	0.111614497	-0.27236339	0.075646007	0.060469105	-0.138564809	1					
9	4 GB	0.097654391	-0.04316721	-0.06331855	0.06538164	-0.009001195	-0.051869866	-0.59824393	1				
10	8 GB	-0.18232971	-0.10591632	0.361405355	-0.186296835	-0.015788849	0.193716103	-0.28609763	-0.56165237	1			
11	320 GB	0.194695858	0.042508385	-0.29387691	0.191267616	-0.203826567	0.04495614	0.013629326	0.282291771	-0.30869598	1		
12	500 GB	-0.13890029	0.223414565	-0.12117953	0.041916288	0.191413544	-0.251770128	0.078031873	-0.09558351	-0.00112979	-0.58382915	1	
13	750 GB	-0.04251455	-0.30002796	0.443258072	-0.246149705	-0.008199201	0.241920528	-0.10352941	-0.18288286	0.32105073	-0.36685601	-0.54108944	1

▲ **Figure 10.15**

PC Purchase Data *Correlation Matrix*

CHECK YOUR UNDERSTANDING

1. What is association rule mining?

2. Explain the concepts of support, confidence, and lift in association rule mining.

Cause-and-Effect Modeling

Managers are always interested in results, such as profit, customer satisfaction and retention, and production yield. **Lagging measures**, or outcomes, tell what has happened and are often external business results, such as profit, market share, and customer satisfaction. **Leading measures** (performance drivers) predict what *will* happen and usually are internal metrics, such as employee satisfaction, productivity, and turnover. For example, customer satisfaction results in regard to sales or service transactions are a lagging measure; employee satisfaction, sales representative behavior, billing accuracy, and so on are examples of leading measures that might influence customer satisfaction. If employees are not satisfied, their behavior toward customers could be negatively affected, and customer satisfaction could be low. If this can be explained using business analytics, managers can take steps to improve employee satisfaction, leading to improved customer satisfaction. Therefore, it is important to understand what controllable factors significantly influence key business performance measures that managers cannot directly control. Correlation analysis can help to identify these influences and lead to the development of cause-and-effect models that can help managers make better decisions today that will influence results tomorrow.

Recall from Chapter 4 that correlation is a measure of the linear relationship between two variables. High values of the correlation coefficient indicate strong relationships between the variables. The following example shows how correlation can be useful in cause-and-effect modeling.

EXAMPLE 10.11   Using Correlation for Cause-and-Effect Modeling

The Excel file *Ten Year Survey* shows the results of 40 quarterly surveys conducted by a major electronics device manufacturer, a portion of which is shown in Figure 10.16.[5] The data provide average scores on a 1–5 scale for customer satisfaction, overall employee satisfaction, employee job satisfaction, employee satisfaction with their supervisor, and employee perception of training and skill improvement. Figure 10.17 shows the correlation matrix. All the correlations except the one between job satisfaction and customer satisfaction are relatively strong, with the highest correlations between overall employee satisfaction and employee job satisfaction, employee

satisfaction with their supervisor, and employee perception of training and skill improvement.

Although correlation analysis does not prove any cause and effect, we can logically infer that a cause-and-effect relationship exists. The data indicate that customer satisfaction, the key external business result, is strongly influenced by internal factors that drive employee satisfaction. Logically, we could propose the model shown in Figure 10.18. This suggests that if managers want to improve customer satisfaction, they need to start by ensuring good relations between supervisors and their employees and focus on improving training and skills.

	A	B	C	D	E	F
1	**Ten Year Survey**					
2						
3	**Survey Sample**	**Customer satisfaction**	**Employee satisfaction**	**Job satisfaction**	**Satisfaction with supervisor**	**Training and skill improvement**
4	1	2.97	3.51	3.92	3.06	3.48
5	2	3.71	3.58	4.13	3.06	2.57
6	3	3.29	3.43	3.62	4.42	3.06
7	4	2.05	3.81	4.12	4.31	3.17
8	5	4.56	4.17	4.25	4.14	4.15
9	6	4.28	4.13	4.13	4.57	3.61
10	7	2.17	2.42	4.19	2.53	2.72
11	8	3.01	2.95	3.95	3.25	2.56

▲ Figure 10.16

Portion of Ten Year Survey *Data*

	A	B	C	D	E	F
1		*Customer satisfaction*	*Employee satisfaction*	*Job satisfaction*	*Satisfaction with supervisor*	*Training and skill improvement*
2	Customer satisfaction	1				
3	Employee satisfaction	0.493345395	1			
4	Job satisfaction	0.151693544	0.840444148	1		
5	Satisfaction with supervisor	0.495977225	0.881324581	0.606796166	1	
6	Training and skill improvement	0.532307756	0.828657884	0.710624973	0.769700425	1

▲ Figure 10.17

Correlation Matrix of Ten Year Survey *Data*

CHECK YOUR UNDERSTANDING

1. What is the difference between a leading and lagging measure?

2. How is correlation used in cause-and-effect modeling?

---

[5]Based on a description of a real application by Steven H. Hoisington and Tse-His Huang, "Customer Satisfaction and Market Share: An Empirical Case Study of IBM's AS/400 Division," in Earl Naumann and Steven H. Hoisington (eds.) *Customer-Centered Six Sigma* (Milwaukee, WI: ASQ Quality Press, 2001). The data used in this example are fictitious, however.

▶ **Figure 10.18**

Cause-and-Effect Model

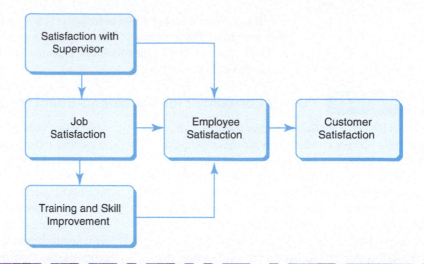

ANALYTICS IN PRACTICE: Successful Business Applications of Data Mining[6]

Many different companies use data mining to segment customers, identify the most profitable types of customers, reduce costs, and enhance customer relationships through improved marketing efforts. Some successful application areas of data mining include the following:

- A pharmaceutical company analyzed sales force activity data to better target high-value physicians and determine which marketing activities will have the greatest impact. Sales representatives can use the results and plan their schedules and promotional activities.

- A credit-card company used data mining to analyze customer transaction data to identify customers most likely to be interested in a new credit product. As a result, costs for mail campaigns decreased by more than 20 times.

- A large transportation company used data mining to segment its customer base and identify the best types of customers for its services. By applying this segmentation to a general business database such as those provided by Dun & Bradstreet, they can develop a prioritized list of prospects for its regional sales force members.

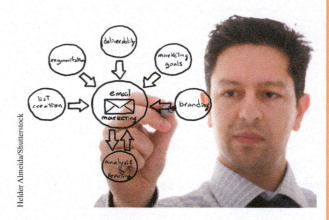

Helder Almeida/Shutterstock

- A large consumer package goods company applied data mining to improve its retail sales process. They used data from consumer panels, shipments, and competitor activity to understand why customers choose different brands and switch stores. Armed with this data, the company can select more effective promotional strategies.

KEY TERMS

Agglomerative clustering methods	Average linkage clustering
Association rule mining	Classification matrix
Average group linkage clustering	Cluster analysis

[6]Based on Kurt Thearling, "An Introduction to Data Mining," White Paper from Thearling.com. http://www.thearling.com/text/dmwhite/dmwhite.htm.

Complete linkage clustering
Cut-off value
Confidence of the (association) rule
Data mining
Dendogram
Discriminant analysis
Discriminant score
Divisive clustering methods
Euclidean distance
Hierarchical clustering

k-nearest neighbors (k-NN) algorithm
Lagging measures
Leading measures
Lift
Market basket analysis
Normalize
Single linkage clustering
Support for the (association) rule
Ward's hierarchical clustering

CHAPTER 10 TECHNOLOGY HELP

Useful Excel Functions

SUMXMY2(*array_x*, *array_y*) Sums the squares of the differences in two corresponding ranges or arrays.

SMALL(*array*, *k*) Finds the *k*th smallest value in an array.

MATCH(*lookup_value*, *lookup_array*, *0*) Finds the position of an exact match to a *lookup_value* in a *lookup_array*.

Analytic Solver

Analytic Solver provides powerful data-mining tools. See the online supplement *Using Data Mining in Analytic Solver*. We suggest that you first read the online supplement *Getting Started with Analytic Solver Basic*. This provides information for both instructors and students on how to register for and access Analytic Solver.

PROBLEMS AND EXERCISES

Cluster Analysis

1. Compute the Euclidean distance between the following sets of points:
   a. (2, 5) and (8, 4)
   b. (12, −1, 32) and (18, 15, −52)

2. For the Excel file *Colleges and Universities Cluster Analysis Worksheet*, compute the normalized Euclidean distances between Berkeley, Cal Tech, UCLA, and UNC, and illustrate the results in a distance matrix.

3. For the three clusters identified in Table 10.3, find the average and standard deviations of each numerical variable for the schools in each cluster and compare them with the average and standard deviation for the entire data set. Does the clustering show distinct differences among these clusters?

4. In Problem 2, you found a normalized distance matrix between Berkeley, Cal Tech, UCLA, and UNC for the Excel file *Colleges and Universities Cluster Analysis Worksheet*. Apply single linkage clustering to these schools and draw a dendogram illustrating the clustering process.

5. Using only Credit Score, Years of Credit History, Revolving Balance, and Revolving Utilization,

apply single linkage cluster analysis to the first six records in the Excel file *Credit Approval Decisions* and draw a dendogram illustrating the clustering process.

6. Apply single linkage cluster analysis to the first five records in the Excel file *Sales Data*, using the variables Percent Gross Profit, Industry Code, and Competitive Rating, and draw a dendogram illustrating the clustering process.

Classification

7. Using the approach described in Example 10.6, classify first record in the worksheet *Records to Classify* in the Excel file *Credit Risk Data* using the k-NN algorithm for $k = 1$ to 5. Use only Checking, Savings, Months Customer, and Months Employed.

8. Use the k-NN algorithm to classify the new records in the Excel file *Credit Approval Decisions Classification Data* using only Credit Score and Years of Credit History for $k = 1$ to 5.

9. Use discriminant analysis to classify the new records in the Excel file *Credit Approval Decisions Discriminant Analysis* using only Credit Score and Years of Credit History as input variables.

10. Extract the records for business loans in the Excel file *Credit Risk Data* and code the non-numerical data. Apply discriminant analysis to classify the credit risk for the business loans in the *Records to Classify* worksheet.

Association

11. The Excel file *Automobile Options* provides data on options ordered together for a particular model of automobile. By examining the correlation matrix, suggest some associations.

12. The Excel file *Automobile Options* provides data on options ordered together for a particular model of automobile. Consider the following rules:

Rule 1: If Fastest Engine, then Traction Control.

Rule 2: If Faster Engine and 16-inch Wheels, then 3 Year Warranty.

Compute the support, confidence, and lift for each of these rules.

Cause and Effect Modeling

13. The Excel file *Myatt Steak House* provides five years of data on key business results for a restaurant. Identify the leading and lagging measures, find the correlation matrix, and propose a cause-and-effect model using the strongest correlations.

CASE: PERFORMANCE LAWN EQUIPMENT

The worksheet *Purchasing Survey* in the *Performance Lawn Care* Database provides data related to predicting the level of business (Usage Level) obtained from a third-party survey of purchasing managers of firms that are customers of Performance Lawn Equipment.[6] The seven PLE attributes rated by each respondent are the following:

Delivery speed—the amount of time it takes to deliver the product once an order is confirmed

Price level—the perceived level of price charged by PLE

Price flexibility—the perceived willingness of PLE representatives to negotiate price on all types of purchases

Manufacturing image—the overall image of the manufacturer

Overall service—the overall level of service necessary for maintaining a satisfactory relationship between PLE and the purchaser

Sales force image—the overall image of PLE's sales force

Product quality—perceived level of quality

Responses to these seven variables were obtained using a graphic rating scale, where a 10-centimeter line was drawn between endpoints labeled "poor" and "excellent." Respondents indicated their perceptions using a mark on the line, which was measured from the left endpoint. The result was a scale from 0 to 10, rounded to one decimal place.

Two measures were obtained that reflected the outcomes of the respondent's purchase relationships with PLE:

Usage level—how much of the firm's total product is purchased from PLE, measured on a 100-point scale, ranging from 0% to 100%

Satisfaction level—how satisfied the purchaser is with past purchases from PLE, measured on the same graphic rating scale as perceptions 1 through 7

The data also include four characteristics of the responding firms:

Size of firm—size relative to others in this market (0 = small; 1 = large)

Purchasing structure—the purchasing method used in a particular company (1 = centralized procurement, 0 = decentralized procurement)

Industry—the industry classification of the purchaser [1 = retail (resale such as Home Depot), 0 = private (nonresale, such as a landscaper)]

Buying type—a variable that has three categories (1 = new purchase, 2 = modified rebuy, 3 = straight rebuy)

Elizabeth Burke would like to understand what she learned from these data.

a. What types of data-mining techniques might provide useful information? Explain.

b. Develop a cause-and-effect model that can provide insights about the drivers of satisfaction and usage level.

Summarize your results in a report to Ms. Burke.

---

[6] The data and description of this case are based on the HATCO example on pages 28–29 in Joseph F. Hair, Jr., Rolph E. Anderson, Ronald L. Tatham, and William C. Black, *Multivariate Analysis*, 5th ed. (Upper Saddle River, NJ: Prentice Hall, 1998).

Spreadsheet Modeling and Analysis

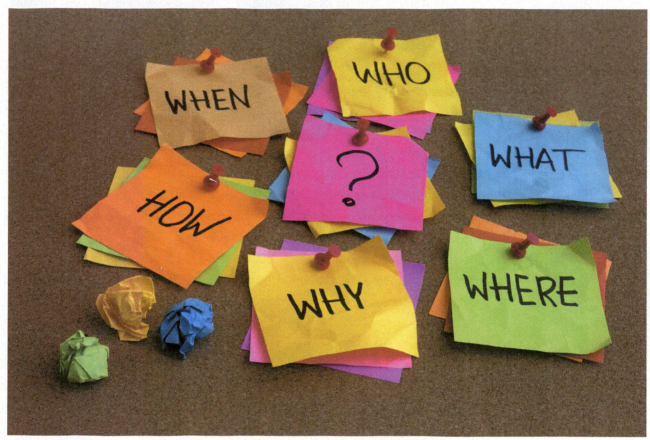

marekuliasz/Shutterstock

LEARNING OBJECTIVES   After studying this chapter, you will be able to:

- Explain how to use logic, business principles, influence diagrams, and historical data to develop analytic decision models.
- Apply principles of spreadsheet engineering to designing and implementing spreadsheet models.

- Use Excel features and spreadsheet engineering to ensure the quality of spreadsheet models.
- Explain how model validity can be assessed.
- Build spreadsheet models for descriptive, predictive, and prescriptive applications.
- Perform what-if analysis on spreadsheet models.

- Construct one- and two-way data tables using Excel.
- Use data tables to analyze uncertainty in decision models.

- Use the Excel *Scenario Manager* to evaluate different model scenarios.
- Apply the Excel *Goal Seek* tool for break-even analysis and other types of models.

Models are essential to understanding decision problems that managers face, predicting the outcomes of decision alternatives, and finding the best decisions to make; therefore, modeling is the heart and soul of business analytics.

We introduced the concept of a decision model in Chapter 1. Decision models transform inputs—data, uncontrollable inputs, and decision options—into outputs, or measures of performance or behavior. When we build a decision model, we are essentially predicting what outputs will occur based on the model inputs. The model itself is simply a set of assumptions that characterize the relationships between the inputs and the outputs. The quality of a model depends on the quality of the assumptions used to create it. This can only be assessed by comparing model outputs to known results, either from the past or in the future. Thus, model assumptions must be based either on sound logic and experience or on the analysis of historical data that may be available.

In this chapter, we focus on building logical and useful decision models using spreadsheets. We also describe approaches for analyzing models to evaluate future scenarios and ask "what-if" questions to facilitate better business decisions.

Being able to build spreadsheet models, understand their logic, and use them to evaluate scenarios are important skills that professionals in all areas of business can use in their daily work. For example, financial analysts use spreadsheet models for predicting cash flow requirements in the future; marketing analysts use them for evaluating the impact of advertising decisions and budget allocations; operations personnel use them for planning production and staffing requirements. Spreadsheet models are also used routinely in managerial accounting, investment banking, consulting activities, and many other areas. We will begin by considering some useful approaches to initiate the model-building process in order to get a better understanding of the scenarios that we wish to model using spreadsheets.

ANALYTICS IN PRACTICE: Using Spreadsheet Modeling and Analysis at Nestlé[1]

Nestlé is a large and well-known global food and beverage company. Nestlé's executive information system (EIS) department gathers data from the firm's subsidiaries (reporting units) to provide top management with operational, financial, and strategic information. The EIS department decided to improve its service by using Excel-based business analytics tools and encourage analysts and controllers to make better use of the information that these tools provide. They developed descriptive, predictive, and prescriptive spreadsheet models for evaluating the economic profitability of investment decisions, setting selling prices, managing cash flows, and evaluating business opportunities and risks.

Through this process, they discovered that spreadsheets were extremely useful for developing small and simple (but not simplistic) quality models. As a result, they made a concerted effort to increase the number of managers accustomed to using spreadsheet models and other analytic tools in their decision making. To do this, they developed and conducted seminars to train employees to use Excel to develop small spreadsheet models devoted to sensitivity analysis, forecasting, simulation, and optimization using mini-case studies. The teaching materials were adapted from real Nestlé data (for example, a forecast of the annual sales of ice cream in the U.S.). The goal was to encourage users to explore new problems by themselves and to apply these small and simple tools. The training courses became popular among Nestlé's employees, increased the number of managers accustomed to analytic decision making, and established new reporting protocols imposing the use of analytic models.

Model-Building Strategies

Building decision models is more of an art than a science. Creating good decision models requires a solid understanding of basic business principles in all functional areas, such as accounting, finance, marketing, and operations, knowledge of business practice and research, and logical skills. Models often evolve from simple to complex and from deterministic to stochastic (see the definitions in Chapter 1), so it is generally best to start simple and enrich models as necessary.

Building Models Using Logic and Business Principles

We introduced the concept of using logic and business principles to develop models in Chapter 1. For instance, Example 1.3 showed how to use the dimensions of terms to help develop logically consistent formulas. Example 1.4 described a decision model for outsourcing. Let us look more closely at the logic of how that model was developed.

EXAMPLE 11.1   A Total Cost Decision Model

From basic business principles, we know that the total cost of producing a fixed volume of a product is composed of fixed cost and variable cost. We also know that the variable cost depends on the unit variable cost as well as the quantity produced. The quantity produced, however, is a decision option because it can be controlled by the manager of the operation.

To develop a mathematical model, we need to specify the precise nature of the relationships among these quantities. For example, we can easily state that

Total Cost = Fixed Cost + Variable Cost   **(11.1)**

(continued)

[1] Adapted from Christophe Oggier, Emmanuel Fragnière, and Jeremy Stuby, "Nestlé Improves Its Financial Reporting with Management Science," *Interfaces*, Vol. 35, No. 4, July–August 2005, 271–280.

Logic also suggests that the variable cost is the unit variable cost times the quantity produced. Thus,

Variable Cost = Unit Variable Cost × Quantity Produced
(11.2)

By substituting this into equation (11.1), we have

Total Cost = Fixed Cost + Variable Cost

= Fixed Cost + Unit Variable Cost
× Quantity Produced   **(11.3)**

Using these relationships, we may develop a mathematical representation by defining symbols for each of these quantities:

C = Total cost
V = Unit variable cost
F = Fixed cost
Q = Quantity produced

This results in the following model:

$$C = F + VQ \qquad \textbf{(11.4)}$$

This descriptive model can be used to evaluate the cost for any input value of the quantity produced.

Building Models Using Influence Diagrams

Although it can be easy to develop a model from basic logic and business principles, as we illustrated in the previous example, most model development requires a more formal approach. A simple descriptive model is a visual representation called an **influence diagram** because it describes how various elements of the model influence, or relate to, others. An influence diagram is a useful approach for conceptualizing the structure of a model and can assist in building a mathematical or spreadsheet model. The elements of the model are represented by circular symbols called *nodes*. Arrows called *branches* connect the nodes and show which elements influence others. Influence diagrams are quite useful in the early stages of model building when we need to understand and characterize key relationships.

| **EXAMPLE** 11.2 | **Developing a Decision Model Using an Influence Diagram** |

We will develop a decision model for predicting profit in the face of uncertain future demand using an influence diagram, which is shown in Figure 11.1. We all know that profit depends on both revenues and costs. Thus, at the top of the influence diagram we see that both revenue and cost influence profit. Using some basic business logic, we can expand the influence diagram by adding more information. Revenue depends on the unit price and the quantity sold, and cost depends on the unit cost, quantity produced, and fixed costs of production. The quantity sold depends on the uncertain demand as well as the quantity produced, since we cannot sell more than the demand or the quantity produced, whichever is less. These facts are reflected in the influence diagram shown in Figure 11.1. In this figure, all the nodes that have no branches pointing into them are inputs to the model. Unit price, unit cost, and fixed cost are data inputs; demand is an uncertain input, and the quantity produced is a decision option because it can be controlled by the manager of the operation. Profit is the output (note that it has no branches pointing out of it) that we wish to calculate. Any nodes that have arrows pointing both in and out of them are intermediate

calculations that link the inputs with the output and can be considered as "building blocks" of the model.

The next step is to translate the influence diagram into a more formal model using the following definitions:

P = profit
R = revenue
C = cost
p = unit price
c = unit cost
F = fixed cost
S = quantity sold
Q = quantity produced
D = demand

A logical approach is to work down in the influence diagram starting from the output. First, profit = revenue - cost; thus,

$$P = R - C$$

Next, revenue equals the unit price (p) multiplied by the quantity sold (S):

$$R = p \times S$$

Note that cost consists of the fixed cost (F) plus the variable cost of producing Q units ($c \times Q$):

$$C = F + c \times Q$$

The quantity sold, however, must be the smaller of the demand (D) and the quantity produced (Q), or

$$S = \min\{D, Q\}$$

Therefore, $R = p \times S = p \times \min\{D, Q\}$. Substituting these results into the basic formula for profit $P = R - C$, we have

$$P = p \times \min\{D, Q\} - (F + c \times Q) \qquad \textbf{(11.5)}$$

Building Models Using Historical Data

Data used in models can come from existing databases and other data sources, analysis of historical data, or surveys, experiments, and other methods of data collection. In the profit model, for example, we might query accounting records for values of the unit cost and fixed costs. Statistical methods that we have studied are often used to estimate data required in predictive models. For instance, we might use historical data to compute the mean demand; we might also use quartiles or percentiles in the model to evaluate different scenarios. The next example shows how we might use empirical data and line-fitting techniques to build a model.

EXAMPLE 11.3   Building a Decision Model Using Historical Data[2]

The DTP Corporation has two major products. Marketing analysts have conducted experiments to gather data on the effect of media advertising on profits. These data are shown in Figure 11.2 (see the Excel file *DTP Corporation*). First, create scatter charts for the data. Then, using the *Add Trendline* feature in Excel, we find that logarithmic functions can adequately fit the data:

Product 1 Profit = 49.699 + 1.1568 ln(X_1)
Product 2 Profit = 19.913 + 0.4177 ln(X_2)

where X_1 and X_2 are the amount spent on advertising products 1 and 2, respectively. The total profit as a function of the amount spent on advertising each product can therefore be expressed as:

Total Profit = 49.699 + 1.1568 ln(X_1) + 19.913 + 0.4177 ln(X_2)

= 69.612 + 1.1568 ln(X_1) + 0.4177 ln(X_2)

▶ **Figure 11.1**

An Influence Diagram for Profit

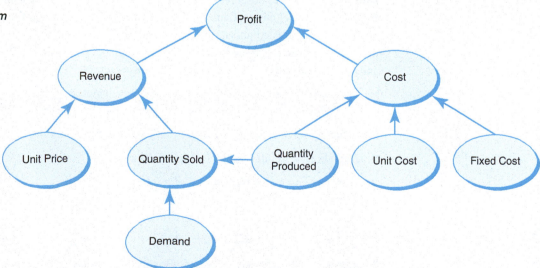

[2] Glen L. Urban, "Building Models for Decision Makers," *Interfaces*, 4, 3 (May 1974): 1–11.

▶ **Figure 11.2**

DTP Corporation
Marketing Data

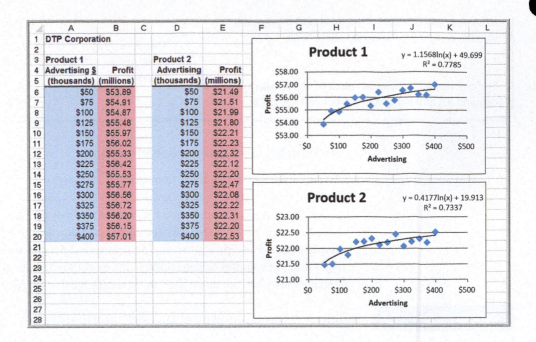

However, even if data are not available, using a good subjective estimate is better than sacrificing the completeness of a model that may be useful to managers.

Model Assumptions, Complexity, and Realism

Models cannot capture every detail of a real problem, and managers must understand the limitations of models and their underlying assumptions. Validity refers to how well a model represents reality. One approach for judging the validity of a model is to identify and examine the assumptions made in a model to see how they agree with our perception of the real world; the closer the agreement, the higher the validity. Another approach is to compare model results to observed results; the closer the agreement, the more valid the model. A "perfect" model corresponds to the real world in every respect; unfortunately, no such model has ever existed and never will, because it is impossible to include every detail of real life in one model. To add more realism to a model generally requires more complexity, and analysts have to know how to balance these factors.

 CHECK YOUR UNDERSTANDING

1. What are three types of strategies that can be used in building decision models?

2. Explain how to interpret an influence diagram.

3. How can you assess the validity of a model?

Implementing Models on Spreadsheets

Spreadsheets have the advantage of allowing you to easily modify the model inputs and calculate the numerical results. We will use both spreadsheets and analytical modeling approaches in our model-building applications—it is important to be able to "speak both languages." We may creatively apply various Excel tools and capabilities to improve the structure and use of spreadsheet models. In this section, we discuss approaches for developing good, useful, and correct spreadsheet models. Good spreadsheet analytic applications should also be user-friendly; that is, it should be easy to input or change data and see key results, particularly for users who may not be as proficient in using spreadsheets. Good design reduces the potential for errors and misinterpretation of information, leading to more insightful decisions and better results.

Spreadsheet Design

Because decision models characterize the relationships between inputs and outputs, it is useful to separate the data, model calculations, and model outputs clearly in designing a spreadsheet. It is particularly important not to use input data in model formulas, but to reference the spreadsheet cells that contain the data. In this way, if the data change or if you want to experiment with the model, you need not change any of the formulas, which can easily result in errors. We illustrate these concepts in the following example.

EXAMPLE 11.4 Spreadsheet Implementation of the Profit Model

The analytical model we developed in Example 11.2 can easily be implemented in an Excel spreadsheet to evaluate profit (Excel file *Profit Model*). Let us assume that unit price = \$40, unit cost = \$24, fixed cost = \$400,000, and demand = 50,000. For the purposes of building a spreadsheet model, we assume that the quantity produced is 40,000 units, although as we have noted, this is a decision option that can be changed. Figure 11.3 shows a spreadsheet implementation of this model. To better understand the model, study the relationships between the spreadsheet formulas, the influence diagram, and the mathematical model. Observe the correspondence between the spreadsheet formulas and the model, for instance:

$$\text{Profit (cell C22)} = \text{Revenue} - \text{Variable Cost} - \text{Fixed Cost}$$
$$= \text{C15} - \text{C19} - \text{C20}$$

Thus, if you can write a spreadsheet formula, you can develop a mathematical model by substituting symbols or numbers into the Excel formulas. Notice also how the data are separated from the model and referenced in the model formulas. A manager might use the spreadsheet to evaluate how profit would be expected to change for different values of the uncertain future demand and/or the quantity produced. We do this later in this chapter.

▶ **Figure 11.3**

Spreadsheet Implementation of Profit Model

	A	B	C
1	Profit Model		
2			
3	Data		
4			
5	Unit Price	\$40.00	
6	Unit Cost	\$24.00	
7	Fixed Cost	\$400,000.00	
8	Demand	50000	
9			
10			
11	Model		
12			
13	Unit Price	\$40.00	
14	Quantity Sold	40000	
15	Revenue		\$1,600,000.00
16			
17	Unit Cost	\$24.00	
18	Quantity Produced	40000	
19	Variable Cost		\$960,000.00
20	Fixed Cost		\$400,000.00
21			
22	Profit		\$240,000.00

	A	B	C
1	Profit Model		
2			
3	Data		
4			
5	Unit Price	40	
6	Unit Cost	24	
7	Fixed Cost	400000	
8	Demand	50000	
9			
10			
11	Model		
12			
13	Unit Price	=B5	
14	Quantity Sold	=MIN(B8,B18)	
15	Revenue		=B13*B14
16			
17	Unit Cost	=B6	
18	Quantity Produced	40000	
19	Variable Cost		=B17*B18
20	Fixed Cost		=B7
21			
22	Profit		=C15-C19-C20

There are often many options when implementing a model on a spreadsheet. It is important to keep the end user in mind and format the spreadsheet model in a form that he or she—who may be a financial manager, for example—can easily interpret and use. In the following example, we show three different ways of implementing a financial model to illustrate this issue.

EXAMPLE 11.5   Modeling Net Income on a Spreadsheet

The calculation of net income is based on the following formulas:

- gross profit = sales − cost of goods sold
- operating expenses = administrative expenses
  + selling expenses
  + depreciation expenses
- net operating income = gross profit −
  − operating expenses
- earnings before taxes = net operating income
  − interest expense
- net income = earnings before taxes − taxes

We could develop a simple model to compute net income using these formulas by substitution:

net income = sales − cost of goods sold − administrative
  expenses − selling expenses − depreciation
  expenses − interest expense − taxes

We can implement this model on a spreadsheet, as shown in Figure 11.4. This spreadsheet provides only the

end result and, from a financial perspective, provides little information to the end user.

An alternative is to break down the model by writing the preceding formulas in separate cells in the spreadsheet using a data-model format, as shown in Figure 11.5. This clearly shows the individual calculations and provides better information. However, although both of these models are technically correct, neither is in the form to which most accounting and financial employees are accustomed.

A third alternative is to express the calculations as a **pro forma income statement** using the structure and formatting that accountants are used to, as shown in Figure 11.6. Although this has the same calculations as in Figure 11.5, note that the use of negative dollar amounts requires a change in the formulas (that is, addition of negative amounts rather than subtraction of positive amounts). The Excel workbook *Net Income Models* contains each of these examples in separate worksheets.

Spreadsheet Quality

Building spreadsheet models, often called **spreadsheet engineering**, is part art and part science. The quality of a spreadsheet can be assessed by both its logical accuracy and its design. Spreadsheets need to be accurate, understandable, and user-friendly.

First and foremost, spreadsheets should be accurate. **Verification** is the process of ensuring that a model is accurate and free from logical errors. Spreadsheet errors can be

▶ **Figure 11.4**

Simple Spreadsheet Model for Net Income

▶ **Figure 11.5**

Data-Model Format for Net Income

	A	B	C
1	Net Income Model - Data Model Format		
2			
3	Data		
4			
5	Sales	$ 5,000,000	
6	Cost of Goods Sold	$ 3,200,000	
7	Administrative Expenses	$ 250,000	
8	Selling Expenses	$ 450,000	
9	Depreciation Expenses	$ 325,000	
10	Interest Expense	$ 35,000	
11	Taxes	$ 296,000	
12			
13	Model		
14			
15	Gross Profit	$ 1,800,000	=B5-B6
16	Operating Expenses	$ 1,025,000	=SUM(B7:B9)
17	Net Operating Income	$ 775,000	=B15-B16
18	Earnings Before Taxes	$ 740,000	=B17-B10
19			
20	Net Income	$ 444,000	=B18-B11

▶ **Figure 11.6**

Pro Forma Income Statement Format

	A	B	C	D
1	Pro Forma Income Statement			
2				
3	Sales		$ 5,000,000	
4	Cost of Goods Sold		$(3,200,000)	
5	Gross Profit		$ 1,800,000	=C3+C4
6				
7	Operating Expenses			
8	Administrative Expenses	$250,000		
9	Selling Expenses	$450,000		
10	Depreciation Expenses	$325,000		
11	Total		$(1,025,000)	=-(SUM(B8:B10))
12				
13	Net Operating Income		$ 775,000	=C5+C11
14	Interest Expense		$ (35,000)	
15				
16	Earnings Before Taxes		$ 740,000	=C13+C14
17	Taxes		$ (296,000)	
18				
19	Net Income		$ 444,000	=C16+C17

disastrous. A large investment company once made a $2.6 billion error. They notified holders of one mutual fund to expect a large dividend; fortunately, they caught the error before sending the checks. One research study of 50 spreadsheets found that fewer than 10% were error free.[3] Significant errors in business have resulted from mistakes in copying and pasting, sorting, numerical input, and spreadsheet formula references. Industry research has found that more than 90% of spreadsheets with more than 150 rows were incorrect by at least 5%.

There are three basic approaches to spreadsheet engineering that can improve spreadsheet quality:

1. *Improve the design and format of the spreadsheet itself.* After the inputs, outputs, and key model relationships are well understood, you should sketch a logical design of the spreadsheet. For example, you might want the spreadsheet to resemble a financial statement to make it easier for managers to read. It is good practice to separate the model inputs from the model itself and to reference the input cells in the model formulas; that way, any changes in the

---

[3] S. Powell, K. Baker, and B. Lawson, "Errors in Operational Spreadsheets," *Journal of End User Computing*, 21 (July–September 2009): 24–36.

inputs will be automatically reflected in the model. We have done this in the examples.

Another useful approach is to break complex formulas into smaller pieces. We also did this in the previous examples. This reduces typographical errors, makes it easier to check your results, and makes the spreadsheet easier to read for the user.

2. *Improve the process used to develop a spreadsheet.* If you sketched out a conceptual design of the spreadsheet, work on each part individually before moving on to the others to ensure that each part is correct. As you enter formulas, check the results with simple numbers (such as 1) to determine if they make sense, or use inputs with known results. Be careful in using the *Copy* and *Paste* commands in Excel, particularly with respect to relative and absolute addresses. Use the Excel function wizard (the f_x button on the formula bar) to ensure that you are entering the correct values in the correct fields of the function.

3. *Inspect your results carefully, and use appropriate tools available in Excel.* For example, the Excel *Formula Auditing* tools (in the *Formulas* tab) help you validate the logic of formulas and check for errors. Using *Trace Precedents* and *Trace Dependents*, you can visually show what cells affect or are affected by the value of a selected cell, similar to an influence diagram. The *Formula Auditing* tools also include *Error Checking*, which checks for common errors that occur when using formulas, and *Evaluate Formula*, which helps to debug a complex formula by evaluating each part of the formula individually. We encourage you to learn how to use these tools.

Other Excel tools that can be used to improve the quality and user-friendliness of spreadsheet models include range names and form controls, both of which were introduced in Chapter 2.

Data Validation

One useful Excel tool is the **data validation** feature, which allows you to define acceptable input values in a spreadsheet and provides an error alert if an invalid entry is made. This can help you avoid inadvertent user errors. This feature can be found in the *Data Tools* group within the *Data* tab on the Excel ribbon. Select the cell range, click on *Data Validation*, and then specify the criteria that Excel will use to flag invalid data.

EXAMPLE 11.6   **Using Data Validation**

Figure 11.7 shows a spreadsheet model for the *Outsourcing Decision Model* we discussed in Example 1.4. Observe how the IF function is used in cell B20 to identify the best decision. If the cost difference is negative or zero, then the function returns "Manufacture" as the best decision; otherwise, it returns "Outsource."

Suppose that you are asked to use the spreadsheet to evaluate the manufacturing and purchase cost options and best decisions for a large number of parts used in an automobile system assembly. You are given lists of data that cost accountants and purchasing managers have compiled and printed and must look up the data and enter them into the spreadsheet. Such a manual process

leaves plenty of opportunity for error. However, suppose that we know that the unit cost of any item is at least $10 but no more than $100. If a cost is $47.50, for instance, a misplaced decimal would result in either $4.75 or $475, which would clearly be out of range. In the *Data Validation* dialog, you can specify that the value must be a decimal number between 10 and 100, as shown in Figure 11.8. On the *Error Alert* tab, you can also create an alert box that pops up when an invalid entry is made (see Figure 11.9). On the *Input Message* tab, you can create a prompt to display a comment in the cell about the correct input format. Data validation has other customizable options that you might want to explore.

▶ **Figure 11.7**

Outsourcing Decision Model
Spreadsheet

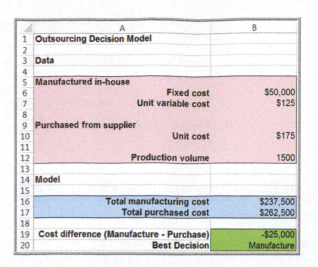

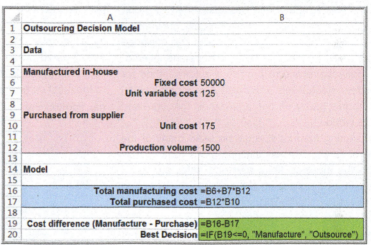

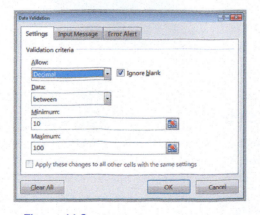

▲ **Figure 11.8**

Data Validation Dialog

▲ **Figure 11.9**

Example of an Error Alert

CHECK YOUR UNDERSTANDING

1. Why is it important to separate data, calculations, and outputs in designing a spreadsheet model?

2. What approaches can be used to improve spreadsheet quality?

3. Explain how Excel's *Data Validation* feature can help avoid user errors in spreadsheet applications.

ANALYTICS IN PRACTICE: Spreadsheet Engineering at Procter & Gamble[4]

In the mid-1980s, Procter & Gamble (P&G) needed an easy and consistent way to manage safety stock inventory. P&G's Western European Business Analysis group created a spreadsheet model that eventually grew into a suite of global inventory models. The model was designed to help supply chain planners better understand inventories in supply chains and to provide a quick method for setting safety stock levels. P&G also developed several spin-off models based on this application that are used around the world.

In designing the model, analysts used many of the principles of spreadsheet engineering. For example, they separated the input sections from the calculation and results sections by grouping the appropriate cells and using different formatting. This speeded up the data entry process. In addition, the spreadsheet was designed to display all relevant data on one screen so the user does not need to switch between different sections of the model.

Analysts used a combination of data validation and conditional formatting to highlight errors in the data input. They also provided a list of warnings and errors that a user should resolve before using the results of the model. The list flags obvious mistakes such as negative transit times, input data that may require checking, and forecast errors that fall outside the boundaries of the model's statistical validity.

At the basic level, all input fields had comments attached; this served as a quick online help function for the planners. For each model, they also provided a user manual that describes every input and result and explains the formulas in detail. The model templates and all documentation were posted on an intranet site that was accessible to all P&G employees. This ensured that all employees had access to the most current versions of the models, supporting material, and training schedules.

Bryan Busovicki/Shutterstock

Descriptive Spreadsheet Models

A wide variety of practical problems in business analytics can be modeled using spreadsheets. In this section, we present several examples of descriptive modeling applications. These models allow one to answer questions such as "How many resources do

---

[4] Based on Ingrid Farasyn, Koray Perkoz, and Wim Van de Velde, "Spreadsheet Models for Inventory Target Setting at Procter & Gamble," *Interfaces*, 38, 4 (July–August 2008): 241–250.

we need to meet estimated demand?" or "What would our profit be under different scenarios?" One thing to note is that a useful spreadsheet model need not be complex; often, simple models can provide managers with the information they need to make good decisions.

Staffing Models

Staffing is an area of any business where making changes can be expensive and time-consuming. Thus, it is important to understand staffing requirements well in advance. The time it takes to hire and train new employees can be 90 to 180 days, so it is not always possible to react quickly to staffing needs. Hence, advance planning is vital so that managers can make good decisions about overtime or reductions in work hours, or adding or reducing temporary or permanent staff. Planning for staffing requirements is an area where analytics can be of tremendous benefit. Example 11.7 is adapted from a real application in the banking industry.

EXAMPLE 11.7   **A Staffing Model for Resource Requirements**[5]

Suppose that the manager of a loan-processing department wants to know how many employees will be needed over the next several months to process a certain number of loan files per month so she can better plan capacity. Let's also suppose that there are different types of products that require processing. A product could be a 30-year fixed rate mortgage, a 7/1 ARM, an FHA loan, or a construction loan. Each of these loan types vary in their complexity, require different levels of documentation, and, consequently, have different times to complete. Assume that the manager forecasts 700 loan applications in May, 750 in June, 800 in July, and 825 in August. Each employee works productively for 6.5 hours each day, and there are 22 working days in May, 20 in June, 22 in July, and 22 in August. The manager also knows, based on historical loan data, the percentage of each product type and how long it takes to process one loan of each type. These data are as follows:

Product	Product Mix (%)	Hours Per File
Product 1	22	3.50
Product 2	17	2.00
Product 3	13	1.50
Product 4	12	5.50
Product 5	9	4.00

Product	Product Mix (%)	Hours Per File
Product 6	9	3.00
Product 7	6	2.00
Product 8	5	2.00
Product 9	3	1.50
Product 10	1	3.50
Miscellaneous	3	3.00
Total	100	

The manager would like to determine the number of full-time equivalent (FTE) staff needed each month to ensure that all loans can be processed.

Figure 11.10 shows a simple descriptive model on a spreadsheet to calculate the FTEs required (Excel file *Staffing Model*). For each month, we take the desired throughput and convert this to the number of files for each product based on the product mix percentages. By multiplying by the hours per file, we then calculate the number of hours required for each product. Finally, we divide the total number of hours required each month by the number of working hours each month (hours worked per day × days in the month). This yields the number of FTEs required.

---

[5] The author is indebted to Mr. Craig Zielazny of BlueNote Analytics, LLC, for providing this example.

▶ Figure 11.10

Staffing Model *Spreadsheet Implementation*

	A	B	C	D	E	F	G	H	I	J	K
1	Staffing Model										
2											
3	Data										
4		May	June	July	August						
5	Desired Throughput	700	750	800	825						
6	Hours Worked Per Day	6.5	6.5	6.5	6.5						
7	Days in Month	22	20	22	22						
8											
9	Model										
10					May		June		July		August
11	Products	Product Mix	Hours Per File	Files/Month	Hours Required	Files/Month	Hours Required	Files/Month	Hours Required	Files/Month	Hours Required
12	Product 1	22%	3.50	154	539.00	165.00	577.50	176.00	616.00	181.50	635.25
13	Product 2	17%	2.00	119	238.00	127.50	255.00	136.00	272.00	140.25	280.50
14	Product 3	13%	1.50	91	136.50	97.50	146.25	104.00	156.00	107.25	160.88
15	Product 4	12%	5.50	84	462.00	90.00	495.00	96.00	528.00	99.00	544.50
16	Product 5	9%	4.00	63	252.00	67.50	270.00	72.00	288.00	74.25	297.00
17	Product 6	9%	3.00	63	189.00	67.50	202.50	72.00	216.00	74.25	222.75
18	Product 7	6%	2.00	42	84.00	45.00	90.00	48.00	96.00	49.50	99.00
19	Product 8	5%	2.00	35	70.00	37.50	75.00	40.00	80.00	41.25	82.50
20	Product 9	3%	1.50	21	31.50	22.50	33.75	24.00	36.00	24.75	37.13
21	Product 10	1%	3.50	7	24.50	7.50	26.25	8.00	28.00	8.25	28.88
22	Misc	3%	3.00	21	63.00	22.50	67.50	24.00	72.00	24.75	74.25
23	Total	100%		700	2089.50	750.00	2238.75	800.00	2388.00	825.00	2462.63
24			FTEs Required		14.61		17.22		16.70		17.22

	A	B	C	D	E
1	Staffing Model				
2					
3	Data				
4		May	June	July	August
5	Desired Througl	700	750	800	825
6	Hours Worked F	6.5	6.5	6.5	6.5
7	Days in Month	22	20	22	22
8					
9	Model				
10					May
11	Products	Product Mix	Hours Per File	Files/Month	Hours Required
12	Product 1	0.22	3.5	=B12*$B$5	=C12*D12
13	Product 2	0.17	2	=B13*$B$5	=C13*D13
14	Product 3	0.13	1.5	=B14*$B$5	=C14*D14
15	Product 4	0.12	5.5	=B15*$B$5	=C15*D15
16	Product 5	0.09	4	=B16*$B$5	=C16*D16
17	Product 6	0.09	3	=B17*$B$5	=C17*D17
18	Product 7	0.06	2	=B18*$B$5	=C18*D18
19	Product 8	0.05	2	=B19*$B$5	=C19*D19
20	Product 9	0.03	1.5	=B20*$B$5	=C20*D20
21	Product 10	0.01	3.5	=B21*$B$5	=C21*D21
22	Misc	=1-SUM(B12:B21)	3	=B22*$B$5	=C22*D22
23	Total	1		=SUM(D12:D22)	=SUM(E12:E22)
24			FTEs Required		=E23/(B6*B7)

Single-Period Purchase Decisions

Banana Republic, a division of Gap, Inc., was trying to build a name for itself in fashion circles as parent company Gap shifted its product line to basics such as cropped pants, jeans, and khakis. One holiday season in the early 2000s, the company bet that blue would be the top-selling color in stretch merino wool sweaters. They were wrong; as the company president noted, "The number 1 seller was moss green. We didn't have enough."[6]

This example describes one of many practical situations in which a one-time purchase decision must be made in the face of uncertain demand. Department store buyers must purchase seasonal clothing well in advance of the buying season, and candy shops must

[6] Louise Lee, "Yes, We Have a New Banana, *BusinessWeek* (May 31, 2004): 70–72.

decide on how many special holiday gift boxes to assemble. The general scenario is commonly known as the **newsvendor problem**: A street newsvendor sells daily newspapers and must make a decision about how many to purchase. Purchasing too few results in lost opportunity to increase profits, but purchasing too many results in a loss since the excess must be discarded at the end of the day.

We first develop a general model for this problem and then illustrate it with an example. Let us assume that each item costs C to purchase and is sold for R. At the end of the period, any unsold items can be disposed of at S each (the salvage value). Clearly, it makes sense to assume that $R > C > S$. Let D be the number of units demanded during the period and Q be the quantity purchased. Note that D is an uncontrollable input, whereas Q is a decision variable. If demand is known, then the optimal decision is obvious: Choose $Q = D$. However, if D is not known in advance, we run the risk of overpurchasing or underpurchasing. If $Q < D$, then we lose the opportunity of realizing additional profit (since we assume that $R > C$), and if $Q > D$, we incur a loss (because $C > S$).

Notice that we cannot sell more than the minimum of the actual demand and the amount produced. Thus, the quantity sold at the regular price is the smaller of D and Q. Also, the surplus quantity is the larger of 0 and $Q - D$. The net profit is calculated as

$$\text{Net Profit} = R \times \text{Quantity Sold} + S \times \text{Surplus Quantity} - C \times Q \quad \textbf{(11.6)}$$

In reality, the demand D is uncertain and can be modeled using a probability distribution based on approaches that we described in Chapter 5. We do not deal with models that involve probability distributions (building the models is enough of a challenge at this point); however, we learn how to deal with them in the next chapter.

EXAMPLE 11.8   A Single-Period Purchase Decision Model

Suppose that a small candy store makes Valentine's Day gift boxes that cost $12.00 and sell for $18.00. In the past, at least 40 boxes have been sold by Valentine's Day, but the actual amount is uncertain, the owner has often run short or made too many. After the holiday, any unsold boxes are discounted 50% and are eventually sold.

The net profit can be calculated using formula (11.6) for any values of Q and D:

$$\text{Net Profit} = \$18.00 \times \min\{D, Q\} + \$9.00 \times \max\{0, Q-D\}$$
$$-\$12.00 \times Q$$

Figure 11.11 shows a spreadsheet that implements this model, assuming a demand of 41 and a purchase quantity of 44 (Excel file *Newsvendor Model*).

▶ **Figure 11.11**

Spreadsheet Implementation of Newsvendor Model

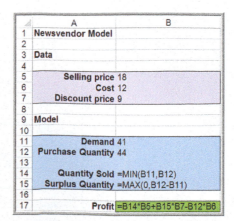

Overbooking Decisions

An important operations decision for service businesses such as hotels, airlines, and car-rental companies is the number of reservations to accept to effectively fill capacity, knowing that some customers may not use their reservations or tell the business. If a hotel, for example, holds rooms for customers who do not show up, they lose revenue opportunities. (Even if they charge a night's lodging as a guarantee, rooms held for additional days may go unused.) A common practice in these industries is **overbooking** reservations; that is, to take more reservations than can be handled, with the expectation that some customers will cancel. When more customers arrive than can be handled, the business usually incurs some cost to satisfy them (by putting them up at another hotel or, for most airlines, providing extra compensation such as ticket vouchers). Therefore, the decision becomes how much to overbook to balance the costs of overbooking against the lost revenue for underuse. The following example illustrates a model to evaluate the net revenue under different scenarios and overbooking policies.

EXAMPLE 11.9   A Hotel Overbooking Model

Figure 11.12 shows a spreadsheet model (Excel file *Hotel Overbooking Model*) for a popular resort hotel that has 300 rooms and is usually fully booked. The hotel charges $120 per room. Reservations may be canceled by the 6:00 p.m. deadline with no penalty. The hotel has estimated that the average overbooking cost is $100.

The logic of the model is straightforward. In the model section of the spreadsheet, cell B11 represents the decision variable of how many reservations to accept, in this case, 310. In this example, we assume that the hotel is willing to accept 310 reservations; that is, to overbook by 10 rooms. Cell B12 represents the actual customer demand (the number of customers who want a reservation). Here we assume that 312 customers tried to make a reservation. The hotel cannot accept more reservations than its predetermined limit; therefore, the number of reservations made in cell B13 is the smaller of the customer demand and the reservation limit. Cell B14 is the number of customers who decide to cancel their reservation. In this example, we assume that only 6 of the 310 reservations are cancelled. Therefore, the actual number of customers who arrive (cell B15) is the difference between the number of reservations made and the number of cancellations. If the actual number of customer arrivals exceeds the room capacity, overbooking occurs. This is modeled by the MAX function in cell B17. Net revenue is computed in cell B18. A manager would probably want to use this model to analyze how the number of overbooked customers and net revenue would be influenced by changes in the reservation limit, customer demand, and cancellations.

As with the newsvendor model, the customer demand and the number of cancellations are random variables that we cannot specify with certainty. We also show how to incorporate randomness into the model in the next chapter.

▶ **Figure 11.12**

Hotel Overbooking Model *Spreadsheet*

ANALYTICS IN PRACTICE: Using an Overbooking Model at a Student Health Clinic

The East Carolina University (ECU) Student Health Service (SHS) provides health care services and wellness education to enrolled students.[7] Patient volume consists almost entirely of scheduled appointments for non-urgent health care needs. In a recent academic year, 35,050 appointments were scheduled. Patients failed to arrive for over 10% of these appointments. The no-show problem is not unique. Various studies report that no-show rates for health service providers often range as high as 30% to 50%.

To address this problem, a quality-improvement (QI) team was formed to analyze an overbooking option. Their efforts resulted in developing a novel overbooking model that included the effects of employee burnout resulting from the need to see more patients than the normal capacity allowed. The model provided strong evidence that a 10% to 15% overbooking level produces the highest value. The overbooking model was also instrumental in alleviating staff concerns about disruption and pressures that result from large numbers of overscheduled patients. At a 5% overbooking rate, the staff was reassured by model results that predicted 95% of the operating days with no patients being overscheduled; in the

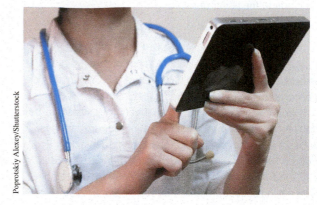

Poprotskiy Alexey/Shutterstock

worst case, eight patients would be overscheduled a few days each month. In addition, at a 10% overbooking rate, the model predicted that during 85% of the operating days per month, no patients would be overscheduled; a maximum of 16 overscheduled patients would rarely occur.

Based on the model, the SHS implemented an overbooking policy and overbooked by 7.3% with plans to increase to 10% in future semesters. The SHS director estimated that the savings from overbooking during the first semester of implementation would be approximately $95,000.

Retail Markdown Decisions

In Example 1.1 in Chapter 1, we described markdown pricing decisions that retail stores must make in managing their inventory. The following example shows how to implement a spreadsheet model for a simple scenario.

EXAMPLE 11.10   Modeling Retail Markdown Pricing Decisions

A chain of department stores is introducing a new brand of bathing suit for $70. The prime selling season is 50 days during the late spring and early summer; after that, the store has a clearance sale around July 4 and marks down the price by 70% (to $21.00), typically selling any remaining inventory at the clearance price. Merchandise buyers have purchased 1,000 units and allocated them to the stores prior to the selling season. After a few weeks, the stores reported an average sales of 7 units/day, and past experience suggests that this constant level of sales will continue over the remainder of the selling season. Thus, over the 50-day selling season,

the stores would be expected to sell 50 × 7 = 350 units at the full retail price and earn a revenue of $70.00 × 350 = $24,500. The remaining 650 units would be sold at $21.00, for a clearance revenue of $13,650. Therefore, the total revenue would be predicted as $24,500 + $13,650 = $38,150.

As an experiment, the store reduced the price to $49 for one weekend and found that the average daily sales were 32.2 units. Assuming a linear trend model for sales as a function of price, as in Example 1.7,

$$\text{Daily Sales} = a - b \times \text{Price}$$

(continued)

[7] Based on John Kros, Scott Dellana, and David West, "Overbooking Increases Patient Access at East Carolina University's Student Health Services Clinic," *Interfaces*, Vol. 39, No. 3 May–June 2009, pp. 271–287.

we can find values for a and b by solving these two equations simultaneously based on the data the store obtained.

$$7 = a - b \times \$70.00$$
$$32.2 = a - b \times \$49.00$$

This leads to the linear demand model:

$$\text{Daily Sales} = 91 - 1.2 \times \text{Price}$$

We may also use Excel's SLOPE and INTERCEPT functions to find the slope and intercept of the straight line between the two points ($70, 7) and ($49, 32.2); this is incorporated into the Excel model that follows.

Because this model suggests that higher sales can be driven by price discounts, the marketing department has the basis for making improved discounting decisions. For instance, suppose they decide to sell at full retail price for d days and then discount the price by $y\%$ for the remainder of the selling season, followed by the clearance sale. What total revenue could they predict?

We can compute this easily. Selling at the full retail price for d days yields revenue of

$$\text{Full Retail Price Revenue} = 7 \text{ units/day} \times d \text{ days}$$
$$\times \$70.00 = \$490.00x$$

The markdown price applies for the remaining $50 - d$ days:

$$\text{Markdown Price} = \$70 \times (100\% - y\%)$$
$$\text{Daily Sales} = a - b \times \text{Markdown Price}$$
$$= 91 - 1.2 \times \$70 \times (100\% - y\%)$$

Units sold at markdown = daily sales $\times (50 - d)$ as long as this is less than or equal to the number of units remaining in inventory from full retail sales. If not, this number needs to be adjusted.

Then we can compute the markdown revenue as

$$\text{Markdown Revenue} = \text{Units Sold} \times \text{Markdown Price}$$

Finally, the remaining inventory after 50 days is

$$\text{Clearance Inventory} = 1000 - \text{Units Sold at Full Retail}$$
$$- \text{Units Sold at Markdown}$$
$$= 1{,}000 - 7 \times d - [91 - 1.2$$
$$\times \$70.00 \times (100\% - y\%)]$$
$$\times (50 - d)$$

This amount is sold at a price of $21.00, resulting in revenue of

$$\text{Clearance Price Revenue} = [1{,}000 - 7 \times d - [91 - 1.2$$
$$\times \$70.00 \times (100\% - y\%)]$$
$$\times (50 - d)] \times \$21.00$$

The total revenue would be found by adding the models developed for full retail price revenue, discounted price revenue, and clearance price revenue.

Figure 11.13 shows a spreadsheet implementation of this model (Excel file *Markdown Pricing Model*). By changing the values in cells B7 and B8, the marketing manager could predict the revenue that could be achieved for different markdown decisions.

▶ **Figure 11.13**

Markdown Pricing Model *Spreadsheet*

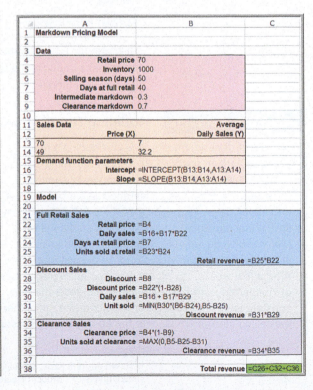

	A	B	C
1	Markdown Pricing Model		
2			
3	Data		
4	Retail price	$70.00	
5	Inventory	1000	
6	Selling season (days)	50	
7	Days at full retail	40	
8	Intermediate markdown	30%	
9	Clearance markdown	70%	
10			
11	Sales Data		Average
12	Price (X)	Daily Sales (Y)	
13	$70	7.00	
14	$49	32.20	
15	Demand function parameters		
16	Intercept	91	
17	Slope	-1.2	
18			
19	Model		
20			
21	Full Retail Sales		
22	Retail price	$70.00	
23	Daily sales	7.00	
24	Days at retail price	40	
25	Units sold at retail	280	
26		Retail revenue	$19,600.00
27	Discount Sales		
28	Discount	30%	
29	Discount price	$49.00	
30	Daily sales	32.20	
31	Unit sold	322	
32		Discount revenue	$15,778.00
33	Clearance Sales		
34	Clearance price	$21.00	
35	Units sold at clearance	398	
36		Clearance revenue	$8,358.00
37			
38		Total revenue	$43,736.00

	A	B	C
1	Markdown Pricing Model		
2			
3	Data		
4	Retail price	70	
5	Inventory	1000	
6	Selling season (days)	50	
7	Days at full retail	40	
8	Intermediate markdown	0.3	
9	Clearance markdown	0.7	
10			
11	Sales Data		Average
12	Price (X)		Daily Sales (Y)
13	70	7	
14	49	32.2	
15	Demand function parameters		
16	Intercept	=INTERCEPT(B13:B14,A13:A14)	
17	Slope	=SLOPE(B13:B14,A13:A14)	
18			
19	Model		
20			
21	Full Retail Sales		
22	Retail price	=B4	
23	Daily sales	=B16+B17*B22	
24	Days at retail price	=B7	
25	Units sold at retail	=B23*B24	
26		Retail revenue	=B25*B22
27	Discount Sales		
28	Discount	=B8	
29	Discount price	=B22*(1-B28)	
30	Daily sales	=B16 + B17*B29	
31	Unit sold	=MIN(B30*(B6-B24),B5-B25)	
32		Discount revenue	=B31*B29
33	Clearance Sales		
34	Clearance price	=B4*(1-B9)	
35	Units sold at clearance	=MAX(0,B5-B25-B31)	
36		Clearance revenue	=B34*B35
37			
38		Total revenue	=C26+C32+C36

CHECK YOUR UNDERSTANDING

1. State some examples of descriptive spreadsheet models that you might use in your work or leisure activities.

2. Explain the logic of modeling single-period purchase decisions.

3. Explain the logic of modeling overbooking decisions on spreadsheets.

4. Discuss the practical implications for building realistic models such as the retail markdown model.

Predictive Spreadsheet Models

Predictive models focus on understanding the future. Practical business models focus on predicting financial performance such as profitability or cash flow, customer retention, product sales, and many other key metrics. Individuals often use spreadsheet models for financial planning, personal budgets, and so on. Such models usually involve multiple time periods, and spreadsheets are an ideal vehicle for capturing this. We will present several examples.

New Product Development Model

Many firms face the decision to launch a new product. In the pharmaceutical industry, for instance, research and development (R&D) is a long and arduous process; total development expenses can approach $1 billion. The following example illustrates a scenario to predict the profitability of a new pharmaceutical venture.

EXAMPLE 11.11   New-Product Development

Suppose that Moore Pharmaceuticals has discovered a potential drug breakthrough in the laboratory and needs to decide whether to conduct clinical trials and seek FDA approval to market the drug. Total R&D costs are expected to reach $700 million, and the cost of clinical trials will be about $150 million. The current market size is estimated to be two million people and is expected to grow at a rate of 3% each year. In the first year, Moore estimates gaining an 8% market share, which is anticipated to grow by 20% each year. It is difficult to estimate beyond five years because new competitors are expected to be entering the market. A monthly prescription is anticipated to generate revenue of $130 while incurring variable costs of $40. A discount rate of 9% is assumed for computing the net present value of the project. The company needs to know how long it will take to recover its fixed expenses and the net present value over the first five years.

Figure 11.14 shows a spreadsheet model for this situation (Excel file *Moore Pharmaceuticals*). The model is based on a variety of known data, estimates, and assumptions. If you examine the model closely, you will see that some of the inputs in the model are easily obtained from corporate accounting (for example, discount rate, unit revenue, and unit cost) using historical data (for example, project costs), forecasts, or judgmental estimates based on preliminary market research or previous experience (for example, market size, market share, and yearly growth rates). The model itself is a straightforward application of accounting and financial logic; you should examine the Excel formulas to see how the model is built.

The assumptions used represent the "most likely" estimates, and the spreadsheet shows that the product will begin to be profitable by the fourth year. However, the model is based on some rather tenuous assumptions about the market size and market-share growth rates. In reality, much of the data used in the model are uncertain, and the corporation would be remiss if it simply used the results of this one scenario. The real value of the model would be in analyzing a variety of scenarios that use different values for these assumptions.

▶ **Figure 11.14**

Spreadsheet Implementation of Moore Pharmaceuticals *Model*

	A	B	C	D	E	F
1	**Moore Pharmaceuticals**					
2						
3	**Data**					
4						
5	Market size	2,000,000				
6	Unit (monthly Rx) revenue	$ 130.00				
7	Unit (monthly Rx) cost	$ 40.00				
8	Discount rate	9%				
9						
10	Project Costs					
11	R&D	$ 700,000,000				
12	Clinical Trials	$ 150,000,000				
13	Total Project Costs	$ 850,000,000				
14						
15	**Model**					
16						
17	Year	1	2	3	4	5
18	Market growth factor		3.00%	3.00%	3.00%	3.00%
19	Market size	2,000,000	2,060,000	2,121,800	2,185,454	2,251,018
20	Market share growth rate		20.00%	20.00%	20.00%	20.00%
21	Market share	8.00%	9.60%	11.52%	13.82%	16.59%
22	Sales	160,000	197,760	244,431	302,117	373,417
23						
24	Annual Revenue	$ 249,600,000	$ 308,505,600	$ 381,312,922	$ 471,302,771	$ 582,530,225
25	Annual Costs	$ 76,800,000	$ 94,924,800	$ 117,327,053	$ 145,016,237	$ 179,240,069
26	Profit	$ 172,800,000	$ 213,580,800	$ 263,985,869	$ 326,286,534	$ 403,290,156
27						
28	Cumulative Net Profit	$ (677,200,000)	$ (463,619,200)	$ (199,633,331)	$ 126,653,203	$ 529,943,358
29						
30	Net Present Value	$ 185,404,860				

	A	B	C	D	E	F
1	**Moore Pharmaceuticals**					
2						
3	**Data**					
4						
5	Market size	2000000				
6	Unit (monthly Rx) revenue	130				
7	Unit (monthly Rx) cost	40				
8	Discount rate	0.09				
9						
10	Project Costs					
11	R&D	700000000				
12	Clinical Trials	150000000				
13	Total Project Costs	=B11+B12				
14						
15	**Model**					
16						
17	Year	1	2	3	4	5
18	Market growth factor		0.03	0.03	0.03	0.03
19	Market size	=B5	=B19*(1+C18)	=C19*(1+D18)	=D19*(1+E18)	=E19*(1+F18)
20	Market share growth rate		0.2	0.2	0.2	0.2
21	Market share	0.08	=B21*(1+C20)	=C21*(1+D20)	=D21*(1+E20)	=E21*(1+F20)
22	Sales	=B19*B21	=C19*C21	=D19*D21	=E19*E21	=F19*F21
23						
24	Annual Revenue	=B22*$B$6*12	=C22*$B$6*12	=D22*$B$6*12	=E22*$B$6*12	=F22*$B$6*12
25	Annual Costs	=B22*$B$7*12	=C22*$B$7*12	=D22*$B$7*12	=E22*$B$7*12	=F22*$B$7*12
26	Profit	=B24-B25	=C24-C25	=D24-D25	=E24-E25	=F24-F25
27						
28	Cumulative Net Profit	=B26-B13	=B28+C26	=C28+D26	=D28+E26	=E28+F26
29						
30	Net Present Value	=NPV(B8,B26:F26)-B13				

Cash Budgeting

Cash budgeting is the process of projecting and summarizing a company's cash inflows and outflows expected during a planning horizon, usually 6 to 12 months.[8] The cash budget also shows the monthly cash balances and any short-term borrowing used to cover cash shortfalls. Positive cash flows can increase cash, reduce outstanding loans, or be used elsewhere in the business; negative cash flows can reduce cash available or be offset with additional borrowing. Most cash budgets are based on sales forecasts.

EXAMPLE 11.12 A Cash Budget Model

Figure 11.15 shows an example of a cash budget spreadsheet (Excel file *Cash Budget Model*). The budget begins in April; thus, the sales for April and subsequent months are forecast values. On average, 20% of sales are collected in the month of sale, 50% in the month following the sale, and 30% in the second month following the sale (see cells B7:B9). For example, in the figure in column E, $120,000 is collected on April sales, $250,000 on March sales, and $120,000 on February sales. Purchases are 60% of sales and are paid for one month prior to the sale. Wages and salaries are 12% of sales and are paid in the same month as the sale. Rent of $10,000 is paid each month. Additional cash operating expenses of $30,000 per month will be incurred for April through July, decreasing to $25,000 for August and September. Tax payments of $20,000 and $30,000 are expected in April and July, respectively. A capital expenditure of $150,000 will occur in June, and the company has a mortgage payment of $60,000 in May. The cash balance at the end of March is $150,000, and managers want to maintain a minimum balance of $100,000 at all times. The company will borrow the amounts necessary to ensure that the minimum balance is achieved. Any cash above the minimum will be used to pay off any loan balance until it is eliminated. The available cash balances in row 25 of the spreadsheet are the outputs the company wants to predict.

	A	B	C	D	E	F	G	H	I	J	K
1	Cash Budget Model										
2											
3	Desired Minimum Balance	$100,000									
4			February	March	April	May	June	July	August	September	October
5		Sales	$400,000	$500,000	$600,000	$700,000	$800,000	$800,000	$700,000	$ 600,000	$500,000
6	Cash Receipts										
7	Collections (current)	20%			$120,000	$140,000	$160,000	$160,000	$140,000	$ 120,000	
8	Collections (previous month)	50%			$250,000	$300,000	$350,000	$400,000	$400,000	$ 350,000	
9	Collections (2nd month previous)	30%			$120,000	$150,000	$180,000	$210,000	$240,000	$ 240,000	
10	Total Cash Receipts				$490,000	$590,000	$690,000	$770,000	$780,000	$ 710,000	
11											
12	Cash Disbursements										
13	Purchases				$420,000	$480,000	$480,000	$420,000	$360,000	$ 300,000	
14	Wages and Salaries				$ 72,000	$ 84,000	$ 96,000	$ 96,000	$ 84,000	$ 72,000	
15	Rent				$ 10,000	$ 10,000	$ 10,000	$ 10,000	$ 10,000	$ 10,000	
16	Cash Operating Expenses				$ 30,000	$ 30,000	$ 30,000	$ 30,000	$ 25,000	$ 25,000	
17	Tax Installments				$ 20,000			$ 30,000			
18	Capital Expenditure						$150,000				
19	Mortgage Payment					$ 60,000					
20	Total Cash Disbursements				$552,000	$664,000	$766,000	$586,000	$479,000	$ 407,000	
21											
22	Ending Cash Balance										
23	Net Cash Flow				$ (62,000)	$ (74,000)	$ (76,000)	$184,000	$301,000	$ 303,000	
24	Beginning Cash Balance				$150,000	$100,000	$100,000	$100,000	$122,000	$ 423,000	
25	Available Balance				$ 88,000	$ 26,000	$ 24,000	$284,000	$423,000	$ 726,000	
26	Monthly Borrowing				$ 12,000	$ 74,000	$ 76,000	$ -	$ -	$ -	
27	Monthly Repayment				$ -	$ -	$ -	$162,000	$ -	$ -	
28	Ending Cash Balance			$150,000	$100,000	$100,000	$100,000	$122,000	$423,000	$ 726,000	
29	Cumulative Loan Balance			$ -	$ 12,000	$ 86,000	$162,000	$ -	$ -	$ -	

▲ Figure 11.15

Cash Budget Model

[8] Douglas R. Emery, John D. Finnerty, and John D. Stowe, *Principles of Financial Management* (Upper Saddle River, NJ: Prentice Hall, 1998): 652–654.

Retirement Planning

Retirement planning is important to every individual. Understanding the impact of compounding on investment growth, saving at an early age, and selecting the right portfolio of investments is key to a successful financial future. Simple spreadsheet models can be used to help develop sound financial plans.

EXAMPLE 11.13   A Retirement-Planning Model

A student who is completing her MBA has accepted a job with a starting salary of $80,000. The company will match up to 5% of pre-tax salary under their 401(k) retirement plan. She expects her salary to increase on average 4% each year, and plans to contribute 15% of her pre-tax income to her 401(k) account, up to the annual cap of $18,000. She has chosen to invest in two aggressive mutual funds: the Vanguard Balanced Index Fund, which has less risk, for her 401(k), and the Boston Trust Asset Management Fund for a Roth IRA. Based on a ten-year average return, she expects the Vanguard fund to earn 6% each year and the Boston Trust fund to earn 6.5% each year. Currently, individuals can contribute up to $18,000 of pre-tax income annually into a 401(k) account; individuals under age 55 can contribute $5,500 to their Roth IRA annually; and individuals 55 and over can contribute $6,500 into a Roth IRA. She hopes to retire at 60 and wants to predict the value of her retirement investments.

Figure 11.16 shows a spreadsheet model of this scenario (Excel file *Retirement Planning Model*). Several key assumptions have been made in developing this model. One, of course, is that the annual salary increase and returns on investment will be the same each year. In reality, these values will be uncertain and vary randomly each year (and investment returns can be negative). A second assumption is how the model calculates the return on investment. The annual returns assume that the investment return is applied to the previous year's balance and not to the current year's contributions (examine the formulas in columns H and I). An alternative would be to calculate the investment return based on the end-of-year balance, including current-year contributions. This will produce a different result. In reality, neither of these are quite correct, since the 401(k) contributions would normally be made on a monthly basis. Reflecting this would require a much larger and more complex spreadsheet model.

Project Management

Project management is concerned with scheduling the activities of a project involving interrelated activities. An important aspect of project management is predicting the expected completion time of the project. To do this, we first define the set of activities that comprise the project, the time it takes for each activity, and the predecessors of each activity (those activities that must immediately precede it). These precedence relationships are

	A	B	C	D	E	F	G	H	I	J
1	Retirement Planning			Age	Salary	401K Contributio	Employer Match	401K Balance	Roth IRA Balance	Final Balance
2	Yearly 401K Contribution	15.0%		24	$80,000	$12,000	$4,000	$16,000	$5,500	$21,500
3	Employer Match of Salary	5.0%		25	$83,200	$12,480	$4,160	$33,600.00	$11,357.50	$44,958
4	Salary increase	4.0%		26	$86,528	$12,979	$4,326	$52,921.60	$17,595.74	$70,517
5				27	$89,989	$13,498	$4,499	$74,094.72	$24,239.46	$98,334
6	Vanguard Balanced Index Fund			28	$93,589	$14,038	$4,679	$97,258.14	$31,315.03	$128,573
7	Expected annual return	6.0%		29	$97,332	$14,600	$4,867	$122,560.08	$38,850.50	$161,411
8				30	$101,226	$15,184	$5,061	$150,158.78	$46,875.78	$197,035
9	Boston Trust Asset Management Fund			31	$105,275	$15,791	$5,264	$180,223.22	$55,422.71	$235,646
10	Expected annual return	6.5%		32	$109,486	$16,423	$5,474	$212,933.72	$64,525.19	$277,459
11				33	$113,865	$17,080	$5,693	$248,482.73	$74,219.32	$322,702
12				34	$118,420	$17,763	$5,921	$287,075.60	$84,543.58	$371,619
13				35	$123,156	$18,000	$6,158	$328,457.95	$95,538.91	$423,997

▲ Figure 11.16

Portion of Retirement Planning Model

usually depicted as a network. The longest path through the network defines the minimum project completion time and is called the **critical path**. To find the critical path, we first compute the earliest time that each activity can start and the earliest time it can finish. The earliest time that the last activity can finish is the minimum project completion time. Then we find the latest time each activity can start and the latest time that it can finish without delaying the project. The difference between the latest finish time and earliest finish time is called the **slack**. If the slack of an activity is zero, then it is on the critical path.

EXAMPLE 11.14   A Project Management Spreadsheet Model

Becker Consulting has been hired to assist in the evaluation of new software. The manager of the information systems department is responsible for coordinating all activities involving consultants and the company's resources. The activities shown in Table 11.1 have been defined for this project, which is depicted graphically in Figure 11.17.

Figure 11.18 shows a spreadsheet designed to calculate the project completion time (Excel file *Becker Consulting Project Management Model*). The model uses Excel MAX, MIN, and IF functions to implement the logic of calculating the project schedule and critical path. We start by finding the earliest start and finish times for each activity. Activities A, B, C, and D have no immediate predecessors and, therefore, have the earliest start times, 0. The earliest start time for each other activity is the *maximum* of the earliest finish times for the activity's

immediate predecessor. Earliest finish times are computed as the early start time plus the activity time. The earliest finish time for the last activity, Q (cell D21, copied to cell F23), represents the earliest time the project can be completed, that is, the minimum project completion time.

To compute latest start and latest finish times, we set the latest finish time of the last activity equal to the project completion time. The latest start time is computed by subtracting the activity time from the latest finish time. The latest finish time for any other activity, say X, is defined as the *minimum* latest start of all activities for which activity X is an immediate predecessor. Slack is computed as the difference between the latest finish and earliest finish times. The critical path consists of activities with zero slack. Based on the expected activity times, the critical path consists of activities B-F-G-H-I-K-M-O-P-Q and has a predicted duration of 159 days.

▼ TABLE 11.1

Activity and Time Estimate List

Activity		Predecessor(s)	Activity Time (days)
A	Select steering committee	—	15
B	Develop requirements list	—	50
C	Develop system size estimates	—	20
D	Determine prospective vendors	—	3
E	Form evaluation team	A	7
F	Issue request for proposal	B, C, D, E	6
G	Bidders conference	F	1
H	Review submissions	G	36
I	Select vendor short list	H	6
J	Check vendor references	I	6
K	Vendor demonstrations	I	32
L	User site visit	I	4
M	Select vendor	J, K, L	3
N	Volume-sensitive test	M	15
O	Negotiate contracts	M	18
P	Cost–benefit analysis	N, O	2
Q	Obtain board of directors' approval	P	5

▶ Figure 11.17

Project Network Structure

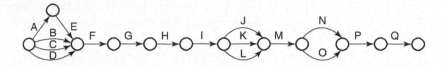

▶ Figure 11.18

Becker Consulting Project
Management Model
Spreadsheet

	A	B	C	D	E	F	G	H
1	Becker Consulting Project Management Model							
2								
3		Activity	Earliest	Earliest	Latest	Latest		On Critical
4	Activity	Time	Start Time	Finish Time	Start Time	Finish Time	Slack	Path?
5	A	15.00	0.00	15.00	28.00	43.00	28.00	
6	B	50.00	0.00	50.00	0.00	50.00	0.00	Yes
7	C	20.00	0.00	20.00	30.00	50.00	30.00	
8	D	3.00	0.00	3.00	47.00	50.00	47.00	
9	E	7.00	15.00	22.00	43.00	50.00	28.00	
10	F	6.00	50.00	56.00	50.00	56.00	0.00	Yes
11	G	1.00	56.00	57.00	56.00	57.00	0.00	Yes
12	H	36.00	57.00	93.00	57.00	93.00	0.00	Yes
13	I	6.00	93.00	99.00	93.00	99.00	0.00	Yes
14	J	6.00	99.00	105.00	125.00	131.00	26.00	
15	K	32.00	99.00	131.00	99.00	131.00	0.00	Yes
16	L	4.00	99.00	103.00	127.00	131.00	28.00	
17	M	3.00	131.00	134.00	131.00	134.00	0.00	Yes
18	N	15.00	134.00	149.00	137.00	152.00	3.00	
19	O	18.00	134.00	152.00	134.00	152.00	0.00	Yes
20	P	2.00	152.00	154.00	152.00	154.00	0.00	Yes
21	Q	5.00	154.00	159.00	154.00	159.00	0.00	Yes
22								
23				Project completion time		159.00		

	A	B	C	D	E	F	G	H
1	Becker Consulting							
2								
3		Activity	Earliest	Earliest	Latest	Latest		On Critical
4	Activity	Time	Start Time	Finish Time	Start Time	Finish Time	Slack	Path?
5	A	15	0	=C5+B5	=F5-B5	=E9	=F5-D5	=IF(G5<0.0001,"Yes","")
6	B	50	0	=C6+B6	=F6-B6	=E10	=F6-D6	=IF(G6<0.0001,"Yes","")
7	C	20	0	=C7+B7	=F7-B7	=E10	=F7-D7	=IF(G7<0.0001,"Yes","")
8	D	3	0	=C8+B8	=F8-B8	=E10	=F8-D8	=IF(G8<0.0001,"Yes","")
9	E	7	=D5	=C9+B9	=F9-B9	=E10	=F9-D9	=IF(G9<0.0001,"Yes","")
10	F	6	=MAX(D6,D7,D8,D9)	=C10+B10	=F10-B10	=E11	=F10-D10	=IF(G10<0.0001,"Yes","")
11	G	1	=D10	=C11+B11	=F11-B11	=E12	=F11-D11	=IF(G11<0.0001,"Yes","")
12	H	36	=D11	=C12+B12	=F12-B12	=E13	=F12-D12	=IF(G12<0.0001,"Yes","")
13	I	6	=D12	=C13+B13	=F13-B13	=MIN(E14,E15,E16)	=F13-D13	=IF(G13<0.0001,"Yes","")
14	J	6	=D13	=C14+B14	=F14-B14	=E17	=F14-D14	=IF(G14<0.0001,"Yes","")
15	K	32	=D13	=C15+B15	=F15-B15	=E17	=F15-D15	=IF(G15<0.0001,"Yes","")
16	L	4	=D13	=C16+B16	=F16-B16	=E17	=F16-D16	=IF(G16<0.0001,"Yes","")
17	M	3	=MAX(D14,D15,D16)	=C17+B17	=F17-B17	=MIN(E18,E19)	=F17-D17	=IF(G17<0.0001,"Yes","")
18	N	15	=D17	=C18+B18	=F18-B18	=E20	=F18-D18	=IF(G18<0.0001,"Yes","")
19	O	18	=D17	=C19+B19	=F19-B19	=E20	=F19-D19	=IF(G19<0.0001,"Yes","")
20	P	2	=MAX(D18,D19)	=C20+B20	=F20-B20	=E21	=F20-D20	=IF(G20<0.0001,"Yes","")
21	Q	5	=D20	=C21+B21	=F21-B21	=D21	=F21-D21	=IF(G21<0.0001,"Yes","")
22								
23				Project completion time	=D21			

▮▮ CHECK YOUR UNDERSTANDING

1. Explain how spreadsheets can be effectively designed to model problems involving multiple time periods.

2. What practical assumptions should be included in retirement planning models?

3. Explain how the critical path is determined in a project management model.

Prescriptive Spreadsheet Models

We introduced prescriptive models in Chapter 1. Recall that a prescriptive decision model helps decision makers to identify the best solution to a decision problem. Prescriptive models are often called *optimization models*. An optimization model is most often formulated mathematically and specifies a set of *decision variables*, numerical quantities that represent the decision options from which to choose; an *objective function* that minimizes or maximizes some quantity of interest, profit, revenue, cost, time, and so on; and *constraints*, which are limitations, requirements, or other restrictions that are imposed on any solution. We will study the mathematical formulation and solution techniques for optimization problems in Chapters 13–15. Here we focus on building spreadsheet models that capture these model elements.

Portfolio Allocation

Selecting investment portfolios is fundamentally a trade-off between return and risk. Clearly, a major source of uncertainty is the annual return of each asset, leaving the decision maker with the risk of not achieving a desired return. In addition, the decision maker faces other risks—for example, unanticipated changes in inflation or industrial production, the spread between high- and low-grade bonds, and the spread between long- and short-term interest rates.

One approach to incorporating such risk factors in a decision model is **arbitrate pricing theory (APT)**.[9] APT provides estimates of the sensitivity of a particular asset to these types of risk factors, leading to a prescriptive model for finding the best portfolio mix.

EXAMPLE 11.15 | **A Portfolio Allocation Model**

An investor has $100,000 to invest in four assets. The expected annual returns and minimum and maximum amounts with which the investor will be comfortable allocating to each investment are as follows:

Investment	Annual Return	Minimum	Maximum
1. Life insurance	5%	$2,500	$5,000
2. Bond mutual funds	7%	$30,000	None
3. Stock mutual funds	11%	$15,000	None
4. Savings account	4%	None	None

(continued)

---

[9]M. Schniederjans, T. Zorn, and R. Johnson, "Allocating Total Wealth: A Goal Programming Approach," *Computers and Operations Research*, 20, 7 (1993): 679–685.

Let us assume that the risk factors per dollar allocated to each asset have been determined as follows:

Investment	Risk Factor/Dollar Invested
1. Life insurance	−0.5
2. Bond mutual funds	1.8
3. Stock mutual funds	2.1
4. Savings account	−0.3

The investor may specify a target level for the weighted risk factor, thus leading to a constraint that limits the risk to the desired level. For example, suppose that our investor will tolerate a weighted risk per dollar invested of at most 1.0. Then the weighted risk for a $100,000 total investment will be limited to 100,000. If our investor allocates $5,000 in life insurance, $50,000 in bond mutual funds, $15,000 in stock mutual funds, and $30,000 in a savings account, the total expected annual return would be

$$\text{Expected Annual Return} = 0.05(\$5,000) + 0.07(\$50,000)$$
$$+ 0.11(\$15,000) + 0.04(\$30,000) = \$6,600$$

However, the total weighted risk associated with this solution is

$$\text{Total Weighted Risk} = -0.5(5,000) + 1.8(50,000)$$
$$+ 2.1(15,000) - 0.3(30,000) = 110,000$$

Because this is greater than the limit of 100,000, this solution could not be chosen. The decision problem, then, is to determine how much to invest in each asset to maximize the total expected annual return, remain within the minimum and maximum limits for each investment, and meet the limitation on the weighted risk.

A spreadsheet for this problem is shown in Figure 11.19 (Excel file *Portfolio Allocation Model*). Problem data are specified in rows 4 through 10. On the bottom half of the spreadsheet, we specify the amounts invested (decision options) in cells B16:B19, all of which must add up to $100,000; the total expected return (objective function) in cell B24, which is the formula =SUMPRODUCT(B6:B9,B16:B19); and the total amount invested and total weighted risk (constraints) in cells B21 and B22. Note that the formulas used in cells B24 and B22 [=SUMPRODUCT(E6:E9,B16:B19)] match the calculations shown above.

Locating Central Facilities

A common problem in designing service systems is to locate a facility in a "central" location with respect to other facilities to minimize some measure of distance from the central location to each of the other facilities. Distances can be measured in several ways. Suppose that the X - and Y-coordinates of two locations are (X_1, Y_1) and (X_2, Y_2). One measure is the straight-line distance between the points. This is computed as

$$\text{Straight-line distance between } (X_1, Y_1) \text{ and } (X_2, Y_2) = \sqrt{(X_1 - X_2)^2 + (Y_1 - Y_2)^2}$$

$$(11.7)$$

▶ Figure 11.19

Portfolio Allocation Model *Spreadsheet*

	A	B	C	D	E
1	Portfolio Allocation Model				
2					
3	Data				
4		Annual			Risk factor
5	Investment	return	Minimum	Maximum	per dollar
6	Life Insurance	5.0%	$2,500.00	$5,000.00	-0.5
7	Bond mutual funds	7.0%	$30,000.00	none	1.8
8	Stock mutual funds	11.0%	$15,000.00	none	2.1
9	Savings Account	4.0%	none	none	-0.3
10	Total amount available	$100,000		Limit	100,000
11					
12	Model				
13					
14		Amount			
15		invested			
16	Life Insurance	$5,000.00			
17	Bond mutual funds	$50,000.00			
18	Stock mutual funds	$15,000.00			
19	Savings Account	$30,000.00			
20					
21	Total amount invested	$100,000.00			
22	Total weighted risk	110,000.00			
23					
24	Total expected return	$6,600.00			

► **Figure 11.20**

Straight-Line Versus Rectilinear Distance

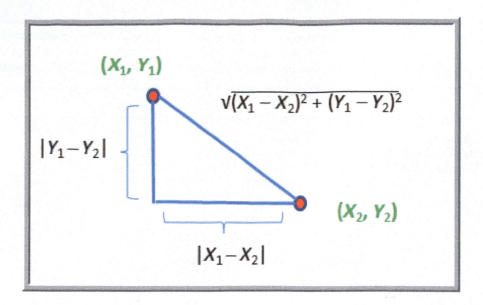

Formula (11.7) is often called **Euclidian distance**. A second measure of distance is called **rectilinear distance**, which is computed as

Rectilinear distance between (X_1, Y_1) and $(X_2, Y_2) = |X_1 - X_2| + |Y_1 - Y_2|$ **(11.8)**

This is often called the "city block" metric. Rectilinear distance moves along the coordinate axes; Figure 11.20 illustrates the difference. Distances are often weighted based on the volume, or frequency, of trips between locations. The goal is to find the location for the central facility that minimizes the distance measure.

EXAMPLE 11.16 **A Location Model for a Medical Laboratory**

A medical testing laboratory needs to collect blood samples from several regional hospitals to perform diagnostic testing. Currently, the laboratory is in the middle of a small town, but several new hospitals have been constructed in outlying areas. The lab wants to relocate to reduce the travel distance required to pick up the samples. The *X*- and *Y*-coordinates for the hospital locations have been found on a grid. The following table shows these, along with the average number of trips per month that the lab must make to each location.

Hospital Location	X-Coordinate	Y-Coordinate	Trips/ Month
1	0	0	5
2	20	80	25
3	60	30	20
4	100	100	35
5	70	110	15

The area is quite rural, so the distances between locations can be measured "as the crow flies"—that is, using the straight-line distance formula (11.7). Suppose that X_C and Y_C represent the coordinates of the laboratory. We want to minimize the weighted distance between the laboratory and all locations, where the weights are the number of trips/month:

$$\text{minimize } 5\sqrt{(X_1 - X_C)^2 + (Y_1 - Y_C)^2}$$
$$+ 25\sqrt{(X_2 - X_C)^2 + (Y_2 - Y_C)^2}$$
$$+ 20\sqrt{(X_3 - X_C)^2 + (Y_3 - Y_C)^2}$$
$$+ 35\sqrt{(X_4 - X_C)^2 + (Y_4 - Y_C)^2}$$
$$+ 15\sqrt{(X_5 - X_C)^2 + (Y_2 - Y_C)^2}$$

Figure 11.21 shows a spreadsheet model (Excel file *Laboratory Location Model*). The locations (decision options) are

(continued)

specified in cells B11:C11, and the total weighted distance (objective function) is calculated in cell C19. This model has no constraints. The spreadsheet also shows the hospital locations and recommended laboratory location visually on a coordinate system using a bubble chart. The sizes of the bubbles represent the numbers of trips/month, and the orange bubble is the location of the laboratory for the specified lab location coordinates.

▶ **Figure 11.21**

Laboratory Location Model Spreadsheet

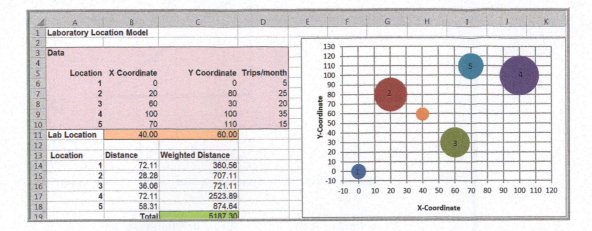

Job Sequencing

A unique application of Excel modeling is for job-sequencing problems. Job-sequencing problems involve finding an optimal sequence, or order, by which to process a set of jobs. For any job sequence, we may compute the completion time for each job by successively adding the processing times of that job and all that were completed before it. We may then compare the completion times with the requested due dates to determine if the job is completed either early or late. For any job i, *lateness* (L_i) is the difference between the completion time (C_i) and the due date (D_i), which can be either positive or negative. *Tardiness* (T_i) is the amount of time by which the completion time exceeds the due date; thus, tardiness is zero if a job is completed early. Hence, for job i,

$$L_i = C_i - D_i \tag{11.9}$$
$$T_i = \max(0, L_i) \tag{11.10}$$

Researchers have shown that sequencing jobs in order of shortest processing time (SPT) first will minimize the average completion time for all jobs. Sequencing by earliest due date (EDD) first will minimize the maximum number of tardy jobs. However, the manager might be interested in minimizing other criteria, such as the average tardiness, total tardiness, or total lateness.

EXAMPLE 11.17   **A Spreadsheet Model for Job Sequencing**

Suppose that a custom manufacturing company has ten jobs waiting to be processed. Each job i has an estimated processing time (P_i) and a due date (D_i) that was requested by the customer, as shown in the following table:

Job	1	2	3	4	5	6	7	8	9	10
Time	8	7	6	4	10	8	10	5	9	5
Due date	20	27	39	28	23	40	25	35	29	30

To develop a spreadsheet model for this problem, we use the Excel function INDEX to identify the processing times and due date for the job assigned to a particular sequence.

Figure 11.22 shows the model (Excel file *Job Sequencing Model*) and a portion of the Excel formulas. A particular job sequence (the decision options) is given in row 10; for this example, we show the sequence for the EDD rule. In rows 11 and 13, we use the INDEX function to identify the processing time and due date associated with a specific job. For example, the formula in cell B11 is =INDEX($B$4:$K$6, 2, B10). This function references the value in the second row of the range B4:K6 corresponding to the job assigned to cell B10, in this case, job 5. Likewise, the formula in cell B13, =INDEX($B$4:$K$6, 3, B10), finds the due date associated with job 5. Any sequence of integers in the decision variable range is called a **permutation**. The goal is to find a permutation that optimizes the chosen criterion.

▶ **Figure 11.22**

Spreadsheet Model for Job Sequencing Model

	A	B	C	D	E	F	G	H	I	J	K
1	Job Sequencing Model										
2											
3	Data										
4	Job	1	2	3	4	5	6	7	8	9	10
5	Time	8	7	6	4	10	8	10	5	9	5
6	Due date	26	27	39	28	23	40	25	35	29	30
7											
8	Model										
9	Sequence	1	2	3	4	5	6	7	8	9	10
10	Job Assigned	5	7	1	2	4	9	10	8	3	6
11	Processing time	10	10	8	7	4	9	5	5	6	8
12	Completion time	10	20	28	35	39	48	53	58	64	72
13	Due Date	23	25	26	27	28	29	30	35	39	40
14	Lateness	-13	-5	2	8	11	19	23	23	25	32
15	Tardiness	0	0	2	8	11	19	23	23	25	32
16											
17	Average Completion Time	42.7									
18	Maximum Number Tardy	8									
19	Total Lateness	125									
20	Average Lateness	12.5									
21	Variance of Lateness	188.85									
22	Total Tardiness	143									
23	Average Tardiness	14.3									
24	Variance of Tardiness	121.21									

	A	B	C	D
1	Job Sequencing Model			
2				
3	Data			
4	Job	1	2	3
5	Time	8	7	6
6	Due date	26	27	39
7				
8	Model			
9	Sequence	1	2	3
10	Job Assigned	5	7	1
11	Processing time	=INDEX($B$4:$K$6,2,B10)	=INDEX($B$4:$K$6,2,C10)	=INDEX($B$4:$K$6,2,D10)
12	Completion time	=B11	=B12+C11	=C12+D11
13	Due Date	=INDEX($B$4:$K$6,3,B10)	=INDEX($B$4:$K$6,3,C10)	=INDEX($B$4:$K$6,3,D10)
14	Lateness	=B12-B13	=C12-C13	=D12-D13
15	Tardiness	=MAX(0,B14)	=MAX(0,C14)	=MAX(0,D14)
16				
17	Average Completion Time	=AVERAGE(B12:K12)		
18	Maximum Number Tardy	=COUNTIF(B15:K15,">0")		
19	Total Lateness	=SUM(B14:K14)		
20	Average Lateness	=AVERAGE(B14:K14)		
21	Variance of Lateness	=VAR.P(B14:K14)		
22	Total Tardiness	=SUM(B15:K15)		
23	Average Tardiness	=AVERAGE(B15:K15)		
24	Variance of Tardiness	=VAR.P(B15:K15)		

CHECK YOUR UNDERSTANDING

1. Summarize the key features of prescriptive spreadsheet models.

2. What is the difference between Euclidian and rectilinear distance in modeling facility location problems?

3. Explain the decision variables and possible objective functions that can be used for job-sequencing models.

Analyzing Uncertainty and Model Assumptions

Because predictive analytical models are based on assumptions and incorporate data that most likely are not known with certainty or are subject to error, it is usually important to investigate how these assumptions and uncertainty affect the model outputs. This is one of the most important and valuable activities for using spreadsheet models to gain insights and make good decisions. In this section, we describe several different approaches for doing this.

What-If Analysis

Spreadsheet models allow you to easily evaluate what-if questions—how specific combinations of inputs that reflect key assumptions will affect model outputs. **What-if analysis** is as easy as changing values in a spreadsheet and recalculating the outputs. However, systematic approaches make this process easier and more useful.

In Example 11.2, we developed a spreadsheet model for profit and suggested how a manager might use the model to change inputs and evaluate different scenarios. A more informative way of evaluating a wider range of scenarios is to build a table in the spreadsheet to vary the input or inputs in which we are interested over some range and calculate the output for this range of values. The following example illustrates this.

EXAMPLE 11.18   **Using Excel for What-If Analysis**

In the profit model used in Example 11.2, we stated that demand is uncertain. A manager might be interested in the following question: For any fixed quantity produced, how will profit change as demand changes? In Figure 11.23, we created a table for varying levels of demand and computed the profit. This shows that a loss is incurred for low levels of demand, whereas profit is limited to $240,000 whenever the demand exceeds the quantity produced, no matter how high it is. Notice that the formula refers to cells in the model; thus, the user could change the quantity produced or any of the other model inputs and still have a correct evaluation of the profit for these values of demand. One of the advantages of evaluating what-if questions for a range of values rather than one at a time is the ability to visualize the results in a chart, as shown in Figure 11.24. This clearly shows that profit increases as demand increases until it hits the value of the quantity produced.

Conducting what-if analysis in this fashion can be quite tedious. Fortunately, Excel provides several tools—*Data Tables*, *Scenario Manager*, and *Goal Seek*—that facilitate what-if and other types of decision model analyses. These can be found within the *What-If Analysis* menu in the *Data* tab.

Data Tables

Data tables summarize the impact of one or two inputs on a specified output. Excel allows you to construct two types of data tables. A **one-way data table** evaluates an output

► Figure 11.23

What-If Table for Uncertain Demand

	A	B	C	D	E	F	G	H	I
1	Profit Model								
2									
3	Data				Demand	Qty. Sold	Revenue	Cost	Profit
4					25000	25000	$1,000,000	$1,360,000	$(360,000)
5	Unit Price	$40.00			30000	30000	$1,200,000	$1,360,000	$(160,000)
6	Unit Cost	$24.00			35000	35000	$1,400,000	$1,360,000	$ 40,000
7	Fixed Cost	$400,000.00			40000	40000	$1,600,000	$1,360,000	$ 240,000
8	Demand	50000			45000	40000	$1,600,000	$1,360,000	$ 240,000
9					50000	40000	$1,600,000	$1,360,000	$ 240,000
10					55000	40000	$1,600,000	$1,360,000	$ 240,000
11	Model				60000	40000	$1,600,000	$1,360,000	$ 240,000
12									
13	Unit Price	$40.00							
14	Quantity Sold	40000							
15	Revenue		$1,600,000.00						
16									
17	Unit Cost	$24.00							
18	Quantity Produced	40000							
19	Variable Cost		$960,000.00						
20	Fixed Cost		$400,000.00						
21									
22	Profit		$240,000.00						

	A	B	C	D	E	F	G	H	I
1	Profit Model								
2									
3	Data				Demand	Qty. Sold	Revenue	Cost	Profit
4					25000	=MIN(E4,$B$18)	=$B$13*F4	=$C$19+$C$20	=G4-H4
5	Unit Price	40			30000	=MIN(E5,$B$18)	=$B$13*F5	=$C$19+$C$20	=G5-H5
6	Unit Cost	24			35000	=MIN(E6,$B$18)	=$B$13*F6	=$C$19+$C$20	=G6-H6
7	Fixed Cost	400000			40000	=MIN(E7,$B$18)	=$B$13*F7	=$C$19+$C$20	=G7-H7
8	Demand	50000			45000	=MIN(E8,$B$18)	=$B$13*F8	=$C$19+$C$20	=G8-H8
9					50000	=MIN(E9,$B$18)	=$B$13*F9	=$C$19+$C$20	=G9-H9
10					55000	=MIN(E10,$B$18)	=$B$13*F10	=$C$19+$C$20	=G10-H10
11	Model				60000	=MIN(E11,$B$18)	=$B$13*F11	=$C$19+$C$20	=G11-H11
12									
13	Unit Price	=B5							
14	Quantity Sold	=MIN(B8,B18)							
15	Revenue		=B13*B14						
16									
17	Unit Cost	=B6							
18	Quantity Produced	40000							
19	Variable Cost		=B17*B18						
20	Fixed Cost		=B7						
21									
22	Profit		=C15-C19-C20						

► Figure 11.24

Chart of What-If Analysis

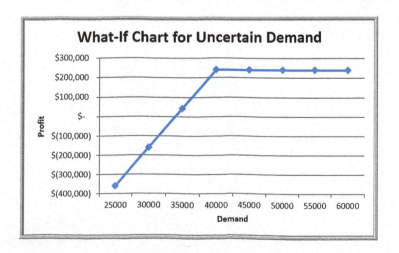

variable over a range of values for a single input variable. **Two-way data tables** evaluate an output variable over a range of values for two different input variables.

To create a one-way data table, first create a range of values for some input cell in your model that you wish to vary. The input values must be listed either down a column (column oriented) or across a row (row oriented). If the input values are column oriented, enter the cell reference for the output variable in your model that you wish to evaluate in the row *above* the first value and one cell to the *right* of the column of input values. Reference any other output variable cells to the right of the first formula. If the input values are listed across a row, enter the cell reference of the output variable in the column to the *left* of the first value and one cell *below* the row of values. Type any additional output cell references below the first one. Next, select the range of cells that contains *both* the formulas and values you want to substitute. From the *Data* tab in Excel, select *Data Table* under the *What-If Analysis* menu. In the dialog box (see Figure 11.25), if the input range is column oriented, type the cell reference for the input cell in your model in the *Column input cell* box. If the input range is row oriented, type the cell reference for the input cell in the *Row input cell* box.

EXAMPLE 11.19   A One-Way Data Table for Uncertain Demand

In this example, we create a one-way data table for profit for varying levels of demand. First, create a column of demand values in column E exactly as we did in Example 11.18. Then in cell F3, enter the formula = C22. This simply references the output of the profit model. Highlight the range E3:F11 (note that this range includes both the column of demand as well as the cell reference to profit), and select *Data Table* from the *What-If Analysis* menu. In the *Column input cell* field, enter B8; this tells the tool that the values in column E are different values of demand in the model. When you click *OK*, the tool produces the results (which we formatted as currency) shown in Figure 11.26.

We may evaluate multiple outputs using one-way data tables.

EXAMPLE 11.20   One-Way Data Tables with Multiple Outputs

Suppose that we want to examine the impact of the uncertain demand on revenue in addition to profit. We simply add another column to the data table. For this case, insert the formula =C15 into cell G3. Also, add the labels "Profit" in F2 and "Revenue" in G2 to identify the results. Then highlight the range E3:G11 and proceed as described in the previous example. This process results in the data table shown in Figure 11.27.

To create a two-way data table, type a list of values for one input variable in a column and a list of input values for the second input variable in a row, starting one row above and one column to the right of the column list. In the cell in the upper left-hand corner immediately above the column list and to the left of the row list, enter the cell reference of the output variable you wish to evaluate. Select the range of cells that contain this cell reference and both the row and column of values. On the *What-If Analysis* menu, click *Data Table*. In the *Row input cell* of the dialog box, enter the reference for the input cell

▶ **Figure 11.25**

Data Table *Dialog*

▶ Figure 11.26

*One-Way Data Table for
Uncertain Demand*

	A	B	C	D	E	F
1	Profit Model					
2						
3	Data				Demand	$240,000.00
4					25000	$ (360,000.00)
5	Unit Price	$40.00			30000	$ (160,000.00)
6	Unit Cost	$24.00			35000	$ 40,000.00
7	Fixed Cost	$400,000.00			40000	$ 240,000.00
8	Demand	50000			45000	$ 240,000.00
9					50000	$ 240,000.00
10					55000	$ 240,000.00
11	Model				60000	$ 240,000.00
12						
13	Unit Price	$40.00				
14	Quantity Sold	40000				
15	Revenue		$1,600,000.00			
16						
17	Unit Cost	$24.00				
18	Quantity Produced	40000				
19	Variable Cost		$960,000.00			
20	Fixed Cost		$400,000.00			
21						
22	Profit		$240,000.00			

▶ Figure 11.27

*One-Way Data Table with
Two Outputs*

E	F	G
	Profit	Revenue
Demand	$240,000	$1,600,000
25000	$(360,000)	$1,000,000
30000	$(160,000)	$1,200,000
35000	$ 40,000	$1,400,000
40000	$ 240,000	$1,600,000
45000	$ 240,000	$1,600,000
50000	$ 240,000	$1,600,000
55000	$ 240,000	$1,600,000
60000	$ 240,000	$1,600,000

in the model that corresponds to the input values in the row. In the *Column input cell* box, enter the reference for the input cell in the model that corresponds to the input values in the column. Then click *OK*.

Two-way data tables can evaluate only one output variable. To evaluate multiple output variables, you must construct multiple two-way tables.

EXAMPLE 11.21   A Two-Way Data Table for the Profit Model

In most models, the assumptions used for the input data are often uncertain. For example, in the profit model, the unit cost might be affected by supplier price changes and inflationary factors. Marketing might be considering price adjustments to meet profit goals. We use a two-way data table to evaluate the impact of changing these assumptions. First, create a column for the unit prices you wish to evaluate and a row for the unit costs in the form of a matrix (see column E and row 2 in Figure 11.28). In the upper left corner (which is cell E2 in Figure 11.28),

enter the formula =C22, which references the profit in the model. Select the range of all the data (not including the descriptive titles; that is, E2:I13 in Figure 11.28) and then select the data table tool in the *What-If Analysis* menu. In the *Data Table* dialog, enter B6 for the *Row input cell* since the unit cost corresponds to cell B6 in the model, and enter B5 for the *Column input cell* since the unit price corresponds to cell B5. Figure 11.28 shows the completed result as a heat map, using conditional formatting color scales (see Chapter 3).

Scenario Manager

The Excel *Scenario Manager* tool allows you to create **scenarios**—sets of values that are saved and can be substituted automatically on your worksheet. Scenarios are useful for conducting what-if analyses when you have more than two output variables (which data tables

▶ **Figure 11.28**

Two-Way Data Table

	D	E	F	G	H	I
1		Profit		Unit Cost		
2		$240,000.00	$22.00	$23.00	$24.00	$25.00
3		$35.00	$120,000.00	$80,000.00	$40,000.00	$0.00
4		$36.00	$160,000.00	$120,000.00	$80,000.00	$40,000.00
5		$37.00	$200,000.00	$160,000.00	$120,000.00	$80,000.00
6		$38.00	$240,000.00	$200,000.00	$160,000.00	$120,000.00
7	Unit	$39.00	$280,000.00	$240,000.00	$200,000.00	$160,000.00
8	Price	$40.00	$320,000.00	$280,000.00	$240,000.00	$200,000.00
9		$41.00	$360,000.00	$320,000.00	$280,000.00	$240,000.00
10		$42.00	$400,000.00	$360,000.00	$320,000.00	$280,000.00
11		$43.00	$440,000.00	$400,000.00	$360,000.00	$320,000.00
12		$44.00	$480,000.00	$440,000.00	$400,000.00	$360,000.00
13		$45.00	$520,000.00	$480,000.00	$440,000.00	$400,000.00

cannot handle). The Excel *Scenario Manager* is found under the *What-If Analysis* menu in the *Data Tools* group on the *Data* tab. When the tool is started, click the *Add* button to open the *Add Scenario* dialog and define a scenario (see Figure 11.29). Enter the name of the scenario in the *Scenario name* box. In the *Changing cells* box, enter the references, separated by commas, for the cells in your model that you want to include in the scenario (or hold down the Ctrl key and click on the cells). In the *Scenario Values* dialog that appears next, enter values for each of the changing cells. If you have put these into your spreadsheet, you can simply reference them. After all scenarios are added, they can be selected by clicking on the name of the scenario and then the *Show* button. Excel will change all values of the cells in your spreadsheet to correspond to those defined by the scenario for you to see the results within the model. When you click the *Summary* button on the *Scenario Manager* dialog, you will be prompted to enter the result cells and choose either a summary or a PivotTable report. The *Scenario Manager* can handle up to 32 variables.

The *Scenario Manager* is a useful tool for **best-case/worst-case analysis**. For example, in the profit model, a best-case scenario would be a high unit price, low unit and fixed cost, and high demand, while a worst-case scenario would have a low unit price, high unit and fixed cost, and low demand.

EXAMPLE 11.22   **Using the Scenario Manager for the Markdown Pricing Model**

In the *Markdown Pricing Model* spreadsheet, suppose that we wish to evaluate four different strategies, which are shown in Figure 11.30. In the *Add Scenario* dialog, enter Ten/ten as the scenario name and specify the changing cells as B7 and B8 (that is, the number of days at full retail price and the intermediate markdown). In the *Scenario Values* dialog, enter the values for these variables in the appropriate fields, or enter the formulas for the cell references; for instance, enter =E2 for the changing cell B7

or =E3 for the changing cell B8. Repeat this process for each scenario. Click the *Summary* button. In the *Scenario Summary* dialog that appears next, enter C33 (the total revenue) as the result cell. The *Scenario Manager* evaluates the model for each combination of values and creates the summary report shown in Figure 11.31. The results indicate that the largest profit can be obtained using the twenty/twenty markdown strategy.

Goal Seek

If you know the result that you want from a formula but are not sure what input value the formula needs to get that result, use the *Goal Seek* feature in Excel. *Goal Seek* works only with one variable input value. If you want to consider more than one input value or wish to maximize or minimize some objective, you must use the *Solver* add-in, which is discussed in other chapters. On the *Data* tab, in the *Data Tools* group, click *What-If Analysis*, and then click *Goal Seek*. The dialog shown in Figure 11.32 will appear. In the *Set cell* box, enter the reference for the cell that contains the formula that you want to resolve. In the *To value* box, type the formula result that you want. In the *By changing cell* box, enter the reference for the cell that contains the value that you want to adjust.

▶ **Figure 11.29**

Add Scenario *Dialog*

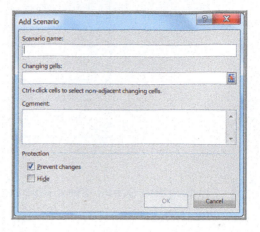

▲ **Figure 11.30**

Markdown Pricing Model with Scenarios

	A	B	C	D	E	F	G	H
1	Markdown Pricing Model				Scenarios	Ten/ten	Twenty/twenty	Thirty/thirty	Forty/forty
2				Days at full retail price	10	20	30	40
3	Data			Intermediate markdown	10%	20%	30%	40%
4	Retail price	$70.00						
5	Inventory	1000						
6	Selling season (days)	50						
7	Days at full retail	40						
8	Intermediate markdown	30%						
9	Clearance markdown	70%						

▶ **Figure 11.31**

Scenario Summary for the Markdown Pricing Model

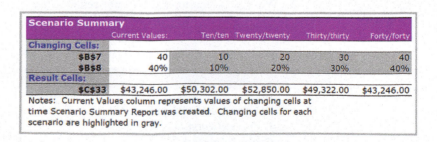

▶ **Figure 11.32**

Goal Seek *Dialog*

▶ **Figure 11.33**

*Break-Even Analysis Using
Goal Seek*

	A	B
1	**Outsourcing Decision Model**	
2		
3	**Data**	
4		
5	**Manufactured in-house**	
6	Fixed cost	$50,000
7	Unit variable cost	$125
8		
9	**Purchased from supplier**	
10	Unit cost	$175
11		
12	Production volume	1000
13		
14	**Model**	
15		
16	Total manufacturing cost	$175,000
17	Total purchased cost	$175,000
18		
19	Cost difference (Manufacture - Purchase)	$0
20	Best Decision	Manufacture

EXAMPLE 11.23   Finding the Break-Even Point in the Outsourcing Model

In the outsourcing decision model we introduced in Chapter 1 (see Example 1.4), we might wish to find the break-even point. The break-even point is the value of demand volume for which total manufacturing cost equals total purchased cost or, equivalently, for which the difference is zero. Therefore, you seek to find the value of production volume in cell B12 that yields a value of zero in cell B19. In the *Goal Seek* dialog, enter B19 for the *Set cell*, enter 0 in the *To value* box, and enter B12 in the *By changing cell* box. The *Goal Seek* tool determines that the break-even volume is 1,000 and enters this value in cell B12 in the model, as shown in Figure 11.33.

CHECK YOUR UNDERSTANDING

1. What Excel tools can be used to perform what-if analyses?

2. Explain the advantages of using Excel data tables.

3. What is a scenario, and why are scenarios useful in what-if analyses?

4. Explain how the Excel *Goal Seek* tool works.

KEY TERMS

Arbitrate pricing theory (APT)
Best-case/worst-case analysis
Critical path
Data table
Data validation
Euclidian distance
Influence diagram
Newsvendor problem
One-way data table
Overbooking

Permutation
Pro forma income statement
Rectilinear distance
Scenarios
Slack
Spreadsheet engineering
Two-way data table
Verification
What-if analysis

CHAPTER 11 TECHNOLOGY HELP

Excel Techniques

Data Validation (Example 11.6):

Select the cell range, click on *Data Validation* in the *Data Tools* group within the *Data* tab on the Excel ribbon, and then specify the criteria that Excel will use to flag invalid data. On the *Error Alert* tab, you can also create an alert box that pops up when an invalid entry is made. On the *Input Message* tab, you can create a prompt to display a comment in the cell about the correct input format

One-Way Data Tables (Example 11.19):

Create a range of values for some input cell in your model that you wish to vary. The input values must be listed either down a column (column oriented) or across a row (row oriented). If the input values are column oriented, enter the cell reference for the output variable in your model that you wish to evaluate in the row above the first value and one cell to the right of the column of input values. Reference any other output variable cells to the right of the first formula. If the input values are listed across a row, enter the cell reference of the output variable in the column to the left of the first value and one cell below the row of values. Type any additional output cell references below the first one. Next, select the range of cells that contains both the formulas and values you want to substitute. From the *Data* tab in Excel, select *Data Table* under the *What-If Analysis* menu. In the dialog box, if the input range is column oriented, type the cell reference for the input cell in your model in the *Column input cell* box. If the input range is row oriented, type the cell reference for the input cell in the *Row input cell* box.

Two-Way Data Tables (Example 11.21):

Type a list of values for one input variable in a column and a list of input values for the second input variable in a row, starting one row above and one column to the right of the column list. In the cell in the upper left-hand corner immediately above the column list and to the left of the row list, enter the cell reference of the output variable you wish to evaluate. Select the range of cells that contain this cell reference and both the row and column of values. On the *What-If Analysis* menu, click *Data Table*. In the *Row input cell* of the dialog box, enter the reference for the input cell in the model that corresponds to the input values in the row. In the *Column input cell* box, enter the reference for the input cell in the model that corresponds to the input values in the column.

Scenario Manager (Example 11.22):

Click the *What-If Analysis* menu in the *Data Tools* group on the *Data* tab. When the tool is started, click the *Add* button to open the *Add Scenario* dialog and define a scenario. Enter the name of the scenario in the *Scenario name* box. In the *Changing cells* box, enter the references, separated by commas, for the cells in your model that you want to include in the scenario (or hold down the Ctrl key and click on the cells). In the *Scenario Values* dialog that appears next, enter values for each of the changing cells. If you have put these into your spreadsheet, you can simply reference them. After all scenarios are added, they can be selected by clicking on the name of the scenario and then the *Show* button. When you click the *Summary* button on the *Scenario Manager* dialog, you will be prompted to enter the result cells and choose either a summary or a PivotTable report.

Goal Seek (Example 11.23):

On the *Data* tab, in the *Data Tools* group, click *What-If Analysis*, and then click *Goal Seek*. In the *Set cell* box, enter the reference for the cell that contains the formula that you want to resolve. In the *To value* box, type the formula result that you want. In the *By changing cell* box, enter the reference for the cell that contains the value that you want to adjust.

Analytic Solver

Analytic Solver provides tools for what-if and sensitivity analysis in spreadsheet models. See the online supplement *Model Analysis in Analytic Solver*. We suggest that you first read the online supplement *Getting Started with Analytic Solver Basic*. This provides information for both instructors and students on how to register for and access Analytic Solver.

PROBLEMS AND EXERCISES

Model-Building Strategies

1. A manufacturer of kitchen appliances is preparing to set the price on a new blender. Demand is thought to depend on the price and is represented by the model

$$D = 2{,}500 - 3P$$

The accounting department estimates that the total cost can be represented by

$$C = 5{,}000 + 5D$$

Develop a mathematical model for the total profit in terms of the price, P.

2. Modern Electronics sells two popular models of wireless headphones, model A and model B. The sales of these products are not independent of each other (in economics, we call these substitutable products because if the price of one increases, sales of the other will increase). The store wishes to establish a pricing policy to maximize revenue from these products. A study of price and sales data shows the following relationships between the quantity sold (N) and prices (P) of each model:

$$N_A = 20 - 0.62P_A + 0.30P_B$$

$$N_B = 29 + 0.10P_A - 0.60P_B$$

   a. Construct a mathematical model for the total revenue.

   b. What is the predicted revenue if $P_A = \$18$ and $P_B = \$30$?

3. Few companies take the time to estimate the value of a good customer (and often spend little effort to keep one). Suppose that a customer at a restaurant spends, on average, R per visit and comes F times each year (for example, if a customer purchases once every two years, then $F = \frac{1}{2} = 0.5$). The restaurant realizes a gross profit margin of M (expressed as a fraction) on the average bill for food and drinks. In addition, the fraction of customers defecting (not returning) each year is D.

   a. Develop a mathematical model to compute the gross profit during a customer's lifetime in doing business with the restaurant (this is often called the *economic value of a customer*).

   b. If the average purchase per visit is $50, the gross profit margin is 0.4 (that is, 40%), customers visit an average of six times each year, and 30% of customers defect each year, what is the economic value of the customer?

4. The demand for airline travel is quite sensitive to price. Typically, there is an inverse relationship between demand and price; when price decreases, demand increases and vice versa. One major airline has found that when the price (P) for a round trip between Chicago and Los Angeles is $600, the demand ($D$) is 500 passengers per day. When the price is reduced to $400, demand is 1,200 passengers per day.

   a. Plot these points on a coordinate system and develop a function that relates demand to price.

   b. Develop a model that will determine the total revenue as a function of the price.

5. A company is trying to predict the long-run market share of a new men's deodorant.[10] Based on initial marketing studies, they believe that 35% of new purchasers in this market will ultimately try this brand, and of these, about 60% will purchase it in the future. Preliminary data also suggest that the brand will attract heavier-than-average buyers, such as those who exercise frequently and participate in sports, and that they will purchase about 20% more than the average buyer.

   a. Calculate the long-run market share that the company can anticipate under these assumptions.

   b. Develop a general model for predicting long-run market share.

6. A pharmaceutical company makes sales calls to physicians to market their drugs. They want to determine the impact of sales calls on the profit it achieves per physician. The profit per physician depends on the number of new prescriptions that are written and the cost of making the sales calls. The cost of making the sales calls depends on the number of sales calls made. The number of sales calls also impacts the number of new prescriptions written. Construct an influence diagram that relates these variables.

7. Construct an influence diagram for profit in the single-period purchase decision (*Newsvendor Model*)

---

[10] Based on an example of the Parfitt-Collins model in Gary L. Lilien, Philip Kotler, and K. Sridhar Moorthy, *Marketing Models* (Englewood Cliffs, NJ: Prentice Hall, 1992): 483.

discussed in this chapter. Use Excel's *Formula Auditing* capability to demonstrate the relationship between the spreadsheet model and the influence diagram.

8. Construct an influence diagram for the net revenue in the overbooking decision model discussed in this chapter. Use Excel's *Formula Auditing* capability to demonstrate the relationship between the spreadsheet model and the influence diagram.

9. Construct an influence diagram for the staffing model in Example 11.7.

10. Construct an influence diagram for the portfolio allocation model in Example 11.15.

11. Return on investment (ROI) is computed in the following manner: ROI is equal to turnover multiplied by earnings as a percent of sales. Turnover is sales divided by total investment. Total investment is current assets (inventories, accounts receivable, and cash) plus fixed assets. Earnings equal sales minus the cost of sales. The cost of sales consists of variable production costs, selling expenses, freight and delivery, and administrative costs.

   a. Construct an influence diagram that relates these variables.

   b. Define symbols and develop a mathematical model.

12. A (greatly) simplified model of the national economy can be described as follows. The national income is the sum of three components: consumption, investment, and government spending. Consumption is related to the total income of all individuals and to the taxes they pay on income. Taxes depend on total income and the tax rate. Investment is also related to the size of the total income.

   a. Use this information to draw an influence diagram by recognizing that the phrase "A is related to B" implies that A influences B in the model.

   b. If we assume that the phrase "A is related to B" can be translated into mathematical terms as $A = kB$, where k is some constant, develop a mathematical model for the information provided.

13. Economists believe that housing starts depend on interest rates and demographic factors such as population size, family income, and the age of the home-buying population. In addition, interest rates depend on inflation, Federal Reserve policies, and government borrowing. Government borrowing depends on

government spending and tax revenues. Develop an influence diagram that illustrates the relationships among these factors.

14. The monthly demand for a digital camera is sensitive to price:

Price	Demand
$150.00	2,317
$160.00	2,068
$170.00	1,839
$180.00	1,708
$190.00	1,542
$200.00	1,421
$210.00	1,314
$220.00	1,293
$230.00	1,195
$240.00	1,150

Find the most appropriate trendline to explain the relationship between demand and price, and develop a model for the monthly revenue.

Implementing Models on Spreadsheets

15. Develop a spreadsheet model for the gasoline usage scenario (Example 1.3) using the data provided. Apply the principles of spreadsheet engineering in developing your model.

16. Develop a spreadsheet model to compute the total revenue for any price in Example 1.6 (prescriptive pricing model). Use the model to create a table for a range of prices to help you identify the price that results in the maximum revenue.

17. Develop a spreadsheet for calculating the total revenue for the Modern Electronics scenario in Problem 2. Design it so that the price-demand function parameters can easily be changed.

Descriptive Spreadsheet Models

18. A clothing company is planning its winter pricing. One popular line of quarter-zip sweatshirts sells for $98. The products hit the stores at the beginning of September and sell through Christmas (assume 120 days). After that time, they are discounted to $39 for clearance and sell out. The planned inventory is 2,500 units. Last year, an average of 12 units per day were sold at the full retail price. A Columbus Day sale (40 days into the selling season) reduced the price to $79 and increased sales to an average of

30 units per day. A Black Friday sale (90 days into the selling season) with a price of $69 increased sales to an average of 40 units per day. For the coming year, the company is considering permanent reductions starting with the Columbus Day sale until Black Friday, and then from Black Friday until the after-Christmas clearance. Develop a spreadsheet model to evaluate their revenue under the following scenarios:

a. No price reductions at all until the after-Christmas clearance

b. A price reduction to $79 starting at Columbus Day until Black Friday and then a sale price of $59 until the after-Christmas clearance

c. A price reduction to $69 starting at Columbus Day until Black Friday and then a sale price of $59 until the after-Christmas clearance

19. Develop a spreadsheet model to determine how much a person or a couple can afford to spend on a house.[11] Lender guidelines suggest that the allowable monthly housing expenditure should be no more than 28% of monthly gross income. From this, you must subtract total nonmortgage housing expenses, which would include insurance and property taxes and any other additional expenses. This defines the affordable monthly mortgage payment. In addition, guidelines also suggest that total affordable monthly debt payments, including housing expenses, should not exceed 36% of gross monthly income. This is calculated by subtracting total nonmortgage housing expenses and any other installment debt, such as car loans, student loans, credit card debt, and so on, from 36% of total monthly gross income. The smaller of the affordable monthly mortgage payment and the total affordable monthly debt payments is the affordable monthly mortgage. To calculate the maximum that can be borrowed, find the monthly payment per $1,000 mortgage based on the current interest rate and duration of the loan. Divide the affordable monthly mortgage amount by this monthly payment to find the affordable mortgage. Assuming a 20% down payment, the maximum price of a house would be the affordable mortgage divided by 0.8. Use the following data to test your model: total monthly gross income = $6,500; nonmortgage housing expenses = $350; monthly installment debt = $500; monthly payment per $1,000 mortgage = $7.25.

20. MasterTech is a new software company that develops and markets productivity software for municipal government applications. In developing their income statement, the following formulas are used:

gross profit = net sales − cost of sales

net operating profit = gross profit − administrative expenses − selling expenses

net income before taxes = net operating profit − interest expense

net income = net income before taxes − taxes

Net sales are expected to be $1,250,000. Cost of sales is estimated to be $300,000. Selling expenses have a fixed component that is estimated to be $90,000 and a variable component that is estimated to be 8% of net sales. Administrative expenses are $50,000. Interest expenses are $8,000. The company is taxed at a 50% rate. Develop a spreadsheet model to calculate the net income. Design your spreadsheet using good spreadsheet-engineering principles.

21. A garage band wants to hold a concert. The expected crowd is 2,500. The average expenditure on concessions is $25. Tickets sell for $20 each, and the band's profit is 80% of the gate and concession sales minus a fixed cost of $18,000. Develop a mathematical model and implement it on a spreadsheet to find the band's expected profit.

22. A stockbroker calls on potential clients from referrals. For each call, there is a 20% chance that the client will decide to invest with the stockbroker's firm. Forty percent of those interested are found not to be qualified, based on the brokerage firm's screening criteria. The remaining are qualified. Of these, half will invest an average of $5,000, 25% will invest an average of $20,000, 15% will invest an average of $50,000, and the remainder will invest $100,000. The commission schedule is as follows:

Transaction Amount	Commission
Up to $25,000	$75 + 0.5% of the amount
$25,001 to $75,000	$100 + 0.4% of the amount
$75,001 to $100,000	$150 + 0.3% of the amount

The broker keeps half the commission. Develop a spreadsheet to calculate the broker's commission based on the number of calls per month made. What is the expected commission based on making 250 calls?

---

[11] Based on Ralph R. Frasca, *Personal Finance*, 8th ed. (Boston: Prentice Hall, 2009).

23. The director of a nonprofit ballet company in a medium-sized U.S. city is planning its next fundraising campaign. In recent years, the program has found the following percentages of donors and gift levels:

Gift Level	Amount	Average Number of Gifts
Benefactor	$10,000	3
Philanthropist	$5,000	10
Producer's Circle	$1,000	25
Director's Circle	$500	50
Principal	$100	7% of solicitations
Soloist	$50	12% of solicitations

Develop a spreadsheet model to calculate the total amount donated based on this information if the company contacts 1,000 potential donors to donate at the $100 level or below.

24. Tanner Park is a small amusement park that provides a variety of rides and outdoor activities for children and teens. In a typical summer season, the number of adult and children's tickets sold is 20,000 and 10,000, respectively. Adult ticket prices are $18 and the children's price is $10. Revenue from food and beverage concessions is estimated to be $60,000, and souvenir revenue is expected to be $25,000. Variable costs per person (adult or child) are $3, and fixed costs amount to $150,000. Determine the profitability of this business.

25. The admissions director of an engineering college has $500,000 in scholarships each year from an endowment to offer to high-achieving applicants. The value of each scholarship offered is $25,000 (thus, 20 scholarships are offered). The benefactor who provided the money would like to see all of it used each year for new students. However, not all students accept the money; some take offers from competing schools. If they wait until the end of the admission deadline to decline the scholarship, it cannot be offered to someone else because any other good students would already have committed to other programs. Consequently, the admissions director offers more money than available in anticipation that a percentage of offers will be declined. If more than 20 students accept the offers, the college is committed to honoring them, and the additional amount has to come out of the dean's budget. Based on prior history, the percentage of applicants that accept the offer is about 70%. Develop a spreadsheet model for this situation to evaluate how much money must be allocated from the dean's budget based on the number of scholarships offered.

26. J&G Bank receives an average of 30,000 credit card applications each month. Approximately 60% of them are approved. Each customer charges an average of $2,000 to his or her credit card each month. Approximately 85% pay off their balances in full, and the remaining incur finance charges. The average finance charge is 3.5% per month. The bank also receives income from fees charged for late payments and annual fees associated with the credit cards. This is a percentage of total monthly charges and is approximately 7%. It costs the bank $20 per application, whether it is approved or not. The monthly maintenance cost for credit card customers is $10. Finally, losses due to charge-offs of customers' accounts average 5% of total charges. Develop a spreadsheet model to calculate the bank's total monthly profit.

Predictive Spreadsheet Models

27. With the growth of digital photography, a young entrepreneur is considering establishing a new business, Cruz Wedding Photography. He believes that the average number of wedding bookings per year is 15. One of the key variables in developing his business plan is the life he can expect from a single digital single-lens reflex (DSLR) camera before it needs to be replaced. Due to heavy usage, the shutter life expectancy is estimated to be 150,000 clicks. For each booking, the average number of photographs taken is assumed to be 2,000. Develop a model to predict the camera life (in years).

28. For a new product, sales volume in the first year is estimated to be 80,000 units and is projected to grow at a rate of 6% per year. The selling price is $12 and will increase by $0.50 each year. Per-unit variable costs are $3, and annual fixed costs are $400,000. Per-unit costs are expected to increase 3% per year. Fixed costs are expected to increase 8% per year. Develop a spreadsheet model to predict the net present value of profit over a three-year period, assuming a 4% discount rate.

29. The Executive Committee of Reder Electric Vehicles is debating whether to replace its original model, the REV-Touring, with a new model, the REV-Sport, which would appeal to a younger audience. Whatever vehicle is chosen will be produced for the next four years, after which time a reevaluation will be necessary. The REV-Sport has passed through the concept and initial design phases and is ready for final design and manufacturing. Final development costs are estimated to be $75 million, and the new fixed costs for tooling and manufacturing are estimated to be $600 million. The REV-Sport is expected to sell for $30,000. The first-year sales for the REV-Sport is estimated to be 60,000, with a sales growth for the subsequent years of 6% per year. The variable cost per vehicle is uncertain until the design and supply-chain decisions are finalized, but is estimated to be $22,000. Next-year sales for the REV-Touring are estimated to be 50,000, but the sales are expected to decrease at a rate of 10% for each of the next three years. The selling price is $28,000. Variable costs per vehicle are $21,000. Since the model has been in production, the fixed costs for development have already been recovered. Develop a four-year model to predict the profitability of each vehicle and recommend the best decision using a net present value discount rate of 5%. How sensitive is the result to the estimated variable cost of the REV-Sport? How might this affect the decision?

30. The Schoch Museum is embarking on a five-year fundraising campaign. As a nonprofit institution, the museum finds it challenging to acquire new donors, as many donors do not contribute every year. Suppose that the museum has identified a pool of 8,000 potential donors. The actual number of donors in the first year of the campaign is estimated to be 60% of this pool. For each subsequent year, the museum expects that 30% of current donors will discontinue their contributions. In addition, the museum expects to attract some percentage of new donors. This is assumed to be 10% of the pool. The average contribution in the first year is assumed to be $50, and will increase at a rate of 2.5%. Develop a model to predict the total funds that will be raised over the five-year period.

31. The Hyde Park Surgery Center specializes in high-risk cardiovascular surgery. The center needs to forecast its profitability over the next three years to plan for capital growth projects. For the first year, the hospital anticipates serving 1,200 patients, which is expected to grow by 8% per year. Based on current reimbursement formulas, each patient provides an average billing of $125,000, which will grow by 3% each year. However, because of managed care, the center collects only 25% of billings. Variable costs for supplies and drugs are calculated to be 10% of billings. Fixed costs for salaries, utilities, and so on will amount to $20,000,000 in the first year and are assumed to increase by 5% per year. Develop a spreadsheet model to predict the net present value of profit over the next three years. Use a discount rate of 4%.

32. Adam is 24 years old and has a 401(k) plan through his employer, a large financial institution. His company matches 50% of his contributions up to 6% of his salary. He currently contributes the maximum amount he can. In his 401(k), he has three funds. Investment A is a large-cap index fund, which has had an average annual growth over the past ten years of 6.63%. Investment B is a mid-cap index fund with a ten-year average annual growth of 9.89%. Finally, Investment C is a small-cap index fund with a ten-year average annual growth rate of 8.55%. Fifty percent of his contribution is directed to Investment A, 25% to Investment B, and 25% to Investment C. His current salary is $48,000, and based on a compensation survey of financial institutions, he expects an average raise of 2.7% each year. Develop a spreadsheet model that predicts his retirement balance at age 65.

33. Develop a realistic retirement planning spreadsheet model for your personal situation. If you are currently employed, use as much information as you can gather for your model, including potential salary increases, promotions, contributions, and rates of return based on the actual funds in which you invest. If you are not employed, try to find information about salaries in the industry in which you plan to work and the retirement benefits that companies in that industry offer for your model. Estimate rates of returns based on popular mutual funds used for retirement or average performance of stock market indexes. Clearly state your assumptions and how you arrived at them.

34. Jennifer Bellin has been put in charge of planning her company's annual leadership conference. The dates of the conference have been determined by her company's executive team. The table that

follows contains information about the activities, predecessors, and activity times (in days):

	Activity	Predecessors	Activity Time (days)
A	Develop conference theme		3
B	Determine attendees		3
C	Contract facility	A	7
D	Choose entertainment	A	10
E	Send announcement	B	5
F	Order gifts	B	5
G	Order materials	B	1
H	Plan schedule of sessions	C	40
I	Design printed materials	B, H	15
J	Schedule session rooms	C	1
K	Print directions	H	10
L	Develop travel memo	E	5
M	Write gift letter	F	5
N	Confirm catering	H	3
O	Communicate with speakers	H	3
P	Track RSVPs and assign roommates	L	30
Q	Print materials	I	3
R	Assign table numbers	P	1
S	Compile packets of materials	G	3
T	Submit audiovisual needs	O	1
U	Put together welcome letter	P	5
V	Confirm arrangements with hotel	P	3
W	Print badges	G, P	5

Develop a spreadsheet model for finding the project completion time and critical path.

Prescriptive Spreadsheet Models

35. Experiment with the portfolio allocation model in Example 11.15 to attempt to find the best solution that maximizes the expected annual return and meets the total weighted risk constraint.

36. A business student has $2,500 available from a summer job and has identified three potential stocks in which to invest. The cost per share and expected return over the next two years is given in the table.

Stock	A	B	C
Price/share	$25	$15	$30
Return/share	$8	$7	$11

Develop a spreadsheet model that computes the total return for any mix of investments. Experiment with the model to attempt to find the best solution with the highest total return and that limits the investment to $2,500.

37. The Gardner Theater, a community playhouse, needs to determine the lowest-cost production budget for an upcoming show. Specifically, they have to determine which set pieces to construct and which, if any, set pieces to rent from another local theater at a predetermined fee. However, the organization has only two weeks to fully construct the set before the play goes into technical rehearsals. The theater has two part-time carpenters who work up to 12 hours a week, each at $10 an hour. Additionally, the theater has a part-time scenic artist who can work 15 hours per week to paint the set and props as needed at a rate of $15 per hour. The set design requires 20 flats (walls), two hanging drops with painted scenery, and three large wooden tables (props). The number of hours required for each piece for carpentry and painting is shown below:

	Carpentry	Painting
Flats	0.5	2.0
Hanging Drops	2.0	12.0
Props	3.0	4.0

Flats, hanging drops, and props can also be rented at a cost of $75, $500, and $350 each, respectively. The theater wants to determine how many of each unit should be built by the theater and how many should be rented to minimize total costs. Develop a spreadsheet model that computes the total cost for any mix of units built and rented, as well as the total hours required for carpenters and painters (which must meet the limited hours available per week). Experiment with the model to attempt to find the best solution that meets the labor availability and the required number of units of each type.

38. A franchise of a chain of Mexican restaurants wants to determine the best location to attract customers from three suburban neighborhoods. The coordinates of the three suburban neighborhoods are as follows:

Neighborhood	X-Coordinate	Y-Coordinate
Liberty	2	12
Jefferson	9	6
Adams	1	1

The population of Adams is four times as large as that of Jefferson, and Jefferson is twice as large as Liberty. The restaurant wants to consider the population in its location decision. Develop a model to find the best location, assuming that straight-line distances can be used between the locations. Experiment with the model to find the best location.

39. ElectroMart wants to identify a location for a warehouse that will ship to five retail stores. The coordinates and annual number of truckloads are given here. Develop a model to find the best location, assuming that straight-line distances can be used between the locations. Experiment with the model to find the best location.

Retail Store	X-Coordinate	Y-Coordinate	Truckloads
A	18	15	12
B	3	4	18
C	20	5	24
D	3	16	12
E	10	20	18

40. An IT support group at Thomson State College has seven projects to complete. The time and project deadlines (both in days) are shown next.

Project	1	2	3	4	5	6	7
Time	4	9	12	16	9	15	8
Deadline	12	24	60	28	24	36	48

a. Develop a spreadsheet model for this situation.

b. Use a spreadsheet model to try to find a sequence that minimizes the average lateness.

c. Use a spreadsheet model to try to find a sequence that minimizes the average tardiness.

d. Compare these solutions to the SPT and EDD rules.

Analyzing Uncertainty and Model Assumptions

41. Implement the model you developed in Problem 1 on a spreadsheet and construct a one-way data table to estimate the price for which profit is maximized.

42. For the stockbroker model you developed in Problem 22, use a one-way data table to show how the broker's commission is a function of the number of calls made.

43. For the nonprofit ballet company fundraising model you developed in Problem 23, use a one-way data table to show how the amount varies based on the number of solicitations.

44. For the Schoch Museum (Problem 30), use data tables to investigate the impacts of the percentage assumptions used in the model on the cumulative funds raised by the fifth year.

45. A gasoline mini-mart orders 25 copies of a monthly magazine. Depending on the cover story, demand for the magazine varies. The mini-mart purchases the magazines for $1.50 and sells them for $4.00. Any magazines left over at the end of the month are donated to hospitals and other health care facilities. Modify the newsvendor example spreadsheet to model this situation. Use what-if analysis to investigate the financial implications of this policy if the demand is expected to vary between 10 and 30 copies each month.

46. The weekly price at an extended-stay hotel (renting by the week for business travelers) is $950. Operating costs average $20,000 per week, regardless of the number of rooms rented. Construct a spreadsheet model to determine the profit if 40 rooms are rented. The manager has observed that the number of rooms rented during any given week varies between 32 and 50 (the total number of rooms available).

a. Use a data table to evaluate the profit for this range of unit rentals.

b. Suppose the manager is considering lowering or increasing the weekly price by $100. Use a data table to evaluate how the profit will be affected.

47. Use the *Markdown Pricing Model* spreadsheet model and a two-way data table to find the total revenue if days at full retail vary from 20 to 40 in increments of 5 and the intermediate markdown varies from 15% to 50% in increments of 5%.

48. For the engineering admissions situation in Problem 25, apply two-way data tables to analyze the impact on extra funds needed and the number of students who accept scholarships as the acceptance rate and number of offers vary.

49. Koehler Vision Associates (KVA) specializes in laser-assisted corrective eye surgery. Prospective patients make appointments for prescreening exams to determine their candidacy for the surgery. If they qualify, a $250 charge is applied as a deposit for the actual procedure. The weekly demand is 150, and about 12% of prospective patients fail to show up or cancel their exam at the last minute. Patients who do not show up are refunded the prescreening fee less a $25 processing fee. KVA can handle 125 patients per week and is considering overbooking its appointments to reduce the lost revenue associated with cancellations. However, any patient who is overbooked may spread unfavorable comments about the company; thus, the overbooking cost is estimated to be $125. Develop a spreadsheet model for calculating net revenue. Use data tables to study how revenue is affected by changes in the number of appointments accepted and patient demand.

50. For the garage band model you developed in Problem 21, use the *Scenario Manager* to evaluate profitability for the following scenarios:

	Likely	Optimistic	Pessimistic
Expected crowd	2,500	4,500	1,500
Concessions Expenditure	$25.00	$40.00	$10.00
Fixed cost	$18,000.00	$10,000.00	$25,000.00

51. Think of any retailer that operates many stores throughout the country, such as Old Navy, Hallmark Cards, or Vineyard Vines, to name just a few. The retailer is often seeking to open new stores and needs to evaluate the profitability of a proposed location that would be leased for five years. An Excel model is provided in the *New Store Financial Model* spreadsheet. Use the *Scenario Manager* to evaluate the cumulative discounted cash flow for the fifth year under the following scenarios:

	Scenario 1	Scenario 2	Scenario 3
Inflation rate	1%	5%	3%
Cost of merchandise (% of sales)	25%	30%	26%
Labor cost	$150,000	$225,000	$200,000
Other expenses	$300,000	$350,000	$325,000
First-year sales revenue	$600,000	$600,000	$800,000
Sales growth year 2	15%	22%	25%
Sales growth year 3	10%	15%	18%
Sales growth year 4	6%	11%	14%
Sales growth year 5	3%	5%	8%

52. For the gasoline usage situation in Problem 15, apply the *Goal Seek* tool to find the fuel economy needed to consume 20 gallons per month, with all other data being constant.

53. For the stockbroker situation in Problem 22, use the *Goal Seek* tool to find the number of calls needed to achieve a broker commission of $5,000.

54. For the ballet company situation in Problem 23, use the *Goal Seek* tool to find the number of solicitations required to achieve $150,000 in total donations.

CASE: PERFORMANCE LAWN EQUIPMENT

Part 1: The *Performance Lawn Equipment Database* contains data needed to develop a pro forma income statement. Dealers selling PLE products all receive 18% of sales revenue for their part of doing business, and this is accounted for as the selling expense. The tax rate is 50%. Develop an Excel worksheet to extract and summarize the data needed to develop the income statement for 2018 and implement an Excel model in the form of a pro forma income statement for the company.

Part 2: The CFO of Performance Lawn Equipment, J. Kenneth Valentine, would like to have a model to predict the net income for the next three years. To do this, you need to determine how the variables in the pro forma income statement will likely change in the future. Using the calculations and worksheet that you developed along with other historical data in the database, estimate the annual rate of change in sales revenue, cost of goods sold, operating expense, and interest expense. Use these rates to modify the pro forma income statement to predict the net income over the next three years.

Because the estimates you derived from the historical data may not hold in the future, conduct appropriate what-if and scenario analyses to investigate how the projections might change if these assumptions don't hold. Summarize your results and conclusions in a report to Mr. Valentine.

Simulation and Risk Analysis

0.00	$2.07	$2.00	$1.50	$0.00	$0.90	8.97	5720.79	6.47	16.52	14.21	30.73	33.69	$1,010.80
2.50	$3.77	$2.50	$1.00	$1.40	$0.00	15.17	5735.96	11.2	17.00	14.02	31.02	32.55	$976.43
1.50	$9.07	$3.50	$0.00	$0.70	$0.30	33.07	5768.73	15.1	17.65	14.09	31.74	33.15	$994.37
9.00	$10.47	$0.50	$1.50	$1.40	$0.00	46.87	5815.60	22.9	18.14	14.38	32.52	34.08	$1,022.43
2.50	$4.44	$2.00	$1.50	$3.15	$0.00	33.84	5849.44	13.6	17.49	14.39	31.88	33.58	$1,007.52
5.50	$13.72	$1.50	$0.50	$4.20	$0.90	44.57	5893.11	26.3	17.76	14.34	32.10	33.45	$1,003.55
2.00	$4.72	$2.50	$0.00	$1.05	$0.00	37.27	5930.38	10.3	16.40	15.13	31.52	33.09	$992.62
2.00	$1.56	$0.00	$0.00	$0.35	$0.60	10.51	5940.29	4.51	16.16	15.23	31.39	31.51	$945.38
1.00	$4.00	$0.00	$1.00	$1.05	$0.00	14.05	5954.34	7.05	16.08	15.27	31.35	29.91	$897.27
3.50	$13.42	$7.50	$0.50	$0.70	$0.30	48.07	6002.11	25.9	16.85	15.71	32.56	31.06	$931.67
8.00	$8.23	$4.00	$2.50	$1.05	$0.00	42.68	6044.79	23.8	17.44	15.65	33.09	32.04	$961.09
4.00	$7.90	$2.00	$1.25	$4.20	$0.30	43.95	6088.44	19.7	17.12	16.17	33.28	32.28	$968.37
3.00	$6.78	$2.50	$2.50	$1.05	$0.00	44.63	6133.07	15.8	16.78	16.94	33.72	32.34	$970.29
2.00	$6.20	$1.50	$1.50	$0.70	$0.00	36.50	6169.57	11.9	15.31	17.55	32.87	31.79	$953.70
1.00	$1.26	$0.00	$0.25	$0.35	$0.00	10.96	6180.53	2.86	15.06	17.95	33.01	30.95	$928.44
0.00	$3.83	$1.50	$2.00	$1.40	$0.90	18.63	6198.26	9.63	14.95	18.31	33.26	30.91	$927.25
8.50	$9.74	$3.00	$0.00	$1.40	$0.00	42.44	6240.70	22.6	15.49	18.44	33.93	31.78	$953.46
3.50	$9.17	$2.00	$1.50	$1.40	$1.20	41.87	6281.37	18.8	15.19	18.38	33.57	32.81	$984.22
3.00	$9.73	$5.50	$0.50	$5.25	$0.00	49.78	6331.15	24	15.94	18.77	34.71	33.70	$1,011.02
4.00	$11.00	$3.00	$2.00	$21.00	$0.00	72.20	6403.35	41	16.99	19.70	36.68	34.91	$1,047.20
1.00	$7.12	$1.00	$1.50	$3.50	$0.00	40.52	6443.87	14.1	17.26	19.65	36.91	34.83	$1,044.93

Stephen Rees/Shutterstock

LEARNING OBJECTIVES

After studying this chapter, you will be able to:

- Explain the concept of Monte Carlo simulation and its importance of analyzing risk in business decisions.
- Use Excel's *Random Number Generation* tool.
- Generate random variates for common probability distributions using Excel functions.

- Apply Monte Carlo simulation to prescriptive, predictive, and prescriptive applications.
- Analyze and interpret simulation results.
- Use data tables to conduct simple Monte Carlo simulations.

Many of the models we developed in Chapter 11, such as the newsvendor, over-booking, and retirement-planning models, incorporated uncontrollable inputs, such as customer demand, hotel cancellations, and annual returns on investments, which exhibit random behavior. We often assume such variables to be constant to simplify the model and the analysis. However, many situations dictate that randomness be explicitly incorporated into our models. This is usually done by specifying probability distributions for the appropriate uncontrollable inputs. As we noted earlier in this book, models that include randomness are called *stochastic*, or *probabilistic*, models.

Nearly every manager deals with risk. **Risk** is the likelihood of an undesirable outcome. It can be assessed by evaluating the probability that the outcome will occur along with the severity of the outcome. For example, an investment that has a high probability of losing money is riskier than one with a lower probability. Similarly, an investment that may result in a $10 million loss is certainly riskier than one that might result in only a $10,000 loss. In assessing risk, we could answer questions such as, What is the probability that we will incur a financial loss? How do the probabilities of different potential losses compare? What is the probability that we will run out of inventory? What are the chances that a project will be completed on time? **Risk analysis** is an approach for developing "a comprehensive understanding and awareness of the risk associated with a particular variable of interest (be it a payoff measure, a cash flow profile, or a macroeconomic forecast)."[1] Hertz and Thomas present a simple scenario to illustrate the concept of risk analysis:

> The executives of a food company must decide whether to launch a new packaged cereal. They have come to the conclusion that five factors are the determining variables: advertising and promotion expense, total cereal market, share of market for this product, operating costs, and new capital investment. On the basis of the "most likely" estimate for each of these variables, the picture looks very bright—a healthy 30% return, indicating a significantly positive expected net present value. This future, however, depends on each of the "most likely" estimates coming true in the actual case. If each of these "educated guesses" has, for example, a 60% chance of being correct, there is only an 8% chance that all five will be correct (0.60 × 0.60 × 0.60 × 0.60 × 0.60) if the factors are assumed to be independent. So the "expected" return, or present value measure, is actually dependent on a rather unlikely coincidence. The decision

---

[1]David B. Hertz and Howard Thomas, *Risk Analysis and Its Applications* (Chichester, UK: John Wiley & Sons, Ltd., 1983): 1.

maker needs to know a great deal more about the other values used to make each of the five estimates and about what he stands to gain or lose from various combinations of these values.[2]

Thus, risk analysis seeks to examine the impacts of uncertainty in the estimates and their potential interaction with one another on the output variable of interest.

In this chapter, we discuss how to build models involving uncertainty and risk, and analyze them using an approach called simulation. We discuss two types of simulation approaches: Monte-Carlo simulation and systems simulation. Monte Carlo simulation is generally focused on risk analysis, particularly for spreadsheet models. Systems simulation models dynamic systems that change over time, such as waiting lines, inventory systems, manufacturing systems, and so on. These techniques have wide applicability in finance and operations, as well as other areas of business.

Monte Carlo Simulation

Monte Carlo simulation is the process of generating random values for uncertain inputs in a model, computing the output variables of interest, and repeating this process for many trials (replications) to understand the distribution of the output results and their statistical properties. For example, in a predictive financial model, we might be interested in the distribution of the cumulative discounted cash flow over several years (the model output) when future sales, sales growth rate, operating expenses, and inflation factors (model inputs) are all uncertain.

In Monte Carlo simulation, we model uncertain inputs using probability distributions. Determining the appropriate probability distributions is crucial to building good simulation models. For many uncertain inputs, empirical data may be available, either in historical records or collected through special efforts. For example, maintenance records might provide data on machine failure rates and repair times, or observers might collect data on service times in a bank or post office. This provides a factual basis for choosing the appropriate probability distribution to model the input variable. We can also identify an appropriate distribution by fitting historical data to a theoretical model, as we illustrated in Chapter 5.

In other situations, historical data are not available, and we can draw upon the properties of common probability distributions and typical applications that we discussed in Chapter 5 to help choose a representative distribution that has the shape that would most reasonably represent the analyst's understanding about the uncertain variable. For example, a normal distribution is symmetric, with a peak in the middle. Exponential data are very positively skewed, with no negative values. A triangular distribution has a limited range and can be skewed in either direction.

In many cases, uniform or triangular distributions are used in the absence of data. These distributions depend on simple parameters that one can easily identify based on

---

[2]Ibid., 24.

	A	B	C
1	Profit Model		
2			
3	Data		
4			
5	Unit Price	$40.00	
6	Unit Cost	$24.00	
7	Fixed Cost	$400,000.00	
8	Demand	50000	
9			
10			
11	Model		
12			
13	Unit Price	$40.00	
14	Quantity Sold	40000	
15	Revenue		$1,600,000.00
16			
17	Unit Cost	$24.00	
18	Quantity Produced	40000	
19	Variable Cost		$960,000.00
20	Fixed Cost		$400,000.00
21			
22	Profit		$240,000.00

managerial knowledge and judgment. For example, to define the uniform distribution, we need to know only the smallest and largest possible values that the variable might assume. For the triangular distribution, we also include the most likely value. In the construction industry, for instance, experienced supervisors can easily tell you the fastest, most likely, and slowest times for performing a task such as framing a house, taking into account possible weather and material delays, labor absences, and so on.

EXAMPLE 12.1   **Profit Model**

In Chapter 11 (Example 11.4), we developed a simple spreadsheet model for computing profit. The model is shown again in Figure 12.1. Suppose that the demand, unit cost, and fixed cost are uncertain. Historical data for the demand might suggest a probability distribution with the following values:

Demand	Probability
40,000	0.1
45,000	0.3
50,000	0.4
55,000	0.15
60,000	0.05

This is a simple discrete probability distribution. For the unit cost, managers might only know that it might vary between $22 and $26; thus, a uniform distribution would be an appropriate distribution to model this. They might also estimate that the fixed cost might be as low as $350,000 or as high as $500,000, with the most likely value being $400,000. In this case, a triangular distribution can be used.

If values from these distributions are generated randomly in the appropriate cells of the spreadsheet, the value of profit will also change. By repeating this many times, we will obtain a probability distribution of profit; this is the essence of Monte Carlo simulation.

Before we learn how to implement Monte Carlo simulation on spreadsheets, we need to introduce some key concepts and methods for generating random values from probability distributions in Excel.

CHECK YOUR UNDERSTANDING

1. What is Monte Carlo simulation?

2. How can uncertain inputs be identified and specified in Monte Carlo simulation models?

Random Sampling from Probability Distributions

The basis for generating random samples from probability distributions—which underlies Monte Carlo simulation—is the concept of a random number. In the context of simulation, **random numbers** are numbers that are uniformly distributed between 0 and 1. Technically speaking, computers cannot generate truly random numbers since they must use a predictable algorithm. However, the algorithms are designed to generate a sequence of numbers that appear to be random. In Excel, we may generate a random number that is greater than or equal to 0 and less than 1 within any cell using the function RAND( ). This function has no arguments; therefore, nothing should be placed within the parentheses (but the parentheses are required). Figure 12.2 shows a table of ten random numbers generated in Excel. You should be aware that unless the automatic recalculation feature is suppressed, whenever any cell in the spreadsheet is modified, the values in any cell containing the RAND( ) function will change. Automatic recalculation can be changed to manual by choosing *Calculation Options* in the *Calculation* group under the *Formulas* tab. Under manual recalculation mode, the worksheet is recalculated only when the F9 key (in Excel for Windows) is pressed, or the *Calculate Now* button in the *Formulas* tab is clicked.

A value randomly generated from a specified probability distribution is called a **random variate**. Most techniques for generating random variates involve transforming random numbers into outcomes from a probability distribution. Excel allows you to generate random variates from discrete distributions and certain others using the *Random Number Generation* option in the *Analysis Toolpak*. ("*Random Number Generation*" is a bit of a misnomer, as the tool generates random variates.) From the *Data* tab in the ribbon, select *Data Analysis* in the *Analysis* group and then *Random Number Generation*. The *Random Number Generation* dialog, shown in Figure 12.3, will appear. From the *Random Number Generation* dialog, you may select from seven distributions: uniform, normal, Bernoulli, binomial, Poisson, patterned, and discrete. (The patterned distribution is characterized by a lower and upper bound, a step, a repetition rate for values, and a repetition rate for the sequence.) If you select the *Output Range* option, you are asked to specify the upper-left cell reference of the output table that will store the outcomes, the number of variables (columns of values you want generated), number of random numbers (the number of data points you want generated for each variable), and the type of distribution. The default distribution is the discrete distribution.

The dialog in Figure 12.3 also allows you the option of specifying a random number seed. A **random number seed** is a value from which a stream of random numbers is generated. By specifying the same seed, you can produce the same random numbers at a later time. This is desirable when we wish to reproduce an identical sequence of "random"

► **Figure 12.2**

A Sample of Random Numbers

	A	B
1	Random Numbers	
2		
3	Sample	Random Number
4	1	0.326510048
5	2	0.743390121
6	3	0.801687688
7	4	0.804777187
8	5	0.848401291
9	6	0.614517898
10	7	0.452136913
11	8	0.600374163
12	9	0.533963502
13	10	0.638112424

EXAMPLE 12.2   Using Excel's *Random Number Generation* Tool

We will generate 100 outcomes from a Poisson distribution with a mean of 12. In the *Random Number Generation* dialog, set the *Number of Variables* to 1 and the *Number of Random Numbers* to 100 and select Poisson from the drop-down *Distribution* box. The dialog will change and prompt you for the value of *Lambda*, the mean of the Poisson distribution; enter 12 in the box and click *OK*. The tool will display the random numbers in a column. Figure 12.4 shows a histogram of the results.

▶ **Figure 12.3**

Excel Random Number Generation *Dialog*

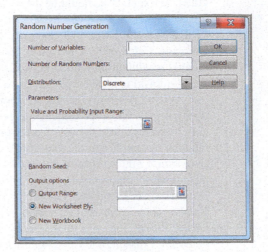

▶ **Figure 12.4**

Histogram of Samples from a Poisson Distribution

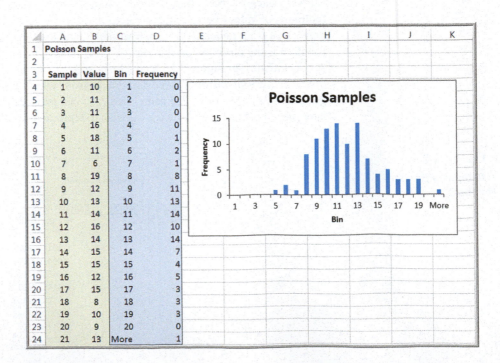

Sample	Value	Bin	Frequency
1	10	1	0
2	11	2	0
3	11	3	0
4	16	4	0
5	18	5	1
6	11	6	2
7	6	7	1
8	19	8	8
9	12	9	11
10	13	10	13
11	14	11	14
12	16	12	10
13	14	13	14
14	15	14	7
15	15	15	4
16	12	16	5
17	15	17	3
18	8	18	3
19	10	19	3
20	9	20	0
21	13	More	1

events in a simulation to test the effects of different policies or decision variables under the same circumstances. However, one disadvantage with using the *Random Number Generation* tool is that you must repeat the process to generate a new set of sample values; pressing the recalculation (F9) key will not change the values. This can make it difficult to use this tool for simulation models.

CHECK YOUR UNDERSTANDING

1. What is a random number, and how can it be generated in Excel?

2. What are the limitations of using the Excel *Random Number Generation* tool for simulation?

Generating Random Variates Using Excel Functions

One disadvantage with using the *Random Number Generation* tool is that it only generates one stream of random values. This can make it difficult to use this tool to perform Monte Carlo simulation in spreadsheet models. We need a method to replace uncertain input data cells in a model with random variates. Fortunately, this can be accomplished using various Excel functions. In this section, we describe how to generate random variates for many common types of probability distributions using Excel functions.

Discrete Probability Distributions

Generating a random variate from discrete probability distributions is quite easy. We will illustrate this process using the probability distribution for rolling two dice.

EXAMPLE 12.3   Sampling from the Distribution of Dice Outcomes

The probability mass function and cumulative distribution in decimal form are as follows:

x	f(x)	F(x)
2	0.0278	0.0278
3	0.0556	0.0833
4	0.0833	0.1667
5	0.1111	0.2778
6	0.1389	0.4167
7	0.1667	0.5833
8	0.1389	0.7222
9	0.1111	0.8333
10	0.0833	0.9167
11	0.0556	0.9722
12	0.0278	1.0000

Notice that the values of F(x) divide the interval from 0 to 1 into smaller intervals that correspond to the probabilities of the outcomes. For example, the interval from 0 and up

to but not including 0.0278 has a probability of 0.0278 and corresponds to the outcome x = 2; the interval from 0.0278 and up to but not including 0.0833 has a probability of 0.0556 and corresponds to the outcome x = 3; and so on. This is summarized as follows:

Interval	Outcome
≥ 0 and < 0.0278	2
≥ 0.0278 and < 0.0833	3
≥ 0.0833 and < 0.1667	4
≥ 0.1667 and < 0.2778	5
≥ 0.2778 and < 0.4167	6
≥ 0.4167 and < 0.5833	7
≥ 0.5833 and < 0.7222	8
≥ 0.7222 and < 0.8323	9
≥ 0.8323 and < 0.9167	10
≥ 0.9167 and < 0.9722	11
≥ 0.9722 and < 1.0000	12

(continued)

Any random number, then, must fall within one of these intervals. Thus, to generate an outcome from this distribution, all we need to do is to select a random number and determine the interval into which it falls. Suppose we use the data in Figure 12.2. The first random number is 0.326510048. This falls in the interval corresponding to the sample outcome of 6. The second random number is 0.743390121. This number falls in the interval corresponding to an outcome of 9. Essentially, we have developed a technique to roll dice on a computer. If this is done repeatedly, the frequency of occurrence of each outcome should be proportional to the size of the random number range (that is, the probability associated with the outcome) because random numbers are uniformly distributed.

We can easily use this approach to generate outcomes from any discrete distribution; the VLOOKUP function in Excel can be used to implement this on a spreadsheet.

EXAMPLE 12.4   **Using the VLOOKUP Function for Random Variate Generation**

Suppose that we want to sample from the probability distribution of the predicted change in the Dow Jones Industrial Average index shown in Chapter 5 in Figure 5.7. We first construct the cumulative distribution $F(x)$. Then assign intervals to the outcomes based on the values of the cumulative distribution, as shown in Figure 12.5. This specifies the table range for the VLOOKUP function, namely, $E$2:$G$10. List the random numbers in a column using the RAND( ) function. The formula in cell J2 is =VLOOKUP(I2, $E$2:$G$10, 3), which is copied down that column. This function takes the value of the random number in cell I2, finds the last number in the first column of the table range that is less than the random number, and returns the value in the third column of the table range. In this case, 0.49 is the last number in column E that is less than 0.530612386, so the function returns 5% as the outcome. An alternative way to generate outcomes in this fashion is simply to embed RAND( ) within the VLOOKUP function, for instance, =VLOOKUP(RAND( ), $E$2:$G$10, 3). This is useful when we need to generate uncertain inputs within spreadsheet models. We will see this used in examples later in this chapter.

Uniform Distributions

It is quite easy to transform a random number into a random variate from a uniform distribution between a and b. Consider the formula

$$U = a + (b - a) \times \text{RAND}( ) \tag{12.1}$$

Note that when RAND( ) = 0, $U = a$, and when RAND( ) approaches 1, U approaches b. For any other value of RAND( ) between 0 and 1, $(b - a) \times$ RAND( ) represents the same proportion of the interval (a, b) as RAND( ) does of the interval $(0, 1)$. For instance, if RAND( ) = 0.5, then $U = a + (b - a)/2 = (a + b)/2$, which is the midpoint of the

▶ **Figure 12.5**

Using the VLOOKUP Function to Sample from a Discrete Distribution

	A	B	C	D	E	F	G	H	I	J
1	**Change in DJIA**	**f(x)**	**F(x)**		**Interval**		**Change in DJIA**		**Random Number**	**Outcome**
2	-20%	0.01	0.01		0	0.01	-20%		0.530612386	5%
3	-15%	0.05	0.06		0.01	0.06	-15%		0.232776591	-5%
4	-10%	0.08	0.14		0.06	0.14	-10%		0.780924503	10%
5	-5%	0.15	0.29		0.14	0.29	-5%		0.363267546	0%
6	0%	0.2	0.49		0.29	0.49	0%		0.489479718	0%
7	5%	0.25	0.74		0.49	0.74	5%		0.062832805	-10%
8	10%	0.18	0.92		0.74	0.92	10%		0.53878251	5%
9	15%	0.06	0.98		0.92	0.98	15%		0.52525315	5%
10	20%	0.02	1		0.98	1	20%		0.99381738	20%
11									0.840872917	10%

interval (a, b). Thus, all real numbers between a and b can occur. Since RAND( ) is uniformly distributed, so also is U.

EXAMPLE 12.5　Modeling Uncertainty with a Uniform Random Variate

The uniform distribution is often used when little is known about the distribution of an uncertain variable. For example, in trying to assess the financial implications of outsourcing manufacturing, uncertainty might exist in the supplier's price prior to designing the manufacturing process. Suppose that the supplier states that the price might be as low as $42 or as high as $50 per unit. Without any other information, a uniform distribution is a reasonable model. Thus, the price can be generated as Price = $42 + ($50 − $42) × RAND( ), or $42 + $8 × RAND( ). If RAND generates the random number 0.6200, then the price would be $42 + $8 × 0.6200 = $46.96.

If you want to generate whole numbers from a uniform distribution between a and b (called a **discrete uniform distribution**), use the Excel function RANDBETWEEN(a, b).

Exponential Distributions

The exponential distribution was introduced in Chapter 5. Exponential random variates can be generated easily using the Excel formula $=-(1/\lambda)*$LN(RAND( )), where $1/\lambda = \mu$ is the mean of the exponential distribution, and LN is the Excel function for the natural logarithm.

Normal Distributions

Normal random variates can be generated in Excel using inverse functions. Inverse functions find the value for a distribution that has a specified cumulative probability. For normal distributions, Excel has two inverse functions:

- NORM.INV(*probability, mean, standard_deviation*)
- NORM.S.INV(*probability*)

For example, the z-value for a standard normal distribution with a cumulative probability of 0.95 is =NORM.S.INV(0.95), which returns 1.645. You may recall using this function to find z-values for confidence intervals and hypothesis tests in Chapters 6 and 7. To use these functions to generate random variates, simply enter *RAND( )* in place of *probability* in the function. Thus,

- NORM.INV(*RAND( ), mean, standard_deviation*) generates a random variate from a normal distribution with a specified mean and standard deviation

- NORM.S.INV(*RAND( )*) generates a random variate from a standard normal distribution (mean = 0 and variance = 1).

For example, NORM.INV(*RAND( )*, 5, 2) will generate a random variate from a normal distribution with mean 5 and standard deviation 2. Each time the worksheet is recalculated, a new random number and, hence, a new random variate are generated. These functions may be embedded in cell formulas and will generate new values whenever the worksheet is recalculated. The following example shows how sampling from probability distributions can provide insights about business decisions that would be difficult to analyze mathematically.

EXAMPLE 12.6   A Monte Carlo Experiment for Evaluating Capital Budgeting Projects

In finance, one way of evaluating capital budgeting projects is to compute a profitability index (PI), which is defined as the ratio of the present value of future cash flows (PV) to the initial investment (I):

$$PI = PV/I \qquad (12.2)$$

Because the cash flow and initial investment that may be required for a particular project are often uncertain, the profitability index is also uncertain. If we can characterize PV and I by some probability distributions, then we would like to know the probability distribution for PI. For example, suppose that PV is estimated to be normally distributed with a mean of $12 million and a standard deviation of $2.5 million, and the initial investment is also estimated to be normal with a mean of $3.0 million and standard deviation of $0.8 million. Intuitively, we might believe that the profitability index is also normally distributed with a mean of $12 million/$3 million = $4 million; however, as we shall

see, this is not the case. We can use a Monte Carlo experiment to identify the probability distribution of PI for these assumptions.

Figure 12.6 shows a simple model from the Excel file *Profitability Index Experiment*. For each experiment, the values of PV and I are sampled from their assumed normal distributions using the NORM.INV function in columns F and G, and PI is calculated in column H. The average value of PI for 1,000 experiments is shown in cell D7. We clearly see that this is not equal to 4 as previously suspected. The histogram in Figure 12.7 also demonstrates that the distribution of PI is not normal but is skewed to the right. This experiment confirms that the ratio of two normal distributions is not normally distributed. We encourage you to create this spreadsheet and replicate this experiment (note that your results will not be exactly the same as these because you are generating random values!).

▶ **Figure 12.6**

Sampling Experiment for Profitability Index

	A	B	C	D	E	F	G	H
1	Profitability Index Analysis				Experiment	PV	I	PI
2					1	11.79045	2.116217	5.571475
3		Mean	Standard Deviation		2	10.62588	2.839064	3.742741
4	PV	12	2.5		3	12.22324	1.049416	11.64765
5	I	3	0.8		4	11.25269	3.947846	2.850337
6					5	11.3254	3.995613	2.83446
7	Mean PI for 1000 Experiments			4.365203	6	15.02659	3.324238	4.52031
8					7	12.79318	3.255405	3.929827
9					8	13.19409	3.000283	4.397616
10					9	12.7466	3.532532	3.608346
11					10	12.5399	3.675463	3.411789

▶ **Figure 12.7**

Frequency Distribution and Histogram of Profitability Index

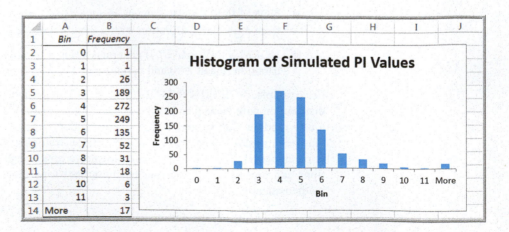

Binomial Distributions

To generate a random variate from a binomial distribution, we may use the Excel function BINOM.INV(*trials, probability, alpha*). In terms of the notation we used in Chapter 5, *trials* represents the number of experiments, n, and *probability* is probability of a success in each trial, p. This function finds the smallest value for which the cumulative binomial distribution is greater than or equal to *alpha*. To randomly generate a binomial random variate, we simply replace *alpha* with RAND( ); that is, BINOM.INV(n, p, RAND()).

EXAMPLE 12.7   **Binomial Random Variates**

Sixty potential customers are called each hour by a telemarketer. The probability that any of them will make a purchase is 0.08. Over a ten-hour period, how many customers might make a purchase? Figure 12.8 shows a spreadsheet for generating ten random samples from a binomial distribution with $n = 60$ and $p = 0.08$ (Excel file

Binomial Random Variates). In column E, we use the formula =BINOM.INV($B$3, $B$4, RAND()). Note that the mean of the binomial (see Chapter 5) is $np = (60)(0.08) = 4.8$. Thus, over a ten-hour period, we would expect 48 successes. Although the ten samples had an average of 6.1, a larger number of samples would average closer to 4.8.

Triangular Distributions

We briefly discussed the triangular distribution in Chapter 5. Recall that this distribution depends on three parameters, $a = $ minimum, $b = $ maximum, and $c = $ most likely. Triangular distributions are often used when one can reasonably estimate these three parameters with little other information. A triangular random variate, X, can be generated using one of the following formulas that depend on the value of R (R is a random number):

$$X = a + \sqrt{R(b - a)(c - a)} \quad \text{for } 0 < R < F(c)$$
$$X = b - \sqrt{(1 - R)(b - a)(b - c)} \quad \text{for } F(c) \leq R < 1 \tag{12.3}$$

where $F(c) = (c - a)/(b - a)$.

To generate a triangular random variate, X, first generate a random number R and compute $F(c)$. If $R < F(c)$, then use the first formula to generate X; otherwise, use the second formula. Here is the Excel logic expressed as an IF function:

$$=IF(R < F(c), a + SQRT(R*(b-a)*(c-a)), b - SQRT((1-R)*(b-a)*(b-c))) \tag{12.4}$$

(You might consider using range names instead of cell references for R, a, b, and c to make this easier to implement in Excel.)

▶ **Figure 12.8**

Random Variates from a Binomial Distribution

	A	B	C	D	E
1	Binomial Random Variates			Sample	Number of purchases/hour
2				1	3
3	Trials (n)	60		2	7
4	Probability (p)	0.08		3	4
5				4	4
6				5	9
7				6	8
8				7	6
9				8	6
10				9	7
11				10	7
12				Total	61

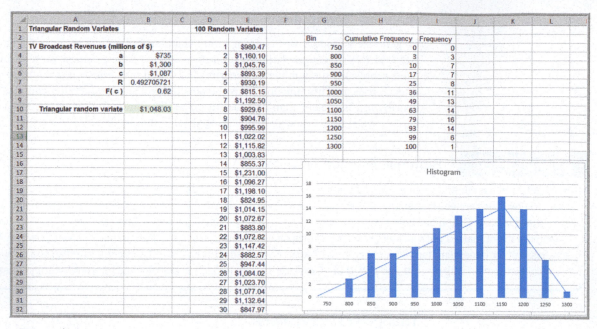

▲ **Figure 12.9**

Triangular Random Variates

Caution: You cannot replace the R's in formula (12.4) with $RAND(\ )$ since they must all be the same random number. Therefore, in an application, R must be referenced from a cell outside of this formula.

| EXAMPLE 12.8 | **Using the Triangular Distribution for a U.S. Olympic Bid Risk Assessment** |

In 1990, Cincinnati was one of six cities in the United States that submitted a proposal to host the 2012 Olympics. The author was involved in assisting the committee to conduct a risk analysis of their budget to predict whether the budget would be met (a condition required by the U.S. Olympic Committee). Because of the uncertainty in estimating many of the financial parameters, such as TV broadcast revenues, ticket sales, and so on, triangular distributions were used to model these uncertainties. Figure 12.9 shows a spreadsheet (Excel file *Triangular Random Variates*) with the parameters used for estimating broadcast revenues. Formula (12.4) was used to generate the random variate in cell B10. The spreadsheet also shows a table with 100 samples from the distribution. The histogram shows that the distribution is roughly triangular; however, with only 100 samples, there will be considerable variation in the outcomes as the spreadsheet is recalculated.

CHECK YOUR UNDERSTANDING

1. Explain how to use the VLOOKUP function for random variate generation of discrete distributions.

2. Explain how Excel is used to generate random variates from common probability distributions.

Monte Carlo Simulation in Excel

We may apply the methods in the previous section for generating random variates to build and implement Monte Carlo simulation models in Excel spreadsheets and conduct risk analyses. In this section, we illustrate several examples of using simulation for descriptive, predictive, and prescriptive analytics applications.

The general process for performing Monte Carlo simulation in Excel is as follows:

1. Develop the spreadsheet model.
2. Determine the probability distributions that describe the uncertain inputs in the model and use the appropriate Excel functions to generate random variates for these uncertain inputs.
3. Identify the model output that you wish to evaluate.
4. Determine the number of trials (replications) for the simulation.
5. Create a data table to summarize the values of the model output for the replications.
6. Compute summary statistics, percentiles, confidence intervals, frequency distributions, and histograms to interpret the results.

Profit Model Simulation

We will illustrate this process for the profit model that we discussed earlier in this chapter.

EXAMPLE 12.9   **Setting Up the Monte Carlo Simulation Model for the *Profit Model***

In Example 12.1, we discussed the probability distributions for the uncertain inputs for the *Profit Model* spreadsheet. Specifically, we have the following:

- Demand—Discrete distribution with the following data:

Demand	Probability
40,000	0.1
45,000	0.3
50,000	0.4
55,000	0.15
60,000	0.05

- Unit cost—Uniform distribution between $22 and $26
- Fixed cost—Triangular distribution with minimum = $350,000, maximum = $500,000, and most likely = $400,000

Figure 12.10 shows the spreadsheet for the simulation (Excel file *Profit Model Monte Carlo Simulation*). Using the approach illustrated in Example 12.4, we constructed a lookup table to generate the demand based on the probability distribution. Then we use the function =VLOOKUP(RAND( ), H3:J7, 3) in cell B8 to generate the random variates for the demand. Next, we use formula (12.1) to generate random variates for the unit cost. In cell B6, we use the formula =F11+(G11−F11)*RAND( ). Finally, we use formula (12.4) to generate triangular random variates for the fixed cost. In cell B7, we enter the formula =IF(G17<G18, G14+SQRT(G17*(G15−G14)*(G16−G14)), G15−SQRT((1−G17)*(G15−G14)*(G15−G16))). This completes step 2 of the Monte Carlo simulation process. If you recalculate the spreadsheet manually, you will see that each of these values changes randomly and leads to a different value for the profit, which is the model output that we wish to evaluate.

To perform the simulation (steps 4 and 5), we need to generate a sufficient number of trials, or replications, to obtain enough data to create a reasonable distribution of the model output values. From statistics, you already know that the larger the sample size, the more accurate the results. Because we are doing the simulation in Excel, we are somewhat limited in the number of replications that we can easily compute. (Professional software has more powerful capabilities.) However, when dealing with relatively small models, it is fairly easy to generate several hundred replications using data tables. The following example illustrates how to do this.

▶ Figure 12.10

*Setting Up
the Simulation
Spreadsheet*

	A	B	C	D	E	F	G	H	I	J
1	Profit Model					Demand Distribution			VLOOKUP Table	
2					Demand	Probability	Cumulative	Interval	Demand	
3	Data				40,000	0.1	0.1	0	0.1	40,000
4					45,000	0.3	0.4	0.1	0.4	45,000
5	Unit Price	$40.00			50,000	0.4	0.8	0.4	0.8	50,000
6	Unit Cost	$22.16			55,000	0.15	0.95	0.8	0.95	55,000
7	Fixed Cost	$402,563.89			60,000	0.05	1	0.95	1	60,000
8	Demand	50000								
9						Unit cost distribution				
10					Uniform	a	b			
11	Model					$22.00	$26.00			
12										
13	Unit Price	$40.00				Fixed cost distribution				
14	Quantity Sold	40000			Triangular	a	$350,000			
15	Revenue		$1,600,000.00			b	$500,000			
16						c	$400,000			
17	Unit Cost	$22.16				R	0.367080337			
18	Quantity Produced	40000				F(c)	0.33			
19	Variable Cost		$886,520.16							
20	Fixed Cost		$402,563.89							
21										
22	Profit		$310,915.95							

EXAMPLE 12.10   Using Data Tables and Analyzing Results for Monte Carlo Simulation in Excel

We may use either one-way or two-way data tables for simulation. For a relatively small number of trials, say 100 or less, one-way tables suffice, but for a larger number of trials it is better to use a two-way table to consolidate the data in a better fashion. In this example, we will use a one-way table; in later examples, we will use two-way tables.

First, construct a one-way data table (see Chapter 11) by listing the number of trials down a column and referencing the cell associated with the profit (cell C22) in the cell above and to the right of the list. In Figure 12.11, we used 100 trials, extending from cell L3 to cell L102, and cell M2 references cell C22. Select the range of the table (L2:M102)—and here's the trick—in the *Column Input Cell* field in the *Data Table* dialog, enter *any blank cell* in the spreadsheet. Make sure that this is one you will not use. When you click *OK* in the *Data Table* dialog, the values in column M will display the simulation results for the profit.

Why does this work? Recall that when you create a data table, the *Column Input Cell* generally refers to some parameter in the model. The data table simply takes these values, replaces them in the model, and then displays the output. Because we used a blank cell for the *Column Input Cell*, the trial numbers do not affect the model. However, for each trial, the spreadsheet is recalculated. Since we used the RAND function to generate random variates, each recalculation uses different values for the uncertain inputs. We may repeat the simulation by recalculating the spreadsheet. Because you may want to preserve the results for subsequent analysis, we suggest setting the *Calculations Options* in the *Formulas* tab to *Automatic Except for Data Tables*.

The last step in the simulation process is to analyze the results using various statistical tools such as summary statistics, percentiles, confidence intervals, and frequency distributions and histograms. We don't recommend using *Data Analysis* tools because if you wish to recalculate the spreadsheet and run a new simulation, the results will not update. Instead, use Excel functions.

Figure 12.12 shows statistical analyses of the simulation results. A point estimate for the mean profit is $228,038.77. Because the simulation was based on only 100 trials, sampling error may be significant. We can assess this by developing a confidence interval. Using formula (6.3), a 95% confidence interval for the mean is

$$\$228,038.77 \pm 1.96(\$53,828.55/10),$$
$$\text{or } [\$217,488.37, 238,589.17]$$

A larger number of trials would decrease the width of this interval.

The standard deviation of profit is quite large, and you can also see that the simulated values range from about $117,000 to $337,000, indicating that considerable variability exists in the predicted profit. The frequency distribution and histogram likewise provide a perspective of this variability. To find the frequency distribution, we used COUNTIF functions to find the cumulative frequencies, and then found the number of observations in each cell by subtraction. The percentiles allow you to make probability statements about the profit and analyze risk. For instance, we see that the 90th percentile is $290,592. This means that there is only a 10% chance that the profit will exceed this value. Percentiles allow you to analyze risk. For instance, if the company needs to make a profit of at least $280,000, we see that the probability of that happening is only about 0.15.

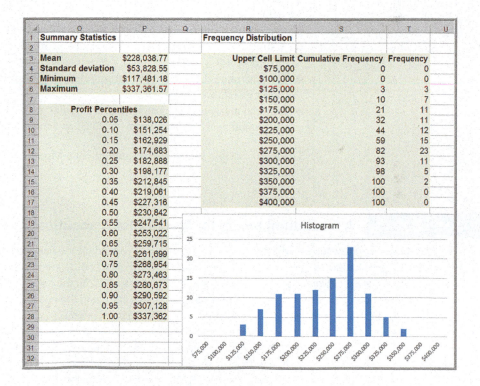

▲ Figure 12.11

Portion of Simulation Data Table

► Figure 12.12

Statistical Analysis of the Profit Model Simulation Results

An easier alternative to compute a frequency distribution is to use the Excel function FREQUENCY(*data array, bin array*). This is an "array" function and returns values in a range of cells. When you enter an array formula, you must first select the range in which to place the results. Then, after entering the formula, you must press Ctrl+Shift+Enter in Windows or Command+Shift+Enter on a Mac simultaneously. For example, in Figure 12.12, instead of computing the cumulative frequencies and frequencies in columns S and T, we could have selected the range S4:S17, then entered =FREQUENCY(M3:M102, R4:R17), and pressed Ctrl (or Command) +Shift+Enter. The range would be filled with the frequencies. In this fashion, we avoid having to compute cumulative frequencies and then convert them into frequencies.

New-Product Development

The *Moore Pharmaceuticals* spreadsheet model to support a new-product development decision was introduced in Chapter 11 (see Figure 11.14). Although the values used in the spreadsheet suggest that the new drug would become profitable by the fourth year, much of the data in this model are uncertain. Thus, we might be interested in evaluating the risk associated with the project. Three questions we might be interested in are as follows:

1. What is the risk that the net present value over the five years will not be positive?
2. What are the chances that the product will show a positive cumulative net profit in the third year?
3. In the fifth year, what minimum amount of cumulative profit are we likely to achieve with a probability of at least 0.90?

Suppose that the project manager of Moore Pharmaceuticals has identified the following uncertain variables in the model and the distributions and parameters that describe them, as follows:

- Market size: normal with mean of 2,000,000 units and standard deviation of 400,000 units
- R&D costs: uniform between $600,000,000 and $800,000,000
- Clinical trial costs: normal with mean of $150,000,000 and standard deviation $30,000,000
- Annual market growth factor: triangular with minimum = 2%, maximum = 6%, and most likely = 3%
- Annual market share growth rate: triangular with minimum = 15%, maximum = 25%, and most likely = 20%

The next example shows how to implement a Monte Carlo simulation for this scenario.

EXAMPLE 12.11   **A Simulation Model for Moore Pharmaceuticals**

Figure 12.13 shows how to set up the *Moore Pharmaceuticals* model for Monte Carlo simulation (Excel file *Moore Pharmaceuticals Excel Simulation Model*). In the profit model example (Example 12.9), we saw how to incorporate uniform and triangular random variates into a simulation model. In this example, we also have uncertain inputs that are normally distributed. We may use the NORM.INVfunction to generate these random variates. Thus, the formula in cell B5 is =NORM.INV(RAND( ), I4, J4), and the formula in cell B12 is =NORM.INV(RAND( ), I12, J12).

In this problem, note that we must use different values for the annual market growth factor and annual market share growth rate in each year of the model, as these values are independent of each other. This means that we have to use a different random number for each of the triangular distributions. Thus, we have created a separate distribution for each year in columns J through M.

Based on the risk analysis questions that we wish to answer, we define the cumulative net profit in the third year (cell D28), the cumulative net profit in the fifth year (cell F28), and the net present value (cell B30) as the model outputs of interest.

Because we have three different model outputs, we need to create three data tables, one for each output to conduct the simulation. In general, you should use a larger number of trials for models that have a larger number of uncertain inputs. In this example, we will use 500 trials. To do this more efficiently, we can use two-way data tables. Figure 12.14 shows a portion of one of these data tables, which uses five columns of 100 trials (a total of 500 trials). See the Excel file for the complete results. In the *Data Table* dialog, enter any blank cell for both the *Row Input Cell* and *Column Input Cell* (being sure not to use the same cells for different data tables).

	A	B	C	D	E	F
1	Moore Pharmaceuticals					
2						
3	Data					
4						
5	Market size	2,166,428				
6	Unit (monthly Rx) revenue	$ 130.00				
7	Unit (monthly Rx) cost	$ 40.00				
8	Discount rate	9%				
9						
10	Project Costs					
11	R&D	$785,947,068.66				
12	Clinical Trials	$142,929,242.22				
13	Total Project Costs	$928,876,310.88				
14						
15	Model					
16						
17	Year	1	2	3	4	5
18	Market growth factor		3.74%	3.04%	4.68%	2.74%
19	Market size	2,166,428	2,247,439	2,315,810	2,424,088	2,490,583
20	Market share growth rate		20.82%	21.69%	17.97%	17.62%
21	Market share	8.00%	9.67%	11.76%	13.88%	16.32%
22	Sales	173,314	217,220	272,386	336,369	406,502
23						
24	Annual Revenue	$270,370,159.07	$338,863,240.62	$424,922,006.63	$524,735,869.63	$634,143,179.68
25	Annual Costs	$83,190,818.17	$104,265,612.50	$130,745,232.81	$161,457,190.66	$195,120,978.36
26	Profit	$187,179,340.89	$234,597,628.12	$294,176,773.82	$363,278,678.98	$439,022,201.32
27						
28	Cumulative Net Profit	$ (741,696,969.99)	$ (507,099,341.87)	$ (212,922,568.05)	$ 150,356,110.93	$ 589,378,312.24
29						
30	Net Present Value	$210,152,512.84				

	H	I	J	K	L	M
2		Market size distribution				
3	Normal	Mean	Standard Deviation			
4		2000000	400000			
6		R&D cost distribution				
7	Uniform	a	b			
8		$600,000,000	$800,000,000			
10		Clinical trial cost distribution				
11	Normal	Mean	Standard Deviation			
12		$150,000,000	$30,000,000			
14			Annual market growth factor distribution			
15			Year 2	Year 3	Year 4	Year 5
16	Triangular	a	2%	2%	2%	2%
17		b	6%	6%	6%	6%
18		c	3%	3%	3%	3%
19		R	0.574149045	0.270921919	0.85383452	0.138044397
20		F(c)	0.25	0.25	0.25	0.25
22			Annual market share growth rate distribution			
23			Year 2	Year 3	Year 4	Year 5
24	Triangular	a	15%	15%	15%	15%
25		b	25%	25%	25%	25%
26		c	20%	20%	20%	20%
27		R	0.649762986	0.781434287	0.176883419	0.137651264
28		F(c)	0.50	0.50	0.50	0.50

▲ **Figure 12.13**

Moore Pharmaceuticals Simulation Model

► **Figure 12.14**

Portion of Data Table Simulation for Moore Pharmaceuticals

	O	P	Q	R	S	T
1		Third Year Cumulative Net Profit				
2						
3	$ (212,922,568.05)	1	2	3	4	5
4	1	$ (245,707,637)	$ (122,911,680)	$ (154,068,080)	$ (403,108,697)	$ (18,839,133)
5	2	$ (257,534,992)	$ (237,813,185)	$ (274,181,051)	$ (94,362,507)	$ (203,086,811)
6	3	$ 23,221,754	$ 139,708,957	$ (492,230,168)	$ (297,159,046)	$ 14,740,621
7	4	$ (204,320,475)	$ (130,193,652)	$ (68,172,798)	$ (513,339,863)	$ (385,647,094)
8	5	$ (216,663,990)	$ (506,101,512)	$ (83,836,995)	$ (342,952,843)	$ (328,388,878)
9	6	$ (335,171,592)	$ (169,566,469)	$ (49,161,857)	$ (36,026,190)	$ (112,620,438)
10	7	$ (514,068,190)	$ (188,460,942)	$ (127,187,172)	$ (329,348,855)	$ (268,227,428)
11	8	$ (339,300,974)	$ 9,840,554	$ (43,071,809)	$ 16,191,428	$ (264,657,524)
12	9	$ (136,935,834)	$ (118,483,642)	$ 29,634,056	$ (250,551,064)	$ (191,919,293)
13	10	$ (29,544,233)	$ (73,754,260)	$ (77,250,633)	$ (120,511,277)	$ (424,535,487)

As with the profit model example, we can compute summary statistics, percentiles, and frequency distributions to analyze the results. Then we may address the risk analysis questions posed earlier using percentile information (see Chapter 4) as the following example illustrates.

EXAMPLE 12.12 Risk Analysis for Moore Pharmaceuticals

1. *What is the probability that the net present value over the five years will not be positive?* We can answer this question using percentiles. Be careful when using Excel tools. For example, the *Rank & Percentile* tool requires the data to be in one column; therefore, you cannot use this tool with a two-way data table. However, we can easily construct a simple table of percentiles using the PERCENTILE.INC(*array*, *k*) function, as shown in Figure 12.15. The *array* is the data table range, and *k* is the percentile value between 0 and 1, which is referenced in column AJ. We see that the probability that the net present value will not be positive is somewhere between 0.15 and 0.20. More refinement can find this more accurately.
2. *What are the chances that the product will show a positive cumulative net profit in the third year?* Again, using percentiles for the third-year value, you should try to verify that the probability of a positive cumulative net profit in the third year is less than 10%.
3. *In the fifth year, what minimum amount of cumulative profit are we likely to achieve with a probability of at least 0.90?* Here, we are looking for the 10th percentile, which, depending on the variation in the simulation results, is about $160 million.

▶ **Figure 12.15**

Percentiles Table for Net Present Value

	AJ	AK
1		
2		
3		
4	**Net Present Value Percentiles**	
5	0.05	$ (159,916,435.10)
6	0.10	$ (107,871,262.73)
7	0.15	$ (31,613,348.27)
8	0.20	$ 13,055,115.98
9	0.25	$ 38,024,192.84
10	0.30	$ 72,790,221.07
11	0.35	$ 109,272,078.11
12	0.40	$ 138,778,622.24
13	0.45	$ 166,968,192.24
14	0.50	$ 197,119,519.55
15	0.55	$ 225,949,641.47
16	0.60	$ 253,676,885.65
17	0.65	$ 278,125,314.43
18	0.70	$ 306,995,325.04
19	0.75	$ 344,355,960.76
20	0.80	$ 387,955,067.81
21	0.85	$ 437,120,942.60
22	0.90	$ 493,273,288.94
23	0.95	$ 577,919,181.93
24	1.00	$ 901,961,366.71

Retirement Planning

In Chapter 11, we developed a model for predicting the amount that an individual might have saved upon retirement. This model, however, was based on some unrealistic assumptions, namely, that the annual salary increase and expected annual investment returns are constant for each year. In reality, these values will typically change each year. More importantly, investment returns may not necessarily have a positive return each year. Thus, using these assumptions can easily overestimate the predicted retirement savings and not provide any information about risk and variability. The following example shows how we can enrich this model to make it more realistic through Monte Carlo simulation.

EXAMPLE 12.13   *Retirement Planning Model* Monte Carlo Simulation

To build more realism into the model, let us assume that the annual salary increase may vary from 2% to 5%, uniformly distributed. Data on the Vanguard Balanced Index Fund show that the average recent return is about 6%, but that it also has a standard deviation of 6.5%. For the Boston Trust Asset Management Fund, the average return is about 6.5% with a standard deviation of 7%.

To build these assumptions into the model, we need to create new columns to generate random values for the annual salary increase and investment returns and make some adjustments in other formulas. Figure 12.16 shows the enhanced model (Excel file *Retirement Planning Excel Simulation Model*). The formulas in the Salary Increase column generate uniform random variates using the Excel formula =2% + 3%*RAND( ). The formulas for the returns of the mutual funds use the NORM.INV(*RAND( )*, *mean*, *standard deviation*) function with the respective means and standard deviations. We assume that these annual

changes apply to the previous year's salary and fund balances.

Now we can use a data table to simulate the final retirement balance. We used a two-way table with 100 rows and 10 columns to simulate 1,000 trials. A portion of the data table is shown in Figure 12.17. Figure 12.18 on page 442 shows the statistical analysis. We see that although the average is around $3.6 million, the standard deviation and range are quite large. The histogram is positively skewed, indicating that very high returns are rather unlikely.

We may also calculate probabilities to address questions of risk. For example, the probability that the retirement balance will be $3 million or less is estimated to be $235/1,000 = 0.235$. The probability that it will be more than $5 million is only $1 - 955/1,000 = 0.045$. The probability that it will be greater than $3 million but less than 4.5 million is $(873 - 235)/1,000 = 0.638$. This should provide reasonable assurance of a comfortable retirement.

	A	B	C	D	E	F	G	H	I	J	K	L	M
1	Retirement Planning Simulation			Age	Salary Increase	Salary	Annual 401K Contribution	Employer Match	Vanguard Return	401K Balance	Boston Trust Return	Roth IRA Balance	Final Balance
2	Yearly 401K Contribution 15.0%			24		$80,000	$12,000	$4,000		$16,000		$5,500	$21,500
3	Employer Match of Salary 5.0%			25	2.47%	$81,979	$12,297	$4,099	-2.02%	$32,072	-1.51%	$10,917	$42,989
4	Salary increase 4.0%			26	4.81%	$85,919	$12,888	$4,296	9.86%	$52,419	-1.30%	$16,275	$68,694
5				27	4.02%	$89,371	$13,406	$4,469	3.90%	$72,339	-3.33%	$21,233	$93,572
6	Vanguard Balanced Index Fund			28	4.61%	$93,492	$14,024	$4,675	13.20%	$100,585	-8.42%	$24,944	$125,529
7	Expected annual return 6.0%			29	4.14%	$97,360	$14,604	$4,868	9.22%	$129,333	-2.09%	$29,924	$159,256
8				30	4.54%	$101,783	$15,267	$5,089	1.84%	$152,072	-1.20%	$35,064	$187,136
9	Boston Trust Asset Management Fund			31	4.44%	$106,297	$15,945	$5,315	13.03%	$193,142	5.70%	$42,562	$235,704
10	Expected annual return 6.5%			32	3.87%	$110,411	$16,562	$5,521	9.53%	$233,623	2.07%	$48,942	$282,565
11				33	2.93%	$113,642	$17,046	$5,682	5.83%	$269,974	-1.03%	$53,940	$323,914
12				34	2.97%	$117,016	$17,552	$5,851	4.17%	$304,635	5.67%	$62,496	$367,131
13				35	2.49%	$119,925	$17,989	$5,996	17.77%	$382,754	12.68%	$75,919	$458,673
14				36	3.89%	$124,585	$18,000	$6,229	9.00%	$441,427	25.20%	$100,550	$541,976
15				37	2.45%	$127,642	$18,000	$6,382	-3.51%	$450,329	7.17%	$113,259	$563,588
16				38	4.28%	$133,109	$18,000	$6,655	9.89%	$519,522	12.07%	$132,427	$651,948
17				39	3.92%	$138,324	$18,000	$6,916	-1.23%	$538,071	-2.51%	$134,608	$672,678
18				40	3.36%	$142,979	$18,000	$7,149	14.05%	$638,823	15.33%	$160,740	$799,563
19				41	2.63%	$146,742	$18,000	$7,337	10.35%	$730,274	9.24%	$181,086	$911,360
20				42	4.17%	$152,861	$18,000	$7,643	-0.46%	$752,548	-0.05%	$186,503	$939,051
21				43	3.18%	$157,721	$18,000	$7,886	3.08%	$801,619	1.88%	$195,502	$997,121
22				44	4.62%	$165,006	$18,000	$8,250	3.76%	$858,050	4.46%	$209,720	$1,067,770
23				45	2.58%	$169,260	$18,000	$8,463	1.74%	$899,409	13.59%	$243,715	$1,143,124
24				46	3.58%	$175,318	$18,000	$8,766	4.33%	$965,124	-6.14%	$234,257	$1,199,381
25				47	4.96%	$184,019	$18,000	$9,201	-0.16%	$990,759	-3.79%	$230,873	$1,221,633
26				48	2.07%	$187,824	$18,000	$9,391	10.59%	$1,123,040	16.15%	$273,670	$1,396,710
27				49	4.72%	$196,686	$18,000	$9,834	-5.04%	$1,094,281	7.08%	$298,539	$1,392,820
28				50	3.12%	$202,824	$18,000	$10,141	11.04%	$1,243,184	9.81%	$333,312	$1,576,495
29				51	2.84%	$208,582	$18,000	$10,429	17.68%	$1,491,413	5.68%	$357,734	$1,849,147
30				52	3.72%	$216,344	$18,000	$10,817	0.74%	$1,531,291	12.66%	$408,531	$1,939,822
31				53	4.96%	$227,069	$18,000	$11,353	6.88%	$1,665,956	3.19%	$427,067	$2,093,023
32				54	2.59%	$232,939	$18,000	$11,647	4.22%	$1,765,916	10.25%	$476,336	$2,242,252
33				55	2.96%	$239,836	$18,000	$11,992	1.20%	$1,817,071	16.21%	$560,032	$2,377,103
34				56	3.10%	$247,264	$18,000	$12,363	12.78%	$2,079,717	9.17%	$617,894	$2,697,611
35				57	4.60%	$258,645	$18,000	$12,932	8.19%	$2,280,909	-4.32%	$597,730	$2,878,639
36				58	4.25%	$269,629	$18,000	$13,481	-0.15%	$2,308,986	6.51%	$643,168	$2,952,154
37				59	2.97%	$277,634	$18,000	$13,882	6.01%	$2,479,714	15.12%	$746,888	$3,226,602
38				60	3.15%	$286,385	$18,000	$14,319	6.40%	$2,670,638	-7.20%	$699,602	$3,370,240

▲ Figure 12.16

Enhanced Retirement Model

	O	P	Q	R	S	T	U	V	W	X	Y
1											
2	$3,370,240	1	2	3	4	5	6	7	8	9	10
3	1	$2,972,612.30	$4,125,489.61	$4,797,783.10	$2,703,771.68	$3,514,270.00	$4,779,412.92	$2,682,145.74	$3,852,707.81	$3,844,107.08	$4,510,170.69
4	2	$3,057,747.32	$3,219,266.60	$2,723,019.27	$3,284,853.71	$2,259,297.61	$3,323,296.97	$3,800,118.62	$4,289,250.60	$2,849,252.49	$3,359,736.32
5	3	$2,769,131.64	$5,093,121.81	$2,654,622.52	$3,924,632.12	$3,029,786.72	$3,524,913.15	$3,390,938.96	$2,616,131.52	$2,324,308.44	$2,844,006.25
6	4	$2,723,876.79	$2,820,133.16	$4,399,788.88	$4,425,761.53	$3,268,764.18	$4,342,841.60	$5,161,633.23	$2,962,830.19	$2,796,172.19	$4,414,897.51
7	5	$3,436,156.84	$4,701,085.14	$3,066,694.62	$2,846,726.34	$3,398,339.90	$4,726,058.86	$4,393,709.87	$3,173,826.53	$3,393,236.90	$3,860,802.62
8	6	$4,642,910.55	$3,387,098.84	$3,873,594.91	$3,469,364.94	$3,601,006.57	$3,279,704.72	$4,502,853.46	$3,080,111.91	$2,750,442.94	$2,163,172.28
9	7	$3,598,555.08	$5,684,821.05	$3,841,224.95	$3,677,723.29	$3,411,599.69	$3,429,350.95	$4,639,879.37	$2,947,605.37	$2,798,555.49	$3,611,197.81
10	8	$2,703,124.44	$3,312,924.65	$3,924,574.28	$3,264,019.96	$4,034,926.58	$3,638,393.93	$4,872,441.91	$4,345,954.79	$2,807,696.09	$5,811,100.66
11	9	$3,906,544.35	$4,013,895.35	$2,440,240.28	$2,891,771.57	$2,770,552.28	$3,128,537.79	$4,274,633.80	$4,770,789.32	$3,202,082.54	$4,130,254.05
12	10	$3,124,797.54	$4,448,443.27	$3,568,510.29	$3,211,756.14	$3,867,210.02	$2,697,401.23	$2,962,349.96	$3,763,642.63	$4,573,512.24	$4,224,896.47

▲ Figure 12.17

Portion of Retirement Model Simulation Data Table

Single-Period Purchase Decisions

In Chapter 11, we introduced a model for single-period purchase decisions: the newsvendor problem. In this model, we calculated the net profit associated with purchasing a quantity Q in advance of a selling season and realizing a demand D in the future. In reality, demand will not be known and can be modeled by some probability distribution. The decision problem that a seller faces is determining the best purchase quantity to maximize the expected net profit. We may use Monte Carlo simulation to find the distribution of the net profit for any purchase quantity, and then use the model to find the best purchase quantity. In this fashion, we are using Monte Carlo simulation as a prescriptive approach.

The following example shows how to use Monte Carlo simulation by resampling from historical data.

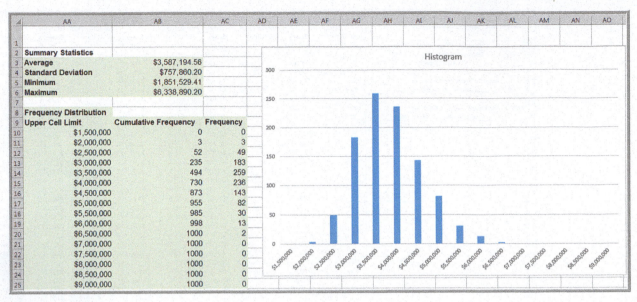

▲ **Figure 12.18**

Statistical Analysis of Retirement Model Simulation Results

EXAMPLE 12.14   **A Prescriptive Simulation Model for a Single-Period Purchase Decision**

Suppose that the candy store in Example 11.8 has 20 years of historical data on Valentine's Day sales of gift boxes, shown in the table below:

Historical Candy Sales	
42	47
45	41
40	41
46	45
43	51
43	43
46	45
42	42
44	44
43	48

We may use these data to construct an empirical probability distribution and use this to simulate the demand. Figure 12.19 (Excel file *Newsvendor Simulation Model*) shows the frequency distribution and histogram of these data, from which we calculated the probability and cumulative probability of each demand value. Using this distribution, we can construct a lookup table to generate random variates for the demand as

shown in columns K through M. We enter the formula =VLOOKUP(RAND( ), K3:M14, 3) in cell B11. Now we may easily simulate the profit using a data table.

However, the objective is to find the purchase quantity that maximizes the expected profit. To do this, we can use a two-way data table where the rows represent the simulation trials and the columns correspond to the purchase quantity. Thus, in the *Data Table* dialog, set the *Row Input Cell* as B12 and the *Column Input Cell* as any blank cell. Figure 12.20 shows a portion of the data table. Each column of the data table provides 200 trials for each purchase quantity. We computed the average and standard deviation of profit above the table. We can observe two things. First, the largest average profit occurs for a purchase quantity of 45 (although this may vary in different simulations). Second, the range and standard deviation generally increase as the purchase quantity increases, suggesting a higher amount of risk. For example, we see that 29 of the 200 profit values for $Q = 45$ fall at or below $240, resulting in a lower profit than if the seller orders 40 boxes. However, the remainder are larger than $240. Thus, a high probability exists of making a higher profit. Similarly, if 51 boxes are ordered, the profit may be as low as $207, but as high as $306—a higher profit potential, but also a higher risk.

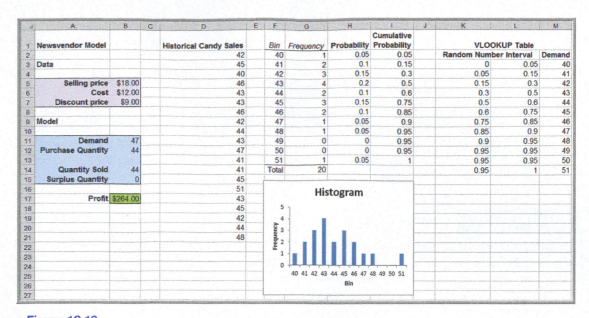

Newsvendor Model			Historical Candy Sales	Bin	Frequency	Probability	Cumulative Probability	VLOOKUP Table		
			42	40	1	0.05	0.05	Random Number Interval		Demand
Data			45	41	2	0.1	0.15	0	0.05	40
			40	42	3	0.15	0.3	0.05	0.15	41
Selling price	$18.00		46	43	4	0.2	0.5	0.15	0.3	42
Cost	$12.00		43	44	2	0.1	0.6	0.3	0.5	43
Discount price	$9.00		43	45	3	0.15	0.75	0.5	0.6	44
			46	46	2	0.1	0.85	0.6	0.75	45
Model			42	47	1	0.05	0.9	0.75	0.85	46
			44	48	1	0.05	0.95	0.85	0.9	47
Demand	47		43	49	0	0	0.95	0.9	0.95	48
Purchase Quantity	44		47	50	0	0	0.95	0.95	0.95	49
			41	51	1	0.05	1	0.95	0.95	50
Quantity Sold	44		41	Total	20			0.95	1	51
Surplus Quantity	0		45							
			51							
Profit	$264.00		43							
			45							
			42							
			44							
			48							

▲ Figure 12.19

Simulation Model Data

	O	P	Q	R	S	T	U	V	W	X	Y	Z	AA				
															Frequency Distribution for Q = 45		
Minimum		$240.00	$237.00	$234.00	$231.00	$228.00	$225.00	$222.00	$219.00	$216.00	$213.00	$210.00	$207.00		Upper Cell Limit	Cumulative Frequency	Frequency
Maximum		$240.00	$246.00	$252.00	$258.00	$264.00	$270.00	$276.00	$282.00	$288.00	$294.00	$300.00	$306.00		200	0	0
Average Profit		$240.00	$245.55	$250.65	$252.96	$255.99	$266.50	$263.32	$251.72	$248.94	$249.99	$245.37	$241.11		210	0	0
Standard Deviation		$0.00	$1.97	$4.11	$8.35	$10.74	$14.18	$16.46	$19.58	$19.96	$20.05	$22.83	$21.59		220	0	0
$264.00		40	41	42	43	44	45	46	47	48	49	50	51		230	7	7
1		$240.00	$246.00	$252.00	$258.00	$264.00	$270.00	$249.00	$228.00	$234.00	$240.00	$210.00	$252.00		240	29	22
2		$240.00	$246.00	$252.00	$258.00	$264.00	$270.00	$276.00	$219.00	$243.00	$285.00	$246.00	$243.00		250	58	29
3		$240.00	$246.00	$252.00	$258.00	$246.00	$270.00	$249.00	$237.00	$261.00	$249.00	$255.00	$234.00		260	91	33
4		$240.00	$246.00	$252.00	$258.00	$264.00	$270.00	$267.00	$219.00	$234.00	$240.00	$219.00	$225.00		270	200	109
5		$240.00	$246.00	$252.00	$258.00	$264.00	$252.00	$267.00	$264.00	$243.00	$249.00	$246.00	$243.00		280	200	0
6		$240.00	$246.00	$252.00	$258.00	$255.00	$261.00	$276.00	$237.00	$225.00	$276.00	$237.00	$234.00		290	200	0
7		$240.00	$246.00	$252.00	$258.00	$246.00	$225.00	$249.00	$246.00	$225.00	$240.00	$228.00	$252.00		300	200	0
8		$240.00	$246.00	$252.00	$258.00	$237.00	$234.00	$231.00	$255.00	$270.00	$276.00	$300.00	$243.00				
9		$240.00	$246.00	$252.00	$258.00	$264.00	$270.00	$240.00	$237.00	$243.00	$258.00	$210.00	$270.00				
10		$240.00	$246.00	$252.00	$258.00	$246.00	$225.00	$276.00	$246.00	$243.00	$249.00	$264.00	$252.00		Histogram for Q = 45		
11		$240.00	$246.00	$252.00	$231.00	$264.00	$270.00	$267.00	$228.00	$234.00	$267.00	$264.00	$225.00				
12		$240.00	$246.00	$252.00	$258.00	$246.00	$261.00	$240.00	$228.00	$243.00	$231.00	$210.00	$252.00				
13		$240.00	$246.00	$252.00	$258.00	$255.00	$252.00	$240.00	$282.00	$234.00	$231.00	$219.00	$234.00				
14		$240.00	$246.00	$252.00	$258.00	$264.00	$270.00	$240.00	$219.00	$279.00	$249.00	$273.00	$270.00				
15		$240.00	$246.00	$243.00	$240.00	$264.00	$252.00	$231.00	$246.00	$243.00	$258.00	$255.00	$207.00				
16		$240.00	$246.00	$252.00	$258.00	$246.00	$270.00	$276.00	$273.00	$261.00	$240.00	$255.00	$234.00				
17		$240.00	$246.00	$252.00	$258.00	$246.00	$243.00	$267.00	$228.00	$243.00	$231.00	$255.00	$261.00				
18		$240.00	$246.00	$252.00	$240.00	$255.00	$270.00	$267.00	$264.00	$243.00	$258.00	$228.00	$306.00				
19		$240.00	$246.00	$252.00	$249.00	$237.00	$234.00	$240.00	$237.00	$216.00	$222.00	$300.00	$243.00				
20		$240.00	$246.00	$252.00	$258.00	$264.00	$270.00	$231.00	$246.00	$261.00	$258.00	$237.00	$234.00				
21		$240.00	$246.00	$252.00	$258.00	$264.00	$270.00	$240.00	$264.00	$234.00	$222.00	$237.00	$234.00				
22		$240.00	$246.00	$252.00	$249.00	$264.00	$270.00	$249.00	$228.00	$261.00	$222.00	$255.00	$243.00				
23		$240.00	$246.00	$252.00	$240.00	$264.00	$234.00	$267.00	$273.00	$243.00	$222.00	$273.00	$261.00				
24		$240.00	$246.00	$252.00	$258.00	$264.00	$261.00	$249.00	$264.00	$243.00	$231.00	$237.00	$207.00				

▲ Figure 12.20

Portion of Data Table and Summary Results for 200 Trials

This example showed how we can use simulation for prescriptive decisions, as well as assess the risk of alternative decisions. We should note that while sampling from empirical data is easy to do, it does have some drawbacks. First, the empirical data may not adequately represent the true underlying population because of sampling error. Second, using an empirical distribution precludes sampling values outside the range of the actual data. Therefore, it is usually advisable to fit a distribution and use it for the uncertain variable. We can do this by fitting a distribution to the data using the techniques we described in Chapter 5.

Overbooking Decisions

We also introduced overbooking decisions in Chapter 11. In any realistic situation, the actual customer demand as well as the number of cancellations would be uncertain. Similar to the newsvendor problem, we may use Monte Carlo simulation to help determine the best overbooking policy. We illustrate this in the following, based on Example 11.9.

EXAMPLE 12.15 Monte Carlo Simulation for the *Hotel Overbooking Model*

In the *Hotel Overbooking Simulation Model* spreadsheet (see Figure 12.21), let us assume that customer demand is normally distributed with a mean of 320 and standard deviation of 15. Because we want this value to be a whole number, we can use the Excel function ROUND (*number, num_digits*); thus using

=ROUND(NORM.INV(RAND( ), 320, 15), 0)

in cell B14 will guarantee this. The number of cancellations depends on the number of reservations made in cell B15. Suppose that the probability that any reservation is cancelled is 0.04. Then the number of cancellations is a binomial random variable, with n = number of reservations made, and $p = 0.04$. We may enter the formula =BINOM.INV(B15, 0.04, RAND( )) into cell B16 to generate a binomial random variate. Note that we are referencing cell B15 in this formula as it will change in each simulated trial because of the random demand.

The decision variable is the reservation limit in cell B13. We may use data tables to simulate the number of overbooked customers and net revenue for each choice of the reservation limit. We use two-way data tables with 100 rows (trials) and 7 columns corresponding to reservation limit decisions, from 300 to 330 in increments of 5. In Figure 12.21, we see that as the reservation limit (shown in the range E4:K4) increases, so does the average number of overbooked customers. We also see that the average net revenue is maximized when 315 reservations are taken; that is, overbooking the hotel by 15 rooms. For 315 reservations, we also see that the average number of overbooked customers is 1.74 (cell H3).

Project Management

In Chapter 11, we developed a project management spreadsheet model for scheduling activities and computing the critical path. For many real projects, activity times are random variables. In most cases, times must be estimated judgmentally, so we often assume that

	A	B	C	D	E	F	G	H	I	J	K
1	Hotel Overbooking Model			Overbooked customers							
2											
3				Average	0.00	0.00	0.32	1.74	4.16	6.22	7.32
4				0	300	305	310	315	320	325	330
5	Data			1	0	0	0	0	2	15	18
6				2	0	0	0	0	8	13	6
7	Rooms available	300		3	0	0	1	0	0	7	0
8	Price	$120		4	0	0	0	3	6	0	0
9	Overbooking cost	$100		5	0	0	0	0	0	0	13
10				6	0	0	0	3	11	1	13
11	Model			7	0	0	0	0	0	5	0
12				8	0	0	1	0	5	12	10
13	Reservation limit	310		9	0	0	0	4	9	10	20
14	Customer demand	305		10	0	0	0	0	9	10	0
15	Reservations made	305		11	0	0	0	3	0	0	5
16	Cancellations	12		12	0	0	0	1	7	5	18
17	Customer arrivals	293		13	0	0	0	0	6	6	19
18				14	0	0	0	5	0	12	0
19	Overbooked customers	0		15	0	0	0	0	5	1	0
20	Net revenue	$35,160		16	0	0	0	0	0	15	18

	L	M	N	O	P	Q	R	S	T
1		Net Revenue							
2									
3		Average	$34,501.20	$34,963.20	$35,138.40	$35,489.00	$35,212.80	$34,864.20	$34,836.20
4		$35,160	300	305	310	315	320	325	330
5		1	$34,920.00	$34,920.00	$35,640.00	$36,000.00	$35,880.00	$35,520.00	$35,900.00
6		2	$35,040.00	$35,280.00	$35,520.00	$35,880.00	$35,640.00	$35,900.00	$35,800.00
7		3	$34,920.00	$35,520.00	$35,520.00	$35,880.00	$34,700.00	$34,900.00	$33,900.00
8		4	$33,840.00	$34,800.00	$35,900.00	$34,800.00	$35,200.00	$35,400.00	$35,760.00
9		5	$34,440.00	$34,680.00	$35,160.00	$35,900.00	$35,200.00	$35,800.00	$34,080.00
10		6	$34,320.00	$34,680.00	$34,560.00	$35,160.00	$35,300.00	$34,900.00	$34,200.00
11		7	$33,840.00	$35,280.00	$34,920.00	$34,900.00	$35,800.00	$35,900.00	$35,600.00
12		8	$34,080.00	$34,440.00	$34,800.00	$36,000.00	$34,320.00	$34,400.00	$34,300.00
13		9	$34,440.00	$34,800.00	$33,960.00	$35,500.00	$35,520.00	$34,000.00	$34,100.00
14		10	$34,800.00	$35,160.00	$34,680.00	$35,760.00	$35,640.00	$34,400.00	$34,400.00
15		11	$34,800.00	$35,280.00	$34,320.00	$35,760.00	$35,640.00	$33,240.00	$32,760.00
16		12	$34,200.00	$35,520.00	$34,800.00	$32,760.00	$35,400.00	$34,800.00	$35,100.00
17		13	$34,560.00	$35,280.00	$35,760.00	$35,400.00	$34,200.00	$34,800.00	$35,700.00
18		14	$34,680.00	$35,040.00	$35,900.00	$35,600.00	$34,700.00	$35,700.00	$33,960.00
19		15	$34,560.00	$34,920.00	$34,920.00	$35,040.00	$34,920.00	$35,600.00	$34,900.00
20		16	$34,560.00	$34,440.00	$35,640.00	$35,640.00	$34,600.00	$35,760.00	$34,200.00

▲ Figure 12.21

Hotel Overbooking Model Monte Carlo Simulation

they have a triangular distribution. Analytical methods, such as the Program Evaluation and Review Technique (PERT), allow us to determine probabilities of project completion times by assuming that the project completion time is normally distributed. However, this assumption may not always be valid. Simulation can provide a more realistic characterization of the project completion time and the associated risks. We will illustrate this with the Becker Consulting project management model used in Chapter 11.

EXAMPLE 12.16   A Spreadsheet Simulation Model for Project Management

Suppose that the information systems manager has determined the most likely time for each activity but, recognizing the uncertainty in the times to complete each task, has estimated the smallest and largest times that the activities might take. These are shown in Table 12.1. With only these estimates, a triangular distribution is an appropriate assumption. Note that the times for activities A, G, M, P, and Q are constant.

Figure 12.22 shows a spreadsheet model designed to simulate the project completion time when the activity times are uncertain (Excel file *Becker Consulting Project Management Simulation Model*). The mean activity times

using the triangular distributions are calculated in column E. If we use these to find the critical path, we would find that the project completion time is 155, with activities B, F, G, H, I, K, M, O, P, and Q on the critical path. Calculations for the triangular random variates are shown in columns N and O. The values of the random variates are referenced in the shaded cells in column F. For this simulation, we see that the completion time is 148.67. Note that as the activity times vary randomly, different project completion times and critical paths may result (manually recalculate the spreadsheet several times to see this). We may use a data table to simulate the project completion time. Figure 12.23 shows the simulation results.

▼ Table 12.1

Uncertain Activity Time Data

Activity	Predecessors	Minimum Time (days)	Most Likely Time (days)	Maximum Time (days)
A Select steering committee	—	15	15	15
B Develop requirements list	—	40	45	60
C Develop system size estimates	—	10	14	30
D Determine prospective vendors	—	2	3	5
E Form evaluation team	A	5	7	9
F Issue request for proposal	B, C, D, E	4	5	8
G Bidders conference	F	1	1	1
H Review submissions	G	25	30	50
I Select vendor short list	H	3	5	10
J Check vendor references	I	3	7	10
K Vendor demonstrations	I	20	30	45
L User site visit	I	3	4	5
M Select vendor	J, K, L	3	3	3
N Volume-sensitive test	M	10	13	20
O Negotiate contracts	M	10	14	28
P Cost–benefit analysis	N, O	2	2	2
Q Obtain board of directors' approval	P	5	5	5

Activity	Minimum Time (a)	Most Likely Time (c)	Maximum Time (b)	Mean	Activity Time (Triangular)	Early Start	Early Finish	Latest Start	Latest Finish	Slack	On Critical Path?	Triangular Random Variates	
A	15	15	15	15.00	15.00	0.00	15.00	19.65	34.65	19.65		Activity B	
B	40	45	60	48.33	42.98	0.00	42.98	0.00	42.98	0.00	Yes	a	40.00
C	10	14	30	18.00	17.93	0.00	17.93	25.05	42.98	25.05		b	60.00
D	2	3	5	3.33	2.34	0.00	2.34	40.64	42.98	40.64		c	45.00
E	5	7	9	7.00	8.33	15.00	23.33	34.65	42.98	19.65		R	0.089
F	4	5	8	5.67	6.01	42.98	48.99	42.98	48.99	0.00	Yes	F(c)	0.2500
G	1	1	1	1.00	1.00	48.99	49.99	48.99	49.99	0.00	Yes	Triangular random variate	42.98
H	25	30	50	35.00	32.54	49.99	82.53	49.99	82.53	0.00	Yes		
I	3	5	10	6.00	7.29	82.53	89.82	82.53	89.82	0.00	Yes	Activity C	
J	3	7	10	6.67	8.20	89.82	98.01	116.21	124.41	26.40		a	10.00
K	20	30	45	31.67	34.59	89.82	124.41	89.82	124.41	0.00	Yes	b	30.00
L	3	4	5	4.00	4.10	89.82	93.91	120.31	124.41	30.49		c	14.00
M	3	3	3	3.00	3.00	124.41	127.41	124.41	127.41	0.00	Yes	R	0.545
N	10	13	20	14.33	12.34	127.41	139.75	129.33	141.67	1.92		F(c)	0.2000
O	10	14	28	17.33	14.26	127.41	141.67	127.41	141.67	0.00	Yes	Triangular random variate	17.93
P	2	2	2	2.00	2.00	141.67	143.67	141.67	143.67	0.00	Yes	Activity D	
Q	5	5	5	5.00	5.00	143.67	148.67	143.67	148.67	0.00	Yes	a	2.00
								Project completion time		148.67		b	5.00

Becker Consulting Project Management Simulation Model

▲ **Figure 12.22**

Portion of Project Management Spreadsheet Model

▶ **Figure 12.23**

Simulation Results

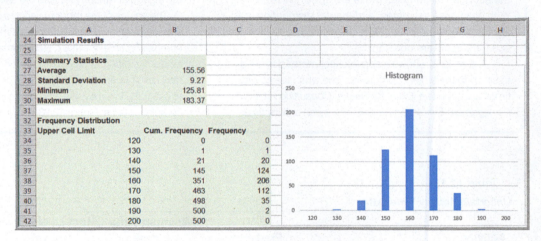

Simulation Results

Summary Statistics

Average	155.56
Standard Deviation	9.27
Minimum	125.81
Maximum	183.37

Frequency Distribution

Upper Cell Limit	Cum. Frequency	Frequency
120	0	0
130	1	1
140	21	20
150	145	124
160	351	206
170	463	112
180	498	35
190	500	2
200	500	0

CHECK YOUR UNDERSTANDING

1. Summarize the general process for performing Monte Carlo simulations in Excel.

2. Explain how to use data tables for Monte Carlo simulations on spreadsheets.

3. What statistical tools can you use to evaluate risk using the results from Monte Carlo simulations?

ANALYTICS IN PRACTICE: Implementing Large-Scale Monte Carlo Spreadsheet Models[3]

Implementing large-scale Monte Carlo models in spreadsheets in practice can be challenging. This example shows how one company used Monte Carlo simulation for commercial real estate credit-risk analysis but had to develop new approaches to effectively implementing spreadsheet analytics across the company.

Based in Stuttgart, Germany, Hypo Real Estate Bank International (Hypo), with a large portfolio in commercial real

[3]Based on Yusuf Jafry, Christopher Marrison, and Ulrike Umkehrer-Neudeck, "Hypo International Strengthens Risk Management with a Large-Scale, *Secure Spreadsheet-Management Framework*," *Interfaces*, 38, 4 (July–August 2008): 281–288.

estate lending, undertakes some of the world's largest real estate transactions. Hypo was faced with the challenge of complying with Basel II banking regulations in Europe. Basel II sets the minimum capital to be held in reserve by internationally active banks. If a bank is able to comply with the more demanding requirements of the regulations, it can potentially save €20–€60 million per year in capital costs. To qualify however, Hypo needed new risk models and reporting systems. The company also wished to upgrade its internal reporting and management framework to provide better analytical tools to its lending officers, who were responsible for structuring new loans, and to provide its managers with better insights into the risks of the overall portfolio.

Monte Carlo simulation is the only practical approach for analyzing the risk models the bank needed. For example, in one commercial real estate application, 200 different macroeconomic and market variables are typically simulated over 20 years. The cash-flow modeling process can be even more complex, particularly if the effects of all the intricate details of the transaction must be quantified. However, the computational process of Monte Carlo simulation is numerically intensive because the entire spreadsheet must be recalculated both for each iteration of the simulation and for each individual asset (or transaction) within the portfolio. This pushes the limits of stand-alone Excel models, even for a single asset. Moreover, because the bank is usually interested in analyzing its entire portfolio of thousands of assets, in practice, it becomes impossible to do so using stand-alone Excel.

Therefore, Hypo needed a way to implement the complex analytics of simulation in a way that its global offices could use on all their thousands of loans. In addition to the computational intensity of simulation analytics, the option to build the entire simulation framework in Excel can lead to human error, which they called *spreadsheet risk*. Spreadsheet risks that Hypo wished to minimize included the following:

- Proliferation of spreadsheet models that are stored on individual users' desktop computers throughout the organization and are untested and lack version

Vladitto/Shutterstock

data, and the unsanctioned manipulation of the results of spreadsheet calculations.

- Potential for serious mistakes resulting from typographical and "cut and copy-and-paste" errors when entering data from other applications or spreadsheets.
- Accidental acceptance of results from incomplete calculations.
- Errors associated with running an insufficient number of Monte Carlo iterations because of data or time constraints.

Given these potential problems, Hypo deemed a pure Excel solution as impractical. Instead, they used a consulting firm's proprietary software, called the Specialized Finance System (SFS), which embeds spreadsheets within a high-performance, server-based system for enterprise applications. This eliminated the spreadsheet risks but allowed users to exploit the flexible programming power that spreadsheets provide, while giving confidence and trust in the results. The new system has improved management reporting and the efficiency of internal processes and has also provided insights into structuring new loans to make them less risky and more profitable.

Dynamic Systems Simulation

Dynamic systems involve processes that consist of interacting events occurring over time. For example, nearly everyone experiences waiting lines, or **queues**, at supermarkets, banks, toll booths, telephone call centers, restaurants, and amusement parks. Many other waiting line systems involve "customers" other than people—for example, messages in communication systems, trucks waiting to be unloaded at a warehouse, work in process at a manufacturing plant, and photocopying machines awaiting repair by a traveling technician. In these systems, customers arrive at random times, and service times are rarely predictable. Managers of these systems would be interested in knowing how long customers have to wait, the length of waiting lines, the utilization of the servers, and other measures of performance. Another example of a dynamic system is an inventory management system. Managers would be interested in knowing inventory levels, numbers of lost sales or backorders incurred, and the costs of operating the system. More complex examples of dynamic systems include entire production systems, which might incorporate aspects of both waiting lines and inventory systems, as well as material movement, information flow, and so on.

Dynamic system simulation models allow us to draw conclusions about the behavior of a real system by studying the behavior of a model of the system, usually with random sequences of events similar to what we might observe in the real system. Simulation has the advantage of being able to incorporate nearly any practical assumption and thus is the most flexible tool for dealing with dynamic systems. In practice, simulating dynamic systems is best accomplished using powerful commercial software; however, we can use Excel quite easily for some simple applications to illustrate the concepts.

Modeling dynamic systems is generally more difficult than just replicating spreadsheets using Monte Carlo simulation, primarily because the logical sequence of events over time must be explicitly taken into account. For instance, key events that occur in a waiting line system are arrivals of customers, completions of services, and possibly line-switching or a customer's decision to leave a line because of excessive waiting. Events in an inventory system include the demand for items, receipt of replenishment orders, and placement of new orders. The next example illustrates a dynamic simulation for a simple production and inventory scenario and also shows how we can incorporate Monte Carlo simulation into the analysis.

EXAMPLE 12.17   A Production/Inventory Simulation

Mantel Manufacturing supplies various engine components to manufacturers of motorcycles on a just-in-time basis. Planned production capacity for one component is 100 units per shift, and the plant operates one shift per day. Because of fluctuations in customers' assembly operations, however, demand fluctuates and is historically between 80 and 130 units per day. To maintain sufficient inventory to meet its just-in-time commitments, the operations manager is considering a policy to run a second shift the next day if inventory falls to 50 or below at the end of a day (after the daily demand is known). For the annual budget planning process, he needs to know how many additional shifts will be needed.

The fundamental equation that governs this process each day is

Ending Inventory = Beginning Inventory
            + Production − Demand   **(12.5)**

Figure 12.24 shows a spreadsheet model (Excel file *Mantel Manufacturing*) to simulate 260 working days (one year) and count the number of additional shifts that

are required. We assume that on day 1, the initial inventory and production are 100 units. The Excel function RANDBETWEEN(80, 130) is used in column C to generate discrete uniform random variates for the demand. In column D, we use an IF function to determine whether an additional production shift is scheduled based on the beginning inventory in column B. For example, the formula in cell D6 is =100+IF(B6<50, 100). Then in column E, the ending inventory is calculated using equation (12.5); this is copied to the beginning inventory for the next day.

The chart shows the ending inventory and production each day. The spikes in the production series show when additional shifts were scheduled. If you recalculate the spreadsheet, the results will change. Since the model provides results only for one year, we use a data table with 100 trials to replicate the spreadsheet in order to determine a distribution for the number of additional shifts (see Figure 12.25). While the average is close to 13, considerable variation exists among simulation runs. To be safe, the budget should plan on a higher number of shifts than the average.

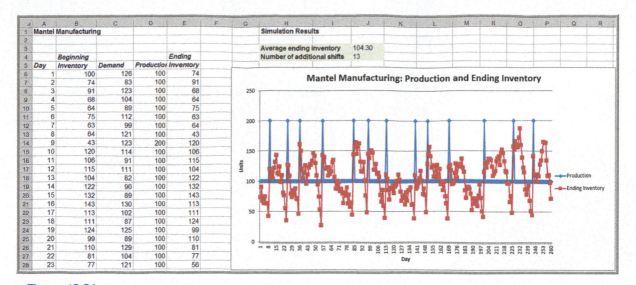

▲ **Figure 12.24**

Spreadsheet Simulation Model for Mantel Manufacturing

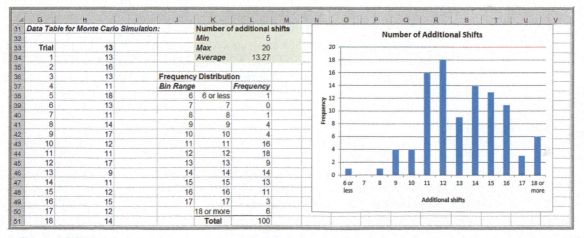

▲ **Figure 12.25**

Monte Carlo Simulation for Number of Additional Shifts

Simulating Waiting Lines

We will develop a simulation model for a waiting line (queueing) system. We assume that there is a single server. Customers arrive at the system and join a waiting line on a first-come, first-served basis if the server is busy; otherwise, they go immediately into service. Then they receive the service and leave the system. Typical performance measures that we would like to evaluate are the average waiting time per customer, the idle time of the server, and the number in the queue.

To understand the logic behind the simulation model, consider the sequence of activities that each customer undergoes:

1. Customer arrives.
2. Customer waits for service if the server is busy.

    3. Customer receives service.

    4. Customer leaves the system.

We can make the following observations:

1. If a customer arrives at time t and the server is not busy, then that customer can begin service immediately upon arrival.

2. If a customer arrives at time t and the server is busy, then that customer will begin service at the time that the *previous* customer completes service (which will be greater than t). The waiting time is the difference between the time service begins and the arrival time.

3. In either case, the time at which a customer completes service is computed as the time that the customer begins service plus the time it takes to perform the service.

4. The server idle time is the difference between the arrival time of the next customer and the time at which the current customer completes service, if the arrival time of the next customer is greater than the time at which the current customer completes service; otherwise, the idle time is zero.

5. To find the number in the queue, we note that when a customer arrives, then all prior customers who have not completed service by that time must still be waiting.

These observations provide all the information we need to run a small manual simulation.

Table 12.2 shows such a simulation. We assume that the system opens at time 0 and that the arrival times and service times have been generated by some random mechanism and are known. We can use the logic above to complete the last five columns. For example, the first customer arrives at time 3.2 (the server is idle from time 0 until this event). Because the queue is empty, customer 1 immediately begins service and ends at time $3.2 + 3.7 = 6.9$. The server is idle until the next customer arrives at time 10.5. Customer 2 then begins service and completes service at time $10.5 + 3.5 = 14.0$. Customer 3 arrives at time 12.8. Because customer 2 is still in service, customer 3 must wait until time 14.0 to begin service, incurring a waiting time of 1.2. You should verify the calculations for the remaining customers in this simulation.

Figure 12.26 shows the formulas for a spreadsheet model that corresponds to Table 12.2. The start time for any customer is the maximum of the arrival time and the time of service of the previous customer. This is implemented using the MAX function. Waiting time is simply the difference between the start time and arrival time; idle time for the server is computed from the arrival time of the current customer and the completion time of the last customer. Finding the length of the queue when a customer arrives is a bit tricky. For example, in Table 12.2, customer 5 arrives at time 17.2. At this time, customer 3 is still in service and

▼ **Table 12.2**

Customer	Arrival Time	Service Time	Number in Queue	Start Time	End Time	Waiting Time	Server Idle Time
1	3.2	3.7	0	3.2	6.9	0	3.2
2	10.5	3.5	0	10.5	14	0	3.6
3	12.8	4.3	1	14	18.3	1.2	0
4	14.5	3	1	18.3	21.3	3.8	0
5	17.2	2.8	2	21.3	24.1	4.1	0
6	19.7	4.2	2	24.1	28.3	4.4	0
7	26.9	2.8	1	28.3	31.1	1.4	3
8	28.7	1.3	1	31.1	32.4	2.4	0
9	32.7	2.1	0	32.7	34.8	0	0.3
10	36.9	4.8	0	36.9	41.7	0	2.1

▶ Figure 12.26

Spreadsheet Formulas for Simulation Model

	A Customer	B Arrival Time	C Service Time	D Number in Queue	E Start Time	F End Time	G Waiting Time	H Server Idle Time
2						0		
3	1	3.2	3.7	0	=B3	=E3+C3	=E3-B3	=E3
4	2	10.5	3.5	=A4-MATCH(B4,$F$2:F3,1)	=MAX(B4,F3)	=E4+C4	=E4-B4	=E4-F3
5	3	12.8	4.3	=A5-MATCH(B5,$F$2:F4,1)	=MAX(B5,F4)	=E5+C5	=E5-B5	=E5-F4
6	4	14.5	3	=A6-MATCH(B6,$F$2:F5,1)	=MAX(B6,F5)	=E6+C6	=E6-B6	=E6-F5
7	5	17.2	2.8	=A7-MATCH(B7,$F$2:F6,1)	=MAX(B7,F6)	=E7+C7	=E7-B7	=E7-F6
8	6	19.7	4.2	=A8-MATCH(B8,$F$2:F7,1)	=MAX(B8,F7)	=E8+C8	=E8-B8	=E8-F7
9	7	26.9	2.8	=A9-MATCH(B9,$F$2:F8,1)	=MAX(B9,F8)	=E9+C9	=E9-B9	=E9-F8
10	8	28.7	1.3	=A10-MATCH(B10,$F$2:F9,1)	=MAX(B10,F9)	=E10+C10	=E10-B10	=E10-F9
11	9	32.7	2.1	=A11-MATCH(B11,$F$2:F10,1)	=MAX(B11,F10)	=E11+C11	=E11-B11	=E11-F10
12	10	36.9	4.8	=A12-MATCH(B12,$F$2:F11,1)	=MAX(B12,F11)	=E12+C12	=E12-B12	=E12-F11

customer 4 is waiting until time 18.3. Thus, the length of the queue is now 2. We can compute this using a MATCH function to determine the last customer whose completion time is less than or equal to the arrival time of the current customer. For example, the value of the MATCH function in cell D7 is 3; therefore, $5 - 3 = 2$ provides the correct value for the number in the queue. To make this work correctly if customer 2 arrives before customer 1 has completed service, we needed to add an end time of 0 in cell F2 for "customer zero" and include this in the MATCH range.

EXAMPLE 12.18　Car Wash Simulation

Mike and Judy operate a small car wash; Judy is in charge of finance, accounting, and marketing and Mike is in charge of operations. On a typical Saturday, customers arrive randomly between three and eight minutes apart. A standard wash takes four minutes (the service time), but may run as high as seven minutes for those cars that purchase extra services. The probabilities of the times between arrivals and service times are estimated from historical data. Although customers may complain a bit, they do not leave if they have to wait.

To simulate this system on a spreadsheet, we need to generate the arrival times of customers and the time required for service. For arrivals, we could specify either the actual *times* that customers arrive for service or the *times between successive arrivals*. The second approach works better for simulation purposes because we need only know the time that the last customer arrived to generate the arrival time of the next customer. Therefore, we set the arrival time for a customer as the arrival time of the previous customer plus a random value. We use discrete distributions in column J–M for the time between arrivals and service time, and we use VLOOKUP functions to find these.

Figure 12.27 shows the spreadsheet model (Excel file *Car Wash Simulation*). The model simulates 100 customers (roughly one day). We calculate the maximum number in the queue, maximum waiting time, average waiting time per customer, and total idle time. It does not make sense to average the number in the queue or idle time, because these are time-dependent, not customer-dependent. Because this model only simulates one day, you could use data tables to find distributions for these measures. We leave this as an exercise.

	A Customer	B Arrival Time	C Service Time	D Number in Queue	E Start Time	F End Time	G Waiting Time	H Server Idle Time	J	K	L	M	O	P
1	Mike & Judy's Carwash													
4						0							Maximum number in queue	1
5	1	5	4	0	5	9	0	5					Maximum waiting time	6
6	2	11	5	0	11	16	0	2		Time between arrivals			Average waiting time per customer	1.18
7	3	16	5	0	16	21	0	0	Probability 0.20	Random Number Intervals 0.00	0.20	Time (min) 3	Total idle time	67
8	4	23	4	0	23	27	0	2	0.20	0.20	0.40	4		
9	5	26	5	1	27	32	1	0	0.30	0.40	0.70	5		
10	6	30	5	2	32	37	2	0	0.15	0.70	0.85	6		
11	7	33	6	1	37	43	4	0	0.10	0.85	0.95	7		
12	8	37	4	1	43	47	6	0	0.05	0.95	1.00	8		
13	9	44	4	1	47	51	3	0						
14	10	49	4	1	51	55	2	0		Service time				
15	11	54	6	1	55	61	1	0	Probability	Random Number Intervals		Time (min)		
16	12	60	6	1	61	67	1	0	0.50	0	0.5	4		
17	13	68	4	0	68	72	0	1	0.25	0.5	0.75	5		
18	14	72	6	0	72	78	0	0	0.15	0.75	0.9	6		
19	15	75	4	1	78	82	3	0	0.10	0.9	1	7		

▲ Figure 12.27

Portion of Spreadsheet Implementation of Car Wash Simulation

CHECK YOUR UNDERSTANDING

1. Explain the characteristics of dynamic systems and why these are more challenging to simulate than Monte Carlo spreadsheet models.

2. Summarize the logic used in developing spreadsheet models to simulate waiting lines.

3. Why is it important to use data tables to replicate simulation models of dynamic systems?

ANALYTICS IN PRACTICE: Using Systems Simulation for Agricultural Product Development[4]

Syngenta, a leading developer of crop varieties (seeds) that provide food for human and livestock consumption, is committed to bringing greater food security to an increasingly populous world by creating a transformational shift in farm productivity. Syngenta Soybean Research and Development (R&D) is leading Syngenta's corporate plant-breeding strategy by developing and implementing a new product development model that is enabling the creation of an efficient and effective soybean-breeding strategy. Their strategy addressed the following objectives:

- Increase the frequency of favorable traits within the population of soybean plant varieties;
- Reduce the time required to develop new soybean plant varieties with favorable traits;
- Build a process to efficiently transfer favorable traits among soybean plant varieties;
- Improve data quality, prediction of variety performance, and characterization of environments;
- Make better decisions to positively impact the probability, cost, and timeline of developing a new soybean plant variety.

The soybean-variety pipeline produces commercial soybean varieties containing traits that enable it to thrive in different environmental conditions and against various diseases and pests, while providing higher yields than the varieties currently available. A project lead makes decisions that improve the parent varieties with a process called trait introgression (TI).

Key to the new strategy is the combination of advanced analytics and plant-breeding knowledge to find opportunities to increase crop productivity and optimize plant-breeding processes. Syngenta uses systems and Monte Carlo simulation models to codify Syngenta Soybean R&D best practices and help create the best soybean-breeding plans and strategically align its research efforts. The TI tool uses systems simulation to model the flow of a process, the biology of creating progeny from mating two plants, and the calculations needed to track genetic segregation according to Mendel's principles of genetics. Although cost and time are key outcomes of the TI tool, knowing both the expected number of seeds with the desired combination of traits and the uncertainty surrounding that expectation is critical to selecting the best breeding plan. Simulation allows a project lead to plan each process step and view the consequences of the plan on cumulative cost, cumulative time, and genetic composition of the progeny at the end of each step in the TI process. If the cost, time, or probability of successfully transferring the desired traits of the planned TI process is unacceptable, the project lead can change one or more decisions and quickly rerun the simulation.

[4]Adapted from Joseph Byrum, Craig Davis, Gregory Doonan, Tracy Doubler, David Foster, Bruce Luzzi, Ronald Mowers, and Chris Zinselmeiers, "Advanced Analytics for Agricultural Product Development," *Interfaces*, Vol. 46, No. 1, January–February 2016, pp. 5–17.

KEY TERMS

Discrete uniform distribution
Dynamic system
Monte Carlo simulation
Queue
Random number

Random number seed
Random variate
Risk
Risk analysis

CHAPTER 12 TECHNOLOGY HELP

Useful Excel Functions

RAND( ) Generates a random number greater than or equal to 0 and less than 1.

RANDBETWEEN(*a*, *b*) Generates a discrete uniform random variate between *a* and *b* inclusive.

NORM.INV(*RAND( ), mean, standard_deviation*) Generates a normal random variate with specified mean and standard deviation.

NORM.S.INV(*RAND( )*) Generates a standard normal random variate.

BINOM.INV(*n, p, RAND( )*) Generates a binomial random variate.

FREQUENCY(*data array, bin array*) Computes a frequency distribution for the *data array* corresponding to the bins defined in the *bin array*. Note that this is an array function and must be implemented using the process discussed in the chapter.

Analytic Solver

Analytic Solver provides a powerful Monte Carlo simulation tool. See the online supplement *Using Monte Carlo Simulation in Analytic Solver*. We suggest that you first read the online supplement *Getting Started with Analytic Solver Basic*. This provides Information for both instructors and students on how to register for and access Analytic Solver.

PROBLEMS AND EXERCISES

(Many of these problems extend problems in Chapter 11 and will be noted accordingly.)

Random Sampling from Probability Distributions

1. Generate 100 random numbers using the RAND function and create a frequency distribution and a histogram with bins of width 0.1. Apply the chi-square goodness of fit test (see Chapter 5) to test the hypothesis that the data are uniformly distributed.

2. A financial consultant has an average of seven customers he consults with each day; assume a Poisson distribution. The consultant's overhead requires that he consult with at least five customers per day for his fees to cover his expenses. Use the Excel *Random Number Generation* tool to generate 100 samples of the number of customers that the financial consultant will have on a daily basis. What percentage of these will meet his target of at least five?

Generating Random Variates Using Excel Functions

3. The weekly demand of a slow-moving product has the following probability mass function:

Demand, *x*	Probability, *f(x)*
0	0.2
1	0.4
2	0.3
3	0.1
4 or more	0

Use VLOOKUP to generate 25 random variates from this distribution.

4. The number and frequency of Atlantic hurricanes annually from 1940 through 2015 are shown on the next page. Use VLOOKUP to generate 25 random variates from this distribution.

Number	Frequency
0	5
1	16
2	20
3	14
4	4
5	5
6	5
7	3
8	2
10	1
12	1

5. The time required to play a game of Battleship™ is uniformly distributed between 15 and 60 minutes. Use formula (12.1) to obtain a sample of 50 outcomes and compute the mean, minimum, maximum, and standard deviation.

6. The distribution of GMAT scores in math for an incoming class of business students has a mean of 620 and standard deviation of 15. Assume that the scores are normally distributed. Generate 25 random variates from this distribution as whole numbers.

7. Historical data show that customers who download music from a popular Web service spend approximately $26 per month, with a standard deviation of $8, normally distributed, but never spend more than $30. Assuming that each customer is independent of the others, predict the average amount of the next ten transactions. (Hint: Develop a method to ensure that the random variate is limited to $30.)

8. A formula in financial analysis is the following: Return on equity = net profit margin × total asset turnover × equity multiplier. Suppose that the equity multiplier is fixed at 4.0, but that the net profit margin is normally distributed with a mean of 3.8% and a standard deviation of 0.4%, and that the total asset turnover is normally distributed with a mean of 1.5 and a standard deviation of 0.2. Set up and conduct a sampling experiment similar to Example 12.6 to find the distribution of the return on equity. Show your results as a histogram to help explain your analysis and conclusions.

9. A government agency is putting a large project out for low bid. Bids are expected from ten contractors

and will have a normal distribution with a mean of $3.5 million and a standard deviation of $0.25 million. Devise and implement a sampling experiment similar to that in Example 12.6 for estimating the distribution of the minimum bid and the expected value of the minimum bid.

10. A cell phone company is conducting a telemarketing campaign to generate new clients. One hundred calls are made, and the probability of successfully gaining a new customer from each call is 0.07. Generate 25 random variates for the number of new customers that would be gained to estimate the minimum and maximum number that might be expected.

11. In the Olympic bid proposal discussed in Example 12.8, ticket sales revenue was estimated to be a triangular distribution with minimum = $560 million, maximum = $600 million, and most likely = $600 million. Generate a random variate from this distribution.

12. To further refine the assumptions of the Olympic bid proposal discussed in Example 12.8, based on additional research, analysts created the following distribution for TV broadcast revenues, which were assumed to be uniform over various intervals, each with a discrete probability (all in millions of dollars):

a	b	Probability
$735	$887	0.1
$887	$930	0.15
$930	$1040	0.25
$1040	$1087	0.25
$1087	$1170	0.15
$1170	$1300	0.1

Develop a spreadsheet to generate ten random variates from this distribution. (Hint: Use Excel lookup functions to choose the distribution parameters based on the probabilities and then apply the uniform random variate formula.)

Monte Carlo Simulation in Excel

13. Use the profit model developed in Example 12.1 to implement a financial simulation model for a new product proposal and determine a distribution of profits using the discrete distributions below for the unit cost, demand, and fixed costs. Price is fixed at $1,000. Unit costs are unknown and follow the distribution

Unit Cost	Probability
$400	0.20
$600	0.40
$700	0.25
$800	0.15

Demand is also variable and follows the following distribution:

Demand	Probability
120	0.25
140	0.50
160	0.25

Fixed costs are estimated to follow the distribution

Fixed Costs	Probability
$45,000	0.20
$50,000	0.50
$55,000	0.30

Simulate this model for 50 trials and a production quantity of 140. What is the average profit?

14. Refer back to the outsourcing decision model in Example 1.4 of Chapter 1. The Excel file *Outsourcing Decision Model* is a spreadsheet implementation of the model. The model calculates the total cost for manufacturing and outsourcing. The key outputs in the model are the difference in these costs and the decision that results in the lowest cost. Note how the IF function is used in cell B20 to identify the best decision. Assume that the production volume is uncertain. Suppose the manufacturer has enough data and information to estimate that the production volume will be normally distributed with a mean of 1,100 and a standard deviation of 100. Use a 100-trial Monte Carlo simulation to find the average cost difference and percent of trials that result in manufacturing or outsourcing as the best decision. (Hint: Your data table should show both the cost difference and decision for each trial.)

15. For the market share model in Problem 5 of Chapter 11, suppose that the estimate of the percentage of new purchasers who will ultimately try the brand is uncertain and assumed to be normally distributed with a mean of 35% and a standard deviation of 4%. Conduct a Monte Carlo simulation with 250 trials and compute summary statistics of the long-run market share.

16. For the garage band model in Problem 21 of Chapter 11, suppose that the expected crowd is normally distributed with a mean of 3,000 and standard deviation of 200, and the average concession expenditure has a triangular distribution with minimum = $15, maximum = $30, and most likely = $25. Conduct a Monte Carlo simulation with 500 trials to find the distribution and summary statistics of the expected profit.

17. A professional football team is preparing its budget for the next year. One component of the budget is the revenue that they can expect from ticket sales. The home venue, Dylan Stadium, has five different seating zones with different prices. Key information is given below. The demands are all assumed to be normally distributed.

Seating Zone	Seats Available	Ticket Price	Mean Demand	Standard Deviation
First Level Sideline	15,000	$100.00	14,500	750
Second Level	5,000	$90.00	4,750	500
First Level End Zone	10,000	$80.00	9,000	1,250
Third Level Sideline	21,000	$70.00	17,000	2,500
Third Level End Zone	14,000	$60.00	8,000	3,000

Determine the distribution of total revenue under these assumptions using 250 trials. Summarize the statistical results.

18. Financial analysts often use the following model to characterize changes in stock prices:

$$P_t = P_0 e^{(\mu - 0.5\sigma^2)t + \sigma Z \sqrt{t}}$$

where

P_0 = current stock price

P_t = price at time t

μ = mean (logarithmic) change of the stock price per unit time

σ = (logarithmic) standard deviation of price change

Z = standard normal random variable

This model assumes that the logarithm of a stock's price is a normally distributed random variable (although we did not discuss it, the lognormal distribution is used in many financial applications). Using historical data, we can estimate values for μ and σ. Suppose that the average daily change for a stock is 0.003227, and the standard deviation is 0.026154. Develop a spreadsheet to simulate the price of the stock over the next 30 days if the current price is $53. Construct a chart showing the movement in the stock price.

19. The owner of Cruz Wedding Photography (see Problem 27 in Chapter 11) believes that the average number of wedding bookings per year can be estimated by triangular distribution with a minimum of 10, maximum of 22, and most likely value of 15. One of the key variables in developing his business plan is the life he can expect from a single digital single-lens reflex (DSLR) camera before it needs to be replaced. Due to heavy usage, the shutter life expectancy is estimated by a normal distribution with a mean of 150,000 clicks with a standard deviation of 10,000. For each booking, the average number of photographs taken is assumed to be normally distributed with a mean of 2,000 and a standard deviation of 300. Develop a simulation model using 500 trials to determine summary statistics and the distribution of the camera life (in years).

20. The manager of the extended-stay hotel in Problem 46 of Chapter 11 believes that the number of rooms rented during any given week has a triangular distribution with minimum 32, most likely 38, and maximum 50. The weekly price is $950, and weekly operating costs follow a normal distribution with mean $20,000 and a standard deviation of $2,500 but with a minimum value of $15,000. (Hint: You cannot embed RAND within the NORM.INV function in order to truncate the value using an IF statement. Use the same idea as generating a triangular random variate.) Run a simulation using 200 trials to answer the following questions.

   a. What is the probability that weekly profit will exceed $20,000?

   b. What is the probability that weekly profit will be less than $10,000?

21. A plant manager is considering investing in a new $35,000 machine. Use of the new machine is expected to generate a cash flow of about $8,000 per year for each of the next five years. However, the cash flow is uncertain, and the manager estimates that the actual cash flow will be normally distributed with a mean of $8,000 and a standard deviation of $500. The discount rate is set at 5% and assumed to remain constant over the next five years. The company evaluates capital investments using net present value. How risky is this investment? Develop and run a simulation model to answer this question using 100 trials.

22. Develop a simulation model for a three-year financial analysis of total profit based on the following data and information. Sales volume in the first year is estimated to be 100,000 units and is projected to grow at a rate that is normally distributed with a mean of 7% per year and a standard deviation of 4%. The selling price is $10, and the price increase is normally distributed with a mean of $0.50 and standard deviation of $0.05 each year. Per-unit variable costs are $3, and annual fixed costs are $200,000. Per-unit costs are expected to increase by an amount normally distributed with a mean of 5% per year and standard deviation of 2%. Fixed costs are expected to increase following a normal distribution with a mean of 10% per year and standard deviation of 3%. Based on 500 simulation trials, compute summary statistics for the average three-year undiscounted cumulative profit.

23. Tanner Park (see Problem 24 in Chapter 11) is a small amusement park that provides a variety of rides and outdoor activities for children and teens. In a typical summer season, the number of adult tickets sold has a normal distribution with a mean of 20,000 and a standard deviation of 2,000. The number of children's tickets sold has a normal distribution with a mean of 10,000 and a standard deviation of 1,000. Adult ticket prices are $18 and the children's price is $10. Revenue from food and beverage concessions is estimated to be between $50,000 and $100,000, with a most likely value of $60,000. Likewise, souvenir revenue has a minimum of $20,000, most likely value of $25,000, and a maximum value of $30,000. Variable costs per person are $3, and fixed costs amount to $150,000. Determine the distribution of profit for this business using 500 trials. What is the probability that the park will make a profit of $250,000 or less?

24. Lily's Gourmet Ice Cream Shop offers a variety of gourmet ice cream and shakes. Although Lily's competes with other ice cream shops and frozen yogurt stores, none of them offer gourmet ice creams with a wide variety of different flavors. The shop is also located in an upscale area and therefore can command higher prices. The owner is a culinary school

graduate without much business experience and has engaged the services of one of her friends who recently obtained an MBA to assist her with financial analysis of the business and evaluation of the profitability of introducing a new product. The shop is open during the spring and summer, with higher sales in the summer season. Based on past observation, Lily has defined three sales scenarios for the new product.

Summer:

- High—3,000 units
- Most likely—2,500 units
- Low—2,100 units

Spring:

- High—2,500 units
- Most likely—1,500 units
- Low—1,000 units

The expected price is $3.00. However, the unit cost is uncertain, and driven by the costs of the ingredients she has to buy for the product. This is estimated to be between $1.40 and $2.00, with a most likely value of $1.50 in the summer, but in the spring, the most likely cost is $2.00 because the ingredients are more difficult to obtain then. Fixed costs are estimated to be $2,600. Find the distribution of profit for each season and the annual profit distribution using 200 trials for each.

25. The Kelly Theater produces plays and musicals for a regional audience. For a typical performance, the theater sells at least 250 tickets and occasionally reaches its capacity of 600 seats. Most often, about 450 tickets are sold. The fixed cost for each performance is normal with a mean of $2,500 and a standard deviation of $250. Ticket prices range from $30 to $70 depending on the location of the seat. Of the 600 seats, 150 are priced at $70, 200 at $55, and the remaining at $30. Of all the tickets sold, the $55 seats sell out first. If the total demand is at least 500, then all the $70 seats sell out. If not, then between 50% and 75% of the $70 seats sell, with the remainder being the $30 seats. If, however, the total demand is less than or equal to 350, then the number of $70 and $30 seats sold are usually split evenly. The theater runs 160 performances per year and incurs an annual fixed cost of $2 million. Develop a simulation model to evaluate the profitability of the theater using 250 trials. What is the distribution of annual net profit and the risk of losing money over a year?

26. J&G Bank receives a large number of credit card applications each month, an average of 30,000 with a standard deviation of 4,000, normally distributed. Approximately 60% of them are approved, but this typically varies between 50% and 70%. Each customer charges a total of $2,000, normally distributed, with a standard deviation of $250, to his or her credit card each month. Approximately 85% pay off their balances in full, and the remaining incur finance charges. The average finance charge has recently ranged from 3% to 4% per month. The bank also receives income from fees charged for late payments and annual fees associated with the credit cards. This is a percentage of total monthly charges and has varied between 6.8% and 7.2%. It costs the bank $20 per application, whether it is approved or not. The monthly maintenance cost for credit card customers is normally distributed with a mean of $10 and standard deviation of $1.50. Finally, losses due to charge-offs of customers' accounts are between 4.6% and 5.4% of total charges. Use Monte Carlo simulation with 500 trials to analyze the profitability of the credit card product.

27. Sturgill Manufacturing Inc. needs to predict the numbers of machines and employees required to produce its planned production for the coming year. The plant runs three shifts continuously during the workweek, for a total of 120 hours of capacity per week. The shop efficiency (the percent of total time available for production), which accounts for setups, changeovers, and maintenance, averages 70% with a standard deviation of 5%, which reduces the weekly capacity. Six key parts are produced, and the plant has three different types of machines to produce each part. The machines are not interchangeable as they each have a specific function. The time to produce each part on each machine varies. The mean time and standard deviation (in hours) to produce each part on each machine are shown below:

Mean Time

Part Type	Machine A	Machine B	Machine C
1	3.5	2.6	8.9
2	3.4	2.5	8
3	1.8	3.5	12.6
4	2.4	5.8	12.5
5	4.2	4.3	28
6	4	4.3	28

Standard Deviation

Part Type	Machine A	Machine B	Machine C
1	0.15	0.12	0.15
2	0.15	0.12	0.15
3	0.1	0.15	0.25
4	0.15	0.15	0.25
5	0.15	0.15	0.5
6	0.15	0.15	0.5

The forecasted demand is shown below.

Part Type	Demand (Parts/Week)
1	42
2	18
3	6
4	6
5	6
6	6

Machines A and B require only one person to run two machines. Machine C requires only one person per machine. Develop a simulation model to determine how many machines of each type and number of employees will be required to meet the forecasted demand. Use 200 trials for each data table and compute summary statistics only.

28. O'Brien Chemicals makes three types of products: industrial cleaning, chemical treatment, and some miscellaneous products. Each is sold in 55-gallon drums. The selling price and unit manufacturing cost are shown below:

Manufacturing

Product Type	Selling Price/drum	Cost/drum
Industrial Cleaning		
Alkaline Cleaner	$700.00	$275.00
Acid Cleaner	$600.00	$225.00
Neutral Cleaner	$450.00	$150.00
Chemical Treatment		
Iron Phosphate	$920.00	$400.00
Zirconium	$1,350.00	$525.00
Zinc Phosphate	$1,400.00	$625.00
Other		
Sealant	$850.00	$350.00
Rust Prevention	$600.00	$260.00

Fixed costs are assumed normal with a mean of $5 million and a standard deviation of $20,000. Demands are all assumed to be normally distributed with the following means and standard deviations:

Product Type	Mean Demand	Standard Deviation
Industrial Cleaning		
Alkaline Cleaner	5,000	100
Acid Cleaner	2,000	500
Neutral Cleaner	5,000	350
Chemical Treatment		
Iron Phosphate	5,500	250
Zirconium	2,800	130
Zinc Phosphate	4,350	300
Other		
Sealant	8,000	350
Rust Prevention	4,250	250

The operations manager has to determine the quantity to produce in the face of uncertain demand. One option is to simply produce the mean demand for each product; depending on the actual demand, this could result in a shortage (lost sales). The other option is to produce at a level equal to the 75th percentile of the demand distribution for each product (that is, find the value so that 75% of the area under the normal distribution is to the left). Using Monte Carlo simulation with 150 trials, evaluate and compare summary statistics for these policies and write a report for the operations manager summarizing your findings. Make sure to round normal variates to whole numbers. What trade-offs does the manager have to make?

29. The Executive Committee of Reder Electric Vehicles (see Problem 29 in Chapter 11) is debating whether to replace its original model, the REV-Touring, with a new model, the REV-Sport, which would appeal to a younger audience. Whatever vehicle is chosen, it will be produced for the next four years, after which time a reevaluation will be necessary. The REV-Sport has passed through the concept and initial design phases and is ready for final design and manufacturing. Final development costs are estimated to be $75 million, and the new fixed costs for tooling and manufacturing are estimated to be $600 million. The REV-Sport is expected to sell for $30,000. The first-year sales for the REV-Sport is estimated to be normally distributed with an average of 60,000/year and standard deviation of 12,000/year. The sales growth

for subsequent years is estimated to be normally distributed with an average of 6% and standard deviation of 2%. The variable cost per vehicle is uncertain until the design and supply-chain decisions are finalized, but is estimated to be between $20,000 and $28,000 with the most likely value being $22,000. Next-year sales for the REV-Touring are estimated to be 50,000 with a standard deviation of 9,000/year, but the sales are expected to decrease at a rate that is normally distributed with a mean of 10% and standard deviation of 3.5% for each of the next three years. The selling price is $28,000. Variable costs are constant at $21,000. Since the model has been in production, the fixed costs for development have already been recovered. Develop a four-year Monte Carlo simulation model using 500 trials to recommend the best decision using a net present value discount rate of 4%.

30. The Schoch Museum (see Problem 30 in Chapter 11) is embarking on a five-year fundraising campaign. As a nonprofit institution, the museum finds it challenging to acquire new donors, as many donors do not contribute every year. Suppose that the museum has identified a pool of 8,000 potential donors. The actual number of donors in the first year of the campaign is estimated to be somewhere between 60% and 75% of this pool. For each subsequent year, the museum expects that a certain percentage of current donors will discontinue their contributions. This is expected to be between 10% and 60%, with a most likely value of 35%. In addition, the museum expects to attract some percentage of new donors. This is assumed to be between 5% and 40% of the current year's donors, with a most likely value of 10%. The average contribution in the first year is assumed to be $50 and will increase at a rate between 0% and 8% each subsequent year, with the most likely increase of 2.5%. Develop and analyze a simulation model to predict the total funds that will be raised over the five-year period using 500 trials.

31. For the Hyde Park Surgery Center scenario described in Problem 31 in Chapter 11, suppose that the following assumptions are made. The number of patients served the first year is uniform between 1,300 and 1,700; the growth rate for subsequent years is triangular with parameters (5%, 8%, 9%), and the growth rate for year 2 is independent of the growth rate for year 3; average billing is normal with mean of $150,000 and standard deviation $10,000; and the annual increase in fixed costs is uniform between 5% and 7% and independent of other years. Find the

distribution of the net present value of profit over the three-year horizon and analyze the summary statistics using 200 trials. Summarize your conclusions.

32. Adam is 24 years old and has a 401(k) plan through his employer, a large financial institution. His company matches 50% of his contributions up to 6% of his salary. He currently contributes the maximum amount he can. In his 401(k), he has three funds. Investment A is a large-cap index fund, which has had an average annual growth over the past ten years of 6.63% with a standard deviation of 13.46%. Investment B is a mid-cap index fund with a ten-year average annual growth of 9.89% and a standard deviation of 15.28%. Finally, Investment C is a small-cap index fund with a ten-year average annual growth rate of 8.55% and a standard deviation of 16.90%. Fifty percent of his contribution is directed to Investment A, 25% to Investment B, and 25% to Investment C. His current salary is $48,000 and based on a compensation survey of financial institutions, he expects an average raise of 2.7% with a standard deviation of 0.4% each year. Develop a simulation model to predict how much he will have available at age 60 using 500 trials.

33. Develop a realistic retirement planning simulation model for your personal situation. If you are currently employed, use as much information as you can gather for your model, including potential salary increases, promotions, contributions, and rates of return based on the actual funds in which you invest. If you are not employed, try to find information about salaries in the industry in which you plan to work and the retirement benefits that companies in that industry offer for your model. Estimate rates of returns based on popular mutual funds used for retirement or average performance of stock market indexes. Clearly state your assumptions and how you arrived at them and fully analyze and explain your model results.

34. Waring Solar Systems provides solar panels and other energy-efficient technologies for buildings. In response to a customer inquiry, the company is conducting a feasibility study to determine if solar panels will provide enough energy to pay for themselves within the payback period. Capacity is measured in MWh/year (1,000 kWh). This figure is determined by the number of panels installed and the amount of sunlight the panels receive each year. Capacity can vary greatly due to weather conditions, especially clouds and snow. Engineers have determined that this client

should use an 80 MWh/year system. The cost of the system and installation is $80,000. The amount of power the system will produce is normally distributed with a standard deviation of 10 MWh/year. The solar panels become less efficient over time mostly due to clouding of their protective cases. The annual loss in efficiency is normally distributed with a mean of 1% and a standard deviation of 0.2% and will apply after the first year. The client currently obtains electricity from its provider at a rate of $0.109/kWh. Based on analysis of previous years' electric bills, the annual cost of electricity is expected to increase following a triangular distribution with most likely value of 3%, min of 2.5%, and max of 4%, beginning with the first year. The cost of capital is estimated to be 5%. Develop a simulation model using 500 trials to find the net present value of the technology over a ten-year period, including the system and installation cost. What is the probability that the system will be economical?

35. SPD Tax Service is a regional tax preparation firm that competes with such national chains as H&R Block. The company is considering expanding and needs a financial model to analyze the decision to open a new store. Key factors affecting this decision include the demographics of the proposed location, price points that can be achieved in the target market, and the availability of funds for marketing and advertising. Capital expenditures will be ignored because unused equipment from other locations can often be shifted to a new store for the first year until they can be replaced periodically through the fixed cost budget. SPD's target markets are communities with populations between 30,000 and 50,000, assumed to be uniformly distributed. Market demand for tax preparation service is directly related to the number of households in the territory; approximately 15% of households are anticipated to use a tax preparation service. Assuming an average of 2.5 people per household, this can be expressed as 0.15*population/2.5. SPD estimates that its first-year demand will have a mean of 5% of the total market demand, and for every dollar of advertising, the mean increases by 2%. The first-year demand is assumed to be normal with a standard deviation of 20% of the mean demand. An advertising budget of $5,000 has been approved but is limited to 10% of annual revenues. Demand grows fairly aggressively in the second and third year and is assumed to have a triangular distribution with a minimum value of 20%, most likely value of 35%, and maximum value of 40%. After year 3, demand growth is between 5% and 15%, with a most likely value of

7%. The average charge for each tax return is $175 and increases each year at a rate that is normally distributed with a mean of 4% and a standard deviation of 1.0%. Variable costs average $15 per customer and increase annually at a rate that is normally distributed with a mean of 3% and a standard deviation of 1.5%. Fixed costs are estimated to be approximately $35,000 for the first year and grow annually at a rate between 1.5% and 3%. Develop a Monte Carlo simulation model to find the distribution of the net present value of the profitability over a five-year period using a discount rate of 3%. Use 500 trials.

36. For the profit model developed in Example 12.1, suppose that the demand is uniform with a minimum of 35,000 and maximum of 60,000; fixed costs are normal with a mean of $400,000 and a standard deviation of $25,000; and unit costs are triangular with a minimum of $22.00, most likely value of $24.00, and maximum value of $30.00. Simulate 100 trials for production quantities from 35,000 to 60,000 in increments of 5,000 and compute the average and standard deviation for each quantity. What is the best quantity to produce?

37. Use the *Newsvendor Model* spreadsheet to set up and run a Monte Carlo simulation assuming that demand is triangular with minimum value = 40, maximum value = 50, and most likely value = 47. Find the distribution of profit for order quantities between 40 and 50 to identify the best order quantity. Use 100 simulation trials.

38. Simulate the mini-mart situation described in Problem 45 of Chapter 11. Identify the best order quantity using a simulation with 100 trials.

39. Develop and analyze a simulation model for Koehler Vision Associates (KVA) in Problem 49 of Chapter 11 with the following assumptions. Assume that the demand is uniform between 110 and 160 per week and that anywhere between 10% and 20% of prospective patients fail to show up or cancel their exam at the last minute. Determine the averages for the net profit (revenue less overbooking costs) and number overbooked for taking appointments between 130 and 150 patients in increments of 2, using a simulation with 100 trials. What would you recommend?

40. Refer back to the college admission director scenario (Problem 25 in Chapter 11). Develop a spreadsheet model and identify uncertain distributions that you believe would be appropriate to conduct a Monte Carlo simulation. Based on your model and simulation, make a recommendation on how many scholarships to offer.

41. In Jennifer Bellin's leadership conference project (Problem 34 in Chapter 11), suppose that the activity times are uncertain. Estimated ranges for these times are shown in the table below.

Activity	Description	Prede-cessors	Minimum Time Estimate	Maximum Time Estimate
A	Develop conference theme		1	5
B	Determine attendees		1	5
C	Contract facility	A	5	15
D	Choose entertain-ment	A	5	20
E	Send announce-ment	B	1	10
F	Order gifts	B	1	10
G	Order materials	B	1	1
H	Plan schedule of sessions	C	30	50
I	Design printed materials	B, H	10	20
J	Schedule session rooms	C	1	1
K	Print directions	H	5	15
L	Develop travel memo	E	1	10
M	Write gift letter	F	1	10
N	Confirm catering	H	1	5
O	Communi-cate with speakers	H	1	5
P	Track RSVPs and assign roommates	L	15	40

Activity	Description	Prede-cessors	Minimum Time Estimate	Maximum Time Estimate
Q	Print materials	I	1	5
R	Assign table numbers	P	1	1
S	Compile packets of materials	G	1	5
T	Submit audio-visual needs	O	1	1
U	Put together welcome letter	P	1	10
V	Confirm arrange-ments with hotel	P	1	5
W	Print badges	G, P	3	7

Develop a spreadsheet simulation model to find the distribution of the project completion time using 100 trials. All times should be expressed as whole numbers of days.

Dynamic Systems Simulation

42. Passengers wait for an airport shuttle service that arrives every 15 minutes. Passengers arrive within each 5-minute period according to the following distribution:

Number of Passengers	Probability
0	0.2
1	0.1
2	0.2
3	0.3
4	0.15
5	0.05

The shuttle holds ten passengers. Develop a simulation model to find the probability that some passengers will not be able to board the shuttle and must wait for the next one. Simulate the arrival of 100 shuttles.

43. Dynamic Research (DR) is a company that produces software designed to enhance company creativity. The current stock price for the company is $9.75 per share. The daily changes in stock price over the last 500 trading days have been analyzed, resulting in the following frequency distribution:

Price Change	Frequency
−$5.00	1
−$1.00	20
−$0.50	55
−$0.38	95
$0.00	140
$0.38	100
$0.50	73
$1.00	10
$2.00	6

Develop a spreadsheet model to simulate the stock performance over the next 90 trading days. Use Monte Carlo simulation with 100 trials to replicate the spreadsheet and find the distribution of the ending stock price.

44. The Miller-Orr model in finance addresses a firm's problem of managing its cash position by purchasing or selling securities at a transaction cost in order to lower or raise its cash position. That is, the firm needs to have enough cash on hand to meet its obligations, but does not want to maintain too high a cash balance because it loses the opportunity for earning higher interest by investing in other securities. The Miller-Orr model assumes that the firm will maintain a minimum cash balance, m, a maximum cash balance, M, and an ideal level, R, called the return point. Cash is managed using a decision rule that states that whenever the cash balance falls to m, $R − m$ securities are sold to bring the balance up to the return point. When the cash balance rises to M, $M − R$ securities are purchased to reduce the cash balance back to the return point. Using some advanced mathematics, the return point and maximum cash balance levels are shown to be

$$R = m + Z$$
$$M = R + 2Z$$

where

$$Z = \left(\frac{3C\sigma^2}{4r} \right)^{1/3}$$

$C =$ fixed transaction cost to sell securities

$\sigma^2 =$ variance of the daily cash flows

$r =$ average daily rate of return corresponding to the premium associated with securities

For example, if the premium is 4%, $r = 0.04/365$. To apply the model, note that we do not need to know the actual demand for cash, only the daily variance. Essentially, the Miller-Orr model determines the decision rule that minimizes the expected costs of making the cash-security transactions and the expected opportunity costs of maintaining the cash balance based on the variance of the cash requirements.

Suppose that the daily requirements are normally distributed with a mean of 0 and variance of $60,000. Assume a transaction cost equal to $35, interest rate premium of 4%, and required minimum balance of $7,500. Develop a spreadsheet implementation for this model. Apply Monte-Carlo simulation to simulate the cash balance over the next year (365 days). Your simulation should apply the decision rule that if the cash balance for the current day is less than or equal to the minimum level, sell securities to bring the balance up to the return point. Otherwise, if the cash balance exceeds the upper limit, buy enough securities (that is, subtract an amount of cash) to bring the balance back down to the return point. If neither of these conditions hold, then there is no transaction and the balance for the next day is simply the current value plus the net requirement. Show the cash balance results on a line chart.

45. For the car wash simulation model (Example 12.18), use data tables to find the distributions for the maximum number in the queue, maximum waiting time, average waiting time per customer, and total idle time. Use 100 trials for each.

46. In many queueing situations, the time between arrivals and service times are assumed to have exponential distributions. Modify the car wash model (Example 12.18) to simulate a queueing system with an exponential arrival rate of $\lambda = 2$ customers/minute (that is, a time between arrivals of 1/2 minute) and an exponential service rate of $\lambda = 3$ customers per minute (that is, a service time of 1/3 minute). Simulate the arrival of 100 customers. Use data tables to find the distributions for the maximum number in the queue, maximum waiting time, average waiting time per customer, and total idle time. Use 100 trials for each. What happens when the arrival rate approaches the service rate?

CASE: PERFORMANCE LAWN EQUIPMENT

In one of PLE's manufacturing facilities, a drill press that has three drill bits is used to fabricate metal parts. Drill bits break occasionally and need to be replaced. The present policy is to replace a drill bit when it breaks or can no longer be used. The operations manager is considering a different policy in which all three drill bits are replaced when any one bit breaks or needs replacement. The rationale is that this would reduce downtime. It costs $200 each time the drill press must be shut down. A drill bit costs $85, and the variable cost of replacing a drill bit is $15 per bit. The company that supplies the drill bits has historical evidence that the reliability of a single drill bit is described by a Poisson probability distribution with the mean number of failures per hour equal to $\lambda = 0.01$. Thus, the time between failures is an exponential distribution with mean $\mu = 1/\lambda = 1/0.01 = 100$ hours. The operations manager at PLE would like to compare the cost of the two replacement policies. Develop spreadsheet simulation models to determine the total cost for each policy over 1,000 hours and make a recommendation.

Linear Optimization

Pinon Road/Shutterstock

LEARNING OBJECTIVES   After studying this chapter, you will be able to:

- Understand the three basic types of optimization models.
- Apply the four-step process to develop a mathematical model for an optimization problem.
- Recognize different types of constraints in problem statements.
- State the properties that characterize linear optimization models.
- Implement linear optimization models on spreadsheets.

- Use the *Solver* add-in to solve linear optimization models in Excel.
- Interpret the *Solver* Answer Report.
- Illustrate and solve two-variable linear optimization problems graphically.
- Explain how *Solver* works.
- List the four possible outcomes when solving a linear optimization model and recognize them from *Solver* messages.
- Formulate and solve linear optimization models for a variety of applications in business.

Up to now, we have concentrated on the role of descriptive analytics and predictive analytics in managerial decisions. In every area of business, managers want to make the best possible decisions. For example, marketing analysts want to choose the best advertising to attract the most customers; finance managers want to set the best prices to maximize profit; operations managers need to determine the best inventory and production policies. In your own life, you might want to find the best route for vacation travel (thank you, Google maps) or determine the best players for a fantasy sports team.

While many decisions involve only a limited number of alternatives and can be addressed using statistical analysis, simple spreadsheet models, or simulation, others have a very large or even an infinite number of possibilities. We introduced optimization—the fundamental tool in prescriptive analytics—in Chapter 1. **Optimization** is the process of selecting values of decision variables that *minimize* or *maximize* some quantity of interest and is the most important tool for prescriptive analytics.

Optimization models have been used extensively in operations and supply chains, finance, marketing, and other disciplines for more than 50 years to help managers allocate resources more efficiently and make lower-cost or more-profitable decisions. Optimization is a very broad and complex topic; in this chapter, we focus on formulating and solving many practical optimization models in business.

Optimization Models

There are three basic types of optimization models: linear, integer, and nonlinear. A **linear optimization model** (often called a **linear program**, or **LP model**) has two basic properties. First, the objective function and all constraints are linear functions of the decision variables. This means that each function is simply a sum of terms, each of which is some constant multiplied by a decision variable, such as $5x + 4y$. Linear optimization models are easy to solve using highly efficient solution algorithms. The second property of a linear optimization model is that all variables are continuous, meaning that they may assume any real value (typically, nonnegative, that is, greater than or equal to zero). Of course, this assumption may not be realistic for a practical business problem (you cannot produce half a refrigerator). However, because this assumption simplifies the solution method and analysis, we often apply it in many situations where the solution would not be seriously affected. For example, in deciding on the optimal number of cases of diapers to produce next month, we could use a linear model, since rounding a value like 5,621.63 would have little impact on the results. However, in a production-planning decision involving low-volume, high-cost items such as airplanes, an optimal value of 10.42 would make little sense, and a difference of one unit (rounded up or down) could have significant economic and production planning consequences.

In an **integer linear optimization model** (also called an **integer program**, or **IP model**), some of or all the variables are restricted to being whole numbers. A special type of integer problem is one in which variables can be only 0 or 1; these are used to model logical yes-or-no decisions. Integer linear optimization models are generally more difficult to solve than pure linear optimization models, but have many important applications in areas such as scheduling and supply chains.

Finally, there are many situations in which the relationship among variables in a model is not linear. Whenever either the objective function or a constraint is not linear, we have a **nonlinear optimization model** (also called a **nonlinear program**, or **NLP model**). In a nonlinear optimization model, the objective function and/or constraint functions are nonlinear functions of the decision variables; that is, terms cannot be written as a constant multiplied by a variable. Some examples of nonlinear terms are $3x^2$, $4/y$, and $6xy$. Building nonlinear optimization models requires more creativity and analytical expertise than linear or integer models; they also require different solution techniques. We will address integer and nonlinear models in the next chapter. Nonlinear models may also include integer restrictions; these are among the most difficult types of optimization models to solve. In this chapter, we will focus exclusively on linear optimization.

Linear optimization models are the most ubiquitous of optimization models used in organizations today. Applications abound in operations, finance, marketing, engineering, and many other disciplines. Table 13.1 summarizes some common types of generic linear optimization models. This list represents but a very small sample of the many practical types of linear optimization models that are used in practice throughout business. We will see examples of many of these later in this chapter.

▶ **Table 13.1**

Generic Examples of Linear Optimization Models

Type of Model	Decisions	Objective	Typical Constraints
Product mix	Quantities of product to produce and sell	Maximize contribution to profit	Resource limitations (for example, production time, labor, material); minimum sales requirements; maximum sales potential
Process selection	Quantities of product to make using alternative processes	Minimize cost	Demand requirements; resource limitations
Blending	Quantity of materials to mix to produce one unit of output	Minimize cost	Specifications on acceptable mixture
Portfolio selection	Proportions to invest in different financial instruments	Maximize future return or minimize risk exposure	Limit on available funds; sector requirements or restrictions; proportional relationships on investment mix
Transportation	Amount to ship between sources of supply and destinations	Minimize total transportation cost	Limited availability at sources; required demands met at destinations
Multiperiod production planning	Quantities of product to produce in each of several time periods; amount of inventory to hold between periods	Minimize total production and inventory costs	Limited production rates; material balance equations
Multiperiod financial management	Amounts to invest in short-term instruments	Maximize cash on hand	Cash balance equations; required cash obligations
Production/marketing	Allocation of advertising expenditures; production quantities	Maximize profit	Budget limitation; production limitations; demand requirements

CHECK YOUR UNDERSTANDING

1. What are the properties of linear, integer, and nonlinear optimization models?
2. State several examples of linear optimization models that are often used in practice.

ANALYTICS IN PRACTICE: Using Optimization Models for Sales Planning at NBC[1]

The National Broadcasting Company (NBC), a subsidiary of General Electric, is primarily in the business of delivering eyeballs (audiences) to advertisers. NBC's television network, cable network, TV stations, and Internet divisions generate billions of dollars in revenues. Of these, the television network business is by far the largest.

The television broadcast year in the United States starts in the third week of September. The broadcast networks announce their programming schedules for the new broadcast year in the middle of May. Shortly after that, the sale of inventory (advertising slots) begins. The broadcast networks sell about 60% to 80% of their airtime inventory during a brief period starting in late May and lasting about two to three weeks. This sales period is known as

the up-front market. During this time, advertising agencies approach the TV networks with requests to purchase time for their clients for the entire season. A typical request consists of the dollar amount, the demographic (for example, adults between 18 and 49 years of age) in which the client is interested, the program mix, weekly weighting, unit-length distribution, and a negotiated cost per 1,000 viewers. NBC must develop a detailed sales plan consisting of the schedule of commercials to be aired to meet the requirements. In addition, the plan should also meet the objectives of NBC's sales management, whose goal is to maximize the revenues for the available fixed amount of inventory.

Traditionally, NBC developed sales plans manually. This process was laborious, taking several hours. Moreover, most

bizoo_n/Fotolia

[1]Based on Srinivas Bollapragada, Hong Cheng, Mary Phillips, Marc Garbiras, Michael Scholes, Tim Gibb, and Mark Humphreville, "NBC's Optimization Systems Increase Revenues and Productivity," *Interfaces*, 32, 1 (January–February 2002): 47–60.

plans required a great deal of rework because, owing to their complexity, they initially met neither management's goals nor the customer's requirements. NBC developed a system using linear optimization that would generate sales plans quickly in a manner that made optimal use of the available inventory. The sales-planning problem was to minimize the amount of premium inventory assigned to a plan and the total penalty incurred in meeting goals, while meeting constraints on inventory, airtime availability, product conflicts, client requirements, budget, show mix, weekly weighting, and unit mix. The decision variables are the numbers of commercials of each spot length requested by the client that are to be placed in the shows and weeks included in the sales plan. The objective function includes a term that represents the total value of inventory assigned to the sales plan and terms that measure the penalties incurred in not meeting the client requirements these systems have provided.

The model and its implementation have saved millions of dollars of good inventory for NBC while meeting all the customer requirements; increased revenues; reduced the time needed to produce a sales plan from three to four hours to about 20 minutes; helped NBC to respond quickly to agencies and secure a greater share of the available money in the market; helped NBC sales managers to resolve deals more quickly than in the past and better read the market, resulting in a more accurate prediction of the upfront outcome; decreased rework on plans by more than 80%; and increased NBC's revenues by at least $50 million a year.

Developing Linear Optimization Models

Any optimization model has the following elements:

1. Decision variables
2. An objective to maximize or minimize
3. Constraints

Decision variables in an optimization model are the unknown values that the model seeks to determine. Depending on the application, decision variables might be the quantities of different products to produce, the amount of money spent on R&D projects, the amount to ship from a warehouse to a customer, the amount of shelf space to devote to a product, and so on. The quantity we seek to minimize or maximize is called the **objective function**; for example, we might wish to maximize profit or revenue, or minimize cost or some measure of risk. **Constraints** are limitations, requirements, or other restrictions that are imposed on any solution, either from practical or technological considerations or by management policy. The presence of constraints along with a large number of variables usually makes identifying an optimal solution considerably more difficult and necessitates the use of powerful software tools. The essence of building an optimization model is to first identify these model components, and then translate the objective function and constraints into mathematical expressions. Managers can generally describe the decisions they have to make, the performance measures they use to evaluate the success of their decisions, and the limitations and requirements they face or must ensure rather easily in plain language. The task of the analyst is to take this information and extract the key elements that form the basis for developing a model.

Developing any optimization model consists of three basic steps:

1. Identify the decision variables, the objective, and all appropriate constraints.
2. Write the objective and constraints as mathematical expressions to create a mathematical model of the problem.
3. Implement the mathematical model on a spreadsheet.

We will begin with a simple scenario to illustrate the development and spreadsheet implementation of a linear optimization model. Sklenka Ski Company (SSC) is a small manufacturer of two types of popular all-terrain snow skis, the Jordanelle and the Deercrest models. The manufacturing process consists of two principal departments: fabrication and finishing. The fabrication department has 12 skilled workers, each of whom works seven hours per day. The finishing department has three workers, who also work a seven-hour shift. Each pair of Jordanelle skis requires 3.5 labor-hours in the fabricating department and 1 labor-hour in finishing. The Deercrest model requires 4 labor-hours in fabricating and 1.5 labor-hours in finishing. The company operates five days per week. SSC makes a net profit of $50 on the Jordanelle model and $65 on the Deercrest model. In anticipation of the next ski-sale season, SSC must plan its production of these two models. Because of the popularity of its products and limited production capacity, its products are in high demand, and SSC can sell all it can produce each season. The company anticipates selling at least twice as many Deercrest models as Jordanelle models. The company wants to determine how many of each model should be produced on a daily basis to maximize net profit.

Identifying Decision Variables, the Objective, and Constraints

The first thing to do is to read the problem statement carefully and identify the decision variables, objective, and constraints in plain language before attempting to develop a mathematical model or a spreadsheet.

EXAMPLE 13.1   **Sklenka Ski Company: Identifying Model Components**

Step 1.   *Identify the decision variables.* SSC makes two different models of skis. The decisions are stated clearly: how many of each model ski should be produced each day? Thus, we may define

Jordanelle = Number of pairs of Jordanelle skis produced/day
Deercrest = Number of pairs of Deercrest skis produced/day

It is very important to clearly specify the dimensions of the variables, for example, "pairs of skis produced/day" rather than simply "Jordanelle skis."

Step 2.   *Identify the objective function.* The problem states that SSC wishes to maximize net profit, and we are given the net profit figures for each type of ski. In some problems, the objective is not explicitly stated, and we must use logic and business experience to identify the appropriate objective.

Step 3.   *Identify the constraints.* To identify constraints, look for clues in the problem statement that

describe limited resources that are available, requirements that must be met, or other restrictions. In this example, we see that both the fabrication and finishing departments have limited numbers of workers, who work only seven hours each day; this limits the amount of production time available in each department. Therefore, we have the following constraints:

Fabrication: Total labor-hours used in fabrication cannot exceed the amount of labor-hours available.
Finishing: Total labor-hours used in finishing cannot exceed the amount of labor-hours available.

In addition, the company anticipates selling at least twice as many Deercrest models as Jordanelle models. Thus, we need a constraint that states

Number of pairs of Deercrest skis must be at least twice the number of parts of Jordanelle skis.

Finally, we must ensure that negative values of the decision variables cannot occur. Nonnegativity constraints are assumed in nearly all optimization models.

Developing a Mathematical Model

The challenging part of developing optimization models is translating the descriptions of the objective and constraints into mathematical expressions. We usually represent decision variables by descriptive names (such as Jordanelle and Deercrest), abbreviations, or subscripted letters such as X_1 and X_2. For mathematical formulations involving many variables, subscripted letters are often more convenient; however, in spreadsheet models, we recommend using descriptive names to make the models and solutions easier to understand. In Example 13.2, we show the importance of specifying the dimension of the decision variables. This is extremely helpful to ensure the accuracy of the model.

EXAMPLE 13.2   **Sklenka Ski Company: Modeling the Objective Function**

The decision variables are the number of pairs of each type of ski to produce each day. Because SSC makes a net profit of $50 on the Jordanelle model and $65 on the Deercrest model, then, for example, if we produce 10 pairs of Jordanelle skis and 20 pairs of Deercrest skis during one day, we would make a profit of ($50/pair of Jordanelle skis)(10 pairs of Jordanelle skis) + ($65/pair of Jordanelle skis)(20 pairs of

Deercrest skis) = $500 + $1,300 = $1,800. Because we don't know how many pairs of skis to produce, we write each term of the objective function by multiplying the unit profit by the decision variables we have defined:

   Maximize Total Profit = $50 Jordanelle + $65 Deercrest

Note how the dimensions verify that the expression is correct: ($/pair of skis)(number of pairs of skis) = $.

Constraints are generally expressed mathematically as algebraic inequalities or equations with all variables on the left side and constant terms on the right (this facilitates solving the model on a spreadsheet, as we will discuss later). To model the constraints, we use a similar approach. First, consider the fabrication and finishing constraints. We expressed these constraints as

Fabrication: Total labor-hours used in fabrication cannot exceed the amount of labor hours available.
Finishing: Total labor-hours used in finishing cannot exceed the amount of labor hours available.

First, note that the phrase "cannot exceed" translates mathematically as "≤." In other constraints, we might find the phrase "at least," which would translate as "≥" or "must contain exactly," which would specify an "=" relationship. All constraints in optimization models must be one of these three forms.

Second, note that "cannot exceed" divides each constraint into two parts—the left-hand side ("total labor-hours used") and the right-hand side ("amount of labor-hours available"). The left-hand side of each of these expressions is called a **constraint function**. A constraint function is a function of the decision variables in the problem. The right-hand sides are numerical values (although occasionally they may be constraint functions as well). All that remains is to translate both the constraint functions and the right-hand sides into mathematical expressions.

EXAMPLE 13.3	Sklenka Ski Company: Modeling the Constraints

The amount of labor available in fabrication is (12 workers) × (7 hours/day) = 84 hours/day, whereas in finishing we have (3 workers) × (7 hours/day) = 21 hours/day. Because each pair of Jordanelle skis requires 3.5 labor-hours and each pair of Deercrest skis requires 4 labor-hours in the fabricating department, the total labor used in fabrication is 3.5 Jordanelle + 4 Deercrest. Note that the dimensions of these terms are (hours/pair of skis)(number of pairs of skis produced per day) = hours. Similarly, for the finishing department, the total labor used is 1 Jordanelle + 1.5 Deercrest. Therefore, the appropriate constraints are:

Fabrication: 3.5 Jordanelle + 4 Deercrest ≤ 84
Finishing: 1 Jordanelle + 1.5 Deercrest ≤ 21

For the market mixture constraint "Number of pairs of Deercrest skis must be at least twice the number of pairs of Jordanelle skis," we have

Deercrest ≥ 2 Jordanelle

It is customary to write all the variables on the left-hand side of the constraint. Thus, an alternative expression for this constraint is

Deercrest − 2 Jordanelle ≥ 0

The difference between the number of Deercrest skis and twice the number of Jordanelle skis can be thought of as the excess number of Deercrest skis produced over the minimum market mixture requirement. Finally, nonnegativity constraints are written as

Deercrest ≥ 0
Jordanelle ≥ 0

The complete mathematical model for the SSC problem is

Maximize *Total Profit* = 50 Jordanelle + 65 Deercrest

3.5 Jordanelle + 4 Deercrest ≤ 84
1 Jordanelle + 1.5 Deercrest ≤ 21
Deercrest − 2 Jordanelle ≥ 0
Deercrest ≥ 0
Jordanelle ≥ 0

More About Constraints

The most challenging aspect of model formulation is identifying constraints. Understanding the different types of constraints can help in proper identification and modeling. Constraints generally fall into one of the following categories:

- *Simple Bounds.* **Simple bounds** constrain the value of a single variable. You can recognize simple bounds in problem statements such as no more than $10,000 may be invested in stock ABC, or we must produce at least 350 units of product Y to meet customer commitments this month.
- *Limitations.* **Limitations** usually involve the allocation of scarce resources. Problem statements such as the amount of material used in production cannot exceed the amount available in inventory, minutes used in assembly cannot exceed the available labor hours, or the amount shipped from the Austin plant in July cannot exceed the plant's capacity are typical of these types of constraints.

- *Requirements*. **Requirements** involve the specification of minimum levels of performance. Such statements as enough cash must be available in February to meet financial obligations, production must be sufficient to meet promised customer orders, or the marketing plan should ensure that at least 400 customers are contacted each month are some examples.
- *Proportional Relationships*. **Proportional relationships** are often found in problems involving mixtures or blends of materials or strategies. Examples include the amount invested in aggressive growth stocks cannot be more than twice the amount invested in equity-income funds, or a mixture of fertilizer must contain exactly 30% nitrogen.
- *Balance Constraints*. **Balance constraints** essentially state that input = output and ensure that the flow of material or money is accounted for at locations or between time periods. Examples include production in June plus any available inventory must equal June's demand plus inventory held to July, the total amount shipped to a distribution center from all plants must equal the amount shipped from the distribution center to all customers, or the total amount of money invested or saved in March must equal the amount of money available at the end of February.

| EXAMPLE 13.4 | **Modeling Constraints** |

We illustrate each of these types of constraints in the following examples.

1. Simple bound: We must produce at least 350 units of product Y to meet customer commitments this month.
2. Limitation: The amount of money spent on research and development projects cannot exceed the assigned budget of $300,000.
3. Requirement: Contractual requirements specify that a total of at least 500 units of product must be shipped from factories in Austin and Atlanta.
4. Proportional relationship: A mixture of fertilizer must contain exactly 30% nitrogen.
5. Balance constraint: Available inventory and production in June must satisfy the demand of 150 units or be held over to July.

To model any constraint, first identify the phrase that corresponds to either ≤, ≥, or = and substitute these into the constraint. Thus, for these examples, we would write the following:

1. Amount of product Y ≥ 350
2. Amount spent on research and development ≤ $300,000
3. Number of units of product shipped from Austin and Atlanta ≥ 500
4. Amount of nitrogen in mixture/total amount in mixture = 0.30

5. Inventory and production in current month = demand and inventory held over to the next month

Then it simply becomes an exercise to translate the words into mathematical expressions using the decision variables in the problem. For instance:

1. Define *Product_Y* to be the number of units of product Y produced. Then the constraint is $Product\_Y \geq 350$
2. Define *R&DExpenses* to be the amount of money spent on research and development projects. Then the constraint is $R\&DExpenses \leq \$300,000$
3. Define X_1 = amount shipped from Austin and X_2 = amount shipped from Atlanta. Then the constraint is $X_1 + X_2 \geq 500$
4. Suppose that two ingredients contain 20% and 33% nitrogen, respectively; then the fraction of nitrogen in a mixture of *x* pounds of the first ingredient and *y* pounds of the second ingredient is expressed by $\frac{0.20x + 0.33y}{x + y}$. If the fraction of nitrogen in the mixture must be 0.30, then we would have $\frac{0.20x + 0.33y}{x + y} = 0.3$. Note that this constraint is actually nonlinear. However, we can convert it to a linear form using simple algebra. This can be rewritten as $0.20x + 0.33y = 0.3(x + y)$ and simplified as $-0.1x + 0.03y = 0$.
5. Define *I_June* = inventory available in June, *I_July* = inventory held over to July, and *P_June* = production in June. Then the constraint is $I\_June + P\_June = 150 + I\_July$.

Constraints in linear optimization models are generally some combination of constraints from these categories. Problem data or verbal clues in a problem statement often help you identify the appropriate constraint. In some situations, all constraints may not be explicitly stated, but are required for the model to represent the real problem accurately. An example of an implicit constraint is nonnegativity of the decision variables.

Implementing Linear Optimization Models on Spreadsheets

We will learn how to solve optimization models using an Excel tool called *Solver*. To facilitate the use of *Solver*, we suggest the following spreadsheet engineering guidelines for designing spreadsheet models for optimization problems:

- *Put the objective function coefficients, constraint coefficients, and right-hand values in a logical format in the spreadsheet.* For example, you might assign the decision variables to columns and the constraints to rows, much like the mathematical formulation of the model, and input the model parameters in a matrix. If you have many more variables than constraints, it might make sense to use rows for the variables and columns for the constraints.
- *Define a set of cells (either rows or columns) for the values of the decision variables.* In some models, it may be necessary to define a matrix to represent the decision variables. The names of the decision variables should be listed directly above the decision variable cells. Use shading or other formatting to distinguish these cells.
- *Define separate cells for the objective function and each constraint function (the left-hand side of a constraint).* Use descriptive labels directly above these cells.

EXAMPLE 13.5   A Spreadsheet Model for Sklenka Skis

Figure 13.1 shows a spreadsheet model for the SSC example. (Excel file *Sklenka Skis* already has the optimal solution. Typically, you would start with all decision variables equal to zero as shown in Figure 13.1.). We use the principles of spreadsheet The engineering that we discussed in Chapter 2 to implement the model. The *Data* portion of the spreadsheet provides the objective function coefficients, constraint coefficients, and right-hand sides of the model. Such data should be kept separate from the actual model so that if any data are changed, the model will automatically be updated. In the *Model* section, the number of each product

to make is given in cells B14 and C14. Also in the *Model* section are calculations for the constraint functions,

3.5 Jordanelle + 4 Deercrest (hours used in fabrication, cell D15)

1 Jordanelle + 1.5 Deercrest (hours used in finishing, cell D16)

Deercrest − 2 Jordanelle (market mixture, cell D19)

and the objective function, 50 Jordanelle + 65 Deercrest (cell D22).

To help you understand the correspondence between the mathematical model and the spreadsheet model more clearly, we will write the model in terms of formulas used in the spreadsheet cells,

$$\text{Maximize Profit} = D22 = B9*B14 + C9*C14$$

subject to the constraints

$$D15 = B6*B14 + C6*C14 \leq D6 \text{ (fabrication)}$$
$$D16 = B7*B14 + C7*C14 \leq D7 \text{ (finishing)}$$
$$D19 = C14 - 2*B14 \geq 0 \text{ (market mixture)}$$
$$B14 \geq 0, C14 \geq 0 \text{ (nonnegativity)}$$

	A	B	C	D
1	Sklenka Skis			
2				
3	Data			
4			Product	
5	Department	Jordanelle	Deercrest	Limitation (hours)
6	Fabrication	3.5	4	84
7	Finishing	1	1.5	21
8				
9	Profit/unit	$ 50.00	$ 65.00	
10				
11				
12	Model			
13		Jordanelle	Deercrest	
14	Quantity Produced	0	0	Hours Used
15	Fabrication	0	0	0
16	Finishing	0	0	0
17				
18				Excess Deercrest
19	Market mixture			0
20				
21				Total Profit
22	Profit Contribution	$ -	$ -	$ -

	A	B	C	D
1	Sklenka Skis			
2				
3	Data			
4			Product	
5	Department	Jordanelle	Deercrest	Limitation (hours)
6	Fabrication	3.5	4	84
7	Finishing	1	1.5	21
8				
9	Profit/unit	50	65	
10				
11				
12	Model			
13		Jordanelle	Deercrest	
14	Quantity Produced	0	0	Hours Used
15	Fabrication	=B6*$B$14	=C6*$C$14	=B15+C15
16	Finishing	=B7*$B$14	=C7*$C$14	=B16+C16
17				
18				Excess Deercrest
19	Market mixture			=C14-2*B14
20				
21				Total Profit
22	Profit Contribution	=B9*$B$14	=C9*$C$14	=B22+C22

▲ **Figure 13.1**

Sklenka Skis Model Spreadsheet Implementation

Observe how the constraint functions and right-hand-side values are stored in separate cells within the spreadsheet.

In Excel, the pairwise sum of products of terms can easily be computed using the SUMPRODUCT function. For example, the objective function

$$=B9*B14 + C9*C14 \text{ is equivalent to } =SUMPRODUCT(B9:C9, B14:C14)$$

Similarly, for the labor limitation constraints

$$=B6*B14 + C6*C14 \text{ is equivalent to } =SUMPRODUCT(B6:C6, B14:C14)$$

$$=B7*B14 + C7*C14 \text{ is equivalent to } =SUMPRODUCT(B7:C7, B14:C14)$$

The SUMPRODUCT function often simplifies the model-building process, particularly when many variables are involved.

We should note that optimization models that we develop can be used in all phases of analytics—descriptive, predictive, and prescriptive. For example, we can use the model to evaluate the profit and utilization of resources in a descriptive setting to answer the question "What are we doing now?" We might use the model in a predictive setting to evaluate forecasted cost increases or the effects of inflation in the future. Finally, we can ask "What is the best we can do with our current resources?" In this way, the model can be used as a prescriptive model.

Excel Functions to Avoid in Linear Optimization

Several common functions in Excel can cause difficulties when attempting to solve linear programs using *Solver* because they are discontinuous (or "nonsmooth") and do not satisfy the conditions of a linear model. For instance, in the formula IF(A12 < 45, 0, 1), the cell value jumps from 0 when the value of cell A12 is less than 45, to 1 when the value of cell A12 is 45 or more. In such situations, the correct solution may not be identified. Common Excel functions to avoid are ABS, MIN, MAX, INT, ROUND, IF, and COUNT. Although these are useful in general modeling tasks with spreadsheets, you should avoid them in linear optimization models.

CHECK YOUR UNDERSTANDING

1. Explain the steps in developing a linear optimization model.

2. List the different categories of constraints that one might find in a linear optimization model.

3. What guidelines should you follow when implementing linear optimization models on spreadsheets?

Solving Linear Optimization Models

To solve an optimization problem, we seek values of the decision variables that maximize or minimize the objective function and also satisfy all constraints. Any solution that satisfies all constraints of a problem is called a **feasible solution**. Finding an optimal solution among the infinite number of possible feasible solutions to a given problem is not an easy task. A simple approach is to try to manipulate the decision variables in the spreadsheet model to find the best solution possible; however, for many problems, it might be very difficult to find a feasible solution, let alone an optimal solution. You might try to find the best solution you can for the Sklenka Ski problem by using the spreadsheet model. With a little experimentation and perhaps a bit of luck, you might be able to zero in on the optimal solution or something close to it. However, to guarantee finding an optimal solution, some type of systematic mathematical solution procedure is necessary. Fortunately, such a procedure is provided by the Excel *Solver* tool, which we discuss next.

Solver is an add-in packaged with Excel that was developed by Frontline Systems, Inc. (www.solver.com), and can be used to solve many different types of optimization problems. *Solver* can be found in the *Analysis* group under the *Data* tab in Excel. When *Solver* is invoked, the *Solver Parameters* dialog appears. You use this dialog to define the objective, decision variables, and constraints from your spreadsheet model within *Solver*.

EXAMPLE 13.6   **Using *Solver* for the SSC Problem**

Figure 13.2 shows the completed *Solver Parameters* dialog for the SSC example. Define the objective function cell in the spreadsheet (D22) in the *Set Objective* field. Either enter the cell reference or click within the field and then in the cell in the spreadsheet. Click the appropriate radio button for *Max* or *Min*. Decision variables (cells B14 and C14) are entered in the field called *By Changing Variable Cells*; click within this field and highlight the range corresponding to the decision variables in your spreadsheet.

To enter a constraint, click the *Add* button. A new dialog, *Add Constraint*, appears (see Figure 13.3). In the left field, *Cell Reference*, enter the cell that contains the constraint function (left-hand side of the constraint). For example, the constraint function for the fabrication constraint is in cell D15. Make sure that you select the correct type of constraint (≤, ≥, or =) in the drop-down box in the middle of the dialog. The other options are

discussed in the next chapter. In the right field, called *Constraint*, enter the numerical value of the right-hand side of the constraint or the cell reference corresponding to it. For the fabrication constraint, this is cell D6. Figure 13.3 shows the completed dialog for the fabrication constraint. To add other constraints, click the *Add* button.

You may also define a group of constraints that all have the same algebraic form (all ≤, all ≥, or all =) and enter them together. For example, the department resource limitation constraints are expressed within the spreadsheet model as

$$D15 \leq D6$$
$$D16 \leq D7$$

Because both constraints are ≤ types, we could define them as a group by entering the range D15:D16 in the *Cell Reference* field and D6:D7 in the *Constraint* field to simplify the input process. When all constraints are added, click *OK* to return to the *Solver Parameters* dialog box. You may add, change, or delete these as necessary by clicking the appropriate buttons. You need not enter nonnegativity constraints explicitly. Just check the box in the dialog *Make Unconstrained Variables Non-Negative*.

For linear optimization problems, it is very important to select the correct solving method. The standard Excel *Solver* provides three options for the solving method:

1. *GRG Nonlinear*—used for solving nonlinear optimization problems
2. *Simplex LP*—used for solving linear and linear integer optimization problems

3. *Evolutionary*—used for solving complex nonlinear and nonlinear integer problems

In the field labeled *Select a Solving Method*, choose *Simplex LP*. Then click the *Solve* button to solve the problem. The *Solver Results* dialog appears, as shown in Figure 13.4, with the message "Solver found a solution." If a solution could not be found, *Solver* would notify you with a message to this effect. This generally means that you have an error in your model or you have included conflicting constraints that no solution can satisfy. In such cases, you need to reexamine your model.

Solver generates three reports, as listed in Figure 13.4: Answer, Sensitivity, and Limits. To add them to your Excel workbook, click on the ones you want and then click *OK*. Do *not* check the box *Outline Reports*; this is an Excel feature that produces the reports in "outlined format." After you press *OK, Solver* will replace the current values of the decision variables and the objective in the spreadsheet with the optimal solution. For the SSC

▶ **Figure 13.2**

Solver Parameters *Dialog*

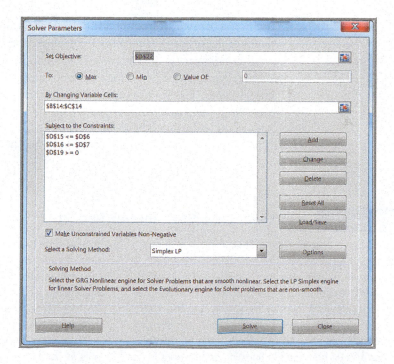

▶ **Figure 13.3**

Add Constraint *Dialog*

▶ **Figure 13.4**

Solver Results *Dialog*

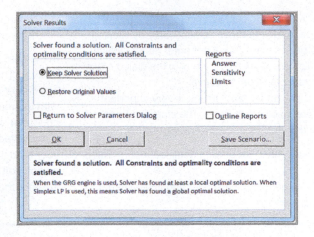

problem, the maximum profit is $945, obtained by producing 5.25 pairs of Jordanelle skis and 10.5 pairs of Deercrest skis per day (remember that linear models allow fractional values for the decision variables). If you save your spreadsheet after setting up a *Solver* model, the *Solver* model will be saved also.

Solver Answer Report

The *Solver* Answer Report provides basic information about the solution, including the values of the original and optimal objective function (in the *Objective Cell* section) and decision variables (in the *Decision Variable Cells* section). In the *Constraints* section, *Cell Value* refers to the value of the constraint function using the optimal values of the decision variables. The *Status* column tells whether each constraint is binding or not binding. A **binding constraint** is one for which the *Cell Value* is equal to the right-hand side of the value of the constraint. *Slack* refers to the difference between the left- and right-hand sides of the constraints for the optimal solution. We discuss the sensitivity and limits reports in Chapter 15.

EXAMPLE 13.7   **Interpreting the SSC Answer Report**

The *Solver* Answer Report for the SSC problem is shown in Figure 13.5. The *Objective Cell* section provides the optimal value of the objective function, $945. The *Decision Variable Cells* section lists the optimal values of the decision variables: 5.25 pairs of Jordanelle skis and 10.5 pairs of Deercrest skis. In the *Constraints* section, the *Cell Values* state that we used 60.375 hours in the fabrication department and 21 hours in the finishing department by producing 5.25 pairs of Jordanelle skis and 10.5 pairs of Deercrest skis. You may easily identify the constraints from the spreadsheet model in the Formulas column. From the Status column, we see that the constraint for fabrication is not binding, although the constraints for finishing and market mixture are binding. This means that there is

excess time that is not used in fabrication; this value is shown in the Slack column as 23.626 hours. For finishing, we used all the time available; hence, the slack value is zero. Because we produced exactly twice the number of Deercrest skis as Jordanelle skis, the market mixture constraint is binding. It would not have been binding if we had produced more than twice the number of Deercrest skis as Jordanelle.

To understand the value of slack better, examine the fabrication constraint:

$$3.5 \text{ Jordanelle} + 4 \text{ Deercrest} \leq 84$$

We interpret this as

Number of Fabrication Hours Used ≤ Hours Available

Note that if the amount used is strictly less than the availability, we have slack, which represents the amount unused; thus,

Number of Fabrication Hours Used $+$ Number of Fabrication Hours Unused $=$ Hours Available

or

Slack $=$ Number of Hours Unused

$=$ Hours Available $-$ Number of Fabrication Hours Used

$= 84 - (3.5 \times 5.25 + 4 \times 10.5) = 23.625$

Slack variables are always nonnegative, so for $\geq$ constraints, slack represents the difference between the left-hand side of the constraint function and the right-hand side of the requirement. The slack on a binding constraint will always be zero.

Graphical Interpretation of Linear Optimization with Two Variables

We can easily illustrate optimization problems with two decision variables graphically. This can help you to better understand the properties of linear optimization models and the interpretation of the *Solver* output. Recall that a feasible solution is a set of values for the decision variables that satisfy all of the constraints. Linear programs generally have an infinite number of feasible solutions. We first characterize the set of feasible solutions, often called the **feasible region**. We use the SSC model to illustrate this graphical approach:

$$\text{Maximize } \textit{Total Profit} = 50 \text{ Jordanelle} + 65 \text{ Deercrest}$$
$$3.5 \text{ Jordanelle} + 4 \text{ Deercrest} \leq 84$$
$$1 \text{ Jordanelle} + 1.5 \text{ Deercrest} \leq 21$$
$$\text{Deercrest} - 2 \text{ Jordanelle} \geq 0$$
$$\text{Deercrest} \geq 0$$
$$\text{Jordanelle} \geq 0$$

For a problem with only two decision variables, x_1 and x_2, we can draw the feasible region on a two-dimensional coordinate system. Let us begin by considering the simplest constraints in a linear optimization model, namely, that the decision variables must be nonnegative. These constraints are $x_1 \geq 0$ and $x_2 \geq 0$. The constraint $x_1 \geq 0$ corresponds to

▶ **Figure 13.5**

Solver *Answer Report*

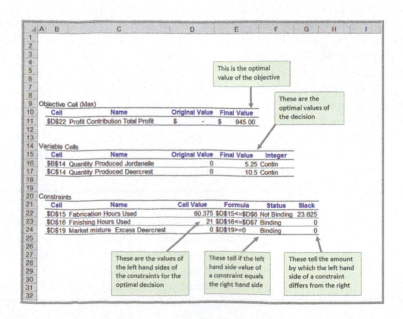

all points on or to the right of the x_2-axis; the constraint $x_2 \geq 0$ corresponds to all points on or above the x_1-axis (see Figure 13.6, where x_1 = Jordanelle and x_2 = Deercrest). Taken together, these nonnegativity restrictions imply that any feasible solution must be restricted to the first (upper-right) quadrant of the coordinate system. This is true for the feasible solutions to the SSC problem.

You are probably very familiar with equations in two dimensions, which define points on a line. An inequality constraint divides the coordinate system into two regions, the set of points that do satisfy the inequality and the set of points that don't. In two dimensions, an equality constraint is simply a line. To graph a line in two dimensions, we need to find two points that lie on the line. As long as the right-hand side term is not zero, the two points that are easiest to find are the x_1- and x_2-intercepts (the points where the line crosses the x_1- and x_2-axes). To find the x_2-intercept, set $x_1 = 0$ and solve for x_2. Likewise, to find the x_1-intercept, set $x_2 = 0$ and solve for x_1.

EXAMPLE 13.8 Graphing the Constraints in the SSC Problem

The fabrication constraint is 3.5 Jordanelle + 4 Deercrest $\leq$ 84. Whenever a constraint is in the form of an inequality (that is, $\geq$ or $\leq$ type), we first graph the equation of the line by replacing the inequality sign with an equal sign. Therefore, we graph the equation 3.5 Jordanelle + 4 Deercrest = 84. If we set Jordanelle = 0, then solving the equation for Deercrest yields Deercrest = 21. Similarly, if we set Deercrest = 0, we find that Jordanelle = 24. This gives us two points, (0, 21) and (24, 0), on the coordinate system and defines the equation of the straight line, as shown in Figure 13.7.

However, the actual constraint is an inequality; therefore, all the points on one side of the line will satisfy the constraint, but points on the other side will not. To identify the proper direction, simply select any point not on the line—the easiest one to choose is the origin, (0, 0)—and determine if that point satisfies the constraint. If it does, then all points on that side of the line will also satisfy the constraint; if not, then all points on the *other* side of the line must satisfy the constraint. Clearly, 3.5(0) + 4(0) = 0 < 84; therefore, all points below the constraint line satisfy the inequality. In mathematical terms, the set of points on one side of a line is called a *half-space*. Only points lying in this half-space can be potential solutions to the optimization model.

To graph the finishing constraint 1 Jordanelle + 1.5 Deercrest $\leq$ 21, we follow the same procedure.

Set Jordanelle = 0 and solve for Deercrest, obtaining Deercrest = 14; set Deercrest = 0 and solve for Jordanelle, obtaining Jordanelle = 21. Choosing the origin again verifies that all points below the line satisfy the inequality constraint. This is shown in Figure 13.8.

The third constraint is the market mix constraint: Deercrest − 2 Jordanelle $\geq$ 0. If we try to set each variable in the equation Deercrest − 2 Jordanelle = 0 to zero and solve for the other, we end up with (0, 0) each time because the equation of the line passes through the origin. When this occurs, we need to select a different value for one of the variables to identify a second point on the line. For example, if we set Jordanelle = 5, then Deercrest = 10. Now we have two points, (0, 0) and (5, 10), which we can use to graph the equation (see Figure 13.9). However, since the line passes through the origin, we cannot determine the proper half-space using the origin (0, 0). Instead, choose any other point not on the line. For example, if we choose the point (2, 10), which is on the left side of the line, we see that Deercrest − 2 Jordanelle = 10 − 2(2) = 6 > 0; therefore, all points to the left of the line satisfy the inequality constraint. Had we chosen a point on the right, say, (5, 2), we would have found that Deercrest − 2 Jordanelle = 2 − 2(5) = −8 < 0, which does not satisfy the inequality.

After graphing each of the constraints, we identify the feasible region. For a linear optimization problem, the feasible region will be some geometric shape that is bounded by straight lines. The points at which the constraint lines intersect along the feasible region are called **corner points**. One of the important properties of linear optimization models is that if an optimal solution exists, then it will occur at a corner point. This makes it easy to identify optimal solutions and is the basis for the computational procedure used by *Solver*.

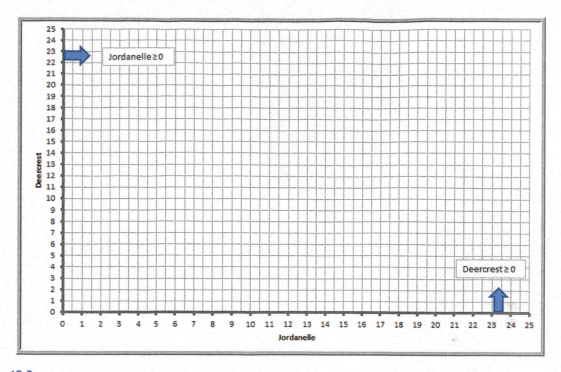

▲ **Figure 13.6**

Feasible Points Satisfying Nonnegativity Constraints

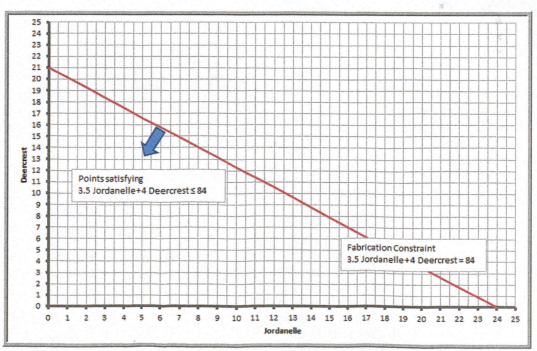

▲ **Figure 13.7**

Graph of the Fabrication Constraint

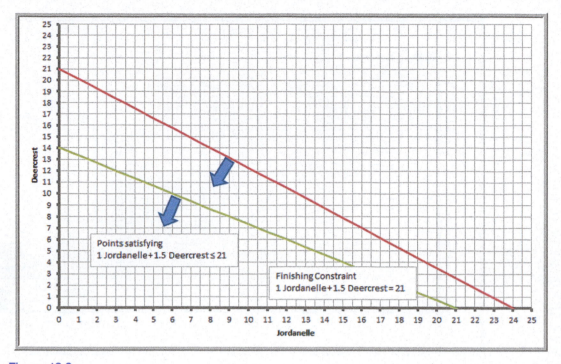

▲ **Figure 13.8**

Graph of the Finishing Constraint

EXAMPLE 13.9	**Identifying the Feasible Region and Optimal Solution**

The feasible region is the set of points that satisfy all constraints simultaneously. From Figure 13.9, we see that the feasible region must be below the fabrication constraint line, below the finishing constraint line, to the left of the market mix constraint line, and, of course, within the first quadrant defined by the nonnegativity constraints. This is shown by the triangular region in Figure 13.10. Notice that every point that satisfies the finishing constraint also satisfies the fabrication constraint. In this case, we say that the fabrication constraint is a *redundant constraint* because it does not impact the feasible region at all.

Because our objective is to maximize profit, we seek a corner point that has the largest value of the objective function total profit = 50 Jordanelle + 65 Deercrest. Note that if we set the objective function to any numerical value, we define a straight line. For example, if we set 50 Jordanelle + 65 Deercrest = 600, then any point on this line will have a total profit of $600. Figure 13.11 shows the dashed-line graphs of the objective function for profit values of $600, $800, and $1,000. Notice that as the profit increases, the graph of the objective function moves in an upward direction. However, for a profit of $1,000, no points

on the line also pass through the feasible region. From the figure, then, we can conclude that the maximum profit must be somewhere between $800 and $1,000.

We also see that as the profit increases, the last point in the feasible region that the profit lines will cross is the corner point on the right side of the triangle, identified by the circle in Figure 13.11. This must be the optimal solution. This point is the intersection of the finishing and market mix constraint lines. We can find this point mathematically by solving these constraint lines simultaneously:

$$1 \text{ Jordanelle} + 1.5 \text{ Deercrest} = 21$$

$$\text{Deercrest} - 2 \text{ Jordanelle} = 0$$

From the second equation, we have Deercrest = 2 Jordanelle; substituting this into the first equation, we obtain

$$1 \text{ Jordanelle} + 1.5(2 \text{ Jordanelle}) = 21$$

$$4 \text{ Jordanelle} = 21$$

$$\text{Jordanelle} = 5.25$$

Then Deercrest = 2(5.25) = 10.5. This is exactly the solution that *Solver* provided.

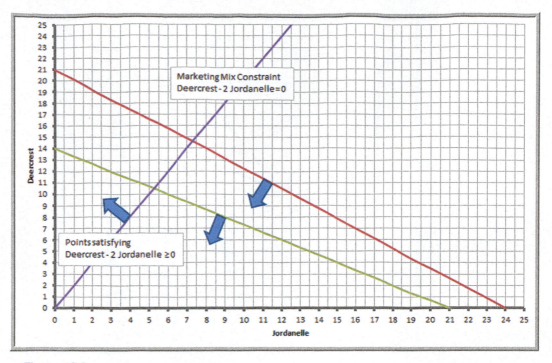

▲ **Figure 13.9**

Graph of the Market Mix Constraint

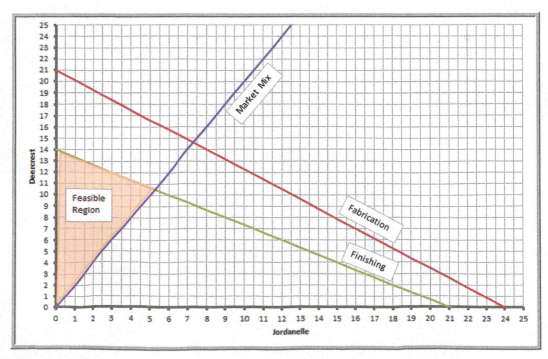

▲ **Figure 13.10**

Identifying the Feasible Region

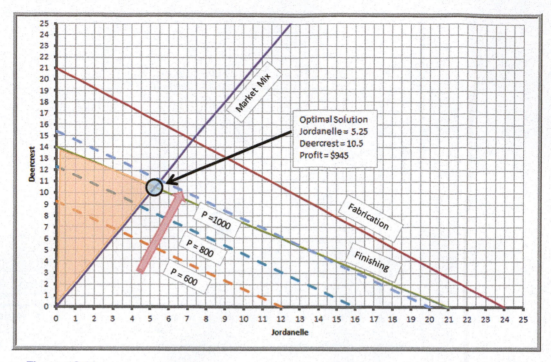

▲ **Figure 13.11**

Identifying the Optimal Solution

Compare the graphical interpretation of the solution to the SSC problem with the *Solver* Answer Report in Figure 13.5. Notice that *Solver* reported that both the finishing constraint and market mix constraint are binding. Graphically, this means that these constraints intersect at the optimal solution. The fabrication constraint, however, is not binding and has a positive value of slack because it does not intersect at the optimal solution. Slack can be interpreted as a measure of the distance from the optimal corner point to the nonbinding constraint.

CHECK YOUR UNDERSTANDING

1. What is a feasible solution?

2. Explain how to use *Solver* to solve a linear optimization model on a spreadsheet.

3. What information is provided in the *Solver* Answer Report?

4. Explain how to graphically visualize linear optimization models with two variables.

How *Solver* Works

Solver uses a mathematical algorithm called the *simplex method*, which was developed in 1947 by the late Dr. George Dantzig. The simplex method characterizes feasible solutions algebraically by solving systems of linear equations. It moves systematically from one corner point to another to improve the objective function until an optimal solution is found (or until the problem is deemed infeasible or unbounded). Because of the linearity of the constraints and objective function, the simplex method is guaranteed to find an optimal solution if one exists and usually does so quickly and efficiently. To gain some intuition into the logic of *Solver*, consider the following example.

EXAMPLE 13.10 **Crebo Manufacturing**

Crebo Manufacturing produces four types of structural support fittings—plugs, rails, rivets, and clips—which are machined on two CNC machining centers. The machining centers have a capacity of 280,000 minutes per year. The gross margin per unit and machining requirements are provided in the table below. How many of each product should be made to maximize gross profit margin?

To formulate this as a linear optimization model, define X_1, X_2, X_3, and X_4 to be the number of plugs, rails, rivets,

and clips, respectively, to produce. The problem is to maximize gross margin = $0.3X_1 + 1.3X_2 + 0.75X_3 + 1.2X_4$ subject to the constraint that limits the machining capacity and nonnegativity of the variables:

$$1X_1 + 2.5X_2 + 1.5X_3 + 2X_4 \leq 280{,}000$$

$$X_1, X_2, X_3, X_4 \geq 0$$

Product	Plugs	Rails	Rivets	Clips
Gross margin/unit	$0.30	$1.30	$0.75	$1.20
Minutes/unit	1	2.5	1.5	2

To solve this problem, your first thought might be to choose the variable with the highest marginal profit. Because X_2 has the highest marginal profit, you might try producing as many rails as possible. Since each rail requires 2.5 minutes, the maximum number that can be produced is $280{,}000/2.5 = 112{,}000$, for a total profit of $\$1.3(112{,}000) = \$145{,}600$. However, notice that each rail uses a lot more machining time than the other products. The best solution isn't necessarily the one with the highest marginal profit, but the one that provides the highest *total* profit. Therefore, more profit might be realized by producing a proportionately larger quantity of a different product having a smaller marginal profit. This is the key insight. What the simplex method essentially does is evaluate the impact of constraints in terms of their contribution to the objective function for each variable. For the simple case of only one constraint, the optimal (maximum) solution is found by simply choosing the variable with the highest ratio of the objective coefficient to the constraint coefficient.

EXAMPLE 13.11 **Solving the Crebo Manufacturing Model**

In the *Crebo Manufacturing Model*, compute the ratio of the gross margin/unit to the minutes per unit of machining capacity used, as shown in row 6 in Figure 13.12 (Excel file *Crebo Manufacturing Model*). These ratios can be interpreted as the marginal profit per unit of resource

consumed. The highest ratio occurs for clips. If we produce the maximum number of clips, $280{,}000/2 = 140{,}000$, the total profit is $\$1.20(140{,}000) = \$168{,}000$. The mathematics gets complicated with more constraints and requires multiple iterations to systematically improve the solution.

▶ **Figure 13.12**

Crebo Manufacturing Model
Analysis

	A	B	C	D	E	F
1	Crebo Manufacturing Model					
2						
3	Product	Plugs (X1)	Rails (X2)	Rivets (X3)	Clips (X4)	Machine Capacity
4	Gross margin/unit	$0.30	$1.30	$0.75	$1.20	
5	Minutes/unit	1	2.5	1.5	2	280,000
6	Gross margin/minute	$0.30	$0.52	$0.50	$0.60	
7	Maximum production	280,000.00	112,000.00	186,666.67	140,000.00	
8	Profit	$84,000	$145,600	$140,000	$168,000	

If we apply similar logic to the SSC problem, we would at first want to produce as many Deercrest skis as possible because they have the largest profit contribution. So for example, if we do, we find that the constraints limit us to the minimum of $84/4 = 21$ units (from the fabrication constraint) or $21/1.5 = 14$ (from the finishing constraint). Note that producing 14 Deercrest skis will also satisfy the market mix constraint. The total profit is $65(14) = 910. However, note that finishing requires 50% more time for Deercrest skis than for Jordanelle skis, so the profit contribution per finishing hour for Deercrest is only $65/1.5 = 43.33, so on a relative basis, the Jordanelle skis are more profitable. Thus, for example, if we produce 1 Jordanelle ski, we can produce $20/1.5 = 13.33$ Deercrest skis, for a total profit of $50(1) + $65(13.33) = 916.67, an increase of $6.67. Similarly, if we produce 2 Jordanelle skis, we can produce 12.67 Deercrest with a total profit of $923.33. If we continue to produce more Jordanelle skis, the profit will continue to increase, but the ratio of Jordanelle to Deercrest also gets larger, and eventually we will violate the market mix constraint. This occurs when more than 5.25 Jordanelle skis are produced. At this point, we have the maximum profit.

Of course, for problems involving many constraints, it is difficult to apply such intuitive logic. The simplex method allows many real business problems involving thousands or even millions of variables—and often hundreds or thousands of constraints—to be solved in reasonable computational time and is the basis for advanced optimization algorithms involving integer variables that we describe in the next chapter.

How *Solver* Creates Names in Reports

How you design your spreadsheet model will affect how *Solver* creates the names used in the output reports. Poor spreadsheet design can make it difficult or confusing to interpret the Answer and other *Solver* reports. Thus, it is important to understand how to do this properly.

Solver assigns names to target cells, changing cells, and constraint function cells by concatenating the text in the first cell containing text to the left of the cell with the first cell containing text above it. For example, in the SSC model in Figure 13.1, the target cell is D22. The first cell containing text to the left of D22 is "Profit Contribution" in A22, and the first cell containing text above D22 is "Total Profit" in cell D21. Concatenating these text strings yields the target cell name "Profit Contribution Total Profit," which is found in the *Solver* reports. The constraint functions are calculated in cells D15 and D16. Note that their report names are "Fabrication Hours Used" and "Finishing Hours Used." Similarly, the changing cells in B14 and C14 have the names "Quantity Produced Jordanelle" and "Quantity Produced Deercrest." These names make it easy to interpret the information in the Answer and Sensitivity Reports. We encourage you to examine each of the target cells, changing variable cells, and constraint function cells in your models carefully so that report names are properly established.

CHECK YOUR UNDERSTANDING

1. Explain the intuitive ideas behind the "simplex method" used by *Solver*.
2. How does *Solver* create range names in its reports?

Solver Outcomes and Solution Messages

Solving a linear optimization model can result in four possible outcomes:

1. a unique optimal solution
2. alternative (multiple) optimal solutions
3. an unbounded solution
4. infeasibility

Unique Optimal Solution

When a model has a **unique optimal solution**, it means that there is exactly one solution that will result in the maximum (or minimum) objective. The solution to the SSC model is unique; there are no solutions other than producing 5.25 pairs of Jordanelle skis and 10.5 pairs of Deercrest skis that result in the maximum profit of $945. We could see this graphically in Figure 13.11 because there is a unique corner point that lies on the objective function line at the optimal value of profit.

Alternative (Multiple) Optimal Solutions

If a model has **alternative optimal solutions**, the objective is maximized (or minimized) by more than one combination of decision variables, all of which have the same objective function value. *Solver* does not tell you when alternative solutions exist and reports only one of the many possible alternative optimal solutions. However, you can use the Sensitivity Report information to identify the existence of alternative optimal solutions. When any of the Allowable Increase or Allowable Decrease values for changing cells are zero, then alternative optimal solutions exist, although *Solver* does not provide an easy way to find them.

EXAMPLE 13.12 **A Model with Alternative Optimal Solutions**

To illustrate a model with alternative optimal solutions, suppose we change the objective function in the SSC model to Max 50 Jordanelle + 75 Deercrest. A solution obtained using *Solver* is shown in Figure 13.13, producing no Jordanelle skis and 14 pairs of Deercrest skis and resulting in a profit of $1,050. However, notice that the original optimal solution also has the same objective function value: profit = $50(5.25) + $75(10.5) = $1,050.

This may be seen graphically in Figure 13.14. The new objective function lines are parallel to the finishing constraint line. Thus, as the profit increases, you can see that the profit line must stop along the top boundary of the feasible region defined by the finishing constraint. Both corner points that are circled are optimal solutions, as is any point connecting them. Therefore, when alternative optimal solutions exist, there actually are an infinite number of them; however, identifying them other than graphically requires some advanced analysis.

Unbounded Solution

A solution is **unbounded** if the value of the objective can be increased or decreased without bound (that is, to infinity for a maximization problem or negative infinity for a minimization problem) without violating any of the constraints. This generally indicates an incorrect model, usually when some constraint or set of constraints have been left out.

▶ **Figure 13.13**

A Solution to the SSC Problem with Modified Objective

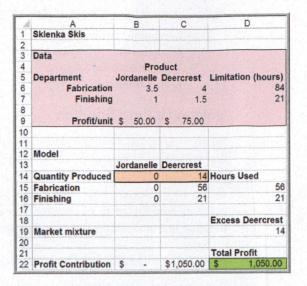

	A	B	C	D
1	Sklenka Skis			
2				
3	Data			
4			Product	
5	Department	Jordanelle	Deercrest	Limitation (hours)
6	Fabrication	3.5	4	84
7	Finishing	1	1.5	21
8				
9	Profit/unit	$ 50.00	$ 75.00	
10				
11				
12	Model			
13		Jordanelle	Deercrest	
14	Quantity Produced	0	14	Hours Used
15	Fabrication	0	56	56
16	Finishing	0	21	21
17				
18				Excess Deercrest
19	Market mixture			14
20				
21				Total Profit
22	Profit Contribution	$ -	$1,050.00	$ 1,050.00

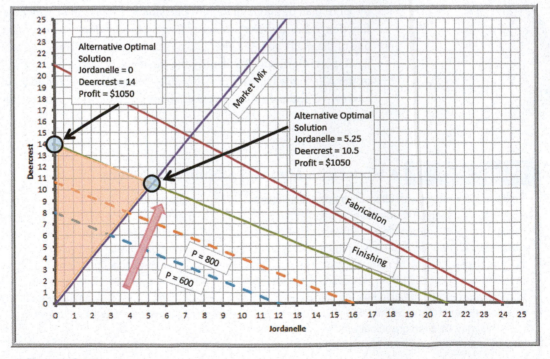

▲ **Figure 13.14**

Graph of Alternative Optimal Solutions

EXAMPLE 13.13   A Model with an Unbounded Solution

Suppose that we solve the SSC model without the fabrication or finishing constraints:

Maximize *Total Profit* = 50 Jordanelle + 65 Deercrest
Deercrest − 2 Jordanelle ≥ 0
Deercrest ≥ 0
Jordanelle ≥ 0

Figure 13.15 shows the *Solver Results* dialog; the message "The objective (Set Cell) values do not converge" is an indication that the solution is unbounded. This can easily be seen graphically in Figure 13.16. Without the finishing and fabrication constraints, the feasible region extends upward in the shaded triangular region with no limit. As the profit values increase, there are no boundary lines to stop the objective function from getting larger and larger. However, it is important to realize that even though the feasible region is unbounded, the problem can have a finite optimal solution if the profit lines move in a different direction.

Infeasibility

Finally, an **infeasible problem** is one for which no feasible solution exists—that is, when there is no solution that satisfies all constraints simultaneously. When a problem is infeasible, *Solver* will report "Solver could not find a feasible solution." Infeasible problems *can* occur in practice—for example, when a demand requirement is higher than available capacity or when managers in different departments have conflicting requirements or limitations. In such cases, the model must be reexamined and modified. Sometimes infeasibility, or unboundedness, is simply a result of a misplaced decimal, an incorrect inequality sign, or other error in the model or spreadsheet implementation, so accuracy checks should be made.

EXAMPLE 13.14   An Infeasible Model

Suppose the modeler for the SSC problem mistakenly reversed the inequality sign for the fabrication constraint:

Maximize *Total Profit* = 50 Jordanelle + 65 Deercrest
3.5 Jordanelle + 4 Deercrest ≥ 84
1 Jordanelle + 1.5 Deercrest ≤ 21
Deercrest − 2 Jordanelle ≥ 0
Deercrest ≥ 0
Jordanelle ≥ 0

Figure 13.17 shows the *Solver Results* dialog for this model. When *Solver* provides the message "Solver could not find a feasible solution," then we know the problem is infeasible. Figure 13.18 shows what happened graphically. The points satisfying the erroneous fabrication constraint lie above the constraint and do not intersect the points that are feasible to the market mix and finishing constraints.

▶ **Figure 13.15**

Solver Results Dialog for *Unbounded Problem*

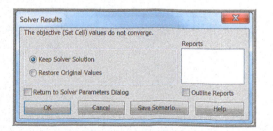

▶ **Figure 13.16**

An Unbounded Feasible Region

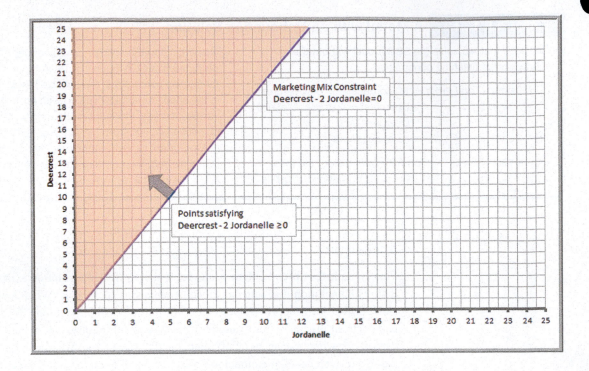

▶ **Figure 13.17**

Solver Results *Dialog for Infeasible Solution*

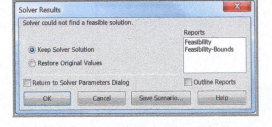

▶ **Figure 13.18**

Graphical Illustration of Infeasibility

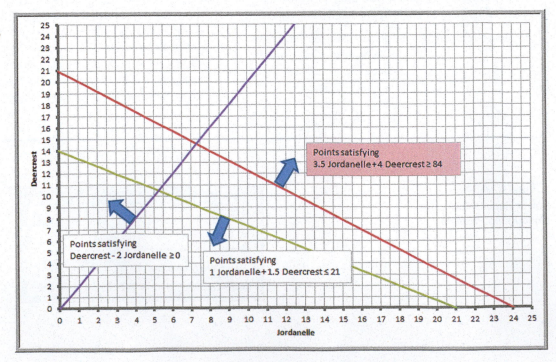

CHECK YOUR UNDERSTANDING

1. Describe the four outcomes that may occur after solving a linear optimization model.

2. What messages does *Solver* provide when a problem is unbounded or infeasible?

Applications of Linear Optimization

Building optimization models is more of an art than a science because there often are several ways of formulating a particular problem. Learning how to build optimization models requires logical thought but can be facilitated by studying examples of different models and observing their characteristics. In this section, we illustrate examples of other types of linear optimization models and describe unique issues associated with formulation, spreadsheet implementation, and interpreting results.

Blending Models

Blending problems involve mixing several raw materials that have different characteristics to make a product that meets certain specifications. Dietary planning, gasoline and oil refining, coal and fertilizer production, and the production of many other types of bulk commodities involve blending. We typically see proportional constraints in blending models.

EXAMPLE 13.15   **BG Seed Company**

The BG Seed Company specializes in food products for birds and other household pets. In developing a new birdseed mix, company nutritionists have specified that the mixture should contain at least 13% protein and 15% fat and no more than 14% fiber. The percentages of each of these nutrients in eight types of ingredients that can be used in the mix are given in Table 13.2, along with the wholesale cost per pound. What is the minimum-cost mixture that meets the stated nutritional requirements?

The decisions are the amount of each ingredient to include in a given quantity—for example, 1 pound—of mix. Define X_i = number of pounds of ingredient i to include in 1 pound of the mix, for $i = 1, \ldots, 8$. Defining the variables in this fashion makes the solution easily scalable to any quantity.

The objective is to minimize total cost, obtained by multiplying the cost per pound by the number of pounds used for each ingredient:

Minimize $0.22X_1 + 0.19X_2 + 0.10X_3 + 0.10X_4 + 0.07X_5$
$+ 0.05X_6 + 0.26X_7 + 0.11X_8$

To ensure that the mix contains the appropriate proportion of ingredients, observe that multiplying the number of pounds of each ingredient by the percentage of nutrient in that ingredient (a dimensionless quantity) specifies the number of pounds of nutrient provided. For

example, sunflower seeds contain 16.9% protein; so $0.169X_1$ represents the number of pounds of protein in X_1 pounds of sunflower seeds. Therefore, the total number of pounds of protein provided by all ingredients is

$0.169X_1 + 0.12X_2 + 0.085X_3 + 0.154X_4 + 0.085X_5$
$+ 0.12X_6 + 0.18X_7 + 0.119X_8$

Because the total number of pounds of ingredients that are mixed together equals $X_1 + X_2 + X_3 + X_4 + X_5 + X_6 + X_7 + X_8$, the proportion of protein in the mix is

$$\left(\frac{0.169X_1 + 0.12X_2 + 0.085X_3 + 0.154X_4 + 0.085X_5 + 0.12X_6 + 0.18X_7 + 0.119X_8}{X_1 + X_2 + X_3 + X_4 + X_5 + X_6 + X_7 + X_8} \right)$$

This proportion must be at least 0.13 and can be converted to a linear form as discussed in the fourth part of Example 13.4. However, we wish to determine the best amount of ingredients to include in *1 pound* of mix; therefore, we add the constraint

$X_1 + X_2 + X_3 + X_4 + X_5 + X_6 + X_7 + X_8 = 1$

Now we can substitute 1 for the denominator in the proportion of protein, simplifying the constraint:

$0.169X_1 + 0.12X_2 + 0.085X_3 + 0.154X_4 + 0.085X_5$
$+ 0.12X_6 + 0.18X_7 + 0.119X_8 \geq 0.13$

(Continued)

This ensures that at least 13% of the mixture will be protein. In a similar fashion, the constraints for the fat and fiber requirements are

$$0.26X_1 + 0.041X_2 + 0.038X_3 + 0.063X_4 + 0.038X_5$$
$$+ 0.017X_6 + 0.179X_7 + 0.04X_8 \geq 0.15$$

$$0.29X_1 + 0.083X_2 + 0.027X_3 + 0.024X_4 + 0.027X_5$$
$$+ 0.023X_6 + 0.288X_7 + 0.109X_8 \leq 0.14$$

Finally, we have nonnegative constraints:

$$X_i \geq 0, \text{ for } i = 1, 2, \ldots, 8$$

The complete model is

Minimize $0.22X_1 + 0.19X_2 + 0.10X_3 + 0.10X_4 + 0.07X_5$
$$+ 0.05X_6 + 0.26X_7 + 0.11X_8$$

Mixture: $X_1 + X_2 + X_3 + X_4 + X_5 + X_6 + X_7 + X_8 = 1$

Protein: $0.169X_1 + 0.12X_2 + 0.085X_3 + 0.154X_4 + 0.085X_5$
$$+ 0.12X_6 + 0.18X_7 + 0.119X_8 \geq 0.13$$

Fat: $0.26X_1 + 0.041X_2 + 0.038X_3 + 0.063X_4 + 0.038X_5$
$$+ 0.017X_6 + 0.179X_7 + 0.04X_8 \geq 0.15$$

Fiber: $0.29X_1 + 0.083X_2 + 0.027X_3 + 0.024X_4 + 0.027X_5$
$$+ 0.023X_6 + 0.288X_7 + 0.109X_8 \leq 0.14$$

Nonnegativity: $X_i \geq 0, \text{ for } i = 1, 2, \ldots, 8$

Dealing with Infeasibility

Figure 13.19 shows an implementation of this model on a spreadsheet (Excel file *BG Seed Model*) and Figure 13.20 shows the *Solver* model. If we solve the model, however, we find that the problem is infeasible. *Solver* provides a report, called the **Feasibility Report**, that can help in understanding why. This is shown in Figure 13.21. From this report, it appears that a conflict exists in trying to meet both the fat and fiber constraints. If you look closely at the data, you can see that only sunflower seeds and safflower seeds have high enough amounts of fat needed to meet the 15% requirement; however, they also have very high amounts of fiber, so including them in the mixture makes it impossible to meet the fiber limitation.

So what should the company owner do? One option is to investigate other potential ingredients to use in the mixture that have different nutritional characteristics and see if a feasible solution can be found. The second option is to either lower the fat requirement or raise the fiber limitation, recognizing that these are not ironclad constraints, but simply nutritional goals that can probably be modified in consultation with the company nutritionists. Figure 13.22 shows *Solver* solutions to two what-if scenarios, where the fat requirement is lowered to 14.5%, and the fiber limitation is raised to 14.5%, with all other data remaining the same in each case. Feasible solutions were found for both cases, and there is little difference in the results.

▶ **TABLE 13.2**

Birdseed Nutrition Data

Ingredient	Protein %	Fat %	Fiber %	Cost/lb
Sunflower seeds	16.9	26.0	29.0	$0.22
White millet	12.0	4.1	8.3	$0.19
Kibble corn	8.5	3.8	2.7	$0.10
Oats	15.4	6.3	2.4	$0.10
Cracked corn	8.5	3.8	2.7	$0.07
Wheat	12.0	1.7	2.3	$0.05
Safflower	18.0	17.9	28.8	$0.26
Canary grass seed	11.9	4.0	10.9	$0.11

▶ **Figure 13.19**

Spreadsheet
Model for BG Seed
Company *Problem*

	A	B	C	D	E	F
1	BG Seed Company					
2						
3	Data					
4		Ingredient	Protein %	Fat %	Fiber %	Cost/lb
5		1 Sunflower seeds	16.90%	26%	29%	$ 0.22
6		2 White millet	12%	4.10%	8.30%	$ 0.19
7		3 Kibble corn	8.50%	3.80%	2.70%	$ 0.10
8		4 Oats	15.40%	6.30%	2.40%	$ 0.10
9		5 Cracked corn	8.50%	3.80%	2.70%	$ 0.07
10		6 Wheat	12%	1.70%	2.30%	$ 0.05
11		7 Safflower	18%	17.90%	28.80%	$ 0.26
12		8 Canary grass seed	11.90%	4%	10.90%	$ 0.11
13		Requirement	13%	15%		
14		Limitation			14%	
15						
16	Model					
17		Ingredient	Pounds			
18		1 Sunflower seeds	0			Total
19		2 White millet	0		Cost/lb.	$ -
20		3 Kibble corn	0		Protein	0.00%
21		4 Oats	0		Fat	0.00%
22		5 Cracked corn	0		Fiber	0.00%
23		6 Wheat	0			
24		7 Safflower	0			
25		8 Canary grass seed	0			
26		Total	0			

	A	B	C	D	E	F
1	BG Seed Company					
2						
3	Data					
4		Ingredient	Protein %	Fat %	Fiber %	Cost/lb
5	1	Sunflower seeds	0.169	0.26	0.29	0.22
6	2	White millet	0.12	0.041	0.083	0.19
7	3	Kibble corn	0.085	0.038	0.027	0.1
8	4	Oats	0.154	0.063	0.024	0.1
9	5	Cracked corn	0.085	0.038	0.027	0.07
10	6	Wheat	0.12	0.017	0.023	0.05
11	7	Safflower	0.18	0.179	0.288	0.26
12	8	Canary grass seed	0.119	0.04	0.109	0.11
13		Requirement	0.13	0.15		
14		Limitation			0.14	
15						
16	Model					
17		Ingredient	Pounds			
18	1	Sunflower seeds	0			Total
19	2	White millet	0		Cost/lb.	=SUMPRODUCT(F5:F12,C18:C25)
20	3	Kibble corn	0		Protein	=SUMPRODUCT(C5:C12,C18:C25)
21	4	Oats	0		Fat	=SUMPRODUCT(D5:D12,C18:C25)
22	5	Cracked corn	0		Fiber	=SUMPRODUCT(E5:E12,C18:C25)
23	6	Wheat	0			
24	7	Safflower	0			
25	8	Canary grass seed	0			
26		Total	=SUM(C18:C25)			

Portfolio Investment Models

Many types of financial investment problems are modeled and solved using linear optimization. Such portfolio investment models have the basic characteristics of blending models.

▶ **Figure 13.20**

Solver *Model for* BG Seed
Company *Problem*

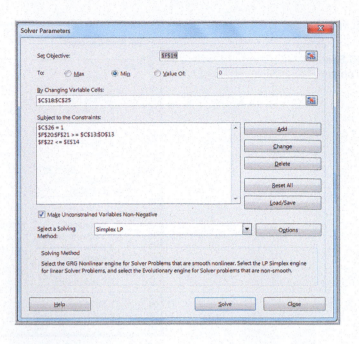

▶ **Figure 13.21**

Feasibility Report for BG
Seed Model

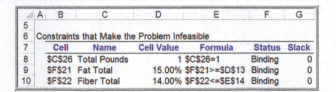

▶ **Figure 13.22**

Model Scenarios for BG
Seed Company *Problem*

	H	I	J	K
1				
2		Scenario	14.5% Fat	14.5% Fiber
3		Ingredient	Pounds	Pounds
4	1	Sunflower seeds	0.434	0.454
5	2	White millet	0.000	0.000
6	3	Kibble corn	0.000	0.000
7	4	Oats	0.422	0.450
8	5	Cracked corn	0.144	0.096
9	6	Wheat	0.000	0.000
10	7	Safflower	0.000	0.000
11	8	Canary grass seed	0.000	0.000
12				
13		Cost/lb.	$0.148	$ 0.152
14		Protein	15.06%	15.42%
15		Fat	14.50%	15.00%
16		Fiber	14.00%	14.50%

EXAMPLE 13.16   Innis Investments

Innis Investments is a small, family-owned business that manages personal financial portfolios. The company manages six mutual funds and has a client who has acquired $500,000 from an inheritance. Characteristics of the funds are given in Table 13.3.

Innis Investments uses a proprietary algorithm to establish a measure of risk for its funds based on the historical volatility of the investments. The higher the volatility, the greater the risk. The company recommends that no more than $200,000 be invested in any individual fund, that at least $50,000 be invested in each of the multinational and balanced funds, and that the total amount invested in income equity and balanced funds be at least 40% of the total investment, or $200,000. The client would like to have an average return of at least 5% but would like to minimize risk. What portfolio would achieve this?

Let X_1 through X_6 represent the dollar amount invested in funds 1 through 6, respectively. The total risk would be measured by the weighted risk of the portfolio, where the weights are the proportion of the total investment in any fund ($X_j/500,000$). Thus, the objective function is

Minimize Total Risk =
$$\frac{10.57X_1 + 13.22X_2 + 14.02X_3 + 2.39X_4 + 9.30X_5 + 7.61X_6}{500,000}$$

The first constraint ensures that $500,000 is invested:

$$X_1 + X_2 + X_3 + X_4 + X_5 + X_6 = 500,000$$

The next constraint ensures that the weighted return is at least 5%:

$$\frac{8.13X_1+9.02X_2+7.56X_3+3.62X_4+7.79X_5+4.40X_6}{500,000} \geq 5.00$$

The next constraint ensures that at least 40% be invested in the income equity and balanced funds:

$$X_5 + X_6 \geq 0.4(500,000)$$

The following constraints specify that at least $50,000 be invested in each of the multinational and balanced funds:

$$X_2 \geq 50,000$$
$$X_6 \geq 50,000$$

Finally, we restrict each investment to a maximum of $200,000 and include nonnegativity:

$$X_j \leq 200,000 \quad \text{for } j = 1,\ldots,6$$
$$X_j \geq 0 \quad \text{for } j = 1,\ldots,6$$

Figure 13.23 shows a spreadsheet implementation of this model (Excel file *Innis Investments*) with the optimal solution. The *Solver* model is given in Figure 13.24. All constraints are met with a minimum risk measure of 6.3073.

Scaling Issues in Using *Solver*

A *poorly scaled* model is one that computes values of the objective, constraints, or intermediate results that differ by several orders of magnitude. Because of the finite precision of computer arithmetic, when these values of very different magnitudes (or others derived from them) are added, subtracted, or compared—in the user's model or in the *Solver*'s own calculations—the result will be accurate to only a few significant digits. As a result, *Solver* may detect or suffer from "numerical instability." The effects of poor scaling in an optimization model can be among the most difficult problems to identify and resolve.

► **Table 13.3**
Mutual Fund Data

Fund	Expected Annual Return	Risk Measure
1. Innis Low-Priced Stock Fund	8.13%	10.57
2. Innis Multinational Fund	9.02%	13.22
3. Innis Mid-Cap Stock Fund	7.56%	14.02
4. Innis Mortgage Fund	3.62%	2.39
5. Innis Income Equity Fund	7.79%	9.30
6. Innis Balanced Fund	4.40%	7.61

▶ **Figure 13.23**

Spreadsheet Model for Innis Investments

	A	B	C	D	E	F
1	Innis Investments					
2						
3	Data					
4			Expected			
5		Fund	Return	Risk Measure	Maximum	Minimum
6		1 Low Priced Stock	8.13%	10.57	$ 200,000	
7		2 Multinational	9.02%	13.22	$ 200,000	$ 50,000
8		3 Mid Cap	7.56%	14.02	$ 200,000	
9		4 Mortgage	3.62%	2.39	$ 200,000	
10		5 Income Equity	7.79%	9.3	$ 200,000	
11		6 Balanced	4.40%	7.61	$ 200,000	$ 50,000
12						
13		Investment =	$ 500,000			
14		Target return ≥	5%			
15		Inc. Eq. + Balanced ≥	$200,000			
16						
17	Model					
18						
19		Fund	Amount Invested			
20		1 Low Priced Stock	$ -			
21		2 Multinational	$ 50,000.00			
22		3 Mid Cap	$ -			
23		4 Mortgage	$ 200,000.00			
24		5 Income Equity	$ 66,371.68			
25		6 Balanced	$ 183,628.32			
26		Total	$ 500,000.00			
27						
28						
29			Total			
30		Risk	6.3073			
31		Weighted Return	5.00%			
32		Inc Eq + Balanced	$250,000			

	A	B	C	D	E	F
1	Innis Investments					
2						
3	Data					
4			Expected			
5		Fund	Return	Risk Measure	Maximum	Minimum
6	1	Low Priced Stock	0.0813	10.57	200000	
7	2	Multinational	0.0902	13.22	200000	50000
8	3	Mid Cap	0.0756	14.02	200000	
9	4	Mortgage	0.0362	2.39	200000	
10	5	Income Equity	0.0779	9.3	200000	
11	6	Balanced	0.044	7.61	200000	50000
12						
13		Investment =	500000			
14		Target return ≥	0.05			
15		Inc. Eq. + Balanced ≥	=0.4*C13			
16						
17	Model					
18						
19		Fund	Amount Invested			
20	1	Low Priced Stock	0			
21	2	Multinational	50000			
22	3	Mid Cap	0			
23	4	Mortgage	200000			
24	5	Income Equity	66371.6814159293			
25	6	Balanced	183628.318584071			
26		Total	=SUM(C20:C25)			
27						
28						
29			Total			
30		Risk	=SUMPRODUCT(D6:D11,C20:C25)/C13			
31		Weighted Return	=SUMPRODUCT(C6:C11,C20:C25)/C13			
32		Inc Eq + Balanced	=C24+C25			

▶ **Figure 13.24**

Solver *Model for* Innis Investments

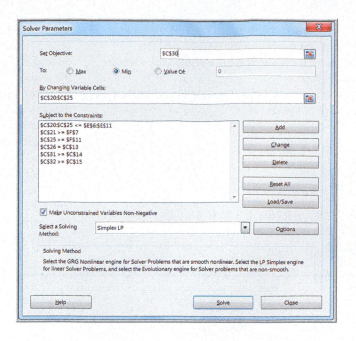

It can cause *Solver* engines to return messages such as "Solver could not find a feasible solution," "Solver could not improve the current solution," or even "The linearity conditions required by this Solver engine are not satisfied," or it may return results that are suboptimal or otherwise very different from your expectations.

In the *Solver* options, you can check the box *Use Automatic Scaling*. When this option is selected, *Solver* rescales the values of the objective and constraint functions internally to minimize the effects of poor scaling. But this can only help with the *Solver*'s own calculations—it may not always work, as it cannot help with poorly scaled results that arise *in the middle of your Excel formulas*. The best way to avoid scaling problems is to carefully choose the "units" implicitly used in your model so that all computed results are within a few orders of magnitude of each other. For example, if you express dollar amounts in units of (say) millions, the actual numbers computed on your worksheet may range from perhaps 1 to 1,000.

EXAMPLE 13.17 **Little Investment Advisors**

Little Investment Advisors is working with a client on determining an optimal portfolio of bond funds. The firm suggests six different funds, each with different expected returns and risk measures (based on historical data):

Bond Portfolio	Expected Return	Risk Measure
1. Ohio National Bond Portfolio	6.11%	4.62
2. PIMCO Global Bond Unhedged Portfolio	7.61%	7.22
3. Federated High Income Bond Portfolio	5.29%	9.75
4. Morgan Stanley UIF Core Plus Fixed Income Portfolio	2.79%	3.95
5. PIMCO Real Return Portfolio	7.37%	6.04
6. PIMCO Total Return Portfolio	5.65%	5.17

(Continued)

The client wants to invest $350,000. Find the optimal investment strategy to achieve the largest weighted percentage return while keeping the weighted risk measure no greater than 5.00.

The model is simple. Let X_1 through X_6 be the amount invested in each of the six funds.

Maximize

$$(6.11X_1 + 7.61X_2 + 5.29X_3 + 2.79X_4 + 7.37X_5 + 5.65X_6)/350,000$$

$$X_1 + X_2 + X_3 + X_4 + X_5 + X_6 = 350,000$$

$$(4.62X_1 + 7.22X_2 + 9.75X_3 + 3.95X_4 + 6.04X_5 + 5.17X_6)/350,000 \leq 5.00$$

$$X_1, \ldots, X_6 \geq 0$$

Figure 13.25 shows the solution without scaling the variables. (Note that in the *Solver Options*, *Automatic Scaling* is not checked.) *Solver* displayed no messages, but the answer is incorrect! This occurs because the objective function (in percent) is several orders of magnitude smaller than the decision variables and investment constraint (in hundreds of thousands of dollars). Figure 13.26 shows the result after *Automatic Scaling* is checked in the *Solver Options*. This is the correct answer. As noted, a better approach would be to scale the investment amount in the model as thousands of dollars (that is, replace cell C11 with $350). Even without automatic scaling, this will yield the optimal solution. So check your models carefully for possible scaling issues!

Transportation Models

Many practical models in supply chain optimization stem from a very simple model called the **transportation problem**. This involves determining how much to ship from a set of sources of supply (factories, warehouses, etc.) to a set of demand locations (warehouses, customers, etc.) at minimum cost.

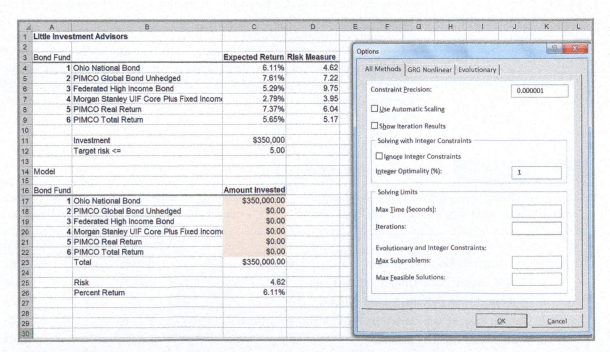

▲ **Figure 13.25**

Solution Without Scaling

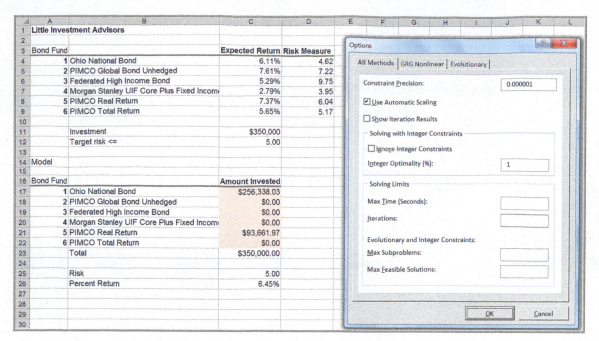

▲ **Figure 13.26**

Solution After Scaling the Model

EXAMPLE 13.18   **General Appliance Corporation**

General Appliance Corporation (GAC) produces refrigerators at two plants: Marietta, Georgia, and Minneapolis, Minnesota. They ship them to major distribution centers in Cleveland, Baltimore, Chicago, and Phoenix. The accounting, production, and marketing departments have provided the information in Table 13.4, which shows the unit cost of shipping between any plant and distribution center, plant capacities over the next planning period, and distribution center demands. GAC's supply chain manager faces the problem of determining how much to ship between each plant and distribution center to minimize the total transportation cost, not exceed available capacity, and meet customer demand.

To develop a linear optimization model, we first define the decision variables as the amount to ship between each plant and distribution center. In this model, we use *double-subscripted variables* to simplify the formulation. Define X_{ij} = amount shipped from plant i to distribution center j, where $i = 1$ represents Marietta, $i = 2$ represents Minneapolis, $j = 1$ represents Cleveland, and so on. Using

the unit-cost data in Table 13.4, the total cost of shipping is equal to the unit cost multiplied by the amount shipped, summed over all combinations of plants and distribution centers. Therefore, the objective function is to minimize total cost:

$$\text{Minimize } 12.60X_{11} + 14.35X_{12} + 11.52X_{13} + 17.58X_{14} + 9.75X_{21} + 16.26X_{22} + 8.11X_{23} + 17.92X_{24}$$

Because capacity is limited, the amount shipped from each plant cannot exceed its capacity. The total amount shipped from Marietta, for example, is $X_{11} + X_{12} + X_{13} + X_{14}$. Therefore, we have the constraint

$$X_{11} + X_{12} + X_{13} + X_{14} \leq 1{,}200$$

Similarly, the capacity limitation at Minneapolis leads to the constraint

$$X_{21} + X_{22} + X_{23} + X_{24} \leq 800$$

Next, we must ensure that the demand at each distribution center is met. This means that the total amount shipped

(Continued)

to any distribution center from both plants must equal the demand. For instance, at Cleveland, we must have

$$X_{11} + X_{21} = 150$$

For the remaining three distribution centers, the constraints are

$$X_{12} + X_{22} = 350$$
$$X_{13} + X_{23} = 500$$
$$X_{14} + X_{24} = 1,000$$

Last, we need nonnegativity, $X_{ij} \geq 0$, for all i and j. The complete model is

$$\text{Minimize } 12.60X_{11} + 14.35X_{12} + 11.52X_{13} + 17.58X_{14}$$
$$+ 9.75X_{21} + 16.26X_{22} + 8.11X_{23} + 17.92X_{24}$$

$$X_{11} + X_{12} + X_{13} + X_{14} \leq 1,200$$
$$X_{21} + X_{22} + X_{23} + X_{24} \leq 800$$
$$X_{11} + X_{21} = 150$$
$$X_{12} + X_{22} = 350$$
$$X_{13} + X_{23} = 500$$
$$X_{14} + X_{24} = 1,000$$
$$X_{ij} \geq 0, \quad \text{for all } i \text{ and } j$$

▶ **Table 13.4**

GAC Cost, Capacity, and Demand Data

	Distribution Center				
Plant	**Cleveland**	**Baltimore**	**Chicago**	**Phoenix**	**Capacity**
Marietta	$12.60	$14.35	$11.52	$17.58	1,200
Minneapolis	$9.75	$16.26	$8.11	$17.92	800
Demand	150	350	500	1,000	

Figure 13.27 shows a spreadsheet implementation for the GAC transportation problem with the optimal solution (Excel file *General Appliance Corporation*), and Figure 13.28 shows the *Solver* model. The Excel model is very simple. In the model section, the decision variables are stored in the plant distribution center matrix. The objective function for total cost in cell B18 can be written in Excel as

$$=B6*B13 + C6*C13 + D6*D13 + E6*E13 + B7*B14$$
$$+ C7*C14 + D7*D14 + E7*E14$$

However, the SUMPRODUCT function is particularly useful for such large expressions; so it is more convenient to express the total cost as

$$=\text{SUMPRODUCT}(B6:E7, B13:E14)$$

▶ **Figure 13.27**

General Appliance Corporation *Model Spreadsheet Implementation and Solution*

▶ **Figure 13.28**

General Appliance Corporation Solver *Model*

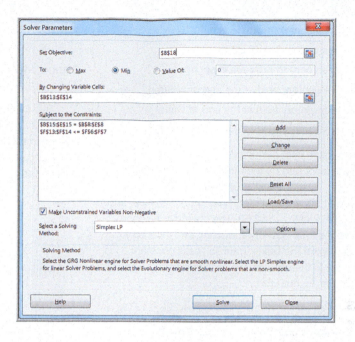

The SUMPRODUCT function can be used for any two arrays as long as the dimensions are the same. Here, the function multiplies pairwise the cost coefficients in the range B6:E7 by the amounts shipped in the range B13:E14 and then adds the terms. In the model, we also use the SUM function in cells F13 and F14 to sum the amount shipped from each plant, and also in cells B15 to E15 to sum the total amount shipped to each distribution center.

Multiperiod Production Planning Models

Many linear optimization problems involve making decisions over a future time horizon. One example is planning production. The basic decision is how much to produce in each time period to meet anticipated demand over each period. Although it might seem obvious to simply produce to the anticipated level of sales, it may be advantageous to produce more than needed in earlier time periods when production costs may be lower and store the excess production as inventory for use in later time periods, thereby letting lower production costs offset the costs of holding the inventory. So the best decision is often not obvious.

EXAMPLE 13.19 **K&L Designs**

K&L Designs is a home-based company that makes hand-painted jewelry boxes for teenage girls. Forecasts of sales for the next year are 150 in the autumn, 400 in the winter, and 50 in the spring. Plain jewelry boxes are purchased from a supplier for $20. The cost of capital is estimated to be 24% per year (or 6% per quarter); thus, the holding cost per item is 0.06($20) = $1.20 per quarter. The company hires art students part-time to craft designs during the autumn, and they earn $5.50 per hour. Because of the high demand for part-time help during the winter holiday season, labor rates are higher in the winter, and workers earn $7.00

per hour. In the spring, labor is more difficult to keep, and the owner must pay $6.25 per hour to retain qualified help. Each jewelry box takes two hours to complete. How should production be planned over the three quarters to minimize the combined production and inventory-holding costs?

The principal decision variables are the number of jewelry boxes to produce during each of the three quarters. However, since we have the option of carrying inventory to other time periods, we must also define decision variables for the number of units to hold in inventory at the end of each quarter. The decision variables are

(Continued)

P_A = amount to produce in autumn

P_W = amount to produce in winter

P_S = amount to produce in spring

I_A = inventory held at the end of autumn

I_W = inventory held at the end of winter

I_S = inventory held at the end of spring

The production cost per unit is computed by multiplying the labor rate by the number of hours required to produce one. Thus, the unit cost in the autumn is ($5.50)(2) = $11.00; in the winter, ($7.00)(2) = $14.00; and in the spring, ($6.25)(2) = $12.50. The objective function is to minimize the total cost of production and inventory. (Because the cost of the boxes themselves is constant, it is not relevant to the problem we are addressing.) The objective function is, therefore,

Minimize $11P_A + 14P_W + 12.50P_S + 1.20I_A + 1.20I_W + 1.20I_S$

The only explicit constraint is that demand must be satisfied. Note that both the production in a quarter as well as the inventory held from the *previous* time quarter can be used to satisfy demand. In addition, any amount in excess of the demand is held to the next quarter. Therefore, the constraints take the form of *inventory balance equations*

that essentially say what is available in any time period must be accounted for somewhere. More formally,

Production + Inventory from the Previous Quarter
= Demand + Inventory Held to the Next Quarter

This can be represented visually using the diagram in Figure 13.29. For each quarter, the sum of the variables coming in must equal the sum of the variables going out. Drawing such a figure is very useful for any type of multiple time period planning model. This results in the constraint set

$$P_A + 0 = 150 + I_A$$
$$P_W + I_A = 400 + I_W$$
$$P_S + I_W = 50 + I_S$$

Moving all variables to the left side results in the model

Minimize $11P_A + 14P_W + 12.50P_S + 1.20I_A + 1.20I_W + 1.20I_S$

subject to

$$P_A - I_A = 150$$
$$P_W + I_A - I_W = 400$$
$$P_S + I_W - I_S = 50$$
$$P_i \geq 0, \quad \text{for all } i$$
$$I_j \geq 0, \quad \text{for all } j$$

Figure 13.30 shows a spreadsheet implementation for the K&L Designs model (Excel file *K&L Designs*); Figure 13.31 shows the associated *Solver* model. For the optimal solution, we produce the demand for the autumn and winter quarters in the autumn and store the excess inventory until the winter. This takes advantage of the lower production cost in the autumn. However, it is not economical to pay the inventory holding cost to carry the spring demand for two quarters.

▶ **Figure 13.29**

Material Balance Constraint Structure

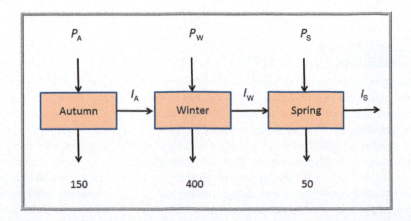

▸ **Figure 13.30**

Spreadsheet Model and Optimal Solution for K&L Designs

	A	B	C	D
1	K&L Designs			
2				
3	Data			
4				
5		Autumn	Winter	Spring
6	Unit Production Cost	$ 11.00	$ 14.00	$ 12.50
7	Unit Inventory Holding Cost	$ 1.20	$ 1.20	$ 1.20
8	Demand	150	400	50
9				
10	Model			
11		Autumn	Winter	Spring
12	Production	550	0	50
13	Inventory	400	0	0
14				
15	Net production	150	400	50
16				
17		Cost		
18	Total	$7,155.00		

	A	B	C	D
1	K&L Designs			
2				
3	Data			
4				
5		Autumn	Winter	Spring
6	Unit Production Cost	11	14	12.5
7	Unit Inventory Holding Cost	1.2	1.2	1.2
8	Demand	150	400	50
9				
10	Model			
11		Autumn	Winter	Spring
12	Production	550	0	50
13	Inventory	400	0	0
14				
15	Net production	=B12-B13	=C12-C13+B13	=D12-D13+C13
16				
17		Cost		
18	Total	=SUMPRODUCT(B6:D7,B12:D13)		

Alternative Models

As we have seen, developing models is more of an art than a science; consequently, there is often more than one way to model a particular problem. Sometimes, alternative models are easier to understand or provide more useful information to the user. Using the ideas presented in the K&L Designs example, we may construct an alternative model involving only the production variables.

EXAMPLE 13.20 An Alternative Optimization Model for K&L Designs

In the *K&L Designs* problem, we simply have to ensure that demand is satisfied. We can do this by guaranteeing that the cumulative production in each quarter is at least as great as the cumulative demand. This is expressed by the following constraints:

$$P_A \geq 150$$
$$P_A + P_W \geq 550$$
$$P_A + P_W + P_S \geq 600$$
$$P_A, P_W, P_S \geq 0$$

The differences between the left- and right-hand sides of these constraints are the ending inventories for each period (and we need to keep track of these amounts because inventory has a cost associated with it). Thus, we use the following objective function:

$$\text{Minimize } 11P_A + 14P_W + 12.50P_S + 1.20(P_A - 150) + 1.20(P_A + P_W - 550) + 1.20(P_A + P_W + P_S - 600)$$

Of course, this function can be simplified algebraically by combining like terms. Although these two models look very different, they are mathematically equivalent and will produce the same solution.

Figure 13.32 shows a spreadsheet implementation of this alternate model (available in the worksheet *Alternate Model* in the *K&L Designs* workbook), and Figure 13.33 shows the *Solver* model. Both have the same optimal solution.

▶ **Figure 13.31**

Solver *Model for* K&L Designs

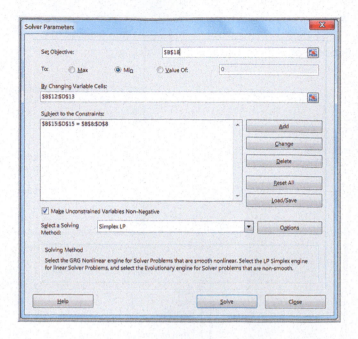

▶ **Figure 13.32**

Alternative Spreadsheet Model for K&L Designs

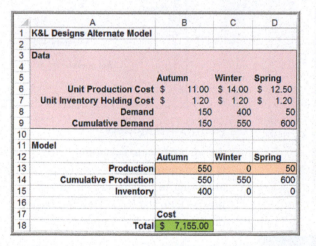

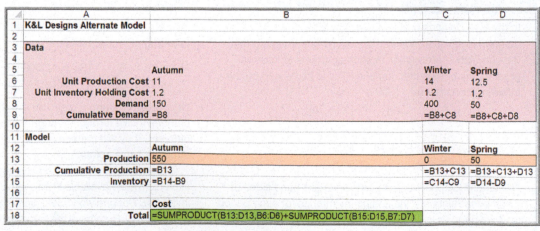

▶ **Figure 13.33**

Solver *Model for Alternative*
K&L Designs *Model*

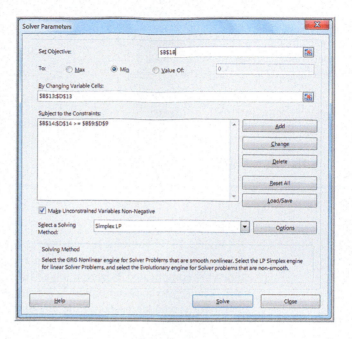

Multiperiod Financial Planning Models

Financial planning often occurs over an extended time horizon. Financial planning models have similar characteristics to multiperiod production planning and can be formulated as multiperiod optimization models.

EXAMPLE 13.21 | **D. A. Branch & Sons**

The financial manager at D. A. Branch & Sons must ensure that funds are available to pay company expenditures in the future but would also like to maximize investment income. Three short-term investment options are available over the next six months: A, a one-month CD that pays 0.25% at maturity each month, and is available each month; B, a three-month CD that pays 1.00% (at maturity), available at the beginning of the first four months; and C, a six-month CD that pays 2.3% (at maturity), available in the first month. The net expenditures for the next six months are forecast as $50,000, ($12,000), $23,000, ($20,000), $41,000, and ($13,000). Amounts in parentheses indicate a net inflow of cash. The company must maintain a cash balance of at least $10,000 at the end of each month. The company currently has $200,000 in cash.

At the beginning of each month, the manager must decide how much to invest in each alternative that may be available. Define the following:

A_i = amount ($) to invest in a one-month CD at the start
of month i

B_i = amount ($) to invest in a three-month CD at the start
of month i

C_i = amount ($) to invest in a six-month CD at the start
of month i

Because the time horizons on these alternatives vary, it is helpful to draw a picture to represent the investments and returns for each year, as shown in Figure 13.34. Each circle represents the beginning of a month. Arrows represent the investments and cash flows. For example, a three-month CD invested in at the start of month 1 (B_1) matures at the beginning of month 4. It is reasonable to assume that all funds available would be invested.

From Figure 13.34, we see that investments A_6, B_4, and C_1 will mature at the end of month 6—that is, at the beginning of month 7. To maximize the amount of cash on hand at the end of the planning period, we have the objective function

Maximize $1.0025A_6 + 1.01B_4 + 1.023C_1$

(Continued)

The only constraints necessary are minimum cash balance equations. For each month, the net cash available, which is equal to the cash in less cash out, must be at least \$10,000. These follow directly from Figure 13.34. The complete model is

$$\text{Maximize } 1.0025A_6 + 1.01B_4 + 1.023C_1$$

subject to

$$200,000 - (A_1 + B_1 + C_1 + 50,000) \geq 10,000 \text{ (month 1)}$$

$$1.0025A_1 + 12,000 - (A_2 + B_2) \geq 10,000 \text{ (month 2)}$$

$$1.0025A_2 - (A_3 + B_3 + 23,000) \geq 10,000 \text{ (month 3)}$$

$$1.0025A_3 + 1.01B_1 + 20,000 - (A_4 + B_4) \geq 10,000 \text{ (month 4)}$$

$$1.0025A_4 + 1.01B_2 - (A_5 + 41,000) \geq 10,000 \text{ (month 5)}$$

$$1.0025A_5 + 1.01B_3 + 13,000 - A_6 \geq 10,000 \text{ (}month\text{ 6)}$$

$$A_i, B_i, C_i \geq 0, \quad \text{for all } i$$

Figure 13.35 shows a spreadsheet model for this problem (Excel file *D. A. Branch & Sons*); the *Solver* model is shown in Figure 13.36. The spreadsheet model may look somewhat complicated; however, it has similar characteristics of a typical financial spreadsheet. The key to constructing the *Solver* model is the summary section. Here we calculate the monthly balance based on the amount of cash available (previous balance plus any investment returns), the net expenditures (remember that a negative expenditure is a cash inflow), and the amount invested as reflected by the decision variables. These balances are a practical interpretation of the constraint functions for each month in the model. In the *Solver* model, these balances simply need to be greater than or equal to the \$10,000 cash-balance requirement for each month.

▶ **Figure 13.34**

Cash Balance Constraint Structure

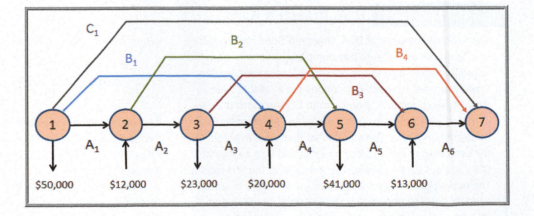

▶ **Figure 13.35**

Spreadsheet Model for D. A. Branch & Sons

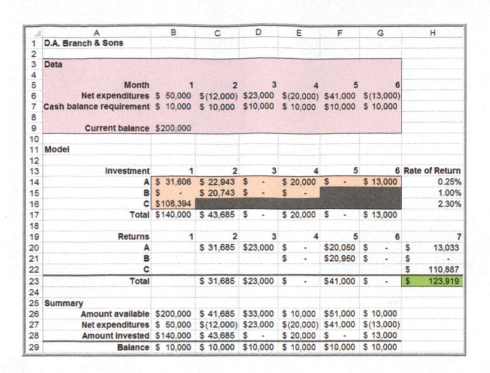

	A	B	C	D	E	F	G	H
1	D.A. Branch & Sons							
2								
3	Data							
4								
5	Month	1	2	3	4	5	6	
6	Net expenditures	$ 50,000	$(12,000)	$23,000	$(20,000)	$41,000	$(13,000)	
7	Cash balance requirement	$ 10,000	$ 10,000	$10,000	$ 10,000	$10,000	$ 10,000	
8								
9	Current balance	$200,000						
10								
11	Model							
12								
13	Investment	1	2	3	4	5	6	Rate of Return
14	A	$ 31,606	$ 22,943	$ -	$ 20,000	$ -	$ 13,000	0.25%
15	B	$ -	$ 20,743	$ -	$ -			1.00%
16	C	$108,394						2.30%
17	Total	$140,000	$ 43,685	$ -	$ 20,000	$ -	$ 13,000	
18								
19	Returns	1	2	3	4	5	6	7
20	A		$ 31,685	$23,000	$ -	$20,050	$ -	$ 13,033
21	B				$ -	$20,950	$ -	$ -
22	C						$ -	$ 110,887
23	Total		$ 31,685	$23,000	$ -	$41,000	$ -	$ 123,919
24								
25	Summary							
26	Amount available	$200,000	$ 41,685	$33,000	$ 10,000	$51,000	$ 10,000	
27	Net expenditures	$ 50,000	$(12,000)	$23,000	$(20,000)	$41,000	$(13,000)	
28	Amount invested	$140,000	$ 43,685	$ -	$ 20,000	$ -	$ 13,000	
29	Balance	$ 10,000	$ 10,000	$10,000	$ 10,000	$10,000	$ 10,000	

	A	B	C	D	E	F	G	H
1	D.A. Branch & Sons							
2								
3	Data							
4								
5	Month	1	2	3	4	5	6	
6	Net expenditures	50000	-12000	23000	-20000	41000	-13000	
7	Cash balance requirement	10000	10000	10000	10000	10000	10000	
8								
9	Current balance	200000						
10								
11	Model							
12								
13	Investment	1	2	3	4	5	6	Rate of Return
14	A	31606.202143588	22942.6433915212	0	20000	0	13000	0.0025
15	B	0	20742.5742574257	0	0			0.01
16	C	108393.797656412						0.023
17	Total	=SUM(B14:B16)	=SUM(C14:C16)	=SUM(D14:D16)	=SUM(E14:E16)	=SUM(F14:F16)	=SUM(G14:G16)	
18								
19	Returns	1	2	3	4	5	6	7
20	A		=(1+$H$14)*B14	=(1+$H$14)*C14	=(1+$H$14)*D14	=(1+$H$14)*E14	=(1+$H$14)*F14	=(1+$H$14)*G14
21	B			=(1+$H$15)*B15	=(1+$H$15)*C15	=(1+$H$15)*D15	=(1+$H$15)*E15	
22	C						=(1+H16)*B16	
23	Total		=SUM(C20:C22)	=SUM(D20:D22)	=SUM(E20:E22)	=SUM(F20:F22)	=SUM(G20:G22)	=SUM(H20:H22)
24								
25	Summary							
26	Amount available	=B9	=B29+C23	=C29+D23	=D29+E23	=E29+F23	=F29+G23	
27	Net expenditures	50000	-12000	23000	-20000	41000	-13000	
28	Amount invested	=B17	=C17	=D17	=E17	=F17	=G17	
29	Balance	=B26-B27-B28	=C26-C27-C28	=D26-D27-D28	=E26-E27-E28	=F26-F27-F28	=G26-G27-G28	

► **Figure 13.36**

Solver *Model for*
D. A. Branch & Sons

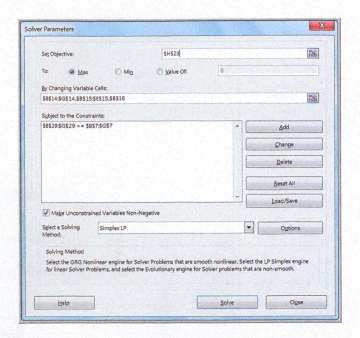

ANALYTICS IN PRACTICE: Linear Optimization in Bank Financial Planning[2]

One of the first applications of linear optimization in banking was developed by Central Carolina Bank and Trust Company (CCB). The bank's management became increasingly concerned with coordinating the activities of the bank to maximize interest rate differentials between sources and uses of funds. To address these concerns, the bank established a financial planning committee comprising all senior bank officers. The committee was charged with the responsibility of integrating the following functions: (1) interest rate forecasting, (2) forecasting demand for bank services, (3) liquidity management policy, and (4) funds allocation. At the same time, CCB's executive committee authorized the development of a balance sheet optimization model using linear programming.

The initial stage in the model's development involved a series of meetings with the financial planning committee to determine how complex the model needed to be. After a thorough discussion of the available options, the group settled on a one-year, single-period model, containing 66 asset and 32 liability and equity categories. Even though a single-period planning model ignores many important time-related linkages, it was felt that a single-period framework would result in a model structure whose output could be readily internalized by management. An integral

part of these discussions involved an attempt to assure senior managers that the resulting model would capture their perceptions of the banking environment.

Next, the model was formulated and its data requirements were clearly identified. The major data inputs needed to implement the model were

- expected yields on all securities and loan categories,
- expected interest rates on deposits and money market liabilities,
- administrative and/or processing costs on major loan and deposit categories,
- expected loan losses, by loan type, as a percentage of outstanding loans,
- maturity structure of all asset and liability categories,
- forecasts of demand for bank services.

The bank's financial records served as a useful database for the required inputs. In those instances where meaningful data did not exist, studies were initiated to fill the gaps.

The decision variables in the model represented different asset categories, such as cash, treasury securities, consumer loans, and commercial loans, among others; other variables represented liabilities and equities such as savings accounts, money market certificates, and certificates of deposit. The

[2]Based on Sheldon D. Balbirer and David Shaw, "An Application of Linear Programming to Bank Financial Planning," *Interfaces*, 11, 5 (October 1981).

2jenn/Shutterstock

objective function was to maximize profits, equaling the difference between net yields and costs. Constraints reflected various operational, legal, and policy considerations, including bounds on various asset or liability categories that represent forecasts of demand for bank services; minimum values of turnover for assets and liabilities; policy constraints that influence the allocation of funds among earning assets or the mix of funds used to finance assets; legal and regulatory constraints; and constraints that prevent the allocation of short-term sources of funds to long-term uses, which gave the model a multiperiod dimension by considering the fund flow characteristics of the target balance sheet beyond the immediate planning horizon. Using the model, CCB successfully structured its assets and liabilities to better determine the bank's future position under different sets of assumptions.

KEY TERMS

Alternative optimal solution
Balance constraint
Binding constraint
Constraint function
Constraints
Corner point
Decision variables
Feasibility Report
Feasible region
Feasible solution
Infeasible problem
Integer linear optimization model
  (integer program, IP model)

Limitation
Linear optimization model (linear
  program, LP model)
Nonlinear optimization model
  (nonlinear program, NLP model)
Objective function
Optimization
Proportional relationship
Requirement
Simple bound
Transportation problem
Unbounded solution
Unique optimal solution

CHAPTER 13 TECHNOLOGY HELP

Excel Techniques

Solver (Example 13.6):

From the *Analysis* group under the *Data* tab in Excel, select *Solver*. Use the *Solver Parameters* dialog to define the objective, decision variables, and constraints. Define the objective function cell in the spreadsheet in the *Set Objective* field. Click the appropriate radio button for *Max* or *Min*. Enter the decision variable cells in the *By Changing Variable Cells* field. To enter a constraint, click the

Add button. In the *Add Constraint* dialog, enter the cell that contains the constraint function (left-hand side of the constraint) in the *Cell Reference* field. In the *Constraint* field, enter the numerical value of the right-hand side of the constraint or the cell reference corresponding to it. Check the box in the dialog *Make Unconstrained Variables Non-Negative*. In the field labeled *Select a Solving Method*, choose *Simplex LP* for a linear optimization model. Then click the *Solve* button. In the *Solver Results* dialog, click on *Answer Report* and then click *OK*.

Analytic Solver

Analytic Solver provides a more powerful optimization tool than the standard *Solver* that comes with Excel. See the online supplement *Using Linear Optimization in* *Analytic Solver*. We suggest that you first read the online supplement *Getting Started with Analytic Solver Basic*. This provides information for both instructors and students on how to register for and access Analytic Solver Basic.

PROBLEMS AND EXERCISES

Note: Data for many problems can be found in the Excel file Chapter 13 Problem Data *to facilitate model development and Excel implementation. Tab names correspond to the problem numbers. These are designated with an asterisk (*).*

Optimization Models

1. Suggest additional generic examples of linear optimization models in the format of Table 13.1 based on your work experience, personal interests (hobbies, etc.), or information found in business articles.

Developing Linear Optimization Models

2. Classify the following descriptions of constraints as bounds, limitations, requirements, proportional relationships, or balance constraints:

   a. Each serving of chili should contain a quarter-pound of beef.

   b. Customer demand for a cereal is not expected to exceed 800 boxes during the next month.

   c. The amount of cash available to invest in March is equal to the accounts receivable in February plus investment yields due on February 28.

   d. A can of premium nuts should have at least twice as many cashews as peanuts.

   e. A warehouse has 3,500 units available to ship to customers.

   f. A call center needs at least 15 service representatives on Monday morning.

   g. An ice cream manufacturer has 40 dozen fresh eggs at the start of the production shift.

3. Review the portfolio allocation model in this chapter. Identify the decision variables, objective function, and constraints in simple verbal statements, and mathematically formulate the linear optimization model.

4. Valencia Products makes automobile radar detectors and assembles two models: LaserStop and Speed-Buster. The firm can sell all it produces. Both models use the same electronic components. Two of these can be obtained only from a single supplier. For the next month, the supply of these is limited to 4,000 of component A and 3,500 of component B. The number of each component required for each product and the profit per unit are given in the table.

	Components Required/Unit		
	A	**B**	**Profit/Unit**
LaserStop	18	6	$124
SpeedBuster	12	8	$136

   a. Identify the decision variables, objective function, and constraints in simple verbal statements.

   b. Mathematically formulate a linear optimization model.

5. A brand manager for ColPal Products must determine how much time to allocate between radio and television advertising during the next month. Market research has provided estimates of the audience exposure for each minute of advertising in each medium, which it would like to maximize. Costs per minute of advertising are also known, and the manager has a limited budget of $25,000. The manager has decided that because television ads have been found to be much more effective than radio ads, at least 75% of the time should be allocated to television.

Suppose that we have the following data:

Type of Ad	Exposure/Minute	Cost/Minute
Radio	350	$400
TV	800	$2,000

   a. Identify the decision variables, objective function, and constraints in simple verbal expressions.

   b. Mathematically formulate a linear optimization model.

6. Burger Office Equipment produces two types of desks, standard and deluxe. Deluxe desks have oak

tops and more-expensive hardware and require additional time for finishing and polishing. Standard desks require 70 board feet of pine and 10 hours of labor, whereas deluxe desks require 50 board feet of pine, 20 square feet of oak, and 18 hours of labor. For the next week, the company has 5,000 board feet of pine, 750 square feet of oak, and 400 hours of labor available. Standard desks net a profit of $250, and deluxe desks net a profit of $350. All desks can be sold to national chains such as Staples or Office Depot.

 a. Identify the decision variables, objective function, and constraints in simple verbal statements.

 b. Mathematically formulate a linear optimization model.

7. A business student has $2,500 available from a summer job and has identified three potential stocks in which to invest. The cost per share and expected return over the next two years is given in the table.

Stock	A	B	C
Price/share	$12	$15	$30
Return/share	$8	$7	$11

 a. Identify the decision variables, objective function, and constraints in simple verbal statements.

 b. Mathematically formulate a linear optimization model.

8. Bangs Leisure Chairs produces three types of handcrafted outdoor chairs that are popular for beach, pool, and patios: sling chairs, Adirondack chairs, and hammocks. The unit profit for these products is $40, $100, and $90, respectively. Each type of chair requires cutting, assembling, and finishing. The owner is retired and is willing to work six hours/day for five days/week, so has 120 hours available each month. He does not want to spend more than 50 hours each month on any one activity (that is, cutting, assembling, and finishing). The retailer he works with is certain that all products he makes can easily be sold. Sling chairs are made up of ten wood pieces for the frame and one piece of cloth. The actual cutting of the wood takes 30 minutes. Assembling includes sewing of the fabric and the attachment of rivets, screws, fabric, and dowel rods, and takes 45 minutes. The finishing stage involves sanding, staining, and varnishing of the various parts and takes one hour. Adirondack chairs take two hours for both the cutting and assembling phases,

and finishing takes one hour. For hammocks, cutting takes 0.4 hour; assembly takes three hours; and finishing also takes one hour. How many of each type of chair should he produce each month to maximize profit?

 a. Identify the decision variables, objective function, and constraints in simple verbal statements.

 b. Mathematically formulate a linear optimization model.

9. The Morton Supply Company produces clothing, footwear, and accessories for dancing and gymnastics. They produce three models of pointe shoes used by ballerinas to balance on the tips of their toes. The shoes are produced from four materials: cardstock, satin, plain fabric, and leather. The number of square inches of each type of material used in each model of shoe, the amount of material available, and the profit/model are shown below:

Material (measured in square inches)	Model 1	Model 2	Model 3	Material Available
Cardstock	12	10	14	1,200
Satin	24	20	15	2,000
Plain fabric	40	40	30	7,500
Leather	11	11	10	1,000
Profit per model	$50	$44	$40	

 a. Identify the decision variables, objective function, and constraints in simple verbal statements.

 b. Mathematically formulate a linear optimization model.

10. Malloy Milling grinds calcined alumina to a standard granular size. The mill produces two different size products from the same raw material. Regular grind can be produced at a rate of 10,000 pounds per hour and has a demand of 400 tons per week with a price per ton of $900. Super grind can be produced at a rate of 6,000 pounds per hour and has demand of 200 tons per week with a price of $1,900 per ton. A minimum of 700 tons has to be ground every week to make room in the raw material storage bins for previously purchased incoming raw material by rail. The mill operates 24/7 for a total of 168 hours/week.

 a. Identify the decision variables, objective function, and constraints in simple verbal statements.

 b. Mathematically formulate a linear optimization model.

Solving Linear Optimization Models

11. Implement the linear optimization model that you developed for Valencia Products in Problem 4 on a spreadsheet and use *Solver* to find an optimal solution. Interpret the *Solver* Answer Report, identify the binding constraints, and verify the values of the slack variables by substituting the optimal solution into the model constraints.

12. Implement the linear optimization model that you developed for ColPal Products in Problem 5 on a spreadsheet and use *Solver* to find an optimal solution. Interpret the Solver Answer Report, identify the binding constraints, and verify the values of the slack variables by substituting the optimal solution into the model constraints.

13. Implement the linear optimization model that you developed for Burger Office Equipment in Problem 6 on a spreadsheet and use *Solver* to find an optimal solution. Interpret the *Solver* Answer Report, identify the binding constraints, and verify the values of the slack variables by substituting the optimal solution into the model constraints.

14. Implement the linear optimization model that you developed for the investment scenario in Problem 7 on a spreadsheet and use *Solver* to find an optimal solution. Interpret the *Solver* Answer Report, identify the binding constraints, and verify the values of the slack variables by substituting the optimal solution into the model constraints.

*15. Implement the linear optimization model that you developed for Bangs Leisure Chairs in Problem 8 on a spreadsheet and use *Solver* to find an optimal solution.

   a. Interpret the *Solver* Answer Report, identify the binding constraints, and verify the values of the slack variables by substituting the optimal solution into the model constraints.

   b. Suppose that Mr. Bangs wants to limit the number of Adirondack chairs to at most 20. Modify and re-solve your model to determine the new solution.

   c. Suppose that Mr. Bangs does not want to spend more than 40 hours each month on any one activity. Modify and re-solve your original model to determine the new solution.

*16. Implement the linear optimization model that you developed for the Morton Supply Company in Problem 9 on a spreadsheet and use *Solver* to find an optimal solution. Interpret the *Solver* Answer Report and identify the binding constraints.

*17. Implement the linear optimization model that you developed for Malloy Milling in Problem 10 on a spreadsheet and use *Solver* to find an optimal solution. Interpret the *Solver* Answer Report and identify the binding constraints.

How *Solver* Works

18. For the Valencia Products model in Problem 4, graph the constraints and identify the feasible region. Then identify each of the corner points and show how increasing the objective function value identifies the optimal solution.

19. For the ColPal model in Problem 5, graph the constraints and identify the feasible region. Then identify each of the corner points and show how increasing the objective function value identifies the optimal solution.

20. For the Burger Office Equipment model in Problem 6, graph the constraints and identify the feasible region. Then identify each of the corner points and show how increasing the objective function value identifies the optimal solution.

Solver Outcomes and Solution Messages

21. For Valencia Products in Problem 4, modify the data in the model to create a problem with each of the following.

   a. alternative optimal solutions

   b. an unbounded solution

   c. infeasibility

22. For ColPal Products in Problem 5, modify the data in the model to create a problem with each of the following.

   a. alternative optimal solutions

   b. an unbounded solution

   c. infeasibility

23. For the investment situation in Problem 7, apply the same logic as we did for the Crebo Manufacturing model in the text to find the optimal solution. Compare your answer with the *Solver* solution.

Applications of Linear Optimization

24. Rosenberg Land Development (RLD) is a developer of condominium properties in the Southwest United States. RLD has recently acquired a 40.625-acre site outside Phoenix, Arizona. Zoning restrictions allow at most eight units per acre. Three types of condominiums are planned: one-, two-, and three-bedroom units. The average construction costs for each type of unit are $450,000, $600,000, and $750,000, respectively. These units will generate a net profit of 10%. The company has equity and loans totaling $180 million dollars for this project. From prior development projects, senior managers have determined that there must be a minimum of 15% one-bedroom units, 25% two-bedroom units, and 25% three-bedroom units.

   a. Develop a mathematical model to determine how many of each type of unit the developer should build.

   b. Implement your model on a spreadsheet and find an optimal solution.

25. Korey is a business student at State U. She has just completed a course in decision models, which had a midterm exam, a final exam, individual assignments, and a class participation grade. She earned a 94% on the midterm, 86% on the final, 93% on the individual assignments, and 85% on participation. The benevolent instructor is allowing his students to determine their own weights for each of the four grade components—of course, with some restrictions:

  ■ The participation weight can be no more than 15%.

  ■ The midterm weight must be at least twice as much as the individual assignment weight.

  ■ The final exam weight must be at least three times as much as the individual assignment weight.

  ■ The weights for each exam must be at least 25%.

  ■ The weights for assignments and participation must be at least 10%.

  ■ The weights must sum to 1.0 and be nonnegative.

   a. Develop a mathematical model that will yield a valid set of weights to maximize Korey's score for the course.

   b. Implement your model on a spreadsheet and find an optimal solution using *Solver*.

*26. The Martinez Model Car Company produces four different radio-controlled model cars based on exotic production models: Ferrari, BMW, Lotus, and Tesla. Each model requires production in five departments:

	Ferrari	BMW	Lotus	Tesla	Minutes Available
Molding	5.00	3.50	1.00	3.00	600
Sanding	4.00	3.20	2.00	3.65	600
Polishing	3.50	2.00	3.00	1.00	480
Painting	3.75	3.25	1.75	2.00	480
Finishing	4.00	1.00	2.00	3.00	480
Prices	$350.00	$330.00	$270.00	$255.00	

   a. How many of each type of car should be produced to maximize profit?

   b. If marketing requires that at least 40 units of each be produced each day, what is the optimal production plan and profit? Before you solve this, how would you expect the profit to compare with your answer to part a?

   c. What happens if marketing requires that at least 50 units of each be produced each day?

27. The International Chef, Inc., markets three blends of oriental tea: premium, Duke Grey, and breakfast. The firm uses tea leaves from India, China, and new domestic California sources.

Quality	Tea Leaves (Percent)		
	Indian	Chinese	California
Premium	40	20	40
Duke Grey	30	50	20
Breakfast	40	40	20

Net profit per pound for each blend is $0.50 for premium, $0.30 for Duke Grey, and $0.35 for breakfast. The firm's regular weekly supplies are 19,000 pounds of Indian tea leaves, 22,000 pounds of Chinese tea leaves, and 16,000 pounds of California tea leaves. Develop and solve a linear optimization model to determine the optimal mix to maximize profit.

28. Young Energy operates a power plant that includes a coal-fired boiler to produce steam to drive a generator. The company can purchase different types of coals and blend them to meet the requirements for burning in the boiler. The following table shows the characteristics of the different types of coals:

Type	BTU/lb	% Ash	% Moisture	Cost ($/lb)
A	11,500	13%	10%	$2.49
B	11,800	10%	8%	$3.04
C	12,200	12%	8%	$2.99
D	12,100	12%	8%	$2.61

The BTU/pound must be at least 11,900. In addition, the ash content can be at most 12.2% and the moisture content at most 9.4%. Develop and solve a linear optimization model to find the best coal blend for Young Energy.

*29. Holcomb Candles, Inc., manufactures decorative candles and has contracted with a national retailer to supply a set of special holiday candles to its 8,500 stores. These include large jars, small jars, large pillars, small pillars, and a package of four votive candles. In negotiating the contract for the display, the manufacturer and retailer agreed that 8 feet would be designated for the display in each store, but that at least 2 feet would be dedicated to large jars and large pillars together, and at least 1 foot to the votive candle packages. At least as many jars as pillars must be provided. The manufacturer has obtained 200,000 pounds of wax, 250,000 feet of wick, and 100,000 ounces of holiday fragrance. The amount of materials and display size required for each product are shown in the table below. How many of each product should be made to maximize the profit?

	Large Jar	Small Jar	Large Pillar	Small Pillar	Votive Pack
Wax	0.5	0.25	0.5	0.25	0.3125
Fragrance	0.24	0.12	0.24	0.12	0.15
Wick	0.43	0.22	0.58	0.33	0.8
Display feet	0.48	0.24	0.23	0.23	0.26
Profit/unit	$0.25	$0.20	$0.24	$0.21	$0.16

30. The Children's Theater Company is a nonprofit corporation managed by Shannon Board. The theater performs in two venues: Kristin Marie Hall and the Lauren Elizabeth Theater. For the upcoming season, seven shows have been chosen. The question

Shannon faces is how many performances of each of the seven shows should be scheduled. A financial analysis has estimated revenues for each performance of the seven shows, and Shannon has set the minimum number of performances of each show based on union agreements with the Actor's Equity Association and the popularity of the shows in other markets. These data are shown in the table below.

Show	Revenue	Cost	Minimum Number of Performances
1	$2,217	$ 968	32
2	$2,330	$1,568	13
3	$1,993	$ 755	23
4	$3,364	$1,148	34
5	$2,868	$1,180	35
6	$3,851	$1,541	16
7	$1,836	$1,359	21

Kristin Marie Hall is available for 60 performances during the season, whereas the Lauren Elizabeth Theater is available for 150 performances. Shows 3 and 7 must be performed in Kristin Marie Hall, and the other shows are performed in either venue. The company wants to achieve revenues of at least $550,000 while minimizing its production costs.

a. Develop and solve a linear optimization model to determine the best way to schedule the shows.

b. Is it possible to achieve revenues of $600,000?

31. Jaycee's department store chain is planning to open a new store. It needs to decide how to allocate the 100,000 square feet of available floor space among seven departments. Data on expected performance of each department per month, in terms of square feet (sf), are shown below.

Department	Investment/sf	Risk as a % of $ Invested	Minimum sf	Maximum sf	Expected Profit per sf
Electronics	$100	24	6,000	30,000	$12.00
Furniture	$50	12	10,000	30,000	$6.00
Men's clothing	$30	5	2,000	5,000	$2.00
Clothing	$600	10	3,000	40,000	$30.00
Jewelry	$900	14	1,000	10,000	$20.00
Books	$50	2	1,000	5,000	$1.00
Appliances	$400	3	12,000	40,000	$13.00

The company has gathered $20 million to invest in floor stock. The risk column is a measure of risk associated with investment in floor stock based on past data from other stores and accounts for outdated inventory, pilferage, breakage, and so on. For instance, electronics loses 24% of its total investment, furniture loses 12% of its total investment, and so on. The amount of risk should be no more than 10% of the total investment.

a. Develop a linear optimization model to maximize profit.

b. If the chain obtains another $1 million of investment capital for stock, what would the new solution be?

*32. A recent MBA graduate, Dara, has gained control over custodial accounts that her parents had established for her. Currently, her money is invested in four funds, but she has identified several other funds as options for investment as shown in the table, *Data for Problem 32*. She has $100,000 to invest with the following restrictions:

■ Keep at least $5,000 in savings.
■ Invest at least 14% in the money market fund.
■ Invest at least 16% in international funds.
■ Keep 35% of funds in current holdings.
■ Do not allocate more than 20% of funds to any one investment except for the money market and savings account.
■ Allocate at least 30% into new investments.

Develop a linear optimization model to maximize the net return.

*33. Janette Douglas is coordinating a bake sale for a non-profit organization. The organization has acquired $2,200 in donations to hold the sale. The table *Data for Problem 33* shows the amounts and costs of ingredients used per batch of each baked good.

Data for Problem 32

	Average Return	Expenses	
1. Large cap blend	17.2%	0.93%	(current holding)
2. Small cap growth	20.4%	0.56%	(current holding)
3. Green fund	26.3%	0.70%	(current holding)
4. Growth and income	15.6%	0.92%	(current holding)
5. Multicap growth	19.8%	0.92%	
6. Midcap index	22.1%	0.22%	
7. Multicap core	27.9%	0.98%	
8. Small cap international	35.0%	0.54%	
9. Emerging international	36.1%	1.17%	
10. Money market fund	4.75%	0	
11. Savings account	1.0%	0	

One batch of each results in 10 brownies, 12 cupcakes, 8 peanut butter cups, and 12 shortbread cookies. Each batch of brownies can be sold for $10.00, cupcakes for $15.00, peanut butter cups for $12.00, and shortbread cookies for $7.50. The organization anticipates that a total of at least 4,000 baked goods must be made. For adequate variety, at least 25 batches of each baked good are required, except for the popular brownies, which require at least 100 batches. In addition, no more than 40 batches of shortbread cookies should be made. How can the organization best use its budget and make the largest amount of money?

Data for Problem 33

Ingredient	Brownies	Cupcakes	Peanut Butter Cups	Shortbread Cookies	Cost/Unit
Butter (cups)	0.67	0.33	1	0.75	$1.44
Flour (cups)	1.5	1.5	1.25	2	$0.09
Sugar (cups)	1.75	1	2	0.25	$0.16
Vanilla (tsp)	2	0.5	0	0	$0.06
Eggs	3	2	1	0	$0.12
Walnuts (cups)	2	0	0	0	$0.31
Milk (cups)	0.5	1	2	0	$0.05
Chocolate (oz)	8	2.5	9	0	$0.10
Baking soda (tsp)	2	1	0	0	$0.07
Frosting (cups)	0.5	1.5	0	1	$2.74
Peanut butter (cups)	0	0	2.5	0	$2.04

*34. Example 13.17 described the Little Investment Advisors problem and illustrated scaling issues. In answering the following questions, be sure to scale the model appropriately.

a. How would the results in Figure 13.26 change if there is a limit of $100,000 in each fund?

b. What if, in addition to the limitation in part a, the client wants to invest at least $50,000 in the Federated High Income Bond fund?

c. What would be the optimal investment strategy if the client wants to minimize risk and achieve a return of at least 6% (with no additional limitations or requirements)?

d. How would your results to part c change if there is a limit of $100,000 in each fund?

e. What if, in addition to the limitation in part d, the client wants to invest at least $50,000 in the Federated High Income Bond fund?

*35. Kelly Foods has two plants and ships canned vegetables to customers in four cities. The cost of shipping one case from a plant to a customer is given in the following table.

Plant/ Customer	Chicago	Cincinnati	Indianapolis	Pittsburgh
Akron	$1.70	$2.30	$2.50	$2.15
Evansville	$1.95	$2.35	$1.65	$2.95

The plant in Akron has a capacity of 3,500 cases per week, and the Evansville plant can produce 4,000 cases per week. Customer orders for the next week are as follows:

Chicago: 1,200 cases

Cincinnati: 2,000 cases

Indianapolis: 2,500 cases

Pittsburgh: 1,400 cases

Find the minimum-cost shipping plan.

*36. Liquid Gold, Inc., transports radioactive waste from nuclear power plants to disposal sites around the country. Each plant has an amount of material that must be moved each period. Each site has a limited capacity per period. The cost of transporting between sites is given in the accompanying table (some combinations of plants and storage sites are not to be used, and no figure is given). Develop and solve a transportation model for this problem.

Plant	Material	Cost to Site S1	S2	S3	S4	Site	Capacity
P1	20,876	$105	$86	–	$23	S1	285,922
P2	50,870	$86	$58	$41	–	S2	308,578
P3	38,652	$93	$46	$65	$38	S3	111,955
P4	28,951	S116	$27	$94	–	S4	208,555
P5	87,423	$88	$56	$82	$89		
P6	76,190	$111	$36	$72	–		
P7	58,237	$169	$65	$48	–		

*37. Shafer Supplies has four distribution centers, located in Atlanta, Lexington, Milwaukee, and Salt Lake City, and ships to 12 retail stores, located in Seattle, San Francisco, Las Vegas, Tucson, Denver, Charlotte, Minneapolis, Fayetteville, Birmingham, Orlando, Cleveland, and Philadelphia. The company wants to minimize the transportation costs of shipping one of its higher-volume products, boxes of standard copy paper. The per-unit shipping cost from each distribution center to each retail location and the amounts currently in inventory and ordered at each retail location are shown in the table below.

Develop and solve an optimization model to minimize the total transportation cost and answer the following questions.

a. What is the minimum cost of shipping?

b. Which distribution centers will operate at capacity in this solution?

Shafer Supplies	Seattle	San Francisco	Las Vegas	Tuscon	Denver	Charlotte	Minneapolis
Atlanta	$2.15	$2.10	$1.75	$1.50	$1.20	$0.65	$0.90
Lexington	$1.95	$2.00	$1.70	$1.53	$1.10	$0.55	$0.60
Milwaukee	$1.70	$1.85	$1.50	$1.41	$0.95	$0.40	$0.40
Salt Lake City	$0.60	$0.55	$0.35	$0.60	$0.40	$0.95	$1.00
Demand	5,000	16,000	4,200	3,700	4,500	7,500	3,000

(Continued)

Shafer Supplies	Fayetteville	Birmingham	Orlando	Cleveland	Philadelphia	Supply
Atlanta	$0.80	$0.35	$0.15	$0.60	$0.50	40,000
Lexington	$1.05	$0.60	$0.50	$0.25	$0.30	35,000
Milwaukee	$0.95	$0.70	$0.70	$0.35	$0.40	15,000
Salt Lake City	$1.10	$1.35	S1.60	$1.60	$1.70	16,000
Demand	9,000	3,300	12,000	9,500	16,000	

*38. Roberto's Honey Farm in Chile makes five types of honey: cream, filtered, pasteurized, mélange (a mixture of several types), and strained, which are sold in 1-kilogram and 0.5-kilogram glass containers, 1-kilogram and 0.75-kilogram plastic containers, and in bulk. Key data are shown in the following tables.

Selling Prices (Chilean pesos)

	0.75-kg Plastic	1-kg Plastic	0.5-kg Glass	1-kg Glass	Bulk/ kg
Cream	744	880	760	990	616
Filtered	635	744	678	840	521
Pasteurized	696	821	711	930	575
Mélange	669	787	683	890	551
Strained	683	804	697	910	563

Minimum Demand

	0.75-kg Plastic	1-kg Plastic	0.5-kg Glass	1-kg Glass
Cream	300	250	350	200
Filtered	250	240	300	180
Pasteurized	230	230	350	300
Mélange	350	300	250	350
Strained	360	350	250	380

Maximum Demand

	0.75-kg Plastic	1-kg Plastic	0.5-kg Glass	1-kg Glass
Cream	550	350	470	310
Filtered	400	380	440	300
Pasteurized	360	390	490	400
Mélange	530	410	390	430
Strained	480	420	380	500

Package Costs (Chilean pesos)

0.75-kg Plastic	1-kg Plastic	0.5-kg Glass	1-kg Glass
91	112	276	351

Harvesting and production costs (in Chilean pesos) for each product per kilogram are as follows:

Cream: 322

Filtered: 275

Pasteurized: 320

Mélange: 300

Strained: 287

Develop a linear optimization model to maximize profit if a total of 10,000 kilograms of honey are available.

*39. Sanford Tile Company makes ceramic and porcelain tile for residential and commercial use. They produce three different grades of tile (for walls, residential flooring, and commercial flooring), each of which requires different amounts of materials and production time and generates different contributions to profit. The following information shows the percentage of materials needed for each grade and the profit per square foot.

	Grade I	Grade II	Grade III
Profit/square foot	$2.50	$4.00	$5.00
Clay	50%	30%	25%
Silica	5%	15%	10%
Sand	20%	15%	15%
Feldspar	25%	40%	50%

Each week, Sanford Tile receives raw material shipments, and the operations manager must schedule the plant to efficiently use the materials to maximize profitability. Currently, inventory consists of 6,000 pounds of clay, 3,000 pounds of silica, 5,000 pounds of sand, and 8,000 pounds of feldspar. Because demand varies for the different grades, marketing estimates that at most 8,000 square feet of grade III tile should be produced and that at least 1,500 square feet of grade I tiles are required. Each square foot of tile weighs approximately 2 pounds.

Develop and solve a linear optimization model to determine how many of each grade of tile the company should make next week to maximize profit contribution.

*40. The Hansel Corporation, located in Bangalore, India, makes plastics materials that are mixed with various additives and reinforcing materials before being melted, extruded, and cut into small pellets for sale to other manufacturers. Four grades of plastic are made, each of which might include up to four different additives. The following table shows the number of pounds of additive per pound of each grade of final product, the weekly availability of the additives, and cost and profitability information.

	Grade 1	Grade 2	Grade 3	Grade 4	Availability
Additive A	0.40	0.37	0.34	0.90	100,000
Additive B	0.30	0.33	0.33		90,000
Additive C	0.20	0.25	0.33		40,000
Additive D	0.10	0.05		0.10	10,000
Profit/lb	$2.00	$1.70	$1.50	$2.80	

Because of marketing considerations, the total amount of grades 1 and 2 should not exceed 75% of the total of all grades produced, and at least 40% of the total product mix should be grade 4. How much of each grade should be produced to maximize profit? Develop and solve a linear optimization model.

*41. Mirza Manufacturing makes four electronic products, each of which comprises three main materials: magnet, wire, and casing. The products are shipped to three distribution centers in North America, Europe, and Asia. Marketing has specified that no location should receive more than the maximum demand and that each location should receive at least the minimum demand. The material costs per unit are magnet—$0.59, wire—$0.29, and casing—$0.31. The following table shows the number of units of each material required in each unit of end product and the production cost per unit.

Product	Production Cost/Unit	Magnet	Wire	Casing
A	$0.25	4	2	2
B	$0.35	3	1	3
C	$0.15	2	2	1
D	$0.10	8	3	2

Additional information is provided next.

Min Demand

Product	NA	EU	Asia
A	850	900	100
B	700	200	500
C	1,100	800	600
D	1,500	3,500	2,000

Max Demand

Product	NA	EU	Asia
A	2,550	2,700	300
B	2,100	600	1,500
C	3,300	2,400	1,800
D	4,500	10,500	6,000

Packaging and Shipping Cost/Unit

Product	NA	EU	Asia
A	$0.20	$0.25	$0.35
B	$0.18	$0.22	$0.30
C	$0.18	$0.22	$0.30
D	$0.17	$0.20	$0.25

Unit Sales Revenue

Product	NA	EU	Asia
A	$4.00	$4.50	$4.55
B	$3.70	$3.90	$3.95
C	$2.70	$2.90	$2.40
D	$6.80	$6.50	$6.90

Available Raw Material

Magnet	120,000
Wire	50,000
Casing	40,000

Develop and solve an appropriate linear optimization model to maximize net profit.

42. Raturi Chemicals Inc., produces four industrial chemicals with variable production costs of $9.00, $6.75, $5.25, and $7.50 per pound, respectively. Because of increasing supplier costs, the variable cost of each of the products will increase by 6% at the beginning of month 3. Demand forecasts are shown in the following table. There are currently 100 pounds of each product on hand, and the company wants to maintain an inventory of 100 pounds of each product at the end of every month. The four products share a common process that operates two shifts of eight hours each per day, seven days per week. Processing requirements are 0.06 hour/pound for product 1, 0.05 hour/pound for product 2, 0.2 hour/pound for product 3, and 0.11 hour/pound for product 4. The per-pound cost of holding inventory each month is estimated to be 12% of the cost of the product. Develop an optimization model to meet demand and minimize the total cost. Assume 30 days per month. Implement your model on a spreadsheet and find an optimal solution.

Product Demand

Product	Month 1	Month 2	Month 3
1	1,000	800	1,000
2	1,000	900	500
3	600	600	500
4	0	200	500

43. Reddy & Rao (R&R) is a small company in India that makes handmade, artistic chairs for commercial businesses. The company makes four models. The time required to make each of the models and cost per chair is given below.

	Model A	Model B	Model C	Model D
Cost per Unit	$900.00	$650.00	$500.00	$750.00
Hours Required per Unit	40	22	12	34

R&R employs four people. Each of them works eight-hour shifts, five days a week (assume four weeks/month). The demand for the next three months is estimated to be as follows:

Demand (Units)	Model A	Model B	Model C	Model D
Month 1	7	4	4	9
Month 2	7	4	5	4
Month 3	6	8	8	6

R&R keeps at most two of each model in inventory each month but wants to have at least one of model D in inventory at all times. The current inventory of each model is two. The cost to hold these finished chairs is 10% of the production cost. Develop and solve an optimization model to determine the optimal number of chairs to produce each month and the monthly inventories to minimize total cost and meet the expected demand.

44. An international graduate student will receive a $28,000 foundation scholarship and reduced tuition. She must pay $1,500 in tuition for each of the autumn, winter, and spring quarters and $500 in the summer. Payments are due on the first day of September, December, March, and May, respectively. Living expenses are estimated to be $1,500 per month, payable on the first day of the month. The foundation will pay her $18,000 on August 1 and the remainder on May 1. To earn as much interest as possible, the student wishes to invest the money. Three types of investments are available at her bank: a three-month CD, earning 0.75% at maturity; a six-month CD, earning 1.9% at maturity; and a 12-month CD, earning 4.2% at maturity. Develop a linear optimization model to determine how she can best invest the money and meet her financial obligations.

*45. Jason Wright is a part-time business student who would like to optimize his financial decisions. Currently, he has $16,000 in his savings account. Based on an analysis of his take-home pay, expected bonuses, and anticipated tax refund, he has estimated his income for each month over the next year. In addition, he has estimated his monthly expenses, which vary because of scheduled payments for insurance, utilities, tuition and books, and so on. The following table summarizes his estimates:

Month	Income	Expenses
1. January	$3,400	$3,360
2. February	$3,400	$2,900
3. March	$3,400	$6,600
4. April	$9,500	$2,750
5. May	$3,400	$2,800
6. June	$5,000	$6,800
7. July	$4,600	$3,200
8. August	$3,400	$3,600
9. September	$3,400	$6,550
10. October	$3,400	$2,800
11. November	$3,400	$2,900
12. December	$5,000	$6,650

Jason has identified several short-term investment opportunities:

- a three-month CD yielding 0.60% at maturity
- a six-month CD yielding 1.42% at maturity
- an 11-month CD yielding 3.08% at maturity
- a savings account yielding 0.0375% per month

To ensure enough cash for emergencies, he would like to maintain at least $2,000 in the savings account. Jason's objective is to maximize his cash balance at the end of the year. Develop a linear optimization model to find the best investment strategy.

*46. Pavlick Products supplies a key component for automobile interiors to U.S. assembly plants. The components can be manufactured in China or Mexico. Unit cost in China is $333 and in Mexico, $350. However, shipping costs per 500 units are $10,000 from China and only $2,000 from Mexico; they are expected to increase 4% each month from China and 1% each month from Mexico. Each unit is sold to the automotive customer for $400. Contracts with the Chinese vendor require that a minimum of 2,500 units be produced each month. Demand for the next 12 months is estimated to be as follows:

	Demand
January	14,000
February	16,000
March	14,000
April	14,000
May	16,000
June	10,500
July	14,000
August	20,000
September	20,000
October	16,000
November	14,000
December	10,500

The Mexican plant is new and is gearing up production; its capacity will increase over the next year as follows:

	Mexican Plant Capacity
January	0
February	2,500
March	5,000
April	7,500
May	10,000
June	12,500
July	15,000
August	15,000
September	15,000

	Mexican Plant Capacity
October	15,000
November	15,000
December	15,000

How should the company source production to maximize total profit?

47. Michelle is a business student who plans to attend medical school. The average state university medical school costs around $35,000 per year, and that cost is escalating rapidly. Michelle created a spreadsheet model to calculate the total expenses for each year of medical school, including both education and living expenses. Her estimates are year 1: $57,067, year 2: $56,572, year 3: $67,846, and year 4: $55,662. She is considering three loan options: the Stafford loan, a 6.8% loan with a cap of $47,167 that does not accrue interest during medical school; the Graduate Plus loan, a 7.9% loan with no cap that does accrue interest during medical school; and a private bank loan, a 5.9% loan with a cap of $30,000, also with accruing interest during medical school. Assume that each loan will be paid over 25 years after graduation. Michelle currently has $39,500 saved from investments, family gifts, and work, and will receive an additional $4,500 in gifts from her grandparents in years 2 through 4. Develop and solve an optimization model to determine how much money to fund from each type of loan to minimize the amount of interest that will have to be paid on the loans. (Hint: Use the Excel function CUMIPMT to find the total interest that will be paid over the life of a loan. For example, if a 30-year loan for $100,000 has an interest rate of 9%, then the formula $= -\text{CUMIPMT}(9\%, 30, 100,000, 1, 30, 0)$ will yield $192,009 cumulative interest paid between years 1 and 30. (Note that this function yields a negative value, so include the minus sign.)

48. Marketing managers have various media alternatives in which to advertise and must determine which to use, the number of insertions in each, and the timing of insertions to maximize advertising effectiveness within a limited budget. Suppose that three media options are available to Kernan Services Corporation: radio, TV, and magazine. The following table provides some information about costs, exposure values, and bounds on the permissible number of ads in each medium desired by the firm. The exposure value is a measure of the number of people exposed to the

advertisement and is derived from market research studies. The company would like to achieve a total exposure value of at least 60,000.

Medium	Cost/Ad	Exposure Value/Ad	Min Units	Max Units
Radio	$500	2,000	0	15
TV	$2,000	4,000	10	no limit
Magazine	$200	2,700	6	12

How many of each type of ad should be placed to minimize the cost of achieving the minimum required total exposure?

49. Klein Industries manufactures three types of portable air compressors: small, medium, and large, which have unit profits of $20.50, $34.00, and $52.00, respectively. The projected monthly sales are as follows:

	Small	Medium	Large
Minimum	14,000	6,200	2,600
Maximum	21,000	12,500	4,200

The production process consists of three primary activities: bending and forming, welding, and painting. The amount of time in minutes needed to process each product in each department is as follows:

	Small	Medium	Large	Available Time
Bending/forming	0.4	0.7	0.8	23,400
Welding	0.6	1.0	1.2	23,400
Painting	1.4	2.6	3.1	46,800

How many of each type of air compressor should the company produce to maximize profit?

*50. Fruity Juices, Inc., produces five different flavors of fruit juice: apple, cherry, pomegranate, orange, and pineapple. Each batch of product requires processing in three departments (blending, straining, and bottling). The relevant data (per 1,000-gallon batch) are shown in the table *Data for Problem 50*. Formulate and solve a linear program to find the amount of each product to produce.

51. Worley Fluid Supplies produces three types of fluid handling equipment: control valves, metering pumps, and hydraulic cylinders. All three products require assembly and testing before they can be shipped to customers. The following data provide the number of minutes that each type requires in assembly and testing, the profit, and sales estimates.

	Control Valve	Metering Pump	Hydraulic Cylinder
Assembly time (min)	45	20	30
Testing time (min)	20	15	25
Profit/unit	$372	$174	$288
Maximum sales	20	50	45
Minimum sales	5	12	22

A total of 3,000 minutes of assembly time and 2,100 minutes of testing time are available next week. Develop and solve a linear optimization model to determine how many pieces of equipment the company should make next week to maximize profit contribution.

*52. MK Manufacturing produces compressor and turbine blades for jet engines. The blades are manufactured from an alloy that is a mix of aluminum and titanium. Sheaths are part of the assembly that keep the compressor blades lightweight and are made from

Data for Problem 50

	Apple	Cherry	Pomegranate	Orange	Pineapple	Minutes Avail.
	Time Required in Minutes/Batch					
Blend	23	22	18	19	19	5,000
Strain	22	40	20	31	28	3,000
Bottle	10	10	10	10	10	5,000

	Apple	Cherry	Pomegranate	Orange	Pineapple
	Profit and Sales Potential				
Profit ($/1,000 gal)	$800	$320	$1,120	$1,440	$800
Max Sales (000)	20	30	50	50	20
Min Sales (000)	10	15	20	40	10

steel. In addition to fabrication, the parts must be put through an acid bath to show any deformities during quality inspection. Compressor blades use 8 lb of aluminum, 2 lb of titanium, and 15 lb of acid and have a profit contribution of $3,000. Sheaths require 3 lb of steel and 5 lb of acid and yield $1,500 in profit. Finally, the turbine blades require 6 lb of aluminum, 1.5 lb of titanium, and 10 lb of acid and yield a profit of $2,000. Warehouse storage is limited, and it can handle a maximum of 14,000 lb of aluminum, 7,000 lb of titanium, 6,000 lb of steel, and 25,000 lb of acid. In the next planning period, orders call for 1,500 compressor blades, 1,500 sheaths, and 2,000 turbine blades; thus, at least this many of each must be produced. Note that compressor blades and sheaths are used together, and the company must produce an equal number of these parts. Formulate and solve a linear optimization model. You will discover that the solution is infeasible. Determine what must be done in order to obtain a feasible solution.

CASE: PERFORMANCE LAWN EQUIPMENT

Elizabeth Burke wants to develop a model to more effectively plan production for the next year. Currently, PLE has a planned capacity of producing 9,100 mowers each month, which is approximately the average monthly demand over the previous year. However, looking at the unit sales figures for the previous year, she observed that the demand for mowers has a seasonal fluctuation, so with this "level" production strategy, there is overproduction in some months, resulting in excess inventory buildup, and underproduction in others, which may result in lost sales during peak demand periods.

Ms. Burke explained that she could change the production rate by using planned overtime or undertime (producing more or less than the average monthly demand), but this incurs additional costs, although it may offset the cost of lost sales or of maintaining excess inventory. Consequently, she believes that the company can save a significant amount of money by optimizing the production plan.

Ms. Burke saw a presentation at a conference about a similar model that another company used but didn't fully understand the approach. The PowerPoint notes didn't have all the details, but they did explain the variables and the types of constraints used in the model. She thought they would be helpful to you in implementing an optimization model. Here are the highlights from the presentation:

Variables:

X_t = planned production in period t
I_t = inventory held at the end of period t
L_t = number of lost sales incurred in period t
O_t = amount of overtime scheduled in period t
U_t = amount of undertime scheduled in period t
R_t = increase in production rate from period $t - 1$ to period t

D_t = decrease in production rate from period $t - 1$ to period t

Material balance constraint:

$X_t + I_{t-1} - I_t + L_t$ = demand in month t

Overtime/undertime constraint:

$O_t - U_t = X_t$ − normal production capacity

Production rate-change constraint:

$X_t - X_{t-1} = R_t - D_t$

Ms. Burke also provided the following data and estimates for the next year: unit production cost = $70.00; inventory-holding cost = $1.40 per unit per month; lost sales cost = $200 per unit; overtime cost = $6.50 per unit; undertime cost = $3.00 per unit; and production-rate-change cost = $5.00 per unit, which applies to any increase or decrease in the production rate from the previous month. Initially, 900 units are expected to be in inventory at the beginning of January, and the production rate for the past December was 9,100 units. She believes that monthly demand will not change substantially from last year, so the mower unit sales figures for the last year in the *Performance Lawn Equipment Database* should be used for the monthly demand forecasts.

Your task is to design a spreadsheet that provides detailed information on monthly production, inventory, lost sales, and the different cost categories and solve a linear optimization model for minimizing the total cost of meeting demand over the next year. Compare your solution with the level production strategy of producing 9,100 units each month. Interpret the Sensitivity Report and conduct an appropriate study of how the solution will be affected by changing the assumption of the lost sales costs. Summarize all your results in a report to Ms. Burke.

Integer and Nonlinear Optimization

Jirsak/Shutterstock

LEARNING OBJECTIVES   After studying this chapter, you will be able to:

- Recognize when to use integer variables in optimization models.
- Incorporate integer variables into *Solver* models.
- Develop integer optimization models for practical applications such as workforce scheduling and location.
- Find alternative optimal solutions to integer optimization models.
- Formulate and solve optimization models with binary variables and logical constraints.
- Recognize when to use nonlinear optimization models.

- Develop and solve nonlinear optimization models for different applications.
- Interpret *Solver* reports for nonlinear optimization.
- Use empirical data and line-fitting techniques in nonlinear optimization.
- Recognize a quadratic optimization model.
- Identify non-smooth optimization models and when to use *Evolutionary Solver*.
- Formulate and solve sequencing and scheduling models using *Solver*'s *alldifferent* constraint.

In the previous chapter, we saw that the variables in linear optimization models can assume any real value. For many practical applications, we need not be concerned with this assumption. For example, in deciding on the optimal number of cases of diapers to produce next month, we could use a linear model, since rounding a value like 5,621.63 would have little impact on the results. However, in a production-planning decision involving low-volume, high-cost items such as airplanes, an optimal value of 10.42 would make little sense, and a difference of one unit (rounded up or down) could have significant economic and production planning consequences. In this situation, we would need to ensure that the solution is integer-valued. Similarly, linear functions may not be appropriate for modeling some objective functions or constraints, so we would need to use nonlinear functions to better reflect the problem. Building nonlinear optimization models requires more creativity and analytical expertise than linear or integer models.

In this chapter, we discuss how to build and solve integer and nonlinear optimization models and illustrate applications in a variety of practical problems. These types of models provide a lot of flexibility to handle many more realistic and difficult situations than do simple linear optimization models. For example, airlines use integer optimization to schedule crews and airline flight segments, and investment firms use nonlinear optimization to create client portfolios.

Integer Linear Optimization Models

Integer linear optimization models are simply linear models with added constraints on variables to ensure that they are integer-valued (that is, whole numbers). Decision variables that we force to be integers are called **general integer variables**. We may model any variable in an ordinary linear program to be a general integer variable simply by specifying it as an integer. For example, if in the Sklenka Ski Company (SSC) model (see Chapter 13) we wish to restrict the number of pairs of skis produced to be whole numbers, we would add the following constraints:

$$\text{Jordanelle} = \text{integer}$$
$$\text{Deercrest} = \text{integer}$$

Many optimization models require **binary variables**, which are variables that are restricted to being either 0 or 1. Mathematically, a binary variable x is simply a general integer variable that is restricted to being between 0 and 1:

$$0 \leq x \leq 1 \text{ and integer} \tag{14.1}$$

However, we usually just write this as $x = 0$ or 1. Binary variables enable us to model logical decisions in optimization problems. For example, binary variables can be used to model decisions such as whether or not to place a facility at a certain location, whether or not to run a production line, or whether or not to invest in a certain stock. For example, we could set $x = 1$ if we place a facility at the location, and $x = 0$ if we don't. Similarly, if we run the production line, we would set $x = 1$; if not, we would set $x = 0$.

Finally, any practical applications of optimization involve a combination of continuous variables, general integer variables, and/or binary variables. This provides the flexibility to model many different types of complex decision problems. However, in this book we will keep things simple and not address these types of models.

Models with General Integer Variables

If we solve the linear optimization model without the integer restrictions (called the **linear program [LP] relaxation**) and the optimal solution happens to have all integer values, then it clearly would have solved the integer model. This is generally not the case, however. The algorithm used to solve integer optimization models begins by solving the LP relaxation and proceeds to enforce the integer restrictions using a systematic search process that involves solving a series of modified linear optimization problems. You need not worry about understanding how this is accomplished, because *Solver* takes care of the algorithmic details.

When using *Solver*, it is important to set a parameter called *Integer Tolerance*. This value specifies when the *Solver* algorithm will terminate. By default, *Integer Tolerance* is set to 0.05 within *Solver*. This means that *Solver* will stop if it finds an integer solution that is within 5% of the optimal solution. With this value, you may end up with a solution that is not the optimum, but is 95% of the way there. It does this for computational efficiency because many practical problems take a very long time to solve, even with today's technology (hours or even days!). If an answer is needed quickly, a manager might be satisfied with a near-optimal solution that is guaranteed to be within a fixed percentage of the best. To find the guaranteed optimal integer solution, *Integer Tolerance* must be set to 0. To do this, click the *Options* button in the *Solver Parameters* dialog and ensure that the value of *Integer Optimality (%)* is 0.

EXAMPLE 14.1   **Sklenka Skis Revisited**

In Chapter 13, we developed a simple linear optimization model for finding the optimal product mix for a ski manufacturer. The model was

Maximize Total Profit = 50 Jordanelle + 65 Deercrest

$$3.5 \text{ Jordanelle} + 4 \text{ Deercrest} \leq 84$$

$$1 \text{ Jordanelle} + 1.5 \text{ Deercrest} \leq 21$$

$$\text{Deercrest} - 2 \text{ Jordanelle} \geq 0$$

$$\text{Deercrest} \geq 0$$

$$\text{Jordanelle} \geq 0$$

We saw that the optimal solution was to produce 5.25 pairs of Jordanelle skis and 10.5 pairs of Deercrest skis. Because the solution involves fractions, it would be beneficial to find the optimal solution for which the decision variables are integers. To do this, we simply add the constraints that Deercrest and Jordanelle must be integers to the model. Figure 14.1 shows the graphical illustration of the set of

feasible values (dark blue dots) that satisfy all constraints as well as the integer restrictions.

To enforce integer restrictions on variables using *Solver*, click the *Add* button to add a constraint. In the *Add Constraint* dialog, enter the variable range in the *Cell Reference* field and choose *int* from the drop-down box, as shown in Figure 14.2. We also need to ensure that we set the *Integer Tolerance* parameter to zero as discussed earlier. Figure 14.3 shows the resulting solution. Notice that the maximum value of the objective function for the model with integer restrictions is smaller than the linear optimization solution. This is expected because we have added an additional constraint (the integer restrictions). Whenever you add a constraint to a model, the value of the objective function can never improve and usually worsens. Figure 14.4 illustrates this graphically. As the profit line increases, the last feasible integer point through which it passes is (3, 12). Notice also that the optimal integer solution is not the same as the solution you would obtain from rounding the optimal solution to the LP relaxation.

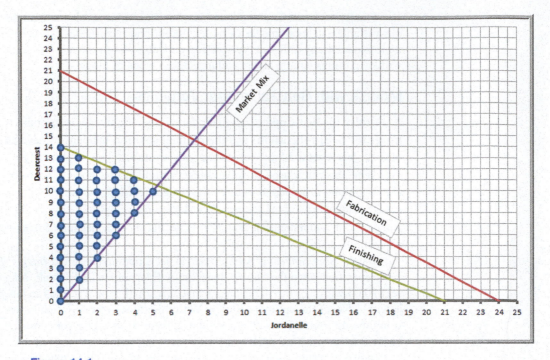

▲ **Figure 14.1**

Graphical Illustration of Feasible Integer Solutions for the Sklenka Ski Problem

▶ **Figure 14.2**

Defining General Integer Variables in Solver

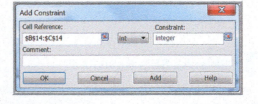

▶ **Figure 14.3**

Optimal Integer Solution to the Sklenka Ski Problem

	A	B	C	D
1	Sklenka Skis			
2				
3	Data			
4			Product	
5	Department	Jordanelle	Deercrest	Limitation (hours)
6	Fabrication	3.5	4	84
7	Finishing	1	1.5	21
8				
9	Profit/unit	$ 50.00	$ 65.00	
10				
11				
12	Model			
13		Jordanelle	Deercrest	
14	Quantity Produced	3	12	Hours Used
15	Fabrication	10.5	48	58.5
16	Finishing	3	18	21
17				
18				Excess Deercrest
19	Market mixture			6
20				
21				Total Profit
22	Profit Contribution	$ 150.00	$ 780.00	$ 930.00

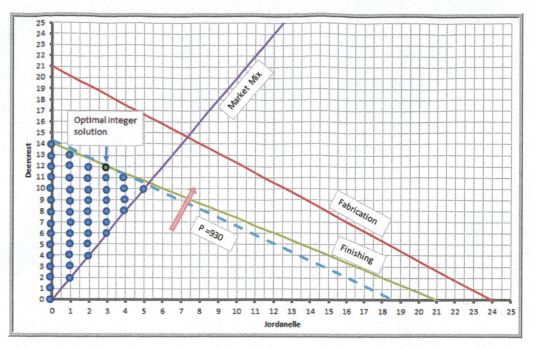

▲ **Figure 14.4**

Graphical Illustration of Optimal Integer Solution

If Sklenka Ski Company were a real company, they would be producing thousands of pairs of skis for the world market. As we noted, it probably would not make much difference if they simply rounded the optimal solution to the linear optimization model. In other types of models, however, it is critical to enforce solution to the integer restrictions. For example, the paper industry needs to find the best mix of cutting patterns to meet demand for various sizes of paper rolls. In a similar fashion, sheet steel producers cut strips of different sizes from rolled coils of thin steel. For these types of problems, fractional values for the decision variables make no sense at all. Finding the best solution for such problems requires integer optimization. Here is one example.

EXAMPLE 14.2   A Cutting-Stock Problem

Suppose that a company makes standard 110-inch-wide rolls of thin sheet metal and slits them into smaller rolls to meet customer orders for widths of 12, 15, and 30 inches. The demands for these widths vary from week to week.

From a 110-inch roll, there are many different ways to slit 12-, 15-, and 30-inch pieces. A *cutting pattern* is a configuration of the number of smaller rolls of each type that are cut from the raw stock. Of course, we would want to use as much of the roll as possible to avoid costly scrap. For example, we could cut seven 15-inch rolls, leaving a 5-inch piece of scrap, or cut three 30-inch rolls and one 12-inch roll, leaving 8 inches of scrap. Finding good cutting patterns for a large set of end products is, in itself, a challenging problem. Suppose that the company has proposed the following cutting patterns:

	Size of End Item			
Pattern	**12 in.**	**15 in.**	**30 in.**	**Scrap**
1	0	7	0	5 in.
2	0	1	3	5 in.
3	1	0	3	8 in.
4	9	0	0	2 in.
5	2	1	2	11 in.
6	7	1	0	11 in.

Demands for the coming week are 500 12-inch rolls, 715 15-inch rolls, and 630 30-inch rolls. The problem is to develop a model that will determine how many 110-inch rolls to cut into each of the six patterns to meet demand and minimize scrap.

(continued)

Define X_i to be the number of 110-inch rolls to cut using cutting pattern i, for $i = 1, \ldots, 6$. Note that X_i needs to be a whole number because each roll that is cut generates a different number of end items. Thus, X_i will be modeled using general integer variables. Because the objective is to minimize scrap, the objective function is

$$\text{Min } 5X_1 + 5X_2 + 8X_3 + 2X_4 + 11X_5 + 11X_6$$

The only constraints are that end-item demand must be met; that is, we must produce at least 500 12-inch rolls, 715 15-inch rolls, and 630 30-inch rolls. The number of end-item rolls produced is found by multiplying the number of end-item rolls produced by each cutting pattern by the number of 110-inch rolls cut using that pattern. Therefore, the constraints are

$$0X_1 + 0X_2 + 1X_3 + 9X_4 + 2X_5 + 7X_6 \geq 500 \quad \text{(12-inch rolls)}$$

$$7X_1 + 1X_2 + 0X_3 + 0X_4 + 1X_5 + 1X_6 \geq 715 \quad \text{(15-inch rolls)}$$

$$0X_1 + 3X_2 + 3X_3 + 0X_4 + 2X_5 + 0X_6 \geq 630 \quad \text{(30-inch rolls)}$$

Finally, we include nonnegativity and integer restrictions:

$$X_i \geq 0 \text{ and integer}$$

Figure 14.5 shows the cutting-stock model implementation on a spreadsheet (Excel file *Cutting-Stock Model*) with the optimal solution. The constraint functions for the number produced in cells B23:D23 and the objective function in cell B26 are SUMPRODUCT functions of the decision variables in B15:B20 and the data in rows 5 through 10. The *Solver* model is shown in Figure 14.6.

Workforce-Scheduling Models

Workforce scheduling is a practical yet highly complex problem that many businesses face. Many fast-food operations hire students who can work in only small chunks of time during the week, resulting in a huge number of possible schedules. In such operations, customer demand varies by day of week and time of day, further complicating the problem of assigning workers to time slots. Similar problems exist in scheduling nurses in hospitals, flight crews in airlines, and many other service operations.

▶ **Figure 14.5**

Spreadsheet Model and Optimal Solution for the Cutting-Stock Model

	A	B	C	D	E
1	Cutting Stock Model				
2					
3	Data				
4	Pattern	12-in rolls	15-in rolls	30-in rolls	Scrap
5	1	0	7	0	5
6	2	0	1	3	5
7	3	1	0	3	8
8	4	9	0	0	2
9	5	2	1	2	11
10	6	7	1	0	11
11	Demand	500	715	630	
12					
13	Model				
14		No. of rolls			
15	Pattern 1	73.00			
16	Pattern 2	210.00			
17	Pattern 3	0.00			
18	Pattern 4	56.00			
19	Pattern 5	0.00			
20	Pattern 6	0.00			
21					
22		12-in rolls	15-in rolls	30-in rolls	
23	Number produced	504	721	630	
24					
25		Total			
26	Scrap	1527			

▶ **Figure 14.6**

Solver Model for Cutting-Stock Problem

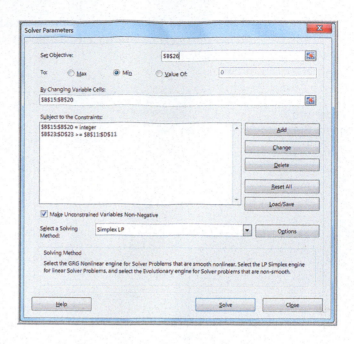

EXAMPLE 14.3   Brewer Services

Brewer Services contracts with outsourcing partners to handle various customer-service functions. The customer-service department is open Monday through Friday from 8 a.m. to 5 p.m. Calls vary over the course of a typical day. Based on a study of call volumes provided by one of the firm's partners, the minimum number of staff needed each hour of the day are as follows:

Hour	Minimum Staff Required
8–9	5
9–10	12
10–11	15
11–Noon	12
Noon–1	11
1–2	18
2–3	17
3–4	19
4–5	14

Mr. Brewer wants to hire some permanent employees and staff the remaining requirements using part-time employees who work four-hour shifts (four consecutive hours starting as early as 8 a.m. or as late as 1 p.m.). Suppose that Mr. Brewer has five permanent employees. What is the minimum number of part-time employees he will need for each four-hour shift to ensure meeting the staffing requirements?

Assuming that the five permanent employees work the full day, the part-time coverage requirements can be calculated by subtracting 5 from the minimum staff required

for each of the time slots in the table. Define X_i to be the number of part-time employees who will work a four-hour shift beginning at hour i, where $i = 1$ corresponds to an 8:00 a.m. start, $i = 2$ corresponds to a 9:00 a.m. start, and so on, with $i = 6$ corresponding to a 1:00 p.m. start as the last part-time shift. The objective is to minimize the total number of part-time employees:

$$Min\ X_1 + X_2 + X_3 + X_4 + X_5 + X_6$$

For each hour, we need to ensure that the total number of part-time employees who work that hour is at least as large as the minimum requirements. For example, only workers starting at 8:00 a.m. will cover the 8:00–9:00 time slot; thus,

$$X_1 \geq 0$$

Workers starting at either 8:00 a.m. or 9:00 a.m. will cover the second time slot; therefore,

$$X_1 + X_2 \geq 7$$

The remaining constraints are

$$X_1 + X_2 + X_3 \geq 10$$
$$X_1 + X_2 + X_3 + X_4 \geq 7$$
$$X_2 + X_3 + X_4 + X_5 \geq 6$$
$$X_3 + X_4 + X_5 + X_6 \geq 13$$
$$X_4 + X_5 + X_6 \geq 12$$
$$X_5 + X_6 \geq 14$$
$$X_6 \geq 9$$

All the variables must also be integers.

Figures 14.7 and 14.8 show the spreadsheet with the optimal solution (Excel file *Brewer Services*) and *Solver* models for this example. The optimal solution is to hire 24 part-time workers.

Figure 14.7

Spreadsheet Model with Optimal Solution for Brewer Services

	A	B	C	D	E
1	Brewer Services				
2					
3	Data				
4		Permanent Employees			
5		5			
6			Part Time Coverage		
7	Hour	Minimum Staff Required	Minimum Requirements		
8	8-9	5	0		
9	9-10	12	7		
10	10-11	15	10		
11	11-noon	12	7		
12	noon-1	11	6		
13	1-2	18	13		
14	2-3	17	12		
15	3-4	19	14		
16	4-5	14	9		
17					
18	Model				
19				Total part-time	
20	Shift	Number of PT employees	Hour	employees	Excess
21	1	7	8-9	7	7
22	2	0	9-10	7	0
23	3	3	10-11	10	0
24	4	0	11-noon	10	3
25	5	5	noon-1	8	2
26	6	9	1-2	17	4
27	Total	24	2-3	14	2
28			3-4	14	0
29			4-5	9	0

	A	B	C	D	E
1	Brewer Services				
2					
3	Data				
4		Permanent Employees			
5		5			
6			Part Time Coverage		
7	Hour	Minimum Staff Required	Minimum Requirements		
8	8-9	5	=B8-$B$5		
9	9-10	12	=B9-$B$5		
10	10-11	15	=B10-$B$5		
11	11-noon	12	=B11-$B$5		
12	noon-1	11	=B12-$B$5		
13	1-2	18	=B13-$B$5		
14	2-3	17	=B14-$B$5		
15	3-4	19	=B15-$B$5		
16	4-5	14	=B16-$B$5		
17					
18	Model				
19				Total part-time	
20	Shift	Number of PT employees	Hour	employees	Excess
21	1	7	8-9	=B21	=D21-C8
22	2	0	9-10	=SUM(B21:B22)	=D22-C9
23	3	3	10-11	=SUM(B21:B23)	=D23-C10
24	4	0	11-noon	=SUM(B21:B24)	=D24-C11
25	5	5	noon-1	=SUM(B22:B25)	=D25-C12
26	6	9	1-2	=SUM(B23:B26)	=D26-C13
27	Total	=SUM(B21:B26)	2-3	=SUM(B24:B26)	=D27-C14
28			3-4	=SUM(B25:B26)	=D28-C15
29			4-5	=B26	=D29-C16

▶ **Figure 14.8**

Solver *Model for* Brewer Services

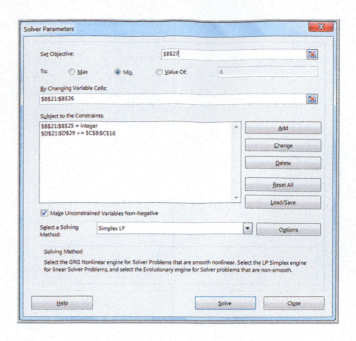

Alternative Optimal Solutions

In looking at the solution, a manager might not be satisfied with the distribution of workers, particularly the fact that there are seven excess employees during the first hour. In most scheduling problems, many alternative optimal solutions usually exist. A little creativity in using the optimization model can help identify these.

| EXAMPLE 14.4 | **Finding Alternative Optimal Solutions for *Brewer Services* Model** |

An easy way to find an alternative optimal solution that reduces the number of excess employees at 8:00 a.m. is to define a constraint setting the objective function equal to its optimal value and then changing the objective function in the model to minimize the number of excess employees during the first hour. Figure 14.9 shows the modified *Solver* model with the constraint

$$X_1 + X_2 + X_3 + X_4 + X_5 + X_6 = 24$$

and the new objective function to minimize the excess number of employees at 8:00 a.m., the value in cell E21. The solution is shown in Figure 14.10. In a "whack-a-mole"

fashion, we now have nine excess employees during the noon hour, a solution which isn't any better than the original one.

A better approach would be to define additional constraints to restrict the excess number of employees in the range E21:E29 to be less than or equal to some maximum number k and then attempt to minimize the original objective function. This *Solver* model is shown in Figure 14.11. If we do this, we find that the smallest value of k that results in a feasible solution is $k = 3$. The result is shown in Figure 14.12. We have achieved a better balance while still maintaining the minimum number of part-time employees.

▶ **Figure 14.9**

Modified Solver Model to Identify an Alternate Optimal Solution

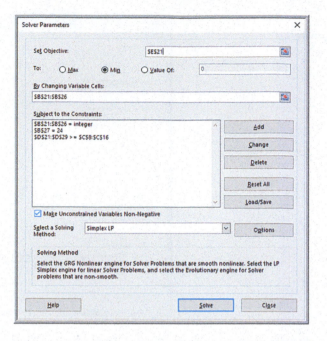

▶ **Figure 14.10**

Alternative Optimal Solution to Brewer Services Problem

	A	B		C	D	E
18	**Model**					
19					**Total part-time**	
20	**Shift**	**Number of PT employees**		**Hour**	**employees**	**Excess**
21	1	0		8-9	0	0
22	2	7		9-10	7	0
23	3	3		10-11	10	0
24	4	0		11-noon	10	3
25	5	5		noon-1	15	9
26	6	9		1-2	17	4
27	**Total**	**24**		2-3	14	2
28				3-4	14	0
29				4-5	9	0

▶ **Figure 14.11**

Solver Model with Constraints on Excess Employees

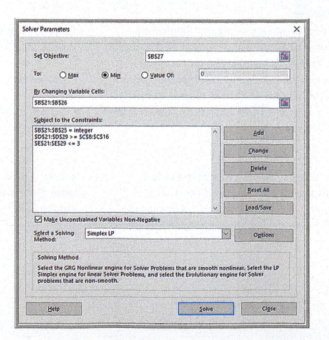

▶ **Figure 14.12**

Improved Alternative Optimal Solution to Brewer Services Problem

	A	B		C	D	E
18	Model					
19					Total part-time	
20	Shift	Number of PT employees		Hour	employees	Excess
21	1	3		8-9	3	3
22	2	5		9-10	8	1
23	3	2		10-11	10	0
24	4	0		11-noon	10	3
25	5	2		noon-1	9	3
26	6	12		1-2	16	3
27	Total	24		2-3	14	2
28				3-4	14	0
29				4-5	12	3

CHECK YOUR UNDERSTANDING

1. Explain the difference between general integer variables and binary variables.

2. What changes must you make in *Solver* to solve integer optimization models?

3. Explain how to find alternate optimal solutions in workforce-scheduling models.

Models with Binary Variables

Binary variables provide an incredible amount of flexibility in optimization modeling. One common example we present next is project selection, in which a subset of potential projects must be selected with limited resource constraints. Capital-budgeting problems in finance have a similar structure.

EXAMPLE 14.5   Hahn Engineering

Hahn Engineering's research and development group has identified five potential new engineering and development projects; however, the firm is constrained by its available budget and human resources. Each project is expected to generate a return (given by the net present value [NPV]) but requires a fixed amount of cash and personnel. Because the resources are limited, all projects cannot be selected. Projects cannot be partially completed; thus, either the project must be undertaken completely or not at all. The data are given in Table 14.1. If a project is selected, it generates the full value of the expected return and requires the full amount of cash and personnel shown in Table 14.1. For example, if we select projects 1 and 3, the total return is $180,000 + $150,000 = $330,000, and these projects require cash totaling $55,000 + $24,000 = $79,000 and 5 + 2 = 7 personnel.

To model this situation, we define the decision variables to be binary, corresponding to either not selecting or selecting each project, respectively. Define $x_i = 1$ if project i is selected and 0 if it is not selected. By multiplying

these binary variables by the expected returns, the objective function is

$$\text{Maximize } \$180{,}000x_1 + \$220{,}000x_2 + \$150{,}000x_3$$
$$+ \$140{,}000x_4 + \$200{,}000x_5$$

Because cash and personnel are limited, we have the following constraints:

$$\$55{,}000x_1 + \$83{,}000x_2 + \$24{,}000x_3 + \$49{,}000x_4$$
$$+ \$61{,}000x_5 \leq \$150{,}000 \quad \text{(cash limitation)}$$
$$5x_1 + 3x_2 + 2x_3 + 5x_4 + 3x_5 \leq 12 \quad \text{(personnel limitation)}$$

Note that if projects 1 and 3 are selected, then $x_1 = 1$ and $x_3 = 1$, and the objective and constraint functions equal

$$\text{Return} = \$180{,}000(1) + \$220{,}000(0) + \$150{,}000(1)$$
$$+ \$140{,}000(0) + \$200{,}000(0) = \$330{,}000$$
$$\text{Cash Required} = \$55{,}000(1) + \$83{,}000(0) + \$24{,}000(1)$$
$$+ \$49{,}000(0) + \$61{,}000(0) = \$79{,}000$$
$$\text{Personnel Required} = 5(1) + 3(0) + 2(1) + 5(0) + 3(0) = 7$$

▼ Table 14.1

Project Selection Data

	Project 1	Project 2	Project 3	Project 4	Project 5	Available Resources
Expected return (NPV)	$180,000	$220,000	$150,000	$140,000	$200,000	
Cash requirements	$55,000	$83,000	$24,000	$49,000	$61,000	$150,000
Personnel requirements	5	3	2	5	3	12

This model is easy to implement on a spreadsheet, as shown in Figure 14.13 (Excel file *Hahn Engineering Project Selection*). The decision variables are defined in cells B11:F11. By multiplying these values by the data for each project in rows 5–7, we can easily compute the total return, cash used, and personnel used for the projects that are selected in rows 12–14. The objective function is computed in cell G12 as the sum of the returns for the selected projects. Similarly, the amounts of cash and personnel used are also summed for the projects selected, representing the constraint functions in cells G13 and G14. The optimal solution is to select projects 1, 3, and 5 for a total return of $530,000.

The *Solver* model is shown in Figure 14.14. To invoke the binary constraints on the variables, use the same process as defining integer variables, but choose *bin* from the drop-down box in the *Add Constraint* dialog. The resulting constraint is $B11:$F11 = binary, as shown in the *Solver* model.

Using Binary Variables to Model Logical Constraints

Binary variables allow us to model a wide variety of logical constraints. For example, suppose that if project 1 is selected, then project 4 must also be selected. Your first thought might be to incorporate an IF function in the Excel model. However, recall that we noted in Chapter 13 that such functions destroy the linearity property of the Excel model; therefore, we need to express such constraints differently. If project 1 is selected, then $x_1 = 1$, and we want to force x_4 to be 1 also. This can be done using the following constraint:

$$x_4 \geq x_1$$

Mathematically, if $x_1 = 1$, then this constraint implies that $x_4 \geq 1$ and, consequently, x_4 must equal 1. If $x_1 = 0$, then $x_4 \geq 0$ and x_4 can be either 0 or 1. Table 14.2 summarizes how to model a variety of logical conditions using binary variables.

▶ Table 14.2

Modeling Logical Conditions Using Binary Variables

Logical Condition	Constraint Model Form
If A, then B	$B \geq A$ or $B - A \geq 0$
If not A, then B	$B \geq 1 - A$ or $A + B \geq 1$
If A, then not B	$B \leq 1 - A$ or $A + B \leq 1$
At most one of A and B	$A + B \leq 1$
If A, then B and C	$(B \geq A$ and $C \geq A)$ or $B + C \geq 2A$
If A and B, then C	$C \geq A + B - 1$ or $A + B - C \leq 1$

▶ **Figure 14.13**

Spreadsheet Model for Project Selection Problem

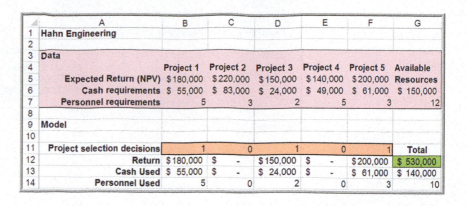

	A	B	C	D	E	F	G
1	Hahn Engineering						
2							
3	**Data**						
4		Project 1	Project 2	Project 3	Project 4	Project 5	Available
5	Expected Return (NPV)	$180,000	$220,000	$150,000	$140,000	$200,000	Resources
6	Cash requirements	$ 55,000	$ 83,000	$ 24,000	$ 49,000	$ 61,000	$ 150,000
7	Personnel requirements	5	3	2	5	3	12
8							
9	**Model**						
10							
11	Project selection decisions	1	0	1	0	1	Total
12	Return	$180,000	$ -	$150,000	$ -	$200,000	$ 530,000
13	Cash Used	$ 55,000	$ -	$ 24,000	$ -	$ 61,000	$ 140,000
14	Personnel Used	5	0	2	0	3	10

▶ **Figure 14.14**

Solver *Model for* Hahn Engineering Project Selection *Problem*

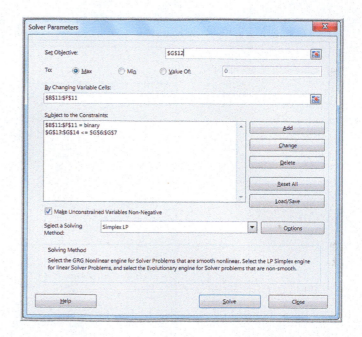

EXAMPLE 14.6   **Adding Logical Constraints into the Project Selection Model**

Suppose that we want to ensure that if project 1 is selected, then project 4 is selected, and that at most one of projects 1 and 3 can be selected in the Hahn Engineering model. To incorporate the constraint $x_4 \geq x_1$, write it as $x_4 - x_1 \geq 0$ by defining a cell for the constraint function $x_4 - x_1$ (cell B17 in Figure 14.15). Similarly, for the constraint $x_1 + x_3 \leq 1$, define a cell for $x_1 + x_3$ (cell

B18 in Figure 14.15). Then add these constraints to the *Solver* model, as shown in Figure 14.16 (Excel file *Hahn Engineering Project Selection with Logical Conditions*). In the optimal solution, we do not select project 1, although project 4 is selected anyway. With the additional constraints, the expected return is smaller than the original solution.

Applications in Supply Chain Optimization

Supply chain optimization is one of the broadest applications of integer optimization and is used extensively today as companies seek to reduce logistics costs and improve customer service in tough economic environments.

▶ **Figure 14.15**

Modified Project Selection Model with Logical Conditions

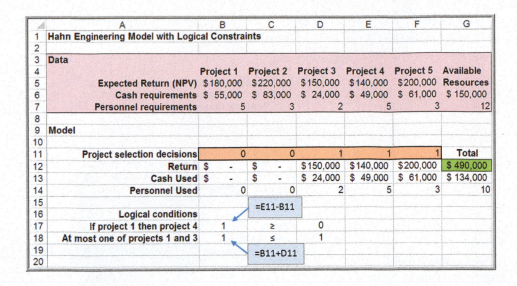

	A	B	C	D	E	F	G
1	Hahn Engineering Model with Logical Constraints						
2							
3	**Data**						
4		Project 1	Project 2	Project 3	Project 4	Project 5	Available
5	Expected Return (NPV)	$180,000	$220,000	$150,000	$140,000	$200,000	Resources
6	Cash requirements	$ 55,000	$ 83,000	$ 24,000	$ 49,000	$ 61,000	$ 150,000
7	Personnel requirements	5	3	2	5	3	12
8							
9	**Model**						
10							
11	Project selection decisions	0	0	1	1	1	Total
12	Return	$ -	$ -	$150,000	$140,000	$200,000	$ 490,000
13	Cash Used	$ -	$ -	$ 24,000	$ 49,000	$ 61,000	$ 134,000
14	Personnel Used	0	0	2	5	3	10
15							
16	Logical conditions		=E11−B11				
17	If project 1 then project 4	1	≥	0			
18	At most one of projects 1 and 3	1	≤	1			
19			=B11+D11				
20							

▶ **Figure 14.16**

Modified Solver Model with Logical Conditions

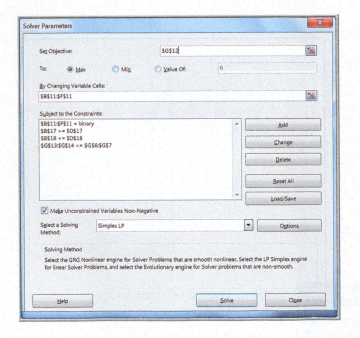

Suppose that a company has numerous potential locations for distribution centers that will ship products to many customers and wants to redesign its supply chain by selecting a fixed number of distribution centers. In an effort to provide exceptional customer service, some companies have a single-sourcing policy—that is, every customer can be supplied from only one distribution center. The problem is to determine how to assign customers to the distribution centers so as to minimize the total cost of shipping to the customers.

Define $X_{ij} = 1$ if customer j is assigned to distribution center i and 0 if it is not; $Y_i = 1$ if distribution center i is chosen from among a set of potential locations and 0 if it is not chosen; and C_{ij} = the total cost of satisfying the demand of customer j from distribution center i. We wish to minimize the total cost, ensure that every customer is assigned to one and only one distribution center, and select k distribution centers from the set of potential locations. This can be accomplished by the following model:

$$\text{Min} \sum_i \sum_j C_{ij} X_{ij}$$

$$\sum_i X_{ij} = 1, \text{ for every } j$$

$$\sum_i Y_i = k$$

$$X_{ij} \leq Y_i, \text{ for every } i \text{ and } j$$

$$X_{ij} \text{ and } Y_i \text{ are binary}$$

The first constraint ensures that each customer is assigned to exactly one distribution center. The next constraint limits the number of distribution centers selected. The final constraint ensures that customer j cannot be assigned to distribution center i unless that distribution center is selected in the supply chain. This is similar to the logical constraints we described in Table 14.2. If $Y_i = 1$, then any customer may be assigned to distribution center i; if $Y_i = 0$, then X_{ij} is forced to be 0 for all customers j because distribution center i is not selected.

EXAMPLE 14.7   Paul & Giovanni Foods

Paul & Giovanni (PG) Foods distributes supplies to restaurants in five major cities: Houston, Las Vegas, New Orleans, Chicago, and San Francisco. In a study to reconfigure their supply chain, they have identified four possible locations for distribution centers: Los Angeles, Denver, Pensacola, and Cincinnati. The costs of supplying each customer city from each possible distribution center are shown in the table below. PG Foods wishes to determine the best supply chain configuration to minimize cost.

Define $X_{ij} = 1$ if customer city j is assigned to distribution center i and 0 if not, and $Y_i = 1$ if distribution center i is chosen from among a set of potential locations. The integer optimization model is

Minimize $\$40{,}000X_{11} + \$11{,}000X_{12} + \$75{,}000X_{13}$

$+ \$70{,}000X_{14} + \$60{,}000X_{15} + \$72{,}000X_{21} + \$77{,}000X_{22}$

$+ \$120{,}000X_{23} + \$30{,}000X_{24} + \$75{,}000X_{25} + \$24{,}000X_{31}$

$+ \$44{,}000X_{32} + \$45{,}000X_{33} + \$80{,}000X_{34} + \$90{,}000X_{35}$

$+ \$32{,}000X_{41} + \$55{,}000X_{42} + \$90{,}000X_{43} + \$20{,}000X_{44}$

$+ \$105{,}000X_{45}$

$$X_{11} + X_{21} + X_{31} + X_{41} = 1$$

$$X_{12} + X_{22} + X_{32} + X_{42} = 1$$

$$X_{13} + X_{23} + X_{33} + X_{43} = 1$$

$$X_{14} + X_{24} + X_{34} + X_{44} = 1$$

$$X_{15} + X_{25} + X_{35} + X_{45} = 1$$

$$Y_1 + Y_2 + Y_3 + Y_4 = k$$

$X_{ij} \leq Y_i$, for every i and j (e.g., $X_{11} \leq Y_1$, $X_{21} \leq Y_1$, and so on)

X_{ij} and Y_i are binary

Figure 14.17 shows a spreadsheet model and the optimal solution for $k = 2$ (Excel file *Paul & Giovanni Foods*); Figure 14.18 shows the *Solver* model. We see that the distribution centers in Los Angeles and Cincinnati should be chosen, with Los Angeles serving Las Vegas, New Orleans, and San Francisco, and Cincinnati serving Houston and Chicago.

This model can easily be used to evaluate alternatives for different values of k. The supply chain manager can use this information to determine the trade-offs associated with opening different numbers of distribution centers.

Sourcing Costs	Houston	Las Vegas	New Orleans	Chicago	San Francisco
Los Angeles	$40,000	$11,000	$75,000	$70,000	$60,000
Denver	$72,000	$77,000	$120,000	$30,000	$75,000
Pensacola	$24,000	$44,000	$45,000	$80,000	$90,000
Cincinnati	$32,000	$55,000	$90,000	$20,000	$105,000

▶ **Figure 14.17**

Spreadsheet Model and Optimal Solution for Paul & Giovanni Foods *for k = 2*

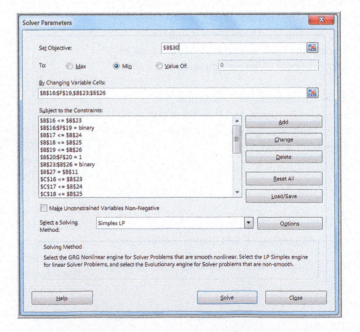

	A	B	C	D	E	F
1	Paul & Giovanni Foods					
2						
3	Data					
4						
5	Sourcing Costs	Houston	Las Vegas	New Orleans	Chicago	San Francisco
6	Los Angeles	$40,000	$11,000	$75,000	$70,000	$60,000
7	Denver	$72,000	$77,000	$120,000	$30,000	$75,000
8	Pensacola	$24,000	$44,000	$45,000	$80,000	$90,000
9	Cincinnati	$32,000	$55,000	$90,000	$20,000	$105,000
10						
11	Number of DCs	2				
12						
13	Model					
14						
15	Customer Assignments	Houston	Las Vegas	New Orleans	Chicago	San Francisco
16	Los Angeles	0	1	1	0	1
17	Denver	0	0	0	0	0
18	Pensacola	0	0	0	0	0
19	Cincinnati	1	0	0	1	0
20	Sum	1	1	1	1	1
21						
22	DCs Chosen					
23	Los Angeles	1				
24	Denver	0				
25	Pensacola	0				
26	Cincinnati	1				
27	Sum	2				
28						
29		Total				
30	Cost	$ 198,000				

▶ **Figure 14.18**

Solver *Model for* Paul & Giovanni Foods
(*Note:* All constraints are not visible in the constraint window.)

Solver Parameters

Se$t Objective: $B$30

To: ◯ Max   ◉ Min   ◯ Value Of: 0

By Changing Variable Cells:
$B$16:$F$19,$B$23:$B$26

Subject to the Constraints:

$B$16 <= $B$23
$B$16:$F$19 = binary
$B$17 <= $B$24
$B$18 <= $B$25
$B$19 <= $B$26
$B$20:$F$20 = 1
$B$23:$B$26 = binary
$B$27 = $B$11
$C$16 <= $B$23
$C$17 <= $B$24
$C$18 <= $B$25

[Add] [Change] [Delete] [Reset All] [Load/Save]

☐ Make Unconstrained Variables Non-Negative

Select a Solving Method: Simplex LP   [Options]

Solving Method

Select the GRG Nonlinear engine for Solver Problems that are smooth nonlinear. Select the LP Simplex engine for linear Solver Problems, and select the Evolutionary engine for Solver problems that are non-smooth.

[Help]   [Solve]   [Close]

CHECK YOUR UNDERSTANDING

1. Why are binary variables needed for project selection models?

2. Explain, using examples, how to use binary variables to model various logical constraints.

3. Explain how the constraint $X_{ij} \leq Y_i$ is used in the supply chain optimization model.

ANALYTICS IN PRACTICE: Supply Chain Optimization at Procter & Gamble

In 1993, Procter & Gamble (P&G) began an effort entitled Strengthening Global Effectiveness (SGE) to streamline work processes, drive out non-value-added costs, and eliminate duplication.[1]

A principal component of SGE was the North American Product Supply Study, designed to reexamine and reengineer P&G's product-sourcing and distribution system for its North American operations, with an emphasis on plant consolidation. Prior to the study, the North American supply chain consisted of hundreds of suppliers, more than 50 product categories, more than 60 plants, 15 distribution centers, and more than 1,000 customers. The need to consolidate plants was driven by the move to global brands and common packaging and the need to reduce manufacturing expense, improve speed to market, avoid major capital investments, and deliver better consumer value.

Kheng Guan Toh/Shutterstock

P&G had a policy of single sourcing; therefore, one of the key submodels in the overall optimization effort was the customer assignment optimization model described in this section to identify optimal distribution center locations in the supply chain and to assign customers to the distribution centers. Customers were aggregated into 150 zones. The parameter *k* was varied by the analysis team to examine the effects of choosing different numbers of locations. This model was used in conjunction with a simple transportation model for each of 30 product categories. Product-strategy teams used these models to specify plant locations and capacity options and optimize the flow of product from plants to distribution centers and customers. In reconfiguring the supply chain, P&G realized annual cost savings of more than $250 million.

Nonlinear Optimization Models

A **nonlinear optimization model** is one in which the objective function and/or at least one constraint is nonlinear. Nonlinear models are generally more difficult to develop than linear models simply because the objective function and constraint functions may not have the familiar linear structure. Thus, building nonlinear optimization models relies on fundamental modeling principles that we introduced in Chapter 11. For example, we may use business logic to identify the appropriate functional relationships among the decision variables or empirical data and line-fitting techniques to characterize the nonlinearities. In addition, nonlinear optimization models are considerably more difficult to solve than either linear or integer models. However, *Solver* provides solution procedures that can solve nonlinear optimization problems quite effectively.

A Nonlinear Pricing Decision Model

In Example 1.6 in Chapter 1, we introduced a simple nonlinear prescriptive model. In this example, a market research study collected data that estimated the expected annual sales

[1]Based on Jeffrey D. Camm, Thomas E. Chorman, Franz A. Dill, James R. Evans, Dennis J. Sweeney, and Glenn W. Wegryn, "Blending OR/MS, Judgment, and GIS: Restructuring P&G's Supply Chain," *Interfaces*, 27, 1 (January–February, 1997): 128–142.

for different levels of pricing. Analysts determined that sales can be expressed by the following model:

$$\text{Sales} = -2.9485 \times \text{Price} + 3{,}240.92$$

Using the fact that revenue equals price times sales, we express total revenue as

$$\text{Total Revenue} = \text{Price} \times \text{Sales}$$
$$= \text{Price} \times (-2.9485 \times \text{Price} + 3{,}240.92)$$
$$= -2.9485 \times \text{Price}^2 + 3{,}240.92 \times \text{Price}$$

Note that the total revenue function contains both a linear and a squared term for price, making the model nonlinear.

EXAMPLE 14.8 **Solving the Pricing Decision Model**

A spreadsheet model for this problem is shown in Figure 14.19. To find the maximum revenue using *Solver*, we identify the objective function cell as B13 and the decision variable cell as B5. The model has no constraints. In *Solver*, you should select "GRG Nonlinear" as the solving method. The *Solver* model is shown in Figure 14.20. The optimal solution is to set the price at about $549 and achieve a total revenue of $890,574.

▶ **Figure 14.19**

Optimal Solver *Solution for Pricing Decision Model*

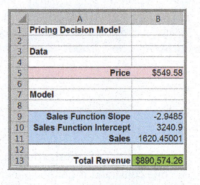

▶ **Figure 14.20**

Solver *Model for Pricing Decision*

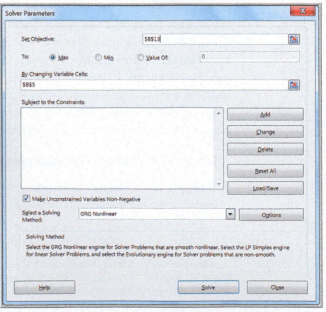

The next example shows a more complex pricing decision model involving multiple decision variables.

EXAMPLE 14.9	**A Hotel Pricing Model**

The Marquis Hotel is considering a major remodeling effort and needs to determine the best rates and room sizes to maximize revenues. Currently, the hotel has 450 rooms with the following history:

Room Type	Rate	Daily Avg. No. Sold	Revenue
Standard	$85	250	$21,250
Gold	$98	100	$9,800
Platinum	$139	50	$6,950
Total Revenue			$38,000

Each market segment has its own price/demand elasticity. Estimates are as follows:

Room Type	Price Elasticity of Demand
Standard	−1.5
Gold	−2.0
Platinum	−1.0

This means, for example, that a 1% *decrease* in the price of a standard room will *increase* the number of rooms sold by 1.5%. Similarly, a 1% increase in the price will decrease the number of rooms sold by 1.5%. For any pricing structure (in $), the projected number of rooms of a given type sold (we allow continuous values for this example) can be found using the formula

Historical Average Number of Rooms Sold +

[(Elasticity) × (New Price − Current Price)
× (Historical Average Number of Rooms Sold)]
Current Price

The hotel owners want to keep the price of a standard room between $70 and $90, a gold room between $90 and $110, and a platinum room between $120 and $149. Define S = price of a standard room, G = price of a gold room, and P = price of a platinum room. Then, for standard rooms, the projected number of rooms sold is $250 − [1.5(S − 85)(250)]/85 = 625 − 4.41176S$. The objective is to set the room prices to maximize total revenue. Total revenue will equal the price multiplied by the projected number of rooms sold, summed over all three types of rooms. Therefore, total revenue will be

Total Revenue
$$= S(625 − 4.41176S) + G(300 − 2.04082G) + P(100 − 0.35971P)$$
$$= 625S + 300G + 100P − 4.41176S^2 − 2.04082G^2 − 0.35971P^2$$
To keep prices within the stated ranges, we need constraints:

$$70 \le S \le 90$$
$$90 \le G \le 110$$
$$120 \le P \le 149$$

Finally, although the rooms may be renovated, there are no plans to expand beyond the current 450-room capacity. Thus, the projected number of total rooms sold cannot exceed 450:

$$(625 − 4.41176S) + (300 − 2.04082G) + (100 − 0.35971P) \le 450$$

or

$$1,025 − 4.41176S − 2.04082G − 0.35971P \le 450$$

The complete model is

Maximize $625S + 300G + 100P − 4.41176S^2 − 2.04082G^2 − 0.35971P^2$
$$70 \le S \le 90$$
$$90 \le G \le 110$$
$$120 \le P \le 149$$
$$1,025 − 4.41176S − 2.04082G − 0.35971P \le 450$$

Figure 14.21 shows a spreadsheet model (Excel file *Hotel Pricing Model*) for this example with the optimal solution. The decision variables, the new prices to charge, are given in cells B15:B17. The projected numbers of rooms sold are computed in cells E15:E17 using the preceding formula. By multiplying the number of rooms sold by the new price for each room type, the projected revenue is calculated, as given in cells F15:F17. The total revenue in cell F18 represents the objective function.

Note that it is easier to formulate this model as a financial spreadsheet than to enter the analytical formulas as they were developed. The *Solver* model is shown in Figure 14.22. The optimal prices predict a demand for all 450 rooms with total revenue of $39,380.65.

▶ **Figure 14.21**

Spreadsheet for Hotel Pric-ing Model

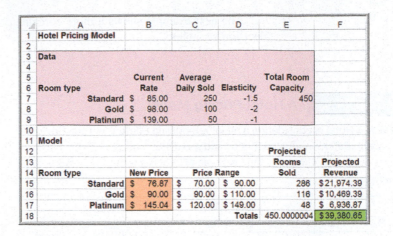

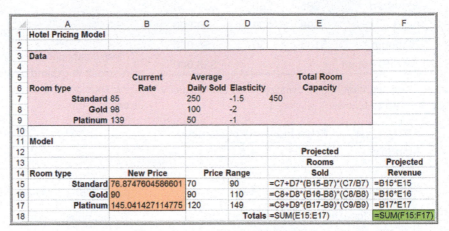

▶ **Figure 14.22**

Hotel Pricing Example Solver Model

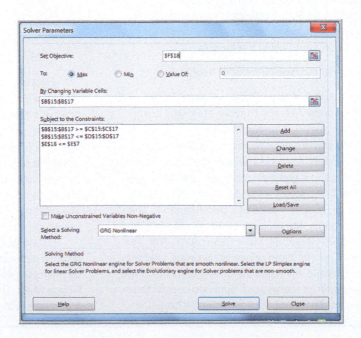

Quadratic Optimization

A special class of nonlinear optimization models is known as a quadratic optimization model, sometimes called a quadratic programming model. A **quadratic optimization model** is one that has a quadratic objective and all linear constraints. Recall from algebra that a quadratic function is $f(x) = ax^2 + bx + c$. In other words, a quadratic function has only constant, linear, and squared terms. Quadratic optimization models can be solved using the *Standard LP/Quadratic* solving method within *Solver*.

The **Markowitz portfolio model** is a classic quadratic optimization model in finance that seeks to minimize the risk of a portfolio of stocks subject to a constraint on the portfolio's expected return.[2] The decision variables are the percent of each stock to allocate to the portfolio. (You might be familiar with the term *asset allocation model* that many financial investment companies suggest to their clients—for example, "maintain 60% equities, 30% bonds, and 10% cash.") Recall from Chapter 4 that we can measure risk by the standard deviation, or, equivalently, the variance. In the Markowitz model, the objective function is to minimize the risk of the portfolio as measured by its variance. Because stock prices are correlated with one another, the variance of the portfolio must reflect not only variances of the stocks in the portfolio but also the covariance between stocks.

Define x_j to be the fraction of the portfolio to invest in stock j. The variance of a portfolio consisting of k stocks is the weighted sum of the variances and covariances:

$$\text{Variance of Portfolio} = \sum_{i=1}^{k} s_i^2 x_i^2 + \sum_{i=1}^{k}\sum_{j>i} 2s_{ij} x_i x_j \tag{14.2}$$

where

$s_i^2 = $ the sample variance in the return of stock i
$s_{ij} = $ the sample covariance between stocks i and j

EXAMPLE 14.10 An Example of the Markowitz Model

Suppose an investor is considering three stocks in which to invest. The expected return for stock 1 is 10%, for stock 2, 12%, and for stock 3, 7%, and she would like an expected return of at least 10%. Clearly one option is to invest everything in stock 1; however, this may not be a good idea because the risk might be too high. Research has found the variance-covariance matrix of the individual stocks to be the following:

	Stock 1	Stock 2	Stock 3
Stock 1	0.025	0.015	−0.002
Stock 2		0.030	0.005
Stock 3			0.004

Using these data and formula (14.2), the objective function is

Minimize Variance $= 0.025x_1^2 + 0.030x_2^2 + 0.004x_3^2$
$+ 2(0.015)x_1x_2 + 2(-0.002)x_1x_3$
$+ 2(0.005)x_2x_3$

The constraints must first ensure that we invest 100% of our budget. Because the variables are defined as fractions, we must have

$x_1 + x_2 + x_3 = 1$

Second, the portfolio must have an expected return of at least 10%. The return on a portfolio is simply the weighted sum of the returns of the stocks in the portfolio. This results in the constraint

$10x_1 + 12x_2 + 7x_3 \geq 10$

Finally, we will assume that we cannot invest negative amounts:

$x_1, x_2, x_3 \geq 0$

The complete model is

Minimize Variance $= 0.025x_1^2 + 0.030x_2^2 + 0.004x_3^2$
$+ 0.03x_1x_2 - 0.004x_1x_3$
$+ 0.010x_2x_3$

$x_1 + x_2 + x_3 = 1$
$10x_1 + 12x_2 + 7x_3 \geq 10$
$x_1, x_2, x_3 \geq 0$

[2]H.M. Markowitz, *Portfolio Selection, Efficient Diversification of Investments* (New York: John Wiley & Sons, 1959).

Figure 14.23 shows a spreadsheet model for this example (Excel file *Markowitz Model*); Figure 14.24 shows the *Solver* model. The minimum variance of the optimal portfolio is 0.012.

Practical Issues Using *Solver* for Nonlinear Optimization

Many nonlinear problems are notoriously difficult to solve. *Solver* cannot guarantee that it will find the absolute best solution (called a global optimal solution) for all problems. A **local optimum solution** is one for which all points close by are no better than the solution (an analogy is being at the top of a mountain when the highest peak is on another mountain). The solution found often depends a great deal on the starting solution in your spreadsheet. For complex problems, it is wise to run *Solver* from different starting points. You should also look carefully at the *Solver* results dialog box when the model has completed running. If it indicates "Solver has found a solution. All constraints and optimality conditions are satisfied," then at least a local optimal solution has been found. If you get the message "Solver has converged to the current solution. All constraints are satisfied," then you should run *Solver* again from the current solution to try to find a better solution.

▶ **Figure 14.23**

Markowitz Portfolio Model *Spreadsheet Implementation*

	A	B	C	D	E	F	G
1	Markowitz Portfolio Model						
2							
3	Data						
4		Expected			Variance-Covariance Matrix		
5		Return			Stock 1	Stock 2	Stock 3
6	Stock 1	10%		Stock 1	0.025	0.015	-0.002
7	Stock 2	12%		Stock 2		0.03	0.005
8	Stock 3	7%		Stock 3			0.004
9	Target Return	10%					
10							
11	Model						
12					Variance Calculations		
13		Allocation			Squared Terms	Cross-Products	
14	Stock 1	0.25			0.001579256	0.003387	
15	Stock 2	0.45			0.006053362	-0.000301067	
16	Stock 3	0.30			0.000358718	0.001345191	
17	Total	1					
18							
19							
20		Return	Variance				
21	Portfolio	10.0%	0.012				

	A	B	C	D	E	F	G
1	Markowitz Portfolio Model						
2							
3	Data						
4		Expected			Variance-Covariance Matrix		
5		Return			Stock 1	Stock 2	Stock 3
6	Stock 1	0.1		Stock 1	0.025	0.015	-0.002
7	Stock 2	0.12		Stock 2		0.03	0.005
8	Stock 3	0.07		Stock 3			0.004
9	Target Return	0.1					
10							
11	Model						
12					Variance Calculations		
13		Allocation			Squared Terms	Cross-Products	
14	Stock 1	0.251336898395722			=B14^2*E6	=F6*B14*B15*2	
15	Stock 2	0.449197860962567			=B15^2*F7	=G6*B14*B16*2	
16	Stock 3	0.299465240641711			=B16^2*G8	=G7*B15*B16*2	
17	Total	=SUM(B14:B16)					
18							
19							
20		Return	Variance				
21	Portfolio	=SUMPRODUCT(B14:B16,B6:B8)	=SUM(E14:E16)+ SUM(F14:F16)				

▶ **Figure 14.24**

Solver *Model for* Markowitz Portfolio Model

ANALYTICS IN PRACTICE: Applying Nonlinear Optimization at Prudential Securities[3]

Prudential Securities Inc. (PSI) created a mortgage-backed securities (MBS) department to develop models for managing complex investments. The department developed a variety of analytical models, including linear, integer, and nonlinear optimization models, to help value, trade, and hedge MBS in inventory and construct portfolios. In one example, nonlinear optimization models are used to construct optimal portfolios for clients, matching the clients' investment performance profile with constraints under a variety of interest rate scenarios. The model inputs required from the client include the portfolio performance target, the securities to consider, diversification restrictions, and view of future interest rates.

Analysts use a scenario analysis to generate an optimal portfolio composed of securities with widely differing characteristics that matches the investor's performance profile under a variety of interest rate scenarios and employs a weighting scheme on the scenarios to reflect the portfolio manager's actual view of the direction of future interest rates. The portfolio manager defines the desired portfolio performance for each scenario. An optimization technique bundles the different scenarios and leads to optimal structured portfolios that meet specified performance targets while taking into account possible interest rate movement. These models have been used hundreds of times per day by PSI personnel.

CHECK YOUR UNDERSTANDING

1. Why are nonlinear optimization models generally harder to model than linear optimization models?

2. What are the characteristics of a quadratic optimization model?

3. State some practical issues that must be considered when solving nonlinear optimization models.

---

[3]Based on Yosi Ben-Dov, Lakhbir Hayre, and Vincent Pica, "Mortgage Valuation Models at Prudential Securities," *Interfaces*, 22, 1 (January–February 1992): 55–71.

Non-Smooth Optimization

As we noted in Chapter 13, such Excel functions as IF, ABS, MIN, and MAX lead to **non-smooth optimization models**. Non-smooth models violate the linearity conditions required for the linear optimization solution method used by *Solver*. Nevertheless, using these Excel functions can simplify the modeling task, especially for nonanalytics professionals.

Evolutionary Solver

Problems that are non-smooth or involve both nonlinear functions and integer variables are usually difficult to solve using conventional techniques. To overcome these limitations, new approaches called *metaheuristics* have been developed by researchers. These approaches have some exotic names, including genetic algorithms, neural networks, and tabu search. Such approaches use heuristics—intelligent rules for systematically searching among solutions—that remember the best solutions they find and then modify or combine them in attempting to find better solutions. *Solver's Evolutionary* algorithm uses such an approach.

Many business problems involve fixed costs; they are either incurred in full or not at all. We may use binary variables to model such situations; however, this approach can be difficult to understand, as it requires advanced optimization logic to incorporate the fixed costs correctly in the model constraints. However, from a spreadsheet modeling perspective, it is quite easy to use IF functions to model fixed costs. This is acceptable when using the evolutionary algorithm.

EXAMPLE 14.11   **Incorporating Fixed Costs into the K&L Designs Model**

Consider the multiperiod production-inventory-planning model for K&L Designs that we developed in Example 13.19. Suppose that the company must rent some equipment, which costs $65 for three months. The equipment can be rented or returned each quarter, so if nothing is produced in a quarter, it makes no sense to incur the rental cost. Thus, if the production in any quarter is positive, we want to include the rental cost in the objective function; otherwise, we don't. We can do this by adding IF functions to the original objective function:

$$\text{Minimize } 11P_A + 14P_W + 12.50P_S + 1.20I_A + 1.20I_W + 1.20I_S$$
$$+ \text{IF}(P_A > 0, 65, 0)$$
$$+ \text{IF}(P_W > 0, 65, 0) + \text{IF}(P_S > 0, 65, 0)$$

The material balance constraints remain the same:
$$P_A - I_A = 150$$
$$P_W + I_A - I_W = 400$$
$$P_S + I_W - I_S = 50$$

We illustrate *Evolutionary Solver* first using this model for K&L Designs and then with some other applications.

EXAMPLE 14.12   **Using *Evolutionary Solver* for the K&L Design Fixed-Cost Problem**

Figure 14.25 shows a simpler, modified spreadsheet for the K&L Designs fixed-cost problem (Excel file *K&L Designs Evolutionary Solver Model*). The objective function in cell B20 is =SUMPRODUCT(B6:D7, B14:D15) + IF(B14>0, B9, 0) + IF(C14>0, C9, 0) + IF(D14>0, D9, 0). Figure 14.26 shows the *Solver* model. The *Evolutionary Solver* algorithm requires that all variables have simple upper and lower

bounds to restrict the search space to a manageable region. Thus, we set upper bounds of 600 (the total demand) and lower bounds of 0 for each of them. *Evolutionary Solver* finds essentially the same optimal solution as the integer optimization problem we solved in Chapter 13, except for a minor rounding issue in cell C14.

▶ **Figure 14.25**

Modified Spreadsheet for
K&L Designs

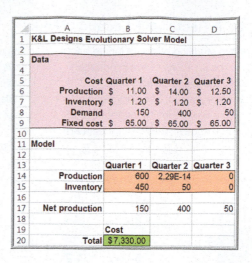

	A	B	C	D
1	K&L Designs Evolutionary Solver Model			
2				
3	Data			
4				
5		Cost	Quarter 1	Quarter 2	Quarter 3
6	Production	$ 11.00	$ 14.00	$ 12.50
7	Inventory	$ 1.20	$ 1.20	$ 1.20
8	Demand		150	400	50
9	Fixed cost	$ 65.00	$ 65.00	$ 65.00
10				
11	Model			
12				
13			Quarter 1	Quarter 2	Quarter 3
14	Production		600	2.29E-14	0
15	Inventory		450	50	0
16				
17	Net production		150	400	50
18				
19		Cost			
20	Total	$ 7,330.00			

▶ **Figure 14.26**

Evolutionary Solver *Model*
for K&L Designs

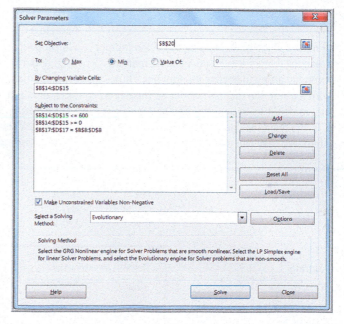

Because *Evolutionary Solver* is a search procedure, it may get "stuck" on an inferior or even an infeasible solution. We suggest that you run the method several times from the previous solution, until you receive a message that the current solution cannot be improved.

The results obtained by *Evolutionary Solver* can depend heavily on the starting values of the decision variables and the amount of time devoted to the search. For simple problems, it usually doesn't make much difference; however, for complex models, different starting values can produce different results. In addition, increasing the maximum search time may improve the solution. Thus, for complex problems, it is wise to run the procedure from different starting points, and we suggest you try this on this example and end-of-chapter problems. The maximum search time and other parameters can be changed from the *Options* button in the dialog; however, this is usually only necessary for advanced users.

Some applications use absolute value functions in the objective function or constraints. Absolute value functions are similar to IF functions and result in non-smooth functions, necessitating the use of *Evolutionary Solver*. The following example draws upon the central facilities location problem introduced in Example 11.16 using the rectilinear distance formula (11.8).

EXAMPLE 14.13 A Rectilinear Location Model

Edwards Manufacturing is studying where to locate a tool bin on the factory floor. The locations of five production cells are expressed as *x*- and *y*-coordinates on a rectangular grid of the factory layout. The daily demand for tools (measured as the number of trips to the tool bin) at each production cell is also known. The relevant data are as follows:

Cell	*x*-Coordinate	*y*-Coordinate	Demand
Fabrication	1	4	12
Paint	1	2	24
Subassembly 1	2.5	2	13
Subassembly 2	3	5	7
Assembly	4	4	17

Because of the nature of the equipment layout in the factory and for safety reasons, workers must travel along marked horizontal and vertical aisles to access the tool bin. Thus, the distance from a cell to the tool bin cannot be measured as a straight line; rather, it must be measured as *rectilinear distance*—that is, the distances parallel to the axes of the coordinate system. We may use the rectilinear distance measure in formula (11.8) to compute the distance between locations. The optimal location should minimize the total weighted distance between the tool bin and all production cells, where the weights are the daily number of trips to the tool bin.

To formulate an optimization model for the best location, define (X, Y) as the location coordinates of the tool bin. The weighted distance between the tool bin and each cell is expressed by the objective function

$$\text{Minimize } 12(|X-1|+|Y-4|)+24(|X-1|+|Y-2|)$$
$$+13(|X-2.5|+|Y-2|)+7(|X-3|+|Y-5|)$$
$$+17(|X-4|+|Y-4|)$$

The absolute value functions used in this objective function create a non-smooth model. Thus, *Evolutionary Solver* is an appropriate solution technique.

Figure 14.27 shows a spreadsheet model with the optimal *Evolutionary Solver* solution for the Edwards Manufacturing example (Excel file *Edwards Manufacturing*). The upper bounds are chosen as the maximum coordinate values and the lower bounds are zero. The *Solver* model is shown in Figure 14.28.

▶ Figure 14.27

Edwards Manufacturing *Spreadsheet*

	A	B	C	D
1	Edwards Manufacturing			
2				
3	Data			
4				
5	Cell	x-coordinate	y-coordinate	Demand
6	Fabrication	1	4	12
7	Paint	1	2	24
8	Subassembly 1	2.5	2	13
9	Subassembly 2	3	5	7
10	Assembly	4	4	17
11	Maximum	4	5	
12				
13	Model			
14				
15	Tool bin location	2.499971843	2.436117169	
16				
17	Cell	Weighted Distance		
18	Fabrication	36.76625608		
19	Paint	46.4661363		
20	Subassembly 1	5.669889242		
21	Subassembly 2	21.44737691		
22	Assembly	52.08648679		
23	Total	162.4361453		

▶ **Figure 14.28**

Evolutionary Solver *Model for* Edwards Manufacturing

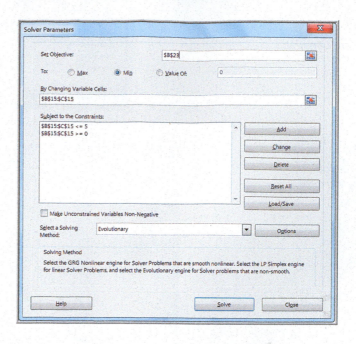

Evolutionary Solver for Sequencing and Scheduling Models

A unique application of Excel modeling and *Evolutionary Solver* is for job-sequencing problems, which we introduced in Chapter 11. We will use the model developed in Example 11.17 to illustrate this. Job-sequencing problems involve finding an optimal sequence, or order, by which to process a set of jobs.

EXAMPLE 14.14   Finding Optimal Job Sequences

In Example 11.17, we developed a spreadsheet model for the following scenario. Suppose that a custom manufacturing company has ten jobs waiting to be processed. Each job i has an estimated processing time (P_i) and a due date (D_i) that was requested by the customer, as shown in the table below.

Any sequence of integers in the decision variable range is called a **permutation**. Our goal is to find a permutation that optimizes the chosen criteria. *Solver* has an option to

define decision variables as a permutation; this is called an *alldifferent* constraint. To do this, open the *Add Constraint* dialog, choose the range of the decision variables, and then choose *dif* from the drop-down box, as shown in Figure 14.29. The final model, shown in Figure 14.30, is quite simple: Minimize the chosen objective cell—in this case, total tardiness—and ensure that the decision variables are a valid permutation of the job numbers. Figure 14.31 shows the *Solver* solution.

Job	1	2	3	4	5	6	7	8	9	10
Time	8	7	6	4	10	8	10	5	9	5
Due date	26	27	39	28	23	40	25	35	29	30

▶ **Figure 14.29**

Solver alldifferent *Constraint Definition*

▶ **Figure 14.30**

Solver *Model for Job Sequencing to Minimize Total Tardiness*

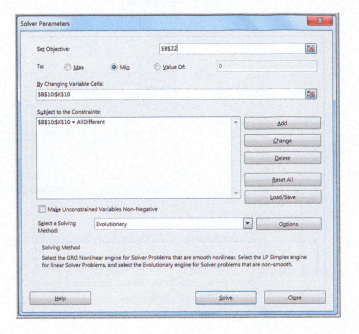

▶ **Figure 14.31**

Evolutionary Solver *Solution for Minimum Total Tardiness*

	A	B	C	D	E	F	G	H	I	J	K
1	**Job Sequencing Model**										
2											
3	**Data**										
4	**Job**	1	2	3	4	5	6	7	8	9	10
5	**Time**	8	7	6	4	10	8	10	5	9	5
6	**Due date**	26	27	39	28	23	40	25	35	29	30
7											
8	**Model**										
9	**Sequence**	1	2	3	4	5	6	7	8	9	10
10	**Job Assigned**	2	5	1	4	10	8	3	6	9	7
11	**Processing time**	7	10	8	4	5	5	6	8	9	10
12	**Completion time**	7	17	25	29	34	39	45	53	62	72
13	**Due Date**	27	23	26	28	30	35	39	40	29	25
14	**Lateness**	-20	-6	-1	1	4	4	6	13	33	47
15	**Tardiness**	0	0	0	1	4	4	6	13	33	47
16											
17	**Average Completion Time**	38.3									
18	**Maximum Number Tardy**	7									
19	**Total Lateness**	81									
20	**Average Lateness**	8.1									
21	**Variance of Lateness**	331.69									
22	**Total Tardiness**	108									
23	**Average Tardiness**	10.8									
24	**Variance of Tardiness**	236.96									

The Traveling Salesperson Problem

The **traveling salesperson problem (TSP)** can be described as follows. A salesperson needs to visit each of *n* different cities and return home in the minimum total distance. A route that visits each city exactly once and returns to the start is called a **tour**. Many practical problems can be formulated as a TSP. For example, drivers for FedEx and UPS must deliver packages to customers and return to their central location. Soft-drink vendors must collect money and replenish bottles for a set of retail locations and then return to the warehouse. Other examples are programming drilling machines to drill holes in circuit boards and picking orders within a warehouse. In all these applications, the goal is to perform the task in minimum total time or distance.

In general, a TSP for *n* cities has $(n - 1)!$ possible tours, making it quite difficult to identify the optimal one. For example, if *n* is just 14, there are more than 6 billion possible tours! The use of algorithms such as *Evolutionary Solver* allow us to find near-optimal solutions efficiently.

EXAMPLE 14.15   **Touring American League Cities**

An avid baseball fan in Detroit would like to plan a trip to visit all 14 ballparks of baseball teams in the American League and minimize the total travel distance. Figure 14.32 shows a spreadsheet model designed for *Evolutionary Solver* (Excel file *American Baseball League*). The data matrix shows the distances between each pair of cities. The model is somewhat tricky to develop and requires some detailed explanation to understand. We number the cities from 0 to 13 (this is necessary to use one of the capabilities of *Solver*, as we soon explain). Any city can be chosen as the start because a feasible sequence requires that the fan visit each city exactly once and return to the starting point, so we arbitrarily choose city 0 as the start. The decision variables are in the range B23:B35; the tour shown in the figure sequences the cities in numerical order. Note that when we get to the $(n - 1)$st city, we must return to city 0, so cell B36 is not a decision variable. We set cell A23 to 0. Whatever city is chosen to move to from city 0 is the decision variable in cell B23. We must ensure that this city becomes the "From" city in the next row; thus, the formula in cell A24 is =B23. In other words, we simply copy the value of cell B23 into cell A24 and do this for the remaining cells in column A. The INDEX function is used to find the distance from one city to another (note that the INDEX function refers to the row and column numbers of the range $D$5:$Q$18

and not the numbering of the cities that we used). Finally, in columns E and F, we use the VLOOKUP function to translate the numerical city values in the tour to their names.

Figure 14.33 shows the *Solver* model. The objective is to minimize the total tour length in cell C37 by changing the decision variables in the range B23:B35. The key to using Solver's *Evolutionary* algorithm is to ensure that the decision variables include each of the remaining 13 cities visited in the tour from city 0 exactly once. This is accomplished by using the *alldifferent* constraint, as we saw in the job-sequencing example. Because the *alldifferent* constraint applies to a set of positive integers from 1 to *n*, we needed to designate the first city as 0 (otherwise, the decision variables would have had to range from 2 to 14, and we would not have been able to use the *alldifferent* constraint).

Figure 14.34 shows the solution that *Evolutionary Solver* finds. To find the tour starting in Detroit, begin at row 34 and cycle around the tour to Milwaukee, Minnesota, Seattle, Oakland, California, Texas, Kansas City, Chicago, Cleveland, Baltimore, New York, Boston, Toronto, and return to Detroit, for a total distance of 6,718 miles. Of course, the real problem should incorporate the game schedules (who wants to visit a ballpark if the team is out of town?), but this would require a much more complicated model.

► **Figure 14.32**

Spreadsheet Model for the American Baseball League TSP

	A	B	C	D	E	F	G	H	I	J	K	L	M	N	O	P	Q
1	**American Baseball League Intercity Distances**																
2																	
3				**0**	**1**	**2**	**3**	**4**	**5**	**6**	**7**	**8**	**9**	**10**	**11**	**12**	**13**
4	**Data**			**SEA**	**OAK**	**CAL**	**TEX**	**KC**	**MIN**	**CHI**	**MIL**	**DET**	**CLE**	**TOR**	**BAL**	**NY**	**BOS**
5		**0**	**SEA**	0	678	954	1670	1489	1399	1720	1694	1932	2023	2124	2334	2421	2469
6		**1**	**OAK**	678	0	337	1467	1498	1589	1846	1845	2079	2161	2286	2457	2586	2704
7		**2**	**CAL**	954	337	0	1246	1363	1536	1745	1756	1979	2053	2175	2329	2475	2611
8		**3**	**TEX**	1670	1467	1246	0	460	853	798	843	982	1015	1186	1209	1383	1555
9		**4**	**KC**	1489	1498	1363	460	0	394	403	483	630	694	822	961	1113	1254
10		**5**	**MIN**	1399	1589	1536	853	394	0	334	297	528	622	780	936	1028	1124
11		**6**	**CHI**	1720	1846	1745	798	403	334	0	74	235	307	430	621	740	867
12		**7**	**MIL**	1694	1845	1756	843	483	297	74	0	237	328	434	641	746	860
13		**8**	**DET**	1932	2079	1979	982	630	528	235	237	0	95	206	409	509	632
14		**9**	**CLE**	2023	2161	2053	1015	694	622	307	328	95	0	193	314	425	563
15		**10**	**TOR**	2124	2286	2175	1186	822	780	430	434	206	193	0	384	366	463
16		**11**	**BAL**	2334	2457	2329	1209	961	936	621	641	409	314	384	0	184	370
17		**12**	**NY**	2421	2586	2475	1383	1113	1028	740	746	509	425	366	184	0	187
18		**13**	**BOS**	2469	2704	2611	1555	1254	1124	867	860	632	563	463	370	187	0
19																	
20	**Model**																
21																	
22	**From**	**To**	**Distance**		**From**	**To**											
23	0	1	678		SEA	OAK											
24	1	2	337		OAK	CAL											
25	2	3	1246		CAL	TEX											
26	3	4	460		TEX	KC											
27	4	5	394		KC	MIN											
28	5	6	334		MIN	CHI											
29	6	7	74		CHI	MIL											
30	7	8	237		MIL	DET											
31	8	9	95		DET	CLE											
32	9	10	193		CLE	TOR											
33	10	11	384		TOR	BAL											
34	11	12	184		BAL	NY											
35	12	13	187		NY	BOS											
36	13	0	2469		BOS	SEA											
37		**Total**	7272														

	A	B	C	D	E	F
20	**Model**					
21						
22	**From**	**To**	**Distance**		**From**	**To**
23	0	1	=INDEX($D$5:$Q$18,A23+1,B23+1)		=VLOOKUP(A23,$B$5:$C$18,2)	=VLOOKUP(B23,$B$5:$C$18,2)
24	=B23	2	=INDEX($D$5:$Q$18,A24+1,B24+1)		=VLOOKUP(A24,$B$5:$C$18,2)	=VLOOKUP(B24,$B$5:$C$18,2)
25	=B24	3	=INDEX($D$5:$Q$18,A25+1,B25+1)		=VLOOKUP(A25,$B$5:$C$18,2)	=VLOOKUP(B25,$B$5:$C$18,2)
26	=B25	4	=INDEX($D$5:$Q$18,A26+1,B26+1)		=VLOOKUP(A26,$B$5:$C$18,2)	=VLOOKUP(B26,$B$5:$C$18,2)
27	=B26	5	=INDEX($D$5:$Q$18,A27+1,B27+1)		=VLOOKUP(A27,$B$5:$C$18,2)	=VLOOKUP(B27,$B$5:$C$18,2)
28	=B27	6	=INDEX($D$5:$Q$18,A28+1,B28+1)		=VLOOKUP(A28,$B$5:$C$18,2)	=VLOOKUP(B28,$B$5:$C$18,2)
29	=B28	7	=INDEX($D$5:$Q$18,A29+1,B29+1)		=VLOOKUP(A29,$B$5:$C$18,2)	=VLOOKUP(B29,$B$5:$C$18,2)
30	=B29	8	=INDEX($D$5:$Q$18,A30+1,B30+1)		=VLOOKUP(A30,$B$5:$C$18,2)	=VLOOKUP(B30,$B$5:$C$18,2)
31	=B30	9	=INDEX($D$5:$Q$18,A31+1,B31+1)		=VLOOKUP(A31,$B$5:$C$18,2)	=VLOOKUP(B31,$B$5:$C$18,2)
32	=B31	10	=INDEX($D$5:$Q$18,A32+1,B32+1)		=VLOOKUP(A32,$B$5:$C$18,2)	=VLOOKUP(B32,$B$5:$C$18,2)
33	=B32	11	=INDEX($D$5:$Q$18,A33+1,B33+1)		=VLOOKUP(A33,$B$5:$C$18,2)	=VLOOKUP(B33,$B$5:$C$18,2)
34	=B33	12	=INDEX($D$5:$Q$18,A34+1,B34+1)		=VLOOKUP(A34,$B$5:$C$18,2)	=VLOOKUP(B34,$B$5:$C$18,2)
35	=B34	13	=INDEX($D$5:$Q$18,A35+1,B35+1)		=VLOOKUP(A35,$B$5:$C$18,2)	=VLOOKUP(B35,$B$5:$C$18,2)
36	=B35	0	=INDEX($D$5:$Q$18,A36+1,B36+1)		=VLOOKUP(A36,$B$5:$C$18,2)	=VLOOKUP(B36,$B$5:$C$18,2)
37		**Total**	=SUM(C23:C36)			

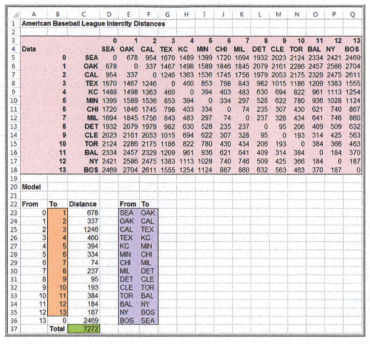

► **Figure 14.33**

Solver Model for the American Baseball League TSP

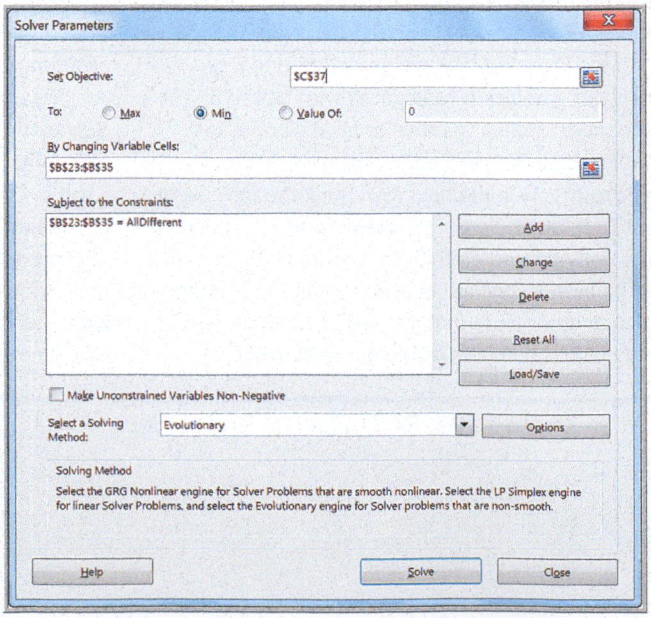

▶ **Figure 14.34**

Solver *Solution for the American Baseball League TSP*

	A	B	C	D	E	F
20	Model					
21						
22	From	To	Distance		From	To
23	0	1	678		SEA	OAK
24	1	2	337		OAK	CAL
25	2	3	1246		CAL	TEX
26	3	4	460		TEX	KC
27	4	6	403		KC	CHI
28	6	9	307		CHI	CLE
29	9	11	314		CLE	BAL
30	11	12	184		BAL	NY
31	12	13	187		NY	BOS
32	13	10	463		BOS	TOR
33	10	8	206		TOR	DET
34	8	7	237		DET	MIL
35	7	5	297		MIL	MIN
36	5	0	1399		MIN	SEA
37		Total	6718			

CHECK YOUR UNDERSTANDING

1. For what types of problems should *Evolutionary Solver* be used?

2. Explain how to use IF functions to model fixed costs on a spreadsheet.

3. What does the *alldifferent* constraint do in *Evolutionary Solver*?

4. Explain the traveling salesperson problem. Can you think of other applications?

KEY TERMS

Binary variable
General integer variable
Integer linear optimization model
Linear program (LP) relaxation
Local optimum solution
Markowitz portfolio model

Non-smooth optimization model
Nonlinear optimization model
Permutation
Quadratic optimization model
Tour
Traveling salesperson problem (TSP)

CHAPTER 14 TECHNOLOGY HELP

Excel Techniques

Solver with general integer variables (Example 14.1):

Set up the *Solver* model using the approach described in Chapter 13. Add a new constraint. In the *Add Constraint* dialog, enter the variable range of integer variables in the *Cell Reference* field and choose *int* from the drop-down box. In the field labeled *Select a Solving Method*, choose *Simplex LP*. Click the *Options* button and set *Integer Tolerance* to 0. Then click the *Solve* button.

Solver with binary variables (Example 14.5):

Set up the *Solver* model using the approach described in Chapter 13. Add a new constraint. In the *Add Constraint*

dialog, enter the variable range of binary variables in the *Cell Reference* field and choose *bin* from the drop-down box. In the field labeled *Select a Solving Method*, choose *Simplex LP*. Click the *Options* button and set *Integer Tolerance* to 0. Then click the *Solve* button.

Solver for nonlinear optimization (Example 14.8):

Set up the *Solver* model using the approach described in Chapter 13. In the field labeled *Select a Solving Method*, choose "GRG Nonlinear." Then click the *Solve* button.

Evolutionary Solver (Example 14.12):

Set up the *Solver* model using the approach described in Chapter 13. Make sure that all variables have upper and lower bounds by adding constraints on the variables.

In the field labeled *Select a Solving Method*, choose *Evolutionary Solver*. Then click the *Solve* button.

Analytic Solver

Analytic Solver provides a more powerful optimization tool than the standard *Solver* that comes with Excel.

See the online supplement *Using Integer and Nonlinear Optimization in Analytic Solver*. We suggest that you first read the online supplement *Getting Started with Analytic Solver Basic*. This provides information for both instructors and students on how to register for and access *Analytic Solver*.

PROBLEMS AND EXERCISES

Note: Data for many of these problems are provided in the Excel file Chapter 14 Problem Data *to facilitate model building. Tab names correspond to the problem numbers. These are designated with an asterisk (*).*

Integer Linear Optimization Models

1. Solve Problem 11 in Chapter 13 (Valencia Products) to ensure that the number of units produced is integer-valued. How much difference is there between the optimal integer solution objective function and the linear optimization solution objective function? Would rounding the continuous solution have provided the optimal integer solution?

2. Solve Problem 12 in Chapter 13 (ColPal Products) to ensure that the number of minutes of radio and TV ads is integer-valued. How much difference is there between the optimal integer solution objective function and the linear optimization solution objective function? Would rounding the continuous solution have provided the optimal integer solution?

*3. Solve Problem 15 in Chapter 13 (Bangs Leisure Chairs) to ensure that the number of units produced is integer-valued. How much difference is there between the optimal integer solution objective function and the linear optimization solution objective function? Would rounding the continuous solution have provided the optimal integer solution?

4. For the Brewer Services scenario described in this chapter, suppose that 11 permanent employees are hired. Find an optimal solution to minimize the number of part-time employees needed.

*5. The Gardner Theater, a community playhouse, needs to determine the lowest-cost production budget for an upcoming show. Specifically, they have to determine which set pieces to construct and which, if any, set

pieces to rent from another local theater at a predetermined fee. However, the organization has only two weeks to fully construct the set before the play goes into technical rehearsals. The theater has two part-time carpenters who work up to 12 hours a week, each at $10 an hour. Additionally, the theater has a part-time scenic artist who can work 15 hours per week to paint the set and props as needed at a rate of $15 per hour. The set design requires 20 flats (walls), two hanging drops with painted scenery, and three large wooden tables (props). The number of hours required for each piece for carpentry and painting is shown below:

	Carpentry	Painting
Flats	0.5	2.0
Hanging drops	2.0	12.0
Props	3.0	4.0

Flats, hanging drops, and props can also be rented at a cost of $75, $500, and $350 each, respectively. How many of each unit should be built by the theater and how many should be rented to minimize total costs?

6. Van Nostrand Hospital must schedule nurses so that the hospital's patients are provided with adequate care. At the same time, in the face of tighter competition in the health care industry, careful attention must be paid to keeping costs down. From historical records, administrators can project the minimum number of nurses to have on hand for the various times of day and days of the week. The nurse-scheduling problem seeks to find the minimum total number of nurses required to provide adequate care. Nurses start work at the beginning of one of the four-hour shifts given next and work for eight hours. Formulate and solve the nurse-scheduling problem as an integer program for one day for the data shown next.

Shift	Time	Minimum Number of Nurses Needed
1	12:00 a.m.–4:00 a.m.	5
2	4:00 a.m.–8:00 a.m.	10
3	8:00 a.m.–12:00 p.m.	14
4	12:00 p.m.–4:00 p.m.	8
5	4:00 p.m.–8:00 p.m.	12
6	8:00 p.m.–12:00 a.m.	10

*7. Joe is an active 26-year-old male who lifts weights six days a week. His rigorous training program requires a diet that will help his body recover efficiently. He is also a graduate student who is looking to minimize the cost of consuming his favorite foods. Joe is trying to gain weight, or at least maintain his current body weight, so he is not concerned about calories. His personal trainer suggests at least 300 grams of protein, 95 grams of fat, 225 grams of carbohydrates, and no more than 110 grams of sodium per day. His favorite foods are all items that he is familiar with preparing, as shown in the table *Data for Problem 7*. He is willing to consume multiple servings of each food per day to meet his requirements, although he cannot eat more than one steak per day and does not want to eat more than three pulled pork sandwiches a day. He needs to consume at least two servings of broccoli and one serving of carrots per day but is willing to eat two servings of carrots if necessary. Joe likes a certain brand of nutrition bars, but he would not eat more than one. Unless previously noted, he does not want more than five servings of any one food. How many servings of each food should he have in an optimal daily diet?

8. Gales Products manufactures ribbon for thermal transfer printing, which transfers ink from a ribbon onto paper through a combination of heat and pressure. Different types of printers use different sizes of ribbons. The company has forecasted demand for seven different ribbon sizes, as shown below.

Ribbon Size	Forecast Demand
60 mm	1,620
83 mm	520
102 mm	840
110 mm	2,640
120 mm	500
130 mm	740
165 mm	680

The rolls from which ribbons are cut are 900 mm in length. Scrap is valued at $0.07 per millimeter. Generate ten different cutting patterns so that each size can be cut from at least one pattern. Use your data to construct and solve an optimization model for finding the number of patterns to cut to meet demand and minimize trim loss.

Models with Binary Variables

*9. Hatch Financial, which recently absorbed another firm, is now downsizing and must relocate five information systems analysts from recently closed locations. Unfortunately, there are only three positions available for five people. Salaries are fairly uniform among this group (those with higher pay were already given the opportunity to begin anew).

Data for Problem 7

Food	Protein (grams)	Fat (grams)	Carbohydrates (grams)	Sodium (grams)	Cost/ Serving	Max Servings
Chicken breast	40	10	2	6	$4.99	5
Steak	49	16	3	11	$8.99	1
Pulled pork sandwich	27	16	27	19	$3.99	3
Salmon filet	39	15.5	1	5	$5.15	5
Rolled oats	9	1	27	9	$0.80	5
Baked potato	4	0	34	18	$1.50	5
Nutrition bar	19	18	17	3	$3.00	1
Serving of broccoli	2	0	6	2	$0.50	5
Serving of carrots	1	1	7	2	$0.50	2

Moving expenses will be used as the means of determining who will be sent where. Estimated moving expenses are as follows:

| Analyst | Moving Cost To | | |
	Gary	Salt Lake City	Fresno
Arlene	$8,500	$6,000	$5,000
Bobby	$5,000	$8,000	$12,000
Charlene	$9,500	$14,000	$17,000
Douglas	$4,000	$8,000	$13,000
Emory	$7,000	$3,500	$4,500

Model this as an integer optimization model to minimize cost and determine which analysts to relocate to the three locations.

*10. Fuller Legal Services wants to determine how much time to allocate to four different services: business consulting, criminal work, nonprofit consulting, and wills/trusts. Mr. Fuller has determined the average hourly fees and the minimum and maximum hours (for consulting and criminal work) and cases (for wills/trusts) that he would like to spend on each. He has no shortage of demand for his services. The relevant data are shown in the table *Data for Problem 10*. Develop and solve an integer optimization model to maximize monthly revenue.

*11. Riesemberg Medical Devices is allocating next year's budget among its divisions. As a result, the R&D Division needs to determine which R&D projects to fund. Each project requires various software and hardware and consulting expenses, along with internal human resources. A budget allocation of $1,300,000 has been approved, and 35 engineers are available to work on the projects. The R&D group has determined that at most one of projects 1 and 2 should be pursued, and that if project 4 is chosen,

then project 2 must also be chosen. Develop a model to select the best projects within the budget.

Project	NPV	Internal Engineers	Additional Costs
1	$600,000	9	$196,000
2	580,000	4	400,000
3	550,000	7	70,000
4	400,000	12	180,000
5	650,000	8	225,000
6	725,000	10	200,000
7	340,000	8	130,000

*12. A software-support division of Blain Information Services has eight projects that can be performed. Each project requires different amounts of development time and testing time. In the coming planning period, 1,150 hours of development time and 900 hours of testing time are available, based on the skill mix of the staff. The internal transfer price (revenue to the support division) and the times required for each project are shown in the table. Which projects should be selected to maximize revenue?

Project	Development Time	Testing Time	Transfer Price
1	80	67	$23,520
2	248	208	$72,912
3	41	180	$62,054
4	10	92	$32,340
5	240	202	$70,560
6	195	164	$57,232
7	269	226	$19,184
8	110	92	$32,340

Data for Problem 10

	Billables/hr	Minimum Hours	Maximum Hours
Business consulting	$200.00	30.00	45.00
Criminal work	$150.00	20.00	100.00
Nonprofit consulting	$100.00	35.00	70.00

	Billables/Client	Minimum Cases	Maximum Cases	Hours/Case	Hours Worked per Month
Wills/Trusts	$3,000.00	2.00	6.00	17	200.00

*13. The Kelmer Performing Arts Center offers a series of four programs that includes jazz, bluegrass, folk, classical, and comedy. The Program Coordinator needs to determine which acts to choose for next year's series. She assigned an "impact" rating to each artist that reflects how well the act meets the center's mission and provides community value. This rating is on a scale from 1 to 4, with 4 being the greatest impact and 1 being the least impact. The theater has 500 seats with an average ticket price of $12. Based on an estimate of the potential sales, the revenue from each artist is calculated. The center has a budget of $20,000 and would like the total impact factor to be at least 12, reflecting an average impact per artist of at least 3. To avoid duplication of genres, at most one of artists 2, 7, and 9 may be chosen, and at most one of artists 3 and 6 may be chosen. Finally, the center wishes to maximize its revenue. Data are shown below.

Artist	Cost	Impact	Ticket Estimate
1	$7,000.00	3	350
2	$975.00	4	500
3	$1,500.00	3	230
4	$5,000.00	3	400
5	$8,000.00	2	400
6	$1,500.00	3	600
7	$6,500.00	4	500
8	$3,000.00	2	350
9	$2,500.00	4	400

Develop and solve an optimization model to find the best program schedule to maximize the total profit.

*14. Dannenfelser Design works with clients in three major project categories: architecture, interior design, and combined. Each type of project requires an estimated number of hours for different categories of employees, as shown in the table *Data for Problem 14*.

In the coming planning period, 184 hours of principal time, 414 hours of senior designer time, 588 hours of drafter time, and 72 hours of administrator time are available. Revenue per project averages $12,900 for architecture, $11,110 for interior design, and $18,780 for combined projects. The firm would like to work on at least one of each type of project for exposure among clients. Assuming that the firm has more demand than they can possibly handle, find the best mix of projects to maximize profit.

*15. Anya is a part-time business student who works full time and is constantly on the run. She recognizes the challenge of eating a balanced diet and wants to minimize cost while meeting some basic nutritional requirements. Based on some research, she found that a very active woman should consume 2,250 calories per day. According to one author's guidelines, the following daily nutritional requirements are recommended.

Source	Recommended Intake (Grams)
Fat	Maximum 70
Carbohydrates	Maximum 225
Fiber	Maximum 30
Protein	At least 160

Anya chose a sample of meals in the table *Data for Problem 15* that could be obtained from healthy quick-service restaurants around town as well as some items that could be purchased at the grocery store. She does not want to eat the same entrée (first six foods) more than once each day but does not mind eating breakfast or side items (last five foods) twice a day and protein powder-based drinks up to four times a day, for convenience. Additional data are given below. Develop an integer linear optimization model to find the number of servings of each food choice in a daily diet to minimize cost and meet Anya's nutritional targets.

Data for Problem 14

	Architecture	Interior Design	Combined	Hourly Rate
Principal	15	5	18	$150
Senior designer	25	35	40	$110
Drafter	40	30	60	$75
Administrator	5	5	8	$50

Data for Problem 15

Food	Cost/Serving	Calories	Fat	Carbs	Fiber	Protein
Turkey sandwich	$4.69	530	14	73	4	28
Baked-potato soup	$3.39	260	16	23	1	6
Whole-grain chicken sandwich	$6.39	750	28	33	10	44
Bacon turkey sandwich	$5.99	770	28	34	5	47
Southwestern chicken wrap	$3.69	220	8	29	15	21
Sesame chicken wrap	$3.69	250	10	26	15	26
Yogurt	$0.75	110	2	19	0	5
Raisin bran with skim milk	$0.40	270	1	58	8	12
Cereal bar	$0.43	110	2	22	0	1
1 cup broccoli	$0.50	25	0.3	4.6	2.6	2.6
1 cup carrots	$0.50	55	0.25	13	3.8	1.3
1 scoop protein powder	$1.29	120	4	5	0	17

*16. Josh Steele manages a professional choir in a major city. His marketing plan is focused on generating additional local demand for concerts and increasing ticket revenue and also gaining attention at the national level to build awareness of the ensemble across the country. He has $20,000 to spend on media advertising. The goal of the advertising campaign is to generate as much local recognition as possible while reaching at least 3,000 units of national exposure. He has set a limit of 100 total ads. Additional information is shown in the table *Data for Problem 16*. The last column sets limits on the number of ads to ensure that the advertising markets do not become saturated. Find the optimal number of ads of each type to run to meet the choir's goals by developing and solving an integer optimization model.

17. Soapbox is a local band that plays classic and contemporary rock. The band members charge $600 for a three-hour gig. They would like to play at least 30 gigs per year but need to determine the best way to promote themselves. The most they are willing to spend on promotion is $2,500. The possible promotion options are as follows:

- Playing free gigs
- Making a demo CD
- Hiring an agent
- Handing out fliers
- Creating a Web site

Each free gig costs them $250 for travel and equipment but generates about three paying gigs. A high-quality studio demo CD should help the band book

Data for Problem 16

Media	Price	Local Exposure	National Exposure	Limit
FM radio spot	$80.00	110	40	30
AM radio spot	$65.00	55	20	30
Cityscape ad	$250.00	80	5	24
MetroWeekly ad	$225.00	65	8	24
Hometown paper ad	$500.00	400	70	10
Neighborhood paper ad	$300.00	220	40	10
Downtown magazine ad	$55.00	35	0	15
Choir journal ad	$350.00	10	75	12
Professional organization magazine ad	$300.00	20	65	12

20 gigs but will cost $1,000. A demo CD made on home recording equipment will cost $400 but may result in only ten bookings. A good agent will get the band 15 gigs but will charge $1,500. The band can create a Web site for $400 and would expect to generate six gigs from this exposure. They also estimate that they may book one gig for every 500 fliers they hand out, which would cost $0.08 each. They don't want to play more than ten free gigs or send out more than 2,500 fliers. Develop and solve an integer optimization model to find the best promotion strategy to maximize their profit.

*18. Cady Industries produces custom induction motors for specific customer applications. Each motor can be configured from different options for horsepower, the driveshaft forming process, spider bar component material, rotor plate process, type of bearings, tophat (a system of channels encased in a box that is placed on top of the motor to reduce airflow velocity both entering and exiting the motor) design, torque direction, and an optional mounting base.

	Cost	Time Requirement (Days)
Horsepower		
1000 HP	$155,000	32
5000 HP	$165,000	36
10000 HP	$180,000	42
15000 HP	$205,000	50
Shaft		
Heat-Rolled	$10,000	10
Oil-Quenched	$5,000	16
Forged	$15,000	8
Spider Bar Material		
Copper	$10,000	4
Aluminum	$2,500	8
Rotor Plates		
Laser-Cut	$12,500	5
Machine-Punched	$7,500	12
Bearings		
Sleeve	$5,000	4
Anti-Friction	$5,000	4
Oil Well	$3,000	2
Oil Guard	$5,000	4

	Cost	Time Requirement (Days)
Tophat Design		
Box	$5,000	15
V-Box	$20,000	15
Torque Direction		
Vertical	$35,000	10
Horizontal	$40,000	6
Optional Base	$75,000	10

a. Develop and solve an optimization model to find the minimum cost configuration of a motor.

b. Develop and solve an optimization model to find the configuration that can be completed in the shortest amount of time.

c. Customer A has a new plant opening in 90 days and needs a motor with at least 5,000 horsepower. The customer has specified that sleeve bearings be installed for easy maintenance and a V-box tophat is required to meet airflow velocity limitations. Find the optimal configuration that can be built within the 90-day requirement.

d. Customer B has a budget of $365,000 and requires a motor with 15,000 horsepower, a heat-rolled shaft, and the optional base. They want the highest-quality product, which implies that they are willing to maximize the cost up to the budget limitation. Find the optimal configuration that will meet these requirements.

*19. For the General Appliance Corporation transportation model discussed in Example 13.18, suppose that the company wants to enforce a single sourcing constraint that each distribution center be served from only one plant. Assume that the capacity at the Marietta plant is 1,500. Set up and solve a model to find the minimum cost solution.

*20. For the Shafer Office Supplies problem (Problem 37 in Chapter 13), suppose that the company wants to enforce a single sourcing constraint that each retail store be served from only one distribution center. Set up and solve a model to find the minimum cost solution.

*21. Tunningley Services is establishing a new business to serve customers in the Ohio, Kentucky, and Indiana region around the Cincinnati, Ohio area. The company has identified 15 key market areas and wants to establish regional offices to meet the goal of being able to travel to all key markets within 60 minutes. The data file *Tunningley* provides travel times in minutes between each pair of cities.

a. Develop and solve an optimization model to find the minimum number of locations required to meet their goal.

b. Suppose they change the goal to 90 minutes. What would be the best solution?

22. Tindall Gifts is a major national retail chain with stores located principally in shopping malls. For many years, the company published a Christmas catalog that was sent to current customers on file. This strategy generated additional e-commerce business, while also attracting customers to the stores. However, the cost-effectiveness of this strategy was never determined. John Harris, vice president of marketing, conducted a major study on the effectiveness of Tindall's Christmas catalog. The results were favorable: Patrons who were catalog recipients spent more, on average, than did comparable non-recipients. These revenue gains more than compensated for the costs of production, handling, and mailing, which had been substantially reduced by cooperative allowances from suppliers. With the continuing interest in direct mail as a vehicle for delivering holiday catalogs, Harris continued to investigate how new customers could most effectively be reached. One of these ideas involved purchasing mailing lists of magazine subscribers through a list broker. To determine which magazines might be appropriate, a mail questionnaire was administered to a sample of current customers to ascertain which magazines they regularly read. Ten magazines were selected for the survey. The assumption behind this strategy is that subscribers of magazines having a high proportion of current customers would be viable targets for future purchases at Tindall stores. The question is which magazine lists should be purchased to maximize the reaching of potential customers in the presence of a limited budget for purchasing lists. Data from the customer survey have begun to trickle in. The information about the ten magazines to which a customer subscribes is provided on the returned questionnaire. Harris has asked you to develop a prototype model, which later can be used to decide which lists to purchase. So far, only 53 surveys have been returned. To keep the prototype model manageable, Harris has instructed you to go ahead with the model development using the data from the 53 returned surveys. These data are shown in the table below. The costs of the first ten lists are given, and your budget is $3,000.

Data for Tindall Gifts Survey

List	1	2	3	4	5	6	7	8	9	10
Cost (000)	$1	$1	$1	$1.5	$1.5	$1.5	$1	$1.2	$0.5	$1.1

Customer	Magazines	Customer	Magazines	Customer	Magazines	Customer	Magazines
1	10	28	4,7	15	8	42	4, 5, 6
2	1, 4	29	6	16	6	43	None
3	1	30	3, 4, 5, 10	17	4, 5	44	5, 10
4	5, 6	31	4	18	7	45	1, 2
5	5	32	8	19	5, 6	46	7
6	10	33	1, 3, 10	20	2, 8	47	1, 5, 10
7	2, 9	34	4, 5	21	7, 9	48	3
8	5, 8	35	1, 5, 6	22	6	49	1, 3, 4
9	1, 5, 10	36	1, 3	23	3, 6, 10	50	None
10	4, 6, 8, 10	37	3, 5, 8	24	None	51	2,6
11	6	38	3	25	5, 8	52	None
12	3	39	2, 7	26	3, 10	53	2, 5, 8, 9, 10
13	5	40	2, 7	27	2, 8		
14	2, 6	41	7				

What magazines should be chosen to maximize overall exposure? (Hint: Define binary variables X_j for whether magazine j is selected or not, and Y_i for whether customer i is reached by any selected magazine. In other words, Y_i cannot be 1 unless customer i is a subscriber to one of the magazines selected.)

Nonlinear Optimization Models

23. Problem 1 in Chapter 11 posed the following situation: A manufacturer of kitchen appliances is preparing to set the price on a new blender. Demand is thought to depend on the price and is represented by the model

$$D = 2,500 - 3P$$

The accounting department estimates that the total costs can be represented by

$$C = 5,000 + 5D$$

You were asked to develop a model for the total profit. Implement the model on a spreadsheet and use nonlinear optimization with *Solver* to find the price that maximizes profit.

24. Problem 4 in Chapter 11 posed the following situation: The demand for airline travel is quite sensitive to price. Typically, there is an inverse relationship between demand and price; when price decreases, demand increases, and vice versa. One major airline has found that when the price (P) for a round trip between Chicago and Los Angeles is $600, the demand ($D$) is 500 passengers per day. When the price is reduced to $400, demand is 1,200 passengers per day. You were asked to develop an appropriate model. Implement the model on a spreadsheet and use nonlinear optimization with *Solver* to find the optimal price to maximize revenue.

25. Problem 2 in Chapter 11 posed the following situation: Modern Electronics sells two popular models of wireless headphones, model A and model B. The sales of these products are not independent of each other (in economics, we call these substitutable products because if the price of one increases, sales of the other will increase). The store wishes to establish a pricing policy to maximize revenue from these products. A study of price and sales data shows the following relationships between the quantity sold (N) and prices (P) of each model:

$$N_A = 20 - 0.62P_A + 0.30P_B$$

$$N_B = 29 + 0.10P_A - 0.60P_B$$

You were asked to construct a model for total revenue. Implement it on a spreadsheet and use nonlinear

optimization with *Solver* to find the optimal prices to maximize revenue.

26. For the pricing decision model in Example 14.8, suppose that the company wants to keep the price at a maximum of $500. Note that the solution in Figure 14.19 will no longer be feasible. Modify the spreadsheet model to include a constraint on the maximum price and solve the model.

27. In the hotel pricing problem in Example 14.9, suppose that the hotel is considering adding suites to its room mix. Based on an analysis of local competitors, suites can sell for a rate of $180, and they expect to sell 20 per day to business travelers. The price elasticity of demand is estimated to be -2.5. The hotel would want to keep the price of suites between $150 and $200. Modify the spreadsheet to include suites and find prices that will maximize total revenue.

28. A franchise of a chain of Mexican restaurants wants to determine the best location to attract customers from three suburban neighborhoods. The coordinates of the three suburban neighborhoods are as follows:

Neighborhood	X-Coordinate	Y-Coordinate
Liberty	2	12
Jefferson	9	6
Adams	1	1

The population of Adams is four times as large as Jefferson, and Jefferson is twice as large as Liberty. The restaurant wants to consider the population in its location decision. Develop and solve a model to find the best location, assuming that straight-line distances can be used between the locations.

29. ElectroMart wants to identify a location for a warehouse that will ship to five retail stores. The coordinates and annual number of truckloads are given here. Develop and solve a model to find the best location, assuming that straight-line distances can be used between the locations.

Retail Store	X-Coordinate	Y-Coordinate	Truckloads
A	18	15	12
B	3	4	18
C	20	5	24
D	3	16	12
E	10	20	18

30. In Chapter 8, we noted that the least-squares coefficients in a regression model are found by minimizing the sums of squares of the errors, as given in equation (8.4). This is a nonlinear optimization problem. Using the *Home Market Value* data, set up a spreadsheet model to find the values for the slope and intercept using nonlinear optimization. Compare your answer with Example 8.5.

*31. Many manufacturing situations, for example, the production of such large and complex items as aircraft or machines, exhibit a learning effect in which the production time per unit decreases as more units are produced. This is often modeled by a power curve, $y = ax^{-b}$, where a and b are constants. Suppose that data on production times for the first ten units produced were collected from a new project at Glasgow Machine Tool:

Unit	Production Hours
1	3,161
2	2,720
3	2,615
4	2,278
5	2,028
6	2,193
7	2,249
8	2,268
9	1,994
10	2,000

Develop a model for estimating the power curve to minimize the sum of the squared deviations of the errors. Use nonlinear optimization to find the parameters.

32. The DTP Corporation has two major products. Marketing analysts have conducted experiments to gather data on the effect of media advertising on profits. These data are available in the Excel file *DTP Corporation*. Suppose that the total advertising budget is $500,000 and that at least $50,000 must be spent on each product. Use the *Add Trendline* feature in Excel to fit logarithmic functions for profit as a function of advertising for each product. Then formulate and solve a nonlinear optimization model to determine how the company should allocate its advertising budget between the two products.

*33. The Hal Chase Investment Planning Agency is in business to help investors optimize their return from investment. Hal deals with three investment mediums: a stock fund, a bond fund, and his own Sports and Casino Investment Plan (SCIP). The stock fund is a mutual fund investing in openly traded stocks. The bond fund focuses on the bond market, which has a more stable, but lower, expected return. SCIP is a high-risk scheme, often resulting in heavy losses but occasionally coming through with spectacular gains. Average returns, their variances, and covariances are given in the table *Data for Problem 33*. Develop and solve a portfolio optimization model for this situation for a target return of 12%.

Non-Smooth Optimization

34. An IT support group at Thomson State College has seven projects to complete. The time each will take and project deadlines (both given in number of days) are shown next.

Project	1	2	3	4	5	6	7
Time	4	9	12	16	9	15	8
Deadline	12	24	60	28	24	36	48

a. Sequence the projects to minimize the average lateness.

b. Sequence the projects to minimize the average tardiness.

c. Compare these solutions to the SPT and EDD rules discussed in the job sequencing application in Chapter 11.

*35. Suppose the distances that a pharmaceutical representative, Tracy Ross, travels between medical offices are as follows:

				To				
From	1	2	3	4	5	6	7	8
1	0	19	57	51	49	4	12	92
2	19	0	51	10	53	25	80	53
3	57	51	0	49	18	30	6	47
4	51	10	49	0	50	11	91	38
5	49	53	18	50	0	68	62	9
6	4	25	30	11	68	0	48	94
7	12	80	6	91	62	48	0	9
8	92	53	47	38	9	94	9	0

Set up and solve a traveling salesperson problem using *Evolutionary Solver*.

Data for Problem 33

	Stock	Bond	SCIP
Average return	0.148	0.060	0.152
Variance	0.014697	0.000155	0.160791
Covariance with stock		0.000468	−0.002222
Covariance with bond			−0.000227

CASE: PERFORMANCE LAWN EQUIPMENT

The CFO at Performance Lawn Equipment, Brian Ferguson, wishes to design a minimum variance portfolio of index funds to invest some of the firm's cash reserves. The funds selected for consideration and their variance-covariance matrix and average returns are given in the table below.

a. The firm would like to achieve a target return of 0.19%. What mix of investments would achieve this?

b. To obtain better diversification, the CFO would like to restrict the percentage of investments in each fund as follows:

- Bond: between 10% and 50%
- S&P 500: between 30% and 50%
- Small cap: no more than 20%
- Mid cap: no more than 20%
- Large cap: no more than 25%
- Emerging market: no more than 10%
- Commodity: no more than 20%

How would the optimal portfolio and objective change? Summarize your findings in a short memo to Mr. Ferguson.

	Bond	S&P 500	Small Cap	Mid Cap	Large Cap	Emerging Market	Commodity
Bond	0.002%						
S&P 500	−0.001%	0.020%					
Small cap	−0.001%	0.027%	0.047%				
Mid cap	−0.001%	0.024%	0.039%	0.033%			
Large cap	−0.001%	0.019%	0.027%	0.023%	0.027%		
Emerging market	0.000%	0.032%	0.050%	0.043%	0.041%	0.085%	
Commodity	0.000%	0.000%	0.005%	0.005%	0.009%	0.015%	0.054%
Average weekly return	0.044%	0.118%	0.256%	0.226%	0.242%	0.447%	0.053%

Optimization Analytics

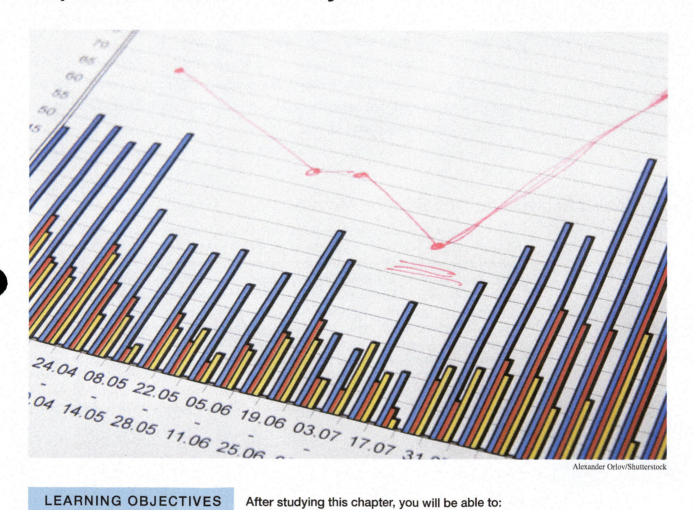

Alexander Orlov/Shutterstock

LEARNING OBJECTIVES   After studying this chapter, you will be able to:

- Interpret the *Solver* Sensitivity Report for both linear and nonlinear optimization models.
- Conduct what-if analysis for optimization models.
- Use auxiliary variables to model bound constraints and obtain more complete sensitivity information.
- Understand and interpret the *Solver* Sensitivity Report for models that have bounded variables.

- Use Excel to evaluate scenarios for integer linear optimization models and gain practical insights into the solutions.
- Visualize *Solver* reports using Excel charts.
- Ensure that assumptions underlying the use of sensitivity information hold when interpreting *Solver* reports.

When the field of optimization was quite young, a wise professor stated that its purpose is "insight, not numbers." In other words, there is much to be gleaned from optimization models beyond just finding the optimal solution. *Solver* provides useful information about the impact of changes and variations in model data on the optimal solution from its Sensitivity Report. In addition, modern analytics provides a rich platform to conduct what-if analysis, provide further insights that can help use optimization in practice, and to visualize results in order to communicate them easily to non-technical managers.

In this chapter, we will focus on using what-if analysis in optimization, interpreting the *Solver* Sensitivity Report, and visualizing optimization results. We will also discuss the special case of models with bounded variables and ensure that sensitivity analysis is used properly.

What-If Analysis for Optimization Models

The principal purpose of formulating and solving an optimization model should never be to just find a "best answer"; rather, the model should be used to provide insight for making better decisions. Thus, it is important to analyze optimization models from a predictive analytics perspective to determine what might happen should the model assumptions change or when the data used in the model are uncertain. For example, managers have some control over pricing but may not be able to control supplier costs. Even though we may have solved a model to find an optimal solution, it would be beneficial to determine what impact a change in a price or cost would have on net profit. Similarly, many constraints represent resource limitations or customer commitments. Limited capacity can be adjusted through overtime, or supplier contracts can be renegotiated. So managers would want to know whether it would be worthwhile to increase capacity or change a contract. With *Solver*, answers to such questions can easily be found by simply changing the data and re-solving the model.

| EXAMPLE 15.1 | Using *Solver* for What-If Analysis |

In the Sklenka Ski Company (SSC) model, managers might wish to answer the following questions:

1. Suppose that the unit profit on Jordanelle skis is increased by $10. How will the optimal solution change? What is the best product mix?

2. Suppose that the unit profit on Jordanelle skis is decreased by $10 because of higher material costs. How will the optimal solution change? What is the best product mix?

3. Suppose that ten additional finishing hours become available through overtime. How will manufacturing plans be affected?

4. What if the number of finishing hours available is decreased by two hours because of planned equipment maintenance? How will manufacturing plans be affected?

Figure 15.1 shows a summary of the solutions for each of these scenarios after re-solving the model.

In the first scenario, when the unit profit of Jordanelle skis is increased to $60, the optimal product mix does not change from the base scenario; however, the total profit increases. You might think that if the profit of Jordanelle skis increases, it would be advantageous to produce more of them. However, doing so would require producing more

Deercrest skis to meet the marketing mix constraint, which would then violate the finishing time constraint. Therefore, the solution is "maxed out," so to speak, because of the constraints. Nevertheless, each pair of Jordanelle skis produced would gain an additional $10 in profit, so the 5.25 pairs we produce increase the profit by 5.25($10) = $52.50 to $997.50. From a practical perspective, a manager might need to consider whether the price increase will still ensure that all the skis can be sold—an implicit assumption in the model.

In the second scenario, the situation is different. If the profit of Jordanelle skis is reduced to $40, it becomes unprofitable to produce any of them. The marketing mix constraint is no longer relevant, and similar to the Crebo Manufacturing example, the profit per unit of finishing time is higher for Deercrest; consequently, it is best to produce only that model. Eliminating a product from the optimal

mix might be a poor marketing decision, or it can offer advantages by simplifying the supply chain.

In the third scenario, we see that we still have a mix of both products. With the additional finishing hours, we are able to produce more of the higher-profit Deercrest skis and use the remaining capacity to produce a smaller amount of the Jordanelle skis. However, you can also see that we have now used all the fabrication hours as well as all the finishing hours, suggesting that the operations manager has no slack in fabrication; any breakdown of equipment or absence of labor will affect the solution.

Finally, in the last scenario, a small reduction in the finishing capacity results in the same two-to-one ratio of Deercrest to Jordanelle skis because of the marketing mix constraint, but the reduction in finishing capacity reduced the amount of each product that can be produced, as well as reducing the overall profit by $90.

For many models, we often want to conduct more systematic what-if analyses to examine how solutions change as input data vary within some reasonable ranges. For example, look back at the Innis Investment problem (Example 13.16). In financial decisions such as these, it is often useful to compare risk versus reward to make an informed decision, particularly since the target return is subjective. We illustrate this in the following example.

EXAMPLE 15.2   **Evaluating Risk Versus Reward in Portfolio Management**

In the Innis Investment problem, we might be interested in comparing how the return, risk, and investment mix would change for different values of the target return. Figure 15.2 shows such an analysis for target returns between 4% and 7%. We see that below 5%, we can obtain a return of 4.89% with a minimum risk. Visualizing these results can provide better insight. The chart on the right shows that as the target

return increases, the risk increases, and at 6%, begins to increase at a faster rate. As the target return increases, the investment mix begins to change to a higher percentage of low-price stock, which is a riskier investment, as shown in the chart on the left. A more conservative client might be willing to take a small amount of additional risk to achieve a 6% return but not venture beyond that value.

Solver Sensitivity Report

In Example 15.1, we evaluated only a few distinct scenarios. Managers might also want to know what would happen if the profit for Jordanelle skis is decreased only by $1, $2, or $5, and so on. We could keep changing the data and re-solving the model, but that would be

▶ **Figure 15.1**

Summary of What-If Scenarios

	G	H	I	J	K	L
1		Quantity Produced		Hours Used		
2	Scenario	Jordanelle	Deercrest	Fabrication	Finishing	Profit
3	Base Case	5.25	10.5	60.375	21	$945.00
4	Jordanelle profit = $60	5.25	10.5	60.375	21	$997.50
5	Jordanelle profit = $40	0	14	56	21	$910.00
6	Finishing hours = 31	1.6	19.6	84	31	$1,354.00
7	Finishing hours = 19	4.75	9.5	54.625	19	$855.00

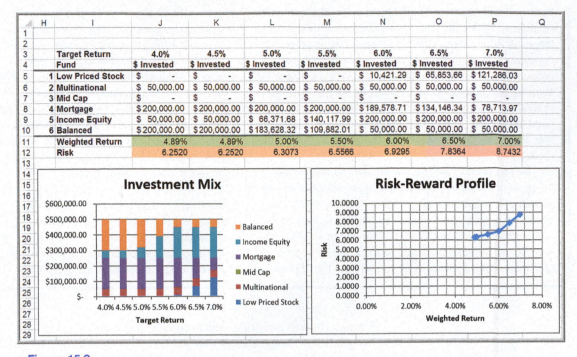

H	I	J	K	L	M	N	O	P	Q
	Target Return	4.0%	4.5%	5.0%	5.5%	6.0%	6.5%	7.0%	
	Fund	$ Invested	$ Invested	$ Invested	$ Invested	$ Invested	$ Invested	$ Invested	
	1 Low Priced Stock	$ -	$ -	$ -	$ -	$ 10,421.29	$ 65,853.66	$121,286.03	
	2 Multinational	$ 50,000.00	$ 50,000.00	$ 50,000.00	$ 50,000.00	$ 50,000.00	$ 50,000.00	$ 50,000.00	
	3 Mid Cap	$ -	$ -	$ -	$ -	$ -	$ -	$ -	
	4 Mortgage	$200,000.00	$200,000.00	$200,000.00	$200,000.00	$189,578.71	$134,146.34	$ 78,713.97	
	5 Income Equity	$ 50,000.00	$ 50,000.00	$ 66,371.68	$140,117.99	$200,000.00	$200,000.00	$200,000.00	
	6 Balanced	$200,000.00	$200,000.00	$183,628.32	$109,882.01	$ 50,000.00	$ 50,000.00	$ 50,000.00	
	Weighted Return	4.89%	4.89%	5.00%	5.50%	6.00%	6.50%	7.00%	
	Risk	6.2520	6.2520	6.3073	6.5566	6.9295	7.8364	8.7432	

▲ **Figure 15.2**

Scenario Analysis for Innis Investments

tedious. Fortunately, we can answer these and other what-if questions more easily by using the Sensitivity Report generated by *Solver*.

The *Solver* Sensitivity Report provides a variety of useful information for managerial interpretation of the solution. Specifically, it allows us to understand how the optimal objective value and optimal decision variables are affected by changes in the objective function coefficients, the impact of forced changes in certain decision variables, or the impact of changes in the constraint resource limitations or requirements. Figure 15.3 shows the Sensitivity Report for the SSC model, with annotations that summarize what the numbers represent. To obtain it, select *Sensitivity Report* in the *Solver Results* dialog. We use this for the examples in this section.

One important caution: The Sensitivity Report information applies to changes in only one of the model parameters at a time; all others are assumed to remain at their original values. In other words, you cannot accumulate or add the effects of sensitivity information if you change the values of multiple parameters in a model simultaneously.

The *Decision Variable Cells* section provides information about the decision variables and objective function coefficients and how changes in their values would affect the optimal solution.

EXAMPLE 15.3 **Interpreting Sensitivity Information for Decision Variables**

The *Decision Variable Cells* section lists the final value for each decision variable, a number called the reduced cost, the coefficients associated with the decision variables from the objective function, and two numbers called allowable increase and allowable decrease. The **reduced cost** tells *how much the objective coefficient needs to* be reduced for a nonnegative variable that is zero in the optimal solution to become positive. If a variable is positive in the optimal solution, as it is for both variables in the SSC example, its reduced cost is always zero. We will see an example later that will help you to understand reduced costs.

The Allowable Increase and Allowable Decrease values tell how much an individual objective function coefficient can change before the optimal values of the decision variables will change (a value listed as "1E + 30" is interpreted as infinity). For example, the Allowable Increase for Deercrest skis is 10, and the Allowable Decrease is 90. This means that if the unit profit for Deercrest skis, $65, either increases by more than 10 or decreases by more than 90, then the optimal values of the decision variables will change (as long as all other objective coefficients stay the same). For instance, if we increase the unit profit by $11 (to $76) and re-solve the model, the new optimal solution will be to produce 14 pairs of Deercrest skis and no Jordanelle skis. However, any increase of less than 10 will keep the current solution optimal. For Jordanelle skis, we can increase the unit profit as much as we wish without affecting the current optimal solution; however, a decrease of at least 6.66 will force a change in the solution.

If the objective coefficient of any one variable that has positive value in the current solution changes but stays within the range specified by the Allowable Increase and Allowable Decrease, the optimal decision variables will stay the same; however, *the objective function value will change*. For example, if the unit profit of Jordanelle skis were changed to $46 (a decrease of $4, which is within the Allowable Increase), then we are guaranteed that the optimal solution will still be to produce 5.25 pairs of Jordanelle skis and 10.5 pairs of Deercrest skis. However, each of the 5.25 pairs of Jordanelle skis produced and sold would realize $4 less profit—a total decrease of 5.25($4) = $21. Thus, the new value of the objective function would be $945 − $21 = $924. If an objective coefficient changes beyond the Allowable Increase or Allowable Decrease, then we must re-solve the problem with the new value to find the new optimal solution and profit.

The range within which the objective function coefficients will not change the optimal solution provides a manager with some confidence about the stability of the solution in the face of uncertainty. If the allowable ranges are large, then reasonable errors in estimating the coefficients will have no effect on the optimal policy (although they will affect the value of the objective function). Tight ranges suggest that more effort might be spent in ensuring that accurate data or estimates are used in the model.

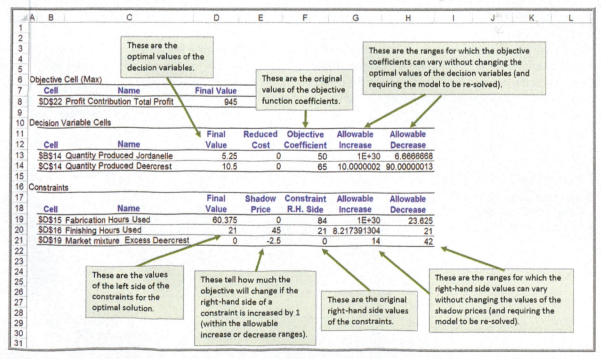

▲ **Figure 15.3**

Solver *Sensitivity Report*

To understand what a nonzero reduced cost means, let us use the second scenario in Example 15.1.

EXAMPLE 15.4 Understanding Nonzero Reduced Costs

Figure 15.4 shows the Sensitivity Report when the unit profit for Jordanelle skis is $40. As before, the reduced cost for Deercrest skis is 0 because the value of the variable is positive. We do not produce any Jordanelle skis in this optimal solution simply because it is not profitable to do so. Using the definition of the reduced cost, *how much the objective coefficient needs to be reduced for a nonnegative variable that is zero in the optimal solution to become positive*, we see that the profit on Jordanelle skis must be reduced by more than −$3.33 (or equivalently, *increased* by more than $3.33) to make it profitable to produce them. If you re-solve the model with the unit profit for Jordanelle as $43.34, you will obtain the original optimal product mix (except that the total profit will be $910.04 because of the different objective function coefficient).

The *Constraints* section of the Sensitivity Report lists the final value of the constraint function (the left-hand side), a number called the shadow price, the original right-hand-side value of the constraint, and an Allowable Increase and Allowable Decrease. The **shadow price** tells *how much the value of the objective function will change as the right-hand side of a constraint is increased by 1*. Whenever a constraint has positive slack (the difference between the left- and right-hand side of the constraint for the optimal solution as defined in Chapter 13), the shadow price is zero. When a constraint involves a limited resource, the shadow price represents the economic value of having an additional unit of that resource.

EXAMPLE 15.5 Interpreting Sensitivity Information for Constraints

In the fabrication constraint (see Figure 15.3), we are using only 60.375 of the 84 available hours in the optimal solution. Thus, having one more hour available will not help us to increase our profit. However, if a constraint is binding, then any change in the right-hand side will cause the optimal values of the decision variables as well as the objective function value to change. We illustrate this with the finishing constraint.

The shadow price of the finishing constraint is 45. This means that if an additional hour of finishing time is available, then the total profit will change by $45. To see this, change the limitation of the number of finishing hours available to 22 and re-solve the problem. The new solution is to produce 5.5 pairs of Jordanelle skis and 11.0 pairs of Deercrest skis, yielding a profit of $990. We see that the total profit increases by $45, as predicted.

Figure 15.4

Solver Sensitivity Report for SSC Objective: Max 40 Jordanelle + 65 Deercrest

Objective Cell (Max)

Cell	Name	Final Value
$D$22	Profit Contribution Total Profit	910

Decision Variable Cells

Cell	Name	Final Value	Reduced Cost	Objective Coefficient	Allowable Increase	Allowable Decrease
$B$14	Quantity Produced Jordanelle	0	-3.333333333	40	3.333333333	1E+30
$C$14	Quantity Produced Deercrest	14	0	65	1E+30	5.00000015

Constraints

Cell	Name	Final Value	Shadow Price	Constraint R.H. Side	Allowable Increase	Allowable Decrease
$D$15	Fabrication Hours Used	56	0	84	1E+30	28
$D$16	Finishing Hours Used	21	43.33333333	21	10.5	21
$D$19	Market mixture Excess Deercrest	14	0	0	14	1E+30

The profit on Jordanelle must be reduced by more than −$3.33 (increased by more than $3.33) to make it economical to produce.

The shadow price is a valid predictor of the change in the objective function value for each unit of increase in the right-hand side of the constraint up to the value of the Allowable Increase. Thus, if up to about 8.2 additional hours of finishing time were available, profit would increase by $45 for each additional hour (but we would have to re-solve the problem to actually find the optimal values of the decision variables). Similarly, a negative of the shadow price predicts the change in the objective function value for each unit the constraint's right-hand side is *decreased*, up to the value of the Allowable Decrease. For example, if one person were ill or injured, resulting in only 14 hours of finishing time available, then profit would decrease by 7($45) = $315, resulting in a total profit of $945 − $315 = $630. This can be predicted because a decrease of 7 hours is within the Allowable Decrease of 21. Beyond these ranges, the shadow price does not predict what will happen, and the problem must be re-solved.

Another way of understanding the shadow price is to break down the impact of a change in the right-hand side of the value. How was the extra hour of finishing time used? After solving the model with 22 hours of finishing time, we see that we were able to produce an additional 0.25 pairs of Jordanelle skis and 0.5 pairs of Deercrest skis as compared to the original solution. Therefore, the profit increased by 0.25($50) + 0.5(65) = $12.50 + 32.50 = $45. In essence, a small change in a binding constraint causes a reallocation of how the resources are used.

Interpreting the shadow price associated with the market mixture constraint is a bit more difficult. If you examine the constraint *Deercrest* −2 *Jordanelle* ≥ 0 closely, an increase in the right-hand side from 0 to 1 results in a change of the constraint to

$$(\text{Deercrest} - 1) - 2\,\text{Jordanelle} \geq 0$$

This means that the number of pairs of Deercrest skis produced would be one short of the requirement that it be at least twice the number of Jordanelle skis. If the problem is re-solved with this constraint, we find the new optimal solution to be 4.875 Jordanelle skis, 10.75 Deercrest skis, and profit = $942.50. The profit changed by the value of the shadow price, and we see that 2 × Jordanelle = 9.75, one short of the requirement.

Shadow prices are useful to a manager because they provide guidance on how to reallocate resources or change values over which the manager may have control. In linear optimization models, the parameters of some constraints cannot be controlled. For instance, the amount of time available for production or physical limitations on machine capacities would clearly be uncontrollable. Other constraints represent policy decisions, which, in essence, are arbitrary. Although it is correct to state that having an additional hour of finishing time will improve profit by $45, does this necessarily mean that the company should spend up to this amount for additional hours? This depends on whether the relevant costs have been included in the objective function coefficients. If the cost of labor *has not* been included in the objective function unit profit coefficients, then the company will benefit by paying less than $45 for additional hours. However, if the cost of labor *has* been included in the profit calculations, the company should be willing to pay up to an *additional* $45 over and above the labor costs that have already been included in the unit profit calculations.

The Limits Report (Figure 15.5) shows the lower limit and upper limit that each variable can assume while satisfying all constraints and holding all the other variables

► **Figure 15.5**

Solver *Limits Report*

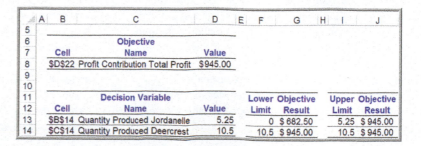

constant. Generally, this report provides little useful information for decision making and can be effectively ignored.

Using the Sensitivity Report

It is easy to use the sensitivity information to evaluate the impact of different scenarios. The following rules summarize how to do this.

    a. If a change in an objective function coefficient remains within the Allowable Increase and Allowable Decrease ranges in the *Decision Variable Cells* section of the report, then the optimal values of the decision variables will not change. However, you must recalculate the value of the objective function using the new value of the coefficient.

    b. If a change in an objective function coefficient exceeds the Allowable Increase or Allowable Decrease limits in the *Decision Variable Cells* section of the report, then you must re-solve the model to find the new optimal values.

    c. If a change in the right-hand side of a constraint remains within the Allowable Increase and Allowable Decrease ranges in the *Constraints* section of the report, then the shadow price allows you to predict how the objective function value will change. Multiply the change in the right-hand side (positive if an increase, negative if a decrease) by the value of the shadow price. However, you must re-solve the model to find the new values of the decision variables.

    d. If a change in the right-hand side of a constraint exceeds the Allowable Increase or Allowable Decrease limits in the *Constraints* section of the report, then you cannot predict how the objective function value will change using the shadow price. You must re-solve the problem to find the new solution.

We will illustrate these rules for the SSC what-if scenarios (see Example 15.1) using the sensitivity report in Figure 15.3.

EXAMPLE 15.6 **Using the Sensitivity Report to Evaluate Scenarios**

1. *Suppose that the unit profit on Jordanelle skis is increased by $10. How will the optimal solution change? What is the best product mix?*
The first thing to do is to determine if the increase in the objective function coefficient is within the range of the Allowable Increase and Allowable Decrease in the *Decision Variable Cells* portion of the report. Because $10 is less than the Allowable Increase of infinity, we can safely conclude that the optimal quantities of the decision variables will not change. However, because the objective function changed, we need to compute the new value of the total profit: 5.25($60) + 10.5($65) = $997.50.

2. *Suppose that the unit profit on Jordanelle skis is decreased by $10 because of higher material costs. How will the optimal solution change? What is the best product mix?*

In this case, the change in the unit profit exceeds the Allowable Decrease ($6.67). We can conclude that the optimal values of the decision variables will change, although we must re-solve the problem to determine what the new values would be.

3. *Suppose that ten additional finishing hours become available through overtime. How will manufacturing plans be affected?*
When the scenario relates to the right-hand side of a constraint, first check if the change in the right-hand-side value is within the range of the Allowable Increase and Allowable Decrease in the *Constraints* section of the report. In this case, ten additional finishing hours exceeds the Allowable Increase. Therefore, we must re-solve the problem to determine the new solution.

4. *What if the number of finishing hours available is decreased by two hours because of planned equipment* *maintenance? How will manufacturing plans be affected?*

In this case, a decrease of two hours in finishing capacity is within the Allowable Decrease. We may conclude that the total profit will decrease by the value of the shadow price for each hour that finishing capacity is decreased. Therefore, we can predict that the total profit will decrease by 2 × $45 = $90 to $855. However, we must re-solve the model to determine the new values of the decision variables.

Degeneracy

A solution is a **degenerate solution** if the right-hand-side value of any constraint has a zero Allowable Increase or Allowable Decrease. A full discussion of the implications of degeneracy is beyond the scope of this book; however, it is important to know that degeneracy can impact the interpretation of sensitivity analysis information. For example, reduced costs and shadow prices may not be unique, and you may have to change objective function coefficients beyond their allowable increases or decreases before the optimal solution will change. Thus, some caution should be exercised when interpreting the information. When in doubt, consult a business analytics expert.

Interpreting *Solver* Reports for Nonlinear Optimization Models

Solver provides Answer, Sensitivity, and Limits Reports for nonlinear optimization models. However, the Sensitivity Report is quite different from that for linear models. We use the hotel pricing example from Chapter 14 to discuss these differences.

EXAMPLE 15.7   **Interpreting *Solver* Reports for the *Hotel Pricing Model***

The Answer Report, shown in Figure 15.6, provides the same basic information as for linear models. The Constraints section provides the value for the left-hand side of each constraint in the Cell Value column, its binding or nonbinding status, and the value of the slack. In this example, we see that the limit of 450 rooms and the lower bound on the price of a gold room are binding. This suggests that we could increase revenue if we could either increase the capacity of the hotel or lower the minimum price for a gold room.

In the Adjustable Cells section of the Sensitivity Report (Figure 15.7), the **Reduced Gradient** is analogous to the *Reduced Cost* in linear models. For this problem, however, the objective function coefficient of each price depends on many parameters; therefore, the reduced gradient is more difficult to interpret in relation to the problem data. Thus, we cannot necessarily conclude that a decrease in the price of a gold room of $42.69 will force a change in the solution.

Lagrange Multipliers in the Constraints section are similar to shadow prices for linear models. However, for nonlinear models, the Lagrange multipliers give only an *approximate* rate of change in the objective function as the right-hand side of a binding constraint is increased by 1 unit. Thus, for this example, if the number of available rooms is increased by 1 to 451, the total revenue would increase by *approximately* $12.08. (For linear models, as we have seen, shadow prices give the *exact* rate of change within the Allowable Increase and Allowable Decrease limits.) Thus, you should be somewhat cautious when interpreting these values and will need to re-solve the models to find the true effect of changes to constraints. For this example, the optimal revenue for a 451-room capacity is $39,392.52, an increase of $39,392.52 − $39,380.65 = $11.87, which is close to, but not exactly, the amount predicted by the Lagrange multiplier value.

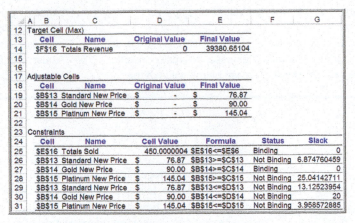

▶ **Figure 15.6**

Hotel Pricing Example Solver *Answer Report*

Cell	Name	Cell Value	Formula	Status	Slack
12 Target Cell (Max)					
13 Cell	Name	Original Value	Final Value		
14 $F$16	Totals Revenue	0	39380.65104		
17 Adjustable Cells					
18 Cell	Name	Original Value	Final Value		
19 $B$13	Standard New Price	$ -	$ 76.87		
20 $B$14	Gold New Price	$ -	$ 90.00		
21 $B$15	Platinum New Price	$ -	$ 145.04		
23 Constraints					
24 Cell	Name	Cell Value	Formula	Status	Slack
25 $E$16	Totals Sold	450.0000004	$E$16<=$E$6	Binding	0
26 $B$13	Standard New Price	$ 76.87	$B$13>=$C$13	Not Binding	6.874760459
27 $B$14	Gold New Price	$ 90.00	$B$14>=$C$14	Binding	0
28 $B$15	Platinum New Price	$ 145.04	$B$15>=$C$15	Not Binding	25.04142711
29 $B$13	Standard New Price	$ 76.87	$B$13<=$D$13	Not Binding	13.12523954
30 $B$14	Gold New Price	$ 90.00	$B$14<=$D$14	Not Binding	20
31 $B$15	Platinum New Price	$ 145.04	$B$15<=$D$15	Not Binding	3.958572885

▶ **Figure 15.7**

Hotel Pricing Example Solver *Sensitivity Report*

	Cell	Name	Final Value	Reduced Gradient
5 Target Cell (Max)				
6	Cell	Name	Final Value	
7	$F$16	Totals Revenue	39380.65104	
9 Adjustable Cells				
10			Final Value	Reduced Gradient
11	Cell	Name		
12	$B$13	Standard New Price	$ 76.87	$ -
13	$B$14	Gold New Price	$ 90.00	$ (42.69)
14	$B$15	Platinum New Price	$ 145.04	$ -
16 Constraints				
17			Final Value	Lagrange Multiplier
18	Cell	Name		
19	$E$16	Totals Sold	450.0000004	12.08293216

CHECK YOUR UNDERSTANDING

1. Explain why what-if analysis is important to apply to optimization models.

2. How can *Solver* be used to conduct what-if analysis for optimization models?

3. What information does the *Solver* Sensitivity Report provide for a linear model?

4. Explain how to interpret and use the *Solver* Sensitivity Report for a linear model.

5. How does the Sensitivity Report differ for nonlinear optimization models?

Models with Bounded Variables

Solver handles simple lower bounds (for example, $C \geq 500$) and upper bounds (for example, $D \leq 1,000$) quite differently from ordinary constraints in the Sensitivity Report. In *Solver*, lower and upper bounds are treated in a manner similar to nonnegativity constraints, which also do not appear explicitly as constraints in the model. *Solver* does this to increase the efficiency of the solution procedure used; for large models, this can represent significant savings in computer-processing time. However, this makes it more difficult to interpret the sensitivity information because we no longer have the shadow prices and allowable increases and decreases associated with these constraints. Actually, this isn't quite true; the shadow prices are there but are hidden in the reduced costs. Fortunately, there is simple approach to provide the missing information that we will describe shortly.

EXAMPLE 15.8   J&M Manufacturing

Suppose that J&M Manufacturing makes four models of gas grills, A, B, C, and D. Each grill must flow through five departments: stamping, painting, assembly, inspection, and packaging. Table 15.1 shows the relevant data. In the second table, for instance, the stamping department can produce 40 units of model A each hour. (Grill A uses imported parts and does not require painting.) J&M wants to determine how many grills to make to maximize monthly profit.

   To formulate this as a linear optimization model, let

A, B, C, and D = number of units of models A, B, C, and
D to produce, respectively

The objective function is to maximize the total net profit:

Maximize $(250 - 210)A + (300 - 240)B + (400 - 300)C$
$$+ (650 - 520)D$$

$$= 40A + 60B + 100C + 130D$$

The constraints include limitations on the amount of production hours available in each department, the minimum sales requirements, and maximum sales potential limits. Here is an example of a situation where you must carefully look at the dimensions of the data. The production rates are given in units/hour, so if you multiply these values by the number of units produced, you will have an expression that makes no sense. Therefore, you must divide the decision

variables by units per hour—or, equivalently, convert these data to hours/unit—and then multiply by the decision variables:

$$A/40 + B/30 + C/10 + D/10 \leq 320 \text{ (stamping)}$$

$$B/20 + C/10 + D/10 \leq 320 \text{ (painting)}$$

$$A/25 + B/15 + C/15 + D/12 \leq 320 \text{ (assembly)}$$

$$A/20 + B/20 + C/25 + D/15 \leq 320 \text{ (inspection)}$$

$$A/50 + B/40 + C/40 + D/30 \leq 320 \text{ (packaging)}$$

   The sales constraints are simple upper and lower bounds on the variables:

$$A \geq 0$$
$$B \geq 0$$
$$C \geq 500$$
$$D \geq 500$$
$$A \leq 4,000$$
$$B \leq 3,000$$
$$C \leq 2,000$$
$$D \leq 1,000$$

Nonnegativity constraints are implied by the lower bounds on the variables and, therefore, do not need to be explicitly stated.

▼ **Table 15.1**

J&M Manufacturing *Data (the second table shows production rates in units/hour)*

Grill Model	Selling Price/Unit	Variable Cost/Unit	Minimum Monthly Sales Requirements	Maximum Monthly Sales Potential
A	$250	$210	0	4,000
B	$300	$240	0	3,000
C	$400	$300	500	2,000
D	$650	$520	500	1,000

Department	A	B	C	D	Hours Available
Stamping	40	30	10	10	320
Painting		20	10	10	320
Assembly	25	15	15	12	320
Inspection	20	20	25	15	320
Packaging	50	40	40	30	320

Figure 15.8 shows a spreadsheet implementation (Excel file *J&M Manufacturing*) with the optimal solution, and Figure 15.9 shows the *Solver* model used to find it. Examine the Answer and Sensitivity Reports for the J&M Manufacturing model in Figures 15.10 and 15.11. In the Answer Report, all constraints are listed along with their status. For example, we see that the upper bound on model D and lower bound on model B are binding. However, none of the bound constraints appear in the Constraints section of the Sensitivity Report.

First, let us interpret the reduced costs. Recall that in an ordinary model with only nonnegativity constraints and no other simple bounds, the reduced cost tells how much the objective coefficient needs to be reduced for a variable to become positive in an optimal solution. For product B, we have the lower bound constraint $B \geq 0$. Note that the optimal solution specifies that we produce only the minimum amount required. Why? It is simply not economical to produce more because the profit contribution of B is too low relative to the other products. How much more would the profit on B have to be for it to be economical to produce anything other than the minimum amount required? The answer is given by the reduced cost. The unit profit on B would have to be reduced by at least −$1.905 (that is, *increased* by at least +$1.905). If a nonzero lower-bound constraint is binding, the interpretation is similar; the reduced cost is the amount the unit profit would have to be reduced to produce more than the minimum amount.

For product D, the reduced cost is $19.29. Note that D is at its upper bound, 1,000. We want to produce as much of D as possible because it generates a large profit. How much would the unit profit have to be *lowered* before it is no longer economical to produce the maximum amount? Again, the answer is the reduced cost, $19.29.

Now, let's ask these questions in a different way. For product B, what would the effect be of increasing the right-hand-side value of the bound constraint, $B \geq 0$, by 1 unit? If we increase the right-hand side of a lower-bound constraint by 1, we are essentially forcing the solution to produce one more than the minimum requirement. How would the objective function change if we do this? It would have to decrease because we would lose money by producing an extra unit of a nonprofitable product. How much? The answer again is the reduced cost. Producing an additional unit of product B will result in a profit reduction of $1.905. Similarly, increasing the right-hand side of the constraint $D \leq 1,000$ by 1 will

▶ Figure 15.8

Spreadsheet Implementation for J&M Manufacturing

	A	B	C	D	E	F
1	**J&M Manufacturing**					
2						
3	**Data**					
4		Grill model	Selling price	Variable cost	Min Sales	Max Sales	
5		A	$ 250.00	$ 210.00	0	4000	
6		B	$ 300.00	$ 240.00	0	3000	
7		C	$ 400.00	$ 300.00	500	2000	
8		D	$ 650.00	$ 520.00	500	1000	
9							
10	Production rates (hours/unit)		A	B	C	D	Hours Available
11		Stamping	40	30	10	10	320
12		Painting		20	10	10	320
13		Assembly	25	15	15	12	320
14		Inspection	20	20	25	15	320
15		Packaging	50	40	40	30	320
16							
17	**Model**					
18		Department	A	B	C	D	Hours Used
19		Stamping	96.429	0.000	123.571	100.000	320.000
20		Painting		0.000	123.571	100.000	223.571
21		Assembly	154.286	0.000	82.381	83.333	320.000
22		Inspection	192.857	0.000	49.429	66.667	308.952
23		Packaging	77.143	0.000	30.893	33.333	141.369
24							
25		Number produced	3857.142857	0	1235.714286	1000	
26		Net profit/unit	$ 40.00	$ 60.00	$ 100.00	$ 130.00	Total Profit
27		Profit contribution	$ 154,285.71	$ -	$ 123,571.43	$ 130,000.00	$ 407,857.14

	A	B	C	D	E	F
1	**J&M Manufacturing**					
2						
3	**Data**					
4		Grill model	Selling price	Variable cost	Min Sales	Max Sales	
5		A	250	210	0	4000	
6		B	300	240	0	3000	
7		C	400	300	500	2000	
8		D	650	520	500	1000	
9							
10	Production rates (hours/unit)		A	B	C	D	Hours Available
11		Stamping	40	30	10	10	320
12		Painting		20	10	10	320
13		Assembly	25	15	15	12	320
14		Inspection	20	20	25	15	320
15		Packaging	50	40	40	30	320
16							
17	**Model**					
18		Department	A	B	C	D	Hours Used
19		Stamping	=B$25/B11	=C$25/C11	=D$25/D11	=E$25/E11	=SUM(B19:E19)
20		Painting		=C$25/C12	=D$25/D12	=E$25/E12	=SUM(B20:E20)
21		Assembly	=B$25/B13	=C$25/C13	=D$25/D13	=E$25/E13	=SUM(B21:E21)
22		Inspection	=B$25/B14	=C$25/C14	=D$25/D14	=E$25/E14	=SUM(B22:E22)
23		Packaging	=B$25/B15	=C$25/C15	=D$25/D15	=E$25/E15	=SUM(B23:E23)
24							
25		Number produced	3857.14285714286	0	1235.71428571429	1000	
26		Net profit/unit	=B5-C5	=B6-C6	=B7-C7	=B8-C8	Total Profit
27		Profit contribution	=B25*B26	=C25*C26	=D25*D26	=E25*E26	=SUM(B27:E27)

increase the profit by $19.29. Thus, *the reduced cost associated with a bounded variable is the same as the shadow price of the bound constraint.* However, we no longer have the allowable range over which we can change the constraint values. (*Important*: The Allowable Increase and Allowable Decrease values in the Sensitivity Report refer to the objective coefficients, not the reduced costs.)

▶ **Figure 15.9**

Solver *Model for* J&M
Manufacturing

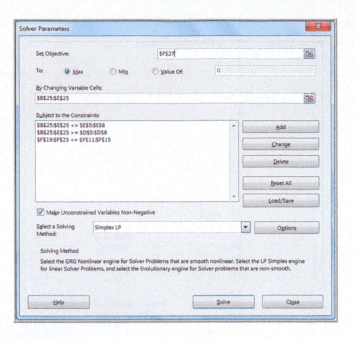

▶ **Figure 15.10**

J&M Manufacturing Solver
Answer Report

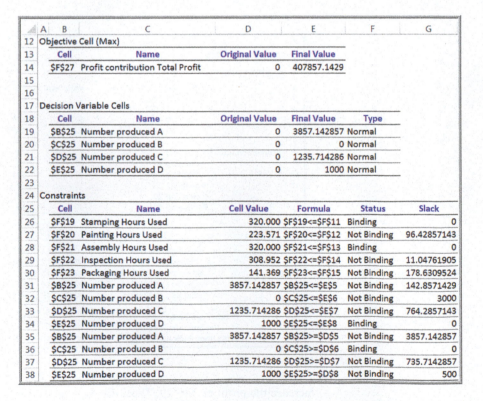

Answer Report table:

Objective Cell (Max)

Cell	Name	Original Value	Final Value		
$F$27	Profit contribution Total Profit	0	407857.1429		

Decision Variable Cells

Cell	Name	Original Value	Final Value	Type	
$B$25	Number produced A	0	3857.142857	Normal	
$C$25	Number produced B	0	0	Normal	
$D$25	Number produced C	0	1235.714286	Normal	
$E$25	Number produced D	0	1000	Normal	

Constraints

Cell	Name	Cell Value	Formula	Status	Slack
$F$19	Stamping Hours Used	320.000	$F$19<=$F$11	Binding	0
$F$20	Painting Hours Used	223.571	$F$20<=$F$12	Not Binding	96.42857143
$F$21	Assembly Hours Used	320.000	$F$21<=$F$13	Binding	0
$F$22	Inspection Hours Used	308.952	$F$22<=$F$14	Not Binding	11.04761905
$F$23	Packaging Hours Used	141.369	$F$23<=$F$15	Not Binding	178.6309524
$B$25	Number produced A	3857.142857	$B$25<=$E$5	Not Binding	142.8571429
$C$25	Number produced B	0	$C$25<=$E$6	Not Binding	3000
$D$25	Number produced C	1235.714286	$D$25<=$E$7	Not Binding	764.2857143
$E$25	Number produced D	1000	$E$25<=$E$8	Binding	0
$B$25	Number produced A	3857.142857	$B$25>=$D$5	Not Binding	3857.142857
$C$25	Number produced B	0	$C$25>=$D$6	Binding	0
$D$25	Number produced C	1235.714286	$D$25>=$D$7	Not Binding	735.7142857
$E$25	Number produced D	1000	$E$25>=$D$8	Not Binding	500

Auxiliary Variables for Bound Constraints

Interpreting reduced costs as shadow prices for bounded variables can be a bit confusing. Fortunately, there is a neat little trick that you can use to eliminate this issue. To recover the missing sensitivity analysis information, define **auxiliary variables**—a new set of cells

▶ **Figure 15.11**

J&M Manufacturing Solver Sensitivity Report

	Cell	Name	Final Value				
5	Objective Cell (Max)						
6	**Cell**	**Name**	**Final Value**				
7	$F$27	Profit contribution Total Profit	407857.1429				
8							
9	Decision Variable Cells						
10			**Final**	**Reduced**	**Objective**	**Allowable**	**Allowable**
11	**Cell**	**Name**	**Value**	**Cost**	**Coefficient**	**Increase**	**Decrease**
12	$B$25	Number produced A	3857.142857	0	40	20.00000004	1.000000042
13	$C$25	Number produced B	0	-1.904761905	60	1.904761905	1E+30
14	$D$25	Number produced C	1235.714286	0	100	13.33333389	33.33333339
15	$E$25	Number produced D	1000	19.28571429	130	1E+30	19.28571429
16							
17	Constraints						
18			**Final**	**Shadow**	**Constraint**	**Allowable**	**Allowable**
19	**Cell**	**Name**	**Value**	**Price**	**R.H. Side**	**Increase**	**Decrease**
20	$F$19	Stamping Hours Used	320.000	571.429	320	44.58333333	5
21	$F$20	Painting Hours Used	223.571	0.000	320	1E+30	96.42857143
22	$F$21	Assembly Hours Used	320.000	642.857	320	3.333333333	71.33333333
23	$F$22	Inspection Hours Used	308.952	0.000	320	1E+30	11.04761905
24	$F$23	Packaging Hours Used	141.369	0.000	320	1E+30	178.6309524

for any decision variables that have upper- or lower-bound constraints by referencing (not copying) the original changing cells. Then in the *Solver* model, use these auxiliary variable cells—*not* the changing variable cells as defined—to define the bound constraints.

EXAMPLE 15.9   Using Auxiliary Variable Cells

Figure 15.12 shows a portion of the J&M Manufacturing model with the inclusion of auxiliary variables in row 29. The formula in cell B29, for example, is =B25. The *Solver* is modified as shown in Figure 15.13 by changing the decision variable cells in the bound constraints to the auxiliary variable cells. The Sensitivity Report for this model is shown in Figure 15.14. We now see that the Constraints section has rows corresponding to the bound constraints and that the shadow prices are the same as the reduced costs in the original Sensitivity Report. Moreover, we now know the allowable increases and decreases for each shadow price, which we did not have before. Thus, we recommend that you use this approach unless solution efficiency is an important issue.

▶ **Figure 15.12**

Auxiliary Variable Cells in J&M Manufacturing Model

	A	B	C	D	E	F
24						
25	Number produced	0	0	0	0	
26	Net profit/unit	$ 40.00	$ 60.00	$ 100.00	$ 130.00	Total Profit
27	Profit contribution	$ -	$ -	$ -	$ -	$ -
28						
29	Auxiliary variable	0	0	0	0	

	A	B	C	D	E	F
24						
25	Number produced	0	0	0	0	
26	Net profit/unit	=B5-C5	=B6-C6	=B7-C7	=B8-C8	Total Profit
27	Profit contribution	=B25*B26	=C25*C26	=D25*D26	=E25*E26	=SUM(B27:E27)
28						
29	Auxiliary variable	=B25	=C25	=D25	=E25	

▶ **Figure 15.13**

Solver *Model for* J&M Manufacturing *with Auxiliary Variables*

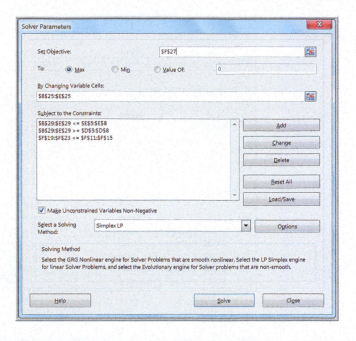

▶ **Figure 15.14**

J&M Manufacturing *Sensitivity Report with Auxiliary Variables*

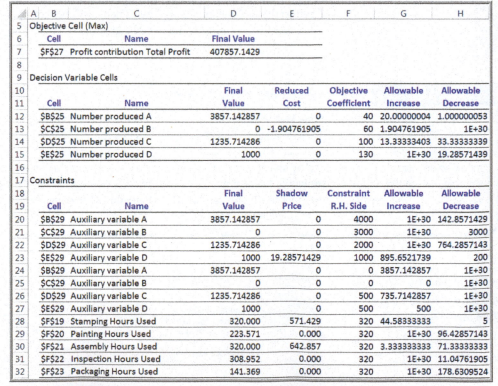

	A	B	C	D	E	F	G	H
5		Objective Cell (Max)						
6		Cell	Name	Final Value				
7		$F$27	Profit contribution Total Profit	407857.1429				
8								
9		Decision Variable Cells						
10				Final	Reduced	Objective	Allowable	Allowable
11		Cell	Name	Value	Cost	Coefficient	Increase	Decrease
12		$B$25	Number produced A	3857.142857	0	40	20.00000004	1.000000053
13		$C$25	Number produced B	0	-1.904761905	60	1.904761905	1E+30
14		$D$25	Number produced C	1235.714286	0	100	13.33333403	33.33333339
15		$E$25	Number produced D	1000	0	130	1E+30	19.28571439
16								
17		Constraints						
18				Final	Shadow	Constraint	Allowable	Allowable
19		Cell	Name	Value	Price	R.H. Side	Increase	Decrease
20		$B$29	Auxiliary variable A	3857.142857	0	4000	1E+30	142.8571429
21		$C$29	Auxiliary variable B	0	0	3000	1E+30	3000
22		$D$29	Auxiliary variable C	1235.714286	0	2000	1E+30	764.2857143
23		$E$29	Auxiliary variable D	1000	19.28571429	1000	895.6521739	200
24		$B$29	Auxiliary variable A	3857.142857	0	0	3857.142857	1E+30
25		$C$29	Auxiliary variable B	0	0	0	0	1E+30
26		$D$29	Auxiliary variable C	1235.714286	0	500	735.7142857	1E+30
27		$E$29	Auxiliary variable D	1000	0	500	500	1E+30
28		$F$19	Stamping Hours Used	320.000	571.429	320	44.58333333	5
29		$F$20	Painting Hours Used	223.571	0.000	320	1E+30	96.42857143
30		$F$21	Assembly Hours Used	320.000	642.857	320	3.333333333	71.33333333
31		$F$22	Inspection Hours Used	308.952	0.000	320	1E+30	11.04761905
32		$F$23	Packaging Hours Used	141.369	0.000	320	1E+30	178.6309524

CHECK YOUR UNDERSTANDING

1. Describe the differences found in the *Solver* Sensitivity Report for models that have bounded variables.

2. How do you incorporate auxiliary variables into *Solver* models with bounded variables?

3. What additional information do auxiliary variables provide in *Solver* Sensitivity Reports?

What-If Analysis for Integer Optimization Models

Because integer models are discontinuous by their very nature, sensitivity information cannot be generated in the same manner as for linear models; therefore, no Sensitivity Report is provided by *Solver*—only the Answer Report is available. To investigate changes in model parameters, it is necessary to re-solve the model. In the following example, we show how integer optimization models can be used in locating facilities and apply what-if analysis to examine trade-offs among different solutions. This example is often called a "covering" problem, because we seek to choose a subset of locations that serve, or cover, all locations in a service area.

EXAMPLE 15.10   **Anderson Village Fire Stations**

Suppose that an unincorporated village wishes to find the best locations for fire stations. Assume that the village is divided into smaller districts, or neighborhoods, and that transportation studies have estimated the response time for emergency vehicles to travel between each pair of districts. The village wants to locate the fire stations so that all districts can be reached within eight minutes. The following table shows the estimated response time in minutes between each pair of districts:

From/To	1	2	3	4	5	6	7
1	0	2	10	6	12	5	8
2	2	0	6	9	11	7	10
3	10	6	0	5	5	12	6
4	6	9	5	0	9	4	3
5	12	11	5	9	0	10	8
6	5	7	12	4	10	0	6
7	8	10	6	3	8	6	0

Define $X_j = 1$ if a fire station is located in district j and 0 if not. The objective is to minimize the number of fire stations that need to be built:

$$\text{Min } X_1 + X_2 + X_3 + X_4 + X_5 + X_6 + X_7$$

Each district must be reachable within eight minutes by some fire station. Thus, from the table, for example, we see that to be able to respond to district 1 in eight minutes or less, a station must be located in district 1, 2, 4, 6, or 7. Therefore, we must have the constraint

$$X_1 + X_2 + X_4 + X_6 + X_7 \geq 1$$

Similar constraints may be formulated for each of the other districts:

$$X_1 + X_2 + X_3 + X_6 \geq 1$$
$$X_2 + X_3 + X_4 + X_5 + X_7 \geq 1$$
$$X_1 + X_3 + X_4 + X_6 + X_7 \geq 1$$
$$X_3 + X_5 + X_7 \geq 1$$
$$X_1 + X_2 + X_4 + X_6 + X_7 \geq 1$$
$$X_1 + X_3 + X_4 + X_5 + X_6 + X_7 \geq 1$$

Figure 15.15 shows a spreadsheet model for this problem (Excel file *Anderson Village Fire Station Location Model*). To develop the constraints in the model, we construct a matrix by converting all response times that are within eight minutes to 1s and those that exceed eight minutes to 0s. Then the constraint functions for each district are simply the SUMPRODUCT of the decision variables and the rows of this matrix, making the *Solver* model, shown in Figure 15.16, easy to define. For instance, the formula in cell I20 is =SUMPRODUCT($B$28:$H$28, B20:H20). For this example, the solution is to site fire stations in districts 3 and 7.

	A	B	C	D	E	F	G	H	I	J
1	Anderson Village Fire Station Model									
2										
3	Data									
4										
5	Response time	8 minutes								
6										
7	Response Times									
8	From/To	1	2	3	4	5	6	7		
9	1	0	2	10	6	12	5	8		
10	2	2	0	6	9	11	7	10		
11	3	10	6	0	5	5	12	6		
12	4	6	9	5	0	9	4	3		
13	5	12	11	5	9	0	10	8		
14	6	5	7	12	4	10	0	6		
15	7	8	10	6	3	8	6	0		
16										
17	Model									
18										
19	From/To	1	2	3	4	5	6	7	Covered?	Requirement
20	1	1	1	0	1	0	1	1	1	1
21	2	1	1	1	0	0	1	0	1	1
22	3	0	1	1	1	1	0	1	2	1
23	4	1	0	1	1	0	1	1	2	1
24	5	0	0	1	0	1	0	1	2	1
25	6	1	1	0	1	0	1	1	1	1
26	7	1	0	1	1	1	1	1	2	1
27									Total	
28	Location	0	0	1	0	0	0	1	2	

▲ **Figure 15.15**

Spreadsheet Model for Anderson Village Fire Station Location Model

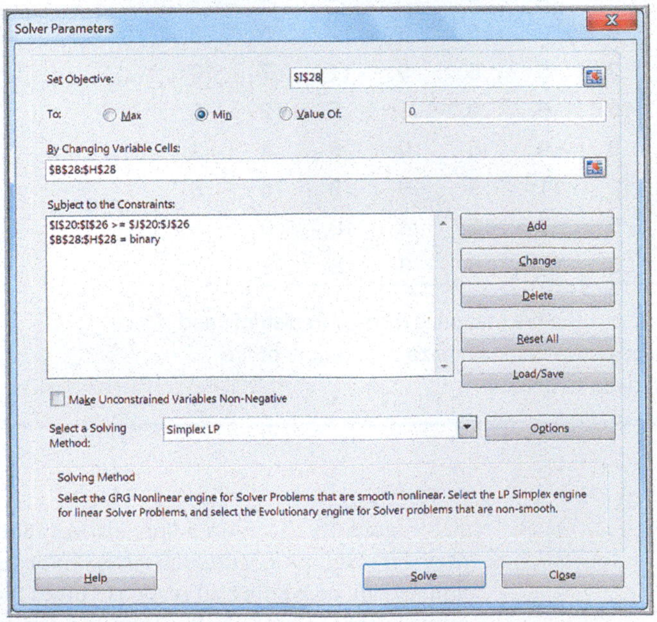

▶ **Figure 15.16**

Solver *Model for Anderson Village Fire Station Location*

Suppose that the Anderson Village township's board of trustees wants to better understand the trade-offs between the response time and minimum number of fire stations needed. We could change the value of the response time in cell B5 and re-solve the model.

EXAMPLE 15.11   What-If Analysis for Response Time

In the Anderson Village example, we changed the response time requirement in cell B8 to vary between 5 and 10 minutes, and re-solved the model for each value. The results are shown in Figure 15.17. In column A are the values of the response time. The 1s in columns B through H show where the fire stations should be located. Column I shows the minimum number of fire stations required.

These results show the maximum response time can be reduced to six minutes while still using only two fire stations (the model solution yields districts 1 and 3). This would clearly be a better alternative. Also, if the response time is increased by only one minute from its original target, the township could save the cost of building a second facility. Of course, such decisions need to be evaluated carefully.

▶ **Figure 15.17**

What-If Analysis Results

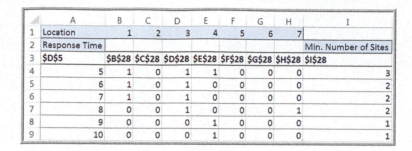

	A	B	C	D	E	F	G	H	I
1	Location	1	2	3	4	5	6	7	
2	Response Time								Min. Number of Sites
3	$D$5	$B$28	$C$28	$D$28	$E$28	$F$28	$G$28	$H$28	$I$28
4	5	1	0	1	1	0	0	0	3
5	6	1	0	1	0	0	0	0	2
6	7	1	0	1	0	0	0	0	2
7	8	0	0	1	0	0	0	1	2
8	9	0	0	0	1	0	0	0	1
9	10	0	0	0	1	0	0	0	1

CHECK YOUR UNDERSTANDING

1. Explain why integer optimization models do not produce *Solver* Sensitivity Reports.

2. How can you conduct what-if analyses for integer optimization models?

Visualization of *Solver* Reports

As you certainly know by now, interpreting the output from *Solver* requires some technical knowledge of linear optimization concepts and terminology, such as reduced costs and shadow prices. Data visualization can help analysts present optimization results in forms that are more understandable and can be easily explained to managers and clients in a report or presentation. We will illustrate this using a process selection model. Process selection models generally involve choosing among different types of processes to produce a good. Make-or-buy decisions are examples of process selection models, whereby we must choose whether to make one or more products in-house or subcontract them out to another firm.

EXAMPLE 15.12   Camm Textiles

Camm Textiles has a mill that produces three types of fabrics on a make-to-order basis. The mill operates on a 24/7 schedule. The key decision facing the plant manager is about the type of loom needed to process each fabric during the coming quarter (13 weeks) to meet demands for the three fabrics and not exceed the capacity of the looms in the mill. Two types of looms are used: dobbie and regular. Dobbie looms can be used to make all fabrics

and are the only looms that can weave certain fabrics, such as plaids. Demands, variable costs for each fabric, and production rates on the looms are given in Table 15.2. The mill has 15 regular looms and 3 dobbie looms. After weaving, fabrics are sent to the finishing department and then sold. Any fabrics that cannot be woven in the mill because of limited capacity will be purchased from an external supplier, finished at the mill, and sold at the selling

(continued)

price. In addition to determining which looms to use to process the fabrics, the manager also needs to determine which fabrics to buy externally.

To formulate a linear optimization model, define D_i = number of yards of fabric i to produce on dobbie looms, $i = 1, 2, 3$. That is, D_1 = number of yards of fabric 1 to produce on dobbie looms, D_2 = number of yards of fabric 2 to produce on dobbie looms, and D_3 = number of yards of fabric 3 to produce on dobbie looms. In a similar fashion, define the following:

R_i = number of yards of fabric i to produce on regular looms, $i = 2, 3$ only

P_i = number of yards of fabric i to purchase from an outside supplier, $i = 1, 2, 3$

Note that we are using *subscripted variables* to simplify their definition rather than defining nine individual variables with unique names.

The objective function is to minimize total cost, found by multiplying the cost per yard based on the mill cost or outsourcing by the number of yards of fabric for each type of decision variable:

$$\text{Min } 0.65D_1 + 0.61D_2 + 0.50D_3 + 0.61R_2 + 0.50R_3$$
$$+ 0.85P_1 + 0.75P_2 + 0.65P_3$$

Constraints to ensure meeting production requirements are

Fabric 1 demand: $D_1 + P_1 = 45{,}000$

This constraint states that the amount of fabric 1 produced on dobbie looms or outsourced must equal the total demand of 45,000 yards. The constraints for the other two fabrics are

Fabric 2 demand: $D_2 + R_2 + P_2 = 76{,}500$

Fabric 3 demand: $D_3 + R_3 + P_3 = 10{,}000$

To specify the constraints on loom capacity, we must convert yards per hour into hours per yard. For example, for fabric 1 on a dobbie loom, 4.7 yards/hour = 0.213 hour/yard. Therefore, the term $0.213D_1$ represents the total time required to produce D_1 yards of fabric 1 on a dobbie loom (hours/yard × yards). The total capacity for dobbie looms is

(24 hours / day)(7 days / week)(13 weeks)(3 looms)
$$= 6{,}552 \text{ hours}$$

Thus, the constraint on available production time on dobbie looms is

$$0.213D_1 + 0.192D_2 + 0.227D_3 \leq 6{,}552$$

For regular looms, we have

$$0.192R_2 + 0.227R_3 \leq 32{,}760$$

Finally, all variables must be nonnegative.

The complete model is

$$\text{Min } 0.65D_1 + 0.61D_2 + 0.50D_3 + 0.61R_2 + 0.50R_3$$
$$+ 0.85P_1 + 0.75P_2 + 0.65P_3$$

Fabric 1 demand: $D_1 + P_1 = 45{,}000$

Fabric 2 demand: $D_2 + R_2 + P_2 = 76{,}500$

Fabric 3 demand: $D_3 + R_3 + P_3 = 10{,}000$

Dobbie loom capacity:

$$0.213D_1 + 0.192D_2 + 0.227D_3 \leq 6{,}552$$

Regular loom capacity:

$$0.192R_2 + 0.227R_3 \leq 32{,}760$$

Nonnegativity: all variables ≥ 0

▼ **Table 15.2**

Textile Production Data

Fabric	Demand (yards)	Dobbie Loom Capacity (yards/hour)	Regular Loom Capacity (yards/hour)	Mill Cost ($/yard)	Outsourcing Cost ($/yard)
1	45,000	4.7	0.0	$0.65	$0.85
2	76,500	5.2	5.2	$0.61	$0.75
3	10,000	4.4	4.4	$0.50	$0.65

Figure 15.18 shows a spreadsheet implementation (Excel file *Camm Textiles*) with the optimal solution to Example 15.12. Observe the design of the spreadsheet and, in particular, the use of labels in the rows and columns in the model section. Using the principles

discussed in the previous chapter, this design makes it easy to read and interpret the Answer and Sensitivity Reports. Figure 15.19 shows the *Solver* model. It is easier to define the decision variables as the range B14:D16; however, because we cannot produce fabric 1 on regular looms, we set cell C14 to zero as a constraint.

Figures 15.20 and 15.21 show the *Solver* Answer and Sensitivity Reports for this problem. The first thing that one might do is to visualize the values of the optimal decision variables and constraints, drawing upon the model output or the information contained in the Answer Report. Figure 15.22 shows a chart of the decision variables, showing the amounts of each fabric produced on each type of loom and outsourced. Figure 15.23 shows the capacity utilization of each type of loom. We can easily see that the utilization of regular looms is approximately half the capacity, while dobbie looms are fully utilized, suggesting that the purchase of additional dobbie looms might be useful, at least under the current demand scenario.

The Sensitivity Report is more challenging to visualize effectively. The reduced costs describe how much the unit production or purchasing cost must be changed to force the

▶ **Figure 15.18**

Spreadsheet Model for Camm Textiles

	A	B	C	D	E	F
1	Camm Textiles					
2						
3	Data					
4		Dobbie	Regular			
5	Fabric	Capacity	Capacity	Mill Cost	Outsourcing Cost	Demand
6	1	4.7	0	$ 0.65	$0.85	45000
7	2	5.2	5.2	$ 0.61	$0.75	76500
8	3	4.4	4.4	$ 0.50	$0.65	10000
9	Hours Available	6552	32760			
10						
11	Model					
12						
13		on Dobbie	on Regular	Purchased	Total Yards Produced	
14	Fabric 1	30794.4	0	14205.6	45000	
15	Fabric 2	0	76500	0	76500	
16	Fabric 3	0	10000	0	10000	
17	Hours Used	6552	16984.26573			
18						
19		Total				
20	Cost	$ 83,756.12				

	A	B	C	D	E	F
1	Camm Textiles					
2						
3	Data					
4		Dobbie	Regular			
5	Fabric	Capacity	Capacity	Mill Cost	Outsourcing Cost	Demand
6	1	4.7	0	0.65	0.85	45000
7	2	5.2	5.2	0.61	0.75	76500
8	3	4.4	4.4	0.5	0.65	10000
9	Hours Available	=24*7*13*3	=24*7*13*15			
10						
11	Model					
12						
13		on Dobbie	on Regular	Purchased	Total Yards Produced	
14	Fabric 1	30794.4	0	14205.6	=SUM(B14:D14)	
15	Fabric 2	0	76500	0	=SUM(B15:D15)	
16	Fabric 3	0	10000	0	=SUM(B16:D16)	
17	Hours Used	=B14/B6+B15/B7+B16/B8	=C15/C7+C16/C8			
18						
19		Total				
20	Cost	=SUMPRODUCT(B14:B16,D6:D8)+SUMPRODUCT(C15:C16,D7:D8)+SUMPRODUCT(D14:D16,E6:E8)				

▶ **Figure 15.19**

Solver *Model for* Camm Textiles

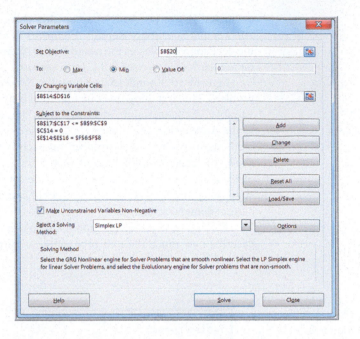

▶ **Figure 15.20**

Solver *Answer Report for* Camm Textiles

	A	B	C	D	E	F	G
11							
12		Objective Cell (Min)					
13		**Cell**	**Name**	**Original Value**	**Final Value**		
14		$B$20	Cost Total	0	83756.12		
15							
16							
17		Decision Variable Cells					
18		**Cell**	**Name**	**Original Value**	**Final Value**	**Type**	
19		$B$14	Fabric 1 on Dobbie	0	30794.4	Normal	
20		$C$14	Fabric 1 on Regular	0	0	Normal	
21		$D$14	Fabric 1 Purchased	0	14205.6	Normal	
22		$B$15	Fabric 2 on Dobbie	0	0	Normal	
23		$C$15	Fabric 2 on Regular	0	76500	Normal	
24		$D$15	Fabric 2 Purchased	0	0	Normal	
25		$B$16	Fabric 3 on Dobbie	0	0	Normal	
26		$C$16	Fabric 3 on Regular	0	10000	Normal	
27		$D$16	Fabric 3 Purchased	0	0	Normal	
28							
29		Constraints					
30		**Cell**	**Name**	**Cell Value**	**Formula**	**Status**	**Slack**
31		$B$17	Hours Used on Dobbie	6552	$B$17<=$B$9	Binding	0
32		$C$17	Hours Used on Regular	16984.26573	$C$17<=$C$9	Not Binding	15775.73427
33		$E$14	Fabric 1 Total Yards Produced	45000	$E$14=$F$6	Binding	0
34		$E$15	Fabric 2 Total Yards Produced	76500	$E$15=$F$7	Binding	0
35		$E$16	Fabric 3 Total Yards Produced	10000	$E$16=$F$8	Binding	0
36		$C$14	Fabric 1 on Regular	0	$C$14=0	Binding	0

value of a variable to become positive in the solution. Figure 15.24 shows a visualization of the reduced cost information. The chart displays the unit cost coefficients for each production or outsourcing decision, and for those not currently utilized, the change in cost required to force that variable to become positive in the solution. Note that since fabric 1 cannot be produced on a regular loom, its reduced cost is meaningless and, therefore, not displayed.

▶ **Figure 15.21**

Solver *Sensitivity Report for Camm Textiles*

	Cell	Name	Final Value				
4							
5	Objective Cell (Min)						
6	**Cell**	**Name**	**Final Value**				
7	$B$20	Cost Total	83756.12				
8							
9	Decision Variable Cells						
10			**Final**	**Reduced**	**Objective**	**Allowable**	**Allowable**
11	**Cell**	**Name**	**Value**	**Cost**	**Coefficient**	**Increase**	**Decrease**
12	$B$14	Fabric 1 on Dobbie	30794.4	0	0.65	0.200000094	1E+30
13	$C$14	Fabric 1 on Regular	0	-0.85	0	1E+30	1E+30
14	$D$14	Fabric 1 Purchased	14205.6	0	0.85	1E+30	0.200000094
15	$B$15	Fabric 2 on Dobbie	0	0.180769231	0.61	1E+30	0.180769231
16	$C$15	Fabric 2 on Regular	76500	0	0.61	0.1400001	1E+30
17	$D$15	Fabric 2 Purchased	0	0.14	0.75	1E+30	0.14
18	$B$16	Fabric 3 on Dobbie	0	0.213636364	0.5	1E+30	0.213636364
19	$C$16	Fabric 3 on Regular	10000	0	0.5	0.1500001	1E+30
20	$D$16	Fabric 3 Purchased	0	0.15	0.65	1E+30	0.15
21							
22	Constraints						
23			**Final**	**Shadow**	**Constraint**	**Allowable**	**Allowable**
24	**Cell**	**Name**	**Value**	**Price**	**R.H. Side**	**Increase**	**Decrease**
25	$B$17	Hours Used on Dobbie	6552	-0.94	6552	3022.468085	6552
26	$C$17	Hours Used on Regular	16984.26573	0	32760	1E+30	15775.73427
27	$E$14	Fabric 1 Total Yards Produced	45000	0.85	45000	1E+30	14205.6
28	$E$15	Fabric 2 Total Yards Produced	76500	0.61	76500	82033.81818	76500
29	$E$16	Fabric 3 Total Yards Produced	10000	0.5	10000	69413.23077	10000

▶ **Figure 15.22**

Summary of Optimal Solution

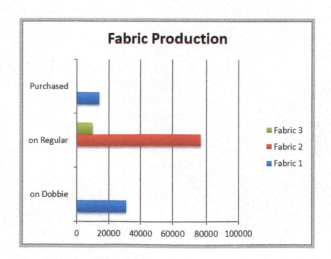

▶ **Figure 15.23**

Chart of Capacity Utilization

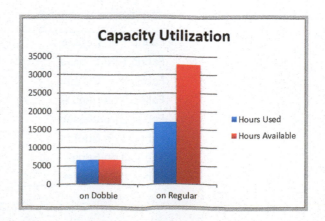

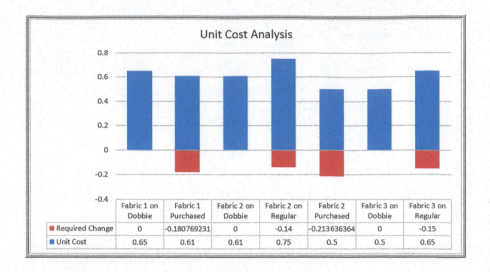

We may also visualize the ranges over which the unit cost coefficients may change without changing the optimal values of the decision variables by using an Excel *Stock Chart*. A stock chart typically shows the "high-low-close" values of daily stock prices; here we can compute the maximum-minimum-current values of the unit cost coefficients. To do this, follow these steps (for Windows; Mac menus are slightly different):

1. Create a table in the worksheet by adding the Allowable Increase values and subtracting the Allowable Decrease values from the cost coefficients, as shown in Table 15.3. Replace $1E + 30$ by #N/A in the worksheet so that infinite values are not displayed. *Note: You must have at least three rows in the table to create a stock chart.*

2. Highlight the range of this table and insert an Excel *Stock Chart* and name the series as Maximum, Minimum, and Current.

3. Click the chart, and in the *Format* tab of *Chart Tools*, go to the *Current Selection* group to the left of the ribbon and click on the drop-down box (it usually says "Chart Area"). Find the series you wish to format and then click *Format Selection*.

► Table 15.3

Data Used to Construct Stock Chart for Cost Coefficient Ranges

	Maximum	Minimum	Current
Fabric 1 on Dobbie	0.85	#N/A	0.65
Fabric 1 Purchased	#N/A	0.65	0.85
Fabric 2 on Dobbie	#N/A	0.429231	0.61
Fabric 2 on Regular	0.75	#N/A	0.61
Fabric 2 Purchased	#N/A	0.61	0.75
Fabric 3 on Dobbie	#N/A	0.286364	0.5
Fabric 3 on Regular	0.65	#N/A	0.5
Fabric 3 Purchased	#N/A	0.5	0.65

4. In the *Format Data Series* pane that appears in the worksheet, click the paint icon and then *Marker*, making sure to expand the *Marker Options* menu.

5. Choose the type of marker you want and increase the width of the markers to make them more visible. We chose the green symbol × for the current value, a red triangle for the minimum value, and a blue dash for the maximum value. This results in the chart shown in Figure 15.25.

Now it is easy to visualize the allowable unit cost ranges. For those lines that have no maximum limit (the blue dash) such as with Fabric 1 Purchased, the unit costs can increase to infinity; for those that have no lower limit (the red triangle) such as Fabric 1 on Dobbie, the unit costs can decrease indefinitely.

Shadow prices show the impact of changing the right-hand side of a binding constraint. Because the plant operates on a 24/7 schedule, changes in loom capacity would require they be in "chunks" (that is, purchasing an additional loom) rather than incremental. However, changes in the demand can easily be assessed using the shadow price information. Figures 15.26 and 15.27 show a simple summary of the shadow prices associated with each product, as well as the ranges based on the Allowable Increase and Allowable Decrease values over which these prices are valid, using a similar approach as described earlier for the cost-coefficient ranges.

▶ **Figure 15.25**

Chart of Allowable Unit Cost Ranges

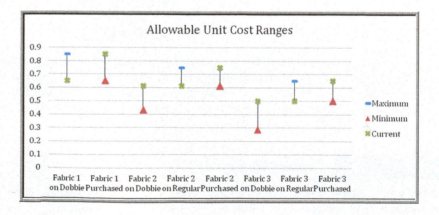

▶ **Figure 15.26**

Summary of Shadow Prices

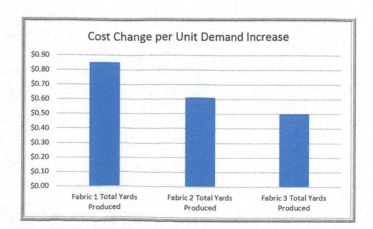

► **Figure 15.27**

Chart of Allowable Demand Ranges for Valid Shadow Prices

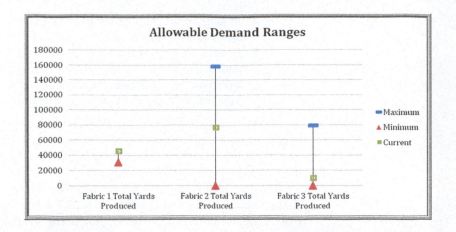

 CHECK YOUR UNDERSTANDING

1. Explain the value of data visualization in communicating information contained in *Solver* reports.

2. What information in *Solver* reports can be visualized using Excel charts? What types of charts are useful?

Using Sensitivity Information Correctly

One crucial assumption in interpreting sensitivity analysis information for changes in model parameters is that all other model parameters are held constant. It is easy to fall into a trap of ignoring this assumption and blindly crunching through the numbers. This is particularly true when using spreadsheet models. We will use the following example to illustrate this.

EXAMPLE 15.13   Walker Wines

A small winery, Walker Wines, buys grapes from local growers and blends the pressings to make two types of wine: Shiraz and merlot.[1] It costs $1.60 to purchase the grapes needed to make a bottle of Shiraz and $1.40 to purchase the grapes needed to make a bottle of merlot. The contract requires that they provide at least 40% but not more than 70% Shiraz. Based on market research, it is estimated that the base demand for Shiraz is 1,000 bottles, but demand increases by five bottles for each $1 spent on advertising; the base demand for merlot is 2,000 bottles and increases by eight bottles for each $1 spent on advertising. Production should not exceed demand. Shiraz sells to retail stores for a wholesale price of $6.25 per bottle and merlot is sold for $5.25 per bottle. Walker Wines has $50,000 available to purchase grapes and advertise its products, with an objective of maximizing profit contribution.

To formulate this model, let

S = number of bottles of Shiraz produced

M = number of bottles of merlot produced

A_s = dollar amount spent on advertising Shiraz

A_m = dollar amount spent on advertising merlot

The objective is to maximize profit (revenue minus costs):

$$= (\$6.25S + \$5.25M) - (\$1.60S + \$1.40M + A_s + A_m)$$
$$= 4.65S + 3.85M - A_s - A_m$$

Constraints are defined as follows:

1. Budget cannot be exceeded:

$$\$1.60S + \$1.40M + A_s + A_m \leq \$50,000$$

2. Contractual requirements must be met:

$$0.4 \leq S/(S + M) \leq 0.7$$

[1]Based on an example in Roger D. Eck, *Operations Research for Business* (Belmont, CA: Wadsworth, 1976): 129–131.

Expressed in linear form,

$$0.6S - 0.4M \geq 0 \text{ and } 0.3S - 0.7M \leq 0$$

3. Production must not exceed demand:

$$S \leq 1{,}000 + 5A_s$$
$$M \leq 2{,}000 + 8A_m$$

4. Nonnegativity

Figure 15.28 shows a spreadsheet implementation of this model (Excel file *Walker Wines*) along with the optimal solution. Figure 15.29 shows the *Solver* model.

▶ Figure 15.28

Walker Wines *Spreadsheet Model*

	A	B	C	D	E
1	Walker Wines				
2					
3	Data				
4		Shiraz	Merlot		
5	Cost/bottle	$ 1.60	$ 1.40		
6	Price/bottle	$ 6.25	$ 5.25		
7					
8	Base demand	1,000.00	2,000.00		
9	Increase/$1 Adv.	5	8		
10	Min. percent requirement	40%			
11	Max. percent limitation	70%			
12					
13	Total Budget	$ 50,000.00			
14					
15	Model				
16					
17		Shiraz	Merlot	Total	
18	Unit profit	$ 4.65	$ 3.85		
19	Advertising dollars	$ 3,912.37	$ 851.53	$ 4,763.90	
20	Demand	20,561.86	8,812.23	29,374.09	
21	Quantity produced	20,561.86	8,812.23	29,374.09	
22					
23	Min. percent requirement	8812.227074	>=	0	
24	Max. percent limitation	0	<=	0	
25					
26				Used	Unused
27	Budget	$ 36,811.35	$ 13,188.65	$ 50,000.00	$ -
28					
29		Total			
30	Profit	$ 124,775.84			

	A	B	C	D	E
1	Walker Wines				
2					
3	Data				
4		Shiraz	Merlot		
5	Cost/bottle	1.6	1.4		
6	Price/bottle	6.25	5.25		
7					
8	Base demand	1000	2000		
9	Increase/$1 Adv.	5	8		
10	Min. percent requirement	0.4			
11	Max. percent limitation	0.7			
12					
13	Total Budget	50000			
14					
15	Model				
16					
17		Shiraz	Merlot	Total	
18	Unit profit	=B6-B5	=C6-C5		
19	Advertising dollars	3912.37263464338	851.528384279476	=SUM(B19:C19)	
20	Demand	=B8+(B9*B19)	=C8+(C9*C19)	=SUM(B20:C20)	
21	Quantity produced	20561.8631732169	8812.22707423581	=SUM(B21:C21)	
22					
23	Min. percent requirement	=(1-B10)*B21-B10*C21	>=	0	
24	Max. percent limitation	=(1-B11)*B21-B11*C21	<=	0	
25					
26				Used	Unused
27	Budget	=B19+(B21*B5)	=C19+(C21*C5)	=SUM(B27:C27)	=B13-D27
28					
29		Total			
30	Profit	=(B18*B21)+(C18*C21)-B19-C19			

▶ **Figure 15.29**

Walker Wines Solver *Model*

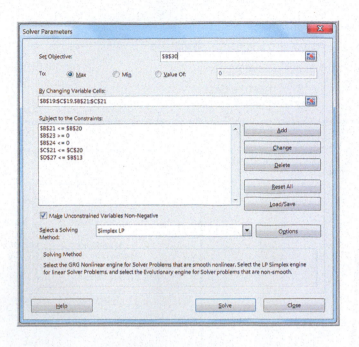

As we noted, interpreting sensitivity analysis information for a change in a model parameter assumes that *all other model parameters are held constant*. Let's see how this applies to the Walker Wines example.

EXAMPLE 15.14   **Evaluating a Cost Increase for Walker Wines**

Figure 15.30 shows the *Solver* Sensitivity Report. A variety of practical questions can be posed around the Sensitivity Report. For example, suppose that the accountant noticed a small error in computing the profit contribution for Shiraz. The cost of Shiraz grapes should have been $1.65 instead of $1.60. How will this affect the solution?

In the model formulation, you can see that a $0.05 increase in cost results in a drop in the unit profit of Shiraz from $4.65 to $4.60. In the Sensitivity Report, however, the change in the profit coefficient is within the allowable decrease of 0.05328; thus, we might conclude that no

change in the optimal solution will result. However, this is *not* the correct interpretation. If the model is re-solved using the new cost parameter, the solution changes dramatically, as shown in Figure 15.31.

Why did this happen? In this case, the unit cost is also reflected in the binding budget constraint. When we change the cost parameter, the constraint also changes. This violates the assumption that all other model parameters are held constant. The change causes the budget constraint to become infeasible, and the solution must be adjusted to maintain feasibility.

This example points out the importance of fully understanding the mathematical model when analyzing sensitivity information. One suggestion to ensure that sensitivity analysis information is interpreted properly in spreadsheet models is to use Excel's formula-auditing capability. If you select the cost of Shiraz (cell B5) and apply the "Trace Dependents" command from the *Formula Auditing* menu, you will see that the unit cost influences both the unit profit (cell B30) *and* the budget constraint function (cell B27).

Figure 15.30

Walker Wines Solver *Sensitivity Report*

	A	B	C	D	E	F	G	H
6	Objective Cell (Max)							
7		Cell	Name	Final Value				
8		$B$30	Profit Total	124775.837				
9								
10	Decision Variable Cells							
11				Final	Reduced	Objective	Allowable	Allowable
12		Cell	Name	Value	Cost	Coefficient	Increase	Decrease
13		$B$19	Advertising dollars Shiraz	$ 3,912.37	$ -	-1	3.771791052	0.266394356
14		$C$19	Advertising dollars Merlot	$ 851.53	$ -	-1	0.36111235	112.8666705
15		$B$21	Quantity produced Shiraz	20,561.86	0.00	4.65	1E+30	0.053278871
16		$C$21	Quantity produced Merlot	8,812.23	0.00	3.85	0.045139044	14.10833381
17								
18	Constraints							
19				Final	Shadow	Constraint	Allowable	Allowable
20		Cell	Name	Value	Price	R.H. Side	Increase	Decrease
21		$B$21	Quantity produced Shiraz	20,561.86	0.69	1000	21297.93978	195000
22		$B$23	Min. percent requirement Shiraz	8812.227074	0	0	8812.227074	1E+30
23		$B$24	Max. percent limitation Shiraz	0	0.047307132	0	6500	9256.880734
24		$C$21	Quantity produced Merlot	8,812.23	0.43	2000	6964.285714	383971.4286
25		$D$27	Budget Used	$ 50,000.00	$ 2.46	50000	1E+30	39000

Figure 15.31

Walker Wines Solver *Solution After Cost Increase*

	A	B	C	D	E
1	Walker Wines				
2					
3	Data				
4		Shiraz	Merlot		
5	Cost/bottle	$ 1.65	$ 1.40		
6	Price/bottle	$ 6.25	$ 5.25		
7					
8	Base demand	1,000.00	2,000.00		
9	Increase/$1 Adv.	5	8		
10	Min. percent requirement	40%			
11	Max. percent limitation	70%			
12					
13	Total Budget	$ 50,000.00			
14					
15	Model				
16					
17		Shiraz	Merlot	Total	
18	Unit profit	$ 4.60	$ 3.85		
19	Advertising dollars	$ 2,238.67	$ 2,036.25	$ 4,274.92	
20	Demand	12,193.35	18,290.03	30,483.38	
21	Quantity produced	12,193.35	18,290.03	30,483.38	
22					
23	Min. percent requirement	0	>=	0	
24	Max. percent limitation	-9145.01511	<=	0	
25					
26				Used	Unused
27	Budget	$ 22,357.70	$27,642.30	$50,000.00	$ -
28					
29		Total			
30	Profit	$122,231.12			

CHECK YOUR UNDERSTANDING

1. What should you look for in models to ensure that sensitivity information is interpreted correctly?
2. How can you use Excel features to ensure the correct interpretation of sensitivity analysis?

KEY TERMS

Auxiliary variables   Reduced cost
Degenerate solution   Reduced gradient
Lagrange multiplier   Shadow price

CHAPTER 15 TECHNOLOGY HELP

Excel Techniques

Obtaining the Solver Sensitivity Report:
After solving a linear or nonlinear optimization model, click on the *Sensitivity Report* in the *Solver Results* dialog, and then click *OK*.

Using auxiliary variables in Solver models (Example 15.9):
Define a new set of cells in the spreadsheet model for any decision variables having upper or lower bounds that reference the original changing cells. In the *Solver* model, use the auxiliary variable cells to define the bound constraints.

Creating a stock chart to visualize allowable increases and allowable decreases in Sensitivity Reports:

1. Create a table in the worksheet by adding the Allowable Increase values and subtracting the Allowable Decrease values from the cost coefficients. Replace any $1E + 30$ values by #N/A in the worksheet so that infinite values are not displayed.

2. Highlight the range of this table and insert an Excel *Stock Chart* and name the series as Maximum, Minimum, and Current.
3. Click the chart, and in the *Format* tab of *Chart Tools*, go to the *Current Selection* group to the left of the ribbon and click on the drop-down box (it usually says "Chart Area"). Find the series you wish to format and then click *Format Selection*.
4. In the *Format Data Series* pane that appears in the worksheet, click the paint icon and then *Marker*, making sure to expand the *Marker Options* menu.
5. Choose the type of marker you want and increase the width of the markers to make them more visible.

Analytic Solver

Analytic Solver provides powerful methods for conducting what-if analysis for optimization. See the online supplement *Using Optimization Parameter Analysis in Analytic Solver*. We suggest that you first read the online supplement *Getting Started with Analytic Solver Basic*. This provides information for both instructors and students on how to register for and access *Analytic Solver*.

PROBLEMS AND EXERCISES

What-If Analysis for Optimization Models

1. For the Valencia Products scenario (Problems 4 and 11 in Chapter 13), use the spreadsheet model to answer the following questions by changing the parameters and re-solving the model. Answer each question independently relative to the original problem.

   a. If the unit profit for SpeedBuster is decreased to $130, how will the optimal solution and profit change?

   b. If the unit profit for LaserStop is increased to $210, how will the optimal solution and profit change?

   c. If an additional 1,500 units of component A are available, can you predict how the optimal solution and profit will be affected?

   d. If a supplier delay results in only 3,000 units of component B being available, can you predict how the optimal solution and profit will be affected? Can you explain the result?

2. For the ColPal Products scenario (Problems 5 and 12 in Chapter 13), use the spreadsheet model to answer the following questions by changing the parameters and re-solving the model. Answer each question independently relative to the original problem.

   a. Suppose that the exposure for TV advertising was incorrectly estimated and should have been 875. How would the optimal solution have been affected?

b. Radio listening has gone down, and new marketing studies have found that the exposure has dropped to 150. How will this affect the optimal solution?

c. The marketing manager has increased the budget by $2,000. How will this affect the solution and total exposure?

3. For the Burger Office Equipment scenario (Problems 6 and 13 in Chapter 13), use the spreadsheet model to answer the following questions by changing the parameters and re-solving the model. Answer each question independently relative to the original problem.

a. If 25% of the pine is deemed to be cosmetically defective, how will the optimal solution be affected?

b. The shop supervisor is suggesting that the workforce be allowed to work an additional 50 hours at an overtime premium of $18/hour. Is this a good suggestion? Why or why not?

c. If the unit profit for standard desks is increased to $280, how will the optimal solution and total profit be affected?

d. If the unit profit of standard desks is only $190, how will the optimal solution and total profit be affected?

4. For the Markowitz model in Example 14.10, determine how the minimum variance and stock allocations change as the target return varies between 8% and 12% (in increments of 1%) by re-solving the model. Summarize your results in a table, and create a chart showing the relationship between the target return and the optimal portfolio variance. Explain what the results mean for an investor.

5. Figure 15.32 shows the *Solver* Sensitivity Report after solving the Crebo Manufacturing problem in Chapter 13 (Example 13.10). Using only the information in the Sensitivity Report, answer the following questions.

a. Explain the value of the reduced cost (-0.3) for the number of plugs to produce.

b. If the gross margin for rails is decreased to $1.05, can you predict what the optimal solution and profit will be?

c. Suppose that the gross margin for rivets is increased to $0.85. Can you predict what the optimal solution and profit will be?

d. If the gross margin for clips is reduced to $1.10, can you predict what the optimal solution and profit will be? What if the gross margin is reduced to $1.00?

e. Suppose that an additional 500 minutes of machine capacity is available. How will the optimal solution and profit change? What if planned maintenance reduces capacity by 300 minutes?

6. Figure 15.33 shows the *Solver* Sensitivity Report for Valencia Products from Problems 4 and 11 in Chapter 13. Using only the information in the Sensitivity Report, answer the following questions, explaining what information you used in the Sensitivity Report.

a. Explain why the reduced cost for SpeedBuster is 0. What does the Allowable Decrease of 53.33 mean?

b. If the unit profit for SpeedBuster is decreased to $130, can you predict how the optimal solution and profit will change?

► **Figure 15.32**

Solver *Sensitivity Report* for Crebo Manufacturing Problem

		Final Value					
Objective Cell (Max)							
Cell	Name	Final Value					
$A$13	Profit	168000					

Decision Variable Cells		Final Value	Reduced Cost	Objective Coefficient	Allowable Increase	Allowable Decrease	
$B$10	Units Produced Plugs (X1)	0	-0.3	0.3	0.3	1E+30	
$C$10	Units Produced Rails (X2)	0	-0.2	1.3	0.2	1E+30	
$D$10	Units Produced Rivets (X3)	0	-0.15	0.75	0.15	1E+30	
$E$10	Units Produced Clips (X4)	140000	0	1.2	1E+30	0.16000008	

Constraints		Final Value	Shadow Price	Constraint R.H. Side	Allowable Increase	Allowable Decrease	
$A$16	Capacity Used	280000	0.6	280000	1E+30	280000	

▶ **Figure 15.33**

Solver *Sensitivity Report for Valencia Products Problem*

			Final	Reduced	Objective	Allowable	Allowable
6	Variable Cells						
8	Cell	Name	Value	Cost	Coefficient	Increase	Decrease
9	$B$5	Numbers Produced LaserStop	0	−80	124	80	1E+30
10	$C$5	Numbers Produced SpeedBuster	333.3333333	0	136	1E+30	53.33333333
12	Constraints		Final	Shadow	Constraint	Allowable	Allowable
14	Cell	Name	Value	Price	R.H. Side	Increase	Decrease
15	$D$8	Component A Used	4000	11.33333333	4000	1250	4000
16	$D$9	Component B Used	2666.666667	0	3500	1E+30	833.3333333

c. If the unit profit for LaserStop is increased to $210, can you predict how the optimal solution and profit will change?

d. If an additional 1,500 units of component A are available, can you predict how the optimal solution and profit will be affected?

e. If a supplier delay results in a shortage of 500 units of component B (that is, only 3,000 are available), can you predict how the optimal solution and profit will be affected?

7. Figure 15.34 shows the *Solver* Sensitivity Report for the ColPal Products scenario from Problems 5 and 12 from Chapter 13. Using only the information in the Sensitivity Report, answer the following questions, explaining what information you used in the Sensitivity Report.

a. Suppose that the exposure for TV advertising was incorrectly estimated and should have been 875. How would the optimal solution have been affected?

b. Radio listening has gone down, and new marketing studies have found that the exposure has dropped to 150. How will this affect the optimal solution?

c. The marketing manager has increased the budget by $2,000. How will this affect the solution and total exposure?

d. The shadow price for the mix constraint (that at least 75% of the time should be allocated to TV) is −237.5. The marketing manager was told that this means that if the percentage of TV advertising is increased to 76%, exposure will fall by 237.5. Explain why this statement is incorrect.

8. Figure 15.35 shows the *Solver* Sensitivity Report for the Burger Office Equipment scenario from Problems 6 and 13 in Chapter 13. Using only the information in the Sensitivity Report, answer the following questions, explaining what information you used in the Sensitivity Report.

a. Explain the reduced cost associated with deluxe desks.

b. If 25% of the pine is deemed to be cosmetically defective, how will the optimal solution be affected?

c. The shop supervisor is suggesting that the workforce be allowed to work an additional 50 hours at an overtime premium of $18/hour. Is this a good suggestion? Why or why not?

d. If the unit profit for standard desks is increased to $280, how will the optimal solution and total profit be affected?

e. If the unit profit of standard desks is only $190, how will the optimal solution and total profit be affected?

▶ **Figure 15.34**

Solver *Sensitivity Report for ColPal Products Problem*

			Final	Reduced	Objective	Allowable	Allowable
6	Variable Cells						
8	Cell	Name	Value	Cost	Coefficient	Increase	Decrease
9	$B$4	Minutes Radio	3.90625	0	350	1E+30	190
10	$C$4	Minutes TV	11.71875	0	800	950	916.6666667
12	Constraints		Final	Shadow	Constraint	Allowable	Allowable
14	Cell	Name	Value	Price	R.H. Side	Increase	Decrease
15	$D$7	Budget	$25,000.00	0.4296875	25000	1E+30	25000
16	$D$8	TV Requirement	0	−237.5	0	3.125	46.875

▶ **Figure 15.35**

Solver *Sensitivity Report for Burger Office Equipment Problem*

Variable Cells							
Cell	Name	Final Value	Reduced Cost	Objective Coefficient	Allowable Increase	Allowable Decrease
$B$4	Number Produced Standard	40	0	250	1E+30	55.55555556
$C$4	Number Produced Deluxe	0	-100	350	100	1E+30

Constraints							
Cell	Name	Final Value	Shadow Price	Constraint R.H. Side	Allowable Increase	Allowable Decrease
$D$7	Pine Used	2800	0	5000	1E+30	2200
$D$8	Oak Used	0	0	750	1E+30	750
$D$9	Labor Used	400	25	400	314.2857143	400

9. Figure 15.36 shows the *Solver* Sensitivity Report for the student investment scenario from Problems 7 and 14 from Chapter 13. Using only the information in the Sensitivity Report, answer the following questions, explaining what information you used in the Sensitivity Report.

   a. How much would the return on stock B have to be in order for the optimal solution to invest fully in that stock?

   b. How much would the return on stock C have to increase in order to invest fully in that stock?

   c. Explain the value of the shadow price for the total investment constraint. If the student could borrow $1,000 at 8% a year to increase her total investment, what would you recommend and why?

10. Obtain the Solver Sensitivity report for the GAC transportation model discussed in Chapter 13 (Example 13.18).

   a. What must the unit shipping cost be to make it attractive to ship from Marietta to Cleveland instead of from Minneapolis?

   b. Why are the Allowable Increases for all demand constraints zero?

   c. Explain why the shadow price for Cleveland makes sense.

d. Explain how the shadow price for Baltimore is calculated by changing the demand at Baltimore to 349, re-solving the model, and showing how the new allocations change the total cost.

11. Use the Sensitivity Report for the Camm Textiles scenario (Figure 15.37) to answer the following:

   a. Explain the reduced cost (0.14) for Fabric 2 Purchased in terms of the original data, and why it makes sense.

   b. Explain the shadow price for the dobbie loom constraint.

   c. Show where the shadow prices for the fabric constraints come from in terms of the original data and why they make sense.

12. The *K&L Designs* workbook contains the Sensitivity Reports for both the original problem formulation (Example 13.19) and the alternative model (Example 13.20). Explain the differences in these reports from a practical perspective. In other words, which model provides more useful information to managers and why?

13. Problems 8 and 15 in Chapter 13 presented a scenario for Bangs Leisure Chairs. The *Solver* Sensitivity Report for the optimal solution (Problem 15a only) is shown in Figure 15.38. Clearly explain the information it provides.

▶ **Figure 15.36**

Solver *Sensitivity Report for Student Investment Problem*

Variable Cells							
Cell	Name	Final Value	Reduced Cost	Objective Coefficient	Allowable Increase	Allowable Decrease
$B$5	Shares Purchased A	208.3333333	0	8	1E+30	2.4
$C$5	Shares Purchased B	0	-3	7	3	1E+30
$D$5	Shares Purchased C	0	-9	11	9	1E+30

Constraints							
Cell	Name	Final Value	Shadow Price	Constraint R.H. Side	Allowable Increase	Allowable Decrease
$E$8	Investment Limit	2500	0.666666667	2500	1E+30	2500

▶ **Figure 15.37**

Solver *Sensitivity Report for Camm Textiles*

Cell	Name	Final Value					
		Objective Cell (Min)					
$B$20	Cost Total	83756.12					

Decision Variable Cells

Cell	Name	Final Value	Reduced Cost	Objective Coefficient	Allowable Increase	Allowable Decrease
$B$14	Fabric 1 on Dobbie	30794.4	0	0.65	0.200000094	1E+30
$C$14	Fabric 1 on Regular	0	-0.85	0	1E+30	1E+30
$D$14	Fabric 1 Purchased	14205.6	0	0.85	1E+30	0.200000094
$B$15	Fabric 2 on Dobbie	0	0.180769231	0.61	1E+30	0.180769231
$C$15	Fabric 2 on Regular	76500	0	0.61	0.1400001	1E+30
$D$15	Fabric 2 Purchased	0	0.14	0.75	1E+30	0.14
$B$16	Fabric 3 on Dobbie	0	0.213636364	0.5	1E+30	0.213636364
$C$16	Fabric 3 on Regular	10000	0	0.5	0.1500001	1E+30
$D$16	Fabric 3 Purchased	0	0.15	0.65	1E+30	0.15

Constraints

Cell	Name	Final Value	Shadow Price	Constraint R.H. Side	Allowable Increase	Allowable Decrease
$B$17	Hours Used on Dobbie	6552	-0.94	6552	3022.468085	6552
$C$17	Hours Used on Regular	16984.26573	0	32760	1E+30	15775.73427
$E$14	Fabric 1 Total Yards Produced	45000	0.85	45000	1E+30	14205.6
$E$15	Fabric 2 Total Yards Produced	76500	0.61	76500	82033.81818	76500
$E$16	Fabric 3 Total Yards Produced	10000	0.5	10000	69413.23077	10000

▶ **Figure 15.38**

Solver *Sensitivity Report for Bangs Leisure Chairs*

Variable Cells

Cell	Name	Final Value	Reduced Cost	Objective Coefficient	Allowable Increase	Allowable Decrease
$B$14	Quantity Produced Sling Chairs	0	-4.758064516	40	4.758064516	1E+30
$C$14	Quantity Produced Adirondack	22.58064516	0	100	2.272727273	8.550724638
$D$14	Quantity Produced Hammocks	1.612903226	0	90	39.33333333	2

Constraints

Cell	Name	Final Value	Shadow Price	Constraint R.H. Side	Allowable Increase	Allowable Decrease
$E$15	Cutting Hours Used	45.80645161	0	50	1E+30	4.193548387
$E$16	Assembling Hours Used	50	1.612903226	50	31.81818182	2
$E$17	Finishing Hours Used	24.19354839	0	50	1E+30	25.80645161
$E$18	Total Hours Used	120	19.35483871	120	5	46.66666667

14. Problems 9 and 16 in Chapter 13 ask you to model and solve an optimization model for the Morton Supply Company. Obtain the *Solver* Sensitivity Report for your solution, and clearly explain all the key information in language that the production manager would understand.

15. Problems 10 and 17 in Chapter 13 ask you to model and solve an optimization model for Malloy Milling. Using the *Solver* Sensitivity Report, answer the following questions, explaining what information you used in the Sensitivity Report.

   a. What impact will changing the required minimum number of tons per week (currently 700) have on the solution?

   b. If the price per ton for regular grind is increased to $1,100, how will the solution be affected?

   c. If the price per ton for super grind is decreased to $1,400 because of low demand, how will the solution change?

16. For the International Chef, Inc. scenario in Problem 13.27, obtain the *Solver* Sensitivity Report and write a short memo to the president, Kathy Chung, explaining the sensitivity information in language that she can understand.

17. For Dara's investment situation (Problem 32 in Chapter 13), obtain the *Solver* Sensitivity Report and interpret the information, making recommendations that Dara might consider for her portfolio.

18. For Problem 13.35 (Kelly Foods), obtain the *Solver* Sensitivity Report and write a short memo to the supply chain director explaining your results.

19. For Problem 13.37 (Shafer Office Supplies), obtain the *Solver* Sensitivity Report and answer the following questions:

    a. Suppose that 500 units of extra supply are available (and that the cost of this extra capacity is a sunk cost). To which distribution center should this extra supply be allocated, and why?

    b. Suppose that the cost of shipping from Atlanta to Birmingham increased to $0.45 per unit. What would happen to the optimal solution?

20. For the Hansel Corporation (Problem 13.40), obtain the *Solver* Sensitivity Report and use it to answer the following:

    a. A labor strike in India leads to a shortage of 20,000 units of additive C. What should the production manager do?

    b. Management is considering raising the price on grade 2 to $2.00 per pound. How will this affect the solution?

21. For the nonlinear pricing decision model in Example 14.8, suppose that the company wants to keep the price at a maximum of $500. Note that the solution in Figure 14.19 will no longer be feasible. Modify the spreadsheet model to include a constraint on the maximum price and solve the model. Interpret the information in the *Solver* Sensitivity Report.

22. Figure 15.39 shows the *Solver* Sensitivity Report for the Markowitz portfolio model (Example 14.10).

    a. Explain how to interpret the Lagrange multiplier value for the target portfolio return.

b. Suppose the target return is increased from 10% to 11%. How much is the minimum portfolio variance predicted to increase using the Lagrange Multiplier value?

c. Re-solve the model with the target return of 11%. How much does the minimum variance actually change?

23. Obtain the *Solver* Sensitivity Report for the Hal Chase portfolio optimization model (Problem 14.33). Explain how to interpret the Lagrange multipliers for the constraints.

Models with Bounded Variables

24. Marketing managers have various media alternatives in which to advertise and must determine which to use, the number of insertions in each, and the timing of insertions to maximize advertising effectiveness within a limited budget. Suppose that three media options are available to Kernan Services Corporation: radio, TV, and magazine. The following table provides some information about costs, exposure values, and bounds on the permissible number of ads in each medium desired by the firm. The exposure value is a measure of the number of people exposed to the advertisement and is derived from market research studies. The company would like to achieve a total exposure value of at least 90,000.

Medium	Cost/Ad	Exposure Value/Ad	Min Units	Max Units
Radio	$500	2,000	0	15
TV	$2,000	4,000	10	
Magazine	$200	2,700	6	12

▶ **Figure 15.39**

Markowitz Model Sensitivity Report

	A B	C	D	E
5	Objective Cell (Min)			
6	Cell	Name	Final Value	
7	$C$21	Portfolio Variance	0.01242246	
8				
9	Decision Variable Cells			
10			Final	Reduced
11	Cell	Name	Value	Gradient
12	$B$14	Stock 1 Allocation	0.25	0.00
13	$B$15	Stock 2 Allocation	0.45	0.00
14	$B$16	Stock 3 Allocation	0.30	0.00
15				
16	Constraints			
17			Final	Lagrange
18	Cell	Name	Value	Multiplier
19	$B$17	Total Allocation	1	-0.038363636
20	$B$21	Portfolio Return	10.0%	63.2%

How many of each type of ad should be placed to minimize the cost of achieving the minimum required total exposure?

a. Formulate and solve a linear optimization model using the auxiliary variable cells method and write a short memo to the marketing manager explaining the solution and sensitivity information using the auxiliary variable Sensitivity Report.

b. Solve the model without the auxiliary variables and explain the relationship between the reduced costs and the shadow prices found in part a.

25. For Problem 13.39 (Sanford Tile Company), use the auxiliary variable technique to handle the bound constraints on grade I and grade III tiles and obtain the *Solver* Sensitivity Report. Answer the following:

a. Explain the sensitivity information for the objective coefficients. What happens if the profit on grade I is increased by $0.05?

b. If an additional 500 pounds of feldspar is available, how will the optimal solution be affected?

c. Suppose that 1,000 pounds of clay are found to be of inferior quality. What should the company do?

26. For Klein Industries (Problem 13.49), formulate and solve a linear optimization model using the auxiliary variable cells method and write a short memo to the production manager explaining the sensitivity information.

27. For the Fruity Juices scenario (Problem 13.50), formulate and solve a linear optimization model using the auxiliary variable cells method and write a short memo explaining the sensitivity information.

28. For Worley Fluid Supplies (Problem 13.51), solve the model using auxiliary variables. Obtain the *Solver* Sensitivity Report, and answer the following:

a. Explain the sensitivity information for the objective coefficients. What happens if the profit on hydraulic cylinders is decreased by $10?

b. Due to scheduled maintenance, the assembly time is expected to be only 2,900 minutes. How will this affect the solution?

c. A worker in the testing department has to take a personal leave because of a death in the family and will miss two days (16 hours of work time). How will this affect the optimal solution?

What-If Analysis for Integer Optimization Models

29. Hahn Engineering (Example 14.5) would like to increase its return, but the project manager knows that this will require more cash and/or personnel. Conduct a what-if analysis of the optimal solution as the cash limitation varies from $150,000 to $270,000 in increments of $20,000, and the personnel limitation varies from 12 to 18. Summarize the returns for these solutions in the form of a heat map. Identify the solutions that represent the lowest amount of resources required to achieve each distinct return value. Write a short memo to the project manager explaining the best options and trade-offs that must be made to increase the returns.

30. For the Paul & Giovanni Foods scenario (Example 14.7), conduct a what-if analysis by varying k, the number of distribution centers, from 1 to 4. Summarize your results and write a short memo to the supply chain manager explaining your findings and recommendations.

31. Analyze the sensitivity of the optimal solution for Riesemberg Medical Devices (Problem 14.11) as the budget and available engineers are varied. Write a short summary memo to the manager in which you outline the results and make recommendations.

32. Conduct a what-if analysis for Tunningley Services (Problem 14.21) as the travel goal varies from 30 to 90 minutes in increments of 10 minutes.

Visualization of *Solver* Reports

33. Use Excel charts to visualize the solution for the Sklenka Skis example that we discussed in Chapter 13 and in this chapter. Use the *Solver* Answer Report (Figure 13.5) and the Sensitivity Report (Figure 15.3). Use the *Sklenka Skis* Excel file to generate these reports and facilitate your analysis.

34. Use Excel charts to visualize the solution for the J&M Manufacturing problem (Example 15.8) using the *Solver* Sensitivity Report with auxiliary variables.

35. Obtain the Answer and Sensitivity Reports for Dara's investment scenario (Problem 13.32). Develop a set of charts that visualize the key information in these reports.

Using Sensitivity Information Correctly

36. Beverly Ann Cosmetics has created two new perfumes: Summer Passion and Ocean Breeze. It costs $5.25 to purchase the fragrance needed for each bottle of Summer Passion and $4.70 for each bottle of Ocean Breeze. The marketing department has stated that at least 30% but no more than 70% of the product mix be Summer Passion; the forecasted monthly demand is 7,000 bottles and is estimated to increase by eight bottles for each $1 spent on advertising. For Ocean Breeze, the demand is forecast to be 12,000 bottles and is expected to increase by 15 bottles for each $1 spent on advertising. Summer Passion sells for $42.00 per bottle and Ocean Breeze for $30.00 per bottle. A monthly budget of $100,000 is available for both advertising and purchase of the fragrances.

a. Develop and solve a linear optimization model to determine how much of each type of perfume should be produced to maximize the net profit.

b. In viewing the *Solver* Sensitivity Report, explain what information is accurate and what information is misleading because it violates the assumptions of sensitivity analysis.

CASE: PERFORMANCE LAWN EQUIPMENT

One of PLE's manufacturing facilities produces metal engine housings from sheet metal for both mowers and tractors. Production of each product consists of five steps: stamping, drilling, assembly, painting, and packaging to ship to its final assembly plant. The production rates in hours per unit and the number of production hours available in each department are given in the following table:

Department	Mower Housings	Tractor Housings	Production Hours Available
Stamping	0.03	0.07	200
Drilling	0.09	0.06	300
Assembly	0.15	0.10	300
Painting	0.04	0.06	220
Packaging	0.02	0.04	100

In addition, mower housings require 1.2 square feet of sheet metal per unit and tractor housings require 1.8 square feet per unit, and 2,500 square feet of sheet metal is available. The company would like to maximize the total number of housings they can produce during the planning period. Formulate and solve a linear optimization model using *Solver* and recommend a production plan. Illustrate the results visually to help explain them in a presentation to Ms. Burke. In addition, conduct whatever what-if analyses you feel are appropriate to include in your presentation (for example, run different scenarios or systematically change model parameters). Summarize your results in a well-written report.

Decision Analysis

marekuliasz/Shutterstock

LEARNING OBJECTIVES   After studying this chapter, you will be able to:

- List the three elements needed to characterize decisions with uncertain consequences.
- Construct a payoff table for a decision situation.
- Apply average, aggressive, conservative, and opportunity-loss decision strategies for problems involving minimization and maximization objectives.
- Assess risk in choosing a decision.
- Apply expected values to a decision problem when probabilities of events are known.

- Find the risk profile for a decision strategy.
- Compute the expected value of perfect information.
- Incorporate sample information in decision trees and apply Bayes's rule to compute conditional probabilities.
- Construct a utility function and use it to make a decision.
- State the properties of different types of utility functions.

Everybody makes decisions, both personal and professional. Managers are continually faced with decisions involving new products, supply chain configurations, new equipment, downsizing, and many others. The ability to make good decisions is the mark of a successful (and promotable) manager. In today's complex business world, intuition alone is not sufficient. This is where analytics plays an important role.

Throughout this book, we have discussed how to analyze data and models using methods of business analytics. Predictive models such as Monte Carlo simulations can provide insight about the impacts of potential decisions, and prescriptive models such as linear optimization provide recommendations as to the best course of action to take. However, the real purpose of such information is to help managers *make decisions*. Their decisions often have significant economic or human resource consequences that cannot always be predicted accurately. For example, in Chapter 12, we analyzed a simulation model for new product development (Example 12.11). In assessing the risks, we observed that the probability that the net present value would not be positive was between 0.15 and 0.20. So what decision (pursue the project or not) should the company make? Similarly, in the Innis Investment example in Chapter 15, we performed a scenario analysis to evaluate the trade-offs between risk and return (Figure 15.2). How should the client make a trade-off between risk and return for their portfolio?

Analytic models and analyses provide decision makers with a wealth of information; however, people make the final decision. Good decisions don't simply implement the results of analytic models; they require an assessment of intangible factors and risk attitudes. **Decision analysis** is the study of how people make decisions, particularly when faced with imperfect or uncertain information, as well as a collection of techniques to support decision choices. Decision analysis differs from other modeling approaches by explicitly considering the individual's preferences and attitudes toward risk and modeling the decision process itself.

Decisions involving uncertainty and risk have been studied for many years. A large body of knowledge has been developed that helps to explain the philosophy associated with making decisions and also provides techniques for incorporating uncertainty and risk in making decisions.

Formulating Decision Problems

Many decisions involve a choice from among a small set of alternatives with uncertain consequences. We may formulate such decision problems by defining three things:

1. the **decision alternatives** that can be chosen,
2. the **uncertain events** that may occur after a decision is made along with their possible **outcomes**, and
3. the consequences associated with each decision and outcome, which are usually expressed as **payoffs**.

The outcomes associated with uncertain events (which are often called **states of nature**), are defined so that one and only one of them will occur. They may be quantitative or qualitative. For instance, in selecting the size of a new factory, the future demand for the product would be an uncertain event. The demand outcomes might be expressed quantitatively in sales units or dollars. On the other hand, suppose that you are planning a spring break vacation to Florida. You might define an uncertain event as the weather; these outcomes might be characterized qualitatively: sunny and warm, sunny and cold, rainy and warm, rainy and cold, and so on. A payoff is a measure of the value of making a decision and having a particular outcome occur. This might be a simple estimate made judgmentally or a value computed from a complex spreadsheet model. Payoffs are often summarized in a **payoff table**, a matrix whose rows correspond to decisions and whose columns correspond to events. The decision maker first selects a decision alternative, after which one of the outcomes of the uncertain event occurs, resulting in the payoff.

EXAMPLE 16.1	Selecting a Mortgage Instrument

Many young families face the decision of choosing a mortgage instrument. Suppose the Durr family is considering purchasing a new home and would like to finance $150,000. Three mortgage options are available: a one-year adjusted-rate mortgage (ARM) at a low interest rate, a three-year ARM at a slightly higher rate, and a 30-year fixed mortgage at the highest rate. However, both ARMs are sensitive to interest rate changes and the rates may change, resulting in either higher or lower interest charges; thus, the potential future change in interest rates represents an uncertain event. Because the family anticipates staying in the home for at least five years, they want to know the total interest costs they might incur; these represent the payoffs associated with their choice and the future change in interest rates

and can easily be calculated using a spreadsheet. The payoff table is given below.

Clearly, no decision is best for each event that may occur. If rates rise, for example, then the 30-year fixed would be the best decision. If rates remain stable or fall, however, then the one-year ARM is best. Of course, you cannot predict the future outcome with certainty, so the question is how to choose one of the options. Not everyone views risk in the same fashion. Most individuals will weigh their potential losses against potential gains. For example, if they choose the one-year ARM mortgage instead of the fixed-rate mortgage, they risk losing money if rates rise; however, they would clearly save a lot if rates remain stable or fall. Would the potential savings be worth the risk? Such questions make decision making a difficult task.

		Outcome	
Decision	Rates Rise	Rates Stable	Rates Fall
1-year ARM	$61,134	$46,443	$40,161
3-year ARM	$56,901	$51,075	$46,721
30-year fixed	$54,658	$54,658	$54,658

CHECK YOUR UNDERSTANDING

1. List the three things that must be specified in formulating a decision problem from among a small set of alternatives with uncertain consequences.

2. Explain the structure of a payoff table.

Decision Strategies Without Outcome Probabilities

We discuss several quantitative approaches that model different risk behaviors for making decisions involving uncertainty when no probabilities can be estimated for the outcomes.

Decision Strategies for a Minimize Objective

Aggressive (Optimistic) Strategy   An aggressive decision maker might seek the option that holds the promise of minimizing the potential loss. This type of decision maker would first ask the question, What is the *best* that could result from each decision? and then choose the decision that corresponds to the "best of the best." For a minimization objective, this strategy is also often called a **minimin strategy**; that is, we choose the decision that minimizes the minimum payoff that can occur among all outcomes for each decision. Aggressive decision makers are often called speculators, particularly in financial arenas, because they increase their exposure to risk in hopes of increasing their return; while a few may be lucky, most will not do very well.

EXAMPLE 16.2   **Mortgage Decision with the Aggressive Strategy**

For the mortgage selection example, we find the best payoff—that is, the lowest-cost outcome—for each decision:

Decision	Outcome			
	Rates Rise	Rates Stable	Rates Fall	Best Payoff
1-year ARM	$61,134	$46,443	$40,161	$40,161
3-year ARM	$56,901	$51,075	$46,721	$46,721
30-year fixed	$54,658	$54,658	$54,658	$54,658

Because the goal is to minimize costs, we would choose the one-year ARM.

Conservative (Pessimistic) Strategy   A conservative decision maker, on the other hand, might take a more pessimistic attitude, asking "What is the worst thing that might result from my decision?" and then select the decision that represents the "best of the worst." Such a strategy is also known as a **minimax strategy** because we seek the decision that minimizes the largest payoff that can occur among all outcomes for each decision. Conservative decision makers are willing to forgo high returns to avoid undesirable losses. This rule typically models the rational behavior of most individuals.

EXAMPLE 16.3   Mortgage Decision with the Conservative Strategy

For the mortgage decision problem, we first find the worst payoff—that is, the largest cost for each option:

	Outcome			
Decision	Rates Rise	Rates Stable	Rates Fall	Worst Payoff
1-year ARM	$61,134	$46,443	$40,161	$61,134
3-year ARM	$56,901	$51,075	$46,721	$56,901
30-year fixed	$54,658	$54,658	$54,658	$54,658

In this case, we want to choose the decision that has the smallest worst payoff, or the 30-year fixed mortgage. Thus, no matter what the future holds, a minimum cost of $54,658 is guaranteed.

Opportunity-Loss Strategy   A third approach that underlies decision choices for many individuals is to consider the *opportunity loss* associated with a decision. Opportunity loss represents the "regret" that people often feel after making a nonoptimal decision (I should have bought that stock years ago!). In general, the opportunity loss associated with any decision and event is the absolute difference between the *best* decision for that particular outcome and the payoff for the decision that was chosen. *Opportunity losses can be only nonnegative values.* If you get a negative number, then you made a mistake. Once opportunity losses are computed, the decision strategy is similar to a conservative strategy. The decision maker would select the decision that minimizes the largest opportunity loss among all outcomes for each decision. For these reasons, this is also called a **minimax regret strategy**.

EXAMPLE 16.4   Mortgage Decision with the Opportunity-Loss Strategy

In our scenario, suppose we chose the 30-year fixed mortgage and later find out that the interest rates had risen. We could not have done any better by selecting a different decision; in this case, the opportunity loss is zero. However, if we had chosen the three-year ARM, we would have paid $56,901 instead of $54,658 with the 30-year fixed instrument, or $56,901 − $54,658 = $2,243 more. This represents the opportunity loss associated with making a nonoptimal decision. Similarly, had we chosen the one-year ARM, we would have incurred an additional

cost (opportunity loss) of $61,134 − $54,658 = $6,476. We repeat this analysis for the other two outcomes and compute the opportunity losses, as summarized in the table below.

Then, we find the maximum opportunity loss that would be incurred for each decision. The best decision is the one with the smallest maximum opportunity loss. Using this strategy, we would choose the one-year ARM. This ensures that, no matter what outcome occurs, we will never be more than $6,476 away from the least cost we could have incurred.

	Outcome			
Decision	Rates Rise	Rates Stable	Rates Fall	Max Opportunity Loss
1-year ARM	$6,476	—	—	$6,476
3-year ARM	$2,243	$4,632	$6,560	$6,560
30-year fixed	—	$8,215	$14,497	$14,497

Different criteria lead to different decisions; there is no "optimal" answer. Which criterion best reflects your personal values?

Decision Strategies for a Maximize Objective

When the objective is to maximize the payoff, we can still apply aggressive, conservative, and opportunity-loss strategies, but we must make some key changes in the analysis.

- For the aggressive strategy, the best payoff for each decision would be the *largest* value among all outcomes, and we would choose the decision corresponding to the largest of these, called a **maximax strategy**.
- For the conservative strategy, the worst payoff for each decision would be the *smallest* value among all outcomes, and we would choose the decision corresponding to the largest of these, called a **maximin strategy**.
- For the opportunity-loss strategy, we need to be careful in calculating the opportunity losses. With a maximize objective, the decision with the largest value for a particular event has an opportunity loss of zero. The opportunity losses associated with other decisions is the absolute difference between their payoff and the largest value. The actual decision is the same as when payoffs are costs: Choose the decision that minimizes the maximum opportunity loss.

Decisions with Conflicting Objectives

Many decisions require some type of trade-off among conflicting objectives, such as risk versus reward. For example, the Innis Investment example in Figure 15.2 showed the results of solving a series of linear optimization models to find the minimum risk that would occur for achieving increasing levels of investment returns. We saw that as the return goes up, the risk begins to increase slowly, and then increases at a faster rate once a 6% investment target is achieved. What decision would be best? Another example we saw was the hotel overbooking model in Chapter 12 (Example 12.15). In this case, we can achieve lower costs but incur a loss in customer satisfaction and goodwill because of higher numbers of overbooked customers.

A simple decision rule can be used whenever one wishes to make an optimal trade-off between any two conflicting objectives, one of which is good, and one of which is bad, that maximizes the ratio of the good objective to the bad (think of this as the "biggest bang for the buck").[1] First, display the trade-offs on a chart with the "good" objective on the *x*-axis and the "bad" objective on the *y*-axis, making sure to scale the axes properly to display the origin (0, 0). Then graph the tangent line to the trade-off curve that goes through the origin. The point at which the tangent line touches the curve (which represents the smallest slope) represents the best return to risk trade-off.

EXAMPLE 16.5 **Risk-Reward Trade-off Decision for Innis Investments Example**

In Figure 15.2, if we take the ratios of the weighted returns to the minimum risk values in the table, we will find that the largest ratio occurs for the target return of 6%. We can visualize this using the risk-reward trade-off curve and a tangent line through the origin, as shown in Figure 16.1.

Note that the tangent line touches the curve at the 6% weighted return value. We can explain this easily from the chart by noting that for any other return, the risk is relatively larger (if all points fell on the tangent line, the risk would increase proportionately with the return).

---

[1]This rule was explained by Dr. Leonard Kleinrock at a lecture at the University of Cincinnati in 2011.

▶ **Figure 16.1**

Innis Investments
Risk-Reward Assessment

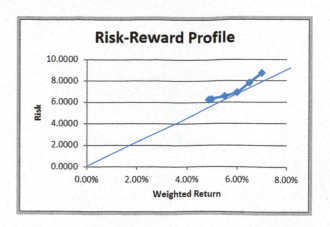

▶ **TABLE 16.1**

Summary of Decision Strategies Under Uncertainty

Objective	Strategy	Aggressive Strategy	Conservative Strategy	Opportunity-Loss Strategy
Minimize objective	Choose the decision with the smallest average payoff.	Find the smallest payoff for each decision among all outcomes and choose the decision with the smallest of these (*minimin*).	Find the largest payoff for each decision among all outcomes and choose the decision with the smallest of these (*minimax*).	For each outcome, compute the opportunity loss for each decision as the absolute difference between its payoff and the *smallest* payoff for that outcome. Find the maximum opportunity loss for each decision and choose the decision with the smallest opportunity loss (*minimax regret*).
Maximize objective	Choose the decision with the largest average payoff.	Find the largest payoff for each decision among all outcomes and choose the decision with the largest of these (*maximax*).	Find the smallest payoff for each decision among all outcomes and choose the decision with the largest of these (*maximin*).	For each outcome, compute the opportunity loss for each decision as the absolute difference between its payoff and the *largest* payoff for that outcome. Find the maximum opportunity loss for each decision and choose the decision with the smallest opportunity loss (*minimax regret*).

Many other analytic techniques are available to deal with more complex multiple objective decisions. These include simple scoring models in which each decision is rated for each criterion (which may also be weighted to reflect the relative importance in comparison with other criteria). The ratings are summed over all criteria to rank the decision alternatives. Other techniques include variations of linear optimization known as *goal programming*, and a pairwise comparison approach known as the *analytic hierarchy process (AHP)*.

Table 16.1 summarizes the decision rules for both minimize and maximize objectives.

CHECK YOUR UNDERSTANDING

1. State the three types of strategies that can be used for decision involving uncertainty when no probabilities can be estimated for the outcomes.

2. Explain how each of these strategies differ for minimizing and maximizing payoffs.

3. Explain how to make an optimal tradeoff between two conflicting objectives, when one is good and the other is bad.

Decision Strategies with Outcome Probabilities

The aggressive, conservative, and opportunity-loss strategies assume no knowledge of the probabilities associated with future outcomes. In many situations, we might have some assessment of these probabilities, through either some method of forecasting or reliance on expert opinions.

Average Payoff Strategy

If we can assess a probability for each outcome, we can choose the best decision based on the expected value using concepts that we introduced in Chapter 5. For any decision, the expected value is the summation of the payoffs multiplied by their probability, summed over all outcomes. The simplest case is to assume that each outcome is equally likely to occur; that is, the probability of each outcome is simply $1/N$, where N is the number of possible outcomes. This is called the **average payoff strategy**. This approach was proposed by the French mathematician Laplace, who stated the *principle of insufficient reason*: If there is no reason for one outcome to be more likely than another, treat them as equally likely. Under this assumption, we evaluate each decision by simply averaging the payoffs. We then select the decision with the best average payoff.

EXAMPLE 16.6   **Mortgage Decision with the Average Payoff Strategy**

For the mortgage selection problem, computing the average payoffs results in the following:

Decision	Outcome			
	Rates Rise	Rates Stable	Rates Fall	Average Payoff
1-year ARM	$61,134	$46,443	$40,161	$49,246
3-year ARM	$56,901	$51,075	$46,721	$51,566
30-year fixed	$54,658	$54,658	$54,658	$54,658

Based on this criterion, we choose the decision having the smallest average payoff, or the one-year ARM.

Expected Value Strategy

A more general case of the average payoff strategy is when the probabilities of the outcomes are not all the same. This is called the **expected value strategy**. We may use the expected value calculation that we introduced in formula (5.12) in Chapter 5.

EXAMPLE 16.7   **Mortgage Decision with the Expected Value Strategy**

Suppose that we can estimate the probabilities of rates rising as 0.6, rates stable as 0.3, and rates falling as 0.1. The following table shows the expected payoffs associated with each decision. The smallest expected payoff, $54,135.20, occurs for the three-year ARM, which represents the best expected value decision.

	Outcome			
Probability	0.6	0.3	0.1	
Decision	**Rates Rise**	**Rates Stable**	**Rates Fall**	**Expected Payoff**
1-year ARM	$61,134	$46,443	$40,161	$54,629.40
3-year ARM	$56,901	$51,075	$46,721	$54,135.20
30-year fixed	$54,658	$54,658	$54,658	$54,658.00

Evaluating Risk

An implicit assumption in using the average payoff or expected value strategy is that the decision is repeated a large number of times. However, for any *one-time* decision (with the trivial exception of equal payoffs), the expected value outcome will *never occur*. In the previous example, for instance, even though the expected value of the three-year ARM (the best decision) is $54,135.20, the actual result would be only one of three possible payoffs, depending on the outcome of the mortgage rate event: $56,901 if rates rise, $51,075 if rates remain stable, or $46,721 if rates fall. Thus, for a one-time decision, we must carefully weigh the risk associated with the decision in lieu of blindly choosing the expected value decision.

EXAMPLE 16.8 **Evaluating Risk in the Mortgage Decision**

In the mortgage selection example, although the average payoffs are fairly similar, note that the one-year ARM has a larger variation in the possible outcomes. We may compute the standard deviation of the outcomes associated with each decision:

Decision	Standard Deviation
1-year ARM	$10,763.80
3-year ARM	$5,107.71
30-year fixed	—

Based solely on the standard deviation, the 30-year fixed mortgage has no risk at all, whereas the one-year ARM appears to be the riskiest. Although based only on three data points, the three-year ARM is fairly symmetric about the mean, whereas the one-year ARM is positively skewed—most of the variation around the average is driven by the upside potential (that is, lower costs), not the downside risk of higher costs. Although none of the formal decision strategies chose the three-year ARM, viewing risk from this perspective might lead to this decision. For instance, a conservative decision maker who is willing to tolerate a moderate amount of risk might choose the three-year ARM over the 30-year fixed because the downside risk is relatively small (and is smaller than the one-year ARM) and the upside potential is much larger. The larger upside potential associated with the one-year ARM might even make this decision attractive.

Thus, it is important to understand that making decisions under uncertainty cannot be done using only simple rules, but by careful evaluation of risk versus rewards. This is why top executives make the big bucks. Evaluating risk in making a decision should also take into account the magnitude of potential gains and losses as well as their probabilities of occurrence, if this can be assessed. For example, a 70% chance of losing $10,000 against a 30% chance of gaining $500,000 might be viewed as an acceptable risk for a company, but a 10% chance of losing $250,000 against a 90% chance of gaining $500,000 might not.

CHECK YOUR UNDERSTANDING

1. How does the average payoff strategy differ from the expected value strategy?
2. Explain the issues involved in using an expected value strategy for one-time decisions.

Decision Trees

A useful approach to structuring a decision problem involving uncertainty is to use a graphical model called a **decision tree**. Decision trees consist of a set of **nodes** and **branches**. Nodes are points in time at which events take place. The event can be a selection of a decision from among several alternatives, represented by a **decision node**, or an outcome over which the decision maker has no control, an **event node**. Event nodes are conventionally depicted by circles, and decision nodes are expressed by squares. Branches are associated with decisions and events. We use a triangle to represent the terminal point of a decision path. Many decision makers find decision trees useful because *sequences* of decisions and outcomes over time can be modeled easily.

EXAMPLE 16.9   **Creating a Decision Tree**

For the mortgage selection problem, we will first create a decision node for the selection of one of the three mortgage instruments. This is shown in Figure 16.2. Although somewhat tedious, you can create this on an Excel worksheet using the *Shapes* button from the *Insert* menu. Next, add an event node at the end of the 1-Year ARM branch, with branches "Rates Rise," "Rates Stable," and "Rates Fall." We assign the probabilities to these outcomes from Example 16.7 above the event branches. This creates the tree shown in Figure 16.3. Repeat this process for the other two mortgage instrument branches.

Finally, enter the payoffs of the outcomes associated with each event in the cells immediately below the branches. Because the payoffs are costs, we enter them as negative values, shown in parentheses using accounting formatting. Sum all payoffs along the paths and place these values next to the terminal nodes. (In this example, there are none associated with the decision branches, but we will see this in another example shortly.) The final decision tree is shown in Figure 16.4.

Next, we need to analyze the decision tree to determine the best strategy that maximizes the expected value of the payoff. We illustrate this in Example 16.10.

EXAMPLE 16.10   **Analyzing a Decision Tree**

To find the best decision strategy in a decision tree, we "roll back" the tree by computing expected values at event nodes and selecting the optimal value of alternative decisions at decision nodes. For example, if the one-year ARM is chosen, the expected value of the chance events is $0.6 \times (-\$61,134) + 0.3 \times (-\$46,443) + 0.1 \times (-\$40,161) = -\$54,629.40$. Enter this value below the one-year ARM decision branch as shown in Figure 16.5 (using accounting format), and repeat this process for each of

the other event nodes. At the decision node, the maximum expected value is chosen from among all decisions; this is −$54,135.20. Write this next to the decision node, as shown in Figure 16.5. Because this corresponds to the three-year ARM, which is branch 2, we can enter a 2 in the decision node, square to indicate the best decision. Therefore, the best strategy is to choose the three-year ARM, having an expected cost of $54,135.20 (the same decision we found in Example 16.7).

▶ **Figure 16.2**

First Partial Decision Tree for Mortgage Selection

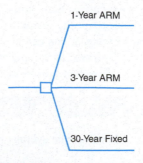

▶ **Figure 16.3**

Second Partial Decision Tree for Mortgage Selection

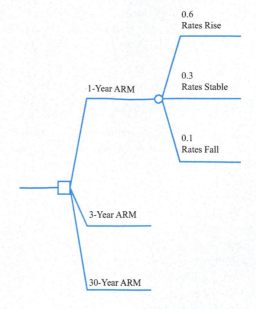

▶ **Figure 16.4**

Final Mortgage Selection Decision Tree

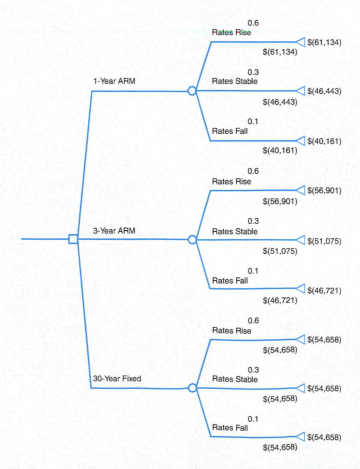

Many decision problems have multiple sequences of decisions and events, as illustrated in the next example. Decision trees are invaluable in helping managers understand the implications of uncertainty and decisions that have to be made over time.

EXAMPLE 16.11　A Pharmaceutical R&D Model

We will consider the R&D process for a new drug (you might recall the basic financial model we developed for the Moore Pharmaceuticals example in Chapter 11). Suppose that the company has spent $300 million to date in research expenses. The first decision is whether or not to proceed with clinical trials. We can decide either to conduct them or to stop development at this point, incurring the $300 million cost already spent on research. The cost of clinical trials is estimated to be $250 million, and the probability of a successful outcome is 0.3. Therefore, if we decide to conduct the trials, we face the chance events that the trials will be either successful or not successful. If they are not successful, then clearly the process stops at this point. If they are successful, the company may seek approval from the Food and Drug Administration or decide to stop the development process. The cost of seeking approval is $25 million, and there is a 60% chance of approval. If the company seeks approval, it faces the chance events that the FDA will approve the drug or not approve it. Finally,

if the drug is approved and is released to the market, the market potential has been identified as large, medium, or small, with the following characteristics:

	Market Potential Expected Revenues (millions of $)	Probability
Large	4,500	0.6
Medium	2,200	0.3
Small	1,500	0.1

A decision tree for this situation is shown in Figure 16.6. When we have sequences of decisions and events, a **decision strategy** is a specification of an initial decision and subsequent decisions to make after knowing what events occur. We can identify the best strategy from the branch number in the decision nodes. For example, the best strategy is to conduct clinical trials and, if successful, seek FDA approval and, if approved, market the drug. The expected net revenue is calculated as $74.3 million.

Decision Trees and Risk

The decision tree approach is an example of expected value decision making. Thus, in the drug development example, if the company's portfolio of drug development projects has similar characteristics, then pursuing further development is justified on an expected value basis. However, this approach does not explicitly consider risk.

From a classical decision analysis perspective, we may summarize the Moore Pharmaceutical's decision as the following payoff table:

	Unsuccessful Clinical Trials	Successful Clinical Trials; No FDA Approval	Successful Trials and Approval; Large Market	Successful Trials and Approval; Medium Market	Successful Trials and Approval; Small Market
Develop drug	($550)	($575)	$3,925	$1,625	$925
Stop development	($300)	($300)	($300)	($300)	($300)

► Figure 16.5

*Rolling Back the
Mortgage-Selection
Decision Tree*

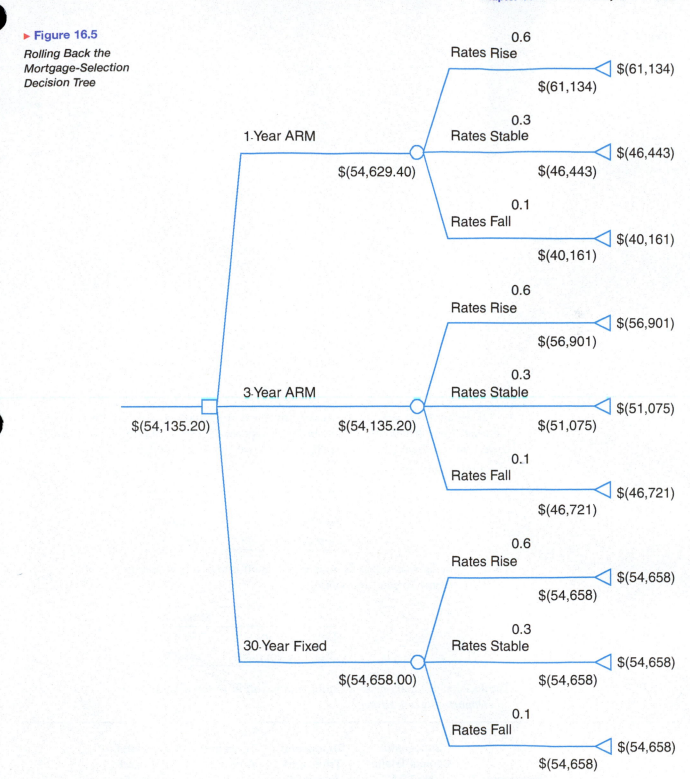

▶ **Figure 16.6**

New-Drug-Development Decision Tree

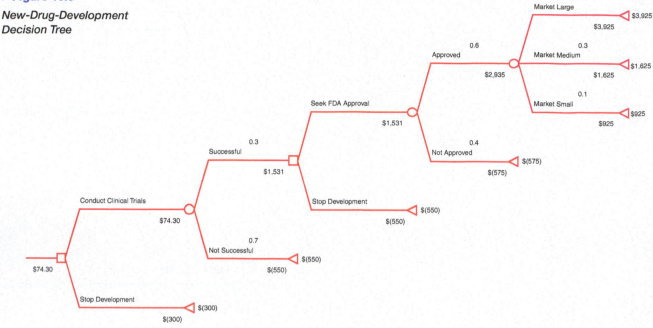

If we apply the aggressive, conservative, and opportunity-loss decision strategies to these data (note that the payoffs are profits as opposed to costs, so it is important to use the correct rule, as discussed earlier in the chapter), we obtain the following.

Aggressive strategy (maximax):

	Maximum
Develop drug	$3,925
Stop development	($300)

The decision that maximizes the maximum payoff is to develop the drug.

Conservative strategy (maximin):

	Minimum
Develop drug	($575)
Stop development	($300)

The decision that maximizes the minimum payoff is to stop development.

Opportunity loss strategy:

	Unsuccessful Clinical Trials	Successful Clinical Trials; No FDA Approval	Successful Trials and Approval; Large Market	Successful Trials and Approval; Medium Market	Successful Trials and Approval; Small Market	Maximum
Develop drug	$250	$275	—	—	—	$275
Stop development	—	—	$4,225	$1,925	$1,225	$4,225

The decision that minimizes the maximum opportunity loss is to develop the drug. However, as we noted, we must evaluate risk by considering both the magnitude of the payoffs and their chances of occurrence. The aggressive, conservative, and opportunity-loss rules do not consider the probabilities of the outcomes.

Each decision strategy has an associated payoff distribution, called a **risk profile**. Risk profiles show the possible payoff values that can occur and their probabilities.

EXAMPLE 16.12   **Constructing a Risk Profile**

In the drug development example, consider the strategy of pursuing development. The possible outcomes that can occur and their probabilities are the following:

Terminal Outcome	Net Revenue	Probability
Market large	$3,925	0.108
Market medium	$1,625	0.054
Market small	$925	0.018
FDA not approved	($575)	0.120
Clinical trials not successful	($550)	0.700

The probabilities are computed by multiplying the probabilities on the event branches along the path to the terminal outcome. For example, the probability of getting to "Market large" is $0.3 \times 0.6 \times 0.6 = 0.108$. Thus, we see that the probability that the drug will not reach the market is $1 - (0.108 + 0.054 + 0.018) = 0.82$, and the company will incur a loss of more than $500 million. On the other hand, if they decide not to pursue clinical trials, the loss would be only $300 million, the cost of research to date. If this were a one-time decision, what decision would you make if you were a top executive of this company?

Sensitivity Analysis in Decision Trees

We may use Excel data tables to investigate the sensitivity of the optimal decision to changes in probabilities or payoff values. We illustrate this using the airline revenue management scenario we discussed in Example 5.26 in Chapter 5.

EXAMPLE 16.13   **Sensitivity Analysis for Airline Revenue Management Decision**

Figure 16.7 shows the decision tree (Excel file *Airline Revenue Management Decision Tree*) for deciding whether or not to discount the fare along with a what-if analysis of the impact of changing the probability that a full-fare ticket sells before the flight. From the data table results, we see that if the probability of selling the full-fare ticket is 0.7 or less, then the best decision is to discount the price.

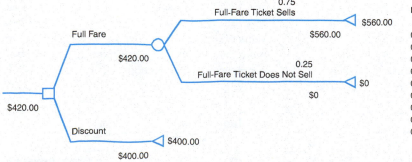

P(Full Fare Ticket Sells)	Expected Value	Decision
	$420.00	Full
0.50	$400.00	Discount
0.55	$400.00	Discount
0.60	$400.00	Discount
0.65	$400.00	Discount
0.70	$400.00	Discount
0.75	$420.00	Full
0.80	$448.00	Full
0.85	$476.00	Full
0.90	$504.00	Full

▶ **Figure 16.7**

Airline Revenue
Management Decision
Tree *and What-If Analysis*

CHECK YOUR UNDERSTANDING

1. Explain the structure and components of decision trees.

2. Describe the process of "rolling back" a decision tree to find the best decision path.

3. Explain what a risk profile is and how to find it.

The Value of Information

When we deal with uncertain outcomes, it is logical to try to obtain better information about their likelihood of occurrence before making a decision. The **value of information** represents the improvement in the expected return that can be achieved if the decision maker is able to acquire—before making a decision—additional information about the future event that will take place. In the ideal case, we would like to have **perfect information**, which tells us with certainty what outcome will occur. Although this will never occur, it is useful to know the value of perfect information because it provides an upper bound on the value of any information that we may acquire. The **expected value of perfect information (EVPI)** is the expected value with perfect information (assumed at no cost) minus the expected value without any information; again, it represents the most you should be willing to pay for perfect information.

The **expected opportunity loss** represents the average additional amount the decision maker would have achieved by making the right decision instead of a wrong one. To find the expected opportunity loss, we create an opportunity-loss table, as discussed earlier in this chapter, and then find the expected value for each decision. *It will always be true that the decision having the best expected value will also have the minimum expected opportunity loss*. The minimum expected opportunity loss is the EVPI.

EXAMPLE 16.14   **Finding EVPI for the Mortgage Selection Decision**

The table below shows the calculations of the expected opportunity losses for each decision (see Example 16.4 for calculation of the opportunity-loss matrix). The minimum expected opportunity loss occurs for the three-year ARM (which was the best expected value decision) and is $3,391.40. This is the value of the EVPI.

Another way to understand this is to use the following logic. Suppose we know that rates will rise. Then we should choose the 30-year fixed mortgage and incur a cost of $54,658. If we know that rates will be stable, then our best decision would be to choose the one-year ARM, with a cost of $46,443. Finally, if we know that rates will fall, we should choose the one-year ARM

with a cost of $40,161. By weighting these values by the probabilities that their associated events will occur, under *perfect information*, our expected cost would be 0.6 × $54,658 + 0.3 × $46,443 + 0.1 × $40,161 = $50,743.80. If we did not have perfect information about the future, then we would choose the three-year ARM no matter what happens and incur an expected cost of $54,135.20. By having perfect information, we would save $54,135.20 − $50,743.80 = $3,391.40. This is the expected value of perfect information. We would never want to pay more than $3,391.40 for any information about the future event, no matter how good.

| | Outcome | | | |
| | 0.6 | 0.3 | 0.1 | |
Decision	Rates Rise	Rates Stable	Rates Fall	Expected Opportunity Loss
1-year ARM	$6,476	—	—	$3,885.60
3-year ARM	$2,243	$4,632	$6,560	$3,391.40
30-year fixed	—	$8,215	$14,497	$3,914.20

Decisions with Sample Information

Sample information is the result of conducting some type of experiment, such as a market research study or interviewing an expert. Sample information is always imperfect. Often, sample information comes at a cost. Thus, it is useful to know how much we should be willing to pay for it. The **expected value of sample information (EVSI)** is the expected value with sample information (assumed at no cost) minus the expected value without sample information; it represents the most you should be willing to pay for the sample information.

EXAMPLE 16.15   **Decisions with Sample Information**

Suppose that a company is developing a new touch-screen cell phone. Historically, 70% of their new phones have resulted in high consumer demand, whereas 30% have resulted in low consumer demand. The company has the decision of choosing between two models with different features that require different amounts of investment and also have different sales potential. Figure 16.8 shows a completed decision tree in which all cash flows are in thousands of dollars. For example, model 1 requires an initial investment for development of $200,000, and model 2 requires an investment of $175,000. If demand is high for model 1, the company will gain $500,000 in revenue, with a net profit of $300,000; it will receive only $160,000 if demand is low, resulting in a net profit of −$40,000. Based on the probabilities of demand, the expected profit is $198,000. For model 2, we see that the expected profit is only $188,000. Therefore, the best decision is to select model 1. Clearly, there is risk in either decision, but on an expected value basis, model 1 is the best decision.

    Now suppose that the firm conducts a market research study to obtain sample information and better understand

the nature of consumer demand. Analysis of past market research studies, conducted prior to introducing similar products, has found that 90% of all products that resulted in high consumer demand had previously received a high survey response, whereas only 20% of all products with ultimately low consumer demand had previously received a high survey response. These probabilities show that the market research is not always accurate and can lead to a false indication of the true market potential. However, we should expect that a high survey response would increase the historical probability of high demand, whereas a low survey response would increase the historical probability of a low demand. Thus, we need to compute the conditional probabilities:

$$P(\text{high demand} \mid \text{high survey response})$$
$$P(\text{high demand} \mid \text{low survey response})$$
$$P(\text{low demand} \mid \text{high survey response})$$
$$P(\text{low demand} \mid \text{low survey response})$$

This can be accomplished using a formula called Bayes's rule.

▶ **Figure 16.8**

Cell Phone Decision Tree

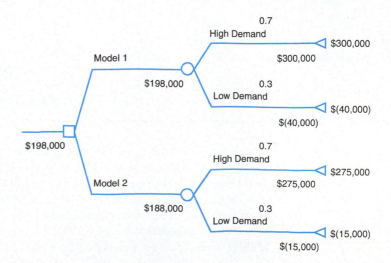

Bayes's Rule

Bayes's rule extends the concept of conditional probability to revise historical probabilities based on sample information. Suppose that $A_1, A_2, \ldots, A_k$ is a set of mutually exclusive and collectively exhaustive events, and we seek the probability that some event A_i occurs given that another event B has occurred. Bayes's rule is stated as follows:

$$P(A_i|B) = \frac{P(B|A_i)\,P(A_i)}{P(B|A_1)\,P(A_1) + P(B|A_2)\,P(A_2) + \ldots + P(B|A_k)\,P(A_k)} \quad \textbf{(16.1)}$$

EXAMPLE 16.16 **Applying Bayes's Rule to Compute Conditional Probabilities**

In the cell phone example, define the following events:

A_1 = high consumer demand
A_2 = low consumer demand
B_1 = high survey response
B_2 = low survey response

We need to compute $P(A_i|B_j)$ for each i and j.

Using these definitions and the information presented in Example 16.15, we have

$$P(A_1) = 0.7$$
$$P(A_2) = 0.3$$
$$P(B_1|A_1) = 0.9$$
$$P(B_1|A_2) = 0.2$$

It is important to carefully distinguish between $P(A|B)$ and $P(B|A)$. As stated, *among all products that resulted in high consumer demand*, 90% received a high market survey response. Thus, the probability of a high survey response *given* high consumer demand is 0.90 and not the other way around. Because the probabilities $P(B_1|A_i) + P(B_2|A_i)$ must add to 1 for each A_i, we have

$$P(B_2|A_1) = 1 - P(B_1|A_1) = 0.1$$
$$P(B_2|A_2) = 1 - P(B_1|A_2) = 0.8$$

Now we may apply Bayes's rule to compute the conditional probabilities of demand given the survey response:

$$P(A_1|B_1) = \frac{P(B_1|A_1)\,P(A_1)}{P(B_1|A_1)\,P(A_1) + P(B_1|A_2)\,P(A_2)}$$

$$= \frac{(0.9)(0.7)}{(0.9)(0.7) + (0.2)(0.3)} = 0.913$$

Therefore, $P(A_2|B_1) = 1 - 0.913 = 0.087$.

$$P(A_1|B_2) = \frac{P(B_2|A_1)\,P(A_1)}{P(B_2|A_1)\,P(A_1) + P(B_2|A_2)\,P(A_2)}$$

$$= \frac{(0.1)(0.7)}{(0.1)(0.7) + (0.8)(0.3)} = 0.226$$

Therefore, $P(A_2|B_2) = 1 - 0.226 = 0.774$.

Although 70% of all previous new models historically had high demand, knowing that the marketing report is favorable increases the likelihood to 91.3%, and if the marketing report is unfavorable, then the probability of low demand increases to 77%.

Finally, we need to compute the nonconditional (marginal) probabilities that the survey response will be either high or low—that is, $P(B_1)$ and $P(B_2)$. These are simply the denominators in Bayes's rule:

$$P(B_1) = P(B_1|A_1)\,P(A_1) + P(B_1|A_2)\,P(A_2)$$
$$= (0.9)(0.7) + (0.2)(0.3) = 0.69$$

$$P(B_2) = P(B_2|A_1)\,P(A_1) + P(B_2|A_2)\,P(A_2)$$
$$= (0.1)(0.7) + (0.8)(0.3) = 0.31$$

The marginal probabilities state that there is a 69% chance that the survey will return a high-demand response, and there is a 31% chance that the survey will result in a low-demand response.

Figure 16.9 shows a decision tree that incorporates the market survey information and the probabilities we calculated in the previous example. The optimal decision strategy is to select model 1 if the survey response is high, and if the response is low, then select model 2. Note that the expected value (which includes the probabilities of obtaining the survey responses) is $202,257. Comparing this to Figure 16.8, we see that the sample information increases the expected value by $202,257 − $198,000 = $4,257. This is the value of EVSI. So we should not pay more than $4,257 to conduct the market survey.

► **Figure 16.9**

*Cell Phone Decision Tree
with Sample Market Survey*

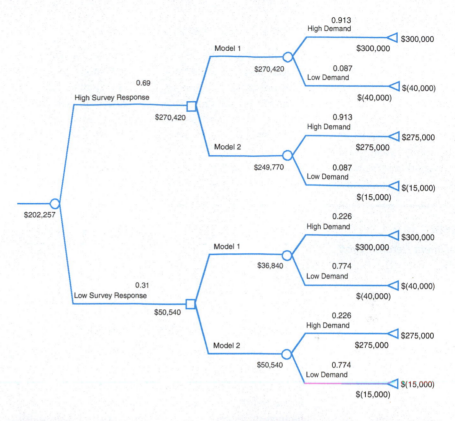

▮▮ **CHECK YOUR UNDERSTANDING**

1. Define the terms value of information, perfect information, and expected value of perfect information.

2. Explain how to find EVPI.

3. What is the expected value of sample information?

4. Explain how Bayes's rule is used in decision trees to find EVSI.

Utility and Decision Making

In Example 5.25 in Chapter 5, we discussed a charity raffle in which 1,000 $50 tickets are sold to win a $25,000 prize. The probability of winning is only 0.001, and the expected payoff is $(-\$50)(0.999) + (\$24,950)(0.001) = -\$25.00$. From a purely economic standpoint, this would be a poor gamble. Nevertheless, many people would take this chance because the financial risk is low (and it's for charity). On the other hand, if only ten tickets were sold at $4,000 with a chance to win $100,000, even though the expected value would be $(-\$4000)(0.9) + (\$96,000)(0.1) = \$6,000$, most people would *not* take the chance because of the higher monetary risk involved.

An approach for assessing risk attitudes quantitatively is called **utility theory**. This approach quantifies a decision maker's relative preferences for particular outcomes. We can determine an individual's utility function by posing a series of decision scenarios. This is best illustrated with an example; we use a personal investment problem to do this.

EXAMPLE 16.17   A Personal Investment Decision

Suppose that you have $10,000 to invest and are expecting to buy a new car in a year, so you can tie the money up for only 12 months. You are considering three options: a bank CD paying 4%, a bond mutual fund, and a stock fund. Both the bond and stock funds are sensitive to changing interest rates. If rates remain the same over the coming year, the share price of the bond fund is expected to remain the same, and you expect to earn $840. The stock fund would return about $600 in dividends and capital gains. However, if interest rates rise, you can anticipate losing about $500 from the bond fund after taking into account the drop in share price and, likewise, expect to lose $900 from the stock fund. If interest rates fall, however, the yield from the bond fund would be $1,000 and the stock fund would net $1,700. Table 16.2 summarizes the payoff table for this decision problem. The decision could result in a variety of payoffs, ranging from a profit of $1,700 to a loss of $900.

▶ **TABLE 16.2**

Investment Return Payoff Table

Decision/Event	Rates Rise	Rates Stable	Rates Fall
Bank CD	$400	$400	$400
Bond fund	$(500)	$840	$1,000
Stock fund	$(900)	$600	$1,700

Constructing a Utility Function

The first step in determining a utility function is to rank-order the payoffs from highest to lowest. We conveniently assign a utility of 1.0 to the highest payoff and a utility of 0 to the lowest. Next, for each payoff between the highest and lowest, consider the following situation: Suppose you have the opportunity of achieving a *guaranteed return of x* or taking a chance of receiving the highest payoff with probability p or the lowest payoff with probability $1-p$. (We use the term **certainty equivalent** to represent the amount that a decision maker feels is equivalent to an uncertain gamble.) What value of p would make you indifferent to these two choices? Then repeat this process for each payoff.

EXAMPLE 16.18   Constructing a Utility Function for the Personal Investment Decision

First, rank the payoffs from highest to lowest; assign a utility of 1.0 to the highest and a utility of 0 to the lowest:

Payoff, x	Utility, U(x)
$1,700	1.0
$1,000	
$840	
$600	
$400	
$(500)	
$(900)	0.0

Let us start with $x = $1,000$. The decision is illustrated in the simple decision tree in Figure 16.10. Because this is a relatively high value, you decide that p would have to be at least 0.9 to take this risk. This represents the utility of a payoff of $1,000, denoted as $U($1,000)$. For example, $1,000 is this decision maker's certainty equivalent for the uncertain situation of receiving $1,700 with probability 0.9 or –$900 with probability 0.1.

Repeating this process for each payoff, suppose we obtain the following utility function:

Payoff, x	Utility, U(x)
$1,700	1.0
$1,000	0.90
$840	0.85
$600	0.80
$400	0.75
$(500)	0.35
$(900)	0.0

▶ **Figure 16.10**

Decision Tree Lottery for Determining the Utility of $1,000

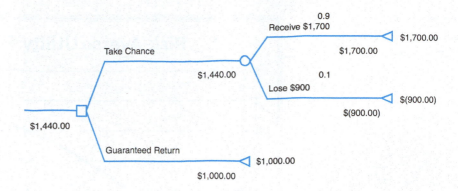

If we compute the expected value of each of the gambles for the chosen values of p, we see that they are higher than the corresponding payoffs. For example, for the payoff of $1,000 and the corresponding $p = 0.9$, the expected value of taking the gamble is

$$0.9(\$1,700) + 0.1(-\$900) = \$1,440$$

This is greater than accepting $1,000 outright. We can interpret this to mean that you require a risk premium of $1,440 − $1,000 = $440 to feel comfortable enough to risk losing $900 if you take the gamble. In general, the **risk premium** is the amount an individual is willing to forgo to avoid risk. This indicates that you are a *risk-averse individual*, that is, relatively conservative.

Another way of viewing this is to find the *break-even probability* at which you would be indifferent to receiving the guaranteed return and taking the gamble. This probability is found by solving the equation

$$1,700p - 900(1 - p) = 1,000$$

resulting in $p = 19/26 = 0.73$. Because you require a higher probability of winning the gamble, it is clear that you are uncomfortable taking the risk.

If we graph the utility versus the payoffs, we can sketch a utility function, as shown in Figure 16.11. This utility function is generally *concave downward*. This type of curve is characteristic of risk-averse individuals. Such decision makers avoid risk, choosing conservative strategies and those with high return-to-risk values. Thus, a gamble must have a higher expected value than a given payoff to be preferable or, equivalently, a higher probability of winning than the break-even value.

Other individuals might be risk takers. What would their utility functions look like? As you might suspect, they are *concave upward*. These individuals would take a gamble that offers higher rewards even if the expected value is less than a certain payoff. An example of a utility function for a risk-taking individual in this situation is as follows:

Payoff, *x*	Utility, *U(x)*
$1,700	1.0
$1,000	0.6
$840	0.55
$600	0.45
$400	0.40
$(500)	0.1
$(900)	0.0

▶ **Figure 16.11**

Example of a Risk-Averse Utility Function

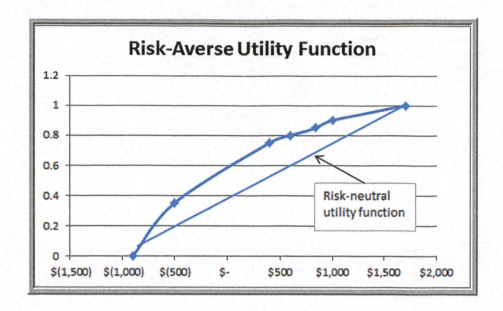

For the payoff of $1,000, this individual would be indifferent between receiving $1,000 and taking a chance at $1,700 with probability 0.6 and losing $900 with probability 0.4. The expected value of this gamble is

$$0.6(\$1,700) + 0.4(-\$900) = \$660$$

Because this is considerably less than $1,000, the individual is taking a larger risk to try to receive $1,700. Note that the probability of winning is less than the break-even value. Risk takers generally prefer more aggressive strategies.

Finally, some individuals are risk neutral; they prefer neither taking risks nor avoiding them. Their utility function is linear and corresponds to the break-even probabilities for each gamble. For example, a payoff of $600 would be equivalent to the gamble if

$$\$600 = p(\$1,700) + (1 - p)(-\$900)$$

Solving for p, we obtain $p = 15/26$, or 0.58, which represents the utility of this payoff. The decision of accepting $600 outright or taking the gamble could be made by flipping a coin. These individuals tend to ignore risk measures and base their decisions on the average payoffs.

A utility function may be used instead of the actual monetary payoffs in a decision analysis by simply replacing the payoffs with their equivalent utilities and then computing expected values. The expected utilities and the corresponding optimal decision strategy then reflect the decision maker's preferences toward risk. For example, if we use the average payoff strategy (because no probabilities of events are given) for the data in Table 16.2, the best decision would be to choose the stock fund. However, if we replace the payoffs in Table 16.2 with the (risk-averse) utilities that we defined and again use the average payoff strategy, the best decision would be to choose the bank CD as opposed to the stock fund, as shown in the following table.

Decision/Event	Rates Rise	Rates Stable	Rates Fall	Average Utility
Bank CD	0.75	0.75	0.75	0.75
Bond fund	0.35	0.85	0.9	0.70
Stock fund	0	0.80	1.0	0.60

If assessments of event probabilities are available, these can be used to compute the expected utility and identify the best decision.

Exponential Utility Functions

It can be rather difficult to compute a utility function, especially for situations involving a large number of payoffs. Because most decision makers typically are risk averse, we may use an exponential utility function to approximate the true utility function. The exponential utility function is

$$U(x) = 1 - e^{-x/R} \qquad (16.2)$$

where e is the base of the natural logarithm (2.71828 . . .) and R is a shape parameter that is a measure of risk tolerance. Figure 16.12 shows several examples of $U(x)$ for different values of R. Notice that all these functions are concave and that as R increases, the functions become flatter, indicating more tendency toward risk neutrality.

One approach to estimating a reasonable value of R is to find the maximum payoff R for which the decision maker is willing to take an equal chance on winning R or losing $R/2$. The smaller the value of R, the more risk averse is the individual. For instance, would you take a bet on winning \$10 versus losing \$5? How about winning \$10,000 versus losing \$5,000? Most people probably would not worry about taking the first gamble but would think twice about the second. Finding one's maximum comfort level establishes the utility function.

▶ **Figure 16.12**
Examples of Exponential Utility Functions

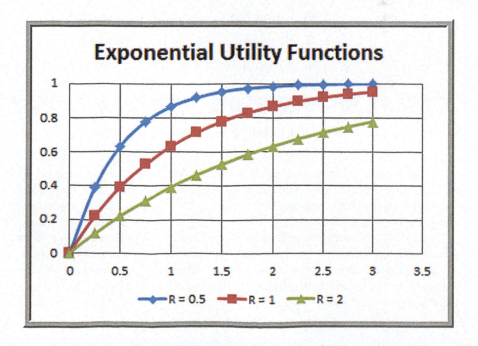

EXAMPLE 16.19 **Using an Exponential Utility Function**

For the personal investment decision example, suppose that $R = \$400$. The utility function is $U(x) = 1 - e^{-x/400}$, resulting in the following utility values:

Payoff, x	Utility, U(x)
$1,700	0.9857
$1,000	0.9179
$840	0.8775
$600	0.7769
$400	0.6321
$(500)	-2.4903
$(900)	-8.4877

Using the utility values in the payoff table, we find that the bank CD remains the best decision, as shown in the following table, as it has the highest average utility.

Decision/Event	Rates Rise	Rates Stable	Rates Fall	Average Utility
Bank CD	0.6321	0.6321	0.6321	0.6321
Bond fund	-2.4903	0.8775	0.9179	-0.2316
Stock fund	-8.4877	0.7769	0.9857	-2.2417

CHECK YOUR UNDERSTANDING

1. What is utility theory, and how does it help to understand decision making?
2. Explain how to construct a utility function.
3. How does a risk-averse utility function compare to a risk-neutral utility function?
4. Why might we use exponential utility functions?

ANALYTICS IN PRACTICE: Using Decision Analysis in Drug Development[2]

Drug development in the United States is time consuming, resource intensive, risky, and heavily regulated. On average, it takes nearly 15 years to research and develop a drug in the United States, with an after-tax cost in 1990 dollars of approximately $200 million.

In July 1999, the biological products leadership committee, composed of the senior managers within Bayer Biological Products (BP), a business unit of Bayer Pharmaceuticals (Pharma), made its newly formed strategic-planning department responsible for the commercial

evaluation of a new blood-clot-busting drug. To ensure that it made the best drug development decisions, Pharma used a structured process based on the principles of decision analysis to evaluate the technical feasibility and market potential of its new drug. Previously, BP had analyzed a few business cases for review by Pharma. This commercial evaluation was BP's first decision analysis project.

Probability distributions of uncertain variables were assessed by estimating the 10th percentile and 90th percentile from experts, who were each asked to review the

---

[2]Based on Jeffrey S. Stonebraker, "How Bayer Makes Decisions to Develop New Drugs," *Interfaces*, 32, 6 (November–December 2002): 77–90.

results to make sure they accurately reflected his or her judgment. Pharma used net present value (NPV) as its decision-making criterion. Given the complexity and inherent structure of decisions concerning new drugs, the new-drug-development decision making was defined as a sequence of six decision points, with identified key market-related and scientific deliverables so senior managers

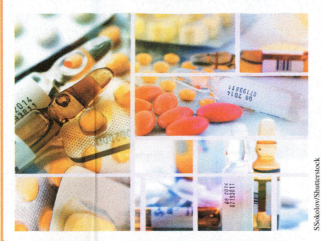

SSokolov/Shutterstock

could assess the likelihood of success versus the company's exposure to risk, costs, and strategic fit. Decision point 1 was whether to begin preclinical development. After successful preclinical animal testing, Bayer can decide (decision point 2) to begin testing the drug in humans. Decision point 3 and decision point 4 are both decisions to invest or not in continuing clinical development. Following successful completion of development, Bayer can choose to file a biological license application with the FDA (decision point 5). If the FDA approves it, Bayer can decide (decision point 6) to launch the new drug in the marketplace.

The project team presented their input assumptions and recommendations for the commercial evaluation of the drug to the three levels of Pharma decision makers, who eventually approved preclinical development. External validation of the data inputs and assumptions demonstrated their rigor and defensibility. Senior managers could compare the evaluation results for the proposed drug with those for other development drugs with confidence. The international committees lauded the project team's effort as top-notch, and the decision-analysis approach set new standards for subsequent BP analyses.

KEY TERMS

Average payoff strategy
Branches
Certainty equivalent
Decision alternatives
Decision making
Decision node
Decision strategy
Decision tree
Event node
Expected opportunity loss
Expected value of perfect information (EVPI)
Expected value of sample information (EVSI)
Expected value strategy
Maximax strategy

Maximin strategy
Minimax regret strategy
Minimax strategy
Minimin strategy
Nodes
Outcomes
Payoffs
Payoff table
Perfect information
Risk premium
Risk profile
Sample information
States of nature
Uncertain events
Utility theory
Value of information

CHAPTER 16 TECHNOLOGY HELP

Analytic Solver

Analytic Solver provides the ability to construct and analyze decision trees in Excel. See the online supplement *Using Decision Trees in Analytic Solver.* We suggest that

you first read the online supplement *Getting Started with Analytic Solver Basic.* This provides information for both instructors and students on how to register for and access *Analytic Solver.*

Note: Data for selected problems can be found in the Excel file Chapter 16 Problem Data *to facilitate problem-solving efforts and Excel Implementation. Worksheet tabs correspond to the problem numbers. These are designated with an asterisk (*).*

Formulating Decision Problems

1. Use the *Outsourcing Decision Model* Excel file to compute the cost of in-house manufacturing and outsourcing for the following levels of demand: 800, 1,000, 1,200, and 1,400. Use this information to set up a payoff table for the decision problem.

Decision Strategies Without Outcome Probabilities

2. For the payoff table you developed in Problem 1, determine the decision using the aggressive, conservative, and opportunity-loss strategies.

*3. The DoorCo Corporation is a leading manufacturer of garage doors. All doors are manufactured in their plant in Carmel, Indiana, and shipped to distribution centers or major customers. DoorCo recently acquired another manufacturer of garage doors, Wisconsin Door, and is considering moving its wood door operations to the Wisconsin plant. Key considerations in this decision are the transportation, labor, and production costs at the two plants. Complicating matters is the fact that marketing is predicting a decline in the demand for wood doors. The company developed three scenarios:

   a. Demand falls slightly, with no noticeable effect on production.

   b. Demand and production decline 20%.

   c. Demand and production decline 40%.

   The following table shows the total costs under each decision and scenario.

	Slight Decline	20% Decline	40% Decline
Stay in Carmel	$1,000,000	$800,000	$840,000
Move to Wisconsin	$1,100,000	$950,000	$750,000

   What decision should DoorCo make using each of the following strategies?

   a. aggressive strategy

   b. conservative strategy

   c. opportunity-loss strategy

4. Suppose that a car-rental agency offers insurance for a week that costs $100. A minor fender bender will cost $3,500, whereas a major accident might cost $16,000 in repairs. Without the insurance, you would be personally liable for any damages. What should you do? Clearly, there are two decision alternatives: take the insurance, or do not take the insurance. The uncertain consequences, or events that might occur, are that you would not be involved in an accident, that you would be involved in a fender bender, or that you would be involved in a major accident. Develop a payoff table for this situation. What decision should you make using each of the following strategies?

   a. aggressive strategy

   b. conservative strategy

   c. opportunity-loss strategy

*5. Slaggert Systems is considering becoming certified to the ISO 9000 series of quality standards. Becoming certified is expensive, but the company could lose a substantial amount of business if its major customers suddenly demand ISO certification and the company does not have it. At a management retreat, the senior executives of the firm developed the following payoff table, indicating the net present value of profits over the next five years.

	Customer Response	
	Standards Required	Standards Not Required
Become certified	$575,000	$525,000
Stay uncertified	$450,000	$675,000

   What decision should the company make using each of the following strategies?

   a. aggressive strategy

   b. conservative strategy

   c. opportunity-loss strategy

*6. For the DoorCo Corporation decision in Problem 3, compute the standard deviation of the payoffs for each decision. What does this tell you about the risk in making the decision?

7. For the car-rental situation in Problem 4, compute the standard deviation of the payoffs for each decision. What does this tell you about the risk in making the decision?

*8. For Slaggert Systems decision in Problem 5, compute the standard deviation of the payoffs for each decision. What does this tell you about the risk in making the decision?

Decision Strategies with Outcome Probabilities

9. What decisions should be made using the average payoff strategy in Problems 3, 4, and 5?

10. For the DoorCo Corporation decision in Problem 2, suppose that the probabilities of the three scenarios are estimated to be 0.15, 0.40, and 0.45, respectively. Find the best expected value decision.

11. For the car-rental situation described in Problem 3, assume that you researched insurance industry statistics and found out that the probability of a major accident is 0.05% and that the probability of a fender bender is 0.16%. What is the expected value decision? Would you choose this? Why or why not?

12. An information systems consultant is bidding on a project that involves some uncertainty. Based on past experience, if all went well (probability 0.1), the project would cost $1.2 million to complete. If moderate debugging were required (probability 0.7), the project would probably cost $1.4 million. If major problems were encountered (probability 0.2), the project could cost $1.8 million. Assume that the consultant is bidding competitively and the expectation of successfully gaining the job at a bid of $2.2 million is 0, at $2.1 million is 0.1, at $2.0 million is 0.2, at $1.9 million is 0.3, at $1.8 million is 0.5, at $1.7 million is 0.8, and at $1.6 million is practically certain.

   a. Calculate the expected value for the given bids.

   b. What is the best bidding decision?

Decision Trees

13. For the DoorCo Corporation decision in Problems 3 and 10, construct a decision tree and compute the rollback values to find the best expected value decision.

14. For the car-rental decision in Problems 4 and 11, construct a decision tree and compute the rollback values to find the best expected value decision.

15. Midwestern Hardware must decide how many snow shovels to order for the coming snow season. Each shovel costs $15.00 and is sold for $29.95. No inventory is carried from one snow season to the next. Shovels unsold after February are sold at a discount price of $10.00. Past data indicate that sales are highly dependent

on the severity of the winter season. Past seasons have been classified as mild or harsh, and the following distribution of regular price demand has been tabulated:

Mild Winter		Harsh Winter	
No. of Shovels	Probability	No. of Shovels	Probability
250	0.5	1,500	0.2
300	0.4	2,500	0.3
350	0.1	3,000	0.5

Shovels must be ordered from the manufacturer in lots of 200; thus, possible order sizes are 200, 400, 1,400, 1,600, 2,400, 2,600, and 3,000 units. Construct a decision tree to illustrate the components of the decision model, and find the optimal quantity for Midwestern to order if the forecast calls for a 40% chance of a harsh winter.

16. Dean Kuroff started a business of rehabbing old homes. He recently purchased a circa-1800 Victorian mansion and converted it into a three-family residence. Yesterday, one of his tenants complained that the refrigerator was not working properly. Dean's cash flow is not extensive, so he was not excited about purchasing a new refrigerator. He is considering two other options: purchase a used refrigerator or repair the current unit. He can purchase a new one for $600, and it will easily last three years. If he repairs the current one, he estimates a repair cost of $150, but he also believes that there is only a 25% chance that it will last a full three years and he will end up purchasing a new one anyway. If he buys a used refrigerator for $200, he estimates that there is a 0.4 probability that it will last at least three years. If it breaks down, he will still have the option of repairing it for $150 or buying a new one. Develop a decision tree for this situation and determine Dean's optimal strategy.

17. Many automobile dealers advertise lease options for new cars. Suppose that you are considering three alternatives:

   1. Purchase a car outright with cash.

   2. Purchase a car with 20% down and a 48-month loan.

   3. Lease a car.

Select an automobile whose leasing contract is advertised in a local paper. Using current interest rates and advertised leasing arrangements, perform a decision analysis of these options. Make, but clearly define, any assumptions that may be required.

18. Perform a sensitivity analysis of the Midwestern Hardware scenario (Problem 15). Find the optimal order quantity and optimal expected profit for probabilities of a harsh winter ranging from 0.2 to 0.8 in increments of 0.2. Plot optimal expected profit as a function of the probability of a harsh winter.

The Value of Information

*19. Mountain Ski Sports, a chain of ski equipment shops in Colorado, purchases skis from a manufacturer each summer for the coming winter season. The most popular intermediate model costs $150 and sells for $275. Any skis left over at the end of the winter are sold at the store's spring sale (for $100). Sales over the years have been quite stable. Gathering data from all its stores, Mountain Ski Sports developed the following probability distribution for demand:

Demand	Probability
150	0.10
175	0.30
200	0.35
225	0.20
250	0.05

The manufacturer will take orders only for multiples of 20, so Mountain Ski is considering the following order sizes: 160, 180, 200, 220, and 240.

a. Construct a payoff table for Mountain Ski's decision problem of how many pairs of skis to order. What is the best decision from an expected value basis?

b. Find the expected value of perfect information.

c. What is the expected demand? What is the expected profit if the shop orders the expected demand? How does this compare with the expected value decision?

20. Bev's Bakery specializes in sourdough bread. Early each morning, Bev must decide how many loaves to bake for the day. Each loaf costs $1.25 to make and sells for $3.50. Bread left over at the end of the day can be sold the next day for $1.00. Past data indicate that demand is distributed as follows:

Number of Loaves	Probability
15	0.02
16	0.05
17	0.11
18	0.15
19	0.27
20	0.21
21	0.15
22	0.04

a. Construct a payoff table and determine the optimal quantity for Bev to bake each morning using expected values.

b. What is the optimal quantity for Bev to bake if the unsold loaves are not sold the next day but are donated to a food bank?

21. A patient arrives at an emergency room complaining of abdominal pain. The ER physician must decide on whether to operate or to place the patient under observation for a non-appendix-related condition. If an appendectomy is performed immediately, the doctor runs the risk that the patient does not have appendicitis. If it is delayed and the patient does indeed have appendicitis, the appendix might perforate, leading to a more severe case and possible complications. However, the patient might recover without the operation.

a. Construct a decision tree for the doctor's dilemma.

b. How might payoffs and probabilities be determined?

c. Would utility be a better measure of payoff than actual costs? If so, how might utilities be derived for each path in the tree?

Decisions with Sample Information

22. Drilling decisions by oil and gas operators involve intensive capital expenditures made in an environment characterized by limited information and high risk. A well site is dry, wet, or gushing. Historically, 50% of all wells have been dry, 30% wet, and 20% gushing. The value (net of drilling costs) for each type of well is as follows:

Dry	−$80.000
Wet	$120,000
Gushing	$200,000

Wildcat operators often investigate oil prospects in areas where deposits are thought to exist by making geological and geophysical examinations of the area before obtaining a lease and drilling permit. This often includes recording shock waves from detonations by a seismograph and using a magnetometer to measure the intensity of Earth's magnetic effect to detect rock formations below the surface. The cost of doing such a study is approximately $15,000. Of course, one may choose to drill in a location based on "gut feel" and avoid the cost of the study. The geological and geo-physical examination classifies an area into one of three categories: no structure (NS), which is a bad sign; open structure (OS), which is an "OK" sign; and closed structure (CS), which is hopeful. Historically, 40% of the tests resulted in NS, 35% resulted in OS, and 25% resulted in CS readings. After the result of the test is known, the company may decide not to drill. The fol-lowing table shows probabilities that the well will actu-ally be dry, wet, or gushing based on the classification provided by the examination (in essence, the examina-tion cannot accurately predict the actual event):

	Dry	Wet	Gushing
NS	0.73	0.22	0.05
OS	0.45	0.32	0.23
CS	0.23	0.35	0.42

a. Construct a decision tree of this problem that includes the decision of whether or not to perform the geological examination.

b. What is the optimal decision under expected value when no experimentation is conducted?

c. Find the overall optimal strategy by rolling back the tree.

23. Hahn Engineering is planning on bidding on a job and often competes against Sweigart and Associates (S&A), as well as other firms. Historically, S&A has bid for the same jobs 80% of the time; thus the prob-ability that S&A will bid on this job is 0.80. If S&A bids on a job, the probability that Hahn Engineering will win it is 0.30. If S&A does not bid on a job, the probability that Hahn will win the bid is 0.60. Apply Bayes's rule to find the probability that Hahn Engi-neering will win the bid. If they do, what is the prob-ability that S&A did bid on it?

24. MJ Logistics has decided to build a new warehouse to support its supply chain activities. They have the option of building either a large warehouse or a small one. Construction costs are $8 million for the large facility versus $3 million for the small facility. The profit (excluding construction cost) depends on the volume of work the company expects to contract for in the future. This is summarized in the following table (in millions of dollars):

	High Volume	Low Volume
Large warehouse	$35	$20
Small warehouse	$25	$15

The company believes that there is a 60% chance that the volume of demand will be high.

a. Construct a decision tree to identify the best choice.

b. Suppose that the company engages an economic expert to provide an opinion about the volume of work based on a forecast of economic condi-tions. Historically, the expert's upside predictions have been 75% accurate, whereas the downside predictions have been 90% accurate. In contrast to the company's assessment, the expert believes that the chance for high demand is 70%. Deter-mine the best strategy if their predictions suggest that the economy will improve or will deteriorate. Given the information, what is the probability that the volume will be high?

Utility and Decision Making

25. Consider the car-rental insurance scenario in Prob-lems 4 and 11. Use the approach described in this chapter to develop your personal utility function for the payoffs associated with this decision. Deter-mine the decision that would result using the utilities instead of the payoffs. Is the decision consistent with your choice?

26. A college football team is trailing 14–0 late in the game. The team just made a touchdown. If they can hold the opponent and score one more time, they can tie or win the game. The coach is wondering whether to go for an extra-point kick or a two-point conver-sion now, and what to do if they score again.

a. Develop a decision tree for the coach's decision.

b. Estimate probabilities for successful kicks or two-point conversions and a last-minute score. (You might want to do this by doing some group brainstorming or by calling on experts, such as your school's coach or a sports journalist.) Using the probabilities from part a, determine the opti-mal strategy.

c. Why would utility theory be a better approach than using the points for making a decision? Pro-pose a utility function and compare your results.

CASE: PERFORMANCE LAWN EQUIPMENT

PLE has developed a prototype for a new snow blower for the consumer market. This can exploit the company's expertise in small-gasoline-engine technology and also balance seasonal demand cycles in the North American and European markets to provide additional revenues during the winter months. Initially, PLE faces two possible decisions: introduce the product globally at a cost of $850,000 or evaluate it in a North American test market at a cost of $200,000. If it introduces the product globally, PLE might find either a high or low response to the product. The probabilities of these events are estimated to be 0.6 and 0.4, respectively. With a high response, gross revenues of $2,000,000 are expected; with a low response, the figure is $450,000. If PLE starts with a North American test market, it might find a low response or a high response, with probabilities of 0.3 and 0.7, respectively. This may or may not reflect the global market potential. In any case, after conducting the marketing research, PLE next needs to decide whether to keep

sales only in North America, market globally, or drop the product. If the North American response is high and PLE stays only in North America, the expected revenue is $1,200,000. If it markets globally (at an additional cost of $200,000), the probability of a high global response is 0.9 with revenues of $2,000,000 ($450,000 if the global response is low). If the North American response is low and it remains in North America, the expected revenue is $200,000. If it markets globally (at an additional cost of $600,000), the probability of a high global response is 0.05, with revenues of $2,000,000 ($450,000 if the global response is low).

Construct a decision tree, determine the optimal strategy, and develop a risk profile associated with the optimal strategy. Evaluate the sensitivity of the optimal strategy to changes in the probability estimates. Summarize all your results, including your recommendation and justification for it, in a formal report to the executive committee, who will ultimately make this decision.

Appendix A: Statistical Tables

▶ Table A.1

The Cumulative Standard Normal Distribution

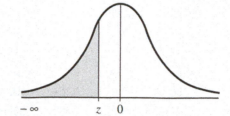

z	.00	.01	.02	.03	.04	.05	.06	.07	.08	.09
−3.9	.00005	.00005	.00004	.00004	.00004	.00004	.00004	.00004	.00003	.00003
−3.8	.00007	.00007	.00007	.00006	.00006	.00006	.00006	.00005	.00005	.00005
−3.7	.00011	.00010	.00010	.00010	.00009	.00009	.00008	.00008	.00008	.00008
−3.6	.00016	.00015	.00015	.00014	.00014	.00013	.00013	.00012	.00012	.00011
−3.5	.00023	.00022	.00022	.00021	.00020	.00019	.00019	.00018	.00017	.00017
−3.4	.00034	.00032	.00031	.00030	.00029	.00028	.00027	.00026	.00025	.00024
−3.3	.00048	.00047	.00045	.00043	.00042	.00040	.00039	.00038	.00036	.00035
−3.2	.00069	.00066	.00064	.00062	.00060	.00058	.00056	.00054	.00052	.00050
−3.1	.00097	.00094	.00090	.00087	.00084	.00082	.00079	.00076	.00074	.00071
−3.0	.00135	.00131	.00126	.00122	.00118	.00114	.00111	.00107	.00103	.00100
−2.9	.0019	.0018	.0018	.0017	.0016	.0016	.0015	.0015	.0014	.0014
−2.8	.0026	.0025	.0024	.0023	.0023	.0022	.0021	.0021	.0020	.0019
−2.7	.0035	.0034	.0033	.0032	.0031	.0030	.0029	.0028	.0027	.0026
−2.6	.0047	.0045	.0044	.0043	.0041	.0040	.0039	.0038	.0037	.0036
−2.5	.0062	.0060	.0059	.0057	.0055	.0054	.0052	.0051	.0049	.0048
−2.4	.0082	.0080	.0078	.0075	.0073	.0071	.0069	.0068	.0066	.0064
−2.3	.0107	.0104	.0102	.0099	.0096	.0094	.0091	.0089	.0087	.0084
−2.2	.0139	.0136	.0132	.0129	.0125	.0122	.0119	.0116	.0113	.0110
−2.1	.0179	.0174	.0170	.0166	.0162	.0158	.0154	.0150	.0146	.0143
−2.0	.0228	.0222	.0217	.0212	.0207	.0202	.0197	.0192	.0188	.0183
−1.9	.0287	.0281	.0274	.0268	.0262	.0256	.0250	.0244	.0239	.0233
−1.8	.0359	.0351	.0344	.0336	.0329	.0322	.0314	.0307	.0301	.0294
−1.7	.0446	.0436	.0427	.0418	.0409	.0401	.0392	.0384	.0375	.0367
−1.6	.0548	.0537	.0526	.0516	.0505	.0495	.0485	.0475	.0465	.0455
−1.5	.0668	.0655	.0643	.0630	.0618	.0606	.0594	.0582	.0571	.0559

(continued)

z	.00	.01	.02	.03	.04	.05	.06	.07	.08	.09
−1.4	.0808	.0793	.0778	.0764	.0749	.0735	.0721	.0708	.0694	.0681
−1.3	.0968	.0951	.0934	.0918	.0901	.0885	.0869	.0853	.0838	.0823
−1.2	.1151	.1131	.1112	.1093	.1075	.1056	.1038	.1020	.1003	.0985
−1.1	.1357	.1335	.1314	.1292	.1271	.1251	.1230	.1210	.1190	.1170
−1.0	.1587	.1562	.1539	.1515	.1492	.1469	.1446	.1423	.1401	.1379
−0.9	.1841	.1814	.1788	.1762	.1736	.1711	.1685	.1660	.1635	.1611
−0.8	.2119	.2090	.2061	.2033	.2005	.1977	.1949	.1922	.1894	.1867
−0.7	.2420	.2388	.2358	.2327	.2296	.2266	.2236	.2006	.2177	.2148
−0.6	.2743	.2709	.2676	.2643	.2611	.2578	.2546	.2514	.2482	.2451
−0.5	.3085	.3050	.3015	.2981	.2946	.2912	.2877	.2843	.2810	.2776
−0.4	.3446	.3409	.3372	.3336	.3300	.3264	.3228	.3192	.3156	.3121
−0.3	.3821	.3783	.3745	.3707	.3669	.3632	.3594	.3557	.3520	.3483
−0.2	.4207	.4168	.4129	.4090	.4052	.4013	.3974	.3936	.3897	.3859
−0.1	.4602	.4562	.4522	.4483	.4443	.4404	.4364	.4325	.4286	.4247
−0.0	.5000	.4960	.4920	.4880	.4840	.4801	.4761	.4721	.4681	.4641

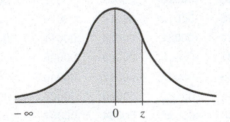

z	.00	.01	.02	.03	.04	.05	.06	.07	.08	.09
0.0	.5000	.5040	.5080	.5120	.5160	.5199	.5239	.5279	.5319	.5359
0.1	.5398	.5438	.5478	.5517	.5557	.5596	.5636	.5675	.5714	.5753
0.2	.5793	.5832	.5871	.5910	.5948	.5987	.6026	.6064	.6103	.6141
0.3	.6179	.6217	.6255	.6293	.6331	.6368	.6406	.6443	.6480	.6517
0.4	.6554	.6591	.6628	.6664	.6700	.6736	.6772	.6808	.6844	.6879
0.5	.6915	.6950	.6985	.7019	.7054	.7088	.7123	.7157	.7190	.7224
0.6	.7257	.7291	.7324	.7357	.7389	.7422	.7454	.7486	.7518	.7549
0.7	.7580	.7612	.7642	.7673	.7704	.7734	.7764	.7794	.7823	.7852
0.8	.7881	.7910	.7939	.7967	.7995	.8023	.8051	.8078	.8106	.8133
0.9	.8159	.8186	.8212	.8238	.8264	.8289	.8315	.8340	.8365	.8389
1.0	.8413	.8438	.8461	.8485	.8508	.8531	.8554	.8577	.8599	.8621
1.1	.8643	.8665	.8686	.8708	.8729	.8749	.8770	.8790	.8810	.8830
1.2	.8849	.8869	.8888	.8907	.8925	.8944	.8962	.8980	.8997	.9015
1.3	.9032	.9089	.9066	.9082	.9099	.9115	.9131	.9147	.9162	.9177
1.4	.9192	.9207	.9222	.9236	.9251	.9265	.9279	.9292	.9306	.9319
1.5	.9332	.9345	.9357	.9370	.9382	.9394	.9406	.9418	.9429	.9441

z	.00	.01	.02	.03	.04	.05	.06	.07	.08	.09
1.6	.9452	.9463	.9474	.9484	.9495	.9505	.9515	.9525	.9535	.9545
1.7	.9554	.9564	.9573	.9582	.9591	.9599	.9608	.9616	.9625	.9633
1.8	.9641	.9649	.9656	.9664	.9671	.9678	.9686	.9693	.9699	.9706
1.9	.9713	.9719	.9726	.9732	.9738	.9744	.9750	.9756	.9761	.9767
2.0	.9772	.9778	.9783	.9788	.9793	.9798	.9803	.9808	.9812	.9817
2.1	.9821	.9826	.9830	.9834	.9838	.9842	.9846	.9850	.9854	.9857
2.2	.9861	.9864	.9868	.9871	.9875	.9878	.9881	.9884	.9887	.9890
2.3	.9893	.9896	.9898	.9901	.9904	.9906	.9909	.9911	.9913	.9916
2.4	.9918	.9920	.9922	.9925	.9927	.9929	.9931	.9932	.9934	.9936
2.5	.9938	.9940	.9941	.9943	.9945	.9946	.9948	.9949	.9951	.9952
2.6	.9953	.9955	.9956	.9957	.9959	.9960	.9961	.9962	.9963	.9964
2.7	.9965	.9966	.9967	.9968	.9969	.9970	.9971	.9972	.9973	.9974
2.8	.9974	.9975	.9976	.9977	.9977	.9978	.9979	.9979	.9980	.9981
2.9	.9981	.9982	.9982	.9983	.9984	.9984	.9985	.9985	.9986	.9986
3.0	.99865	.99869	.99874	.99878	.99882	.99886	.99889	.99893	.99897	.99900
3.1	.99903	.99906	.99910	.99913	.99916	.99918	.99921	.99924	.99926	.99929
3.2	.99931	.99934	.99936	.99938	.99940	.99942	.99944	.99946	.99948	.99950
3.3	.99952	.99953	.99955	.99957	.99958	.99960	.99961	.99962	.99964	.99965
3.4	.99966	.99968	.99969	.99970	.99971	.99972	.99973	.99974	.99975	.99976
3.5	.99977	.99978	.99978	.99979	.99980	.99981	.99981	.99982	.99983	.99983
3.6	.99984	.99985	.99985	.99986	.99986	.99987	.99987	.99988	.99988	.99989
3.7	.99989	.99990	.99990	.99990	.99991	.99991	.99992	.99992	.99992	.99992
3.8	.99993	.99993	.99993	.99994	.99994	.99994	.99994	.99995	.99995	.99995
3.9	.99995	.99995	.99996	.99996	.99996	.99996	.99996	.99996	.99997	.99997

Entry represents area under the cumulative standardized normal distribution from $-\infty$ to z.

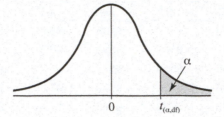

▶ **TABLE A.2**

Critical Values of t

Degrees of Freedom	Upper Tail Areas					
	.25	.10	.05	.025	.01	.005
1	1.0000	3.0777	6.3138	12.7062	31.8207	63.6574
2	0.8165	1.8856	2.9200	4.3027	6.9646	9.9248
3	0.7649	1.6377	2.3534	3.1824	4.5407	5.8409
4	0.7407	1.5332	2.1318	2.7764	3.7469	4.6041
5	0.7267	1.4759	2.0150	2.5706	3.3649	4.0322
6	0.7176	1.4398	1.9432	2.4469	3.1427	3.7074
7	0.7111	1.4149	1.8946	2.3646	2.9980	3.4995
8	0.7064	1.3968	1.8595	2.3060	2.8965	3.3554
9	0.7027	1.3830	1.8331	2.2622	2.8214	3.2498
10	0.6998	1.3722	1.8125	2.2281	2.7638	3.1693
11	0.6974	1.3634	1.7959	2.2010	2.7181	3.1058
12	0.6955	1.3562	1.7823	2.1788	2.6810	3.0545
13	0.6938	1.3502	1.7709	2.1604	2.6503	3.0123
14	0.6924	1.3450	1.7613	2.1448	2.6245	2.9768
15	0.6912	1.3406	1.7531	2.1315	2.6025	2.9467
16	0.6901	1.3368	1.7459	2.1199	2.5835	2.9208
17	0.6892	1.3334	1.7396	2.1098	2.5669	2.8982
18	0.6884	1.3304	1.7341	2.1009	2.5524	2.8784
19	0.6876	1.3277	1.7291	2.0930	2.5395	2.8609
20	0.6870	1.3253	1.7247	2.0860	2.5280	2.8453
21	0.6864	1.3232	1.7207	2.0796	2.5177	2.8314
22	0.6858	1.3212	1.7171	2.0739	2.5083	2.8188
23	0.6853	1.3195	1.7139	2.0687	2.4999	2.8073
24	0.6848	1.3178	1.7109	2.0639	2.4922	2.7969
25	0.6844	1.3163	1.7081	2.0595	2.4851	2.7874
26	0.6840	1.3150	1.7056	2.0555	2.4786	2.7787
27	0.6837	1.3137	1.7033	2.0518	2.4727	2.7707
28	0.6834	1.3125	1.7011	2.0484	2.4671	2.7633
29	0.6830	1.3114	1.6991	2.0452	2.4620	2.7564
30	0.6828	1.3104	1.6973	2.0423	2.4573	2.7500
31	0.6825	1.3095	1.6955	2.0395	2.4528	2.7440
32	0.6822	1.3086	1.6939	2.0369	2.4487	2.7385

Degrees of Freedom	Upper Tail Areas					
	.25	.10	.05	.025	.01	.005
33	0.6820	1.3077	1.6924	2.0345	2.4448	2.7333
34	0.6818	1.3070	1.6909	2.0322	2.4411	2.7284
35	0.6816	1.3062	1.6896	2.0301	2.4377	2.7238
36	0.6814	1.3055	1.6883	2.0281	2.4345	2.7195
37	0.6812	1.3049	1.6871	2.0262	2.4314	2.7154
38	0.6810	1.3042	1.6860	2.0244	2.4286	2.7116
39	0.6808	1.3036	1.6849	2.0227	2.4258	2.7079
40	0.6807	1.3031	1.6839	2.0211	2.4233	2.7045
41	0.6805	1.3025	1.6829	2.0195	2.4208	2.7012
42	0.6804	1.3020	1.6820	2.0181	2.4185	2.6981
43	0.6802	1.3016	1.6811	2.0167	2.4163	2.6951
44	0.6801	1.3011	1.6802	2.0154	2.4141	2.6923
45	0.6800	1.3006	1.6794	2.0141	2.4121	2.6896
46	0.6799	1.3002	1.6787	2.0129	2.4102	2.6870
47	0.6797	1.2998	1.6779	2.0117	2.4083	2.6846
48	0.6796	1.2994	1.6772	2.0106	2.4066	2.6822
49	0.6795	1.2991	1.6766	2.0096	2.4049	2.6800
50	0.6794	1.2987	1.6759	2.0086	2.4033	2.6778
51	0.6793	1.2984	1.6753	2.0076	2.4017	2.6757
52	0.6792	1.2980	1.6747	2.0066	2.4002	2.6737
53	0.6791	1.2977	1.6741	2.0057	2.3988	2.6718
54	0.6791	1.2974	1.6736	2.0049	2.3974	2.6700
55	0.6790	1.2971	1.6730	2.0040	2.3961	2.6682
56	0.6789	1.2969	1.6725	2.0032	2.3948	2.6665
57	0.6788	1.2966	1.6720	2.0025	2.3936	2.6649
58	0.6787	1.2963	1.6716	2.0017	2.3924	2.6633
59	0.6787	1.2961	1.6711	2.0010	2.3912	2.6618
60	0.6786	1.2958	1.6706	2.0003	2.3901	2.6603
61	0.6785	1.2956	1.6702	1.9996	2.3890	2.6589
62	0.6785	1.2954	1.6698	1.9990	2.3880	2.6575
63	0.6784	1.2951	1.6694	1.9983	2.3870	2.6561
64	0.6783	1.2949	1.6690	1.9977	2.3860	2.6549
65	0.6783	1.2947	1.6686	1.9971	2.3851	2.6536
66	0.6782	1.2945	1.6683	1.9966	2.3842	2.6524
67	0.6782	1.2943	1.6679	1.9960	2.3833	2.6512
68	0.6781	1.2941	1.6676	1.9955	2.3824	2.6501
69	0.6781	1.2939	1.6672	1.9949	2.3816	2.6490
70	0.6780	1.2938	1.6669	1.9944	2.3808	2.6479

(continued)

Degrees of Freedom	Upper Tail Areas					
	.25	.10	.05	.025	.01	.005
71	0.6780	1.2936	1.6666	1.9939	2.3800	2.6469
72	0.6779	1.2934	1.6663	1.9935	2.3793	2.6459
73	0.6779	1.2933	1.6660	1.9930	2.3785	2.6449
74	0.6778	1.2931	1.6657	1.9925	2.3778	2.6439
75	0.6778	1.2929	1.6654	1.9921	2.3771	2.6430
76	0.6777	1.2928	1.6652	1.9917	2.3764	2.6421
77	0.6777	1.2926	1.6649	1.9913	2.3758	2.6412
78	0.6776	1.2925	1.6646	1.9908	2.3751	2.6403
79	0.6776	1.2924	1.6644	1.9905	2.3745	2.6395
80	0.6776	1.2922	1.6641	1.9901	2.3739	2.6387
81	0.6775	1.2921	1.6639	1.9897	2.3733	2.6379
82	0.6775	1.2920	1.6636	1.9893	2.3727	2.6371
83	0.6775	1.2918	1.6634	1.9890	2.3721	2.6364
84	0.6774	1.2917	1.6632	1.9886	2.3716	2.6356
85	0.6774	1.2916	1.6630	1.9883	2.3710	2.6349
86	0.6774	1.2915	1.6628	1.9879	2.3705	2.6342
87	0.6773	1.2914	1.6626	1.9876	2.3700	2.6335
88	0.6773	1.2912	1.6624	1.9873	2.3695	2.6329
89	0.6773	1.2911	1.6622	1.9870	2.3690	2.6322
90	0.6772	1.2910	1.6620	1.9867	2.3685	2.6316
91	0.6772	1.2909	1.6618	1.9864	2.3680	2.6309
92	0.6772	1.2908	1.6616	1.9861	2.3676	2.6303
93	0.6771	1.2907	1.6614	1.9858	2.3671	2.6297
94	0.6771	1.2906	1.6612	1.9855	2.3667	2.6291
95	0.6771	1.2905	1.6611	1.9853	2.3662	2.6286
96	0.6771	1.2904	1.6609	1.9850	2.3658	2.6280
97	0.6770	1.2903	1.6607	1.9847	2.3654	2.6275
98	0.6770	1.2902	1.6606	1.9845	2.3650	2.6269
99	0.6770	1.2902	1.6604	1.9842	2.3646	2.6264
100	0.6770	1.2901	1.6602	1.9840	2.3642	2.6259
110	0.6767	1.2893	1.6588	1.9818	2.3607	2.6213
120	0.6765	1.2886	1.6577	1.9799	2.3578	2.6174
∞	0.6745	1.2816	1.6449	1.9600	2.3263	2.5758

For particular number of degrees of freedom, entry represents the critical value of t corresponding to a specified upper tail area (α).

▶ **Table A.3**

Critical Values of χ^2

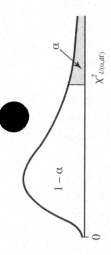

Degrees of Freedom	Upper Tail Areas (α)												
	.995	.99	.975	.95	.90	.75	.25	.10	.05	.025	.01	.005
1			0.001	0.004	0.016	0.102	1.323	2.706	3.841	5.024	6.635	7.879
2	0.010	0.020	0.051	0.103	0.211	0.575	2.773	4.605	5.991	7.378	9.210	10.597
3	0.072	0.115	0.216	0.352	0.584	1.213	4.108	6.251	7.815	9.348	11.345	12.838
4	0.207	0.297	0.484	0.711	1.064	1.923	5.385	7.779	9.488	11.143	13.277	14.860
5	0.412	0.554	0.831	1.145	1.610	2.675	6.626	9.236	11.071	12.833	15.086	16.750
6	0.676	0.872	1.237	1.635	2.204	3.455	7.841	10.645	12.592	14.449	16.812	18.548
7	0.989	1.239	1.690	2.167	2.833	4.255	9.037	12.017	14.067	16.013	18.475	20.278
8	1.344	1.646	2.180	2.733	3.490	5.071	10.219	13.362	15.507	17.535	20.090	21.955
9	1.735	2.088	2.700	3.325	4.168	5.899	11.389	14.684	16.919	19.023	21.666	23.589
10	2.156	2.558	3.247	3.940	4.865	6.737	12.549	15.987	18.307	20.483	23.209	25.188
11	2.603	3.053	3.816	4.575	5.578	7.584	13.701	17.275	19.675	21.920	24.725	26.757
12	3.074	3.571	4.404	5.226	6.304	8.438	14.845	18.549	21.026	23.337	26.217	28.299
13	3.565	4.107	5.009	5.892	7.042	9.299	15.984	19.812	22.362	24.736	27.688	29.819
14	4.075	4.660	5.629	6.571	7.790	10.165	17.117	21.064	23.685	26.119	29.141	31.319
15	4.601	5.229	6.262	7.261	8.547	11.037	18.245	22.307	24.996	27.488	30.578	32.801
16	5.142	5.812	6.908	7.962	9.312	11.912	19.369	23.542	26.296	28.845	32.000	34.267
17	5.697	6.408	7.564	8.672	10.085	12.792	20.489	24.769	27.587	30.191	33.409	35.718
18	6.265	7.015	8.231	9.390	10.865	13.675	21.605	25.989	28.869	31.526	34.805	37.156
19	6.844	7.633	8.907	10.117	11.651	14.562	22.718	27.204	30.144	32.852	36.191	38.582
20	7.434	8.260	9.591	10.851	12.443	15.452	23.828	28.412	31.410	34.170	37.566	39.997
21	8.034	8.897	10.283	11.591	13.240	16.344	24.935	29.615	32.671	35.479	38.932	41.401
22	8.643	9.542	10.982	12.338	14.042	17.240	26.039	30.813	33.924	36.781	40.289	42.796
23	9.260	10.196	11.689	13.091	14.848	18.137	27.141	32.007	35.172	38.076	41.638	44.181
24	9.886	10.856	12.401	13.848	15.659	19.037	28.241	33.196	36.415	39.364	42.980	45.559
25	10.520	11.524	13.120	14.611	16.473	19.939	29.339	34.382	37.652	40.646	44.314	46.928
26	11.160	12.198	13.844	15.379	17.292	20.843	30.435	35.563	38.885	41.923	45.642	48.290
27	11.808	12.879	14.573	16.151	18.114	21.749	31.528	36.741	40.113	43.194	46.963	49.645
28	12.461	13.565	15.308	16.928	18.939	22.657	32.620	37.916	41.337	44.461	48.278	50.993
29	13.121	14.257	16.047	17.708	19.768	23.567	33.711	39.087	42.557	45.722	49.588	52.336
30	13.787	14.954	16.791	18.493	20.599	24.478	34.800	40.256	43.773	46.979	50.892	53.672

For a particular number of degrees of freedom, entry represents the critical value of χ^2 corresponding to a specified upper tail area (α).

For larger values of degrees of freedom (df) the expression $Z = \sqrt{2\chi^2} - \sqrt{2(df)} - 1$ may be used, and the resulting upper tail area can be obtained from the table of the standard normal distribution (Table A.1).

► Table A.4

Critical values of the F distribution

**Upper critical values of the *F* distribution for numerator degrees
of freedom ν_1 and denominator degrees of freedom ν_2, 5% significance level**

ν_2 \ ν_1	1	2	3	4	5	6	7	8	9	10
1	161.448	199.500	215.707	224.583	230.162	233.986	236.768	238.882	240.543	241.882
2	18.513	19.000	19.164	19.247	19.296	19.330	19.353	19.371	19.385	19.396
3	10.128	9.552	9.277	9.117	9.013	8.941	8.887	8.845	8.812	8.786
4	7.709	6.944	6.591	6.388	6.256	6.163	6.094	6.041	5.999	5.964
5	6.608	5.786	5.409	5.192	5.050	4.950	4.876	4.818	4.772	4.735
6	5.987	5.143	4.757	4.534	4.387	4.284	4.207	4.147	4.099	4.060
7	5.591	4.737	4.347	4.120	3.972	3.866	3.787	3.726	3.677	3.637
8	5.318	4.459	4.066	3.838	3.687	3.581	3.500	3.438	3.388	3.347
9	5.117	4.256	3.863	3.633	3.482	3.374	3.293	3.230	3.179	3.137
10	4.965	4.103	3.708	3.478	3.326	3.217	3.135	3.072	3.020	2.978
11	4.844	3.982	3.587	3.357	3.204	3.095	3.012	2.948	2.896	2.854
12	4.747	3.885	3.490	3.259	3.106	2.996	2.913	2.849	2.796	2.753
13	4.667	3.806	3.411	3.179	3.025	2.915	2.832	2.767	2.714	2.671
14	4.600	3.739	3.344	3.112	2.958	2.848	2.764	2.699	2.646	2.602
15	4.543	3.682	3.287	3.056	2.901	2.790	2.707	2.641	2.588	2.544
16	4.494	3.634	3.239	3.007	2.852	2.741	2.657	2.591	2.538	2.494
17	4.451	3.592	3.197	2.965	2.810	2.699	2.614	2.548	2.494	2.450
18	4.414	3.555	3.160	2.928	2.773	2.661	2.577	2.510	2.456	2.412
19	4.381	3.522	3.127	2.895	2.740	2.628	2.544	2.477	2.423	2.378
20	4.351	3.493	3.098	2.866	2.711	2.599	2.514	2.447	2.393	2.348
21	4.325	3.467	3.072	2.840	2.685	2.573	2.488	2.420	2.366	2.321
22	4.301	3.443	3.049	2.817	2.661	2.549	2.464	2.397	2.342	2.297
23	4.279	3.422	3.028	2.796	2.640	2.528	2.442	2.375	2.320	2.275
24	4.260	3.403	3.009	2.776	2.621	2.508	2.423	2.355	2.300	2.255
25	4.242	3.385	2.991	2.759	2.603	2.490	2.405	2.337	2.282	2.236
26	4.225	3.369	2.975	2.743	2.587	2.474	2.388	2.321	2.265	2.220
27	4.210	3.354	2.960	2.728	2.572	2.459	2.373	2.305	2.250	2.204
28	4.196	3.340	2.947	2.714	2.558	2.445	2.359	2.291	2.236	2.190
29	4.183	3.328	2.934	2.701	2.545	2.432	2.346	2.278	2.223	2.177
30	4.171	3.316	2.922	2.690	2.534	2.421	2.334	2.266	2.211	2.165
31	4.160	3.305	2.911	2.679	2.523	2.409	2.323	2.255	2.199	2.153
32	4.149	3.295	2.901	2.668	2.512	2.399	2.313	2.244	2.189	2.142
33	4.139	3.285	2.892	2.659	2.503	2.389	2.303	2.235	2.179	2.133
34	4.130	3.276	2.883	2.650	2.494	2.380	2.294	2.225	2.170	2.123
35	4.121	3.267	2.874	2.641	2.485	2.372	2.285	2.217	2.161	2.114
36	4.113	3.259	2.866	2.634	2.477	2.364	2.277	2.209	2.153	2.106
37	4.105	3.252	2.859	2.626	2.470	2.356	2.270	2.201	2.145	2.098

ν_2 \ ν_1	1	2	3	4	5	6	7	8	9	10
38	4.098	3.245	2.852	2.619	2.463	2.349	2.262	2.194	2.138	2.091
39	4.091	3.238	2.845	2.612	2.456	2.342	2.255	2.187	2.131	2.084
40	4.085	3.232	2.839	2.606	2.449	2.336	2.249	2.180	2.124	2.077
41	4.079	3.226	2.833	2.600	2.443	2.330	2.243	2.174	2.118	2.071
42	4.073	3.220	2.827	2.594	2.438	2.324	2.237	2.168	2.112	2.065
43	4.067	3.214	2.822	2.589	2.432	2.318	2.232	2.163	2.106	2.059
44	4.062	3.209	2.816	2.584	2.427	2.313	2.226	2.157	2.101	2.054
45	4.057	3.204	2.812	2.579	2.422	2.308	2.221	2.152	2.096	2.049
46	4.052	3.200	2.807	2.574	2.417	2.304	2.216	2.147	2.091	2.044
47	4.047	3.195	2.802	2.570	2.413	2.299	2.212	2.143	2.086	2.039
48	4.043	3.191	2.798	2.565	2.409	2.295	2.207	2.138	2.082	2.035
49	4.038	3.187	2.794	2.561	2.404	2.290	2.203	2.134	2.077	2.030
50	4.034	3.183	2.790	2.557	2.400	2.286	2.199	2.130	2.073	2.026
51	4.030	3.179	2.786	2.553	2.397	2.283	2.195	2.126	2.069	2.022
52	4.027	3.175	2.783	2.550	2.393	2.279	2.192	2.122	2.066	2.018
53	4.023	3.172	2.779	2.546	2.389	2.275	2.188	2.119	2.062	2.015
54	4.020	3.168	2.776	2.543	2.386	2.272	2.185	2.115	2.059	2.011
55	4.016	3.165	2.773	2.540	2.383	2.269	2.181	2.112	2.055	2.008
56	4.013	3.162	2.769	2.537	2.380	2.266	2.178	2.109	2.052	2.005
57	4.010	3.159	2.766	2.534	2.377	2.263	2.175	2.106	2.049	2.001
58	4.007	3.156	2.764	2.531	2.374	2.260	2.172	2.103	2.046	1.998
59	4.004	3.153	2.761	2.528	2.371	2.257	2.169	2.100	2.043	1.995
60	4.001	3.150	2.758	2.525	2.368	2.254	2.167	2.097	2.040	1.993
61	3.998	3.148	2.755	2.523	2.366	2.251	2.164	2.094	2.037	1.990
62	3.996	3.145	2.753	2.520	2.363	2.249	2.161	2.092	2.035	1.987
63	3.993	3.143	2.751	2.518	2.361	2.246	2.159	2.089	2.032	1.985
64	3.991	3.140	2.748	2.515	2.358	2.244	2.156	2.087	2.030	1.982
65	3.989	3.138	2.746	2.513	2.356	2.242	2.154	2.084	2.027	1.980
66	3.986	3.136	2.744	2.511	2.354	2.239	2.152	2.082	2.025	1.977
67	3.984	3.134	2.742	2.509	2.352	2.237	2.150	2.080	2.023	1.975
68	3.982	3.132	2.740	2.507	2.350	2.235	2.148	2.078	2.021	1.973
69	3.980	3.130	2.737	2.505	2.348	2.233	2.145	2.076	2.019	1.971
70	3.978	3.128	2.736	2.503	2.346	2.231	2.143	2.074	2.017	1.969
71	3.976	3.126	2.734	2.501	2.344	2.229	2.142	2.072	2.015	1.967
72	3.974	3.124	2.732	2.499	2.342	2.227	2.140	2.070	2.013	1.965
73	3.972	3.122	2.730	2.497	2.340	2.226	2.138	2.068	2.011	1.963
74	3.970	3.120	2.728	2.495	2.338	2.224	2.136	2.066	2.009	1.961
75	3.968	3.119	2.727	2.494	2.337	2.222	2.134	2.064	2.007	1.959
76	3.967	3.117	2.725	2.492	2.335	2.220	2.133	2.063	2.006	1.958
77	3.965	3.115	2.723	2.490	2.333	2.219	2.131	2.061	2.004	1.956

(continued)

ν_2 \ ν_1	1	2	3	4	5	6	7	8	9	10
78	3.963	3.114	2.722	2.489	2.332	2.217	2.129	2.059	2.002	1.954
79	3.962	3.112	2.720	2.487	2.330	2.216	2.128	2.058	2.001	1.953
80	3.960	3.111	2.719	2.486	2.329	2.214	2.126	2.056	1.999	1.951
81	3.959	3.109	2.717	2.484	2.327	2.213	2.125	2.055	1.998	1.950
82	3.957	3.108	2.716	2.483	2.326	2.211	2.123	2.053	1.996	1.948
83	3.956	3.107	2.715	2.482	2.324	2.210	2.122	2.052	1.995	1.947
84	3.955	3.105	2.713	2.480	2.323	2.209	2.121	2.051	1.993	1.945
85	3.953	3.104	2.712	2.479	2.322	2.207	2.119	2.049	1.992	1.944
86	3.952	3.103	2.711	2.478	2.321	2.206	2.118	2.048	1.991	1.943
87	3.951	3.101	2.709	2.476	2.319	2.205	2.117	2.047	1.989	1.941
88	3.949	3.100	2.708	2.475	2.318	2.203	2.115	2.045	1.988	1.940
89	3.948	3.099	2.707	2.474	2.317	2.202	2.114	2.044	1.987	1.939
90	3.947	3.098	2.706	2.473	2.316	2.201	2.113	2.043	1.986	1.938
91	3.946	3.097	2.705	2.472	2.315	2.200	2.112	2.042	1.984	1.936
92	3.945	3.095	2.704	2.471	2.313	2.199	2.111	2.041	1.983	1.935
93	3.943	3.094	2.703	2.470	2.312	2.198	2.110	2.040	1.982	1.934
94	3.942	3.093	2.701	2.469	2.311	2.197	2.109	2.038	1.981	1.933
95	3.941	3.092	2.700	2.467	2.310	2.196	2.108	2.037	1.980	1.932
96	3.940	3.091	2.699	2.466	2.309	2.195	2.106	2.036	1.979	1.931
97	3.939	3.090	2.698	2.465	2.308	2.194	2.105	2.035	1.978	1.930
98	3.938	3.089	2.697	2.465	2.307	2.193	2.104	2.034	1.977	1.929
99	3.937	3.088	2.696	2.464	2.306	2.192	2.103	2.033	1.976	1.928
100	3.936	3.087	2.696	2.463	2.305	2.191	2.103	2.032	1.975	1.927

ν_2 \ ν_1	11	12	13	14	15	16	17	18	19	20
1	242.983	243.906	244.690	245.364	245.950	246.464	246.918	247.323	247.686	248.013
2	19.405	19.413	19.419	19.424	19.429	19.433	19.437	19.440	19.443	19.446
3	8.763	8.745	8.729	8.715	8.703	8.692	8.683	8.675	8.667	8.660
4	5.936	5.912	5.891	5.873	5.858	5.844	5.832	5.821	5.811	5.803
5	4.704	4.678	4.655	4.636	4.619	4.604	4.590	4.579	4.568	4.558
6	4.027	4.000	3.976	3.956	3.938	3.922	3.908	3.896	3.884	3.874
7	3.603	3.575	3.550	3.529	3.511	3.494	3.480	3.467	3.455	3.445
8	3.313	3.284	3.259	3.237	3.218	3.202	3.187	3.173	3.161	3.150
9	3.102	3.073	3.048	3.025	3.006	2.989	2.974	2.960	2.948	2.936
10	2.943	2.913	2.887	2.865	2.845	2.828	2.812	2.798	2.785	2.774
11	2.818	2.788	2.761	2.739	2.719	2.701	2.685	2.671	2.658	2.646
12	2.717	2.687	2.660	2.637	2.617	2.599	2.583	2.568	2.555	2.544
13	2.635	2.604	2.577	2.554	2.533	2.515	2.499	2.484	2.471	2.459
14	2.565	2.534	2.507	2.484	2.463	2.445	2.428	2.413	2.400	2.388
15	2.507	2.475	2.448	2.424	2.403	2.385	2.368	2.353	2.340	2.328

ν_2 \ ν_1	11	12	13	14	15	16	17	18	19	20
16	2.456	2.425	2.397	2.373	2.352	2.333	2.317	2.302	2.288	2.276
17	2.413	2.381	2.353	2.329	2.308	2.289	2.272	2.257	2.243	2.230
18	2.374	2.342	2.314	2.290	2.269	2.250	2.233	2.217	2.203	2.191
19	2.340	2.308	2.280	2.256	2.234	2.215	2.198	2.182	2.168	2.155
20	2.310	2.278	2.250	2.225	2.203	2.184	2.167	2.151	2.137	2.124
21	2.283	2.250	2.222	2.197	2.176	2.156	2.139	2.123	2.109	2.096
22	2.259	2.226	2.198	2.173	2.151	2.131	2.114	2.098	2.084	2.071
23	2.236	2.204	2.175	2.150	2.128	2.109	2.091	2.075	2.061	2.048
24	2.216	2.183	2.155	2.130	2.108	2.088	2.070	2.054	2.040	2.027
25	2.198	2.165	2.136	2.111	2.089	2.069	2.051	2.035	2.021	2.007
26	2.181	2.148	2.119	2.094	2.072	2.052	2.034	2.018	2.003	1.990
27	2.166	2.132	2.103	2.078	2.056	2.036	2.018	2.002	1.987	1.974
28	2.151	2.118	2.089	2.064	2.041	2.021	2.003	1.987	1.972	1.959
29	2.138	2.104	2.075	2.050	2.027	2.007	1.989	1.973	1.958	1.945
30	2.126	2.092	2.063	2.037	2.015	1.995	1.976	1.960	1.945	1.932
31	2.114	2.080	2.051	2.026	2.003	1.983	1.965	1.948	1.933	1.920
32	2.103	2.070	2.040	2.015	1.992	1.972	1.953	1.937	1.922	1.908
33	2.093	2.060	2.030	2.004	1.982	1.961	1.943	1.926	1.911	1.898
34	2.084	2.050	2.021	1.995	1.972	1.952	1.933	1.917	1.902	1.888
35	2.075	2.041	2.012	1.986	1.963	1.942	1.924	1.907	1.892	1.878
36	2.067	2.033	2.003	1.977	1.954	1.934	1.915	1.899	1.883	1.870
37	2.059	2.025	1.995	1.969	1.946	1.926	1.907	1.890	1.875	1.861
38	2.051	2.017	1.988	1.962	1.939	1.918	1.899	1.883	1.867	1.853
39	2.044	2.010	1.981	1.954	1.931	1.911	1.892	1.875	1.860	1.846
40	2.038	2.003	1.974	1.948	1.924	1.904	1.885	1.868	1.853	1.839
41	2.031	1.997	1.967	1.941	1.918	1.897	1.879	1.862	1.846	1.832
42	2.025	1.991	1.961	1.935	1.912	1.891	1.872	1.855	1.840	1.826
43	2.020	1.985	1.955	1.929	1.906	1.885	1.866	1.849	1.834	1.820
44	2.014	1.980	1.950	1.924	1.900	1.879	1.861	1.844	1.828	1.814
45	2.009	1.974	1.945	1.918	1.895	1.874	1.855	1.838	1.823	1.808
46	2.004	1.969	1.940	1.913	1.890	1.869	1.850	1.833	1.817	1.803
47	1.999	1.965	1.935	1.908	1.885	1.864	1.845	1.828	1.812	1.798
48	1.995	1.960	1.930	1.904	1.880	1.859	1.840	1.823	1.807	1.793
49	1.990	1.956	1.926	1.899	1.876	1.855	1.836	1.819	1.803	1.789
50	1.986	1.952	1.921	1.895	1.871	1.850	1.831	1.814	1.798	1.784
51	1.982	1.947	1.917	1.891	1.867	1.846	1.827	1.810	1.794	1.780
52	1.978	1.944	1.913	1.887	1.863	1.842	1.823	1.806	1.790	1.776
53	1.975	1.940	1.910	1.883	1.859	1.838	1.819	1.802	1.786	1.772
54	1.971	1.936	1.906	1.879	1.856	1.835	1.816	1.798	1.782	1.768
55	1.968	1.933	1.903	1.876	1.852	1.831	1.812	1.795	1.779	1.764

(continued)

ν_2 \ ν_1	11	12	13	14	15	16	17	18	19	20
56	1.964	1.930	1.899	1.873	1.849	1.828	1.809	1.791	1.775	1.761
57	1.961	1.926	1.896	1.869	1.846	1.824	1.805	1.788	1.772	1.757
58	1.958	1.923	1.893	1.866	1.842	1.821	1.802	1.785	1.769	1.754
59	1.955	1.920	1.890	1.863	1.839	1.818	1.799	1.781	1.766	1.751
60	1.952	1.917	1.887	1.860	1.836	1.815	1.796	1.778	1.763	1.748
61	1.949	1.915	1.884	1.857	1.834	1.812	1.793	1.776	1.760	1.745
62	1.947	1.912	1.882	1.855	1.831	1.809	1.790	1.773	1.757	1.742
63	1.944	1.909	1.879	1.852	1.828	1.807	1.787	1.770	1.754	1.739
64	1.942	1.907	1.876	1.849	1.826	1.804	1.785	1.767	1.751	1.737
65	1.939	1.904	1.874	1.847	1.823	1.802	1.782	1.765	1.749	1.734
66	1.937	1.902	1.871	1.845	1.821	1.799	1.780	1.762	1.746	1.732
67	1.935	1.900	1.869	1.842	1.818	1.797	1.777	1.760	1.744	1.729
68	1.932	1.897	1.867	1.840	1.816	1.795	1.775	1.758	1.742	1.727
69	1.930	1.895	1.865	1.838	1.814	1.792	1.773	1.755	1.739	1.725
70	1.928	1.893	1.863	1.836	1.812	1.790	1.771	1.753	1.737	1.722
71	1.926	1.891	1.861	1.834	1.810	1.788	1.769	1.751	1.735	1.720
72	1.924	1.889	1.859	1.832	1.808	1.786	1.767	1.749	1.733	1.718
73	1.922	1.887	1.857	1.830	1.806	1.784	1.765	1.747	1.731	1.716
74	1.921	1.885	1.855	1.828	1.804	1.782	1.763	1.745	1.729	1.714
75	1.919	1.884	1.853	1.826	1.802	1.780	1.761	1.743	1.727	1.712
76	1.917	1.882	1.851	1.824	1.800	1.778	1.759	1.741	1.725	1.710
77	1.915	1.880	1.849	1.822	1.798	1.777	1.757	1.739	1.723	1.708
78	1.914	1.878	1.848	1.821	1.797	1.775	1.755	1.738	1.721	1.707
79	1.912	1.877	1.846	1.819	1.795	1.773	1.754	1.736	1.720	1.705
80	1.910	1.875	1.845	1.817	1.793	1.772	1.752	1.734	1.718	1.703
81	1.909	1.874	1.843	1.816	1.792	1.770	1.750	1.733	1.716	1.702
82	1.907	1.872	1.841	1.814	1.790	1.768	1.749	1.731	1.715	1.700
83	1.906	1.871	1.840	1.813	1.789	1.767	1.747	1.729	1.713	1.698
84	1.905	1.869	1.838	1.811	1.787	1.765	1.746	1.728	1.712	1.697
85	1.903	1.868	1.837	1.810	1.786	1.764	1.744	1.726	1.710	1.695
86	1.902	1.867	1.836	1.808	1.784	1.762	1.743	1.725	1.709	1.694
87	1.900	1.865	1.834	1.807	1.783	1.761	1.741	1.724	1.707	1.692
88	1.899	1.864	1.833	1.806	1.782	1.760	1.740	1.722	1.706	1.691
89	1.898	1.863	1.832	1.804	1.780	1.758	1.739	1.721	1.705	1.690
90	1.897	1.861	1.830	1.803	1.779	1.757	1.737	1.720	1.703	1.688
91	1.895	1.860	1.829	1.802	1.778	1.756	1.736	1.718	1.702	1.687
92	1.894	1.859	1.828	1.801	1.776	1.755	1.735	1.717	1.701	1.686
93	1.893	1.858	1.827	1.800	1.775	1.753	1.734	1.716	1.699	1.684
94	1.892	1.857	1.826	1.798	1.774	1.752	1.733	1.715	1.698	1.683
95	1.891	1.856	1.825	1.797	1.773	1.751	1.731	1.713	1.697	1.682

$\nu_2 \backslash \nu_1$	11	12	13	14	15	16	17	18	19	20
96	1.890	1.854	1.823	1.796	1.772	1.750	1.730	1.712	1.696	1.681
97	1.889	1.853	1.822	1.795	1.771	1.749	1.729	1.711	1.695	1.680
98	1.888	1.852	1.821	1.794	1.770	1.748	1.728	1.710	1.694	1.679
99	1.887	1.851	1.820	1.793	1.769	1.747	1.727	1.709	1.693	1.678
100	1.886	1.850	1.819	1.792	1.768	1.746	1.726	1.708	1.691	1.676

Upper critical values of the *F* distribution for numerator degrees of freedom ν_1 and denominator degrees of freedom ν_2, 10% significance level

$\nu_2 \backslash \nu_1$	1	2	3	4	5	6	7	8	9	10
1	39.863	49.500	53.593	55.833	57.240	58.204	58.906	59.439	59.858	60.195
2	8.526	9.000	9.162	9.243	9.293	9.326	9.349	9.367	9.381	9.392
3	5.538	5.462	5.391	5.343	5.309	5.285	5.266	5.252	5.240	5.230
4	4.545	4.325	4.191	4.107	4.051	4.010	3.979	3.955	3.936	3.920
5	4.060	3.780	3.619	3.520	3.453	3.405	3.368	3.339	3.316	3.297
6	3.776	3.463	3.289	3.181	3.108	3.055	3.014	2.983	2.958	2.937
7	3.589	3.257	3.074	2.961	2.883	2.827	2.785	2.752	2.725	2.703
8	3.458	3.113	2.924	2.806	2.726	2.668	2.624	2.589	2.561	2.538
9	3.360	3.006	2.813	2.693	2.611	2.551	2.505	2.469	2.440	2.416
10	3.285	2.924	2.728	2.605	2.522	2.461	2.414	2.377	2.347	2.323
11	3.225	2.860	2.660	2.536	2.451	2.389	2.342	2.304	2.274	2.248
12	3.177	2.807	2.606	2.480	2.394	2.331	2.283	2.245	2.214	2.188
13	3.136	2.763	2.560	2.434	2.347	2.283	2.234	2.195	2.164	2.138
14	3.102	2.726	2.522	2.395	2.307	2.243	2.193	2.154	2.122	2.095
15	3.073	2.695	2.490	2.361	2.273	2.208	2.158	2.119	2.086	2.059
16	3.048	2.668	2.462	2.333	2.244	2.178	2.128	2.088	2.055	2.028
17	3.026	2.645	2.437	2.308	2.218	2.152	2.102	2.061	2.028	2.001
18	3.007	2.624	2.416	2.286	2.196	2.130	2.079	2.038	2.005	1.977
19	2.990	2.606	2.397	2.266	2.176	2.109	2.058	2.017	1.984	1.956
20	2.975	2.589	2.380	2.249	2.158	2.091	2.040	1.999	1.965	1.937
21	2.961	2.575	2.365	2.233	2.142	2.075	2.023	1.982	1.948	1.920
22	2.949	2.561	2.351	2.219	2.128	2.060	2.008	1.967	1.933	1.904
23	2.937	2.549	2.339	2.207	2.115	2.047	1.995	1.953	1.919	1.890
24	2.927	2.538	2.327	2.195	2.103	2.035	1.983	1.941	1.906	1.877
25	2.918	2.528	2.317	2.184	2.092	2.024	1.971	1.929	1.895	1.866
26	2.909	2.519	2.307	2.174	2.082	2.014	1.961	1.919	1.884	1.855
27	2.901	2.511	2.299	2.165	2.073	2.005	1.952	1.909	1.874	1.845
28	2.894	2.503	2.291	2.157	2.064	1.996	1.943	1.900	1.865	1.836
29	2.887	2.495	2.283	2.149	2.057	1.988	1.935	1.892	1.857	1.827
30	2.881	2.489	2.276	2.142	2.049	1.980	1.927	1.884	1.849	1.819

(continued)

ν_2 \ ν_1	1	2	3	4	5	6	7	8	9	10
31	2.875	2.482	2.270	2.136	2.042	1.973	1.920	1.877	1.842	1.812
32	2.869	2.477	2.263	2.129	2.036	1.967	1.913	1.870	1.835	1.805
33	2.864	2.471	2.258	2.123	2.030	1.961	1.907	1.864	1.828	1.799
34	2.859	2.466	2.252	2.118	2.024	1.955	1.901	1.858	1.822	1.793
35	2.855	2.461	2.247	2.113	2.019	1.950	1.896	1.852	1.817	1.787
36	2.850	2.456	2.243	2.108	2.014	1.945	1.891	1.847	1.811	1.781
37	2.846	2.452	2.238	2.103	2.009	1.940	1.886	1.842	1.806	1.776
38	2.842	2.448	2.234	2.099	2.005	1.935	1.881	1.838	1.802	1.772
39	2.839	2.444	2.230	2.095	2.001	1.931	1.877	1.833	1.797	1.767
40	2.835	2.440	2.226	2.091	1.997	1.927	1.873	1.829	1.793	1.763
41	2.832	2.437	2.222	2.087	1.993	1.923	1.869	1.825	1.789	1.759
42	2.829	2.434	2.219	2.084	1.989	1.919	1.865	1.821	1.785	1.755
43	2.826	2.430	2.216	2.080	1.986	1.916	1.861	1.817	1.781	1.751
44	2.823	2.427	2.213	2.077	1.983	1.913	1.858	1.814	1.778	1.747
45	2.820	2.425	2.210	2.074	1.980	1.909	1.855	1.811	1.774	1.744
46	2.818	2.422	2.207	2.071	1.977	1.906	1.852	1.808	1.771	1.741
47	2.815	2.419	2.204	2.068	1.974	1.903	1.849	1.805	1.768	1.738
48	2.813	2.417	2.202	2.066	1.971	1.901	1.846	1.802	1.765	1.735
49	2.811	2.414	2.199	2.063	1.968	1.898	1.843	1.799	1.763	1.732
50	2.809	2.412	2.197	2.061	1.966	1.895	1.840	1.796	1.760	1.729
51	2.807	2.410	2.194	2.058	1.964	1.893	1.838	1.794	1.757	1.727
52	2.805	2.408	2.192	2.056	1.961	1.891	1.836	1.791	1.755	1.724
53	2.803	2.406	2.190	2.054	1.959	1.888	1.833	1.789	1.752	1.722
54	2.801	2.404	2.188	2.052	1.957	1.886	1.831	1.787	1.750	1.719
55	2.799	2.402	2.186	2.050	1.955	1.884	1.829	1.785	1.748	1.717
56	2.797	2.400	2.184	2.048	1.953	1.882	1.827	1.782	1.746	1.715
57	2.796	2.398	2.182	2.046	1.951	1.880	1.825	1.780	1.744	1.713
58	2.794	2.396	2.181	2.044	1.949	1.878	1.823	1.779	1.742	1.711
59	2.793	2.395	2.179	2.043	1.947	1.876	1.821	1.777	1.740	1.709
60	2.791	2.393	2.177	2.041	1.946	1.875	1.819	1.775	1.738	1.707
61	2.790	2.392	2.176	2.039	1.944	1.873	1.818	1.773	1.736	1.705
62	2.788	2.390	2.174	2.038	1.942	1.871	1.816	1.771	1.735	1.703
63	2.787	2.389	2.173	2.036	1.941	1.870	1.814	1.770	1.733	1.702
64	2.786	2.387	2.171	2.035	1.939	1.868	1.813	1.768	1.731	1.700
65	2.784	2.386	2.170	2.033	1.938	1.867	1.811	1.767	1.730	1.699
66	2.783	2.385	2.169	2.032	1.937	1.865	1.810	1.765	1.728	1.697
67	2.782	2.384	2.167	2.031	1.935	1.864	1.808	1.764	1.727	1.696
68	2.781	2.382	2.166	2.029	1.934	1.863	1.807	1.762	1.725	1.694
69	2.780	2.381	2.165	2.028	1.933	1.861	1.806	1.761	1.724	1.693
70	2.779	2.380	2.164	2.027	1.931	1.860	1.804	1.760	1.723	1.691
71	2.778	2.379	2.163	2.026	1.930	1.859	1.803	1.758	1.721	1.690

ν_2 \ ν_1	1	2	3	4	5	6	7	8	9	10
72	2.777	2.378	2.161	2.025	1.929	1.858	1.802	1.757	1.720	1.689
73	2.776	2.377	2.160	2.024	1.928	1.856	1.801	1.756	1.719	1.687
74	2.775	2.376	2.159	2.022	1.927	1.855	1.800	1.755	1.718	1.686
75	2.774	2.375	2.158	2.021	1.926	1.854	1.798	1.754	1.716	1.685
76	2.773	2.374	2.157	2.020	1.925	1.853	1.797	1.752	1.715	1.684
77	2.772	2.373	2.156	2.019	1.924	1.852	1.796	1.751	1.714	1.683
78	2.771	2.372	2.155	2.018	1.923	1.851	1.795	1.750	1.713	1.682
79	2.770	2.371	2.154	2.017	1.922	1.850	1.794	1.749	1.712	1.681
80	2.769	2.370	2.154	2.016	1.921	1.849	1.793	1.748	1.711	1.680
81	2.769	2.369	2.153	2.016	1.920	1.848	1.792	1.747	1.710	1.679
82	2.768	2.368	2.152	2.015	1.919	1.847	1.791	1.746	1.709	1.678
83	2.767	2.368	2.151	2.014	1.918	1.846	1.790	1.745	1.708	1.677
84	2.766	2.367	2.150	2.013	1.917	1.845	1.790	1.744	1.707	1.676
85	2.765	2.366	2.149	2.012	1.916	1.845	1.789	1.744	1.706	1.675
86	2.765	2.365	2.149	2.011	1.915	1.844	1.788	1.743	1.705	1.674
87	2.764	2.365	2.148	2.011	1.915	1.843	1.787	1.742	1.705	1.673
88	2.763	2.364	2.147	2.010	1.914	1.842	1.786	1.741	1.704	1.672
89	2.763	2.363	2.146	2.009	1.913	1.841	1.785	1.740	1.703	1.671
90	2.762	2.363	2.146	2.008	1.912	1.841	1.785	1.739	1.702	1.670
91	2.761	2.362	2.145	2.008	1.912	1.840	1.784	1.739	1.701	1.670
92	2.761	2.361	2.144	2.007	1.911	1.839	1.783	1.738	1.701	1.669
93	2.760	2.361	2.144	2.006	1.910	1.838	1.782	1.737	1.700	1.668
94	2.760	2.360	2.143	2.006	1.910	1.838	1.782	1.736	1.699	1.667
95	2.759	2.359	2.142	2.005	1.909	1.837	1.781	1.736	1.698	1.667
96	2.759	2.359	2.142	2.004	1.908	1.836	1.780	1.735	1.698	1.666
97	2.758	2.358	2.141	2.004	1.908	1.836	1.780	1.734	1.697	1.665
98	2.757	2.358	2.141	2.003	1.907	1.835	1.779	1.734	1.696	1.665
99	2.757	2.357	2.140	2.003	1.906	1.835	1.778	1.733	1.696	1.664
100	2.756	2.356	2.139	2.002	1.906	1.834	1.778	1.732	1.695	1.663

ν_2 \ ν_1	11	12	13	14	15	16	17	18	19	20
1	60.473	60.705	60.903	61.073	61.220	61.350	61.464	61.566	61.658	61.740
2	9.401	9.408	9.415	9.420	9.425	9.429	9.433	9.436	9.439	9.441
3	5.222	5.216	5.210	5.205	5.200	5.196	5.193	5.190	5.187	5.184
4	3.907	3.896	3.886	3.878	3.870	3.864	3.858	3.853	3.849	3.844
5	3.282	3.268	3.257	3.247	3.238	3.230	3.223	3.217	3.212	3.207
6	2.920	2.905	2.892	2.881	2.871	2.863	2.855	2.848	2.842	2.836
7	2.684	2.668	2.654	2.643	2.632	2.623	2.615	2.607	2.601	2.595
8	2.519	2.502	2.488	2.475	2.464	2.455	2.446	2.438	2.431	2.425

(continued)

$\nu_2 \backslash \nu_1$	11	12	13	14	15	16	17	18	19	20
9	2.396	2.379	2.364	2.351	2.340	2.329	2.320	2.312	2.305	2.298
10	2.302	2.284	2.269	2.255	2.244	2.233	2.224	2.215	2.208	2.201
11	2.227	2.209	2.193	2.179	2.167	2.156	2.147	2.138	2.130	2.123
12	2.166	2.147	2.131	2.117	2.105	2.094	2.084	2.075	2.067	2.060
13	2.116	2.097	2.080	2.066	2.053	2.042	2.032	2.023	2.014	2.007
14	2.073	2.054	2.037	2.022	2.010	1.998	1.988	1.978	1.970	1.962
15	2.037	2.017	2.000	1.985	1.972	1.961	1.950	1.941	1.932	1.924
16	2.005	1.985	1.968	1.953	1.940	1.928	1.917	1.908	1.899	1.891
17	1.978	1.958	1.940	1.925	1.912	1.900	1.889	1.879	1.870	1.862
18	1.954	1.933	1.916	1.900	1.887	1.875	1.864	1.854	1.845	1.837
19	1.932	1.912	1.894	1.878	1.865	1.852	1.841	1.831	1.822	1.814
20	1.913	1.892	1.875	1.859	1.845	1.833	1.821	1.811	1.802	1.794
21	1.896	1.875	1.857	1.841	1.827	1.815	1.803	1.793	1.784	1.776
22	1.880	1.859	1.841	1.825	1.811	1.798	1.787	1.777	1.768	1.759
23	1.866	1.845	1.827	1.811	1.796	1.784	1.772	1.762	1.753	1.744
24	1.853	1.832	1.814	1.797	1.783	1.770	1.759	1.748	1.739	1.730
25	1.841	1.820	1.802	1.785	1.771	1.758	1.746	1.736	1.726	1.718
26	1.830	1.809	1.790	1.774	1.760	1.747	1.735	1.724	1.715	1.706
27	1.820	1.799	1.780	1.764	1.749	1.736	1.724	1.714	1.704	1.695
28	1.811	1.790	1.771	1.754	1.740	1.726	1.715	1.704	1.694	1.685
29	1.802	1.781	1.762	1.745	1.731	1.717	1.705	1.695	1.685	1.676
30	1.794	1.773	1.754	1.737	1.722	1.709	1.697	1.686	1.676	1.667
31	1.787	1.765	1.746	1.729	1.714	1.701	1.689	1.678	1.668	1.659
32	1.780	1.758	1.739	1.722	1.707	1.694	1.682	1.671	1.661	1.652
33	1.773	1.751	1.732	1.715	1.700	1.687	1.675	1.664	1.654	1.645
34	1.767	1.745	1.726	1.709	1.694	1.680	1.668	1.657	1.647	1.638
35	1.761	1.739	1.720	1.703	1.688	1.674	1.662	1.651	1.641	1.632
36	1.756	1.734	1.715	1.697	1.682	1.669	1.656	1.645	1.635	1.626
37	1.751	1.729	1.709	1.692	1.677	1.663	1.651	1.640	1.630	1.620
38	1.746	1.724	1.704	1.687	1.672	1.658	1.646	1.635	1.624	1.615
39	1.741	1.719	1.700	1.682	1.667	1.653	1.641	1.630	1.619	1.610
40	1.737	1.715	1.695	1.678	1.662	1.649	1.636	1.625	1.615	1.605
41	1.733	1.710	1.691	1.673	1.658	1.644	1.632	1.620	1.610	1.601
42	1.729	1.706	1.687	1.669	1.654	1.640	1.628	1.616	1.606	1.596
43	1.725	1.703	1.683	1.665	1.650	1.636	1.624	1.612	1.602	1.592
44	1.721	1.699	1.679	1.662	1.646	1.632	1.620	1.608	1.598	1.588
45	1.718	1.695	1.676	1.658	1.643	1.629	1.616	1.605	1.594	1.585
46	1.715	1.692	1.672	1.655	1.639	1.625	1.613	1.601	1.591	1.581
47	1.712	1.689	1.669	1.652	1.636	1.622	1.609	1.598	1.587	1.578
48	1.709	1.686	1.666	1.648	1.633	1.619	1.606	1.594	1.584	1.574
49	1.706	1.683	1.663	1.645	1.630	1.616	1.603	1.591	1.581	1.571

ν_2 \ ν_1	11	12	13	14	15	16	17	18	19	20
50	1.703	1.680	1.660	1.643	1.627	1.613	1.600	1.588	1.578	1.568
51	1.700	1.677	1.658	1.640	1.624	1.610	1.597	1.586	1.575	1.565
52	1.698	1.675	1.655	1.637	1.621	1.607	1.594	1.583	1.572	1.562
53	1.695	1.672	1.652	1.635	1.619	1.605	1.592	1.580	1.570	1.560
54	1.693	1.670	1.650	1.632	1.616	1.602	1.589	1.578	1.567	1.557
55	1.691	1.668	1.648	1.630	1.614	1.600	1.587	1.575	1.564	1.555
56	1.688	1.666	1.645	1.628	1.612	1.597	1.585	1.573	1.562	1.552
57	1.686	1.663	1.643	1.625	1.610	1.595	1.582	1.571	1.560	1.550
58	1.684	1.661	1.641	1.623	1.607	1.593	1.580	1.568	1.558	1.548
59	1.682	1.659	1.639	1.621	1.605	1.591	1.578	1.566	1.555	1.546
60	1.680	1.657	1.637	1.619	1.603	1.589	1.576	1.564	1.553	1.543
61	1.679	1.656	1.635	1.617	1.601	1.587	1.574	1.562	1.551	1.541
62	1.677	1.654	1.634	1.616	1.600	1.585	1.572	1.560	1.549	1.540
63	1.675	1.652	1.632	1.614	1.598	1.583	1.570	1.558	1.548	1.538
64	1.673	1.650	1.630	1.612	1.596	1.582	1.569	1.557	1.546	1.536
65	1.672	1.649	1.628	1.610	1.594	1.580	1.567	1.555	1.544	1.534
66	1.670	1.647	1.627	1.609	1.593	1.578	1.565	1.553	1.542	1.532
67	1.669	1.646	1.625	1.607	1.591	1.577	1.564	1.552	1.541	1.531
68	1.667	1.644	1.624	1.606	1.590	1.575	1.562	1.550	1.539	1.529
69	1.666	1.643	1.622	1.604	1.588	1.574	1.560	1.548	1.538	1.527
70	1.665	1.641	1.621	1.603	1.587	1.572	1.559	1.547	1.536	1.526
71	1.663	1.640	1.619	1.601	1.585	1.571	1.557	1.545	1.535	1.524
72	1.662	1.639	1.618	1.600	1.584	1.569	1.556	1.544	1.533	1.523
73	1.661	1.637	1.617	1.599	1.583	1.568	1.555	1.543	1.532	1.522
74	1.659	1.636	1.616	1.597	1.581	1.567	1.553	1.541	1.530	1.520
75	1.658	1.635	1.614	1.596	1.580	1.565	1.552	1.540	1.529	1.519
76	1.657	1.634	1.613	1.595	1.579	1.564	1.551	1.539	1.528	1.518
77	1.656	1.632	1.612	1.594	1.578	1.563	1.550	1.538	1.527	1.516
78	1.655	1.631	1.611	1.593	1.576	1.562	1.548	1.536	1.525	1.515
79	1.654	1.630	1.610	1.592	1.575	1.561	1.547	1.535	1.524	1.514
80	1.653	1.629	1.609	1.590	1.574	1.559	1.546	1.534	1.523	1.513
81	1.652	1.628	1.608	1.589	1.573	1.558	1.545	1.533	1.522	1.512
82	1.651	1.627	1.607	1.588	1.572	1.557	1.544	1.532	1.521	1.511
83	1.650	1.626	1.606	1.587	1.571	1.556	1.543	1.531	1.520	1.509
84	1.649	1.625	1.605	1.586	1.570	1.555	1.542	1.530	1.519	1.508
85	1.648	1.624	1.604	1.585	1.569	1.554	1.541	1.529	1.518	1.507
86	1.647	1.623	1.603	1.584	1.568	1.553	1.540	1.528	1.517	1.506
87	1.646	1.622	1.602	1.583	1.567	1.552	1.539	1.527	1.516	1.505
88	1.645	1.622	1.601	1.583	1.566	1.551	1.538	1.526	1.515	1.504
89	1.644	1.621	1.600	1.582	1.565	1.550	1.537	1.525	1.514	1.503

(continued)

ν_2 \ ν_1	11	12	13	14	15	16	17	18	19	20
90	1.643	1.620	1.599	1.581	1.564	1.550	1.536	1.524	1.513	1.503
91	1.643	1.619	1.598	1.580	1.564	1.549	1.535	1.523	1.512	1.502
92	1.642	1.618	1.598	1.579	1.563	1.548	1.534	1.522	1.511	1.501
93	1.641	1.617	1.597	1.578	1.562	1.547	1.534	1.521	1.510	1.500
94	1.640	1.617	1.596	1.578	1.561	1.546	1.533	1.521	1.509	1.499
95	1.640	1.616	1.595	1.577	1.560	1.545	1.532	1.520	1.509	1.498
96	1.639	1.615	1.594	1.576	1.560	1.545	1.531	1.519	1.508	1.497
97	1.638	1.614	1.594	1.575	1.559	1.544	1.530	1.518	1.507	1.497
98	1.637	1.614	1.593	1.575	1.558	1.543	1.530	1.517	1.506	1.496
99	1.637	1.613	1.592	1.574	1.557	1.542	1.529	1.517	1.505	1.495
100	1.636	1.612	1.592	1.573	1.557	1.542	1.528	1.516	1.505	1.494

Upper critical values of the F distribution for numerator degrees of freedom ν_1 and denominator degrees of freedom ν_2, 1% significance level

ν_2 \ ν_1	1	2	3	4	5	6	7	8	9	10
1	4052.19	4999.52	5403.34	5624.62	5763.65	5858.97	5928.33	5981.10	6022.50	6055.85
2	98.502	99.000	99.166	99.249	99.300	99.333	99.356	99.374	99.388	99.399
3	34.116	30.816	29.457	28.710	28.237	27.911	27.672	27.489	27.345	27.229
4	21.198	18.000	16.694	15.977	15.522	15.207	14.976	14.799	14.659	14.546
5	16.258	13.274	12.060	11.392	10.967	10.672	10.456	10.289	10.158	10.051
6	13.745	10.925	9.780	9.148	8.746	8.466	8.260	8.102	7.976	7.874
7	12.246	9.547	8.451	7.847	7.460	7.191	6.993	6.840	6.719	6.620
8	11.259	8.649	7.591	7.006	6.632	6.371	6.178	6.029	5.911	5.814
9	10.561	8.022	6.992	6.422	6.057	5.802	5.613	5.467	5.351	5.257
10	10.044	7.559	6.552	5.994	5.636	5.386	5.200	5.057	4.942	4.849
11	9.646	7.206	6.217	5.668	5.316	5.069	4.886	4.744	4.632	4.539
12	9.330	6.927	5.953	5.412	5.064	4.821	4.640	4.499	4.388	4.296
13	9.074	6.701	5.739	5.205	4.862	4.620	4.441	4.302	4.191	4.100
14	8.862	6.515	5.564	5.035	4.695	4.456	4.278	4.140	4.030	3.939
15	8.683	6.359	5.417	4.893	4.556	4.318	4.142	4.004	3.895	3.805
16	8.531	6.226	5.292	4.773	4.437	4.202	4.026	3.890	3.780	3.691
17	8.400	6.112	5.185	4.669	4.336	4.102	3.927	3.791	3.682	3.593
18	8.285	6.013	5.092	4.579	4.248	4.015	3.841	3.705	3.597	3.508
19	8.185	5.926	5.010	4.500	4.171	3.939	3.765	3.631	3.523	3.434
20	8.096	5.849	4.938	4.431	4.103	3.871	3.699	3.564	3.457	3.368
21	8.017	5.780	4.874	4.369	4.042	3.812	3.640	3.506	3.398	3.310
22	7.945	5.719	4.817	4.313	3.988	3.758	3.587	3.453	3.346	3.258
23	7.881	5.664	4.765	4.264	3.939	3.710	3.539	3.406	3.299	3.211
24	7.823	5.614	4.718	4.218	3.895	3.667	3.496	3.363	3.256	3.168
25	7.770	5.568	4.675	4.177	3.855	3.627	3.457	3.324	3.217	3.129

ν_2 \ ν_1	1	2	3	4	5	6	7	8	9	10
26	7.721	5.526	4.637	4.140	3.818	3.591	3.421	3.288	3.182	3.094
27	7.677	5.488	4.601	4.106	3.785	3.558	3.388	3.256	3.149	3.062
28	7.636	5.453	4.568	4.074	3.754	3.528	3.358	3.226	3.120	3.032
29	7.598	5.420	4.538	4.045	3.725	3.499	3.330	3.198	3.092	3.005
30	7.562	5.390	4.510	4.018	3.699	3.473	3.305	3.173	3.067	2.979
31	7.530	5.362	4.484	3.993	3.675	3.449	3.281	3.149	3.043	2.955
32	7.499	5.336	4.459	3.969	3.652	3.427	3.258	3.127	3.021	2.934
33	7.471	5.312	4.437	3.948	3.630	3.406	3.238	3.106	3.000	2.913
34	7.444	5.289	4.416	3.927	3.611	3.386	3.218	3.087	2.981	2.894
35	7.419	5.268	4.396	3.908	3.592	3.368	3.200	3.069	2.963	2.876
36	7.396	5.248	4.377	3.890	3.574	3.351	3.183	3.052	2.946	2.859
37	7.373	5.229	4.360	3.873	3.558	3.334	3.167	3.036	2.930	2.843
38	7.353	5.211	4.343	3.858	3.542	3.319	3.152	3.021	2.915	2.828
39	7.333	5.194	4.327	3.843	3.528	3.305	3.137	3.006	2.901	2.814
40	7.314	5.179	4.313	3.828	3.514	3.291	3.124	2.993	2.888	2.801
41	7.296	5.163	4.299	3.815	3.501	3.278	3.111	2.980	2.875	2.788
42	7.280	5.149	4.285	3.802	3.488	3.266	3.099	2.968	2.863	2.776
43	7.264	5.136	4.273	3.790	3.476	3.254	3.087	2.957	2.851	2.764
44	7.248	5.123	4.261	3.778	3.465	3.243	3.076	2.946	2.840	2.754
45	7.234	5.110	4.249	3.767	3.454	3.232	3.066	2.935	2.830	2.743
46	7.220	5.099	4.238	3.757	3.444	3.222	3.056	2.925	2.820	2.733
47	7.207	5.087	4.228	3.747	3.434	3.213	3.046	2.916	2.811	2.724
48	7.194	5.077	4.218	3.737	3.425	3.204	3.037	2.907	2.802	2.715
49	7.182	5.066	4.208	3.728	3.416	3.195	3.028	2.898	2.793	2.706
50	7.171	5.057	4.199	3.720	3.408	3.186	3.020	2.890	2.785	2.698
51	7.159	5.047	4.191	3.711	3.400	3.178	3.012	2.882	2.777	2.690
52	7.149	5.038	4.182	3.703	3.392	3.171	3.005	2.874	2.769	2.683
53	7.139	5.030	4.174	3.695	3.384	3.163	2.997	2.867	2.762	2.675
54	7.129	5.021	4.167	3.688	3.377	3.156	2.990	2.860	2.755	2.668
55	7.119	5.013	4.159	3.681	3.370	3.149	2.983	2.853	2.748	2.662
56	7.110	5.006	4.152	3.674	3.363	3.143	2.977	2.847	2.742	2.655
57	7.102	4.998	4.145	3.667	3.357	3.136	2.971	2.841	2.736	2.649
58	7.093	4.991	4.138	3.661	3.351	3.130	2.965	2.835	2.730	2.643
59	7.085	4.984	4.132	3.655	3.345	3.124	2.959	2.829	2.724	2.637
60	7.077	4.977	4.126	3.649	3.339	3.119	2.953	2.823	2.718	2.632
61	7.070	4.971	4.120	3.643	3.333	3.113	2.948	2.818	2.713	2.626
62	7.062	4.965	4.114	3.638	3.328	3.108	2.942	2.813	2.708	2.621
63	7.055	4.959	4.109	3.632	3.323	3.103	2.937	2.808	2.703	2.616
64	7.048	4.953	4.103	3.627	3.318	3.098	2.932	2.803	2.698	2.611
65	7.042	4.947	4.098	3.622	3.313	3.093	2.928	2.798	2.693	2.607

(continued)

ν_2 \ ν_1	1	2	3	4	5	6	7	8	9	10
66	7.035	4.942	4.093	3.618	3.308	3.088	2.923	2.793	2.689	2.602
67	7.029	4.937	4.088	3.613	3.304	3.084	2.919	2.789	2.684	2.598
68	7.023	4.932	4.083	3.608	3.299	3.080	2.914	2.785	2.680	2.593
69	7.017	4.927	4.079	3.604	3.295	3.075	2.910	2.781	2.676	2.589
70	7.011	4.922	4.074	3.600	3.291	3.071	2.906	2.777	2.672	2.585
71	7.006	4.917	4.070	3.596	3.287	3.067	2.902	2.773	2.668	2.581
72	7.001	4.913	4.066	3.591	3.283	3.063	2.898	2.769	2.664	2.578
73	6.995	4.908	4.062	3.588	3.279	3.060	2.895	2.765	2.660	2.574
74	6.990	4.904	4.058	3.584	3.275	3.056	2.891	2.762	2.657	2.570
75	6.985	4.900	4.054	3.580	3.272	3.052	2.887	2.758	2.653	2.567
76	6.981	4.896	4.050	3.577	3.268	3.049	2.884	2.755	2.650	2.563
77	6.976	4.892	4.047	3.573	3.265	3.046	2.881	2.751	2.647	2.560
78	6.971	4.888	4.043	3.570	3.261	3.042	2.877	2.748	2.644	2.557
79	6.967	4.884	4.040	3.566	3.258	3.039	2.874	2.745	2.640	2.554
80	6.963	4.881	4.036	3.563	3.255	3.036	2.871	2.742	2.637	2.551
81	6.958	4.877	4.033	3.560	3.252	3.033	2.868	2.739	2.634	2.548
82	6.954	4.874	4.030	3.557	3.249	3.030	2.865	2.736	2.632	2.545
83	6.950	4.870	4.027	3.554	3.246	3.027	2.863	2.733	2.629	2.542
84	6.947	4.867	4.024	3.551	3.243	3.025	2.860	2.731	2.626	2.539
85	6.943	4.864	4.021	3.548	3.240	3.022	2.857	2.728	2.623	2.537
86	6.939	4.861	4.018	3.545	3.238	3.019	2.854	2.725	2.621	2.534
87	6.935	4.858	4.015	3.543	3.235	3.017	2.852	2.723	2.618	2.532
88	6.932	4.855	4.012	3.540	3.233	3.014	2.849	2.720	2.616	2.529
89	6.928	4.852	4.010	3.538	3.230	3.012	2.847	2.718	2.613	2.527
90	6.925	4.849	4.007	3.535	3.228	3.009	2.845	2.715	2.611	2.524
91	6.922	4.846	4.004	3.533	3.225	3.007	2.842	2.713	2.609	2.522
92	6.919	4.844	4.002	3.530	3.223	3.004	2.840	2.711	2.606	2.520
93	6.915	4.841	3.999	3.528	3.221	3.002	2.838	2.709	2.604	2.518
94	6.912	4.838	3.997	3.525	3.218	3.000	2.835	2.706	2.602	2.515
95	6.909	4.836	3.995	3.523	3.216	2.998	2.833	2.704	2.600	2.513
96	6.906	4.833	3.992	3.521	3.214	2.996	2.831	2.702	2.598	2.511
97	6.904	4.831	3.990	3.519	3.212	2.994	2.829	2.700	2.596	2.509
98	6.901	4.829	3.988	3.517	3.210	2.992	2.827	2.698	2.594	2.507
99	6.898	4.826	3.986	3.515	3.208	2.990	2.825	2.696	2.592	2.505
100	6.895	4.824	3.984	3.513	3.206	2.988	2.823	2.694	2.590	2.503

ν_2 \ ν_1	11	12	13	14	15	16	17	18	19	20
1	6083.35	6106.35	6125.86	6142.70	6157.28	6170.12	6181.42	6191.52	6200.58	6208.74
2	99.408	99.416	99.422	99.428	99.432	99.437	99.440	99.444	99.447	99.449
3	27.133	27.052	26.983	26.924	26.872	26.827	26.787	26.751	26.719	26.690
4	14.452	14.374	14.307	14.249	14.198	14.154	14.115	14.080	14.048	14.020
5	9.963	9.888	9.825	9.770	9.722	9.680	9.643	9.610	9.580	9.553
6	7.790	7.718	7.657	7.605	7.559	7.519	7.483	7.451	7.422	7.396
7	6.538	6.469	6.410	6.359	6.314	6.275	6.240	6.209	6.181	6.155
8	5.734	5.667	5.609	5.559	5.515	5.477	5.442	5.412	5.384	5.359
9	5.178	5.111	5.055	5.005	4.962	4.924	4.890	4.860	4.833	4.808
10	4.772	4.706	4.650	4.601	4.558	4.520	4.487	4.457	4.430	4.405
11	4.462	4.397	4.342	4.293	4.251	4.213	4.180	4.150	4.123	4.099
12	4.220	4.155	4.100	4.052	4.010	3.972	3.939	3.909	3.883	3.858
13	4.025	3.960	3.905	3.857	3.815	3.778	3.745	3.716	3.689	3.665
14	3.864	3.800	3.745	3.698	3.656	3.619	3.586	3.556	3.529	3.505
15	3.730	3.666	3.612	3.564	3.522	3.485	3.452	3.423	3.396	3.372
16	3.616	3.553	3.498	3.451	3.409	3.372	3.339	3.310	3.283	3.259
17	3.519	3.455	3.401	3.353	3.312	3.275	3.242	3.212	3.186	3.162
18	3.434	3.371	3.316	3.269	3.227	3.190	3.158	3.128	3.101	3.077
19	3.360	3.297	3.242	3.195	3.153	3.116	3.084	3.054	3.027	3.003
20	3.294	3.231	3.177	3.130	3.088	3.051	3.018	2.989	2.962	2.938
21	3.236	3.173	3.119	3.072	3.030	2.993	2.960	2.931	2.904	2.880
22	3.184	3.121	3.067	3.019	2.978	2.941	2.908	2.879	2.852	2.827
23	3.137	3.074	3.020	2.973	2.931	2.894	2.861	2.832	2.805	2.781
24	3.094	3.032	2.977	2.930	2.889	2.852	2.819	2.789	2.762	2.738
25	3.056	2.993	2.939	2.892	2.850	2.813	2.780	2.751	2.724	2.699
26	3.021	2.958	2.904	2.857	2.815	2.778	2.745	2.715	2.688	2.664
27	2.988	2.926	2.871	2.824	2.783	2.746	2.713	2.683	2.656	2.632
28	2.959	2.896	2.842	2.795	2.753	2.716	2.683	2.653	2.626	2.602
29	2.931	2.868	2.814	2.767	2.726	2.689	2.656	2.626	2.599	2.574
30	2.906	2.843	2.789	2.742	2.700	2.663	2.630	2.600	2.573	2.549
31	2.882	2.820	2.765	2.718	2.677	2.640	2.606	2.577	2.550	2.525
32	2.860	2.798	2.744	2.696	2.655	2.618	2.584	2.555	2.527	2.503
33	2.840	2.777	2.723	2.676	2.634	2.597	2.564	2.534	2.507	2.482
34	2.821	2.758	2.704	2.657	2.615	2.578	2.545	2.515	2.488	2.463
35	2.803	2.740	2.686	2.639	2.597	2.560	2.527	2.497	2.470	2.445
36	2.786	2.723	2.669	2.622	2.580	2.543	2.510	2.480	2.453	2.428
37	2.770	2.707	2.653	2.606	2.564	2.527	2.494	2.464	2.437	2.412
38	2.755	2.692	2.638	2.591	2.549	2.512	2.479	2.449	2.421	2.397

(continued)

ν_2 \ ν_1	11	12	13	14	15	16	17	18	19	20
39	2.741	2.678	2.624	2.577	2.535	2.498	2.465	2.434	2.407	2.382
40	2.727	2.665	2.611	2.563	2.522	2.484	2.451	2.421	2.394	2.369
41	2.715	2.652	2.598	2.551	2.509	2.472	2.438	2.408	2.381	2.356
42	2.703	2.640	2.586	2.539	2.497	2.460	2.426	2.396	2.369	2.344
43	2.691	2.629	2.575	2.527	2.485	2.448	2.415	2.385	2.357	2.332
44	2.680	2.618	2.564	2.516	2.475	2.437	2.404	2.374	2.346	2.321
45	2.670	2.608	2.553	2.506	2.464	2.427	2.393	2.363	2.336	2.311
46	2.660	2.598	2.544	2.496	2.454	2.417	2.384	2.353	2.326	2.301
47	2.651	2.588	2.534	2.487	2.445	2.408	2.374	2.344	2.316	2.291
48	2.642	2.579	2.525	2.478	2.436	2.399	2.365	2.335	2.307	2.282
49	2.633	2.571	2.517	2.469	2.427	2.390	2.356	2.326	2.299	2.274
50	2.625	2.562	2.508	2.461	2.419	2.382	2.348	2.318	2.290	2.265
51	2.617	2.555	2.500	2.453	2.411	2.374	2.340	2.310	2.282	2.257
52	2.610	2.547	2.493	2.445	2.403	2.366	2.333	2.302	2.275	2.250
53	2.602	2.540	2.486	2.438	2.396	2.359	2.325	2.295	2.267	2.242
54	2.595	2.533	2.479	2.431	2.389	2.352	2.318	2.288	2.260	2.235
55	2.589	2.526	2.472	2.424	2.382	2.345	2.311	2.281	2.253	2.228
56	2.582	2.520	2.465	2.418	2.376	2.339	2.305	2.275	2.247	2.222
57	2.576	2.513	2.459	2.412	2.370	2.332	2.299	2.268	2.241	2.215
58	2.570	2.507	2.453	2.406	2.364	2.326	2.293	2.262	2.235	2.209
59	2.564	2.502	2.447	2.400	2.358	2.320	2.287	2.256	2.229	2.203
60	2.559	2.496	2.442	2.394	2.352	2.315	2.281	2.251	2.223	2.198
61	2.553	2.491	2.436	2.389	2.347	2.309	2.276	2.245	2.218	2.192
62	2.548	2.486	2.431	2.384	2.342	2.304	2.270	2.240	2.212	2.187
63	2.543	2.481	2.426	2.379	2.337	2.299	2.265	2.235	2.207	2.182
64	2.538	2.476	2.421	2.374	2.332	2.294	2.260	2.230	2.202	2.177
65	2.534	2.471	2.417	2.369	2.327	2.289	2.256	2.225	2.198	2.172
66	2.529	2.466	2.412	2.365	2.322	2.285	2.251	2.221	2.193	2.168
67	2.525	2.462	2.408	2.360	2.318	2.280	2.247	2.216	2.188	2.163
68	2.520	2.458	2.403	2.356	2.314	2.276	2.242	2.212	2.184	2.159
69	2.516	2.454	2.399	2.352	2.310	2.272	2.238	2.208	2.180	2.155
70	2.512	2.450	2.395	2.348	2.306	2.268	2.234	2.204	2.176	2.150
71	2.508	2.446	2.391	2.344	2.302	2.264	2.230	2.200	2.172	2.146
72	2.504	2.442	2.388	2.340	2.298	2.260	2.226	2.196	2.168	2.143
73	2.501	2.438	2.384	2.336	2.294	2.256	2.223	2.192	2.164	2.139
74	2.497	2.435	2.380	2.333	2.290	2.253	2.219	2.188	2.161	2.135
75	2.494	2.431	2.377	2.329	2.287	2.249	2.215	2.185	2.157	2.132
76	2.490	2.428	2.373	2.326	2.284	2.246	2.212	2.181	2.154	2.128
77	2.487	2.424	2.370	2.322	2.280	2.243	2.209	2.178	2.150	2.125
78	2.484	2.421	2.367	2.319	2.277	2.239	2.206	2.175	2.147	2.122

ν_2 \ ν_1	11	12	13	14	15	16	17	18	19	20
79	2.481	2.418	2.364	2.316	2.274	2.236	2.202	2.172	2.144	2.118
80	2.478	2.415	2.361	2.313	2.271	2.233	2.199	2.169	2.141	2.115
81	2.475	2.412	2.358	2.310	2.268	2.230	2.196	2.166	2.138	2.112
82	2.472	2.409	2.355	2.307	2.265	2.227	2.193	2.163	2.135	2.109
83	2.469	2.406	2.352	2.304	2.262	2.224	2.191	2.160	2.132	2.106
84	2.466	2.404	2.349	2.302	2.259	2.222	2.188	2.157	2.129	2.104
85	2.464	2.401	2.347	2.299	2.257	2.219	2.185	2.154	2.126	2.101
86	2.461	2.398	2.344	2.296	2.254	2.216	2.182	2.152	2.124	2.098
87	2.459	2.396	2.342	2.294	2.252	2.214	2.180	2.149	2.121	2.096
88	2.456	2.393	2.339	2.291	2.249	2.211	2.177	2.147	2.119	2.093
89	2.454	2.391	2.337	2.289	2.247	2.209	2.175	2.144	2.116	2.091
90	2.451	2.389	2.334	2.286	2.244	2.206	2.172	2.142	2.114	2.088
91	2.449	2.386	2.332	2.284	2.242	2.204	2.170	2.139	2.111	2.086
92	2.447	2.384	2.330	2.282	2.240	2.202	2.168	2.137	2.109	2.083
93	2.444	2.382	2.327	2.280	2.237	2.200	2.166	2.135	2.107	2.081
94	2.442	2.380	2.325	2.277	2.235	2.197	2.163	2.133	2.105	2.079
95	2.440	2.378	2.323	2.275	2.233	2.195	2.161	2.130	2.102	2.077
96	2.438	2.375	2.321	2.273	2.231	2.193	2.159	2.128	2.100	2.075
97	2.436	2.373	2.319	2.271	2.229	2.191	2.157	2.126	2.098	2.073
98	2.434	2.371	2.317	2.269	2.227	2.189	2.155	2.124	2.096	2.071
99	2.432	2.369	2.315	2.267	2.225	2.187	2.153	2.122	2.094	2.069
100	2.430	2.368	2.313	2.265	2.223	2.185	2.151	2.120	2.092	2.067

Source: National Institute of Standards and Technology

Glossary

Absolute address. Use of a dollar sign ($) before either the row or column label or both.

Agglomerative clustering methods. A series of partitions takes place from a single cluster containing all objects to n clusters, which proceed by a series of fusions of the n objects into groups.

Algorithm. A systematic procedure that finds a solution to a problem.

Alternative hypothesis. The complement of the null hypothesis; it must be true if the null hypothesis is false. The alternative hypothesis is denoted by H_1.

Alternative optimal solution. A solution that results in maximizing (or minimizing) the objective by more than one combination of decision variables, all of which have the same objective function value.

Analysis of variance (ANOVA). A tool that analyzes variance in the data and examines a test statistic that is the ratio of measures.

Area chart. A chart that combines the features of a pie chart with those of line charts.

Arithmetic mean (mean). The average, which is the sum of the observations divided by the number of observations.

Association rule mining. A tool used to uncover interesting associations and/or correlation relationships among large sets of data. The rules identify attributes that occur frequently together in a given data set.

Autocorrelation. Correlation among successive observations over time and identified by residual plots having clusters of residuals with the same sign. Autocorrelation can be evaluated more formally using a statistical test based on the measure, Durbin–Watson statistic.

Auxiliary variables. The variables used to define the bound constraints and obtain more complete sensitivity information.

Average group linkage clustering. A method that uses the mean values for each variable to compute distances between clusters.

Average linkage clustering. Defines the distance between two clusters as the average of distances between all pairs of objects where each pair is made up of one object from each group.

Average payoff strategy. French mathematician Laplace proposed this approach. For any decision, the expected value is the summation of the payoffs multiplied by their probability, summed over all outcomes. The simplest case is to assume that each outcome is equally likely to occur; that is, the probability of each outcome is simply 1/N, where N is the number of possible outcomes.

Balance constraints. Balance constraints ensure that the flow of material or money is accounted for at locations or between time periods. Example: The total amount shipped to a distribution center from all plants must equal the amount shipped from the distribution center to all customers.

Bar chart. A horizontal bar chart.

Bernoulli distribution. The probability distribution of a random variable with two possible outcomes, each with a constant probability of occurrence.

Best-subsets regression. A tool that evaluates either all possible regression models for a set of independent variables or the best subsets of models for a fixed number of independent variables.

Big data. Massive amounts of business data from a wide variety of sources, much of which is available in real time and much of which is uncertain or unpredictable.

Bimodal. Histograms with exactly two peaks.

Binding constraint. A constraint for which the *Cell Value* is equal to the right-hand side of the value of the constraint.

Binary variable. The variable restricted to being either 0 or 1 and enables to model logical decisions in optimization models. The variable is usually written as $x = 0$ or 1.

Binomial distribution. The distribution that models n independent replications of a Bernoulli experiment, each with a probability p of success.

Boxplot. Graphically displays five key statistics of a data set—the minimum, first quartile, median, third quartile, and maximum—and identifies the shape of a distribution and outliers in the data.

Box-whisker chart. A chart that shows the minimum, first quartile, median, third quartile, and maximum values in a data set graphically.

Branches. Each branch of the decision tree represents an event or a decision.

Bubble chart. A type of scatter chart in which the size of the data marker corresponds to the value of a third variable—a way to plot three variables in two dimensions.

Business analytics (analytics). The use of data, information technology, statistical analysis, quantitative methods, and mathematical or computer-based models to help managers gain improved insight about their business operations and make better, fact-based decisions; a process of transforming data into actions through analysis and insights in the context of organizational decision making and problem solving.

Business intelligence (BI). The collection, management, analysis, and reporting of data.

Categorical (nominal) data. Data that are sorted into categories according to specified characteristics.

Central limit theorem. A theory that states that if the population is normally distributed, then the sampling distribution of the mean will be normal for any sample size.

Certainty equivalent. The term represents the amount that a decision maker feels is equivalent to an uncertain gamble.

Chebyshev's theorem. The theorem that states that for any set of data, the proportion of values that lie within k standard deviations ($k > 1$) of the mean is at least $1 - 1/k^2$.

Chi-square distribution. Distribution of Chi-square statistics characterized by degrees of freedom.

Chi-square statistic. The sum of squares of the differences between observed frequency, fo, and expected frequency, fe, divided by the expected frequency in each cell.

Classification matrix. A tool that shows the number of cases that were classified either correctly or incorrectly.

Cluster analysis. A collection of techniques that seek to group or segment a collection of objects into subsets or clusters such that objects within each cluster are more closely related to one another than objects assigned to different clusters. The objects within clusters exhibit a high amount of similarity.

Cluster sampling. A theory based on dividing a population into subgroups (clusters), sampling a set of clusters, and (usually) conducting a complete census within the clusters sampled.

Conditional probability. The probability of occurrence of one event A, given that another event B is known to be true or has already occurred.

Coefficient of determination (R^2). The tool gives the proportion of variation in the dependent variable that is explained by the independent variable of the regression model and has the value between 0 and 1.

Coefficient of kurtosis (CK). A measure of the degree of kurtosis of a population; "excess kurtosis" is computed using the Excel function KURT (data range).

Coefficient of multiple determination. Similar to simple linear regression, the tool explains the percentage of variation in the dependent variable. The coefficient of multiple determination in the context of multiple regression indicates the strength of association between the dependent and independent variables.

Coefficient of skewness (CS). A measure of the degree of asymmetry of observations around the mean.

Coefficient of variation (CV). Relative measure of the dispersion in data relative to the mean.

Confidence interval. A range of values between which the value of the population parameter is believed to be along with a probability that the interval correctly estimates the true (unknown) population parameter.

Confidence coefficient. The probability of correctly failing to reject the null hypothesis, or P(not rejecting $H_0 | H_0$ is true), and is calculated as $1 - \alpha$.

Confidence of the (association) rule. The conditional probability that a randomly selected transaction will include all the items in the consequent given that the transaction includes all the items in the antecedent.

Constraint function. A function of the decision variables in the problem.

Constraints. Limitations, requirements, or other restrictions that are imposed on any solution, either from practical or technological considerations or by management policy.

Contingency table. A cross-tabulation table.

Continuous metric. A metric that is based on a continuous scale of measurement.

Continuous random variable. A random variable that has outcomes over one or more continuous intervals of real numbers.

Convenience sampling. A method in which samples are selected based on the ease with which the data can be collected.

Column chart. A vertical bar chart.

Complete linkage clustering. The distance between groups is defined as the distance between the most distant pair of objects, one from each group.

Complement. The set of all outcomes in the sample space that is not included in the event.

Corner point. The point at which the constraint lines intersect along the feasible region.

Correlation. A measure of the linear relationship between two variables, X and Y, which does not depend on the units of measurement.

Correlation coefficient (Pearson product moment correlation coefficient). The value obtained by dividing the covariance of the two variables by the product of their standard deviations.

Covariance. A measure of the linear association between two variables, X and Y.

Cross-tabulation. A tabular method that displays the number of observations in a data set for different subcategories of two categorical variables.

Cross-validation. A process of using two sets of sample data; one to build the model (the training set), and the second to assess the model's performance (the validation set).

Cumulative distribution function. A specification of the probability that the random variable X assumes a value less than or equal to a specified value x.

Cumulative relative frequency. The proportion of the total number of observations that fall at or below the upper limit of each group.

Cumulative relative frequency distribution. A tabular summary of cumulative relative frequencies.

Curvilinear regression model. The model is used in forecasting when the independent variable is time.

Cyclical effect. Characteristic of a time series that describes ups and downs over a much longer time frame, such as several years.

Dashboard. A visual representation of a set of key business measures.

Database. A collection of related files containing records on people, places, or things.

Data mining. A rapidly growing field of business analytics that is focused on better understanding characteristics and patterns among variables in large databases using a variety of statistical and analytical tools.

Data profile (fractile). A measure of dividing data into sets.

Data set. A collection of data.

Data table. A table that summarizes the impact of one or two inputs on a specified output.

Data validation. A tool that allows defining acceptable input values in a spreadsheet and providing an error alert if an invalid entry is made.

Data visualization. The process of displaying data (often in large quantities) in a meaningful fashion to provide insights that will support better decisions.

Decision alternatives. Decisions that involve a choice from among a small set of alternatives with uncertain consequences.

Decision making. The study of how people make decisions, particularly when faced with imperfect or uncertain information, as well as a collection of techniques to support decision choices.

Decision model. A logical or mathematical representation of a problem or business situation that can be used to understand, analyze, or facilitate making a decision.

Decision node. A decision node is expressed by a square, and it represents an event of a selected decision from among several alternatives.

Decision strategy. A decision strategy is a specification of an initial decision and subsequent decisions to make after knowing what events occur.

Decision support systems (DSS). A combination of business intelligence concepts and OR/MS models to create analytical-based computer systems to support decision making.

Decision tree. An approach to structuring a decision problem involving uncertainty to use a graphical model.

Decision variables. The unknown values that an optimization model seeks to determine.

Degenerate solution. A solution is a degenerate solution if the right-hand-side value of any constraint has a zero allowable increase or allowable decrease.

Delphi method. A forecasting approach that uses a panel of experts, whose identities are typically kept confidential from one another, to respond to a sequence of questionnaires to converge to an opinion of a future forecast.

Dendrogram. Hierarchical clustering represented by a two-dimensional diagram that illustrates the fusions or divisions made at each successive stage of analysis.

Degrees of freedom (df). An additional parameter used to distinguish different t-distributions.

Descriptive analytics. The use of data to understand past and current business performance and make informed decisions; the most commonly used and most well-understood type of analytics.

Descriptive statistics. Methods of describing and summarizing data using tabular, visual, and quantitative techniques.

Deterministic model. A prescriptive decision model in which all model input information is either known or assumed to be known with certainty.

Discriminant analysis. A technique for classifying a set of observations into predefined classes; the purpose is to determine the class of an observation based on a set of predictor variables.

Discount rate. The opportunity costs of spending funds now versus achieving a return through another investment, as well as the risks associated with not receiving returns until a later time.

Discrete metric. A metric derived from counting something.

Discrete random variable. A random variable for which the number of possible outcomes can be counted.

Discrete uniform distribution. A variation of the uniform distribution for which the random variable is restricted to integer values between a and b (also integers).

Dispersion. The degree of variation in the data, that is, the numerical spread (or compactness) of the data.

Divisive clustering methods. A series of partitions takes place from a single cluster containing all objects to n clusters, which separate n objects successively into finer groupings.

Double exponential smoothing. A forecasting approach similar to simple exponential smoothing used for time series with a linear trend and no significant seasonal components.

Double moving average. A forecasting approach similar to a simple moving average used for time series with a linear trend and no significant seasonal components.

Doughnut chart. A chart that is similar to a pie chart but can contain more than one data series.

Dummy variables. A numerical variable used in regression analysis to represent subgroups of the sample in the study.

Econometric models. Explanatory/causal models that seek to identify factors that explain statistically the patterns observed in the variable being forecast.

Empirical probability distribution. An approximation of the probability distribution of the associated random variable.

Empirical rules. For a normal distribution, all data will fall within three standard deviations of the mean. Depending on the data and the shape of the frequency distribution, the actual percentages may be higher or lower.

Estimation. A method used to assess the value of an unknown population parameter such as a population mean, population proportion, or population variance using sample data.

Estimators. Measures used to estimate population parameters.

Expected opportunity loss. The expected opportunity loss represents the average additional amount the decision maker would have achieved by making the right decision instead of a wrong one.

Expected value. The notion of the mean or average of a random variable; the weighted average of all possible outcomes, where the weights are the probabilities.

Expected value of perfect information (EVPI). The expected value with perfect information (assumed at no cost) minus the expected value without any information.

Expected value of sample information (EVSI). The expected value with sample information (assumed at no cost) minus the expected value without sample information. It represents the most one should be willing to pay for the sample information.

Expected value strategy. A more general case of the average payoff strategy is when the probabilities of the outcomes are not all the same.

Experiment. A process that results in an outcome.

Exponential distribution. A continuous distribution that models the time between randomly occurring events.

Exponential function, $y = ab^x$. Exponential functions have the property that y rises or falls at constantly increasing rates.

Euclidean distance. The most commonly used measure of distance between objects in which the distance between two points on a plane is computed as the hypotenuse of a right triangle.

Event. A collection of one or more outcomes from a sample space.

Event node. An event node is an outcome over which the decision maker has no control.

Factor. The variable of interest in statistics terminology.

Feasibility report. The report analyzes limits on variables and the constraints that make the problem infeasible.

Feasible region. The set of feasible solutions to an optimization problem.

Feasible solution. Any solution that satisfies all constraints of an optimization problem.

Frequency distribution. A table that shows the number of observations in each of several non-overlapping groups.

General integer variables. Any variable in an ordinary linear optimization model.

Goodness of fit. A procedures that attempts to draw a conclusion about the nature of a distribution.

Heat map. Color-coding of quantitative data.

Hierarchical clustering. The data are not partitioned into a particular cluster in a single step but a series of partitions takes place, which may run from a single cluster containing all objects to n clusters, each containing a single object.

Histogram. A graphical depiction of a frequency distribution for numerical data in the form of a column chart.

Historical analogy. A forecasting approach in which a forecast is obtained through a comparative analysis with a previous situation.

Holt-Winters additive model. A forecasting model that applies to time series with relatively stable seasonality.

Holt-Winters models. Forecasting models similar to exponential smoothing models in that smoothing constants are used to smooth out variations in the level and seasonal patterns over time.

Holt-Winters multiplicative model. A forecasting model that applies to time series whose amplitude increases or decreases over time.

Homoscedasticity. The assumption means that the variation about the regression line is constant for all values of the independent variable. The data is evaluated by examining the residual plot and looking for large differences in the variances at different values of the independent variable.

Hypothesis. A proposed explanation made on the basis of limited evidence to interpret certain events or phenomena.

Hypothesis testing. Involves drawing inferences about two contrasting propositions relating to the value of one or more population parameters, such as the mean, proportion, standard deviation, or variance.

Independent events. Events that do not affect the occurrence of each other.

Index. A single measure that weights multiple indicators, thus providing a measure of overall expectation.

Indicators. Measures that are believed to influence the behavior of a variable we wish to forecast.

Infeasible problem. A problem for which no feasible solution exists.

Influence diagram. A visual representation that describes how various elements of a model influence, or relate to, others.

Information systems (IS). The modern discipline evolved from business intelligence (BI).

Integer linear optimization model (integer program). In an integer linear optimization model (integer program), some of or all the variables are restricted to being *whole numbers*.

Interaction. Occurs when the effect of one variable (i.e., the slope) is dependent on another variable.

Interquartile range (IQR, or midspread). The difference between the first and third quartiles, $Q_3 - Q_1$.

Interval estimate. A method that provides a range for a population characteristic based on a sample.

Intersection. A composition with all outcomes belonging to both events.

Interval data. Data that are ordinal but have constant differences between observations and have arbitrary zero points.

Joint probability. The probability of the intersection of two events.

Joint probability table. A table that summarizes joint probabilities.

Judgment sampling. A plan in which expert judgment is used to select the sample.

***k*-nearest neighbors (*k*-NN) algorithm.** A classification scheme that attempts to find records in a database that are similar to one that is to be classified.

***k*th percentile.** A value at or below which at least *k* percent of the observations lie.

Kurtosis. The peakedness (i.e., high, narrow) or flatness (i.e., short, flat-topped) of a histogram.

Lagging measures. Outcomes that tell what happened and are often external business results, such as profit, market share, or customer satisfaction.

Laplace or average payoff strategy. *See* Average payoff strategy.

Leading measures. Performance drivers that predict what *will* happen and usually are internal metrics, such as employee satisfaction, productivity, turnover, and so on.

Least-squares regression. The mathematical basis for the best-fitting regression line.

Level of confidence. A range of values between which the value of the population parameter is believed to be along with a probability that the interval correctly estimates the true (unknown) population parameter.

Level of significance. The probability of making Type 1 error, that is, $P(\text{rejecting } H_0 \mid H_0 \text{ is true})$, is denoted by α.

Lift. Defined as the ratio of confidence to expected confidence. Lift provides information about the increase in probability of the 'then' (consequent) given the 'if' (antecedent) part.

Line chart. A chart that provides a useful means for displaying data over time.

Linear function, $y = a + bx$. Linear functions show steady increase or decrease over the range of x and used in predictive models.

Linear optimization model (linear program, LP). A model with two basic properties: i) The objective function and all constraints are linear functions of the decision variables and ii) all variables are continuous.

Linear program (LP) relaxation. A problem that arises by replacing the constraint that each variable must be 0 or 1.

Logarithmic function, $y = \ln(x)$. Logarithmic functions are used when the rate of change in a variable increases or decreases quickly and then levels out, such as with diminishing returns to scale.

Limitations. Limitations usually involve the allocation of scarce resources. Example: Problem statements such as the amount of material used in production cannot exceed the amount available in inventory.

Marginal probability. The probability of an event irrespective of the outcome of the other joint event.

Marker line. The red line that divides the regions in a "probability of a negative cost difference" chart.

Market basket analysis. A typical and widely used example of association rule mining. The transaction data routinely collected using bar-code scanners are used to make recommendations for promotions, for cross-selling, catalog design and so on.

Maximax strategy. For the aggressive strategy, the best payoff for each decision would be the *largest* value among all outcomes, and one would choose the decision corresponding to the largest of these.

Maximin strategy. For the conservative strategy, the worst payoff for each decision would be the *smallest* value among all outcomes, and one would choose the decision corresponding to the largest of these.

Mean absolute deviation (MAD). The absolute difference between the actual value and the forecast, averaged over a range of forecasted values.

Mean absolute percentage error (MAPE). The average of absolute errors divided by actual observation values.

Mean square error (MSE). The average of the square of the difference s between the actual value and the forecast.

Measure. Numerical value associated with a metric.

Measurement. The act of obtaining data associated with a metric.

Median. The measure of location that specifies the middle value when the data are arranged from the least to greatest.

Metric. A unit of measurement that provides a way to objectively quantify performance.

Midrange. The average of the greatest and least values in the data set.

Minimax regret strategy. The decision maker selects the decision that minimizes the largest opportunity loss among all outcomes for each decision.

Minimax strategy. One seeks the decision that minimizes the largest payoff that can occur among all outcomes for each decision. Conservative decision makers are willing to forgo high returns to avoid undesirable losses.

Mixed-integer linear optimization model. If only a subset of variables is restricted to being integer while others are continuous, we call this a mixed integer linear optimization model.

Mode. The observation that occurs most frequently.

Model. An abstraction or representation of a real system, idea, or object.

Modeling and optimization. Techniques for translating real problems into mathematics, spreadsheets, or other computer languages, and using them to find the best ("optimal") solutions and decisions.

Monte Carlo simulation. The process of generating random values for uncertain inputs in a model, computing the output variables of interest, and repeating this process for many trials to understand the distribution of the output results.

Multicollinearity. A condition occurring when two or more independent variables in the same regression model contain high levels of the same information and, consequently, are strongly correlated with one another and can predict each other better than the dependent variable.

Multiple correlation coefficient. *Multiple R* and *R Square* (or R^2) in the context of multiple regression indicate the strength of association between the dependent and independent variables.

Multiple linear regression. A linear regression model with more than one independent variable. Simple linear regression is just a special case of multiple linear regression.

Multiplication law of probability. The probability of two events A and B is the product of the probability of A given B, and the probability of B (or) the product of the probability of B given A, and the probability of A.

Mutually exclusive. Events with no outcomes in common.

Net present value (discounted cash flow). The sum of the present values of all cash flows over a stated time horizon; a measure of the worth of a stream of cash flows, that takes into account the time value of money.

Newsvendor problem. A practical situation in which a one-time purchase decision must be made in the face of uncertain demand.

Nodes. Nodes are points in time at which events take place.

Nonsampling error. An error that occurs when the sample does not represent the target population adequately.

Normal distribution. A continuous distribution described by the familiar bell-shaped curve and is perhaps the most important distribution used in statistics.

Null hypothesis. Describes the existing theory or a belief that is accepted as valid unless strong statistical evidence exists to the contrary.

Objective function. The quantity that is to be minimized or maximized; minimizing or maximizing some quantity of interest—profit, revenue, cost, time, and so on—by optimization.

Ogive. A chart that displays the cumulative relative frequency.

One-sample hypothesis test. A test that involves a single population parameter, such as the mean, proportion, standard deviation, and a single sample of data from the population is used to conduct the test.

One-tailed test of hypothesis. The hypothesis test that specify a direction of relationship where H_0 is either $\geq$ or $\leq$.

One-way data table. A data table that evaluates an output variable over a range of values for a single input variable.

Overfitting. If too many terms are added to the model, then the model may not adequately predict other values from the population. Overfitting can be mitigated by using good logic, intuition, physical or behavioral theory, and parsimony.

Operations Research/Management Science (OR/MS). The analysis and solution of complex decision problems using mathematical or computer-based models.

Optimal solution. Any set of decision variables that optimizes the objective function.

Optimization. The process of finding a set of values for decision variables that minimize or maximize some quantity of interest and the most important tool for prescriptive analytics.

Ordinal data. Data that can be ordered or ranked according to some relationship to one another.

Outcome. A result that can be observed.

Outcomes. Possible results of a decision or a strategy.

Outlier. The observation that is radically different from the rest.

Overbook. To accept reservations in excess of the number that can be accommodated.

Overlay chart. A feature for superimposition of the frequency distributions from selected forecasts, when a simulation has multiple related forecasts, on one chart to compare differences and similarities that might not be apparent.

Point estimate. A single number derived from sample data that is used to estimate the value of a population parameter.

Population frame. A listing of all elements in the population from which the sample is drawn.

Prediction interval. Provides a range for predicting the value of a new observation from the same population.

Probability interval. In general, a $100(1 - \alpha)\%$ is any interval [A, B] such that the probability of falling between A and B is $1 - \alpha$. Probability intervals are often centered on the mean or median.

p-Value (observed significance level). An alternative approach to find the probability of obtaining a test statistic value equal to or more extreme than that obtained from the sample data when the null hypothesis is true.

Power of the test. Represents the probability of correctly rejecting the null hypothesis when it is indeed false, or P(rejecting $H_0 | H_0$ is false).

Parsimony. A model with the fewest number of explanatory variables that will provide an adequate interpretation of the dependent variable.

Partial regression coefficient. The partial regression coefficients represent the expected change in the dependent variable when the associated independent variable is increased by one unit while the values of all other independent variables are held constant.

Polynomial function. $y = ax^2 + bx + c$ (second order—quadratic function), $y = ax^3 + bx^2 + dx + e$ (third order—cubic function), and so on. A second order polynomial is parabolic in nature and has only one hill or valley; a third order polynomial has one or two hills or valleys. Revenue models that incorporate price elasticity are often polynomial functions.

Power function. $y = ax^b$. Power functions define phenomena that increase at a specific rate. Learning curves that express improving times in performing a task are often modeled with power functions having $a > 0$ and $b < 0$.

Parallel coordinates chart. The chart consists of a set of vertical axes, one for each variable selected and creates a "multivariate profile," that helps an analyst to explore the data and draw basic conclusions. For each observation, a line is drawn connecting the vertical axes. The point at which the line crosses an axis represents the value for that variable.

Proportional relationships. Proportional relationships are often found in problems involving mixtures or blends of materials or strategies.

Payoffs. The decision maker first selects a decision alternative, after which one of the outcomes of the uncertain event occurs, resulting in the payoff.

Payoff table. Payoffs are often summarized in a payoff table, a matrix whose rows correspond to decisions and whose columns correspond to events.

Perfect information. The information that tells us with certainty what outcome will occur and it provides an upper bound on the value of any information that one may acquire.

Parameter analysis. An approach provided by *Analytic Solver Platform* for automatically running multiple optimizations with varying model parameters within predefined ranges.

Parametric sensitivity analysis. The term used by Analytic Solver Platform for systematic methods of what-if analysis.

Pareto analysis. The analysis that uses the Pareto principle, the 80–20 rule, that refers to the generic situation in which 80% of some output comes from 20% of some input.

Pie chart. A chart that partitions a circle into pie-shaped areas showing the relative proportion of each data source to the total.

PivotChart. A data analysis tool provided by Microsoft Excel, which enables visualizing data in PivotTables.

PivotTables. A powerful tool, provided by Excel, for distilling a complex data set into meaningful information.

Poisson distribution. A discrete distribution used to model the number of occurrences in some unit of measure.

Population. Gathering of all items of interest for a particular decision or investigation.

Predictive analytics. A component of business analytics that seeks to predict the future by examining historical data, detecting patterns or relationships in these data, and then extrapolating these relationships forward in time.

Prescriptive analytics. A component of business analytics that uses optimization to identify the best alternatives to minimize or maximize some objective.

Price elasticity. The ratio of the percentage change in demand to the percentage change in price.

Pro forma income statement. A calculation of net income using the structure and formatting that accountants are used to.

Probability. The likelihood that an outcome occurs.

Probability density function. The distribution that characterizes outcomes of a continuous random variable.

Probability distribution. The characterization of the possible values that a random variable may assume along with the probability of assuming these values.

Probability mass function. The probability distribution of the discrete outcomes for a discrete random variable X.

Problem solving. The activity associated with defining, analyzing, and solving a problem and selecting an appropriate solution that solves a problem.

Process capability index. The value obtained by dividing the specification range by the total variation; index used to evaluate the quality of the products and determine the requirement of process improvements.

Proportion. Formal statistical measure; key descriptive statistics for categorical data, such as defects or errors in quality control applications or consumer preferences in market research.

Quartile. The value that breaks data into four parts.

Radar chart. A chart that allows plotting of multiple dimensions of several data series.

Random number. A number that is uniformly distributed between 0 and 1.

Random number seed. A value from which a stream of random numbers is generated.

Random variable. A numerical description of the outcome of an experiment.

Random variate. A value randomly generated from a specified probability distribution.

Range. The difference between the maximum value and the minimum value in the data set.

Ratio data. Data that are continuous and have a natural zero.

Reduced cost. A number that tells how much the objective coefficient needs to be reduced for a nonnegative variable that is zero in the optimal solution to become positive.

Requirements. Requirements involve the specification of minimum levels of performance. Example: Production must be sufficient to meet promised customer orders.

Regression analysis. A tool for building mathematical and statistical models that characterize relationships between a dependent variable and one or more independent, or explanatory, variables, all of which are numerical.

Relative address. Use of just the row and column label in the cell reference.

Relative frequency. Expression of frequency as a fraction, or proportion, of the total.

Relative frequency distribution. A tabular summary of the relative frequencies of all categories.

Reliability. A term that refers to accuracy and consistency of data.

Return to risk. The reciprocal of the coefficient of variation.

R^2 (R-squared). A measure of the "fit" of the line to the data; the value of R^2 will be between 0 and 1. The larger the value of R^2, the better the fit.

Residuals. Observed errors which are the differences between the actual values and the estimated values of the dependent variable using the regression equation.

Risk. The likelihood of an undesirable outcome; a condition associated with the consequences and likelihood of what might happen.

Risk analysis. An approach for developing a comprehensive understanding and awareness of the risk associated with a particular variable of interest.

Risk premium. The amount an individual is willing to forgo to avoid risk, and this indicates that the person is a *risk-averse individual* (relatively conservative).

Risk profile. Risk profiles show the possible payoff values that can occur and their probabilities. Each decision strategy has an associated payoff distribution called a risk profile.

Root mean square error (RMSE). The square root of mean square error (MSE).

Sample. A subset of a population.

Sample correlation coefficient. The value obtained by dividing the covariance of the two variables by the product of their sample standard deviations.

Sample information. The information is a result of conducting some type of experiment, such as a market research study, or interviewing an expert. Sample information is always imperfect and comes at a cost.

Sample proportion. An unbiased estimator of a population proportion where x is the number in the sample having the desired characteristic and n is the sample size.

Sample space. The collection of all possible outcomes of an experiment.

Sampling distribution of the mean. The means of all possible samples of a fixed size n from some population will form a distribution.

Sampling plan. A description of the approach that is used to obtain samples from a population prior to any data collection activity.

Sampling (statistical) error. This occurs for samples are only a subset of the total population. Sampling error is inherent in any sampling process, and although it can be minimized, it cannot be totally avoided.

Scatter chart. A chart that shows the relationship between two variables.

Scenarios. Sets of values that are saved and can be substituted automatically on a worksheet.

Search algorithm. Solution procedure that generally finds good solutions without guarantees of finding the best one.

Seasonal effect. Characteristic of a time series that repeats at fixed intervals of time, typically a year, month, week, or day.

Sensitivity chart. A feature that allows determination of the influence that each uncertain model input has individually on an output variable based on its correlation with the output variable.

Shadow price. A number that tells how much the value of the objective function will change as the right-hand side of a constraint is increased by 1.

Single linkage clustering. The distance between two clusters is given by the value of the shortest link between the clusters. The distance between groups is defined as the distance between the closest pair of objects, where only pairs consisting of one object from each group are considered.

Simple bounds. Simple bounds constrain the value of a single variable. Example: Problem statements such as no more than $10,000 may be invested in stock ABC.

Simple exponential smoothing. An approach for short-range forecasting that is a weighted average of the most recent forecast and actual value.

Simple moving average. A smoothing method based on the idea of averaging random fluctuations in the time series to identify the underlying direction in which the time series is changing.

Simple random sampling. The plan involves selecting items from a population so that every subset of a given size has an equal chance of being selected.

Significance of regression. A simple hypothesis test checks whether the regression coefficient is zero.

Simple linear regression. A tool used to find a linear relationship between one independent variable, X, and one dependent variable, Y.

Simulation and risk analysis. A methodology that relies on spreadsheet models and statistical analysis to examine the impact of uncertainty in the estimates and their potential interaction with one another on the output variable of interest.

Skewness. Lacking symmetry of data.

Slicers. A tool for drilling down to "slice" a PivotTable and display a subset of data.

Smoothing constant. A value between 0 and 1 used to weight exponential smoothing forecasts.

Sparklines. Graphics that summarize a row or column of data in a single cell.

Spreadsheet engineering. Building spreadsheet models.

Standard deviation. The square root of the variance.

Standard error of the estimate, S_{YX}. The variability of the observed Y-values from the predicted values.

Standard residuals. Residuals divided by their standard deviation. Standard residuals describe how far each residual is from its mean in units of standard deviations.

Standard error of the mean. The standard deviation of the sampling distribution of the mean.

Standard normal distribution. A normal distribution with mean 0 and standard deviation 1.

Standardized value (z-score). A relative measure of the distance an observation is from the mean, which is independent of the units of measurement.

States of nature. The outcomes associated with uncertain events are defined so that one and only one of them will occur. They may be quantitative or qualitative.

Stationary time series. A time series that does not have trend, seasonal, or cyclical effects but is relatively constant and exhibits only random behavior.

Statistic. A summary measure of data.

Statistics. The science of uncertainty and the technology of extracting information from data; an important element of business, driven to a large extent by the massive growth of data.

Statistical inference. The estimation of population parameters and hypothesis testing which involves drawing conclusions about the value of the parameters of one or more populations based on sample data.

Statistical thinking. A philosophy of learning and action for improvement that is based on the principles that i) all work occurs in a system of interconnected processes, ii) variation exists in all processes, and iii) better performance results from understanding and reducing variation.

Stratified sampling. A plan that applies to populations that are divided into natural subsets (called strata) and allocates the appropriate proportion of samples to each stratum.

Stochastic model. A prescriptive decision model in which some of the model input information is uncertain.

Stock chart. A chart that allows plotting of stock prices, such as the daily high, low, and close.

Support for the (association) rule. The number of transactions that include all items in the antecedent and consequent parts of the rule; shows probability that a randomly selected transaction from the database will contain all items in the antecedent and the consequent.

Surface chart. A chart that shows 3-D data.

Systematic (or periodic) sampling. A sampling plan that selects every nth item from the population.

Tag cloud. A visualization of text that shows words that appears more frequently using larger fonts.

t-Distribution. The t-distribution is actually a family of probability distributions with a shape similar to the standard normal distribution.

Time series. A stream of historical data.

Transportation problem. The problem involves determining how much to ship from a set of sources of supply (factories, warehouses, etc.) to a set of demand locations (warehouses, customers, etc.) at minimum cost.

Trend. A gradual upward or downward movement of a time series over time.

Trend chart. The single chart that shows the distributions of all output variables, when a simulation has multiple output variables that are related to one another.

Tornado chart. A tool that graphically shows the impact that variation in a model input has on some output while holding all other inputs constant.

Type I error. The null hypothesis is actually true, but the hypothesis test incorrectly rejects it.

Type II error. The null hypothesis is actually false, but the hypothesis test incorrectly fails to reject it.

Two-tailed test of hypothesis. The rejection region occurs in both the upper and lower tail of the distribution

Two-way data table. A data table that evaluates an output variable over a range of values for two different input variables.

Unbounded solution. A solution that has the value of the objective to be increased or decreased without bound (i.e., to infinity for a maximization problem or negative infinity for a minimization problem) without violating any of the constraints.

Uncertain function. A cell referred, by *Analytic Solver Platform*, for which prediction and creation of a distribution of output values from the model is carried out.

Uncertain events. An event that occurs after a decision is made along with its possible outcome.

Uncertainty. Imperfect knowledge of what will happen.

Utility theory. An approach for assessing risk attitudes quantitatively.

Uniform distribution. A function that characterizes a continuous random variable for which all outcomes between some minimum and maximum value are equal likely.

Unimodal. Histograms with only one peak.

Union. A composition of all outcomes that belongs to either of two events.

Unique optimal solution. The exact single solution that will result in the maximum (or minimum) objective.

Value of information. Represents the improvement in the expected return that can be achieved if the decision maker is able to acquire—before making a decision—additional information about the future event that will take place.

Validity. An estimate of whether the data correctly measure what they are supposed to measure; a term that refers to how well a model represents reality.

Variance. The average of the squared deviations of the observations from the mean; a common measure of dispersion.

Verification. The process of ensuring that a model is accurate and free from logical errors.

Visualization. The most useful component of business analytics that is truly unique.

Ward's hierarchical clustering. The clustering method uses a sum-of-squares criterion.

What-if analysis. The analysis shows how specific combinations of inputs that reflect key assumptions will affect model outputs.

Index